# Pest Management
*in*
# Horticulture Crops
*Principles and Practices*

# Pest Management
*in*
# Horticulture Crops

## *Principles and Practices*

*Editors*

**L.R. Verma**
*Vice-Chancellor*
*Himachal Pradesh University*
*Shimla, 171 005, India*

**A.K. Verma**
*Department of Entomology and Apiculture*
*Dr. Y.S. Parmar University of Horticulture & Forestry*
*Nauni, Solan, 173 230, India*

**D.C. Gautam**
*Department of Biosciences*
*Himachal Pradesh University*
*Shimla, 171 005, India*

**ASIATECH PUBLISHERS INC.**
**New Delhi**

*First Edition*, February, 2004

**ASIATECH PUBLISHERS INC.**
New Delhi

**ISBN: 81-87680-10-5**

Published by N.K. Muraleedharan for Asiatech Publishers Inc.
E-20, Arya Nagar, CGHS, Plot No. 91, I.P. Extension, Patparganj,
Delhi 110 092. *Phone*: +91-11-22724601 • *Fax*: +91-11-22727254
*e-mail*: asiatech@bol.net.in

Typeset by Jayanti Graphics Pvt. Ltd., H-296, Uday Niketan,
Narain Vihar, New Delhi 110 028. *Phone*: +91-11-25791026
*e-mail*: jgpl@bol.net.in

Printed at Saras Graphics, A-49/1,
Mayapuri Industrial Area, Phase I, New Delhi.
*Phone:* +91-11-28115318

# Preface

Horticulture crops contribute significantly not only in strengthening the economy of a country but also provide a broad range of products with high nutritive value and thus play a vital role in shaping the health of human populations. But these crops are attacked by a large number of pests, mainly the insects, which cause huge economic losses. Over the years, man has been striving hard to protect these crops and in this struggle, his main emphasis has been to minimize the economic damage without any adverse effects on human populations and environment. Different methods were devised and tried from time to time which yielded varied results reducing the damaging potential of almost all the pests. Appropriate use of different methods resulted in satisfactory control of some of the serious pests.

The present book provides detailed information on the management of pests of almost all the tropical, subtropical and temperate horticulture crops. This book is divided in two parts: Part I explains the principles while Part II deals with the practices followed in the pest management of horticulture crops.

First nine chapters are devoted to the principles of pest management. Ecological approaches and strategies of pest management which help in developing efficient and reliable techniques and tools required for pest management are discussed in detail in these chapters. Different aspects of biological, cultural, mechanical and regulatory control are also appropriately highlighted. Use of insecticides, status and scope of insect growth regulators and the role of semiochemicals for pest control are also dealt with in detail. Keeping in view the recent scientific developments, indigenous technology knowledge and the genetic engineering techniques used for pest management have also been discussed.

Pest control practices used in pest management form the subject matter from chapter 10 through 27. Detailed information on the biological and ecological aspects and diverse strategies practised for effective control of pests of horticulture crops are given in different chapters.

This book is intended to provide a clear overview of the principles and practices of pest management in horticulture crops and should serve as a useful source of information to the scientists as well as to orchardists and farmers alike. It will also serve as a text/reference book to the teachers, postgraduate students and researchers.

The editors wish to thank all the contributors for putting the exhaustive information in a systematic way in the form of chapters.

L.R. VERMA
A.K. VERMA
D.C. GAUTAM

# Contents


*Preface* . . . . . . . . . . . . *v*

*Contributors* . . . . . . . . . . . . *ix*

HORTICULTURAL PEST MANAGEMENT: PRINCIPLES

*Chapter* 1 Ecological Basis of Insect Pest Management: Issues and Strategies . . 3

*Chapter* 2 Population Monitoring, Dispersion, Modelling and Life Table Analysis in Insect Pest Management . . . . . . . . . 26

*Chapter* 3 Biological Control of Insect Pests . . . . . . . 90

*Chapter* 4 Cultural, Mechanical and Regulatory Methods in Pest Management . . 134

*Chapter* 5 Insecticides in Pest Management . . . . . . . 154

*Chapter* 6 Insect Growth Regulators: Current Status and Scope in the Management of Fruit Crop Pests . . . . . . . 171

*Chapter* 7 Role of Semiochemicals in the Management of Insect Pests of Fruit Crops 184

*Chapter* 8 Indigenous Technical Knowledge for Pest Management . . . 202

*Chapter* 9 Genetic Engineering Techniques for Insect Pest Management . . . 215

HORTICULTURAL PEST MANAGEMENT: PRACTICES

*Chapter 10* Insect and Mite Pests of Apple and Pear . . . . . . 237

*Chapter 11* Aphids as Pests of Horticultural Crops . . . . . . 259

*Chapter 12* Insect Pests of Peach, Plum and Apricot . . . . . . 279

*Chapter 13* Pests of Almond, Cherry and Walnut . . . . . . 295

*Chapter 14* Pests of Mango . . . . . . . . . 310

*Chapter 15* Pests of Citrus . . . . . . . . . 325

*Chapter 16* Pests of Grapes . . . . . . . . . 339

*Chapter 17* Insect Pests of Fruits of Arid and Semi-Arid Regions of India . . 347

*Chapter 18* Pests of Potato . . . 387

*Chapter 19* Insect Pests of Beans, Pea and Okra . . . 410

*Chapter 20* Insect Pests of Radish and Turnip . . . 427

*Chapter 21* Insect Pests of Cucurbit Vegetables . . . 437

*Chapter 22* Insect and Mite Pests of Ornamental Plants . . . 449

*Chapter 23* Nematode Pests of Fruit Crops . . . 469

*Chapter 24* Nematode Pests of Vegetable Crops and Ornamental Plants . . . 491

*Chapter 25* Rodent Pests of Horticultutal Crops . . . 506

*Chapter 26* Birds as Pests of Horticultural Crops . . . 519

*Chapter 27* Pesticide-Honeybee Interactions . . . 527

*Subject Index* . . . 541

# Contributors

**Arora, Ramesh** Department of Entomology, Panjab Agricultural University, Ludhiana 141 004, India.

**Chandla, VK** Division of Entomology-Nematology, Central Potato Research Institute, Shimla 171 001, India.

**Chauhan, Usha** Department of Entomology & Apiculture, Dr. Y.S. Parmar University of Horticulture & Forestry, Nauni, Solan 173 230, India.

**Devi, Nirmala** Department of Entomology, CSK Himachal Pradesh Agricultural University, Palampur 176 062, India.

**Dhaliwal, GS** Department of Entomology, Panjab Agricultural University, Ludhiana 141 004, India.

**Gautam, DC** Department of Biosciences, Himachal Pradesh University, Shimla 171 005, India.

**Gupta, Divender** Department of Entomology & Apiculture, Dr. Y.S. Parmar University of Horticulture & Forestry, Nauni, Solan 173 230, India.

**Gupta, PR** Department of Entomology & Apiculture, Dr. Y.S. Parmar University of Horticulture & Forestry, Nauni, Solan 173 230, India.

**Kashyap, NP** Department of Entomology, CSK Himachal Pradesh Agricultural University, Palampur 176 062, India.

**Khajuria, DR** Regional Horticultural Research Station, Dr. Y.S. Parmar University of Horticulture and Forestry, Seobagh, PO Neoli, Kullu 175 138, India.

**Lakra, RK** Department of Entomology, CCS Haryana Agricultural University, Hisar 125 004, India.

**Mehta, PK** Department of Entomology, CSK Himachal Pradesh Agricultural University, Palampur 176 062, India.

**Narang, ML** Department of Entomology & Apiculture, Dr. Y.S. Parmar University of Horticulture & Forestry, Nauni, Solan 173 230, India.

**Nath, A** Department of Entomology & Apiculture, College of Horticulture, Dr. Y.S. Parmar University of Horticulture & Forestry, Nauni, Solan 173 230, India.

**Raj, Des** Department of Entomology, CSK Himachal Pradesh Agricultural University, Palampur 176 062, India.

**Sachan, GC** Department of Entomology, G.B. University of Agriculture and Technology, Pantnagar 263 145, India.

**Sharma, HC** International Crops Research Institute for the Semi-Arid Tropics (ICRISAT), Patancheru 502 324, India.

**Sharma, JP** Regional Horticultural Research station, Dr. Y.S. Parmar University of Horticulture & Forestry, Seobagh, PO Neoli, Kullu 175 138, India.

**Sharma, KC** Department of Entomology & Apiculture, Dr. Y.S. Parmar University of Horticulture & Forestry, Nauni, Solan 173 230, India.

**Sharma, NK** Department of Entomology & Apiculture, Dr. Y.S. Parmar University of Horticulture & Forestry, Nauni, Solan 173 230, India.

**Sheikher, Chander** Department of Entomology & Apiculture, Dr. Y.S. Parmar University of Horticulture & Forestry, Nauni, Solan 173 230, India.

**Singh, Darshan** Department of Entomology, Punjab Agricultural University, Ludhiana 141 004, India.

**Singh, Gursharan** Department of Entomology, Punjab Agricultural University, Ludhiana 141 004, India.

**Srivastava, RP** Division of Entomology, Central Institute of Horticulture for Northern Plains, Indira Nagar, Lucknow 226 016, India.

**Thakur, JR** Regional Horticultural Research Station, Dr. Y.S. Parmar University of Horticulture & Forestry, Bajaura, Kullu 175125, India.

**Verghese Abraham** Institute of Horticultural Research, Hessarghatta Lake Post, Bangalore 560 089, India

**Verma, KD** Division of Entomology-Nematology, Central Potato Research Institute, Shimla 171 001, India.

**Verma, LR** Vice-Chancellor, Himachal Pradesh University, Shimla 171 005, India.

**Walia, RK** Department of Nematology, CCS Haryana Agricultural University, Hisar 125 004, India.

Part

# I

# Horticultural Pest Management: Principles

*Pest Management in Horticulture Crops*
L.R. Verma, A.K. Verma and D.C. Gautam (eds)


*Chapter* 1

# Ecological Basis of Insect Pest Management: Issues and Strategies

*Ramesh Arora and G.S. Dhaliwal*

## INTRODUCTION

INSECT PEST PROBLEMS in agriculture are probably as old as agriculture itself. For thousands of years, humans could do nothing about these pests except to appeal to the power of magic and a variety of Gods. Gradually, farmers developed a number of physical, mechanical and cultural control practices largely through trial and error experiments (Smith *et al.*, 1976). These control practices, still useful and valid today, include sanitation, tillage, crop rotations, sowing dates to avoid high damage prone periods, use of pest free seeds, collection and mechanical destruction of gregarious stages of pests, barriers to prevent entry of insects, etc. The appropriate use of these methods reduced the damage potential of essentially all pests and provided satisfactory control of some. However, many of the highly damaging pests could not be adequately controlled with any combination of the known methods. As biological knowledge grew during the eighteenth and nineteenth centuries and pest problems became more severe due to intensification of agriculture and introduction of pests into new areas, human beings tried to look for more effective pest control measures (Smith, 1978).

Synthetic organic insecticides developed during the mid-twentieth century provided spectacular control of insect pests and resulted in the abandonment of traditional pest control practices. Ecological and environmental problems caused by large scale indiscriminate use of these pesticides led to the ecologically-based concept of integrated pest management (IPM). The agricultural pest control has, thus, passed through three distinct phases, viz., the era of traditional approaches, the era of pesticides and the era of IPM (Dhaliwal *et al.*, 1998).

### Era of traditional approaches (Ancient-1938)

Cultural and mechanical practices like crop rotation, field sanitation, deep ploughing, flooding, collection and destruction of damaging insects/insect infested plants, etc., developed by farmers through experience were among the oldest methods developed by humans to minimize the damage caused by insect pests (Smith *et al.*, 1976). This was followed by the use of plant products from neem, chrysanthemum, rotenone, tobacco and several other lesser known plants in different parts of the world. The Chinese were probably the pioneers in the use of botanical pesticides as well as biological control methods for the management of insect pests of stored grains and field crops (Dhaliwal and Arora, 1994). However, systematized work on many important tactics of pest control including the use of resistant varieties, biological control agents and botanical and inorganic insecticides was done in the USA from the end of the 18th to the end of 19th century. Remarkable success was achieved in the management of grape phylloxera caused by *Vitis vitifoliae* (Fitch) by grafting of European grape vine scions to resistant North American rootstocks during the 1880s. At around the same time, cottony cushion scale, *Icerya purchasi* Maskell which was causing havoc to the citrus industry in California (USA) was successfully controlled by release of the Vedalia beetle, *Rodolia cardinalis* (Mulsant) imported from Australia (Dhaliwal and Arora, 1996).

A number of synthetic inorganic insecticides containing arsenic, mercury, tin and copper were also developed towards the end of the 19th and the beginning of twentieth century. With the development of these insecticides, the focus of research in entomology slowly shifted from ecological and cultural control to chemical control, even before the development of synthetic organic insecticides (Perkins, 1980).

**Era of pesticides (1939-1975)**

The inorganic synthetic insecticides were broad spectrum biocides and were highly toxic to all living organisms. These were followed, in due course, by the synthetic organic insecticides like alkyl thiocyanates, lethane, etc. The era of pesticides, however, began with the discovery of DDT [2,2-(p-chlorophenyl)-1,1,1-trichloroethane] by Paul Muller in 1939. The impact of DDT on pest control is perhaps unmatched by any other synthetic substance. Muller was awarded Nobel Prize for this work in 1948.

DDT was soon followed by a number of other insecticides like HCH, chlordane, aldrin, dieldrin, heptachlor, toxaphene (organochlorine group); parathion, schradan, EPN (organophosphorus group) and allethrin (synthetic pyrethroid) during the 1950s and a large number of other popularly used organophosphates and carbamates in the ensuing decade.

Due to their efficacy, convenience, flexibility and economy, these pesticides played a major role in increasing crop production. The success of high yielding varieties of wheat and rice that ushered in the 'green revolution' was partially due to the protection umbrella of pesticides (Pradhan, 1983). The spectacular success of these pesticides masked their limitations. The intensive and extensive use, misuse and abuse of pesticides during the ensuing decades caused widespread damage to the environment. In addition, insect pest problems in some crops increased following the continuous application of pesticides. This, in turn, further increased the consumption of pesticides resulting in the phenomenon of the pesticide treadmill (Altieri, 1995). The combined impact of all these problems together with the rising cost of pesticides provided the necessary feedback for limiting the use of chemical control strategy and led to the development of the IPM concept.

### Era of IPM (1976 onwards)

To overcome increasing problems associated with the strategy of exclusive and indiscriminate use of pesticides, Stern *et al.* (1959) propounded the concept of integrated control. The use of term 'pest management' was advocated by Geier and Clarke (1961). But the concept of IPM really came of age at the XV International Congress of Entomology during 1976 held at Washington, DC. IPM represents ecologically sound approaches to pest control and automatically rejects the usage of broad spectrum and persistent biocides (Metcalf, 1980).

## CONCEPT OF PEST MANAGEMENT

Though it is not generally recognized, evolution of the concept of pest management spans a period of almost a century. Many components of integrated pest management were developed in the late 19th and 20th century. The rapidly developing technologies and changing social values had their impact on the pest control tactics also (Dhaliwal and Arora, 1994). IPM is now acclaimed as a means for the development of ecologically sound and sustainable agricultural practices to mitigate crop production losses and for strengthening farming communities' participation in that development (Heinrichs, 1998).

Stephen A. Forbes, State Entomologist of Illinois (USA) was probably the pioneer in the application of ecological studies for dealing with insect problems in agriculture (Metcalf, 1930). A number of others concerned with crop protection also advocated this fundamental approach. By the early 1920s, a highly complex and sophisticated system involving the use of multiple component suppression techniques, viz., resistant varieties, sanitation practices and chemical treatments with calcium arsenate at fixed population levels, was already developed for cotton boll weevil control in the USA (Hunter and Coad, 1923).

The discontent with the purely insecticidal approach and early success of biological control method resulted in the origin of the concept of integrated control during 1950s (Michelbacher and Bacon, 1952; Smith and Allen, 1954; Stern *et al.*, 1959). The integrated pest control (IPC) emphasized the selective use of insecticides, so that natural enemies were conserved in the agroecosystem. This integration was expanded in later years to include other control techniques (Smith and Reynolds, 1966). The idea of managing insect pest populations was proposed by Geier and Clarke (1961) who called this concept 'protective management of noxious species' or 'pest management'. IPM refers to an ecological approach in pest management in which all available techniques are consolidated in a unified programme, so that pest populations can be managed in such a manner that economic damage is avoided and adverse side effects are minimized (NAS, 1969). IPM is a systematic approach to crop protection that uses increased information and improved decision-making paradigms to reduce purchased inputs and improve economic, social and environmental conditions on the farm and in society (Allen and Rajotte, 1990).

### Ecological factors regulating pest numbers

Pest populations like all animal populations are governed by their innate capacity to increase as influenced by various abiotic and biotic factors. The abiotic factors include temperature, moisture, light, etc., while biotic factors are comprised of food, intra- and inter-specific competitions as well as natural enemies of pests. The biotic factors are greatly influenced by human activities like agriculture and industrialization. The intensification of agriculture utilizing high yielding varieties grown under new agronomic practices with increased irrigation facilities and extensive use of agro-chemicals, has also resulted in increasing incidence and outbreaks of a number of insect pests during the last three decades (Pimentel, 1997; Bramble, 1989; Dhaliwal and Arora, 1994).

### *Biotic factors*

*Food:* Phytophagous insects compete with humans for food supply from our agroecosystem and are, therefore, labelled as pests. Survival, development rate and multiplication capacity of insect pests are all determined by the quantity and quality of food. An insect population would increase in numbers and attain the status of a pest in case sufficient supply of a suitable host is available. The differences in varieties/species as regards their suitability for a particular insect population are governed by their recognition and acceptability as host, nutritional adequacy and absence of inhibitors/toxicants for that pest species (Atwal and Singh, 1990). The modern agriculture utilizing large scale monocultures of a few high yielding varieties of important crops provides an almost inexhaustable source of food supply for insect pests feeding on these crops and has, therefore, greatly contributed to increasing outbreaks of insect pests (Arora and Dhaliwal, 1996).

*Natural enemies:* The role of natural enemies (parasitoids and predators) in the regulation of population of forest defoliators, gypsy moth, *Porthetria dispar* (Linnaeus), and brown-tail moth, *Euproctis chrysorrhoea* (Linnaeus), was emphasized by Howard and Fiske (1911). They termed these factors as facultative, i.e., those causing increased percentage mortality with increase in insect density. The importance of entomophagous insects and pathogens in regulation of pest populations was also emphasized by Smith (1935), who called these factors as density-dependent factors.

Later on, Nicholson in a series of articles (1927 to 1958) propounded his population theory, which emphasized the role of intra-specific competition and natural enemies in regulation of pest numbers (Clark *et al.*, 1967). Populations of most insects are generally kept in check by their natural enemies. Indiscriminate use of broad spectrum conventional insecticides results in mortality of these 'friends of farmers'. This creates congenial conditions for resurgence of target pests or outbreaks of secondary/potential pests.

### *Abiotic factors*

Climatic or abiotic factors have a dominating influence on the survival, development and reproductive capacity of insect pests and were considered to be the most important cause of fluctuations in animal populations by Elton (1927). The role of climatic factors was further highlighted by Andrewartha and Birch (1954). On the basis of detailed investigations into population dynamics of a swarm-forming grasshopper, *Austroicetes cruciata* Sauss, and the apple blossom thrips, *Thrips imaginis* Bagnall, these authors found that the populations of these and probably other species were almost entirely governed by climatic factors.

*Temperature:* All life survives within a certain narrow range of temperature. Departure from this optimum range on both sides is tolerated to some extent, depending upon the physiological adaptations of the concerned species or populations. Temperatures above or below these limits prove lethal. Exposure to lethal high or low temperature may result in instant killing or failure to grow and reproduce normally. Harmful effects of exposure to sub-lethal temperatures may be manifested at some later critical stage like ecdysis or pupation. Some of the insect species have developed dormancy mechanisms like diapause, hibernation and aestivation to tide over periods of unfavourable temperatures in their life cycles (Atwal and Singh, 1990).

*Moisture:* Most terrestrial insects live in an environment, which is dry. The only source of water for phytophagous insects is the water obtained with food material from their host plants. These insects have, therefore, developed a variety of mechanisms to conserve water. Inspite of these mechanisms, exceptionally dry air may prove lethal to most insects. Likewise, excessive moisture may also adversely affect many insects by encouraging disease outbreaks, affecting normal development and by lowering their capacity to withstand low temperature. The reproductive capacity of most insects is also affected by moisture but there are great differences in the capacity of different insects to tolerate conditions ranging from extreme dryness to near saturated environments. Besides temperature and moisture, a number of other abiotic factors like light, atmospheric pressure, air current and carbon dioxide concentration also influence insect abundance to a lesser extent.

Both biotic and abiotic factors generally act simultaneously on the pest. According to Milne (1962), the insect populations increase due to the combined action of density independent (abiotic) and imperfectly density dependent (interspecific competition, natural enemies, etc.) environmental factors. Only in rare cases, intraspecific competition for food, space, etc., checks the increase in population to the point of collective suicide. It has been reported that abiotic factors are limiting towards the margins of the distribution range of a pest species. In the middle of the distribution range, where climatic factors are favourable, the population increase is checked by the biotic factors (Price, 1984).

### *Genetic factors*

Chitty (1960), on the basis of his studies on voles, *Microtus agrestis* proposed that all species were capable of regulating their own population densities without destroying the renewable resources of their environment or without requiring natural enemies and unfavourable weather to keep them from doing so. In the case of western tent caterpillar, *Malacosoma californicum pluviale* (Dyar), the activity and vitality of larvae decreased as the abundance of individuals in the colony increased. During harsh climatic conditions, individuals in marginal habitats were killed and those in favourable habitats survived to recolonize all the vacant areas as conditions permitted (Wellington *et al.*, 1975).

## MODERN AGRICULTURE, AGROECOLOGICAL CHANGES AND PEST PROBLEMS

Rapidly increasing human population during the last century has necessitated intensification of agriculture through expansion of irrigation facilities, introduction of high yielding varieties, and application of increased amounts of agrochemicals. Besides, the cultural practices like spacing, crop rotation, sowing times and tillage methods have also been modified to achieve maximum productivity per unit time from the available land. While undoubtedly contributing to higher yields, this technology package has resulted in aggravation of insect pest outbreaks in agricultural crops (Bramble, 1989; Dhaliwal and Arora, 1993; Sharma and Veerbhadra Rao, 1993; Arora and Dhaliwal, 1996).

**High yielding varieties**

When altering the genetic make up of the crop plant to increase yield with little or no attention given to pest attack, natural resistance may be lost or greatly reduced. Efforts to breed plants more palatable to human taste by elimination of such factors as bitterness or hairiness may also result in the development of more susceptible varieties (Pimentel, 1997). The fields planted with modern varieties, which are short and heavy tillering, develop a distinctly different microclimate conducive for proliferation of insect pests (Pathak and Dhaliwal, 1981).

**Monoculture**

In agriculture, the natural plant community is removed, destroyed and usually replaced by a single crop species. Larger the area that is planted with a single crop, the greater is the potential for pest problems. Monocropping of vegetable crops in selected production centres especially along the periphery of large cities has been an important factor in the increasing outbreaks of many insect pests (Srinivasan and Krishnamoorty, 1993).

Even within the same crop, modern plant breeding with emphasis on inbred, uniform strains has fostered a widespread trend towards genetic erosion, which is depleting the genetic base of many crop plants. In paddy, where we have the world's most diverse genetic base with thousands of traditional varieties, two thirds of the area is under a few high yielding varieties, which in turn are derived from a fewer number of parents. All these changes have made our agroecosystem very susceptible to insect pests.

**Fertilizers**

Excessive use of inorganic nitrogenous fertilizers generally creates congenial conditions for rapid multiplication and subsequent outbreaks of many insect pests. A significant increase in the incidence of most of the insect pests, including yellow stem borer, leaf folder, gall midge, green leafhopper, whitebacked planthopper, earhead bug, rice hispa, whorl maggot, etc., in rice crop, has been recorded with increasing application of nitrogenous fertilizers. A similar situation prevails in other crops also. On the other hand, application of potash and sometimes phosphorus also, either singly or in combination with nitrogen, results in lower incidence of many insect pests. Aphid and thrips incidence on chillies decline with increase in potash application. Higher dosages of phosphorus and potash reduce the attack of *Empoasca* sp., *Therioaphis trifoii* f. *maculata* (Buckton) and *Aphis craccivora* Koch on lucerne (Dhaliwal and Arora, 1993).

**Pesticides**

Extensive and indiscriminate use of synthetic organic pesticides has also resulted in aggravation of insect pest problems in a number of crops. This, in turn, has further increased the consumption of pesticides. Taking the example of cotton in Punjab, the cost of insecticides as percentage of total cost of cultivation of the crop increased from 2.1 per cent in 1974-75 to 4.6 per cent in 1979-80, 11.9 per cent in 1984-85 and 15.5 per cent in 1989-90. While the cost of cultivation has increased little over three times from 1974-75 to 1989-90, the cost of insecticides used has multiplied over 21-fold during this period. Inspite of this phenomenal incease in the use of insecticides, the losses due to insect pests have been increasing steadily. This is due to killing of natural enemies, resurgence of insect pests, development of insecticide resistance and outbreaks of secondary pests, etc. (Arora and Dhaliwal, 1996).

**Irrigation**

With the expansion of surface - and ground-water irrigation facilities, the share of irrigated area in total cultivated area has been increasing slowly. While only about one-third of the cultivated area in the country is irrigated, the states of Punjab and Haryana have more than 80 per cent area under irrigation. The expansion of irrigation facilities has also contributed to an aggravation of insect pest problems in many crops. In rice, it has been reported that draining the fields and maintaining irrigation water to field capacity tends to inhibit the multiplication of brown planthopper and gall midge (Chelliah, 1987). Water-logging enhances the multiplication of a number of borers including stalk borer, internode borer, Plassey borer and the whitefly in sugarcane crop. One of the important factors in the recent heavy build up of top borer in Uttar Pradesh and Bihar, has been the increased irrigation facilities (Varma *et al*., 1993). Similarly, the multiplication of armyworm on wheat has been found to be positively correlated with the number of irrigations (1 to 12) per plot (Deol *et al*., 1987).

The amount of moisture in the soil has a great effect on the longevity and fecundity of the mustard aphid. maximum fecundity was obtained when the water level was maintained continuously above the field capacity (Arora and Sidhu, 1982).

**Cultural practices**

New agronomic recommendations, which are essentially meant for exploiting the full genetic yield potential of high yielding varieties, are simultaneously favourable to most of the insect pests, diseases and weeds for their attack, multiplication and spread (Rajak, 1986).

The close plant spacings currently recommended in many crops help in the build up of many insect pests. In rice, close spacing resulted in higher build up of brown planthopper, gall midge, leaf folder, stem borer and whitebacked planthopper (Chelliah and Bharathi, 1985; Velusamy, 1987; Yein and Das, 1988). Larval population of pod borer on chickpea was 4 times as large at closest (33 plants/m2) as at the widest (3 plants/m2) spacing. There was little advantage of more than 8 plants/m2 in chemically protected crop (Lal *et al*., 1985). Similarly, higher incidence of a number of insect pests has been reported in closely planted crops of cotton, sugarcane, soybean, etc. Another fallout of intensive agriculture is that in large tracts cereals are followed by cereals and pulses by pulses. Paddy-wheat rotation has resulted in increased incidence of stem borer, pink borer and shootfly. Cotton-sunflower rotation is conducive for the multiplication of the dreaded *Helicoverpa armigera* (Hubner) as well as many of the sucking pests, which attack both these crops.

Modification or abandoning of many other cultural practices like time and manner of planting, growing of intercrops, trenching, harvesting time and procedures, has also contributed to an aggravation of insect pest problems in modern agriculture.

## CONCEPT OF INJURY LEVELS

Most crops harbour a wide variety of insects. Crops like cotton and sugarcane have been reported to be attacked by more than 1000 species of insects worldwide. Potentially, most of these insects could cause havoc but actually only a few of them damage the crop. The critical factor which determines the damaging capacity or otherwise of an insect is its population level. The concept of injury levels was propounded to enable us to identify the population level at which an insect could cause damage to a crop. Most popular terms used in this connection are the economic injury level (EIL) and the economic threshold level (ETL). But a number of other terms like action threshold, action level, threshold level, inaction threshold, control threshold, insect injury threshold, critical injury threshold and critical population threshold have been suggested as alternatives to EIL and ETL (Pedigo *et al*., 1986).

### Economic injury level

Insect colonization and feeding often cause– injury to the plants. The injury does not necessarily result in damage. The latter refers to a measurable loss of host ability most often including yield quantity, quality or aesthetics. The lowest level of injury where the damage can be measured is called the damage boundary (DB), while the lowest number of insects that will cause economic damage is referred to as economic injury level (EIL) (Pedigo, 1996).

Even before the advent of synthetic organic insecticides, questions were raised regarding the level of damage at which control measures should be initiated against insects. However, the concept of EIL was developed by Stern *et al*. (1959) to overcome problems, viz., insecticide resistance, pest resurgence, insecticide residues and non-target effects, caused by indiscriminate use of DDT and other broad-spectrum synthetic organic insecticides.

Although expressed as number of insects per unit area, as the name implies, the EIL is, in reality, a level of injury. Because injury is difficult to measure in a field situation, therefore, number of insects is used as an index of that injury. There have been suggestions that the name should be changed to critical population density. Alternatively, it may be more useful to express the EIL in standard units of injury. The standard units of injury are the injury equivalents, i.e., amount of injury that could be produced by one pest through its complete life cycle; and equivalency, i.e., total injury equivalents (for a population) at a point of time (Pedigo *et al*., 1986). If management action (insect suppression) can be taken quickly and loss averted completely, the EIL may be expressed as follows (Pedigo, 1996):

$$EIL = C/VID \tag{1}$$

EIL = No. of injury equivalents per production unit
C = Cost of management activity per unit of production
V = Market value per unit of product
I = Crop injury per pest density
D = Damage per unit injury

These primary variables are affected by a number of complex variables. In instances, where some loss from the insect is unavoidable, the relationship becomes:

$$EIL = \frac{C}{V \times I \times D \times K} \qquad (2)$$

where K represents proportionate reduction in injury (e.g., 0.6 for 60%)

**Economic threshold level**

Economic threshold level (ETL) is the best known and most widely used index in making pest management decisions. It is defined as the population density at which control measures should be initiated against an increasing pest population to prevent economic damage. Although expressed in insect numbers, ETL is, in fact, a time parameter with pest numbers being used as an index for when to implement management strategies. Just as with EILs, ETLs also can be expressed in insect equivalents (Pedigo, 1996). In economic terms, ETL is defined as the level to which a given pest population should be reduced to achieve the point where marginal revenue just exceeds marginal costs (Mumford and Norton, 1984).

ETL is a complex value based on the EIL, population dynamics of the pest, weather forecasting and the pest's potential for injury. When no action is taken at ETL, the population exceeds EIL, whereas, when management steps for pest suppression are taken as the population crosses ETL, the population is forced down before it could reach EIL.

***Determination of economic threshold***

The economic threshold may be determined experimentally as outlined below (Reichelderfer *et al.*, 1984):

- For a range of pest densities, including zero pests, measure yield and quality of the crop by means of controlled experiments.
- For each management practice to be analysed, measure yields and total crop revenues in the same type of experiments as above except that management actions are taken at each of the possible pest densities.
- Total crop revenue is computed for each management action at each pest density by multiplying yield by price per unit of output.
- Substract cost of each management action from crop revenue for that action at each of the initial pest densities to obtain net revenues.
- Beginning at very high pest densities and moving to lower densities, compare net revenues of taking a management action with taking no action. The pest density where the net revenues under controlled and uncontrolled conditions are equal is the economic threshold level.
- Alternatively, the cost of management action is compared with the marginal crop revenues obtained when progressively smaller pest densities are subjected to treatment. The pest density where marginal crop revenue is equal to the management action cost, is also the economic threshold. Marginal crop revenue is the difference in crop revenue from taking the action minus the crop revenue of not taking the action of each pest density.

The concept of EIL and ETL gained wide acceptability from the time it was presented. However, implementation of the concept in practice presents several problems. This is due to a number of serious limitations in the concept. These limitations and some of the strategies suggested to overcome them are discussed below (Pedigo, 1996):

- The names EIL and ETL are themselves misleading because both are defined in terms of population densities, while former represents an injury level and the latter the time for taking control measures. The limitation may be overcome by defining these levels in terms of injury equivalents (Pedigo *et al.*, 1986). Moreover, it would then be possible to describe the same type of injury for many pest species.
- There is a lack of rigorous definition of economic damage (the amount of injury that will justify the cost of control). Because economic damage was not described mathematically in terms of its components, it could not be assessed solely on the basis of definition given by Stern *et al.* (1959). EIL had to be calculated from ED and it also could not be established. The inadequate definition of EIL delayed the acceptance and precise calculation of EILs.
- The EIL concept overlooks the influence of other production factors that can affect the crop/pest system. The externalities left out include interseasonal dynamics, biological relationships with other pests and natural enemies, environmental contamination by pesticides, resistance to pesticides, effect of control in neighbouring fields and health problems relating to pesticides.
- Decision levels for management of some types of pests cannot be determined with EILs. Besides medical and veterinary pests, it includes most vectors. It is very difficult to place a monetary value on the reduction in aesthetic value associated with a given type of injury. A similar problem exists with respect to forest pests. Almost all components of EILs are difficult to estimate for forest pests; accurate market values are a problem; management costs may vary greatly; frequently include mere environmental and social costs, and the injury-crop response relationships may be difficult to determine because the growth of the crop spans many years.
- The concept is unsuitable in case of attack of multiple pests on a single crop at the same stage.

However, in spite of limitations, the EIL concept continues to offer a practical approach to pest-related decision making in a broad sense.

## Environmental economic injury levels

The challenge of attempting to decrease pesticide inputs further can be met by developing environmentally based EILs and their concomitant ETLs. An environmental EIL is an EIL that evaluates a management tactic based on not only its direct costs and benefits to the user but also its effect on the environment. The EIL equation (equation 2) integrates many management elements, each of which may have a role in making pest management most environmentally sustainable (Pedigo and Higley, 1992).

### *Assigning realistic management costs (C)*

Component C of the EIL equation represents costs associated with taking management action against a pest population, and increased costs cause EIL to increase proportionately. Generally, C does not take into account the environmental costs associated with environmental risks; it is possible to include these costs in variable C of EIL. Methods for computing the environmental costs (EC) of pesticides are discussed in elsewhere. These are utilized for calculating the environmental economic injury levels (En. EIL).

$$\text{En. EIL} = \frac{PC + EC}{VDIK} \tag{3}$$

***Manipulating crop market value (V)***

This could be achieved by putting a higher market value for a pesticide-free produce. The extent of increase would depend on the consumers willingness to pay for a safer product.

***Reducing damage per pest (D)***

Reducing D implies that less loss of yield occurs for a given amount of injury. This is possible if plant is able to tolerate and compensate for injury. Plants that can tolerate or compensate for injury do not place selection pressure on pest populations. Therefore, the benefits of tolerance and compensation in plant are sustainable and permanent. Even partial tolerance will increase EILs (by decreasing D). The need for pesticides and the risk to environment will be reduced correspondingly.

***Developing environmentally responsible K value***

Modified K is the proportion of total pest injury averted by timely application of a management tactic. Increasing the EIL to improve environmental quality implies that we are willing to tolerate more pests. But this is not always the case. By reducing D or K, EIL can be increased even without causing increased losses or costs.

## INTEGRATION OF PEST MANAGEMENT TACTICS

The pest management tactics are either preventive or therapeutic. Preventive practices utilize tactics to lower environmental carrying capacity (reduce the general equilibrium position) or increase tolerance of the host to pest injury. Prevention relies on an intimate understanding of the pest's life cycle, behaviour and ecology. The preventive tactics involve natural enemies, host resistance and cultural practices. In addition, quarantines are also an important component of preventive tactics. Therapeutic tactics are applied as a correction to the system when necessary. The objective of therapy is to dampen pest population below the economic injury level. The only widely used therapeutic tactic is the use of conventional insecticides but other approaches like microbial agents, augmentation of natural enemies, use of growth regulators, etc., may also play a vital role (Pedigo, 1996).

Actual integration involves proper choice of compatible tactics and blending them so that each component potentiates or complements the other. Probably, the earliest example of integration of techniques was the use of a combination of resistant varieties and sanitation practices as prophylactic measures combined with application of calcium arsenate at high population level in case of boll weevil on cotton in the USA during first quarter of the twentieth century (Hunter and Coad,

1923). Similar programmes were being developed for other pests also but the advent of synthetic organic insecticides intervened and these techniques were relegated to the background. The misuses and abuses of insecticides have again focused our attention on integrated control measures. Some of the possible ways in which different tactics may be integrated are illustrated below with suitale examples.

**Cultural control**

The manipulation of cultural practices for reducing or avoiding pest damage to crops is known as cultural control. Cultural practices include all the crop production and management techniques which are utilized by the farmers to maximize their crop productivity and/or farm income. It includes decisions on crops/varieties to be grown, time and manner of planting, tillage, field and crop sanitation, application of irrigation and fertilizers, harvesting times and procedures, and even off-season operations in fallow/cropped fields. Many of these practices can be fitted into the IPM programme.

High plant nitrogen levels are generally favourable to pest insect populations due to greater survival, increased feeding rate and increased fecundity resulting in faster population growth. A study on the effect of nitrogen fertilizer on the development of brown planthopper on rice indicated that nitrogen favoured population growth, regardless of the level of resistance in the host plant; but the increase was limited at the higher levels of resistance as indicated by the highly resistant variety IR60 (Heinrichs, 1994). This indicates the need for planting resistant cultivars when high nitrogen levels are needed to maximize production.

The efficacy of insecticides against insect pests can be increased by manipulation of cultural practices. The mustard aphid, *Lipaphis erysimi* (Kaltenbach), feeding on potted *Brassica* plants grown at lower soil-water regimes were found to be more susceptible to insecticides than those from higher soil-water regime (Arora and Sidhu, 1991). In field studies, lower dosages (250 and 500 ml/ha) of oxydemeton methyl were at par with higher ones (750 ml and 1 l/ha) after 2 and 15 days of spraying, in case of Indian mustard crop growing under water stress conditions. There was no difference in yield of crop under water stress and non-stress conditions (Brar *et al*., 1991). In another study, normal sown crop of *B. juncea* (RLM 619) and *B. campestris* (BSH 1) required one spray for the control of mustard aphid under no water stress conditions. But under moderate and high water stress conditions, no spray was required for aphid control on these crops (Narang *et al*., 1993). Similarly, application of potash has also been reported to increase the efficacy of insecticides against mustard aphid. In case of potted *B. campestris* and *B. juncea* plants, the relative susceptibility of mustard aphid to dilmethoate, endosulfan, carbaryl and fenvalerate increased with increase in rate of application from 50 to 100 and 200 ppm. Application of 200 ppm potash increased the susceptibility of the aphid to dimethoate by 2.83 to 3.27 times on *B. compestris* and *B. juncea* plants, respectively (Singh, 1987).

One of the greatest successes achieved in recent times in the development of IPM has been the integrated use of trap cropping and botanical pesticides for the control of diamondback moth (DBM) and other pests of cabbage. Growing of Indian mustard in paired rows at the beginning and after every 25 rows attracted more than 80 per cent of DBM infestation besides almost entire population of leafwebber, stem borer, bugs and aphids. To control the remaining attack of DBM, a 4 per cent neem seed kernel extract (NSKE) is applied at primordial or head initiation stage of the crop (Srinivasan, 1994). Two or three post-heading sprays of NSKE may also be applied at 10-15 days intervals depending on population pressure. Spraying of NSKE is also safe to *Cotesia plutellae* (Kurdyumov) which is the dominant natural enemy found in the cabbage ecosystem.

The entomologists at the Indian Institute of Horticultural Research, Bangalore have also been successful in developing an IPM system for the control of tomato fruit borer, *Helicoverpa armigera* (Hubner). Trap cropping of marigold after every 8 rows of tomato attracts most of the ovipositing moths to the former crop. The use of conventional insecticides on the trap crop reduces their attractiveness to the pest. Therefore, the pest on the trap crop has to be removed mechanically. The residual pest population on both the crops is controlled by sprays of *H. armigera* nuclear polyhedrosis virus (HaNPV) at 500 larval equivalents (LE) per hectare.

### Host plant resistance

Host plant resistance (HPR) represents the ability of crop plants to restrict, retard or overcome pest infestations (Kumar, 1984). HPR is usually compatible with other tactics of pest management. There are many examples of host plant resistance enhancing the efficacy of insecticides. Evaluations of rice insecticides indicate that they cause higher pest mortality of planthoppers and leafhoppers feeding on resistant than on susceptible rice varieties (Heinrichs, 1994). Mortality of brown planthopper when reared on either a moderately resistant ASD7 or a highly resistant cultivar Sinna Sivappu was higher than when feeding on a susceptible TN1 cultivar. The integration of host plant resistance and insecticides has cumulative effect on *Nephotettix virescens* (Distant), the vector of rice tungro virus. There was no tungro virus infection on the resistant cultivar IR28 even without the application of insecticides. On the moderately resistant cultivar, IR36, tungro virus infected plants decreased from 42 per cent in the control treatment to 10 per cent in the 0.5 kg a.i./ha rate. On the susceptible cultivar, IR22, virus infection was high at all insecticide rates decreasing from 92 per cent in the control to 74 per cent in 1 kg a.i./ha rate.

Pest resistant varieties are usually compatible with biological control, but may sometimes have an adverse effect on natural enemies. By reducing pest numbers, resistant varieties help to shift the pest: predator ratios in favour of biological control (Heinrichs, 1994). In field studies at the International Rice Research Institute, Philippines, the brown planthopper, *Nilaparvata lugens* (Stal): spider, *Lycosa pseudoannulata* (Bosenberg and Strand) ratios increased with the level of susceptibility from ASD7 and IR36, both highly resistant rice cultivars to IR42 and Triveni, moderately resistant cultivars to IR8 and TN1, susceptible cultivars.

The host plant resistance may also enhance the predatory activity. Predation rate of the mirid bug, *Cyrtorhinus lividipennis* Reuter when feeding on the first instar *N. lugens* nymphs increased on the resistant cultivar, IR36 as compared to the susceptible IR8. Combinations of host plant resistance and predation by the mirid bug, *C. lividipennis* have a cumulative effect on the population increase of the green leafhopper, *Nephotettix virescens* (Distant). In cage studies, the number of green leafhopper reached only 6 on IR29 (resistant) with the predator and 31 without the predator, while there were 91 and 220 hoppers, respectively, on susceptible IR22.

In case of sugarcane, tightness of leafsheath and increased hardness of stalk tissues conferring resistance to the borer, *Diatraea saccharalis* (Fabricius) prolong the time during which the pest remains in an exposed situation, vulnerable to the action of natural control agents.

Another successful example of integration of resistant varieties and biological control has been reported from Texas, USA for the management of Rhodes grass scale, *Antonina graminis* (Maskell), on range grasses. A parasitoid, *Neodusmetia sangwani* (Rao), was successfully established against the pest in Texas. However, during the June peak population of the pest, natural enemy population was lower due to general host unsuitabililty during May. The resistant Rhodes grass variety 'Bell' was found to tolerate pest attack without appreciable damage. This variety, thus, provided relief until the parasitoid effectively reduced scale populations (Schuster and Dean, 1973). The predation rate was highest on resistant cultivars and this was attributed to the greater movement of hoppers in search of suitable feeding sites (Kartohardjono and Heinrichs, 1984).

**Biological control**

Biological control refers to the use of natural enemies (parasitoids, predators, pathogens) for the management of insect pests. Control by natural enemies can be greatly strengthened by the use of a large number of cultural practices like intercropping, trap cropping, strip harvesting, etc. Modification of the crop environment by manipulation of irrigation or fertilizer application, row spacing, seed rate and tillage operations, etc., may also lead to enhanced biological control.

In Pakistan, an integration of cultural control practices and natural enemies is used in over 30,000 ha of Habib Sugar Mills and other companies. Field sanitation is used to minimize the carryover of stem borer population from the leftover canes. Egg parasitoids, *Trichogramma chilonis* Ishii are released from March to June. White grubs, *Pachnessa nicobarica* are controlled by early (May) harvesting of heavily infested fields, followed by 2-3 deep ploughings. This practice causes heavy mortality of white grub pupae. Pyrilla is controlled by release of *Epiricania melanoleuca* (Fletcher) in heavily infested fields. Interestingly, other natural enemies also appear to be attracted by the presence of *E. melanoleuca*. About 95 per cent parasitism by *Parachryrocharis javensis* was observed in fields where *E. melanoleuca* was released as compared to only 20 per cent in the fields where no parasitoids were released (Mohyuddin, 1994).

An IPM system based on a combination of biological control and cultural practices has been developed for the management of rhinoceros beetle, *Oryctes rhinoceros* (Linnaeus), one of the major pests of coconut palm. The beetle suffers from natural epizootics produced by a fungus, *Metarhizium anisopliae* (Metchnikoff) and a baculovirus. Release of baculovirus infected beetles has been successfully used for the biosuppression of the beetle in several countries including South Pacific Islands, Fiji, Mauritius, Seychelles and Papua New Guinea. However, there have been reports of resurgence of the beetle in some other regions including Kerala (India) and Western Samoa. There is also the possibility of development of resistance to baculovirus in India, East Jawa and South Sulawese. To overcome these problems, a set of cultural practices has been recommended for strengthening natural control of the beetle. These include planting legumes as cover crops to conceal the potential breeding sites of the beetles and leaving some dead standing palms for promoting the spread of the virus. In addition, breeding places of the beetle are treated with the fungus, *M. anisopliae*. Virus-infected beetles are released in case the pest population shows an upward trend (Pillai *et al.*, 1993).

In sunflower, early sowing favours the build up of important predators, chrysopa and ladybird beetles. It results in lowering the population of a number of insect pests including green jassid, cabbage semilooper and head bug (Arora *et al.*, 1998).

## Botanical pesticides

The use of pesticides of plant origin for the control of agricultural pests has a long history but has assumed greater importance in recent years due to environmental deterioration and health hazards associated with the use of synthetic pesticides. Botanical pesticides exert a range of behavioral and physiological effects on the colonization, development, growth, survival and multiplication of insects. In view of their environmental safety, these pesticides offer an attractive alternative to synthetic pestcides for use in IPM.

The application of neem seed kernel extract for the control of *Spodoptera litura* (Fabr.) did not affect the emergence of the egg parasitoid, *Telenomus remus* Nixon, and predator, *Brinkochrysa scelestes* (Banks). Topical application of seed oil of neem, Chinaberry and custard apple showed no adverse effect on predatory spider, *Lycosa pseudoannulata* (Bosenberg & Strand) and was only slightly toxic to the mirid bug predator, *Crytorhinus lividipennis* Reuter, at and above 10 ug/female in rice crop. An active neem seed kernel fraction evaluated against sorghum pests was found to be safe to midge, *Contarinia sorghicola* (Coquillett); parasitoid, *Tetrastichus* sp. and predator, *Orius* sp. It was, however, toxic to *Apanteles ruficrus* (Haliday), a larval parasite of *Mythimna separata* Walker. Field trials with neem oil for the control of sorghum aphid, *Melanaphis sacchari* (Zehntner) did not show any adverse effect on syrphids and coccinellids.

The plant extracts of some rice varieties were found to be highly toxic to important rice pests including *Chilo suppressalis* (Walker), *Nilaparvata lugens* (Stal and *Sogatella* furcifera (Horvath) but were safe to predators of hopper complex, viz., *Lycosa pseudoannulata* (Bosenberg & Strand) and *Cyrtorhinus lividipennis* Reuter (Arora and Dhaliwal, 1994). In the rice field trials, application of neem seed bitters as ULV sprays had no detrimental effect on a number of natural enemies including the black cricket predator, *Mettioche vittaticollis*; ground beetle predator, *Ophionea nigrofasciata* (Schmidt-Goebel), and lady beetle predator, *Conocephalus longipennis* (deHann). Predation of leaf folder, *Marasmia patnalis* Bradley eggs by *M. vittaticollis* was unaffected in treated fields. The parasitism of *Cnaphalocrocis medinalis* (Guenee), larvae was relatively unaffected in rice fields sprayed with 50 per cent neem oil (ULV application) than in control fields.

Normal male and female *Macrocentrus philippinensis* Ashmead emerged from parasitized leaf folder, *C. medinalis* larvae which had been confined to rice leaf cuts treated with partially purified neem fractions. Similarly, emergence of braconid predator, *Lysiphlebus testacelpes* (Cresson) from parasitized aphids, *Aphis gossypii* Glover was unaffected by application of Margosan-0 to *Hibiscus* leaf discs in laboratory studies. Neem oil 50 per cent sprayed at rates up to 20 ml of spray fluid (recommended rate 3 ml/l) has no adverse effect on *Chrysoperla carnea* (Stephens) and *Tetrastichus howardi* (Lim and Bottrell, 1994).

Neem and other plant products have also been found to increase the efficacy of microbial control gents like baculoviruses, *Bacillus thuringiensis* Berliner and fungi.

## Chemical control

Pesticides will continue to play an important role in IPM programmes. But their use must be made compatible with other components of pest management. Physiologically selective insecticides are presently available only for a limited number of insect pests and field situations. However, even broad spectrum insecticides may be used in an ecologically selective manner. Systemic insecticides applied as granules, seed dressings, root zone application, stem injection and leaf-axil application avoid damage to natural enemies moving on the plant surface. Separation in time could be achieved by timing of sprays to avoid period of maximum adult activity of natural enemies (Jayaraj *et al.*, 1994).

Sublethal concentrations of a number of insecticides alongwith respective nuclear polyhedrosis viruses have been found highly effective for the control of a number of important pests including *Helicoverpa* spp., *Spodoptera* spp. and *Spilosoma obliqua* (Walker) (Battu *et al.*, 1993). Fenvalerate, endosulfan and chlorpyriphos at one third of the recommended dosage when combined with entomopathogenic fungi *Beauveria bassiana* (Balsamo) Vullemin or *Metarhizium anisopliae* (Metchnikoff) were found highly effective against *S. obliqua* (Patel and Kanaujia, 1998).

## IMPLEMENTATION OF IPM

Although IPM has been accepted in principle as the most attractive option for the protection of agricultural crops from the ravages of insect and non-insect pests yet implementation at the farmers' level has been rather limited. Some of the important constraints to wider adoption of IPM and suggested measures to overcome them are discussed as under:

### Constraints in IPM implementation

The Consultant Group of the IPM Task Force has conducted an indepth study of the constraints on the implementation of IPM in developing countries (NRI, 1992).

#### *Institutional constraints*

IPM requires an interdisciplinary, multifunctional approach to solving pest problems. Fragmentation of disciplines between research, extension and implementation, and between institutes, all lead to a lack of institutional integration. Secondly, both the national programmes of developing countries and the donor agencies have lacked a policy commitment to IPM in the context of national economic planning and agricultural development. This has resulted in a low priority for IPM from national programmes and donors alike. Thirdly, the traditional top-down research in many cases does not address the real needs of farmers, who eventually are the end-users and who select to adopt or reject the technology based on its appropriateness. Institutional barriers to research scientists in national programmes conducting on-farm research in developing countries are real, and need to be addressed.

*Informational constraints:* The lack of IPM information which could be used by the farmers and by extension workers is a major constraint in implementation. While the individual control techniques are well known, little knowledge is available on using these in an integrated fashion under farm conditions. The lack of training materials, curricula and experienced teachers on the principles and practice of IPM is another major constraint.

*Sociological constraints:* The conditioning of most farmers and farm level extension workers by the pesticide industry has created a situation where chemicals are presented as highly effective and simple to apply. This acts as a major constraint in IPM implementation. There is a direct conflict between industry's objective of more sales and the IPM message of rational pesticide use. There is a need for private industry and the public sector extension to work in a more complementary manner.

*Economic constraints:* A major constraint, even if IPM is adopted in principle, is the funding for research, extension and farmer training needed for an accelerated programme. IPM must be viewed as an investment and as with other forms of investment, requires an outlay. In the long run, IPM programmes may become self-generating due to savings on inputs for production.

***Political constraints***

The relatively low status of plant protection workers in the administrative hierarchy is a constraint to general improvement in plant protection. Associated with the above are the morale and financial standing of these workers.

The continuance of pesticide subsidies by the government for political reasons and its tie up with the government-provided credit for crop production act as major constraints to farmers' acceptance of IPM. Various vested interests associated with the pesticide trade also act as a political constraint on the implementation of IPM. Indeed, the illicit financial rewards for licensing the import, formulation or distribution of agricultural chemicals are one of the major obstacles limiting the expansion of IPM activities (Jiggins, 1996).

**Measures for improving implementation**

Acceleration of IPM implementation in developing countries requires farmers' participation, increased government support, legislative measures, improved institutional infrastructure and a favourable environment.

***Farmers' participation***

It is not an exaggeration to say that the dawn of civilization started with farmer innovation. Ever since that day, farmers have improved ways of growing crops through successive innovations. Prior to the emergence of crop protection sciences and even before the broad outlines of the biology of pests were understood, farmers evolved many cultural, mechanical and physical control practices for the protection of their crops from insect and non-insect pests (Smith *et al.*, 1976). Farmers' innovations were the only source of improvements in crop production and protection technology until formal research by scientists started complementing it during the late eighteenth and nineteenth century (Haverskort *et al.*, 1991). Unfortunately, with the advent of modern high-tech agriculture comprised of HYVs, fertilizers and pesticides, farmers have been completely displaced from the research and development process. Instead this role has been usurped by the private industry and the government agencies. The technology generated by scientists is being transferred through the extension agencies to the farmers. The new technology package has created a number of ecological and environmental problems. The alternative path of sustainable agriculture requires farmers' participation at every step of the research and development process in order to draw on their understanding of the local conditions and constraints, their innovativeness and skills at making the best possible use of limited resources (van de Fliert, 1998).

Placing the farmer at the centre of development process is wholly consistent with the IPM goal of making the farmer a confident manager and decision maker, free from dependence on a constant stream of pest control instructions from outside. The role of researchers, extension and non-government organizations (NGOs) and the field workers is to act as consultants, facilitators and collaborators, stimulating and empowering the farmers to analyze their own situation, to experiment and to make constructive choices. A number of terms have been proposed for the new apprach. These include: 'farmer-first-and-last', 'farmer participatory research', 'people-centred technology development (PCTD)' and 'participatory technology development (PTD)' (Chambers *et al.*, 1991; Haverskort *et al.*, 1991). PTD serves to improve the experimental capacity of farmers and helps in development of locally adapted improved technologies.

The approach has been used in the development of farmers' field schools (FFS) in IPM in Indonesia and the Philippines. The Field Schools are designed to build up farmers' understanding of agro-ecological relationships and to improve their capacities to systematically observe, document and interpret these. Experimental learning is based on the principles of adult learning, and follows a cycle of focused experience, systematic observation and measurement, critical reflection followed by planning and taking action. The action, in turn, begins a new cycle of experience-based learning and critical reflection. Thousands of farmers have been trained utilizing this approach and it is being tried on a pilot scale (Dhaliwal and Heinrichs, 1998).

***Government support***

Both the national programmes of developing countries and the donor agencies must have a policy commitment to IPM in the context of national economic planning and agricultural development. The costs of developing countries of not bringing their policies in line with the objectives of IPM are relatively greater than the costs to developed countries. National policies to promote IPM require close regulation at all stages related to the importance and/or manufacture, distribution, use and disposal of pesticides. In the case of pesticides that do not meet prescribed standards for safety, persistence, etc., import and manufacturing bans should be enacted. At a minimum, the conditions laid out by the FAO Code of Conduct on the Regulation, Distribution and Use of Pesticides should be adopted. Pesticide subsidies need to be eliminated in order to make IPM an attractive alternative.

The funds so saved may be utilized for the implementation of IPM. Funds may also be diverted from some of the current research programmes to IPM-oriented plant protection programmes. Additional monetary resources may be generated through cooperation with bilateral/multilateral agencies willing to support such programmes (NRI, 1992).

***Legislative measures***

IPM is an information system and its adoption reduces pest control costs. The alternative to IPM is the indiscriminate use of broad spectrum synthetic organic pesticides. Unfortunately, while pesticide manufactures and users (farmers) derive the full benefits from the use of these chemicals, they pass on the environmental and ecoloigcal costs of their use to society as a whole. If they are made to bear the full cost of the use of these toxicants, they may find IPM a more economical and attractive alternative. This could be achieved by enforcing suitable legislative measures.

Secondly, the success of an IPM programme in any geographical region depends upon its implementation by all farmers in the area. Ideally, farmers may voluntarily adopt an IPM programme but some farmers may hold out. Such farmers called 'spoiler holdouts' may impair the success of a programme by failing to adopt a necessary practice thus causing damage to adjacent areas. Besides these, some farmers may free-ride and thus shift the costs of implementation and managing a programme to a group of participating farmers. To overcome 'spoiler holdouts' and 'free riders', it may be necessary to impose a programme upon an unwilling minority through suitable legislative measures (Tarlock, 1980).

***Improved institutional infrastructure***

IPM can not be implemented unless there is a basic infrastructure for plant protection in a country. There is a need to develop and support national programme capabilities for on-farm testing and technology extrapolation. At the international level establishment of the Global IPM Facility to coordinate and monitor funding of IPM projects is needed to provide impetus to the implementation of IPM.

IPM is predominantly knowledge technology, the use of which requires training of the many groups involved. There is currently little training material for most of these groups including farmers, extension personnel and researchers. If IPM is to become the major approach for pest management in the developing world, this deficiency must be remedied (NRI, 1992). Another aspect requiring greater attention is coordinated effort within and between national research, training and implementation institutes/programmes and amongst international development agencies.

Lack of a reliable data-base has also hampered progress of IPM programmes. A reliable source of accurate information on the status of crops and pests in farmers' fields is necessary for many IPM activities. Most of the successful IPM programmes, both in developed and developing countries, have a reasonably accurate system of monitoring and evaluating various biological and environmental parameters in the agroecosystem. A reliable data-base on crop yield and pest losses is required for planning and resource allocation at the national and international level. Systems analysis has been used as a problem diagnosis tool for IPM in developed-country cropping systems and may be used in the developing countries as well.

***Improved awareness***

Increased education and awareness regarding the objectives, techniques and impact of IPM programmes is required at all levels including policy makers, planners, farmers, consumers and the general public. The importance and benefits of pesticides are being over emphasized by a multibillion dollar industry utilizing the services of not only their salesmen but also agricultural scientists, administrators and planners. There is not yet a strong market for IPM information. Policy makers and planners need to be convinced that without IPM current agricultural production systems are not sustainable. Similarly, much important information that might induce a farmer to adopt IPM is not immediately observable and is, therefore, not sought by him. A manufacturer has no incentive to recommend a programme that uses less pesticides, or even selective pesticides that kill a limited range of pests (Tarlock, 1980).

Consumer groups and the general public may also be able to support the implementation of IPM programmes by demanding residue-free commodities. There is now a distinct market for organically produced food and other products. Non-Government Organizations and consumer groups need to be strengthened especially in developing countries, so that there is a public-oriented movement for implementation of IPM.

## CONCLUSIONS

The burgeoning human population requires increased amounts of food and fibre from a shrinking agricultural land base. To date, modern agriculture has largely achieved these objectives but the intensification of agriculture through expansion of irrigation facilities, introduction of high yielding varieties and application of increased amount of agrochemicals during the last three decades has also increased the losses due to insect pests, diseases and weeds. The strategy of exclusive reliance on broad spectrum, synthetic organic pesticides has caused irreparable damage to the environment. As we approach the third millennium, increasing intensification of agriculture will put further strain on our fragile agroecosystems. Integrated pest management appears to be the only viable option to manage future insect pest problems as part of a holistic crop production technology. IPM provides ecological sustainability, social stability and economic sustainability. Therefore, IPM would play a key role in sustainable agriculture production systems in the 21st century.

## REFERENCES


Allen, W.A. and Rajotte, E.G. 1990. The changing role of extension entomology in the IPM era. *Annual Review of Entomology,* **35**: 379-397.

Altieri, M. 1995. Escaping the treadmill. *Ceres* 27(4): 15-23.

Andrewartha, H.G. and Birch, L.C. 1954. *The distribution and abundance of animals*. University of Chicago Press, Chicago.

Arora, R. and Dhaliwal, G.S. 1994. Botanical pesticides in insect pest management: Ecological perspectives. *In:* G.S. Dhaliwal and B.D. Kansal (eds). *Management of Agricultural Pollution in India.* Commonwealth Publishers, New Delhi, pp. 213-245.

Arora, R. and Dhaliwal, G.S. 1996. Agroecological changes and insect pest problems in Indian agriculture. *Indian Journal of Ecology,* **23**: 109-122.

Arora, R., Kaur, S. and Dhillon, A.S. 1998. Manipulation of date of sowing for natural control of insect pest in sunflower agroecosystem. *In*: *Ecological Agriculture and Sustainable Development*. G.S. Dhaliwal, N.S. Randhawa, R. Arora and A.K. Dhawan (eds). Vol. 2. Indian Ecological Society and Centre for Rural and Industrial Development, Chandigarh.

Arora, R. and Sidhu, H.S. 1982. Effect of plant water deficit on the fecundity of the mustard aphid, *Lipaphis erysimi* (Kaltenbach). *Indian Journal of Ecology,* **9**: 346-347.

Arora, R. and Sidhu, H.A. 1991. Efficacy of dimethoate against mustard aphid, *Lipaphis erysimi* (Kalt.) in relation to water stress in the host plants. *Journal of Insect Science,* **4**: 138-140.

Atwal, A.S. and Singh, B. 1990. *Pest Population and Assessment of Crop Losses*. Indian Council of Agricultural Research, New Delhi.

Battu, G.S., Ramakrishnan, N. and Dhaliwal, G.S. 1993. Microbial pesticides in developing countries: Current status and future trends. *In:* G.S. Dhaliwal and B. Singh (eds.). *Pesticides: Their Ecological Impact in Developing Countries*. Commonwealth Publishers, New Delhi, pp.417-468.

Bramble, B.J. 1989. An environmentalists' view of pest management and the green revolution. *Tropical Pest Management,* **35**: 228-230.

Brar, K.S., Singh, I. and Aujla, T.S. 1991. Water stress induced effects of *raya* crop on the efficacy of xoydemeton methyl for the control of mustard aphid. *Journal of Insect Science,* **4**: 141-145.

Chambers, R., Pacey, A. and Thrupp, L.A. (eds) 1991. *Farmer First: Farmer Innovation and Agricultural Research*. Intermediate Technology Publications, London.

Chelliah, S. 1987. Insecticide-induced resurgence of rice brown plant hopper, *Nilaparvata lugens* (Stal). *In:* S. Jayaraj (ed). *Resurgence of Sucking Pests*. Tamil Nadu Agricultural University, Coimbatore, pp.1-10.

Chelliah, S. and Bharathi, M. 1985. Rice brown plant hopper, *Nilparvata lugens* (Stal) -Status and strategies in management. *In:* S. Jayaraj (ed). *Integrated pest and disease management*. Tamil Nadu Agricultural University, Coimbatore, pp. 34-53.

Chitty, D. 1960. Population process in the vole and their relevence to general theory. *Canadian Journal of Zoology,* **38**: 99-113.

Clark, L.R., Geier, P.W., Hughs, R.D. and Morris, R.F. 1967. *The Ecology of Insect Populations in Theory and Practice.* The English Language Book Society and Methuen & Co. Ltd., London.

Deol, G.S., Singh, D. and Sarmam, J.S. 1987. Effect of some crop management practices on population of armyworm (*Mythimna separata* Walker) in wheat. *Journal of Research Punjab Agricultural University, Ludhiana*, **24**: 363-367.

Dhaliwal, G.S. and Arora, R. 1993. Changing status of insect pests and their management strategies. *In:* K.S. Gill, G.S. Dhaliwal and B.S. Hansra (eds). *Changing Scenario of Indian Agriculture*. Commonwealth Publishers, New Delhi, pp. 98-145.

Dhaliwal, G.S. and Arora, R. 1994. Components of insect pest management: a critique. *In:* G.S. Dhaliwal and Ramesh Arora (eds). *Trends in Agricultural Insect Pest Management*. Commonwealth Publishers, New Delhi, pp. 1-55.

Dhaliwal, G.S. and Arora, R. 1996. *Principles of Insect Pest Management*. National Agricultural Technology Information Centre, Ludhiana.

Dhaliwal, G.S. and Heinrichs, E.A. (eds). 1998. *Critical Issues in Insect Pest Management*. National Agricultural Technology Information Centre, Ludhiana.

Dhaliwal, G.S., Arora, R. and Heinrichs, E.A. 1998. Insect pest management: From traditional sustainable approach. *In:* G.S. Dhaliwal and E.A. Heinrichs (eds). *Critical Issues in Insect Pest Management*. National Agricultural Technology Information Centre, Ludhiana, pp. 1-29.

Elton, C.S. 1927. *Animal Ecology*. Sidgwick and Jackson, London.

Geier, P.W. and Clarke, L.R. 1961. An ecological approach to pest control. *In: Proceedings of the 8th Technical Meeting 1960*. International Union of Conservation of Nature and Natural Resources, Warsaw, pp. 10-18.

Haverskort, B., Kamp, Johan van der and Waters-Bayer, A. 1991. *Joining Farmer's Experiments-Experiences in Participatory Technology Development.* Intermediate Technology Publications, London.

Heinrichs, E.A. 1994. Development of multiple pest resistant crop cultivars. *Journal of Agricultural Entomology,* **11**: 325-353.

Heinrichs, E.A. 1998. IPM in the 21st century. Challenges and opportunities. *In*: G.S. Dhaliwal and E.A. Heinrichs (eds). *Critical Issues in Insect Pest Management.* National Agricultural Technology Information Centre, Ludhiana, pp. 307-322.

Howard, L.O. and Fiske, W.F. 1911. The importation into the United States of the parasites of the gypsy moth and the browntail moth. *US Department of Agriculture Bulletin Bureaux of Entomology,* **91**: 29, 33-34.

Hunter, W.D. and Coad, B.R. 1923. The boll weevil problem. *USDA Farmer's Bulletin* **1329**.

Jayaraj, S., Ananthakrishnan, T.N. and Veeresh, G.K. 1994. *Biological Pest Control in India: Progress and Perspectives*. RGICS Project No.2, Rajiv Gandhi Institute of Contemporary Studies, New Delhi.

Jiggins, J. 1996. Women and the re-making of civil society. *Forest Trees People Newsletter,* **30**: 18-22.

Kartohardjono, A. and Heinrichs, E.A. 1984. Population of the brown plant hopper *Nilaparvata lugens* (Stal) (Homoptera : Delphacidae) and its predation on rice varieties with different levels of resistance. *Environmental Entomology,* **13**: 359-365.

Kumar, A. 1984. *Insect Pest Control with Special Reference to African Agriculture*. Edward Arnold, London.

Lal, S.S., Sachan, J.N. and Chandra, S. 1985. Cultural and varietal tools for integrated pest management for increasing pulse production. *Plant Protection Bulletin,* **37**: 13-23.

Lim, G.S. and Bottrell, D.G. 1994. *Neem Pesticides in Rice: Potential and Limitations*. International Rice Research Institute, Manila, Philippines.

Metcalf, C.L. 1930. Obituary, Stephen Alfred Forbes, May 29, 1844-March 13, 1930. *Entomology News*, **41**(5): 175-178.

Metcalf, R.L. 1980. Changing role of insecticides in crop protection. *Annual Review of Entomology*, **25**: 215-256.

Michelbacher, A.E. and Bacon, O.G. 1952. Walnut insect and spider mite control in northern California. *Journal of Economic Entomology*, **45**: 1020-1027.

Milne, A. 1962. On a theory of natural control of insect populations. *Journal of Theoretical Biology*, **3**: 19-50.

Mohyuddin, A.I. 1994. Commercial sugarcane. *IPM Working Division*, **2**: 4-5.

Mumford, J.D. and Norton, G.A. 1984. Economics of decision making in pest management. *Annual Review of Entomology*, **29**: 157-174.

Narang, D.D., Joshan, V.S. and Dilawari, V.K. 1993. Control of mustard aphid, *Lipaphis erysimi* (Kalt.) on *Brassica* crops grown in field under stress conditions. *Journal of Insect Science*, **6**: 195-196.

NAS, 1969. *Principles of Plant and Animal Pest Control. Vol.3, Insect Management and Control*. National Academy of Sciences. Washington DC.

NRI, 1992. A Synopsis of *Integrated Pest Management in Developing Countries in the Tropics. Experience and Prospects*. Natural Resource Institute, Chatham, UK.

Patel, P.S. and Kanaujia, K.R. 1998. Mass multiplication techniques, pathogenicity and feasibility of mycopesticides in integrated pest management. *In:* G.S. Dhaliwal, N.S. Randhawa, R. Arora and A.K.Dhawan (eds). *Ecological Agriculture and Sustainable Development*. Vol. 2. Indian Ecological Society and Centre for Rural and Industrial Development, Chandigarh.

Pathak, M.D. and Dhaliwal, G.S. 1981. *Trends Strategies for Rice Insect Problems in Tropical Asia*. IRRI Research Paper Series No. 64, International Rice Research Institute, Philippines.

Pedigo, L.P. 1996. *Entomology and Pest Management*. Prentice Hall of India Pvt. Ltd., New Delhi, Eastern Indian Edition.

Pedigo, L.P. and Higley, L.G. 1992. The economic injury level concept and environmental quality. *American Entomologist*, **38**: 1,2-21

Pedigo, L.P., Hutchins, S.H. and Higley, L.G. 1986. Economic injury levels in theory and practice. *Annual Review of Entomology*, **31**: 341-368.

Perkins, J.H. 1980. The quest for innovation in agricultural entomology. *In:* D.Pimentel and J.H. Perkins (eds). *Pest Control: Cultural and Environment Aspects*. AAAS Selected Symposium 43, Westview Press, Colorado, pp. 23-80.

Pillai, G.B., Sathiamma, B. and Dangar, T.K. 1993. Integrated control of rhinoceros beetle. *In:* M.K. Nair, H.H. Khan, P. Gopalsundaram, and E.V.V. Bhaskara Rao (eds). *Advances in Coconut Research and development*. Oxford and IBH Publishing Co. Pvt. Ltd., New Delhi, pp.455-463.

Pimentel, D. 1997. Ecological basis of insect pests and weed problems. *In*: J.M.Charret and G.R. Sagar (eds). *Origin of Pest, Parasite, Disease and Weed Problems*. Blackwell Scientific Publications, Oxford, pp.3-31.

Pradhan, S. 1983. *Agricultural Entomology and Pest Control*. Indian Council of Agricultural Research, New Delhi.

Price, P.W. 1984. *Insect Ecology*. John Wiley and Sons. New York.

Rajak, R.L. 1986. Integrated pest management in rice crop-prospects and retrospects. *Plant Protection Bulletin, India*, **38**: 1-4.

Reichelderfer, K.H., Carlson, G.A. and Norton, G.A. 1984. *Economic Guidelines for Crop Pest Control*. FAO Plant Production and Protection Paper 58, Food and Agriculture Organization of the United Nations, Rome.

Schuster, M.F. and Dean, H.A. 1973. Rhodesgrass scale resistance studies in rhodesgrass scale resistance studies in rhodesgrass. *Journal of Economic Entomology,* **66**: 467-469.

Sharma, H.C. and Veerbhadra Rao, M. (eds). 1993. *Pests and Pest Management in India -The Changing Scenario*. Plant Protection Association of India, Hyderabad.

Singh, G. 1987. *Toxicity of insecticides against mustard aphid, Lipaphis erysimi (Kaltenbach) in relation to supply of potassium to host plants.* M.Sc. Dissertation, Punjab Agricultural University, Ludhiana.

Smith, H.S. 1935. The role of biotic factors in the determination of population densities. *Journal of Economic Entomology* **12**: 282-292.

Smith; R.F. 1978. History and complexity of integrated pest management. *In:* E.H. Smith and D. Pimentel (eds). *Pest Control Strategies*. Academic Press, New York, pp. 41-54.

Smith, R.F. and Allen, W.W. 1954. Insect control and the balance of nature. *Scientific American,* **190**(6): 38-42.

Smith, R.F., Apple, J.L. and Bottrell, D.G. 1976. The origin of integrated pest management concepts for agricultural crop. *In:* J.L. Apple and R.F. Smith (eds). *Integrated Pest Management.* Plenum Press, New York, pp. 1-16.

Smith, R.F. and Reynolds, H.T. 1966. Principles, definitions and scope of integrated pest control. *In: Proceedings of FAO Symposium on Integrated Pest Control,* Vol. I, Rome, pp. 11-17.

Srinivasan, K. 1994. Recent trends in insect pest management in vegetable crops. *In:* G.S. Dhaliwal and R. Arora (eds). *Trends in Agricultural Insect Pest Management.* Commonwealth Publishers, New Delhi, India, pp. 345-372.

Srinivasan, K. and Krishnamoorty, P.N. 1993. Changing scenario of pest problems in solanaceous, malvaceous and leguminous vegetables and their management. *In:* H.C. Sharma and M. Veerbhadra Rao (eds). *Pests and Pest Management in India-The Changing Scenario*. Plant Protection Association of India, Hyderabad, pp. 103-105.

Stern, V.M., Smith, R.F., Van den Bosch, R. and Hagen, K.S. 1959. The integrated control concept. *Hilgardia,* **27**: 81-101.

Tarlock, A.D. 1980. Legal aspects of integrated pest management. *In:* D. Pimentel and J.H. Perkins (eds). Pest Control: Cultural and Environmental Aspects. AAAS selected Symposium 43, Westview Press, Colorado, pp. 217-236.

van den Fliert, E. 1998. Integrated pest management: Springboard to sustainable agriculture. *In:* G.S. Dhaliwal and E.A. Heinrichs (eds). *Critical Issues in Insect Pest Management.* National Agricultural Technology Information Centre, Ludhiana, pp. 287-305.

Varma, A., Tiwari, N.K. and Tanwar, R.K. 1993. Changing scenario of insect pests in sugarcane and their management. *In:* H.C. Sharma and M. Veerbhadra Rao (eds). *Pests and Pest Management in India - The Changing Scenario*. Plant Protection Association of India, Hyderabad, pp. 88-95.

Velusamy, M. 1987. Studies on resurgence of rice green leafhopper and brown planthopper. *In:* S. Jayaraj (ed). *Resurgence of Sucking Pests.* Tamil Nadu Agricultural University, Coimbatore, pp. 19-22.

Wellington, W.G., Cameron, P.J., Thompson, W.A., Kertinsky, I.B. and Landsberg, A.S. 1975. A stochastic model for assessing the effects of external and internal heterogeneity of an insect population. *Research in Population Ecology,* **17**: 1-28.

Yein, B.R. and Das, G.R. 1988. Effect of spacing and nitrogen levels on the incidence of insect pests in rice. *Pesticides,* **22**(10): 37-40.

*Pest Management in Horticulture Crops*
L.R. Verma, A.K. Verma and D.C. Gautam (eds)


## *Chapter 2*

# Population Monitoring, Dispersion, Modelling and Life Table Analysis in Insect Pest Management

*H.C. Sharma, J.P. Sharma, P.R. Gupta and J.R. Thakur*

## INTRODUCTION

IN INTEGRATED PEST MANAGEMENT, it is important to know whether an insect found on a crop is really a pest, and its potential to inflict damage on the crop. However, it is difficult to get an estimate of absolute population density of an organism. Hence, most biological studies are based on a relatively small sample of the total population, which should be representative of the population so as to enable us to extract most accurate information. Insect damage on crops and their products can be detected through pest surveillance, which determines if the species exits, estimates its distribution, population density and dynamics. These are basic inputs in decision making in pest management. A survey is a detailed collection of information about an insect at a particular time and in a geographical area, which can be qualitative, or quantitative. In agricultural systems, qualitative surveys precede quantitative ones. The latter are commonly employed in pest management. Such surveys establish the numerical abundance of insects in space and time to predict future population density trends, and extent of damage on the crops. Various sampling techniques are employed to establish population density of different insects. The development of efficient and reliable sampling techniques, surveillance, and forecasting system requires ecological information on insect pests and appropriate sampling and data analytical tools.

Minimizing the losses due to insect pests is one of the means to stabilize and increase the production and productivity of crops. Hence, it is important to develop and implement efficient pest control measures based on pest population dynamics and crop phenology. In this regard, pest occurrence, modelling and forecasting play a central role. A proper and timely forecast enables us to have sufficient time to carry out the control measures against the insect pests. It also helps us to determine the most efficient method of pest control. Forecasts are based on collection and processing of information about the ecology and spread of insect pests, and the environment in which the crops are grown. The forecasting and warning system depends on systematic observations on appearance, development and spread of crop pests. It is useful for:

- Providing information about timely application of appropriate prophylactic crop protection methods.
- Determining the effectiveness of crop protection measures.
- Development of long- and short-term forecasts of emergence and duration of insect attack.
- Provide information to farmers and pest control agencies for planning and implementation of efficient pest control strategies.

Various aspects of insect population monitoring, distribution, and prediction models have been discussed by Southwood (1978), Ruesink and Kogan (1985), Taylor (1984), Pawar *et al.* (1985), Horn (1988), Kuno (1991), Karel (1996), and Sharma *et al.* (1996). An attempt has been made in this chapter to discuss various techniques employed for insect population monitoring, sampling, distribution patterns, population models, life table analysis and use of day-degrees in pest forecasting.

## POPULATION MONITORING

To develop an appropriate pest forecasting system, we need to have a proper insect monitoring system, appropriate sampling techniques, understanding of the relationships between weather (temperature) and insect development, life table analysis and key mortality factors. Pest management programmes can not operate without an accurate estimate of insect and natural enemy population densities or assessment of plant damage and its effect on crop yield. Estimates of population density fall into three main categories; absolute methods, relative methods, and population indices. Absolute methods estimate densities per unit of area in the habitat. Relative methods estimate densities per unit other than area, and cannot be converted to absolute estimates without correcting for the insect behaviour and effect on habitat, e.g., sweep net. Population indices are based on insect products/effects, e.g., frass, nests, plant damage, etc. Relative methods are used in qualitative monitoring programmes, e.g., quarantine. Assessment of insect damage depends on whether the insect is a direct pest (which affect the desired product of the crop), e.g., pod borers, sorghum midge, gypsy moth, etc. or indirect damage (when insects feed on plant parts which may or may not affect the yield), e.g., stem borers in cereals, armyworms, leaf feeders, etc. The effect of indirect damage on crop yields is more difficult to assess.

### Absolute methods

Absolute methods can be used for calculating oviposition and mortality rates and in validation studies of descriptive population models. Four distinct approaches have been used in entomology to estimate absolute population densities: 1) the distance to nearest neighbour approach, 2) sampling a unit of habitat, 3) the recapture of marked individuals, and 4) removal trapping (Ruesink and Kogan, 1985; Southwood, 1978).

Absolute methods provide more useful data often desired by population ecologists. Successive estimates of the number of insects per unit of land area are necessary for constructing life tables and for nearly all other studies of population dynamics of uncaged field populations. Number of insects present, type and stage of the crop, biology, stage, sex, generation, feeding behaviour, effect of weather factors on insect activity, sampling time, position of the attractant/trap, and lack of uniformity in sampling techniques influence the effectiveness of direct sampling methods.

### *Nearest neighbour approach*

The distance to nearest neighbour approach has been applied much more by botanists than by entomologists. The method is based on randomly selecting an individual and then measuring the distance to its nearest neighbour. The mobility of insects and the risk that one will fail to find the nearest neighbour have limited its application in entomology. Furthermore, the theory apparently has been developed only for species that are randomly distributed in space. Nevertheless, the approach can be used in entomology, especially for sessile insects such as scales or for well-marked nests or colonies (e.g., tent caterpillars). The insect density per unit area m can be estimated by:

$$m = (pr)^2 \tag{1}$$

where r = mean distance between nearest neighbours, and p = index of aggregation; that takes the value of 2 for randomness, and exceeds 2 for aggregated distributions (Clark and Evans, 1954; Waloff and Blackith, 1962). A modification of the above formula that uses the sum of $r^2$ rather than the mean r squared is more accurate when p differs greatly from 2 (Southwood, 1978). Other modifications are available that utilize the information contained in the distances to the second, third,..., and nth nearest neighbours (Morisita, 1954; Thompson, 1956). These approaches not only allow more accurate density determinations to be made but also dispersion pattern to be detected over a large area. Waloff and Blackith (1962) indicated that competition effects detectable between nearest neighbours may be marked at greater distances because of the heterogeneity of the habitat. Density per unit area can also be assessed (Clark and Evans, 1954) by:

$$m = 1/4r^2 \tag{2}$$

where m = density per unit area and r = mean distance between the nearest neighbours.

### *Sampling a unit of habitat*

Nearest-neighbour approach is useful only for individuals inhabiting a substrate and works best for those on the soil surface. The method of sampling a unit habitat applies equally well to a broad range of habitats, including soil, litter, soil surface, on or in vegetation and even in the air. The main advantage of this approach is its broad applicability, as it makes no assumptions about the spatial distribution of the species nor about its activity or behaviour. In general, there is no concern about the efficiency of the sampling method, for it is assumed that by bringing in an entire unit of habitat, no individuals are left behind.

*Sampling from air*: Wind traps have been used to sample insects from air. These traps are useful to sample airborne migratory insects. The number of insects trapped will depend on the wind speed and the route of migration. However, they can give an indication of the onset of insect infestation/arrival and possible outbreaks.

The techniques used for sampling from air are better mechanized than for other habitats; suction traps and rotary traps are the two basic methods used. The advantage of the rotary trap is that the catch is independent of wind speed. Taylor (1962) calculated efficiency factors for suction traps that correct for wind speed, trap design, and insect size. Efficiency factors are not available for rotary traps, but it is known that even the best ones miss about 15% of the population (Taylor, 1962). Taylor and Palmer (1972) compared the merits of the suction trap with the rotary trap and concluded that rotary traps have no advantages and are more difficult to make.

Since the catch with these traps is almost exclusively insects, there is no problem in counting the number per sample. But there is a problem in converting the catch, typically in insects per hour, into aerial density. Even after correcting for the efficiency of the trap, one must divide by the volume of air sampled per hour to estimate insect density at the trap height. Johnson (1957) has shown that density of insects at a particular height z is given by:

$$f_z = c(z - z_e)^r \tag{3}$$

where $f_z$ = density at height z, c = a scale factor dependent on the general size of the population in the air, $z_e$ = a constant added to the actual height, the height at which the insects' speed is exceeded by wind speed, and r = an index of the profile of aerial diffusion processes.

*Sampling from vegetation:* The most direct method is to sample all plant matter over a unit area of soil surface. The difficulty in this approach is that in some habitats it is difficult to determine what vegetation is inside the sample. For example, in a crop that has lodged, should parts be taken from plants lying across the sample area? In orchard or forest situations, it is practically impossible to sample this way.

Direct counts of insects on plants do not require sophisticated equipment but depend on keen observations, and is the most widely used method of sampling insects on plants. Insects can be sampled by direct counts in a number of ways, e.g., number of leaf folder larvae per square meter, length of a crop row, weight of crop, numbers per plant, tillers, shoot, stem internode, leaf, panicle, fruit, grain, etc. The unit should be converted to a unit area. Various insects are counted on different plant parts, e.g., aphids are counted per leaf, bugs per panicle, leafhoppers per stem, bollworm eggs per leaf, scale insects per sugarcane stem, etc. Plants may be split to count the number of stem borer larvae or pupae, bean fly larvae and pupae per pod, etc. Insects such as aphids, midges, mites, etc. can be counted by crushing the plant parts on a glossy paper or filter paper. Sorghum spikelets containing midge larvae can be crushed on a ninhydrin paper. The number of insects on a whole plant or plant part can be counted by beating the plant over a sheet of cloth or an inverted umbrella. Insects such as aphids and mites can be brushed off the leaves and preserved in alcohol. A non-persistent knockdown agent such as $CO_2$ or insecticide such as dichlorvos, methyl isobutyl-ketone (for aphids), turpentine vapors (for aphids), and pyrethrin may be used on a plant or panicle in a bag, box, or a sheet of cloth to collect the insects. Eggs of insects can be removed with a solvent or detergent, washed off, and their numbers recorded as volume. Insects inside plant parts such as sorghum midge larvae, lepidopterous larvae in stem, mites in galls, or beetles in stored grain can be detected by X-rays (Karel, 1996).

*Sampling from soil and litter:* Soil insects are sampled by digging a quadrat or drilling the soil with a soil core sampler to a standard depth to sample constant volume of soil. A knowledge of biology and ecology of the insect in question is important to decide on the sampling procedure. The

Colorado potato beetles overwinter as deep as 120 cm in sandy soil (Mail and Salt, 1933). However, bean leaf beetle eggs in soybeans have not been found below 10 cm depth (Waldbauer and Kogan, 1973). Insects can be separated from soil by hot water jacketed Berlese funnel, or the Tullgren funnel heated by an electric bulb or oil heater (Southwood, 1978; Karel, 1996). In the wet extraction methods, such as Salt-Hollick method, the soil samples are soaked, shaken with a detergent, and the insects float off a salt solution such as magnesium sulphate and than separated by centrifugation. Some insects may be driven off the soil by repellents such as formalin or with an insecticide. In case of termites, the insects can be attracted by placing a piece of paper on the soil. If soil samples are taken along the row length, the pest densities can be expressed on area basis.

***Recapturing marked individuals***

Recapturing marked individuals has been extensively used as an alternative method to those based on animal counts within a fixed unit of habitat (Southwood, 1978) and does not depend on the number of sampling units in the habitat. When recapturing marked individuals to estimate the absolute population size, the most serious limitation is in satisfying the assumption that the marked individuals become thoroughly mixed into the population. The simplest way to use recapture data has been proposed by Lincoln (1930) and is called the *Lincoln index*. This method requires the additional assumption that no births, deaths, immigration, or emigrations occur between the time of releasing the marked animals and the subsequent recapture sampling. This assumption applies to both the natural population and the marked individuals. The total population size P can be computed by:

$$P = an/r \qquad (4)$$

where a = number marked and released, n = total number recaptured in a sample, and r = number of marked animals recaptured in a sample. The best estimate of P is obtained when a and n are approximately equal. Modifications of the simple *Lincoln index* include methods based on the successive release of groups of marked animals with interspersed recapture sampling, and a single release with multiple recaptures. Many of these modifications are designed to eliminate the assumption of a stable population without migrations. Southwood (1978) has described the details on modified versions of *Lincoln index*.

Insects can be marked by paints, fluorescent dyes, labels, mutilation (without affecting behaviour of the insects), internal injection, mutant genes, and radioactive isotopes. A basic pre-requisite of these methods is that the insects should be released unharmed and unaffected into the wild, and recognized again on recapture.

***Removal trapping***

Removal trapping is based on the fact that as the population is reduced by removing members from it, the catch per trap per unit time decreases. The initial population size can be estimated from the rate by which the catch decreases, using one of three mathematical techniques described by Southwood (1978). The simplest method is to plot the number caught in the *n*th sampling against the previous total catch and to draw a straight line through these points. An estimate of the initial population size is then obtained by reading from the graph the value at which the line intersects the total-catch axis. Removal trapping is very rarely used in entomology because of the assumptions that must be satisfied and the number of individuals that must be trapped and removed from the population. The assumptions of the method are; no births, deaths or migrations occur during the sampling programme, and the probability of being caught must be the same for each member of the population and does not change during the sampling programme.

## Relative methods

With absolute methods, the intention is to capture and count every insect over a unit area, while the relative methods aim to capture a more or less consistent portion of those present. The sweep net and the black-light trap are two of the most common methods used to sample insect population density. Relative methods have one distinct advantage over absolute methods: a given amount of labour and equipment yields much more data. By wisely choosing which relative method to use, a person can usually capture at least 100 times as many insects as with the same effort using an absolute method. Southwood (1978) listed five factors that may simultaneously affect the catch of any relative method, viz.,

- The actual density or population size.
- The number of animals in a particular phase.
- The level of activity.
- The efficiency of the relative method being used.
- The responsiveness of the particular sex and species to the trap stimulus (includes attraction and avoidance).

### *Visual searches/direct counts*

This is the simplest approach. The observer counts all the insects he can see in an area or per unit time. Searching a fixed area, if completely efficient, provides an absolute estimate of population density. Visually searching an area may be considered an absolute measure of population size but in general less than 100% of the insects present are found. The proportion found may vary with conditions, thus making the relationship between absolute population and catch unpredictable. A variation of this method is called *flushing;* for example, a vehicle driven across the grasslands of Africa has been used to flush locusts (Scheepers and Gunn, 1958), and the number flushed per unit distance can be used as a relative measure of population density. A 5-minute visual search has been used in corn to obtain a relative estimate of the adult corn rootworm density. The use of these estimates must be approached with caution because of changes in behaviour with the weather and the age of the insects, and the differences among observers in their ability to spot and identify the insects.

### *Sweep net*

The sweep net is the most widely used method for sampling insects from small grain, forage, and many row crops. No other method can capture as many insects from vegetation per man-hour without both costing more for the equipment and doing more damage to the crop. The efficiency of the sweep net usually varies with; 1) species, 2) habitats, particularly the height of the vegetation, 3) weather, particularly wind speed, air temperature, and intensity of solar radiation, 4) time of day, reflecting diel cycles of behaviour of the species, and 5) styles of sweeping. The usefulness of sweep net data can be greatly improved by recording data on steps 1-5 for each sample taken. The difference in catch among several persons can be minimized by standardizing the style of sweeping.

Use a sweep net having a 45 cm diameter opening as a pendulum swing, as if sweeping the sidewalk with a broom. In short vegetation, swing the net as deep as possible without taking too much dirt into the net, while in tall vegetation, sweep only deep enough to keep the upper edge of the sweep net opening even with the top of the vegetation. Sweep one stroke per step while walking at a casual pace. The number of sweeps necessary to obtain a mean that is within 25% of the true value has been investigated for many insects (Southwood, 1978). For collections from alfalfa, an average of 25 units of 25 sweeps are adequate, while 10 units of 25 sweeps are adequate for grassland spiders. For cotton, the net is swept across the plants about 20 cm deep. For legumes such as cowpea, a 180° sweep is taken sideways 20 cm deep into the plants. Ten sweeps are taken for leafhoppers in upper part of rice plant for each step forward (Heinrichs *et al.*, 1982). Sweeping is not suitable for insects normally found near the base of the plant.

### *Ground cloth*

The ground cloth is the most widely used method in row crops such as soybeans or sorghum but it is also adequate for sampling the insect fauna of bushy weeds. This method consists of spreading a piece of cloth on the ground along the stems of the plants to be sampled. The plants are bent over the cloth and vigorously shaken or beaten with a wooden stick. The insects dislodged from the plant fall on the cloth and are collected and counted. In sampling for fast-moving insects, it is recommended that an aspirator be used for collecting.

### *Sampling bags*

Insects from plants can also be sampled by using polyethylene, paper or cloth bags (30 x 45 cm or of any other suitable size to enclose the plant part being sampled) containing a swab of cotton wool soaked in chloroform or ethyl acetate (Sharma *et al.*, 1992). The size of the bag may vary with the size of the sampling unit (such as a sorghum panicle, a branch of any crop plant, an inflorescence or a whole plant). The number of insects can be collected from 5 to 10 plants or sampling units.

### *Suction traps*

Trapping or sampling from air by suction is useful when the insects are not attracted to chemicals or light. Continuous sampling at different levels above the crop can give valuable indications of which and how many pests are likely to attack the crop. Suction traps utilize electric power from mains or batteries, and are of two types: 1) fixed, and 2) portable. Flying insects can be sampled by fixed type of suction traps. They usually consist of a metal sieve cone into which air is sucked by an electric fan. A control mechanism is used to separate the catch into time intervals. A portable suction trap is used to sample vertically downward or sideways or in restricted habitats. They are hand held, battery driven, or petrol driven car vacuum cleaner type samplers (Carino *et al.*,

1979; Thorhill, 1978). For larger areas, a variety of electric suction traps are available. The Johnson Southwood suction sampler, Univac sampler, D-vac suction sampler, and FRRMCOP are the commonly used portable suction samplers (Carino *et al.*, 1979; Perfect *et al.*, 1983). A device with the trade name D-Vac® captures insects by sucking them into a fine mesh net held open inside a rigid enclosure. A portable gasoline motor propels a blower that generates the suction. Two models are available; a smaller one which is held in front of the operator, and a heavier, backpack model. The D-Vac has been described by Dietrick *et al.* (1959). This device is most useful for small light-bodied species such as leafhopper adults and parasitic Hymenoptera. Catches in the suction traps are influenced by wind speed, and the size of the insect. Correlations of wind speed with insect catches and size of the insect have been worked out as a logarithm, which is added to the log catch/hour. Correlations have also been made for trap height above the ground level and the volume of air sampled.

### *Water traps*

Water traps are shallow plastic or metal trays, 5 to 8 cm deep, and 0.1 $m^2$, filled with water, containing a detergent or mineral oil film. The traps are placed in or near the crop. They can also incorporate a colour, a pheromone, or an attractant. Water traps are more effective if raised above the ground level. The catch need to be removed daily and the trap re-filled with water. Water traps have been used in combination with yellow colour for rice pests, aphids, and with fishmeal for sorghum shoot fly (Sharma *et al.*, 1992). They can be used for several insects in combination with sex pheromones. Relationship between wind speed and insect catch is less simple than for sticky traps (Southwood, 1978). Water traps need to be attended more frequently. They may overflow in heavy rain or dry out in hot sun. However, insects are caught in good condition for identification, and the catch can be easily separated by forceps or by straining.

### *Interception traps*

These traps intercept insects at random as they move through the habitat. Indices of absolute population estimates can also be obtained from these types of traps than from other types. These traps are particularly useful for determining direction of flight and can also provide data as to when dispersal flights occur. One of the most important uses of stationary traps is to determine the direction of flight. The direction of migration can be determined by a number of stationary flight traps arranged so as to sample from four fixed directions.

***Malaise traps***: A Malaise trap essentially consists of a tent made of netting, with one side open, into which insects either fly or crawl. The net may be black or green in colour. It was developed as a collectors tool by Malaise (1937) and has been modified by other workers. Malaise traps should be placed along the flight paths such as woodland paths but cannot be used in windy situations. Most insects crawl up the netting, and once inside, they can be trapped in vials of preservative placed in the upper corners or the peak of the tent. Since these traps depend upon the insect to enter them accidentally, they work best for highly active species such as adult Diptera and Hymenoptera.

***Air-flight traps***: These are stationary flight traps that are not attractive to insects. One of the simplest type is the suspended cone used by Johnson (1950). These may have a windwane attachment to ensure that it swivels around to face into the wind (Plate 1). Its efficiency can be determined at different wind speeds.

*Windowpane traps:* Flying Coleoptera can be sampled using windowpane traps, which consist of a vertical pane of glass or Plexiglas with a trough of preservative beneath it. Any insect that hits the glass and reacts by falling is caught (Chapman and Kinghorn, 1955).

Plate 1. A wind trap.

### *Sticky traps*

Sticky traps are of several basic designs. Sticky traps (Plate 2) are conveniently thought of as a variation on windowpane traps in which glass, screening wire or some other surface is covered with some sticky substance to hold the insect where it lands rather than allowing it to fall into the preservative. The shape of sticky traps may be flat, cylindrical, triangular or spherical depending on the behaviour and size of the insects. Cylindrical and spherical traps are efficient for passive collection, while triangular or flat structures are used in combination with pheromones. Sticky red-spheres have been used for estimating adult fly populations in orchards. Cylindrical traps have been used for aphids (Phadke and Prasad, 1984) and rice brown plant hopper. Protection of sticky traps from dust and rain is essential to maintain their efficiency. They can also be combined with colour, kairomones, or pheromones. Catch can be washed off with kerosene or any other suitable solvent. Sticky traps can be used for many species that cannot be sampled with windowpane traps. Since the insects are caught in place, the sticky substance can be put on almost anything and anywhere. Often these traps are combined with baits or attractants to enhance the catch of certain species. These traps are most often used with pheromone or kairomone baits to monitor insect populations.

**Plate 2.** A sticky trap.

### *Pitfall traps*

Pitfall traps have been used extensively for species that roam on the soil surface, such as ground beetles, spiders, and Collembola. They consist of smooth sided containers such as glass or plastic jars sunk into the ground with the top at the ground level. They catch insects falling accidentally into the trap. When used without baits, they catch whatever accidentally falls in but when used with artificial or natural baits, draw individuals from considerable distances. The catch need to be protected from predators such as ants and birds. Pitfall traps can be emptied by hand operated or by suction aspirators. Traps baited with beer are an often recommended control practice for slugs in home gardens. The efficiency of these traps is influenced by changes in weather conditions.

### *Light traps*

Light traps are the most widely used visual traps for estimating populations of insect pests in agriculture. They are particularly important in surveillance programmes and the seasonal appearance of many insect species (Bowden, 1982). A light trap basically consists of a light source above a funnel and a container below to collect the insects. The light source can be either an oil lamp, an electric bulb or a fluorescent tube. Because many insects are attracted to short wave lengths of the light spectrum, the black light emitting ultraviolet light is widely used in most traps (Plate 3). Some light traps may also have a fan to suck the insects into the trap. Once inside the trap, the insects may be collected and counted or killed by an insecticide such as dichlorvos.

**Plate 3.** Light trap (Robinson type).

A commonly used trap is the Robinson trap. This trap uses a mercury discharge ultraviolet electric bulb. Other traps such as Entech, Heistands, Rothamsted, New Jersey, and Pennsylvania have also been used (Karel, 1996). Trap catches are influenced by the density of insect population in the environment, responsiveness of the insect toward the light trap and efficiency of the light trap used. Insect catch in one light trap may not be the same as from another light trap nearby or from the same trap placed elsewhere.

As with other relative sampling methods, black-light traps do not yield data that allow the comparison of densities among species. The results are useful for comparing densities of a species from one year to the next and for monitoring the emergence dates of many pests. These traps have been used mostly for moths and mosquitoes, and are restricted to species that fly at night or at twilight. A light trap catch must be interpreted with the same caution as any other relative measure of abundance. With sufficient corroborative data on absolute population size, it may be possible to convert light trap catch to absolute density.

### *Colour traps*

Insects are attracted to various colors, and this behavioural characteristic has been used to monitor various insects. Some insects are attracted to yellow while others prefer white. Yellow traps are effective in attracting various species of aphids, white flies, sorghum midge (Sharma *et al.*, 1990), and jassids, while white traps are useful against pigeonpea pod fly, *Melanagromyza obtusa* M., sorghum head bug, *Calocoris angustatus* Lethiery and groundnut thrips, *Frankliniella schultzei* (Trybom) (Pawar *et al.*, 1985). Colour traps are generally made of white or yellow galvanized sheets, plastic jars, or a tin box suspended by a support. The colour, size, and shape of the trap is determined through experimentation. Colour can also be combined with water traps, sticky traps or pheromone traps.

### *Attractant traps*

Species that are not attracted to light may be trapped by using a sticky board in combination with an attractant or bait. Pheromones and kairomones are often used for this purpose. Crop extracts, fruits or fruit extracts, fishmeal, alcohols, aldehydes, etc. are commonly used in assessing insect populations, e.g., banana weevil, *Cosmopolitus sordidus* (Germ.), is attracted to cut pieces of banana stem, *Rhinoceros beetle to coconut stem, Drosophila* to pieces of tomato; moths and ants to sugar molasses; sorghum shoot fly, *Atherigona soccata* Rondani to fishmeal; fruitflies to fruits and esters; housefly *Musca domestica* to sugar and propionate; diamondback moth, *Plutella xylostella* (L.) to sinigrin and elaterids and scarabeid beetles to ethyl propionate and phenols (Karel, 1996).

Pheromones exploit chemical sex attractants emitted by the insects to attract the opposite sex. Sex pheromones are commonly produced by the females. Several sex pheromones have now been synthesized commercially for a wide range of insects. In cases where synthetic sex pheromones are not available, the traps can be baited by virgin females in screened cages.

Traps of various designs have been used for trapping insects using sex pheromones. These include flat, cylindrical and triangular shapes, cartons, funnels, plastic bags, sticky surfaces, and water baths depending on the behaviour of the insect. Water (in a square pan trap, Plate 4), sticky materials such as tanglefoot, grease or vaseline (Plate 5), or a polyethylene bag (Plate 6) can be used to collect the insects. Some of these traps can also be modified to collect the insects alive (Plate 7), which can either be released again in nature or used for other entomological studies. The data must be interpreted cautiously. The fact that some pheromones are highly specific and attract individuals from great distances makes pheromone traps a useful source of data.

Plate 4. Square pan trap. Water with a emulsifier or mineral oils is used to catch the insects. An attractant (a kairomone or pheromone) is placed in the center of the trap. Several modifications of this trap have been used for different insects.

Plate 5. A pheromone trap using a sticky material on a platform for catching the insects.

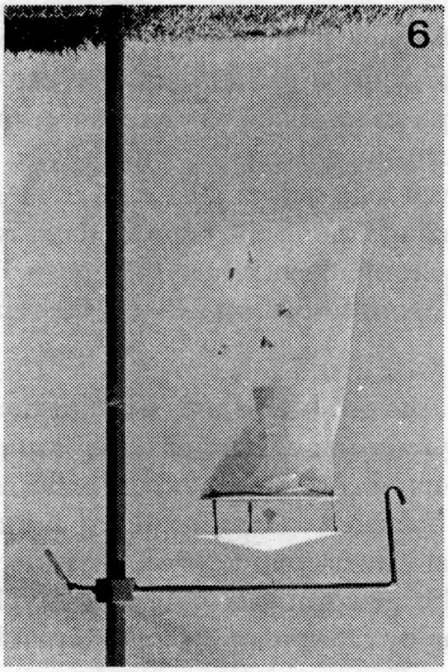

Plate 6. A pheromone trap using a polyethylene bag to catch the insects (mostly Lepidoptera).

Plate 7. A modified funnel trap, which can be used to catch the insects alive (this trap has been used to catch sorghum shoot flies for other entomological studies).

## Indirect Sampling Methods

### *Insect products*

In some cases, a species that is difficult to sample directly, produces products that can be easily sampled by absolute methods. The insect product most often sampled is the frass produced by lepidopterous defoliators or honey dew by aphids and plant hoppers. The rate at which frass is produced by defoliators can be estimated from the amount falling into a box or funnel placed under the trees. The size and shape of the frass pellets are rather constant for a given species and instar. This allows one to identify the species and age composition of the defoliators. Density estimates are more difficult but work done by Waldbauer (1964), Gosswald (1935), Green and DeFreitas (1955), Pond (1961), and many others helps to convert frass catch to absolute population density. Insect webs are another product that can be sampled rather easily. The webs of caterpillars can be counted to obtain a fast yet reasonably accurate estimate of larval density over large areas (Morris, 1964). Insect exuviae (larval or pupal) are often left in conspicuous positions, where it is possible to count them and these can also provide a measure of the population density.

### *Assessing plant damage*

The nature and amount of damage caused are also important in studies in insect ecology. Measures of absolute population can be obtained from plant feeders when they enter the plant, e.g., for leaf miners, it is possible to determine the number of larvae that commenced mining, the number that completed development, age and cause of death of those that failed. The amount of damage caused by arthropods to crop plants is a function of the pest density, the characteristic feeding or oviposition behaviour of the arthropod species, and the biological characteristics of the plants. Each of these factors is differentially affected by environmental and other biotic factors, and a correlation between population levels and plant damage is frequently difficult to establish. Although correlation of a certain type of injury with the causal organism presents no difficulty in most cases, there may be instances in which symptoms are observed long before the causal organism is identified. When we find an ear of corn with a worm tunneling among the kernels, there is little doubt about the cause-effect relationship. In the development of adventitious roots on soybeans stimulated by the feeding of three-cornered alfalfa hopper, *Spissistilus festinus* (Say), the relationship is less obvious.

*Measuring losses by insects causing direct damage:* In this case, the produce is the object of the sampling procedure and techniques to evaluate damage usually refer to absolute or relative numbers of damaged units, for example, ears of corn per 10 plants, number of apples per tree, bean pods per 4 m row, or cabbage heads per ha. Different levels of damage can be recognized and this is sometimes necessary for the conversion of damage to actual value. Prasad (1963) classified cabbage heads as nonmarketable if feeding by the imported cabbageworm was more than two leaves deep or the heads were less than 7.5 cm in diameter. Feeding by caterpillars on the external leaves was, therefore, not enough to disqualify a cabbage head.

Damage to sorghum by the sorghum midge, *Stenodiplosis sorghicola* (Coquillett) can be estimated directly as a percentage loss in terms of number of chaffy spikelets (Sharma *et al.*, 1992). Ten primary branches can be selected at random from sorghum panicles at the milk stage or at maturity. All spikelets are removed from the primary branches, and 250 spikelets are picked up at random. The percentage of damaged spikelets (chaffy spikelets without grain) correspond to percentage yield loss.

Number or percentage of plants damaged can also be recorded as an indication of the population density of the pest. It is a quick and simple method for stem borers, shoot flies, termites, cutworms, beetles, and other soil insects. The number of internodes bored at various growth stages can also be used to assess some insect populations. Leaves having holes, mines, rolls, leaf scrapping or lesions, are measured for caterpillars, semiloopers, caseworms, leaf folders, leaf miners, leaf beetles and bugs, mites etc. The actual leaf area damaged can be measured by graphical or photographical methods, planimeter or by leaf area meter. Damage to roots can be measured in terms of amount of roots damaged, percentage roots pruned below the soil surface, length of roots damaged, weight or volume of the roots damaged. Holes, lesions or other type of damage Rating scales can either be descriptive (nil, light, medium or high) or numeric. The grades of rating scale can represent arithmetic increases (e.g., 1, 2, 3, 4, and 5 being 1-10, 11-20, 21-30, 31-40, and 41-50% damage, respectively). Experienced entomologists are often capable of visually estimating with reasonable accuracy the percentage of defoliation of a plant stand. This is a subjective estimation and often estimates made by different persons in a field may differ significantly.

*Measuring losses by insects causing indirect damage:* Indirect damage to crops includes damage due to defoliation by insects with chewing mouthparts (e.g., grasshoppers, lepidopterous caterpillars, leaf beetles, etc.), insect feeding on the root system, tunneling by stem borers, damage by sap sucking insects, and damage by vectors of plant pathogens. In such cases, one has to determine the relationship between insect numbers or extent of damage with loss in crop yield.

Extensive studies on the effect of insect feeding on roots have been carried out with the corn rootworm larvae feeding on corn (Chiang, 1973). The criteria used to evaluate damage to the root system include; 1) the amount of damage to the entire root system, 2) the percentage of nodes below the soil surface having injury and 3) the percentage of roots pruned on the first and second nodes below the soil surface. Indirect measurements use 1) the force needed to pull a plant from the soil, and 2) ratings of damage based on plant lodging (Chiang, 1973). The vertical-pull technique was developed by Ortman *et al.* (1968), it uses a grip that is attached to a cut corn plant and to a 453.6 kg-capacity recording dynamometer. The pulling weight is highly correlated with visual ratings of corn root damage, but soil factors can greatly influence the results of these measurements. The technique offers the advantage of providing quantitative data free of bias by the observer.

Correlation of damage level with yield loss is usually simpler in case of pests causing direct damage. If one counts the number of damaged apples in a barrel, one can easily calculate the percentage of damage and the consequent loss in yield. Grades of damage can be established to account for losses that result in unmarketable fruit or those with reduced quality but still marketable fruits. The need for sound crop loss evaluations in pest management programmes is generally recognized (Chiarappa *et al.*, 1972). Chiarappa (1971) compiled extremely valuable information on crop loss methodology and Judenko (1973) described some analytical methods for assessing yield losses caused by pests in cereal crops.

**Remote sensing**

Photographs or transmitted images taken from air by an aeroplane or a satellite can also give an indication of the extent of insect damage. These techniques are based on the fact that insect damage influences the water balance of plants, and changes the crop reflectance which can be measured by normal or infra-red photography. Remote sensing can be used for detection of insects, extent of damage, and the environment conducive for their development (Riley, 1989). Aerial photography with films sensitive to infra-red provides accurate delineation and recording of damaged vegetation. Sooty moulds which grow on honey dew can also be detected by this method, e.g., sooty moulds on cotton produced by whitefly can be detected at 3 km from ground level by infrared photography. Multispectral sensing (MSS) can form images over a wide range of electromagnetic wavelengths than phototactic methods. Satellite based MSS has been used to detect changes in vegetation due to insects or diseases (Vegelmann and Rock, 1989). Remote sensing is a useful tool for investigating environmental factors controlling insect and disease development. Meteosite satellite data have been used to study concentration of moths of African armyworm, *Spodoptera exempta* Walk. and locating the outbreaks of the pest (Garland, 1985). Landsat satellite data have been used to identify areas capable of supporting desert locusts. Damage by rice brown planthopper has been identified by SPOT and Landsat TM. Radars have also been used successfully for studying long distance migration and flight behaviour of windborne insects. The Indian Council of Agricultural Research (ICAR) in collaboration with Indian Space Research Organization (ISRO) is involved in studies on plant-soil-atmosphere-continuum (Kamat and Sinha, 1984). Remote sensing for plant protection is being carried out by the Indian Remote Sensing Satellite Utilization Programme (IRSUP) and Remote Sensing Application Mission (RSAM) of the Department of Space.

## DISPERSION

The dispersion of a population, the description of distribution pattern of insects in space, is of considerable ecological importance. It is important for devising an appropriate sampling plan, data analysis and a description of the size and condition of the population, e.g., mortality factors affecting clumping of a sessile population indicate that it acts at the most dense populations or if the dispersion of a population becomes more regular, then intensification of competition may be expected. Several mathematical models have been used to describe the spatial distribution of insects.

### Normal distribution

The normal or Gaussian distribution is the term applied to a continuous variable. When frequency is plotted against magnitude, it gives a symmetrical bell-shaped curve. The dispersion pattern of animals is seldom normal. The normal distribution is rarely of interest to ecologists, except as a means of describing dispersion. Its importance arises solely from the fact that for most statistical methods to be applied to a set of data, the frequency distribution must be normal, and it should possess the associated properties that variance is not dependent on mean, and its components are additive. If the distribution is skewed, the data can be transformed to normalize the data or stabilize the variance. The distribution of individuals in a natural population is such that the variance is not independent of mean.

### Poisson distribution

It is a mathematical distribution describing a truncated random distribution with variance independent of mean (Southwood, 1978). Poisson series gives a curve which is described completely by one parameter and the variance is equal to mean. The probability (P) of finding a certain number (x) of animals in a sample from a population with a given mean (m) in a Poisson distribution is given by:

$$Pn = e^m (m^n/n) \quad (5)$$

Where e = base of natural logarithms (= 2.71828), n = the expected number of insects per unit area, and m = the mean number of insects. One simply plugs the values of 1, 2, 3 and so forth into the formula to determine subsequent probabilities. Value of $e^m$ may be found by using a table of these logs backwards. The goodness of fit of a set of data to Poisson distribution may be tested by chi-square on the observed and expected values or by index of dispersion. An expected value should not be less than 5 for a valid application of chi-square test of goodness of fit.

### Negative binomial

Negative binomial is one of the distributions that describes contagious patterns in which the variance is dependent on mean. This distribution is described by two parameters, the mean m and the exponent k, the latter is a measure of amount of clumping, and is referred to as dispersion parameter. Generally, the values of k are close to 2, and as they become larger, the distribution tends to be identical with Poisson. The value of k increases with the mean. To compute the expected negative binomial frequencies, one needs to estimate:

$$q = S^2/m \quad (6)$$

$$k = m^2/(S^2 - m) \quad (7)$$

where $S^2$ = the variance, k = is the index of aggregation, and m = the mean number of insects per unit area. These formulae provide a reasonable estimate of q and k, but are valid only for data with a low mean, and include a high proportion of zeros. The probability $P_{(0)}$ of finding no insects (n = 0) is:

$$P_{(0)} = q^k \quad (8)$$

Each subsequent probability $P_{(n)}$ is calculated from the previous one $P_{(n-1)}$ by:

$$P(n) = P(n-1)[k-(n-1)]/n[(1-q)/q] \quad (9)$$

A table of probabilities is calculated to arrive at the expected frequencies. If repeated sampling over a range of densities yields data conforming to negative binomial, this strengthens the validity of the general model and its application. Not all contagious populations conform to negative binomial. Value of k is influenced by the size of the sampling unit and the number of samples, because it influences the variance which is used to calculate k.

**Logarithmic and other contagious models**

A number of other models have been developed to describe various non-random distributions. These have been reviewed by Anscombe (1950) and Evans (1953). Logarithmic models describe situation in which negative binomial would give a very small value of k. The logarithmic series, which is derived from negative binomial with values of k tending to be zero, with zero values neglected and the discrete and truncated log normal distributions have been found to be useful in describing relationship between number of species and number of individuals.

**Transforming data**

Use of untransformed data based on normal distribution is limited to insect populations distributed randomly. However, the dispersion pattern of a population is rarely normal. For most statistical methods to be applied to a set of data, the frequency distribution must be normal and the data should possess associated properties that the variance is independent of mean and its components additive. The analysis of variance is more robust than chi-square test (Reimer, 1959). The data whose frequency distribution is more skewed with the variance more closely related to mean cannot be analyzed without the risk of an error (Beall, 1940). In order to overcome these problems, the data are transformed, i.e., the actual numbers are replaced by a function which normalizes the data or stabilizes the variance.

The first step is to compute the mean and the variance. The next step is to check whether the insects are distributed at random. To choose an appropriate transformation, the value of b may be calculated (see under Taylor's Power Law) and hence that of P.

$$P = 1 - 1/2b \quad (10)$$

If P = 0 - a logarithmic transformation should be used; if P = 0.05 — use square root transformation, reciprocal square root and if P = 1.0. Healy and Taylor (1962) have given the tables for

P = 0.2, 0.4, 0.8, and for negative power. In practice, when sampling and other errors are large, use squares for insects with a regular dispersion pattern. For insects with a contagious dispersion, use square roots, and for distinctly aggregated or contagious population use logarithms. To overcome the problem of zeros in logarithm transformation, a control (normally 1) is added to the original count (x) and expressed as log (x + 1). It is customary to transform percentages into angles (Arcsin) but Reimer (1959) suggested that it is only of value when the predictability of finding the individual bearing the attribute is uniform within each area, but varies considerably between different classes. Adequate transformation should eliminate or considerably reduce; 1) the skewness of the frequency curve and 2) dependence of the variance on the mean (instability of variance). An appropriate transformation should reduce the skewness of the frequency curve, and the dependence of the variance on the mean. The value of any transformation can be tested by the use of probability paper (Southwood, 1978). However, the non-normality must be extreme to invalidate the analysis of variance. It is better to concentrate on stabilizing the variance of the samples.

### Indices of aggregation/crowding

Because the biological causes of aggregation are behavioural, its mode of action is unpredictable, as also the biological mechanism appropriate to a distribution. The index of aggregation is a more feasible approach to analyses than the frequency distribution because it makes less demands on the data. Several indices of aggregation, including those of Green (1966) and Lefkovitch (1966), based on the mean and the variance have been used.

#### *Taylor's power law*

Taylor (1961) proposed an exponential function to describe the overall relationship between variance and mean for contagious populations in the absence of fit to the negative binomial. It is known as the Taylor's power law. It is a measure of aggregation (b), is constant for a species, and is of value in interspecific comparisons.

$$S^2 = am^b \tag{11}$$

where a = $S^2$. When m = 1, and b = the slope of plot log $S^2$ on log m. The series of means and variances necessary to calculate b may be obtained from several sets of samples of different sizes or by combining samples from different sized sampling units. The mean and variance are calculated from the raw data by the usual method or by means of probability for a less exact approach (Southwood, 1978). The value of m and $S^2$ are plotted on log to log scale and the value of a is read off on the $S^2$ axis at the value of m = 1. The value of b can be calculated from the following equation:

$$\text{Log } S^2 = \text{Log } a + b \log m \tag{12}$$

Taylor (1984) suggested that a varies with sampling technique while b seems to assume a typical constant value for each species. As such, it may be more appropriate index of aggregation than k of the negative binomial. Value of k in the negative binomial is related to the Taylor's power law by:

$$k = m/[am^{(b-1)} - 1] \tag{13}$$

where a and b are constants. If a = b = 1, then variance is equal to mean, and the data can be fitted into Poisson distribution.

### *Mean crowding index*

Mean crowding is the overall effect of each individual of the crowd, averaged over the local density. Its main aim was to devise a method that measures the effects of others on the behaviour of the individual. Mean crowding and patchiness only have a meaning when the population is loosely aggregated and the quadrat size is not too large as compared with the "ambit" of the individuals. Iwao (1968) indicated that these parameters can be used for any organism or quadrat, without reference to biological interactions. Llyod (1967) and Iwao (1968) devised a mean crowding index M*, when the negative binomial was found to be inadequate.

$$M^* = m + [S^2/m - 1] \tag{14}$$

where m = mean and $S^2$ = the variance. The individual, who is the center of observation, is omitted from consideration in the mean crowding index. This is the sum of mean density representing the population and the variance ratio representing the crowding effect, less one, which is the individual at the center. The other measure introduced by Llyod (1967) was patchiness, the ratio of mean crowding to mean density, i.e., mean crowding ratio equivalent to the variance ratio.

### *Index of dispersion*

Departure of distribution from randomness is often referred to as coefficient of dispersion. The extent to which the distribution satisfies the Poisson distribution is tested by $\chi^2$:

$$\chi^2 = S^2(N - 1)/m \tag{15}$$

where $S^2$ = the variance, N = the number of samples, and m = mean. In Poisson distribution, the value of $\chi^2$ will not be outside the limits, normally 0.95 and 0.05 of chi-square for N – 1 in tables. If the distribution conforms to Poisson distribution, the index of dispersion (I) will approximate to unity.

$$I = \chi^2/(N - 1) \text{ or } I = S^2/m \tag{16}$$

A value of zero would imply that the insects are regularly distributed, and a value significantly greater than one implies aggregation. In contagious distributions, the value of this index is influenced both by the size of the sampling unit as well as by the density.

### *Morisita's indices of dispersion*

Morisita (1954, 1959) developed an index intended to be free from frequency distributions. The basic premise is that all spatial distributions become random or regular when quadrant size is sufficiently small. The parameter $I_a$, equated with Simpson's (1949) diversity index, is independent of total count, mean diversity and size and number of quadrants provided each sample is taken from each of the q groups of a infinite population where q is the number of quadrats. Patil and Stiteler (1974) argued that the basis of this hypothesis is consistent. Morisita (1959) considered its relationship to binomial distribution and derived parameters for simple and stratified sampling (Morisita, 1962), but with some reservations about their application (Morisita, 1964).

It is a measure of dispersion pattern, and is independent of the type of distribution pattern, number of samples and mean. Morisita's index ($I_a$) is given by:

$$I_a = N[\{\Box x^2 - \Box x\}/\{(\Box x)^2 - \Box x\}] \qquad (17)$$

where N = the total number of samples and $\Box x$ = sum of numbers of individuals found in all the samples. When the distribution is Poisson, this index gives a value of unity and greater than one for a negative binomial, and for a regular distribution (binomial), a value of less than one. Another index devised by Morisita (1962) can be used to compare distribution of insects on different parts of a tree or habitat. It allows comparison of populations with different number of samples. Unlike $I_a$, it allows the direct comparison of populations with different number of samples, and is given by:

$$I_b = I_a\{(G - 1)/N\}/\ (G - 1) \qquad (18)$$

where G = number of groups, periods or sub-areas, N = total number of samples in the whole series. $I_a$ is calculated for each group.

***Breder's equation***

It is a measure of cohesion of aggregation (Breder, 1954). When the individuals are normal distance apart d, the attractive force between individuals a can be equal to the repulsive force r, the measure of cohesion c is zero as given by:

$$c = (a - r)/d^2 \qquad (19)$$

Negative value of c will indicate overcrowding, and there will be a tendency among the individuals to move apart. Positive value will indicate that a normal aggregation has not been reached, and this will approach +1 if the attractive force is taken as one. If $r_1$ and $r_2$ are the repulsive forces of two individuals on each other, then:

$$c = (1 - r_1 r_2)d^2 \quad \text{or} \quad r_1 r_2 = ds^2 \qquad (20)$$

where ds = mean normal distance between individuals in the aggregation. Under these conditions c = 0. Breder's equation can be used to compare insects such as locusts and armyworms, which form swarms. Smaller the mean distance(in terms of total length) between individuals, the more stable the aggregation.

***Nearest neighbour***

This has been discussed partly under monitoring insect populations. It is a measure of population size or of the departure from randomness of the distribution. One may either select an individual at random and measure the distance between it and the nearest neighbour or select a point and measure the distance between the point and the nearest or nth nearest individuals. From this information, inference may be drawn about the departure of the distribution from random or if the distribution is assumed to be random, its density can be estimated (Morisita, 1954). This technique is used with stationary, discrete and easily mapped organisms such as deadheart production by stem borers, scales, gall insects or imprinting method for estimating mite numbers on leaves. It can be estimated by (Clark and Evans, 1954):

$$m = 1/4r^2 \quad (2)$$

where m = density per unit area, and r = mean distance between the nearest neighbours.

*Deevey's coefficient of crowding*

This measure is used to express the extent of crowding (Deevey, 1947). It is given by:

$$C_c = 2\pi r^2 N^2 \quad (21)$$

where r = radius of the fully grown animal, $\pi$ = 22/7, and N = density of the animals per unit area. This coefficient is used with sedentary, approximately circular animals (circular galls); when a realistic value can be given to r, it can also be used for mobile forms.

*The $S^2/m$ ratio*

It is based on a concept of aggregation as an absolute property, independent of sampling and density for comparison of distribution events (individuals) within periods for sub-areas of an event (area, plant, etc.). Patil and Stiteler (1974) proposed this as a ratio of two variances in the form of:

$$S^2/m = S^2/S^2_r \quad (22)$$

where $S^2_r$ = is the random variance equivalent to the mean. Considered this way, the variance mean ratio is a measure of deviation from randomness. The size and shape of the sample affects the ratio. In practice, $S^2/m$ changes with m in a characteristic way within a species, and in a different way between species. Llyod (1967) indicated that $S^2/m$ is a misleading measure of aggregation.

*The negative binomial*

Bartlett (1947) gave an explanation of variance greater than mean as a variance law based on the assumption that the mean itself varies leading to quadrics in m.

$$S^2 = m + m^2/k \quad (23)$$

where $S^2$ = variance, m = mean and k = a general, reciprocal index of dispersion which varies as the parameter of the negative binomial. If negative binomial can be fitted into the data the value of k gives the measure of dispersion. The smaller the value of k, the greater the extent of aggregation. A large value of k (>8) indicates that the distribution is approaching Poisson. Unfortunately, the value of k is influenced by the size of the sampling unit, and therefore comparisons can only be made for the same sized sampling units. The degree of aggregation of a population estimated by k-expression could influence predation and parasitism (Waters, 1959). The value of k can be computed by several methods (Anscombe, 1949, 1950; Bliss and Fisher, 1953).

$$k = m^2/(S^2 - m) \quad (24)$$

where m = mean and $S^2$ = variance. The efficiency of this method in relation to various values of k and m can be seen in Fig. 1. Unless the mean is low, it is not reliable to use this method with populations that show a moderate degree of clumping (i.e., k of 3 or less). This limits the use of this method to low population densities. Another approach is described by the following equation:

$$\log (N/n_0) = k \log (1 + m/k) \qquad (25)$$

where N = total number of samples, $n_0$ = number of samples containing no insects. The easiest method to find unknown k is by substituting various values of k into the equation until two sides are equal. A reasonable values of k to start with is 2.

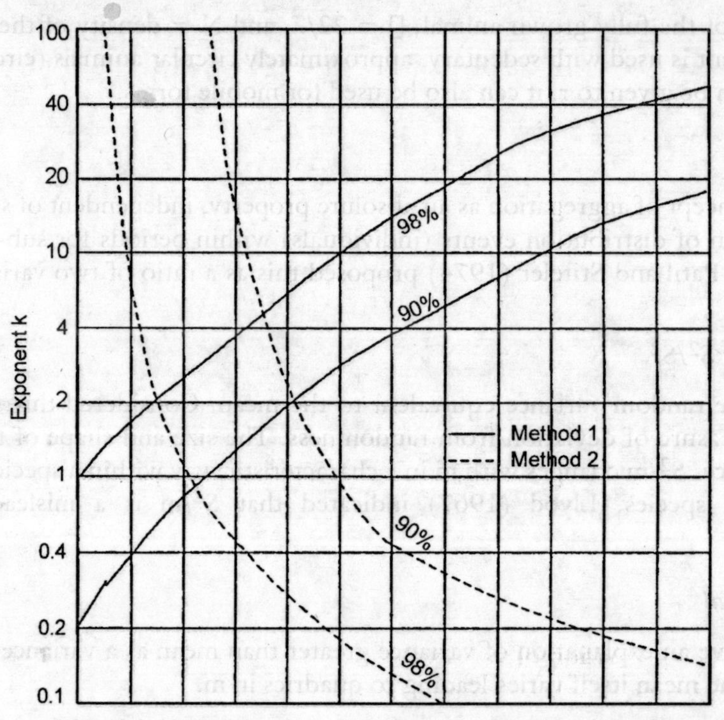


**Fig. 1.** Efficiencies of estimation of k of the negative binomial distribution by methods I and II (for details on methods I and II, see Anscombe, 1950; Southwood, 1978).

The aggregation recognized by negative binomial may be either due to active aggregation or some heterogeneity of the environment (microclimate, soil, plant, natural enemies, etc.). With high infestation, k may be considerably larger. Anscombe (1948) gave a formula for the Poisson, binomial and negative binomial, and tried to fit a constant k into different data sets. Again, k was not constant but increased with m. Kuno (1968) derived 1/k, and showed that random removal of individuals did not change 1/k. If the mortality is presumed to be spatially random, to account for density changes, aggregation as measured by 1/k will remain constant for survivors.

## Relationship between K, $I_a$, and M*

The parameter k derives directly from the assumption of a negative binomial distribution with an apparently unlimited number of potential statistical models (Patil and Joshi, 1968). In contrast $I_a$ has been devised specifically to avoid the constraints of a statistical distribution in describing the nonrandom component of spatial biological distributions (Morisita, 1954). Mean crowding M* is a completely novel biological model, devised within strict biological constraints (Llyod, 1967). The constraints may not be binding depending on the relationship of M* to mean density (Iwao, 1968), but if the relationship does not hold, the original constraints on M* must be applied. These three parameters are closely related despite their different origins (Morisita, 1962; Llyod, 1967).

$$1/k + 1 = I_a = M^*/m \qquad (26)$$

They are equivalents, not equal. There is a close relationship between them for practical purposes. It is therefore likely that one of them is dependent on density and possibly the others as well (Taylor, 1984). If density dependence of 1/k is well established in real biological populations, $I_a$, which differs by unity from 1/k, must have the same functional form, and M*/m a similar one. The close relationship with negative binomial appears when M* is converted back to mean and variance from which it is derived.

According to Llyod (1967):

$$M^* = m - (S^2/m - 1) \qquad (27)$$

and Iwao (1968):

$$M^* = \square + \square \qquad (28)$$

Hence, according to Taylor (1971):

$$S^2 = (\square + 1)\, m + (\square - 1)\, m^2 \qquad (29)$$

The similarity between this variance law and the negative binomial suggests that despite different biological perceptions of these authors, these models are examples of ubiquity of the negative binomial, where:

$$S^2 = m\,(1 + \square\, m) \text{ (Bartlett, 1936)} \qquad (30)$$

$$S^2 = \square^2\, m + \square^2\, m^2 \text{ (Bartlett, 1947)} \qquad (31)$$

$$S^2 = m + m^2/k \text{ (Anscombe, 1949)} \qquad (32)$$

Pieters and Sterling (1974) measured 1/k and $I_a$ in several insects. All taxa fit closely to a linear regression of 1/k on ($I_a - 1$), although a factor of 2 appears to have been lost in computing (Taylor, 1984). Kogan *et al.* (1974) illustrated the linear relationship between 1/k and $I_a$ estimated by different methods; 1/k and $I_a$ are almost identical to each other over most of the range but these parameters are not considered in relation to density. It is also necessary to consider these indices in relation to density over space and time. An index must take account of density in order to find parameters that are preferable and species specific.

## THE SAMPLING PROGRAMME

Reliable estimation of population density is also an essential pre-requisite to any meaningful research into factors determining distribution and abundance of insects. It is almost impossible to count every individual in a population of free-ranging insects. Therefore, we must devise a scheme to sample a proportion of the population and to use such samples along with appropriate statistical techniques to infer generalizations about the population as a whole. Statements about insect populations are only as much accurate as the sampling methods employed, together with validity of statistical analyses performed on the resulting data. At the outset, the investigator must know as much as possible about the biology of the target insect population. Information about activity periods and critical thresholds of survival must be sought for temperature, humidity, light, and colour response, e.g., alate aphids fly primarily during morning and late afternoon hours and are more likely to be collected in yellow pan traps at these times rather than at mid day. Parasitic Hymenoptera rarely fly at temperatures below $10^{o}$C, particularly when sun is obscured. The crepuscular activity of many Diptera is well known. Most moths are attracted to light and careful studies have revealed species-specific differences in flight times. There may be considerable differences in insect response to stimuli, and this becomes highly important when traps are used to monitor insect populations. Individual insects also vary in their ability to avoid traps and an attractant may hold greater or lesser appeal according to age or mating status. Statistical techniques for analysis of data often assume that insects behave identically or that the variation in response is random, whereas neither condition may be satisfied in the real world.

Preliminary sampling is the first step in designing a sampling programme to estimate pest density. An intensive investigation on the effects of environmental factors on an insect might concentrate on a single location and demands a high degree of precision, whereas forecasting pest outbreaks for growers may extend over entire region, state or nation, and high precision may be of secondary importance to timely information. Before statistical inference can be legitimately used, the relationship of an insects distribution to that of an appropriate statistical population should be determined. Many statistical techniques are based on normal distribution (variation is random, and mean and variance are independent of each other). Before any statistical analyses are carried out, we must determine appropriate corresponding statistical distribution whose relation to the normal is known, whereupon the observed data can be normalized.

The distribution of insects in two dimensional space assumes certain patterns (Fig. 2) that correspond broadly to three types. In a normal distribution, individuals are spaced evenly at regular intervals; in a random pattern, the occurrence of each individual is independent of the occurrence of the other; and in a clumped or contagious pattern, most typical of insect populations, the presence of any individual increases the probability of another being found nearby. Apparent dispersion pattern changes with a change in insect density or sampling unit (Taylor, 1984). Alate aphids settle randomly on plants early in the season, but their distribution becomes clumped soon after they commence reproduction (Horn, 1981). Also if the area sampled is expanded, the distribution pattern may change. Plant phenology changes during the cropping season and this may change the apparent distribution of phytophagous insects (Stanton, 1983).

Hutchinson (1953) recognized five kinds of spatial distribution patterns: vectorial — caused by environmental gradients; reproductive — producing density gradients as a result of reproductive behaviour of an insect; social — caused by foraging and social behaviour of a species; coactive

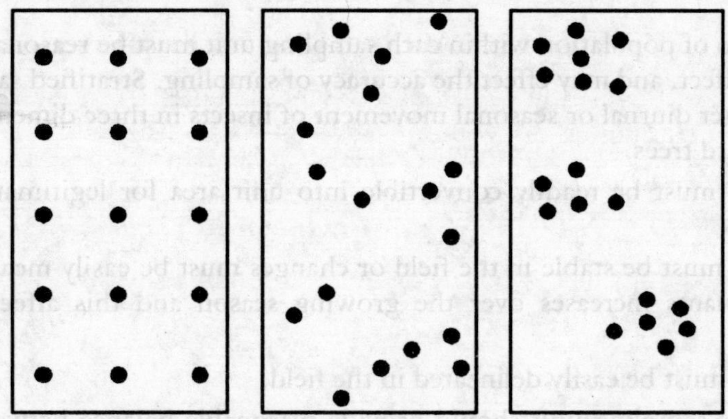


**Fig. 2.** Dispersion patterns of insects. a) regular — the insects are equidistant from each other and under these conditions, the variance is less than the mean; b) random — the probability of occurrence of one insect is independent of other and the variance is approximately equal to mean; and c) contagious or clumped — occurrence of an insect at a point increases the probability of occurrence of other insects nearby and the variance exceeds the mean.

— behavioural interactions between species, and stochastic — result of randomizing environmental processes. It is futile to discuss the spatial structure without empirical experience. The following points may be taken into consideration for deciding on an appropriate sampling plan for an insect.

### Sub-divisions within a habitat

If the distribution of the population is not normal in certain sub-divisions, and the samples are taken randomly, systematic errors will arise. This can be overcome by sampling so that the differential number of samples from each sub-division reproduce in the samples the gradient in the habitat or by recording each part separately and correcting at the end. There may be differences in insect numbers at different levels of the same plant. Sometimes, it may be important to sample at different levels of the same plant. The aspect of the plant to be sampled may also be important for some insects. Sometimes, more insects are found on the south-east aspect than in the north-west aspect. Taking a certain number of samples at random within a habitat, which itself is selected at random from a larger area, e.g., in a field, is often referred to as nested sampling.

### Sampling unit and size

A valid sampling technique should take into account the following factors:

- All units within the habitat must have an equal chance of selection. Random selection satisfies this condition, though one must guard against any systematic, but unconscious bias. Such methods include dice, playing cards, or random numbers.

- The proportion of population within each sampling unit must be reasonably constant. Often there is edge effect, and may effect the accuracy of sampling. Stratified sampling may be necessary to counter diurnal or seasonal movement of insects in three dimensional habitats such as forest/orchard trees.
- Sampling unit must be readily convertible into unit area for legitimate comparison with other studies.
- Sampling unit must be stable in the field or changes must be easily measured, e.g., the biomass of the plants increases over the growing season and this affects the of the areas sampled.
- Sampling unit must be easily delineated in the field.
- The size of the sampling unit should provide reasonable balance between sample variance and cost of sampling.

To compare various sampling units in respect to variance and cost, it is generally convenient to keep one or the other constant. The same method of sampling should be used throughout. From preliminary sampling, variances of different units can be calculated and then computed to a common basis which is often the size of the smallest unit, e.g., if the smallest unit is 1 $m^2$, then the variance of 2 $m^2$ is divided by 2 and that of 3 $m^2$ by 3. The costs can be similarly reduced to a common basis. The relative net cost for the same precision for each unit will be proportional to:

$$C_u S_u^2 \tag{33}$$

where $S_u^2$ = is the variance per unit on a common basis, and $C_u$ = is the cost per unit on a common basis. The relative net precision of each will thus be proportional to $1/C_u S_u^2$. The higher the value of $1/C_u S_u^2$, the greater the precision for the same cost. As the population density and the variance is always fluctuating, too much emphasis should not be placed on the precise determination of optimum size of the sampling unit.

**Number of samples**

The number of samples to be taken depends on the degree of precision required which might be specified as standard error (SE) of a certain size or as confidence limits of a certain half-width as a percentage of the mean. For many purposes, a standard error of 10% of mean is reasonable. With normally distributed continuous data within an ecologically homogenous habitat, the number of samples N required is given by:

$$N = (tS/Dm)^2 \tag{34}$$

where S = standard deviation, D = required level of accuracy as a decimal of the mean (usually 0.1), m = mean and t = value of Student's t-variables, which depends on the number of samples. For more than 10 samples, t approximates to 2 at P=0.05. The larger the value of S with respect to D, the larger is the number of samples that are necessary.

If the dispersion of the population is well described by negative binomial, the desired number N of samples is given by:

$$N = t^2 pq/D^2 \tag{35}$$

where p = probability of occurrence, q = 1 - p and t and D are as above. To determine the optimum number of samples N per plant or part of it, the variance $S^2$ can be compared with the variance of between plant samples $S_p^2$ and set against the cost of sampling within the same plant $C_s$ or moving to another plant $C_p$ and sampling within it.

$$N_s = \sqrt{(S_s^2/S_p^2)(C_p/C_s)} \quad (36)$$

If the interplant variance is the major source of variation, and if the cost of moving from plant to plant is very high, N will be in the order of one or less.

## Pattern of sampling

If the objective is to estimate mean density for precise estimates, e.g., life tables, it is important to keep the variance minimum and if the objective is to get an idea of dispersion, then variance is of less importance. To get an unbiased estimate of the population, the samples should be collected at random. The samples are obtained by the use of random numbers from the whole area being studied. The method of stratified sampling is preferred for most ecological studies, where the area is divided into a number of equal sized sub-divisions or strata, and one sample is taken from each sub-division at random. Taking one sample at random, and the other at a fixed distance from it has been suggested for mapping aggregations. In systematic sampling, the samples are taken at a fixed interval in space or time. Such sampling generally cannot be analyzed statistically. Milne (1959) has shown that if centric-systematic area sample is analyzed as if it were random, the resulting statistics are as good if not better, than those obtained from random sampling. The centric-systematic sample is one which is drawn from the center of each area or stratum. It suffers from the weakness that it might coincide with some unsuspected systematic distribution pattern.

## Timing of sampling

Timing of sampling is determined by the life cycle of the insect. When only one stage of the insect is being sampled, it is important that this operation coincides with the maximum abundance of the insect. The faster the developmental rate of the insect, the more critical is the timing of sampling. Time of the day when samples are taken is also important. The diurnal rhythms of insects may cause them to move from one habitat to another, e.g., larvae of the Oriental armyworm, *Mythimna separata* Walk. hide in weeds, grasses, and leaf whorls during the day, but feed at night on the leaves. Many insects move up and down the vegetation in relation to weather changes. To overcome many of these problems, it is important to sample at dusk or dawn rather than during day time.

## Sequential sampling

The negative binomial distribution, when appropriate, can be used as a basis for devising a sequential sampling scheme to determine the statistical validity of a particular economic injury level (EIL). Sequential sampling has won wide acceptance as a means of providing cost effective technique for population estimation by pest management experts. The first step in formulating a sequential sampling plan is to set up lower ($x_1$) and upper ($x_2$) limits for EIL, and choose probabilities for Type I and Type II errors and if the data conform to negative binomial, calculate:

$$P_1 = x_1/k \quad (37)$$

$$P_2 = x_2/k \quad (38)$$

where $x_1$ and $x_2$ are the minimum and maximum EILs, respectively, chosen previously. The sequential sampling lines are determined by:

$$m = k[\log\{(1 + P_2)/(1 + P_1)\}]/\log[\{P_2(1 + P_1)\}/\{P_1(1 + P_2)\}] \quad (39)$$

$$y_0 = \log[\square/(1 + \square)]/\log[\{P_2(1 + P_1)\}/\{P_1(1 + P_2)\}] \quad (40)$$

$$y_1 = \log[(1 - \square)/\square]/\log[\{P_2(1 + P_1)\}/\{P_1(1 + P_2)\}] \quad (41)$$

With this, we can calculate the sequential sampling lines with the formula for a straight line:

$$y = bx + y_0 \quad (42)$$

$$y = bx + y_1 \quad (43)$$

where $\square$ = probability of Type I error, $\square$ = is the probability of Type II error, b = slope of sequential sampling lines, and $y_0$ and $y_1$ are the intercepts of sequential sampling lines.

The sequential sampling lines are then plotted (Fig. 3) and a pest management scout may use the graph in the field to sample the plants. The cumulative number of insects (y) is plotted against the cumulative number of plants sampled (x) until the plotted line enters either of the areas of

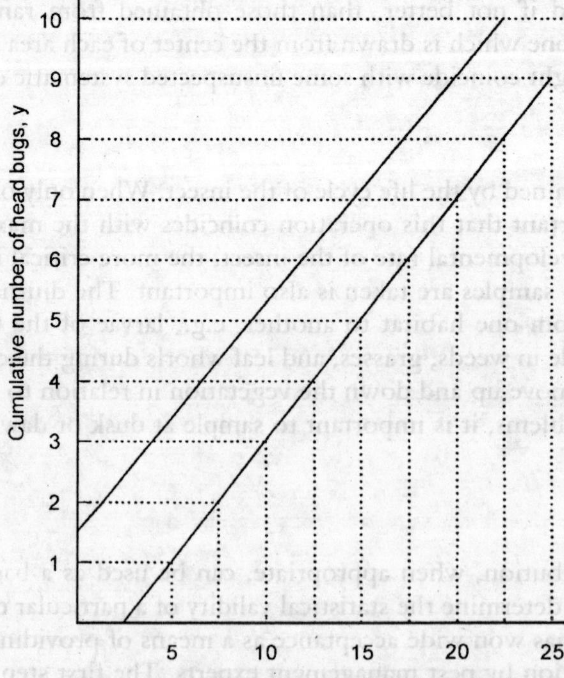


**Fig. 3.** Sequential sampling lines for sorghum head bug, *Calocoris angustatus* on sorghum (a hypothetical example) (modified after Horn, 1988).

decision. If a scout entering Field 1 counts on successive plants 5, 4, 2, 7 and 4 insects, the sampling ceases as soon as the insect population enters the area above EIL, and then the scout can be sure with 90% certainty that insect population in the field has exceeded EIL. In another field, if the insect counts on successive plants are 2, 1, 3, 2 and 1 whereupon the area entered is below EIL, a decision can be made with 90% certainty that the insect population is below EIL. If a scout finds a mean of 3 insects per plant with little variation, the cumulative total will be within the area of uncertainty even if several successive samples are taken. In practice, an arbitrary cut off may be established (e.g., say 20 samples or plants), and if the accumulated total remains in the area of uncertainty, the location can be classified as uncertain, to be re-visited early. Onsager (1976) and Southwood (1978) have given the formulae for computing decision making lines for populations that conform to Poisson or binomial distributions.

Sequential sampling lines can also be generated from regression of log variance on log mean after Taylor's power law (Green 1970) or from the mean crowding index (Iwao, 1975). Grothusen (1984) developed a programme to adjust sequential sampling lines in the field according to prevailing conditions. Pedigo and van Schaik (1984) suggested the use of sequential sampling over time (rather than space) to detect the timing of green cloverworm outbreaks. This method can be applied to such data as light trap surveys, though this has not been attempted so far. Numerous studies have indicated that sequential sampling is an improvement in accuracy, efficiency, and tedium over other techniques, because fixed sample size is likely to be imprecise at low insect densities, and inefficient at high densities. Bellinger *et al.* (1981) showed time savings of 30% - 68% with an accuracy of 82–91% in a sequential sampling scheme. Hoy *et al.* (1983) developed a pattern of sampling throughout the field, varying the number of plants sampled in each spot, and this pattern may be suitable for sampling insect populations displaying high variance and uneven distribution in the field.

## POPULATION PREDICTION MODELS AND PEST MANAGEMENT

Predictive models provide the primary mechanism of information flow in decision making in pest management. The factors influencing the decision making are:

- Time span of the crop.
- Crop phenology in relation to life cycle and activity of the insect in question.
- Number of generations, and population dynamics.
- Dispersal capability.
- Nature and rate of damage on the crop.
- Nature and time sequence of the control treatments to be used.

### *Scope of the pest management programme*

In this, both short and long term predictive modelling is involved, and the models include both, the insect and the host plant. There are three types of predictive models:

- Models that relate to the spatial parameters of the insect and plant populations and provide prognoses of where the phenomena of interest will occur.

- Models that provide estimates of when critical events will occur.
- Models concerning population numbers or quantities and express expectations of how much or how many.

All the models are linked in each system according to the biological, physical, and economic relationships. Predictive models take a variety of mathematical forms, depending on the relationships or phenomena concerned and degree of resolution, accuracy and precision required. In some cases, simple or curvilinear regressions may suffice. In many cases, however the purpose is to forecast when or where a particular event, e.g., egg laying by the insect or flowering in the plant will begin; or when the maximum numbers of the insects will occur and when a specified level of threshold will be reached. For such purposes, differential equation models are more applicable. However, these are deterministic and experience has shown that chance and uncertainty in prediction may be better expressed through stochastic models. It is implicit in the system structure that the research aimed at a particular pest should be carried out with full awareness of other destructive insects, diseases and weeds. The pest management research should be conceptually integrated with other research aimed at optimizing the productivity.

Biological population is a group of individuals actually or potentially capable of interbreeding and thereby producing fully fertile offspring. In practice, both temporal and spatial boundaries are often determined arbitrarily. Simple population models designate the density of an insect population by a single variable (**N**), which conveys the impression that all members of the population are identical. However, insects vary according to age, behaviour, morphology, genotypes, etc. It is, therefore, appropriate to assume that **N** represents a range of individuals assumed to be identical only for the purpose of analysis.

To gain an insight even in a qualitative way into the many interrelationships of the insects with their environment is a difficult task (Clark *et al.*, 1967; van Emden and Williams, 1974). Attempts to mathematically model insect populations from real life data have been limited by the fragmented nature of the data that are available. These generally represent times and locations where the insect is abundant (Campbell, 1967; Fye, 1974). Simulation models have been helpful in filling some void (Stinner *et al.*, 1974a; Barryman, 1974). To understand the basic population dynamics of an insect, we must have the basic tools to estimate the insect numbers in discrete units of space and time within specified limits of error. To ignore the errors of sampling in developing and testing a model is quite risky. The model must be able to handle the spatial variations, anticipate the insect occurrence in different places and know the conditional probabilities of occurrence associated with specific characteristics of the location, which can be quantified, at least on a relative scale. We also must have insight into the long term differences in average population densities within and across locations, to develop models that will forecast average population densities in a given set of conditions in different locations. The best predictors are generally easily measured variables that encompass and reflect a sufficient number of really critical factors to be accurate and consistent over a wide range of conditions. In some cases, a few or specific variables related to the location, host plant, the insect, or its natural enemy may account for a large proportion of the variation in average pest densities to serve as a reliable predictor.

We must be able to deal with short and long term changes in population densities. Short term changes are variations within a season or a given year. The problem is simple for insects having a single generation a year. In this case, we need to model only the developmental rates to forecast the appearance of critical stages and then to model survivorship over the generation beginning with whatever stage appears first in the field. For insects having more than one generation in a year, we need to account for potential changes in pest numbers and differences in survivor rates in successive generations. The purpose is to be able to predict the insect populations for a given location when the damaging stages will appear, and what the pest densities will be at a given point of time. The predictor variables can be any one or a combination of initial counts of the insect and the measures of any of biotic or abiotic factors that serve the purpose.

Long term changes include a time period greater than one year. Such changes encompass changes in insect numbers from one year to the next in the current gradation or cycle and changes in the form and magnitude of the next gradation as were as of successive gradations. Prediction models of this type are needed to forecast; 1) initial population densities or the densities of a critical stage (e.g., egg, larvae, or adult in flight) at 1, 2,. ......n years hence, 2) peak population level and the year in which it will occur in the current gradation, 3) the duration of low or endemic phase of the gradation, and 4) the year and the potential rate of increase from the start of next gradation. In long term, we need to know the probable predictivity and amplitude of the major fluctuations in insect populations over time under different conditions.

Since the insects are mobile, predictive models of both spatial distribution and temporal change in density must include a dispersal component. Insect dispersal is complex, so a particular mathematical model and the manner in which it is coupled to the general model will be unique for each pest (Edmonds *et al.*, 1973; Ewing *et al.*, 1974).

Models used for pest management can be grouped into: statistical, mechanistic, optimization and bioclimatic models. Apart from these, expert or knowledge based systems can also be employed for decision making in pest management. Models can also be based on regression, infection periods, phenology, simulation and analytical (Shoemaker, 1980; Welch *et al.*, 1978; Sharma *et al.*, 1996). These models have been used, both for research and operational purposes.

## Population models

Various aspects of decision making in pest management, and modelling of insect populations have been discussed by Barryman (1981), Horn 1988, Singh (1996), and Karel (1996).

### *Statistical models*

These are regression and multivariate models which are based on purely empirical observations and do not include processes. They are predictive models such as regression models used to predict pest incidence. Most commonly used model in crop protection is EPIPRE - Epidemic Prediction and Prevention Model for wheat (Zadoks, 1981). A range of simulation models have been developed for research purposes. However, for practical crop protection, linear regression models that assess likely insect infestation have been found to be more suitable.

### *Mechanistic models*

These models deal with biological and ecological mechanisms or processes that underline pest population dynamics, damage and/or control. These are more suitable for expressing hypotheses about the reasons for pest outbreaks or why certain biological control agents are likely to be more successful (Norton *et al.*, 1993). Analytical models, simulation models, inferential models, spreadsheets and rule based models fall into this category.

***Analytical models***

These are mathematical models, based on differential equations. They provide general information about system by describing the characteristics in a handful of relationships and parameters (Jones, 1993). Analytical models have been used for developing host-parasitoid relationships for sustainable biocontrol strategies and foı determining population changes. Analytical models are those for which explicit formulae are derived for predicted values or distribution. These include regression and multivariate models, and standard theoretical distributions. These models have been used as tools for developing pest management principles and ecological theory. Differential models are appropriate for organisms with overlapping generations and continuous births and deaths.

*Exponential model:* The simplest population model (Fig. 4) is that regards density (N) as a function of births (B) and deaths (D).

$$N = B - D \tag{44}$$

Since births and deaths occur only in pre-existing population ($N_0$), the density ($N_t$) after time t is:

$$N_t = N_0 + B_t - D_t \tag{45}$$

where $B_t$ and $D_t$ are the number of births and deaths, respectively, during the time interval t. Normally, births and deaths are expressed as birth (b) and death (d) rates as a function of the population density.

$$N_t = N_0 + N_0(b - d), \text{ and generally for any time interval:} \tag{46}$$

$$N_{t+1} = N_t + N_t(b - d) \tag{47}$$

This model describes the exponential population growth but is oversimplified and unrealistic, yet adequately describes changes in density following initial colonization of a favorable habitat. Different equations are useful for simulating population dynamics of different insects since population events are discrete phenomena. Population models are often described in differential equation form. A differential equation describes a continuous function, which many insect population curves approximate as densities increase and intervals between successive samples shorten. The exponential model can be expressed in differential equation form as:

$$N_t = N_0 e^{(b - d)t} \tag{48}$$

where e is the base of natural logarithms, and t = is the time interval. The quantity b - d is usually expressed as r, the instantaneous rate of increase:

$$N_t = N_0 e^{rt} \tag{49}$$

*Logistic model:* The exponential model is of limited practical use in describing population events, except under unusual circumstances over short time intervals. Environmental constraints prevent the world from filling with flies, aphids or grasshoppers. Verhulst (1838) and Lotka (1920) developed a simple mathematical model recognizing the tendency of populations to be regulated about an upper limit set by extrinsic and intrinsic factors. In the simplest form of logistic model, K (the environmental carrying capacity) is incorporated into the exponential model to act as a brake on population growth.

$$N_{t+1} - N_t = N_t(b - d)[(K - N_t)/K] \quad (50)$$

This equation gives a sigmoid curve, and the model is often expressed in differential equation form as:

$$dN/dt = rN[(K - N)/K] \quad (51)$$

or in integral form as:

$$N = K/(1 + e^{rt}) \quad (52)$$

where e and r are the natural logarithm and instantaneous rate of increase, respectively. Despite its simplicity, this model is incapable of a large array of outputs. Most importantly, if b increases relative to d, there is a tendency for greater oscillations about K, alongwith greater instability. This has been discussed in detail by May (1975) and Barryman (1981). A general conclusion is that insect populations with short generation time and high fecundity may fluctuate unpredictably and may exhibit spectacular instability and may reach EIL much more quickly than those with lower r.

*Logistic with a minimum:* Further realism can be added to the logistic model (Fig. 5) by considering a minimum population below which the population is unable to maintain itself. This is added to the logistic in a term creating an unstable equilibrium at an extinction threshold E.

$$N_{t+1} - N_t = N_t(b - d)[(K - N_t)/K)][(N_t - E)/N_t] \quad (53)$$

If E exceeds $N_t$, the right side of the equation becomes substractive, and the population declines to extinction. The population declines if the initial $N_0$ is less than E. This models events such as the failure of biological control to become established following initial release. Also, density of a population with a very high r might enter the extinction zone ($N_t$ less than E) and decline to extinction.

*Stochastic models:* The models described above, like many theoretical population models and applied models in pest management are deterministic. They assume a single value for each input, and devise a single value for $N_t + 1$, after each time interval. A more realistic characterization of population events assigns a probability to each demographic event. Models using probabilistic functions to define variables are called stochastic models. A realistic characterization of actual population events may be made by assigning a probability to each demographic measurement. Birth rate, death rate, carrying capacity, and extinction threshold might be defined by probability functions that describe fluctuations about a mean value in the from:

$$N_{t+1} - N_t = N_t[P(b) - P(d)] \quad (54)$$

where P(b) and P(d) are the probability functions describing birth rate and death rate, respectively. In a simplified example, P(b) and P(d) might be permitted to vary randomly (random numbers assigned to successive calculations of $N_t + 1$. Fig. 4 shows that simulations using such a model produce an array of meaningless results but with one common element, i.e., all become extinct. Biologically, the demography of some local ephemeral populations sometimes approximates this situation. From a practical standpoint, modelling of such populations to predict pest outbreaks is futile because they are inherently unpredictable. Models using probabilistic functions to define variables are said to the stochastic. They are usually less tractable mathematically than are deterministic models, though in many cases, stochastic models produce greater realism in describing actual population events and are thus of some utility in insect pest management. Use of computers has greatly enhanced the use of stochastic models.

***Simulation models***

Application of simulation models in pest management involves simulation of population dynamics of the pest, crop, natural enemies or combination of these. Simulation models are helpful in understanding insect population dynamics, damage, crop research and development strategies, and provide a basis for taking control decisions. Simulation is the process of designing a model of a real system by conducting experiments with the model for the purpose of understanding the be-

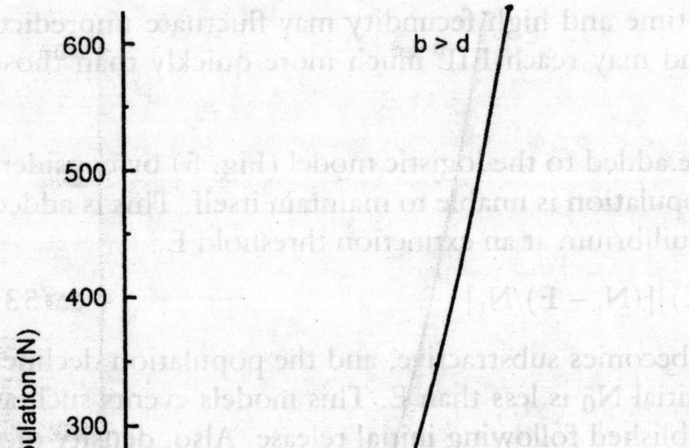


**Fig. 4.** Exponential population growth when b > d, b = d, and b < d (b = number of births and d = number of deaths over time in a population).

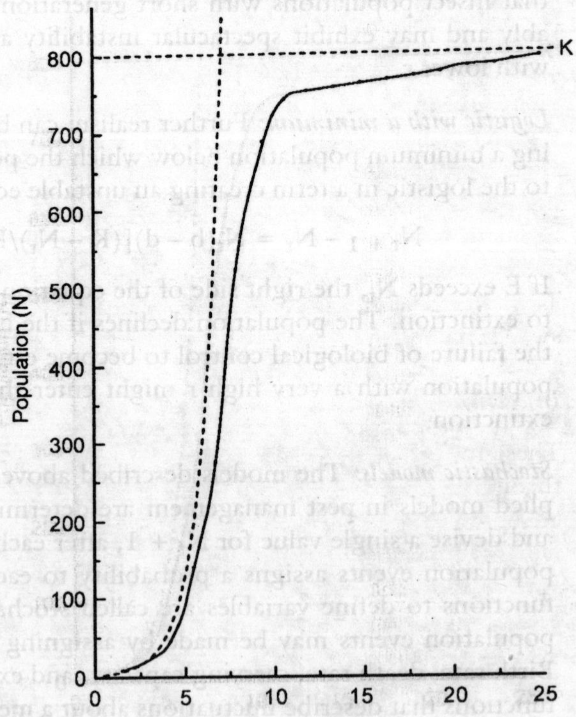


**Fig. 5.** Logistic population growth curve (a hypothetical example).

haviour of the system or of evaluating various strategies for operation of the system (Shannon, 1975). Simulation modelling is used in crop management particularly in relation to population dynamics of the crop and natural enemies. In pest management, decision making simulation models are used for forecasting pest population changes, and for testing a wide range of management options of the model pest population. An example is simulation models for rice brown plant hopper (Holt *et al.*, 1987). It has also been used for timely application of insecticides. Several models have been developed for disease control.The simulation models predicting time events in the population dynamics of the pest are termed phenology models. They are also referred to as day-degree models, because population development is based on accumulation of heat. PETE (Pro Extension Timing Estimator) is an example of such a model. PETE has been applied for several pests of fruit crops in several countries (Croft and Welch, 1984; Welch *et al.*, 1981).

Simulation models of crops coupled with pest models are called crop-pest coupled models. They are useful when insect density damage relationships are complicated or when other factors such as drought interact with crop damage (Boote *et al.*, 1983; Rabbinge and Rijsdijk, 1983). Simple Universal Crop Growth Simulator - SUCROS has been used for cereal crops damaged by aphids (Sutherst *et al.*, 1994). SIRATAC is another pest management system employed to control cotton bollworm, *Helicoverpa armigera* (Hub.) in irrigated cotton growing area in Australia (Hearn and Brook, 1983; Ives *et al.*, 1984). The grower enters into the computer the date of sowing, and then at regular intervals, information on number of stem internodes, number of fruiting points. the number of *Helicoverpa* are classified according to stage of development, and number of predators. The weather data are either entered manually into the central computer or logged and automatically transmitted through the telephone network. The system calculates in advance the plant and insect development, and estimates empirically whether the projected pest population would result in a reduction in yield. This system also has data on the efficiency of different insecticides. If insect resistance to a particular insecticide is detected, recommendations for its use can be discontinued. PMEX - Pest Management Executive System was developed for efficient pest management in apple (Croft *et al.,* 1976), later improved by Whalon and Croft (1984) and applied to other field and vegetable crops. This system consists of software other than computer terminals. The software has features that make it something more than just a collection of programmes. The system manager coordinates different programmes and software.

### *Inferential models*

These models have been proposed for modelling development of an insect population that passes through a series of discrete developmental stages. The relationship between an insect, its hosts, and predators is often dependent on the relative phenologies of each of the interacting components (Hudes and Shoemaker, 1988). Insect control can be made more effective if control measures can be timed with occurrence of most susceptible stage of the insect. This has been successfully used to describe the phenology of spruce budworm, *Choristoneura fumiferana* (Clemens).

### *Rule-based models*

A rule-based approach provides a means by which subjective knowledge about a system can be used to build models. In these models, different components are represented by a small set of discrete states and changes are described by 'if then' type rules. These can be built in any computer language which includes logical operations (and/or, greater than, etc.), and 'if then' statements for interaction. Rule-based models can be used when the level of description of the system provided by the rule-based model is sufficient to analyze the problem. A quantitative model would be useful when the information to build the model is not available, and the problem is such that some useful conclusions may be drawn from a rule-based approach.

#### *Optimization models*

These types of models are used with prescriptive decision problems in pest management. The model is constructed in such a way that optimal strategy of pest management can be determined, either by a mathematical solution or by numerical optimization procedures. Dynamic programming is adopted in multistage decision-making problems and has the unusual feature of starting at the latest point in time, say at harvest, and working backwards to the earliest decision point. The advantage of this model is that at each stage, once the optimal action plan for different levels of pest attack has been worked out through the end of season, all other paths for that point can be ignored. This technique has been applied to a number of pest problems (Holling *et al.*, 1976).

#### *Bioclimatic models*

These are climate driven models used for various purposes of pest management (Mukhopadhyay, 1996). CLIMAX and BIOCLIM are examples of such models. CLIMAX is a climate driven simulation model that enables the prediction of a plant or animal's potential relative abundance and geographical distribution as determined by climate (Sutherst and Maywald, 1985). CLIMEX can be used to assess the risk of introduction of a pest into new areas (Sutherst *et al.*, 1989), appraising likely changes in pest status associated with changes in climate (Sutherst, 1991) and selecting appropriate areas for potential biocontrol agents during the exploratory phase of a classical biological control.

BIOCLIM is a bioclimatic model used for climatic analysis and mapping the potential distribution of any biological entity. It generates a profile of selected meteorological variables which describe the climatic limits of a species. An example of successful use of BIOCLIM is in the prediction and potential distribution of leadbeakers possum *(Gymnobelideus leadbeateri) in* Australia (Lindmyer *et al.*, 1991).

### Systems modeling in pest management

#### *Expert systems*

Expert systems mimic precesses employed by a human expert in diagnosing a problem and prescribing control measures. Expert systems are computer based but techniques employed in expert system can be used to develop better decision making rules even without computers. Expert systems are a set of knowledge-based systems, programmes that store, process, and disseminate knowledge. Knowledge based systems can perform four main roles in pest management: problem solving, information processing, problem structuring, and training. Expert systems can be employed in pest identification, extent of crop losses, determination of available options and their effectiveness, cost/benefit ratio, and constraints involved in decision making in pest management

(Compton *et al.*, 1992; Jones *et al.*, 1990; Knight *et al.*, 1992). Expert systems can be attached to data based systems.

### *Data-based systems*

Data base is a large collection of data organized for a particular use. Data-based systems are used for reference sources, historical archives, geographical information system, and expert systems. Most of the decision tools such as historical and seasonal profiles emphasize the temporal dimensions of pest problems. Data-based systems have been used to investigate migratory pest problems and incorporate both temporal and spatial dimensions. The major use of a data base system is to provide the users with an abstract view of the data as well as the environment. Data-based systems can be handled by any standard data management system.

### *Geographic information system (GIS)*

GIS has been designed to handle spatial aspects of agricultural and forest management problems. It has also been used to investigate and forecast grasshopper and locust outbreaks. GIS technology is used for three main purposes: organization and storage of data, monitoring and management of resources, and as modelling and research tool. GIS utilizes computer based methods of storing, summarizing, and analyzing data with reference to geographic position. Most GIS software is capable of some sort of modelling that allows the user to build equations in which maps, such as pest density, rainfall and crop stage can be used as independent variables, this can be combined according to specific relationships to allow calculation of a predictive map. In case of pest management, it can serve as a tool to provide a data base for crop pest modeling, and improve upon existing constraints pertaining to analysis methodologies.

## Insect population models and host-plant resistance

Plant resistance has been used to suppress pest populations since times immemorial. However, only a few pest species can be controlled by the use of resistant varieties alone. Adequate levels of resistance are present against a few pests only, e.g., against midge and greenbug in sorghum; leaf hoppers and gall midge in rice; jassids in cotton; jassids and thrips in groundnut; borers, scale insects and whitefly in sugarcane, etc. However, varieties with low to moderate levels of resistance can be very useful for pest suppression over a period of time. The impact of resistant varieties on pest populations can be demonstrated by making use of simple insect models of Knipling (1964) as adopted by Adkisson and Dyck (1980), Teetes (1985) and Sharma (1993).

The impact of growing a resistant, a moderately resistant, and a susceptible variety on insect populations over a period of time has been explained for sorghum midge, *Stenodiplosis sorghicola* Coq. in Fig. 6. The effect of resistant cultivars would be similar for other insects depending on the level of resistance and the mechanism of resistance. The rate of multiplication of sorghum midge is six times on a susceptible cultivar (CSH 1), and three times on a moderately resistant cultivar, IS 12664C compared with the resistant cultivar, DJ 6514 (Sharma, 1985). Based on the rates of insect multiplication, there would be 1296 and 81 times as many insects in areas planted with CSH 1 and IS 12664C, respectively, compared with DJ 6514, where the insect population would remain constant (Table 1). Thus, plant resistance has a great influence on insect populations, which is cumulative

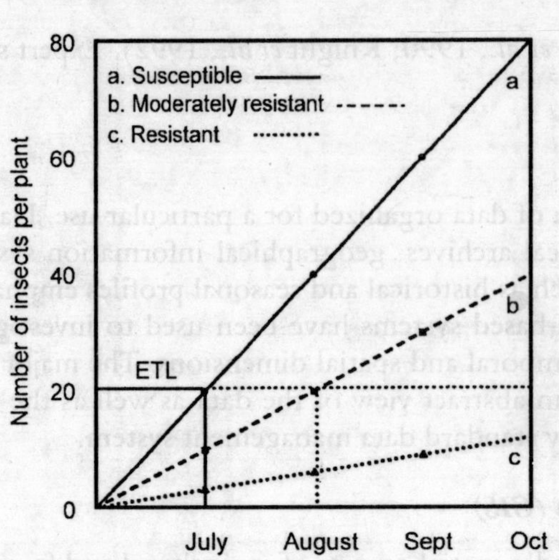


**Fig. 6.** **Effect of host plant resistance on population dynamics of insects and the time taken by insects to reach economic threshold levels when an area is planted to a susceptible, a moderately resistant or a resistant cultivar (Sharma, 1993).**

**over time. These models can also explain the situations where minor pests become very serious with the introduction of newly developed high yielding susceptible cultivars.**

Table 1. Population increase of sorghum midge on CSH 1 (susceptible,) IS 12664C (moderately resistant) cultivars and DJ 6514 (resistant). A hypothetical example based on Knipling (1964) and Sharma *et al.* (1990).

| | *Generation* | *No. of midge flies/ha* | | |
|---|---|---|---|---|
| | | *CSH 1** | *IS 12664C** | *DJ 6514** |
| | $P_1$ | 100** | 100* | 100* |
| First year | $F_1$ | 600 | 300 | 100 |
| | $F_2$ | 3600 | 900 | 100 |
| | $F_3$ | 21600 | 2700 | 100 |
| | $F_4$ | 129600 | 8100 | 100 |
| Diapause population (1%)*** | | 1554 | 120 | 4 |
| | $P_2$ | 1554 | 120 | 4 |
| Second year | $F_1$ | 9324 | 360 | 4 |
| | $F_2$ | 55944 | 1080 | 4 |
| | $F_3$ | 335664 | 3240 | 4 |
| | $F_4$ | 2113984 | 9720 | 4 |
| Diapause population (6%) | | 25149 | 142 | 1 |

$P_1$ and $P_2$ are the parent populations in first and second year, respectively.

$F_1$ to $F_4$ are the number of offspring generations in a cropping season.

* Midge population multiplies by 6 times on CSH 1, 3 times on IS 12664C and once on DJ 6514 (Sharma *et al.*, 1990).

** The midge population at the beginning of the season is assumed to be 100 flies/ha.

*** In each generation, 1% of the total populations is assumed to enter diapause.

The levels of resistance quite often are not adequate, and in such situations, host-plant resistance in combination with insecticides or natural enemies (Table 2) can reduce pest densities over time. One application of an insecticide (with 90% mortality) will lead to 810 insects in an area planted with a moderately resistant cultivar (e.g., IS 12664C) compared with 12960 insects in an area planted with a susceptible cultivar (CSH 1). Similarly, moderate levels of plant resistance in combination with natural enemies (e.g., *Tetrastichus diplosidis* Craw. on sorghum midge resulting in 50% parasitization) can reduce the insect numbers to 506 and 8100 on IS 12664C and CSH 1, respectively. Thus, it is important to include a component of host-plant resistance in insect population modelling in different crops. In addition to affecting the population dynamics of insects, plant resistance to insects either increases the EILs or the time required by the insect populations to reach EILs (Sharma and Lopez, 1994; Sharma *et al.*, 1993).

Table 2. Population increase of sorghum midge on CSH 1 (susceptible) and IS 12664C (moderately resistant) cultivar + one insecticidal spray or parasite release in first generation — A hypothetical example based on Knipling (1964) and Sharma *et al.* (1990).

| | Generation | No. of midge flies/ha | |
|---|---|---|---|
| | | *CSH 1** | *IS 12664C** |
| First year | $P_1$ | **100 | **100 |
| Insecticide/parasite*** | $F_1$ | 60 | 30 |
| | $F_2$ | 360 | 90 |
| | $F_3$ | 2160 | 270 |
| | $F_4$ | 12960 | 810 |
| Diapause**** population | | 155 | 12 |
| Second year | $P_2$ | 155 | 12 |
| Insecticide/parasite*** | $F_1$ | 93 | 4 |
| | $F_2$ | 558 | 12 |
| | $F_3$ | 3348 | 36 |
| | $F_4$ | 20088 | 108 |
| Diapause**** population | | 241 | 2 |

$P_1$ and $P_2$ are the parent populations in first and second year, respectively.

$F_1$ to $F_4$ are the number of offspring generations in a cropping season.

* The population of midge increase by 6 times on CSH 1 and 3 times on IS 12664C (Sharma *et al.*, 1990).

** The midge population at the beginning of the season is assumed to be 100 flies/ha.

*** The insecticide/parasite is applied on $F_1$ and is presumed to kill 90% of the population.

**** In each generation, 1% of the total population is assumed to enter diapause.

A component of host-plant resistance has been incorporated in cotton simulation models. Several morphological characters in cotton affect different insect pests, and sometimes in opposing ways, e.g., smooth leaf character increases resistance to bollworms but increases susceptibility to jassids. A cotton crop model boll weevil and *Heliothis* models have been interfaced. The use of plant and insect models is just beginning in host-plant resistance research. The amount of information on plant growth and daily insect population structure indicates that use of plant insect simulation models, along with traditional research approaches, should enhance the ability to develop hypotheses, to advance the development of insect resistant plant types, and to determine ways of growing such plants.

**Effect of temperature on insect development and insect prediction models**

***Relationship of temperature with rate of insect development***

Pradhan (1946) described the relationship between insect development and temperature by:

$$Y = Y_0 e - ax^2 \quad (55)$$

where $Y_0$ = highest value of developmental index, i.e., at the temperature of quickest development (T), Y = developmental index at temperature t, x = T – t, and e = 2.71828. This is based on the assumption that organic development takes place with the law of compound interest (speed of vital activity acting as the rate of interest), the amount of development from birth to death are constant for a species, and the amount of activity per unit time increases with an increase in temperature with a velocity (which does not remain constant but is continuously retarded with a rise in temperature; the rate of retardation however being uniform). The velocity (v) of change is given by:

$$v = u - at \quad (56)$$

$$r = ut - 1/2at^2 + c \quad (57)$$

v = velocity of change in the value of r, u = constant, i.e., value of v when t = 0, a = retardation constant (i.e., negative acceleration), t = temperature, and c = constant, i.e., value of r when t = 0.

This equation has been found to fit well for observed data for several insects (Pradhan and Bhatia, 1967; Harukawa, 1956; Matsazawa *et al.*, 1957; Chaudhari, 1960). However, Harukawa (1956) observed that this equation cannot be applied to insects such as *Bombyx mori* L. and *Collasobruchus chinensis* L., because for these insects, the velocity curves are a straight line in a fairly wide range on medial temperatures. Also, the equation does not fit in experimental data accurately when rearing temperatures are beyond the temperature of maximum rate of development, and needs further investigation. This equation can be used to calculate the developmental periods corresponding to variable temperatures in nature.

This equation also provides an explanation of the variation in insect development under constant and variable temperatures. Below a certain critical temperature, development may be quicker under variable temperatures, than under constant temperature. Above the critical temperature, the relative effects may be reversed. When the temperature fluctuations take place in the higher region Z, the increase in developmental index due to increase in temperature is less than the decrease in the value of the developmental index when the temperature goes down, and the average of the fluctuations between these two limits is less than the value of developmental index if the temperature stays constant at the mid-point Z. The reverse is the case in the lower regions of temperature at X (Fig. 7). In general, acceleration due to fluctuating temperature in the lower region, and retardation due to fluctuating temperature in the higher region are more apparent than real (Pradhan and Bhatia, 1967). Under variable temperatures, the rate of development can be calculated by:

$$Y(t_1 - t_2) = Y_0 l_{x1}{}^{x2}/(t_1 - t_2)_e \, ax^2 dx \qquad (58)$$

where $Y(t_1 - t_2)$ = average value of developmental index corresponding to temperature fluctuation between $t_1$ (maximum) and $t_2$ (minimum), at which the corresponding values of x are $x_1$ and $x_2$. For more details refer to Pradhan (1945).

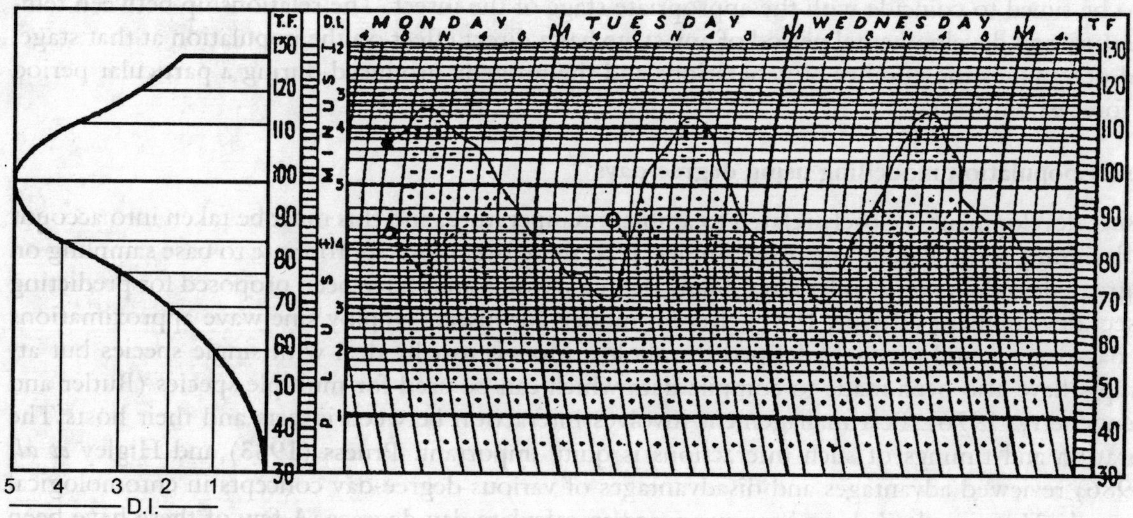

**Fig. 7.** Diagram illustrating the principle of biometer and the method of estimating development periods through biometer. The wavy line represents the temperature tracing as seen through the transparent sheet of the biometer. DI = developmental index, and TF temperature in °F (Pradhan and Bhatia, 1967).

*Biometer*

Pradhan (1945) developed a biometer to estimate the rate of insect development at any temperature, and the amount of development or number of generations in any given period across a range of temperature fluctuations (Fig. 7). The curve **b** represents the curve on the basis of equation given above (equation 58), showing the relation between temperature and rate of insect development. The temperature on this graph when superimposed at **A** over the temperature line of thermograph chart in which the equidistant horizontal lines showing temperature have been removed (except the narrow vertical columns on either side) and replaced by horizontal lines representing rate of development **DI** corresponding to different temperatures. This modified chart when printed on a transparent sheet, is superimposed over a thermograph tracing, it is possible to read the rate of development at different points of the time scale. The ordinate of this line represents the rate of development **DI**, and abscissa - the time. The area below the temperature tracing gives the amount of development. Different areas of this tracing are not equal as the relationship between temperature and development is not linear. This difficulty can be avoided by counting the number of cells formed by horizontal and vertical lines. In this technique, there is no erroneous assumption of a linear relationship between temperature and development. The area above the line showing the maximum value of development index has a minus value, and in case the temperature rises above this line, the area enclosed above **5DI** is subtracted from the area below **5DI**. The biometer can be developed after establishing the curve on the basis of experimental data.

The practical implications of biometer include: chemical control against an insect can be timed to coincide with a particular stage of development, and the release of parasites and predators can also be timed to coincide with the appropriate stage of the insect. The relationship between temperature and developmental period of any stage has a direct effect on the population at that stage, and determines the number of generations and the progeny produced during a particular period to forecast outbreaks.

## Insect population modelling using degree-days

Insect development is closely related to ambient temperature, and this must be taken into account in designing routine sampling programmes. It is, therefore, more appropriate to base sampling on degree - days than on calendar days. Many degree-day methods have been proposed for predicting insect development (Taylor, 1981). Recent models for insects employ sine wave approximations of day-degrees and precisely defined thresholds. Models usually deal with single species but attempts have also been made to compile data which can be used for multiple species (Butler and Henneberry, 1976). Pest management involves interaction between insects and their hosts The duration and timings of such interactions is quite important. Pruess (1983) and Higley *et al.* (1986) reviewed advantages and disadvantages of various degree-day concepts in entomological research. Many methods have been proposed to calculate day-degrees. A few of these have been discussed below.

### *Day-degree methods*

*Mean minus base method:* Day-degrees accumulated on a single day are computed as the difference between the arithmetic mean temperature [(max + min)/2], and the threshold temperature. Most practical applications use two thresholds; 4.4°C for cool season, and 10°C for warm season crops. Development occurs when the temperatures are above the assumed threshold. In this situation, the computed day-degrees are similar to sine wave estimates.

*Upper and lower threshold method:* This modification of the minus base method recognizes both the upper and lower threshold for insect development, and assumes that the insect development occurs when the mean temperature is below the threshold as long it is below the maximum above. The limits are set at 10° and 30°C, and the computation is similar to the minus base method. It overestimates the day-degrees when the daily temperature curve intersects the assumed threshold.

It recognizes the inhibition of the insect to grow in hot weather. Moisture or other stress factors under high temperature may cause this inhibition.

*Sine wave:* This method assumes that the daily temperatures approximate a sine wave, and recognizes that development may occur if only a portion of day lies above the threshold temperature. It estimates the day-degrees closely if minimum temperature is below the threshold. If temperatures are between the thresholds, the estimates are identical to the other methods (Allen, 1976). This method is of importance to insects responding to temperature during spring in the temperate regions. The computations are not easy without a computer, but tables have been published for specific thresholds (Wedberg *et al.*, 1977).

*Bias-corrected sine wave:* Sine wave is only an approximation of actual day-degrees, and bias in such estimates varies geographically (Allen, 1976). A linear method can be used to correct the local bias in sine wave (Allen, 1976). Bias correction may be necessary because presently used day-degrees need to be transformed to actual day-degrees to make use of predictions developed using past technology.

*Actual day-degrees:* The methods discussed above only estimate day-degree accumulations. With the ability to integrate temperatures at frequent intervals above several thresholds from the automated weather stations, the actual day-degrees can be estimated for many locations. These actual values will be more useful both for the crop and insect models. Allen (1976) proposed a regression method to estimate actual day-degrees from the sine wave approximations.

***Degree-day model***

In general, developmental rates above the threshold are proportional to the temperature. This assumption may not be always true and models need to be constructed for nonlinear responses (Logan *et al.*, 1976). However good predictions can be made by the use of linear models by appropriate choice of effective development thresholds. As additional factors that affect crop growth are determined, the practical solutions become more difficult, e.g., some crops (corn and soybean) differ in growth during photophase and scotophase. Insect models which use observations on one active growth stage to predict some future stage are quite useful. Many models of potential use lack information as to when development after dormancy begins. Thus, in many cases, arbitrary dates are selected from which the estimation of accumulation of day-degrees begins. At times, the accumulation of day-degrees before such a date may be negligible. An error in choice of starting date may be the major source of error in some models.

The regression method proposed by Allen (1976) is a good technique to estimate actual day-degrees from sine wave approximations. Actual day-degrees can also be estimated by triangulation method of Sevacherian *et al.* (1977). The triangulation method slightly underestimates day-degrees during the warmer part of the day when the temperature curve is convex, but overestimates when the curve is concave. These errors tend to cancel when the minimum temperatures are above the threshold. Some improvement in estimating actual day-degrees can be achieved by computing day-degrees as the sum of two half-days, having different minimum temperatures. To use a standard, assumption may be made that each half day was of equal duration. However, the exact duration varies daily, seasonally and geographically. Parton and Logan (1981) used a truncated sine wave to estimate day temperatures, and an exponential function for night temperatures to overcome some of these problems. However, Pruess (1983) argued that bias correction applied to the whole day, rather than its component parts is a satisfactory procedure. Bias correction is not useful for individual days. For accumulated day-degrees over a longer period, bias correction may be useful to estimate actual day-degrees. All day-degree methods with a potential application in pest management should be reported to standardized thresholds of 5, 10, and 15°C. Either sine wave or actual day-degrees may be used, and bias correction is not of much value unless data are obtained from instruments recording actual day-degrees. Temperatures should be recorded from air, and state the date on which the accumulation began. A comparison of predictions with the calendar dates should indicate deviations in actual days from the mean calendar days. Upper threshold, if used, should be indicated, and if not resulting in better predictability, would be consistent with the current agronomic practice of 30°C.

Degree-day models have now become established tool in understanding insect and plant phenology, developing computer simulation models and predicting pest incidence. One of the assumptions of day-degree method is that development is directly related to temperature and time. The relationship between temperature and development is not always linear but depends on the various biochemical reactions. A number of these enzyme catalyzed reactions are rate-limiting for growth. The rates of reactions depend on the availability of substrates, enzymes, and temperature. Development may be retarded when any of these are limiting. Fluctuating temperatures can also change the rate of biochemical reactions. Thus, development under a fluctuating temperature regime is different than that under constant temperatures (Howe, 1967). Magnitude of differences under constant and fluctuating temperatures depends on the average temperature, amplitude, and frequency of fluctuations (Campbell *et al.*, 1974). Also, several other factors are important for development of insects under field conditions. The relationship between temperature and growth is often calculated as linear, while in most of the cases it is curvilinear. Algorithms have been developed using a variety of nonlinear functions that characterize the temperature growth relationships (Stinner *et al.*, 1974b; Logan *et al.*, 1976; Wagner *et al.*, 1984). For most species, linear function is acceptable so long as the temperatures considered are in the linear range. For each species, a developmental minimum is calculated, at which the rate limiting reactions will occur. This is determined by calculating the rate of development over a range of temperatures, fitting the growth rates to a line, and extrapolating to zero. This gives only an estimate of developmental minimum.

Other approaches to calculate the developmental minimum have been described by Arnold (1959) and Kirk and Aliniazee (1981). An analogous threshold occurs at the upper end of the scale called developmental maximum temperature. This is the temperature at which the development rate reaches the maximum (where kinetic constraints limit enzyme and substrate diffusion, beyond which the temperatures are sufficiently high so as to produce structural changes in control enzymes to impair general enzyme function, or actually denature some proteins, and developmental maximum often approaches the lethal temperatures for a species). Estimation of developmental maxima is often difficult, but is estimated by choosing the upper temperature beyond which growth rate plateaus or declines. This point can be estimated (Funderburk *et al.*, 1984) or calculated iteratively (Nowierski *et al.*, 1983). Most often, developmental maxima are not calculated, and no upper limit is placed on the temperature used to calculate degree-days. The developmental maximum and minimum are usually calculated as single values but the actual minimum and maximum temperatures may vary with the stage of the insect (Sanborn *et al.*, 1982). However, despite the differences in the threshold temperatures across stages, a single value is used to avoid complexity. Also, differences in micro-habitat and ambient temperature also produce some inaccuracies in degree-day estimations. The other problem in using degree-days is determining when to begin accumulations. Normally, accumulations begin when the temperature exceeds developmental minimum, but the development may not begin until later. Normally, it is related to the termination of diapause. For migratory insects, accumulations may correspond to a phenological event, e.g., capture of mated or un-mated females in light traps. This is a sound approach so long as the temperature is the major factor limiting development. One method to avoid the problems associated with this approach is to use curvilinear functions to describe growth rates rather than degree-days. Sharpe and DeMichele (1977) described a model to predict developmental times, and have developed a computer programme for using this model. When degree-days are used in computer simulation models, many factors affecting degree-days such as different thresholds for different stages, influence of host plant/nutrition, and fluctuating temperatures can be used. A programme, DEGDAY for computing degree-days using microcomputers has been developed by Higley *et al.* (1986) which calculates heating degree-days with rectangle, triangle, and sine wave methods, and cooling degree-days with sine wave method.

Pruess (1983) suggested that all day-degree methods with a potential for application in pest management should be reported to standardized thresholds of 5, 10, and 15$^{o}$C. Either actual day-degrees of sine wave estimates may be reported or air temperature obtained with equipment comparable with instruments used in weather stations. The models should state when the accumulations began. If upper threshold is employed, then it should be clearly stated. If the use of upper threshold is not resulting in decreased predictability, use of 30$^{o}$C as an upper threshold would be consistent with current agronomic practices.

## LIFE TABLES

Life tables may be either age-specific (horizontal) or time-specific (vertical). In age-specific life tables, the fate of a cohort is documented from its collective birth to the death of the last survivor, and survivorship is determined from direct observation. This information is often difficult to obtain for insect populations because egg, larval, pupal, and adult populations may be sampled using different methods, with varying degrees of accuracy. Age-specific life tables are statistically valid. The cohort usually represents individuals from a single population and a single generation. Under field conditions, it may be possible to sample the fate of a cohort in those situations where the oviposition occurs over a short period of time, and the cohort develops through the immature stages more or less in unison. Many life tables of insects are age-specific as most insect populations in temperate regions are seasonal. Age-specific life tables are less feasible when there is an overlap in insect generations or where there is a substantial delay in development, so that there is an overlap in age classes.

All the models discussed above assume that populations contain identical individuals, the term N gives no indication of differences in age, reproductive capacity, genetic makeup, or any other characteristic that contributes to the heterogeneity within biological populations. The dynamics of populations differ vastly, depending upon a number of characteristics. The age structure is most influential. In most of these, immature insects eat more than the adults of the same species, so that EIL may be lower for immature insects than for adults. On the other hand, only adults reproduce, and therefore the birth rate of a population is a direct function of the number of adult females. Immature insects may inhabit entirely different environments and they may feed on a totally different food. It is therefore useful to sub-divide a population into age classes, and to analyze mortality separately for each class. This can be accomplished by intensive sampling and expressing results in the form of a life table — a tabular summary of inter-generation survival and mortality.

## Sampling for life tables

Sampling insect populations in order to construct life tables is subject to all constraints discussed earlier under sampling. The more accurate the sampling technique, the better it is. In studies on life tables of insects, it is important to restrict the error to about 10% of the mean. A life table for a single generation at a locality is of limited value in understanding the population dynamics of a species. It is more appropriate to sample insect populations more intensively over several generations, across locations, and more ideally over the species' entire geographical range. The life table studies are long term, and have been undertaken for a few species of insects.

Time-specific life tables are constructed from direct samples of the age classes of an entire population, and survivorship is inferred from shrinkage between successive age classes. If a population exhibits a stable age distribution, one would expect the relationship, eggs>larvae>pupae>adults. The population is sampled (several times during the season), and the life table is inferred from proportion alive within each age class. A time-specific life table may be constructed from directly estimated mortality (as due to parasites and predators), and the survivorship is inferred from this. The assumption of a stable age distribution is invalid for highly seasonal species (Taylor, 1979), though populations of insects with multiple generations may approach a stable age distribution once developmental stages overlap. Generally, insect populations in seasonal environments are reset to a single developmental stage annually at the outset of the favorable season, and reach a stable age distribution only gradually, if at all. Thus, a time-specific life table may be valid only late in the season, when the pests are well established, whereas the pest control managers are interested in the pest population before the EIL is reached. Southwood (1978) has suggested that more time should be spent in developing time-specific life tables for long lived adult insects whose fecundity may be an important factor in demography.

## Estimation of population

### *Graphical method*

It is assumed that the mortality is constant throughout the life of an individual in a particular stage. It provides an estimate that approximates to the total population at the median age of that stage. If mortality is both constant and heavy, then the value will differ markedly from the actual numbers entering a stage, but if the mortality is heavy, and occurs at the end of the stage, then the population estimate will approximate to the total population entering a stage. After joining the points, and calculating the number of squares under the curve (each square representing one individual), it is divided by the mean developmental time under field conditions (Southwood and Jepson, 1962) (Fig. 8). This method may give an underestimate of the population. So the individuals that are destroyed by the predators are present for less than the full developmental time.

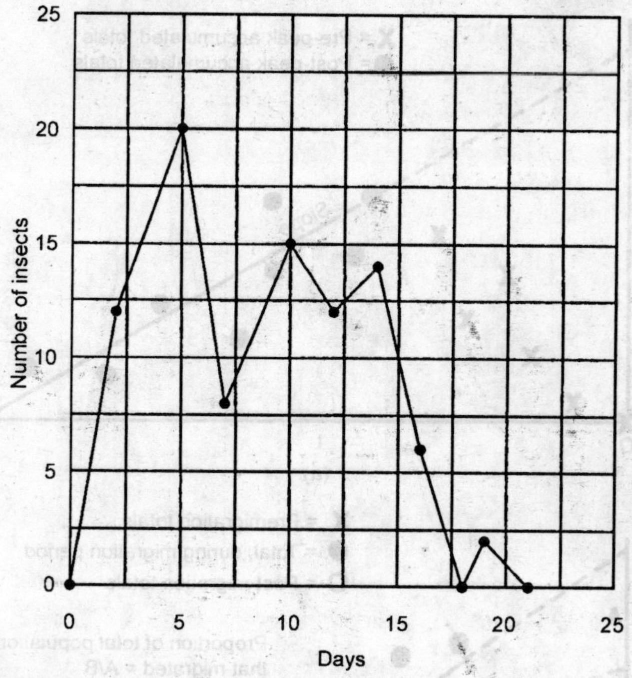


**Fig. 8.** Determination of total number of individuals in a stage from a series of estimates by graphical method — a hypothetical example (number of squares under the curve gives the total number of insects in that stage during the observation period and the number of insects attaining the mean age at that stage is given by the number of insects divided by the developmental period for that stage).

### *Measuring change in a population*

Richards and Waloff (1954) described a method to record the number of individuals entering a stage (Fig. 9 a & b). This is applicable to a stage with a well marked peak, and an approximately steady mortality rate. It consists of plotting the linear regression of fall in numbers with time after the peak, and then extending them back to the time at which the stage was first found ($t_0$); the population corresponding to $t_0$ is the number of individuals at that stage. It is expressed as:

$$Y_t = P_0 S^t \tag{59}$$

where $Y_t$ = population on day t, $P_o$ = peak population (the number hatched), and S = is the function of the population which survived to the end of unit time.

Hence, $$\log Y_t = \log P_o + t \log S \tag{60}$$

After the peak, this will describe a straight line which is conveniently obtained from the regression of log $Y_t$ on t. The regression coefficient will be log S, the logarithm of average, supposed to be constant survival rate. The number entering the stage is given by the value of $\mathbf{Y_t}$, found by inserting into the equation a value of t corresponding to the start of the stage. If t is sequentially numbered from the start of the stage in question, log $P_o$ = number entering the stage. Reasonable results are obtained if $Y_t$ is taken as the accumulated population, i.e., after the peak is obtained for

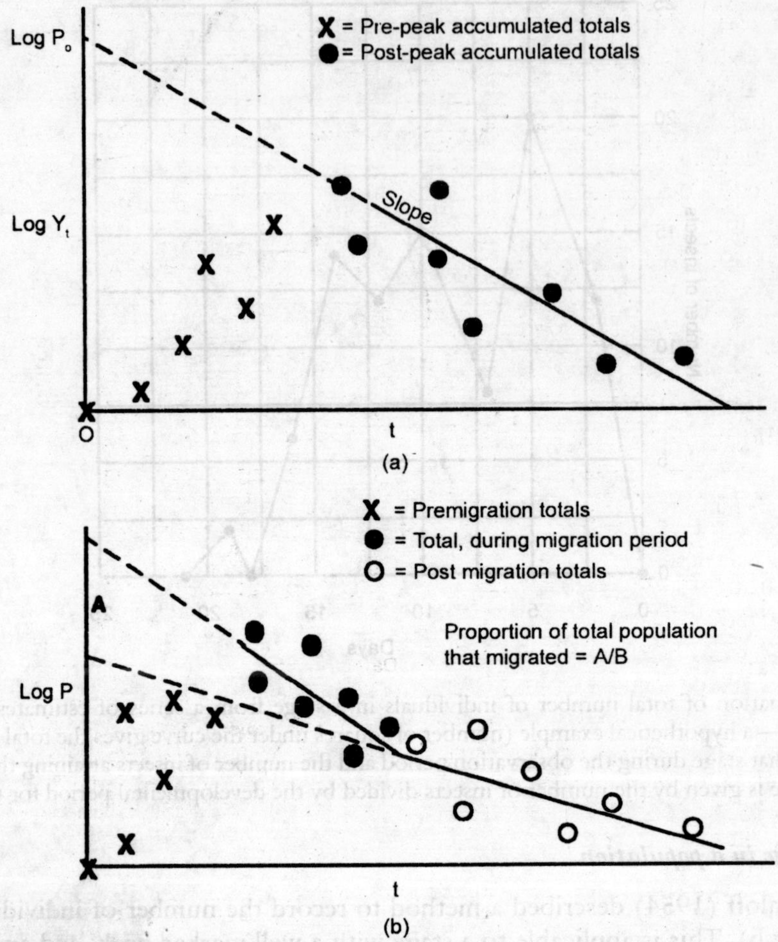


**Fig. 9.** Determination of change in a population (when a factor acts in the middle of the stage with a well marked peak) -- above. a) actual estimate of the stage under observation (egg, larvae, or adult) and the previous stage; b) increase or decrease in the stage as a percentage of N plotted as Probits and projected to determine the change between $N_1$ and $N_2$ — below (redrawn after Waloff and Bakker, 1963). For more details, see the text.

a particular instar plus any individuals that have passed through that stage, e.g., $Y_{tx}$ for the second instar would consist of individuals in the second, third, and fourth instar.

To determine change in a population Waloff and Bakker (1963) developed a method where a stage with a well marked peak is divided into three phases, and migration particularly is restricted to the second phase (Fig. 10). For this, determine:

- The date immediately before the adult (or other stage) appearance = a
- The number of previous stage (say larvae) on this date = N
- The date of adult appearance = b

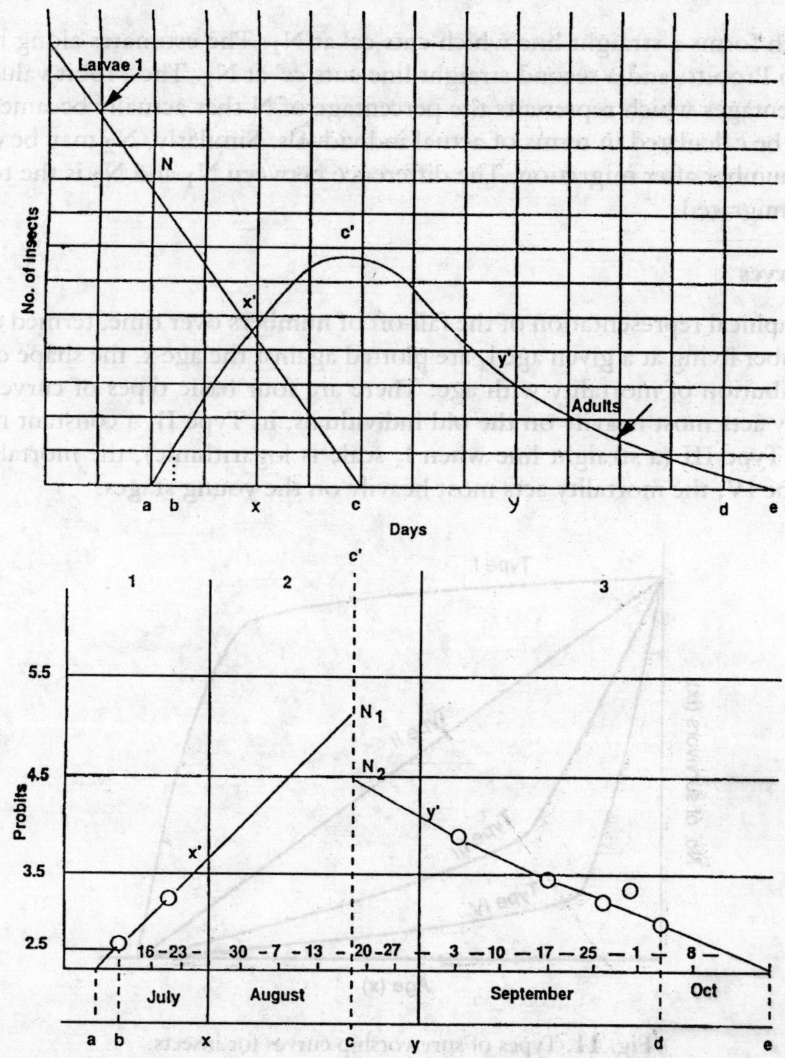


**Fig. 10.** Determination of change in a population; a) by numbers entering a stage, and b) by regression method to determine change due to migration — a hypothetical example (redrawn after Richards and Waloff, 1954).

- Date when all the larvae have become adults = c
- Date when the last adult is recorded = d
- The date of onset of migration = x
- The date of termination of migration = y

Thus: bx = the pre-migration period and yd = the post-migration period. The actual population estimates (Fig. 10a) along bx' are transformed into percentages of N, and then into Probits

(Fig. 10b), which forms a straight line which cuts cc' at $N_1$. The estimates along the y'd are also transformed into Probits, and a second straight line cuts cc' at $N_2$. The Probit value of $N_1$ is converted into percentages which represents the percentage of N that actually became adults. If N is known, $N_1$ may be calculated in terms of actual individuals. Similarly, $N_2$ may be calculated, and this is the total number after migration. The difference between $N_1$ and $N_2$ is the total number of individuals that migrated.

## Survivorship curves

A budget is a graphical representation of the fall-off of numbers over time, termed as survivorship curves. The number living at a given age $l_x$ are plotted against the age x, the shape of the curve describes the distribution of mortality with age. There are four basic types of curves (Fig. 11). In Type I, mortality acts most heavily on the old individuals, in Type II, a constant number die per unit of time, in Type III (a straight line when $l_x$ scale is logarithmic), the mortality rate is constant, and in Type IV, the mortality acts most heavily on the young stages.

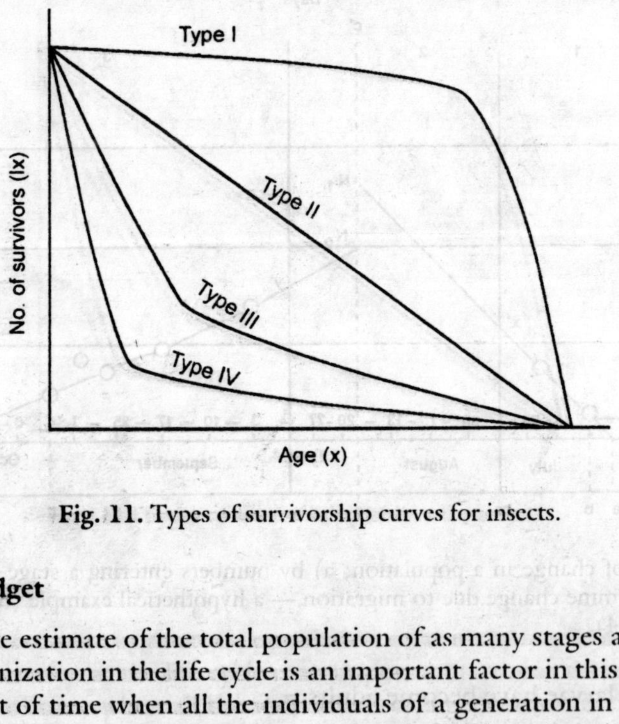


**Fig. 11.** Types of survivorship curves for insects.

## Construction of a budget

A budget is an absolute estimate of the total population of as many stages as possible of a species. The degree of synchronization in the life cycle is an important factor in this regard. In an ideal situation, there is a point of time when all the individuals of a generation in a stage provide a peak estimate, which can be used in the life tables. As the overlap in the successive stages increases, it is necessary to integrate a number of estimates to obtain the total population. When there is a complete overlap of different stages, methods based on the age structure of the population are most appropriate. Once a series of estimates of the total population for each stage have been developed, it can be assumed that the differences between them represent mortality and/or dispersal. It may be possible to check this assumption by direct measurement of these factors. The following methods can be used to determine total population at a stage.

Technically, a life table refers to the fate of a cohort of 1000 individuals. Richards (1940) proposed 'life budget' as a more appropriate term for entomological studies based on actual field census rather than a uniform cohort of 1000 individuals. A typical life table consists of following columns:

Age class interval = x
Survivorship (= number of survivors entering an age interval) = $l_x$
Mortality within an age interval = $d_x$

Mortality rate within an age interval

$$(q_x) = d_x/l_x \tag{61}$$

The expectation of life remaining for individuals of age x = $e_x$
The number of animals alive between age x and x + 1, is found by:

$$L_x = \int_x^{x+1} l_x d_x \tag{62}$$

If the age interval is very small, then

$$L_x = (l_x + l_{x+1})/2 \tag{63}$$

The number of animals in x age units beyond the age x, is given by:

$$T_x = L_x + L_{x+1} + .......... + L_w \tag{64}$$

Where w is the last age. In practice, it can be calculated by summing up the $L_x$ column from the bottom upwards. The expectation of life is:

$$e_x = \int_x^w l_x d_x/l_x \tag{65}$$

It is calculated by:

$$e_x = T_x/l_x \tag{66}$$

When the survivorship curve is of type I, $e_x$ increases with age. It is constant for Type II, and decreases for Type III and IV. The mortality rate per age interval ($q_x$) is usually expressed as rate per 1000 alive at the start of the class interval, i.e.,

$$1000q_x = 1000\ d_x/l_x \tag{67}$$

A typical insect life table omits life expectancy, and gives greater weight to individual mortality factors. Table 3 shows a hypothetical life table for the sorghum head bug, *Calocoris angustatus* Lithiery. Age intervals correspond to easily delineated life stages, and therefore may not be equal in calendar time. Survivorship and mortality are subdivided into components ($d_x$f and d'$_x$) to separate the impact of different causes of mortality. Insect life tables list an additional factor called 'killing power' ($k_x$), calculated from the difference between the log($l_x$) entering the age class and log ($l_{x+1}$), the survivorship at the outset of the subsequent age interval. These k-values are used to determine key-factors relating to density changes.

A valid life table can be used to determine whether a population is growing, declining or remaining stable. A life and fertility table may be constructed by preparing a life table with x and $l_x$ columns, but the $l_x$ column refers only to number of females alive during a given age interval, as a

**Table 3.** Life table for sorghum head bug, *Calocoris angustatus* (a hypothetical example)

| $x$ | $l_x$ | $d_x$ | $L_x$ | $T_x$ | $e_x$ | $1000q_x$ |
|---|---|---|---|---|---|---|
| 5 | 500 | 200 | 400 | 1000 | 2.00 | 400 |
| 10 | 300 | 100 | 250 | 600 | 2.00 | 333 |
| 15 | 200 | 100 | 175 | 350 | 1.75 | 500 |
| 20 | 150 | 50 | 125 | 175 | 1.17 | 333 |
| 25 | 100 | 50 | 50 | 50 | 0.50 | 500 |

x = pivotal age for the age class (days), $l_x$ = number surviving at the beginning of the age class x (generally out of 1000 originally born), $d_x$ = number dying during the age interval x, $L_x$ = number of insects alive between age x and x + 1, $T_x$ = number of insect x age units beyond the age x, $e_x$ = expectation of life and $1000q_x$ = mortality rate per age interval per 1000 insects.

function of initial population of one. It is expressed as life expectancy at birth to age x as a fraction of one (Birch, 1948). Age-specific fertility ($m_x$) indicates the number of living females born per female in an age interval. In practice, we can assume that sex ratio is 50: 50; then:

$$m_x = N_x/2 \tag{68}$$

$N_x$ = the total natality per female at age x. At each age interval, the total number of female births (female eggs laid) in each age interval (the pivotal age being x) is given by $l_x m_x$ (Table 4). The number of times a population will multiply per generation is denoted by net reproductive rate ($R_0$). The replacement rate $R_0$, is calculated from the survivorship of the adult females and the average fecundity of females ($m_x$, fecundity per female).

$$R_0 = \square l_x m_x \tag{69}$$

**Table 4.** Life and fertility table for sorghum head bug, *Calocoris angustatus* (a hypothetical example)

| $x$ *(in days)* | $l_x$ | $m_x$ | $l_x m_x$ $(= V_x)$ |
|---|---|---|---|
| 5 | 2.00 | — | Immature |
| 10 | 1.00 | — | stages |
| 15 | 0.80 | 1.0 | 0.80 |
| 20 | 0.70 | 5.0 | 3.50 |
| 25 | 0.50 | 8.0 | 4.00 |
| 30 | 0.25 | 10.0 | 2.50 |
| 35 | 0.05 | 1.0 | 0.50 |

x = pivotal age for the age class x, $l_x$ = number surviving at the beginning of age class x, $m_x$ = age specific fertility — the number of living females born per female in an age interval and $V_x$ = total number of female births in an age interval. Summation of $l_x m_x$ in a generation will give the number of times a population will multiply in a generation called net reproductive rate ($R_0$).

It equals the rate at which each egg in the cohort at the outset is replaced by an egg entering the subsequent generation. If the sex ratio is 50 : 50, and if each female produces 100 eggs, $R_0$ = 10 x 100/100 = 10, and the population size remains constant, neither increasing nor decreasing. It can also be expressed as a ratio of individuals in a population at the end of a generation to the numbers at the beginning of the generation as:

$$R_0 = N_T/N_0 \tag{70}$$

where T = the generation time. Values of $R_0$ in excess of one imply an increasing population, and less than one — a decreasing population.

**Intrinsic rate of natural increase ($r_m$)**

In developing models of population dynamics, one may wish to know the rate at which population is capable of increasing in a favourable environment. This is an instantaneous growth coefficient expressed when the population is growing in an unlimited environment and the age structure is stable. This is a crucial parameter in temperature dependent models, wherein we intend to know the speed at which the EIL is reached at various temperatures. For a population with a stable age distribution, $r_m$ may be calculated from the life table by:

$$\square_e - r_m x * l_x m_x = 1 \tag{71}$$

In this case, it is assumed that the developmental rates of different individuals are normally distributed around the mean. A series of life tables constructed from laboratory populations reared under several different temperatures can be used to determine the optimum temperature for maximum population growth. Messenger (1964) and Herbert (1981) used this to calculate $r_m$ for spotted alfalfa aphid and European red mites, respectively. When a stable age distribution cannot be assumed, especially among the seasonal insects, $r_m$ may be approximated by the capacity for increase ($r_c$):

$$r_c = \log_e R_0/T_c \tag{72}$$

where $T_c$ = the mean cohort generation time from egg to mean adult longevity. Antilog$_e r_c$ (the finite capacity for increase) will give the number of times a population multiplies per unit of time. A species capacity for increase will be close to intrinsic rate of increase when the generations are distinct, but if the generations are overlapping and the reproductive period is long, the intrinsic rate will be greater than the capacity for increase. The finite rate of natural increase, the number of times population increases per unit of time is:

$$\square = (N_t + 1)/N_t = {}_e r^m = \text{antilog}_e r_m;\ \text{as } N_T/N_0 = {}_e r_m\, T \tag{73}$$

where T = the generation time, which may be defined as:

$$T = \log_e R_0/r_m \tag{74}$$

If $R_0$ = 1.1, $r_c = \log_e 1.1/1$ = 9.5% annual increase. The statistics $r_m$ and $r_c$ are characteristic for each insect species under the appropriate environmental conditions, and are useful indicators of the speed at which the pest may be expected to reach EIL. Cave and Gutierrez (1983) illustrated the utility of this concept by calculating the $r_m$ for *Lygus* bugs on alfalfa and cotton and indicated that alfalfa may be better host (because of greater $r_m$) than cotton, so it can be used as trap crop to draw the bugs away from cotton. This can also be used to quantify the effect of host plant resistance on the population dynamics of insects.

## Analysis of life table data

Analysis of life table data is not only of theoretical interest, but should eventually provide a rational and reliable basis for pest control, enabling both the forecasting of climatically induced outbreaks, and also making the prognoses of the effects of changes in cultural or control practices (Table 5).

**Table 5.** Mortality factors for sorghum head bug, *Calocoris angustatus* (a hypothetical example)

| *Stage* | $l_x$ | $d_x$ | *% apparent mortality* | *% real mortality* | *% indispensable mortality* | *Mortality/ survivor ratio* |
|---|---|---|---|---|---|---|
| Egg | 100 | 20 | 20.0 | 20 | 5 | 0.25 |
| Larva | 80 | 30 | 37.5 | 30 | 5 | 0.60 |
| Pupa | 50 | 25 | 50.0 | 10 | 25 | 1.00 |
| Adult | 25 | — | — | — | — | — |

$l_x$ = number of survivors entering the age class x and $d_x$ = mortality rate within the age class x.

### *Apparent mortality*

The numbers dying as a percentage of the numbers entering a stage $d_x$ as a percentage of $l_x$. It can be used for comparing either with independent factors or with the same factor in different parts of the habitat.

### *Real mortality*

It is calculated from the population density at the beginning of a generation, i.e., 100 $(d_i/l_c)$; where $d_i$ = deaths in the age interval and $l_c$ is the size of the cohort at the commencement of the generation.

### *Indispensable or irreplaceable mortality*

It is a part of the generation mortality that would not occur, should the mortality factor in question be removed from the life system after allowance is made for the action of subsequent mortality factors. It is presumed that these will destroy the same percentage independent of the change in prey density, which may not always be justified. This indispensable component of a factor may be used for assessing its value in control programmes. If the exact density relationship of the subsequent factors is known, then a corrected and more realistic, indispensable mortality can be calculated. Factors acting contemporarily also affect the indispensable factor. The interaction of two factors acting contemporarily can be expressed by (Morris, 1965):

$$S = 1 - (m_1 + m_2) + v(m_1 m_2) \tag{75}$$

where S = proportion surviving, $m_1 = M_1/N$, $M_1$ = the numbers killed by factor 1, and N = the number alive at the beginning of age interval, $m_2 = M_2/N$, and v = the interaction coefficient, an index of vulnerability. If v = 1, insects affected by one factor will be as vulnerable as those unaffected. If $v < 1$, the vulnerability to one factor decreases the vulnerability to other factor, and $v > 1$, means that vulnerability to one factor increases the vulnerability to the other factor. The relation between changes in various terms in the equation may be seen by plotting S against $m_2$, the intercept on the S-axis will represent $1 - m_1$, and that on the $m_2$-axis $(1 - m_1)/(1 - vm_1)$, and the slope $vm_1 - 1$. In pest control programmes, when immediate population reduction is desired, a factor with the two last mentioned properties will be ideal.

*Mortality survivor ratio*

This represents the increase in population that would have occurred if the factor had been absent (Bess, 1945). If the final population is multiplied by the factor, then the resulting value represents the indispensable mortality due to that factor.

*The response of a natural enemy to prey density*

The response of a natural enemy to the density of a prey or host in different parts of the habitat may be determined by the regression of apparent mortality (as a percentage) in different areas of the habitat to the prey density in these areas. If the regression coefficient is significant, the action of the natural enemy is density dependent, either directly (+b) or inversely (-b). This response has two components, functional and numerical (Solomon, 1949). Holling (1959) defined a functional response as one when there are changes in the number of prey taken or killed by each predator or parasite, and a numerical response when the number of predators or parasites per unit area changes with a change in pest density.

**Survivorship analysis and model construction**

The index of population trend (I) (Balch and Bird, 1944) is calculated as:

$$I = P_{n+1}/P_n \quad (76)$$

where P = total population of a stage and n = any given generation. The stage chosen for the determination of I is important. Watt (1961) recommended adult stage, while Morris (1963) considered egg stage as preferable, and is probably the most uniform stage with respect to quality. If the same number of females survive to adulthood, and stress leads to a decrease in fecundity, then this can only be accounted for if I is taken from egg to egg. Morris (1963) computed this by:

$$I = (P_{En+1})/P_{En} = S_E S_L S_P S_A p_o p_F \pm P_{Dn}/ \quad (77)$$

where $S_E$ = survival of eggs, $S_L$ = survival of larvae, $S_P$ = survival of pupae, $S_A$ = survival of adults up to and including the period of oviposition, $p_o$ = the proportion of adults that are females and $p_F$ = the proportion of mean maximal fecundity that can be obtained by population of females because of their quality (e.g., fecundity can be indicated by size), $P_{Dn}$ = population of eggs lost or added to the population as a result of migration during generation n, and $P_{En}$ = the population of eggs of generation n.

**Determination of key factors**

Key factor analysis (Morris, 1959, 1963) is a valuable tool in identifying environmental factors most closely related to intergenerational population trends. Key factor analysis begins with a series of life tables extending over several generations of a population, usually from a single location. Initially, Morris (1959) used the relationship between each $l_x$ and total survivorship (calculated from the product of all the $l_x$s'). Varley and Gradwell's (1968) graphical factor analysis is considered better (Southwood, 1978), in which $k_x$ values, summed over each generation, are used to arrive at total K. Life tables from adult to adult are preferred for this analysis. Each $k_x$ is plotted over several generations, alongwith the total K and the relative importance of each factor is evident from its similarity to the plot of total K. If the curve for immigration ($k_5$) bears the greatest similarity to the overall mortality curve, the key factor should be sought in the $5^{th}$ age interval. Podoler and Rogers (1975) plotted common K against $k_x$. The closer the apparent relationship, the greater is the contribution of $k_x$ to the total K. There are three cautions in using this analysis; 1) individual k's are included in the calculation of total K, so the statistical correlation and regression are invalid. Total K is dependent variable by definition, and statistical correlation is valid only for independent variables, 2) a relationship between $K_x$ and k is only an indicator. Key factors are useful predictors of population trends but may not be actual causes of the observed trends and 3) a key factor at one location may not be the key factor at another location. The effect of some environmental factors (e.g., predation, parasitism, disease, competition, etc.) may be largely density dependent, while other factors (weather, insecticides, soil tillage, burning of crop residue, etc.) are independent, and incapable of regulating density about an equilibrium. Laboratory populations exhibit more density related mortality than the field populations. Strong (1984) indicated that regulation of field populations is density 'vague' due to high variance and time lags. Each population is influenced by a number of complex factors varying in their impact on density. However, key factor analysis can reveal the extent of density dependence. Each $k_x$ is plotted against survivorship ($l_x$) at the start of age interval. Positive slope indicates density dependence, the closer the slope is to 1, the more likely is that the factor capable of stabilizing the population. Slopes greater than 1 indicates overcompensating density dependence, whereas slopes less than 1 indicate under compensating density dependence. A negative slope is an indication of inverse density dependence. Density independent mortality is characterized by lack of any relationship between $k_x$ and $l_x$. Southwood (1978) has given further details of methodology to separate density independent from density dependent factors in life table analysis.

Life table analysis is only as valid as the accuracy of sampling techniques used to obtain the data. Therefore, one must ascertain the validity of assumptions underlying sampling techniques, as well as the resulting calculations. The relationships between various mortality factors are critical. For simplicity, mortality is considered constant within each age class and mortality factors are considered sequential and mutually exclusive.

## CONCLUSIONS

Whatever analysis are adopted, the dependence of descriptive parameters must be well founded in sound data and experience (Taylor, 1984). Negative binomial can be useful over a limited range of **b**-values and mean densities. Power Law provides the most consistent fit, and the most reliable parameters between continents, crops, and potential density changes; once the functional regression is established.

Extension of key-factor analysis for use in detailed studies of within population variations of reproductive or survival performance among different types of groups or individuals can be highly useful. The significance of studying qualitative heterogeneity or individual variation in insect ecology is increasing, particularly in response to evolutionary process. Use of multivariate regression for measuring the components of reproductive success to compare the opportunity for natural selection suggests a promising approach to the use of key factor analysis (Kuno, 1991). Considering plant resistance and key factor analysis as components in insect models will rationalize decision making in pest management.

Conventional computer science techniques equipped with file and information management systems, and decision support systems can serve as a decision making tool. Technical information on insect pests damaging the crops, expert systems and simulation models can be used to mimic the deductive and inductive reasoning processes employed by human beings in crop management. Simulation models and stand-alone systems have been developed as a part of research conducted on IPM and can be adapted for decision making. A timely decision by the farmers can either result in pesticide application by the farmers or non-application of pesticides which can save the costs associated with pesticide use; saving the environment, besides preserving biodiversity, including beneficial insects. Decision making in crop protection involves intricacies and historical experience. A number of pests having latent infestation in conducive environmental conditions and suitable hosts, can result in heavy crop losses. Thus, there is a need for constant monitoring of crops and the pests associated with them. These data need to be fed to computers linked to a central computer having the data bank, including weather. The insect pest information will activate the subroutine on pest models and suggest the decisions to be undertaken for pest management.

Many systems have been developed in advanced countries for high value crops are now in use as important components of integrated pest management to increase productivity, and minimize the application of pesticides. In developing countries, cost intensive knowledge-based expert and modelling systems in agriculture can be applied in high value horticultural and plantation crops, but these systems may not be applicable in low-value crops under marginal farming systems. Under these situations, low cost tactical models may be more rewarding.

## ACKNOWLEDGEMENTS

We thank Mr. V.V. Rao for his help in preparing the figures, and Dr. O. Youm and Dr. S. Chandra for reviewing the manuscript. Approved as JA/2272 by the International Crops Research Institute for the Semi-Arid Tropics.

## REFERENCES

Adkisson, P.L. and Dyck, V.A. 1980. Resistant varieties in pest management systems. *In: Breeding Plants Resistant to Insects.* F.G. Maxwell and P.R. Jennings (eds). John Wiley and Sons, New York, USA, pp. 233-251.

Allen, J.C. 1976. A modified sine wave method for calculating degree-days. *Environmental Entomology,* **5**: 388-396.

Anscombe, F.J. 1948. The transformation of Poisson, binomial and negative binomial data. *Biometrika,* **35**: 246-254.

Anscombe, F.J. 1949. The statistical analysis of insect counts based on negative binomial distribution. *Biometrics,* **5**: 165-173.

Anscombe, F.J. 1950. Sampling theory of negative binomial and logarithmic series distributions. *Biometrika,* **37**: 358-382.

Arnold, C.Y. 1959. The determination and significance of base temperature in linear heat unit system. *Proceedings of the American Society of Horticulture Science,* **74**: 430-435.

Balch, R.E. and Bird, F.T. 1944. A disease of the European spruce sawfly, *Gilpinia hercyniae* (Htg.) and its place in natural control. *Science of Agriculture,* **25**: 65-80.

Barryman, A.A. 1974. Dynamics of bark beetle populations; towards a general predictivity model. *Environmental Entomology,* **3**: 579-585.

Barryman, A.A. 1981. *Population Systems: A General Introduction. Plenum Press, New York, USA*. pp. 222.

Bartlett, M.S. 1936. The square root transformation in analysis of variance. *Journal of Research of Statistical Society (Supplement),* **3**: 68-78.

Bartlett, M.S. 1947. The use of transformations. *Biometrics,* **3**: 39-52.

Beall, G. 1940. The fit and significance of contagious distributions when applied to observations on larval insects. *Ecology,* **21**: 460-474.

Bellinger, R.C., Dively, GP and Douglass, L.W. 1981. Spatial distribution and sequential sampling of Mexican bean beetle defoliation on soybeans. *Environmental Entomology,* **10**: 835-841.

Bess, H.A. 1945. A measure of the influence of natural mortality factors on insect survival. *Annals of the Entomological Society of America,* **38**: 472-482.

Birch, L.C. 1948. The intrinsic rate of increase of an insect population. *Journal of Animal Ecology,* **17**: 15-26.

Bliss, C.I. and Fisher, R.A. 1953. Fitting the negative binomial to biological data and note on efficient fitting of negative binomial. *Biometrics,* **9**: 176-200.

Boote, KJ., Mishoe, J.W. and Berger, R.D. 1983. Coupling pests to crop growth simulators to predict yield reductions. *Phytopathology,* **73**: 1581-1587.

Bowden, J. 1982. An analysis of factors affecting catches of insects in light traps. *Bulletin of Entomological Research,* **72**: 535-556.

Breder, C.M. 1954. Equations descriptive of fish schools and other animal aggregations. *Ecology,* **35**: 361-370.

Butler, G.D. and Henneberry, T.J. 1976. Temperature-dependent development rate tables for insects associated with cotton in the southwest. *United States Department of Agriculture — Agriculture Research Service W* - **38**. P 36.

Campbell, A. 1967. The analysis of numerical change in gypsy moth populations. *Forest Science Monograph* **15**. P 33.

Campbell, A., Fraser, B.D., Gilbert, N., Gutierrez, A.P. and Mackauer, M. 1974. Temperature requirements of some aphids and their parasites. *Journal of Applied Ecology,* **11**: 431-438.

Carino, F.O., Kenmore, P.E. and Dyck, V.A. 1979. The FARMCOP suction sampler for hoppers and predators in flooded rice. *Rice Research Newsletter,* **4**: 21-22.

Cave, R.D. and Gutierrez, A.P. 1983. *Lugus hespersus* field life table studies on cotton and alfalfa (Heteroptera: Miridae). *Canadian Entomologist,* **115**: 649-654.

Chapman, J.A. and Kinghorn, K.M. 1955. Window-trap for flying insects. *Canadian Entomologis,* **82**: 46-47.

Chaudhari, D.K. 1960. Influence of temperature on development of *Onychiurus furciferus* (Barnes) and its mathematical representation. *Proceedings of the Zoological Society of Bengal,* **13**(2): 123-128.

Chiang, H.C. 1973. Bionomics of northern and western corn rootworm. *Annual Review of Entomology,* **18**: 47-72.

Chiarappa, L, (ed.). 1971. *Crop Loss Assessment Methods: FAO Manual on the Evaluation and Prevention of Losses by Pests, Diseases and Weeds.* Commonwealth Agricultural Bureaux, Farnham Royal, Slough, UK, p 162.

Chiarappa, L., Chiang, H.C. and Smith, R.F. 1972. Plant pests and diseases: Assessment of crop losses. *Science,* **176**: 769-773.

Clark, L.R., Geier, P.W., Hughes, R.D. and Morris, R.F. 1967. *The Ecology of Insect Population in Theory and Practice. Methuen and Co. Ltd.*, London, UK. pp 232.

Clark, P.J. and Evans, F.C. 1954. Distance to nearest neighbour as a measure of spatial relationships in populations. *Ecology,* **35**: 445-453.

Croft, B.A., Howes, J.L. and Welch, S.M. 1976. A computer-based extension pest management delivery system. *Environmental Entomology,* **5**: 20-34.

Croft, B.A. and Welch, S.M. 1984. Use of online control systems to implement pest control models. *In: Pest and Pathogen Control: Strategic, Tactical, and Policy Models.* G.R. Conway (ed). John Wiley and Sons, Chichester, UK. pp352-380.

Compton, J.A.F., Tyler, P.S., Mumford, J.D., Norton, G.A., Jones, T.H. and Hindmarsh, P.S. 1992. Potential of an expert system for pest control in tropical grain stores. *Tropical Science,* **32**: 295-303.

Deevey, E.S. 1947. Life tables for natural populations of animals. *Quarterly Review of Biology,* **22**: 283-314.

Dietrick, E.J., Schlinger, E.I. and van den Bosch, R. 1959. A new method for sampling arthropods using a suction collecting machine and modified Burlese funnel. *Journal of Economic Entomology,* **52**: 1085-1091.

Edmonds, R.L., VonLindern, I., Mason, C.J., McManus, M.L. and Wallner, W.E. 1973. A dispersal model for larvae of gypsy moth, *Porthetria dispar* L. - Report of the entomology group. *In*: *Proceedings, Workshop/Conference III on Ecological Systems Approaches to Aerobiology, US/IBP Aerobiogy Program Handbook no. 4.* R.L. Edmonds and W.S. Benninghoff (eds). University of Michigan, Ann Arbor, Michigan, USA. pp 57-74.

Evans, D.A. 1953. Experimental evidence concerning contagious distribution in ecology. *Biometrika,* **40**: 186-211.

Ewing, B., Rauch, P.A. and Barbieri, J.F. 1974. *Simulating the Dynamics and Structure of Populations.* UCRL 76046. University of California, Livermore Laboratory, Lawrence, California, USA. pp 58.

Funderburk, J.E.; Higley, L.G. and Pedigo, L.P. 1984. Seedcorn maggot (Diptera: Anthomyiidae) phenology in central Iowa and examination of a thermal unit system to predict development under field conditions. *Environmental Entomology,* **13**: 105-109.

Fye, R.E. 1974. *Modelling of Cotton Insect Populations. United States Department of Agriculture — Agriculture Research Service (USDA-ARS)* W 14. pp 4.

Garland, A.C. 1985. *The use of Meteosat in Analysis of African Armyworm Outbreaks in East Africa.* Ph.D. Thesis. University of Bristol, United Kingdom. pp 289.

Gosswald, K. 1935. Uber die Frasstatigkeit von Forstschadlingen unter Einfluss von Altersunterscheiden und der Einwirkung verschiedner Temperatur and Luftfeuchtigkeit and ihre praktische und physiologische Bedeutung I. *Zeitschrift fur Angewandte Entomologie,* **21**: 183-201.

Green, G.W. and DeFreitas, A.S. 1955. Frass-drop studies of larvae of *Neodiprion americanus banksianae* Roh. and *N. lecontei* (Fitch.) (Hymenoptera: Diprionidae). *Canadian Entomologist,* **21**: 183-201.

Green, R.H. 1966. Measurement of non-randomness in spatial distributions. *Research in Population Ecology,* **8**: 1-7.

Green, R.H. 1970. On fixed precision level sequential sampling. *Research in Population Ecology,* **12**: 249-251.

Grothusen, J. 1984. A simple calculator programme for flexible sequential sampling of insects. *Bulletin of Entomological Society of America,* 30(3): 35-37.

Harukawa, C. 1956. Examination of Pradhan's theory relating to the development of insects. *Science Report of the Faculty of Agriculture*, Okayama University No. **8**: 1-7.

Healy, M.J.R. and Taylor, L.R. 1962. Tables for power law transformations. *Biometrika,* **49**: 557-559.

Hearn, A.B. and Brook, K.D. 1983. SIRATAC - Case study in pest management in cotton. In: D.E. Byth, M.A. Foale,, V.E. Montgomery, and E.S. Wallis (eds). *New Technology in Field Crop Production,* Australian Institute of Agricultural Science, Brisbane, Queensland, Australia, pp. 199-211.

Heinrichs, E.A.; Chelliah, S.; Pickin, S.R.; Vaencia, S.L.; Fabeller, L.T.; Arceo, M.B. and Aquino, G.B. 1982. *A Manual of Methods in Insecticide Evaluation in Rice.* International Rice Research Institute, Los Banos, Philippines.

Herbert, H.J. 1981. Biology, life tables and intrinsic rate of increase of the European red mite, *Panonychus ulmi* (Acarina: Tetranychidae). *Canadian Entomologist,* **113**: 65-71.

Higley, L.G., Pedigo, L.P. and Ostlie, K.R. 1986. DEGDAY: A programme to calculate degree-days, and assumptions behind the degree day approach. *Environmental Entomology,* **15**: 999-1016.

Holling, C.S. 1959. The components of predation as revealed by a study of small mammal predators on the European pine sawfly. *Canadian Entomologist,* **91**: 293-320.

Holling, C.S.; Jones, D.D. and Clark, W.C. 1976. Ecological policy design: A case study of forest and pest management. *In: Pest Management: Proceedings of an International Conference.* G.A. Norton, and C.S. Holling (eds). Pergamon Press, Oxford, UK. pp 13-90.

Holt, J.; Wareing, D.R.; and Norton, G.A. 1987. Simulation analysis of brown planthopper population dynamics in rice in Philippines. *Journal of Applied Ecology,* **24**: 87-102.

Horn, D.J. 1981. Effect of weedy backgrounds on colonization of collards by green peach aphid, *Myzus persicae*, and its major predators. *Environmental Entomology,* **10**: 285-289.

Horn, D.J. 1988. *Ecological Approaches to Pest Management.* Elsevier Applied Science Publishers, London, UK, pp 285.

Hoy, C.W.; Jennisen, C.; Shelton, A.M. and Andaloro, J.T. 1983. Variable intensity sampling—a new technique for decision making in cabbage pest management. *Journal of Economic Entomology,* **76**: 139-143.

Howe, R.W. 1967. Temperature effects on embryonic development in insects. *Annual Review of Entomology,* **12**: 15-42.

Hudes, E.S. and Shoemaker, C.A. 1988. Inferential method modelling insect phenology and its application to the Budworm (Noctuidae: Lepidoptera). *Environmental Entomology,* **17**: 97-108.

Hutchinson, G.E. 1953. The concept of pattern in ecology. *Proceedings of the Academy of Natural Sciences, Philadelphia,* **105**: 1-12.

Ives, P.M., Wilson, L.T., Cull, P.O., Palmer, W.A., and Haywood, C. 1984. Field use of SIRATAC: An Australian computer-based pest management system for cotton. *Protection Ecology,* **6**: 1-21.

Iwao, S. 1968. A new regression method for analyzing aggregation pattern of insect populations. *Research in Population Ecology,* **10**: 1-20.

Iwao, S. 1975. A new method of sequential sampling to classify populations relative to a critical density. *Research in Population Ecology,* **16**: 281-288.

Johnson, C.G. 1950. The comparison of suction trap, sticky trap and tow-net for the quantification sampling of small airborne insects. *Annals of Applied Biology,* **37**: 268-285.

Johnson, C.G. 1957. The distribution of insects in the air and the empirical relation of density to height. *Journal of Animal Ecology,* **26**: 479-494.

Jones, T.H. 1993. Analytical models. *In: Decision Tools for Pest Management.* C.A. Norton and J.D. Mumford (eds). CAB International, Wallingford, UK. pp 101-107.

Jones, T.H.; Young, J.E.B.; Norton, G.A. and Mumford, J.D. 1990. An expert system for the management of wheat bulb fly, *Delia coarctata* in the United kingdom. *Journal of Economic Entomology,* **83**: 2065-2072.

Judenko, E. 1973. Analytical method for assessing yield losses caused by pests on cereal crops with and without pesticides. *Tropical Pest Bulletin,* **2**: 1-31.

Kamat, D.S. and Sinha, S.K. (eds). 1984. *Proceedings of the Seminar on Crop Growth Conditions and Remote Sensing*, 22-23 June 1974. Indian Agriculture Research Institute, New Delhi, India.

Karel, A.K. 1996. Insect pest sampling, surveillance and forecasting in IPM. *In: IPM System in Agriculture.* R.K. Upadhyay, K.G. Mukerji and R.L. Razak (eds). Aditya Books Private Limited, New Delhi, India, pp 133-174.

Kirk, J.R. and Aliniazee, M.T. 1981. Determining low temperature threshold for pupal development of western cherry fruit fly for use in phenology models. *Environmental Entomology,* **10**: 968-971.

Knight, J.D., Tatchell, G.M., Norton, G.A. and Harrington, R. 1992. FLYPAST: An information management system for the Rothamsted aphid database to aid pest control research and advice. *Crop Protection,* **11**: 419-426.

Knipling, E.F. 1964. *The Potential Role of Sterility Method for Insect Population Control With Special Reference to Combining this Method With Conventional Methods*. United States Department of Agriculture. Agriculture Research Service 33-88.p 54.

Kogan, M., Ruesink, W.G. and McDowell, K. 1974. Spatial and temporal distribution patterns of the bean leaf beetle, *Cerotoma trifurcata* (Foster) in soybeans in Illinois. *Environmental Entomology,* **3**: 607-617.

Kuno, E. 1968. Studies on the population dynamics of rice leafhoppers in a paddy field. *Bulletin of the Kyushu Agriculture Experiment Station,* **14**: 131-246.

Kuno, E. 1991. Sampling and analysis of insect populations. *Annual Review of Entomology,* **36**: 285-304.

Lefkovitch, L.P. 1966. An index of spatial distribution. *Research in Population Ecology,* **8**: 89-92.

Lindmyer, D.B., Nix, H.A., McMohana, J.P., Hutchinson, M.F. and Tanton, M.T. 1991. The conservation of Leadbeakers possum, *Gymnobelideus leadbeateri*, a case. *Biogeography,* **18**: 371-383.

Lincoln, F.C. 1930. *Calculating Waterfowl Abundance on the basis of Banding Returns.* United States Department of Agriculture Circular, **118**: pp4.

Llyod, M. 1967. Mean crowding. *Journal of Animal Ecology,* **36**: 1-30.

Logan, J., Wollkind, D.J., Hoyt, S.C. and Tanigoshi, L.K. 1976. An analytic method for description of temperature dependent rate phenomena in arthropods. *Environmental Entomology,* **5**: 1133-1140.

Lotka, A.J. 1920. Analytical notes on certain rhythmic relations in organic systems. *Proceedings of the National Academy of Sciences,* **7**: 410-415.

Mail, G.A. and Salt, R.W. 1933. Temperature as a possible limiting factor in the northern spread of Colourado potato beetle. *Journal of Economic Entomology,* **26**: 1068-1075.

Malaise, R. 1937. A new insect trap. *Entomologica Tidskr.* **58**: 148-160.

Matsazawa, H., Okamoto, H. and Mizamoto, Y. 1957. Application of Pradhan's formula to the data of development of cabbage butterfly, *Pieris rapae crucivora* and it parasite, *Apanteles glomeratus*. Kontyu, **25**: 89-93.

May, R.M. 1975. *Stability and Complexity of Model Ecosystems.* Princeton University Press, Princeton, New Jersey, USA, pp 265.

Messenger, P.S. 1964. Use of life tables in a bioclimatic study of an experimental aphid-braconid wasp host-parasite system. *Ecology,* **45**: 119-131.

Milne, A. 1959. The centric-systematic area sample treated as a random sample. *Biometrics,* **15**: 270-297.

Morisita, M. 1954. Estimation of population density by spacing method. *Memoirs of the Faculty of Science, Kyushu University Series E* **1**: 187-197.

Morisita, M. 1959. Measuring the distribution of individuals and analysis of the distributional patterns. *Memoirs of the Faculty of Science, Kyushu University Series E* **2**: 215-235.

Morisita, M. 1962. Ia-index, a measure of dispersion of individuals. *Research in Population Ecology,* **4**: 1-7.

Morisita, M. 1964. Application of Ia index to sampling techniques. *Research in Population Ecology,* **6**: 43-53.

Morris, R.F. 1959. Single-factor analysis in population dynamics. *Ecology,* **45**: 119-131.

Morris, R.F. 1963. Predictive population equations based on key factors. *Memoirs of the Entomological Society of Canada,* **31**: 30-37, 116-129.

Morris, R.F. 1964. The value of historical data in population research, with particular reference to *Hyphantria cunea* Drurry. *Canadian Entomologist,* **96**: 356-368.

Morris, R.F. 1965. Contemporaneous mortality factors in population dynamics. *Canadian Entomologist,* **97**: 1173-1184.

Mukhopadhyay, S. 1996. Meteorology in integrated pest management systems. *In: IPM System in Agriculture. R.K. Upadhyay, K.G. Mukerji and R.L. Razak (eds)*. Aditya Books Private Limited, New Delhi, India. 175-206.

Norton, G.A.; Holt, J. and Mumford, J.D. 1993. Introduction to pest models. *In: Decision Tools for Pest Management*. G.A. Norton and J.D. Mumford (eds). CAB International, Wallingford, UK, pp. 89-99.

Nowierski, R.M., Gutierrez, A.P. and Yaninek, J.S. 1983. Estimation of thermal threshold and age-specific life table parameters for the walnut aphid (Homoptera: Aphididae) under field conditions. *Environmental Entomology,* **12**: 680-686.

Ortman, E.E., Peters, D.C. and Fitzgerald, P.J. 1968. Vertical-pull technique for evaluating tolerance of corn root systems to northern and western corn rootworm. *Journal of Economic Entomology,* **61**: 373-375.

Onsager, J.A. 1976. *The Rationale of Sequential Sampling with Emphasis on its use in Pest Management.* United States Department of Agriculture, Technical Bulletin 1526, pp 16.

Parton, W.J. and Logan, J.A. 1981. A model for diurnal variation in soil and air temperature. *Agricultural Meteorology,* **23**: 205-216.

Patil, G.P. and Joshi, S.W. 1968. *A dictionary and bibliography of discrete distributions.* Oliver and Boyd, Edinburgh, England, pp 268.

Patil G.P. and Stiteler, W.M. 1974. Concepts of aggregation and their quantification: a critical review with some new results and applications. *Research in Population Ecology,* **15**: 238-254.

Pawar, C.S.; Sithanantham, S.; Sharma, H.C.; Taneja, S.L.; Amin, P.W.; Leuschner, K. and Reed, W. 1985. Use and development of insect traps at ICRISAT. *In: Use of Traps in Pest/Vector Research and Control: Proceedings of the National Seminar.* S. Mukhopadhaya and M.R. Ghosh (eds). Bidhan Chandra Agricultural University, Kalyani, West Bengal, India, pp133-151.

Pedigo, L.P. and van Schaik, J.W. 1984. Time-sequential sampling. A new use of the sequential probability ratio test for pest management decisions. *Bulletin of the Entomological Society of America,* **30**: 32-36.

Perfect, T.J., Cook, A.G. and Ferrer, E.R. 1983. Population sampling of plant hoppers, leafhoppers (Hemiptera: Delphacidae and Cicadellidae) and their predation in flood rice. *Bulletin of Entomological Research,* **73**: 345-355.

Phadke, K.K. and Prasad, S.K. 1984. Adhesive traps in monitoring aphids/aphid vectors. *In: Proceedings of the National Seminar on use of Traps for Pest/Vector Research and Control.* March 1984. Bidhan Chandra Krishi Vishva Vidalaya, Mohanpur, West Bengal, India.pp 108 - 112

Pieters, E.P. and Sterling, W.L. 1974. Aggregation indices of cotton arthropods in Texas. *Environmental Entomology,* **3**: 598-600.

Podoler, H. and Rogers, D. 1975. A new method for the identification of key factors from life-table data. *Journal of Animal Ecology,* **45**: 85-114.

Pond, D.D. 1961. Frass studies of the armyworm, *Pseudaletia unipuncta*. *Annals of the Entomological Society of America,* **54**: 133-140.

Pradhan, S. 1945. Rate of insect development under variable temperature of the field. *Proceedings of the National Institute of Science, India,* **11**: 74-80.

Pradhan, S. 1946. Dynamics of temperature effect on insect development. *Proceedings of the National Institute of Science, India,* **12**: 385-404.

Pradhan, S. and Bhatia, S.K. 1967 (eds). Effect of temperature and humidity on the life economy of insects. In: *Agricultural Entomology – Reviews.* Indian Council of Agricultural Research, New Delhi, India, pp. 55-95.

Prasad, S.K. 1963. Quantitative estimation of damage to crucifers caused by cabbageworms, cabbage looper, diamondback moth and cabbage aphids. *Indian Journal of Entomology,* **25**: 242-259.

Pruess, K.P. 1983. Day-degree methods for pest management. *Environmental Entomology,* **12**: 613-619.

Rabbinge, R. and Rijsdijk, F.H. 1983. EPIPRE: a disease and pest management system for winter wheat, taking account of micrometeorological factors. *EPPO Bulletin,* **13**: 297-305.

Reimer, C. 1959. Statistical analysis of percentages based on unequal numbers, with examples from entomological research. *Canadian Entomologist,* **91**: 88-92.

Richards, O.W. 1940. Biology of the small white butterfly (*Pieris rapae*) with reference to the factors controlling its abundance. *Journal of Animal Ecology,* **9**: 243-288.

Richards, O.W. and Waloff, N. 1954. Studies on the biology and population dynamics of British grasshoppers. *Anti-Locust Bulletin,* **17**: 1-182.

Riley, J.R. 1989. Remote sensing in entomology. *Annual Review of Entomology,* **34**: 247-271.

Ruesink, W.G. and Kogan, M. 1985. The quantitative basis of pest management: sampling and measuring. *In: Introduction to Insect Pest Management.* R.L. Metcalf and W.H. Luckmann (eds). John Wiley and Sons, New York, USA. pp315-352.

Sanborn, S.M. Wayman J.A. and Chapman R.K. 1982. Threshold temperature and heat unit summations for seedcorn maggot development under controlled temperature. *Annals of the Entomological Society of America,* **75**: 103-106.

Scheepers, C.C. and Gunn, D.L. 1958. Enumerating populations of adults of the red locust, *Nomadacris septemfasciata* (Serville), in its outbreak areas in east and central Africa. *Bulletin of Entomological Research,* **49**: 273-285.

Sevacherian, V., Stern, V.M. and Mueller, A.J. 1977. Heat accumulation for timing *Lugus* control measures in a safflower-cotton complex. *Journal of Economic Entomology,* **70**: 399-402.

Shannon, R.E. 1975. *Systems Simulation. The Art and Science.* Prentice Hall, Englewood Cliffs, New Jersey, USA. pp 387.

Sharma, H.C. 1985. Future strategies for pest control in sorghum in India. *Tropical Pest Management,* **31**: 167-185.

Sharma, H.C. 1993. Host plant resistance to insects in sorghum and its role in integrated pest management. *Crop Protection,* **12**: 11-34.

Sharma, H.C., Leuschner, K. and Vidyasagar, P. 1990. Factors influencing oviposition behaviour of sorghum midge, *Contarinia sorghicola* Coq. *Annals of Applied Biology,* **116**: 431-439.

Sharma, H.C. and Lopez, V.F. 1994. Comparison of economic injury levels for sorghum head bug, *Calocoris angustatus*, on resistant and susceptible genotypes at different stages of panicle development. *Crop protection,* **12**: 259-266.

Sharma, H.C.; Taneja, S.L.; Leuschner, K. and Nwanze, K.F. 1992. *Techniques to Screen Sorghums for Resistance to Insects.* Information Bulletin no 32. International Crops Research Institute for the Semi-Arid Tropics, Patancheru 502 324, Andhra Pradesh, India. pp 48.

Sharma, H.C.; Vidyasagar, P. and Nwanze, K.F. 1993. Effect of host plant resistance on economic injury levels for the sorghum midge, *Contarinia sorghicola. International Journal Pest Management,* **39**: 435-444.

Sharma, O.P.; Upadhyay, R.K.; Jalali, B.L. and Mukerji, K.G. 1996. Decision making tools and their role in successful implementation of integrated pest management. *In: IPM System in Agriculture. R.K. Upadhyay, K.G. Mukerji and R.L. Razak (eds).* Aditya Books Private Limited, New Delhi, India. pp 95-117.

Sharpe, J.H. and DeMichele, D.W. 1977. Reaction kinetic of poikilotherm development. *Journal of Theoretical Biology,* **64**: 649-670.

Shoemaker, C.A. 1980. The role of systems analysis in integrated pest management. In: C.B. Huffaker (ed). *The New Technology for Pest Control.* John Wiley, New York, USA. pp. 25-29.

Simpson, E.H. 1949. Measurement of diversity. *Nature,* **163**: 688.

Singh, S. 1996. Computer modelling in IPM decision making. In: R.K. Upadhyay, K.G. Mukerji and R.L. Razak (eds). *IPM System in Agriculture.* Aditya Books Private Limited, New Delhi, India. pp. 118-132.

Solomon, M.E. 1949. The natural control of animal populations. *Journal of Animal Ecology,* **31**: 1-35.

Southwood, T.R.E. 1978. *Ecological Methods with Particular Reference to Study of Insect Populations,* 2nd Edition. Chapman and Hall, London, UK, pp 524.

Southwood, T.R.E. and Jepson, W.F. 1962. Studies on the populations of *Oscinella frit* L. (Diptera: Chloropidae) in the oat crop. *Journal of Animal Ecology,* **31**: 481-495.

Stanton, M.L. 1983. Spatial patterns in the plant community and their effect on insect search. In: *Herbivorous Insects: Host Seeking Behaviour and Mechanisms.* Academic Press, New York, USA. pp. 125-157.

Stinner, R.E., Rabb, R.L. and Bradley, J.R. 1974a. Population dynamics of *Heliothis zea* (Boddie) and *H. virescens* (F.) in North Carolina: a simulation model. *Environmental Entomology,* **3**: 163-168.

Stinner, R, E.; Gutierrez, A.P. and Butler, G.D. Jr. 1974b. An alogritham for temperature-dependent growth rate simulation model. *Canadian Entomologist,* **106**: 519-524.

Strong, D.R. Jr. 1984. Density-vague ecology and liberal population regulation in insects. *In:*). *A New Ecology: Novel Approaches to Interactive Systems.* P.W. Price, G.N. Slobodchikoff and W.S. Gaud (eds). John Wiley, New York, USA. pp. 313-327.

Sutherst, R.W. 1991. Predicting the survival of immigrant insect pests in new environments. *Crop Protection,* **10**: 331-333.

Sutherst, R.W. and Maywald, G.F. 1985. A computerized system for matching climates in ecology. *Agriculture, Ecosystem, and Environment,* **13**: 281-229.

Sutherst, R.W.; Maywald, G.F. and Skarrat, B. 1994. Predicting insect distributions in changed climate. In: *Insects in a Changing Environment.* Proceedings of the , 17th Symposium of Royal Entomological Society, London, UK.

Sutherst, R.W.; Spreadbery, J.P. and Maywald, G.F. 1989. The potential geographic distribution of old screw-wormfly, *Chrysomia bezziana. Medical and Veterinary Entomology,* **3**: 273-280.

Taylor, F. 1979. Convergence to the stable age distribution in populations of insects. *American Naturalist,* **113**: 511-530.

Taylor, F. 1981. Ecology and evolution of physiological time in insects. *American Naturalist,* **117**: 1-23.

Taylor, L.R. 1961. Aggregation, variance and the mean. *Nature,* **189**: 732-735.

Taylor, L.R. 1962. The absolute efficiency of insect suction traps. *Annals of Applied Biology,* **50**: 405-421.

Taylor, L.R. 1971. Aggregation as a species characteristic. *Statistical Ecology,* **1**: 357-377.

Taylor, L.R. 1979. The Rothamsted insect survey: an approach to the theory and practice of synaptic pest forecasting in agriculture. *In: Movement of Highly Mobile Insects: Concepts and Methodology in Research.*

G.G. Kennedy and R.L. Rabb (eds). North Carolina State University, Raleigh, North Carolina, USA. pp. 148-183.

Taylor, L.R. 1984. Assessing and interpreting spatial distributions of insect populations. *Annual Review of Entomology,* **29**: 321-357.

Taylor, L.R. and Palmer, J.P.M. 1972. Aerial sampling. *In*: *Aphid Technology*. van Emden (ed). Academic Press, London, UK. pp. 189-234.

Teetes, G.L. 1985. Insect resistant sorghums in pest management. *Insect Science and its Application,* **6**: 443-451.

Thompson, H.R. 1956. Distribution of distance to n[th] nearest neighbour in a population of randomly distributed individuals. *Ecology,* **37**: 391-394.

Thorhill, E.W. 1978. A motorized insect sampler. *PANS* **24**: 205-207.

van Emden, H.F. and Williams, G.F. 1974. Insect stability and diversity in agro-ecosystems. *Annual Review of Entomology,* **19**: 455-475.

Varley, G.C. and Gradwell, G.R. 1968. Population models for the winter moth. *In: Insect Abundance,* T.R.E. Southwood, (ed). 4[th] Symposium of Royal Entomological Society. Balckwell, London, UK. pp132-141.

Vegelmann, J.E. and Rock, B.N. 1989. Use of thematic mapper data for the detection of forest damage caused by pear thrips. *Remote Sensing Environment,* **30**: 217-225.

Verhulst, J.P. 1838. Notice sur la loi la population suite dans son accroissement. *Correspondance in Mathematical Physic,* **10**: 113-121.

Wagner, T.L., Wu, H., Sharpe, P.J.H., Schoolfield, R.M. and Coulson, R.N. 1984. Modelling insect development rates: a literature review and application of a biophysical model. *Annals of Entomological Society of America,* **77**: 208-225.

Waldbauer, G.P. 1964. Quantitative relationship between the number of fecal pellets, fecal weights and the weight of food eaten by tobacco hornworm, *Protoparce sexta* (Johan) (Lepidoptera: Sphingidae). *Entomologia Experimentalis et Applicata*, **7**: 310-314.

Waldbauer, G.P. and Kogan, M. 1973. Sampling for bean leaf beetle eggs: Extraction from soil and location in relation to soybean plants. *Environmental Entomology,* **2**: 441-446.

Waloff, N. and Bakker, K. 1963. The flight activity of Miridae (Heteroptera) living on broom, *Sarothamnus scoparius* (L.) Wimn. *Journal of Animal Ecology* **32**: 461-480.

Waloff, N. and Blackith, R.E. 1962. The growth and distribution of mounds of *Lasius flavus* (Fabricius) (Hym: Formicidae) in Silwood Park, Berkshire. *Journal of Animal Ecology,* **31**: 421-437.

Waters, W.E. 1959. A quantitative measure of aggregation in insects. *Journal of Economic Entomology,* **52**: 1180-1189.

Whalon M.E. and Croft, B.A. 1984. Apple IPM implementation in North America. *Annual Review of Entomology,* **29**: 435-470.

Watt, K.E.F. 1961. Mathematical models for use in insect pest control. *Canadian Entomologist,* 93: Supplement 19, pp 82.

Wedberg, J.L., Ruesink, W.G., Armbrust, E.J. and Bartell, D.P. 1977. Alfalfa weevil pest management programme. University of Illinois Extension Circular 1136. University of Illinois, Urbana, Illinois, USA. pp 8.

Welch, S.M.; Croft, B.A., Brunner, J.F. and Michels, M.F. 1978. PETE: An extension phenology modelling system for management of multi-species pest complex. *Environmental Entomology,* **10**: 425-432.

Welch, S.M.; Croft, B.A. and Michels, M.F. 1981. Validation of pest management models. *Environmental Entomology,* **10**: 425-432.

Zadoks, J.C. 1981. EPIPRE — A disease and pest management system for winter wheat developed in Netherlands. *EPPO Bulletin,* **11**: 365-396.

*Pest Management in Horticulture Crops*
L.R. Verma, A.K. Verma and D.C. Gautam (eds)


*Chapter* 3

# Biological Control of Insect Pests

*P.R. Gupta*

## INTRODUCTION

LONG BEFORE THE APPEARANCE OF human being on the earth, lower forms of life, primarily insects and other invertebrates were existing and interacting with other organisms and the environment. This interaction for survival may be beneficial in the form of mutualism (beneficial for both), phoresy (for transportation), inquilism (commensalism in arthropods, when one thrives on surplus food or waste of others), social symbiosis (one species feeds on food storage of the other thus indirectly injurying the latter), etc., or detrimental in the form of predatism, parasitism, diseases and competition. This interaction has resulted in confining a species within certain limits in a given environment consisting of biotic and abiotic factors. Delineation of processes governing sustained stable pest populations reflects the significance of detrimental processes which put a curb on population growth and besides diseases, insectivores or entomophages hold a key control. Although significance of these phenomena is being experienced since commencement of early historical development of mankind, especially after taking up crop husbandry, yet it is only recently that the pioneer U.S. entomologist H.S. Smith used the term biological control for the first time in 1919 in context with exploitation of natural enemies to control insect pests.

The number of animal species is estimated to about 50 million but so far about 1.7 million have been described. Three-fourths of these are believed to be insects and those feeding on green plants comprise about half of the total number of insect species (Strong *et al.*, 1984), most of these occurring in two orders, Coleoptera and Lepidoptera. Strange enough, according to various estimates only 500–10000 herbivorous insects are existing as pests, obviously due to action of indigenous natural enemies which have potential to keep populations of insect species at such a low profile that it causes no or a little serious damage. Thus, nature is exercising its check by a phenomenon known as natural control and without human intervention keeps the potentially injurious pests at such low levels so that they fail to cause economic damage (DeBach and Rosen, 1991; Huffaker and Messenger, 1976). The natural regulative processes which may be additive, conditioning or subtractive in an undisturbed environment tend to keep plant or animal populations to neither become extinct nor rise to infinite density and this is known as Balance of Nature. Host plant, pests and their natural enemies constitute tritrophic level which interact with one another and when harmful effects of an organism (pest) are mitigated through another organism which is neither the host plant, the pest/pathogen or man it becomes biological control which is mediated by a third party. DeBach (1974) has aptly defined biological control as regulation of one organism by its natural enemies at a low average population density than would otherwise occur. Some would include host plant resistance, autocidal control, pheromones, hormonal analogues including insect growth regulators, etc. under the ambit of biological control. While these biorational approaches to pest suppression have a biological basis, these should not be considered as component of biological control as these pertain to the host or pest itself and not constitute the third party. In view of recent advances in biological control, Singh (1997) has defined biological control as use of natural or modified organisms, genes or gene products to reduce the undesirable organism (pest) and favour the desirable crops, trees, animals, beneficial insects and micro-organisms. Some consider natural control as naturally occurring biological control in which man does not actively manipulate natural enemies and the biological control as applied biological control which involves use and manipulation of natural enemies by man. However, sometimes biological control occurs without human manipulation as a result of the accidental immigration and establishment (ecesis) of an exotic natural enemy or ecesis of an exotic pest that is subsequently controlled by indigenous organism, it is known as fortuitous biological control.

## BRIEF HISTORY

Although significance of natural enemies (particularly predators) in pest suppression has been realized since ancient times, yet their use in true sense got an impetus after the introduction of vedalia beetle from Australia to California for the control of the cottony cushion scale on citrus in 1888 and its successful suppression by this coccinellid. Hence, in true sense history of biological control is over a century old. Another sea-change in historical development occurred in the forties of this century following discovery of insecticidal properties of DDT which ushered in the era of synthetic pesticides and interest in biological control got faded away. It was after publication of the book 'Silent Spring' by Carson (1962) when again a need was felt to search for non-agrochemical based methods of pest control because of harm done by excessive use of pesticides in pest control. Following it commenced the recent history of the biological control. The history of biological control, therefore, can be considered under the headings ancient history (prior to 1888), middle history (pre-DDT era and DDT era) and recent history (after 1962) (for details see DeBach, 1964; Hagen and Franz, 1973; Coppel and Mertins, 1977).

### Ancient history

Amongst natural enemies, predatory species have been used in biological control since ancient times. In Asia (China and Arabia), beneficial species of ants (*Oecophylla*) were used to combat menace of foliage feeders in citrus and palm orchards. Creating bamboo run-ways between citrus trees for movement of ants was a practice followed in Burma in 1770s. Use of spiders for pest control in China also dates back 2000 years. Perhaps Carl Linnaeus in 1752 was the first to suggest the use of natural enemies for getting rid of harmful insects. The first successful introduction of natural enemy from one country to another occurred in 1762 when Indian mynah bird, *Acridotheres tristis* (L.) was imported to Mauritius for the control of red locust, *Nomadacris septemfasciata* (Surville) and by 1770, the red locust was no more a serious pest. The pentatomid, *Pentatoma* (*Picromerus*) *bidens* was recommended against bed bugs in Europe as early as 1776. Some more examples in respect of use of predators in biological control in the 19th century were the suggestions of Darwin in 1800 and Kirby and Spence in 1815 to distribute coccinellids on plants and in green houses to destroy aphids, collection and release of the carabid beetle, *Calosoma sycophanta* (L.) on poplars against the gypsy moth in France by Boisgiraurd in 1840 and use of giant toad, *Bufo marinus* (L.), introduced from Jamaica into Barbados in 1844 for control of insect pests.

The concept of parasitism was perhaps first conceived by Vallisnieri in 1706 who described parasitism of the common cabbage butterfly, *Pieris rapae* (L.), by its gregarious internal parasite *Cotesia* (*Apanteles*) *glomeratus* Cameron. Hartig, a German entomologist, recommended in 1827 holding of parasitized caterpillars in cages for subsequent emergence and dispersal of parasitoids. Ratzeburg, another German entomologist provided the first comprehensive report on value of parasites in his book on forest entomology in 1844. Asa Fitch, a New York state entomologist, was the first to suggest in 1854, the international transfer of beneficial insects for pest suppression by transferring European parasites of the wheat midge, *Sitodiplosis mosellana* (Gehin), to the United States but the first successful intercontinental transfer of a parasite from England to the USA was made by Riley in 1883 to control common cabbage butterfly by *C. glomeratus*. He was also the first to purposefully move parasites from one locality to another.

Although the Greek philosopher Aristotle (384–322 BC) had mentioned diseases in bees and some silk worm diseases were described in 1000 AD, yet the Italian Agostino Bassi was the first to suggest employment of microbes for pest suppression in 1836 (Steinhaus, 1956). The concept of using insect disease to control crop pests was ushered in by the Russian Elia Metchnikoff in 1879 by infecting grubs of *Anisoplia austriaca* Herbst. with green muscardine fungus *Metarhizium anisopliae* (Metchnikoff) Sorokin (Steinhaus, 1963).

**Middle history**

***Pre-DDT era (1888–1940)***

Middle history commenced in 1888 with a remarkable success obtained in the control of the cottony cushion scale, *Icerya purchasi* Maskell in citrus orchards in California following importation of the fly *Cryptochaetum iceryae* (Williston) and vedalia beetle, *Rodolia cardinalis* (Mulsant) by C.V. Riley of the USDA from Australia in collaboration with his field agent Albert Koebele. Koebele was again sent to Australia in 1891 from where he sent 46 coccinellid species, of which only 4 got established in California and only *Cryptolaemus mountrouzieri* Mulsant proved as useful as vedalia beetle against mealy bugs. Both these beetles are still being used in biological control. Another successful biocontrol programme was the control of the Australian leafhopper, *Perkinsiella saccharacida* Kirkaldy in Hawaii on sugarcane by concerted efforts of Perkins and Koebele through importation of large number of natural enemies from Australia. Berlese initiated a successful programme in Italy in the year 1906 against the peach white scale, *Pseudaulacaspis pentagona* (Targioni-Tozetti) on mulberry by importing *Encarsia berlesei* from the USA. This parasitoid was later found effective in many other countries. Another massive programme of beneficial insect importation began in 1919 by the USDA against the European corn borer, *Ostrinia nubilalis* (Hubner) but only partial success was obtained. In the thirties, a few biocontrol accomplishments were: complete control of the Comstock mealy bug and citrus blackfly, substantial control of larch case bearer and partial control of the San Jose scale, besides work on *Bacillus popilliae* Dutky and *B. thuringiensis* Berliner continued to be carried on.

In 1927, the British Imperial Bureau of Entomology organized the Farnham House Laboratory for biological control in the countries of British Empire. Consequently many parasitoids were also sent to India. The co-operative effort of this laboratory with US Bureau of Entomology resulted in the control of European spruce sawfly, *Diprion hercyniae* (Hartig) in North America by introduced parasitoids and fortuitously introduced virus. In 1940, it was shifted to Ontario in Canada after World War II broke out.

### *DDT era (1940–1960)*

Due to shift in interest and prime focus on use of insecticides in this period, the biological control programme received some set back. Nevertheless, work on some programmes of importance continued to be persued. As a result milky disease causing bacterium, *Bacillus popilliae* was registered for the control of the Japanese beetle in the USA around 1950 when first registration of any insect pathogen as a microbial control agent was made. Subsequently commercial production of *B. thuringiensis* began in late fifties (1958). There was growing emphasis on conservation and encouragement of native beneficial organisms through environmental manipulation. Some other worth mentioning developments were the establishment of well equipped insect pathology laboratory by 1950 in Sault Ste. Marie (Canada) for work on forest-insects; creation of largest centre of research on biological control in form of Entomology Research Institute for Biological Control by 1955 after renaming Devonian Parasite Laboratory at Belleville, Ontario, Canada; Commonwealth Institute of Biological Control (CIBC) (the former Farnham House Laboratory) in 1951—a unit of the Commonwealth Agricultural Bureaux with its headquarters in Belleville (1940–46), Ottawa (1946–61) and at present in Trinidad, West Indies; International Organization for Biological Control of Noxious Animals and Plants—a reorganization of International Commission on Biological Control with its headquarters in Zurich, Switzerland (established in 1952), which has been publishing a prestegious periodical 'Entomophaga' since 1956. CIBC has been doing marvellous work since its establishment by maintaining insectaries and supplying cultures of beneficial organisms to interested parties on a contract basis. It is a publisher of 'A Catalogue of the Parasitoids and Predators of Terrestrial Arthropods' and a variety of technical bulletins and books on biological control. IOBC in its original form was composed of 22 members (government agencies, private organisations, etc.) from 16 countries of Europe and the Mediterranean area with a major objective of dissemination of information, co-ordination and promotion of research and application of biocontrol works. In the USA, successful work was carried out against coccids, aphids, spider mites and control of Klamath weed. In 1959, publication of the 'Journal of Insect Pathology' commenced in New York (now named as Journal of Invertebrate Pathology).

By the close of this era, information started pouring on harmful effects of pesticides like development of resistance to much used insecticides, large scale destruction of natural enemies and other non-target organisms due to broad-spectral nature and consequently the pest-resurgence and secondary pest outbreaks. In many areas it became impossible to grow the crops economically.

**Recent history**

As mentioned earlier, the impulse behind searching for non-agrochemical based methods set in after realization of harmful effects of pesticides on the environment and human health. Need was felt for biologically sound and ecologically safe practices in the pest control and consequently the concept of integrated pest management got originated which emphasises upon integration of chemical control methods with biological and other non-chemical based approaches. Renewed interest in biological and ecological studies of agroecosystem in totality led to development of new concepts like life tables, population models, systems analysis, etc. which visualized the complex biological problems of agroecosystem as a workable unit. The biorational approaches (utilization of safe and specific chemicals that affect the insect behaviour, growth or reproduction) have resulted into formulation of selective insecticides that attack or affect insect cuticle, endocrine and chemical communication systems in tritrophic interaction, e.g. chitin synthesis inhibitors, insect growth regulators, pheromones and other semiochemicals, and genetic control (sterile male release technique which prevents the progeny production by the female). The non-chemical methods like cultural techniques, and development of plants resistant to insects, nematodes and pathogens are now in practical use. The genetic techniques based on genetic engineering, by which lethal genes are cloned and implanted in a desired variety, has led to formation of transgenic plants which are at present being tested world-over and await their practical use on large scale. Yet biological control is more applicable and sound than biorational control and other innovative approaches and it can be blended with other (chemical or non-chemical) methods in a cost-effective way to optimize yield and quality while minimizing environmental hazards due to pesticides. Intensive and extensive research efforts have yielded fruitful results and gradually pesticidal use is declining with proper pest management strategy.

## ECOLOGICAL AND THEORETICAL ASPECTS

**Ecological base**

The inherent capacity or innate ability ($r$) of an individual to survive and reproduce in isolation from its natural enemies (including diseases) or other inhibitory factors is known as biotic potential. Species have a tendency to increase in number in geometric progression. Such unlimited growth is known as exponential growth and rate of growth at any time is proportional to the size of population. But it does not happen in nature as it has certain carrying capacity ($K$) which results in sigmoid or logistic growth. Initially it is slower, then occurs at a rapid growth rate, and finally declines, as population growth curve tends to approach the environment's carrying capacity. Interaction of biotic potential and environmental resistance tends to bring a state of balance or dynamic equilibrium and population fluctuates between certain limits. Environmental factors (natural enemies, weather, food, competition, etc.) may affect rate of birth (natality) and survival in density-independent and perfectly or imperfectly density-dependent manner. Semantically, density-independent factors cause mortality unrelated to the density of a population (e.g., harsh weather conditions) while density-dependent factors change in intensity with change in population density (e.g., food, natural enemies, etc.). Some factors (e.g., birds and predators) act in a manner opposite to density-dependent factors, thereby destroying a certain number of individuals every year irrespective of abundance of the prey such that the percentage of prey destroyed by them will decrease with the increase in population of the latter. These are known as inversely density dependent factors. Thus both density-dependent and independent processes may contribute significantly for changes in population density. It is only the density-dependent factors that would be able to regulate population growth, though density-independent factors may also modify the performance of a regulatory factor (natural enemy).

### *r and K strategists*

The concept of intrinsic rate of increase (*r*) and environment's carrying capacity (*K*) selection given by MacArthur and Wilson (1967) has also been applied to biological control. When a natural enemy is introduced into a new environment it has its prey or host in abundance. The *r*-selected natural enemy species which are small sized, attain maturity at an early age, have high intrinsic rate of increase (large reproductive capacity) and low parental care, high dispersal power and low interspecific competitive ability would be a better choice. For pest species which are nearly *K* (with large body size, delayed reproduection and low fecundity and high interspecifiic competivness), *K*-strategist (specialist) natural enemy would dominate keeping the pest population at low level. Horen (1988) has suggested to release a sequence of agents having *r*- and *K*-continuum such that the most active, voracious, fecund and highly dispersible species be released first and the most efficient, long lived be released last.

### *Functional and numerical response*

The number of pests destroyed depends upon the capacity of an entomophage to kill and their relative number. Changes in density of the pest may affect either of these two elements. With increase in pest density, if the entomophage gives quick behavioural response and destroys more pests through, e.g., reduced handling time and lowered rate of search, it is known as functional response. But it is rarely responsible for pest regulation. High functional response usually leads to greater numerical response, as increased functional response would result in more consumption of the prey, better growth, survival and reproductive output and greater increase in population. The numerical response is an increase in population of the entomophage with increase in density of the pest. For expression of numerical response on a target pest, an entomophage requires a period to complete at least three generations. The numerical response is of more interest in pest regulation. Hence an entomophage having higher functional response is expected to have greater numerical response.

## Desirable attributes of an effective natural enemy

It is practically difficult to forecast success of a biocontrol agent to be introduced in the new environment. However, on the basis of experience gained through previous biocontrol programmes, and on the basis of ideas gathered through concepts of population dynamics and natural control some characteristics which an entomophage should possess have been identified. These are descibed below.

***Searching and dispersal capacity***

High searching capacity is considered as one of the most desirable attributes of effective natural enemy though it is very difficult to measure. An entomophage should be able to find its prey even when the latter is scarce. An effective searcher is one with a tendency to exploit high density patch of the pest, exhibiting high mobility after population has been decimated to a desired low level, and has fast dispersal rate to another niche where pest population is existing. Ability to search is affected by power of locomotion, perception of the prey, survival, aggressiveness and persistence.

Searching behaviour leads to prey finding which involves four sequential steps, viz., habitat selection, prey finding, prey acceptance and finally suitability for development of the natural enemy. Host habitat is usually found firstly by random movement but when an entomophage enters the arena of food plant of the host, it is more strongly attracted to it than the prey, primarily by visual cues and due to secondary volatile chemicals produced by plants termed as infochemicals i.e. information conveying chemicals (Vet and Dicke, 1992) which act as synomone (chemical of advantage to both, the producer and recipient), e.g., *Trichogramma chilonis* Ishii, an important egg parasite of *Helicoverpa armigera* (Hubner), is attracted to crops like tomato, maize, cotton, etc., but not to the eggs on chickpea. Once the natural enemy has entered the host's habitat, it searches the prey by making directed or non-directed movements under the influence of physical and chemical cues, the latter emitted as volatiles contained in excreta, exudates, secretions, etc., from the victim which act as kairomones (trans-specific chemical of advantage to the recipient rather than the emitter). The natural enemy detects the host by tactile (touch) and/or olfactory (smell) sensilla on antennae, ovipositor and other body parts and distinguishes the preferred stage. Kairomones present in moth scales left near the moth eggs attract the egg parasitoids (e.g., *Trichogramma*). Upon finding the prey, and after receiving proper stimuli, the host is thoroughly examined to ascertain whether to accept or reject it. An effective natural enemy should be able to discriminate an already parasitized host from the unparasitized one. Finally, the selected prey should be suitable for development of the natural enemy.

***Prey specificity***

Monophagous or oligophagous entomophages with high degree of prey specificity exhibit direct density-dependence with respect to prey, give high numerical response to increased pest population and are capable of utilizing the prey at low population density. Such a specialist is better adept to behaviour, phenology, habitat and dispersal of its prey than a polyphage which is a generalist. Such specific feeders are reliable biocontrol agents as they can regulate the prey population effectively. Polyphage may behave as a specialist under particular situation and can be a better control factor (Gupta and Chander, 1992), yet they are poorer regulators. A general feeder exhibits better environmental adaptation and maintains its population by effective spatial distribution in case of periodic population collapse of their hosts. Most of the successful biocontrol programmes have been with mono- or oligo-phagous natural enemies.

*Reproductive capacity*

Natural enemy with high innate potential for increase with respect to its prey is a desired one for its efficacy. The high reproductive potential may be due to short duration of the generation or high fecundity or both. Besides, such factors as searching capacity and adaptability to a particular situation can also affect reproductive capacity, the former may often be negatively correlated with the fecundity.

*Adaptability*

The introduced natural enemy should preferably be pre-adapted to the new habitat, that is why introductions are made from the regions or countries having climatic conditions akin to the place where it is to be introduced. Even ecological requirements should be similar to those of the target pest so that the natural enemy is able to tolerate prevailing climatic conditions and is able to synchronize its activity with life cycle and phenology of the prey. It should be adapted to all ecological niches of the target pest so that it may effectively exploit pest inhabited regimes. Adaptation to man-made hazards like pesticidal application is also important. Under such circumstances, pesticide resistant species should be considered as a desirable biocontrol agent.

*Other attributes*

A desirable entomophage must be free from its natural enemies and every care must be taken to eliminate hyperparasitoid(s) from the stock being introduced, so that there is not a restraint on its population build-up in the new environment presumably free from its hyperparasitoids. The new habitat should be able to meet the food requirements of the introduced natural enemy and suitable alternate hosts be available. The most important consideration is its easy and cost effective culturability. Cost-effective multiplication procedures are known only for limited number of species and such species are frequently being used for biocontrol.

All these attributes may not be present in a candidate natural enemy but these merit consideration for selecting a good and effective one. Even sometimes environmental conditions (climate) or pesticidal usage may make an effective natural enemy as ineffective or modify otherwise expected results.

**Applicability of biological control**

Biological control is not a panacea for all pest problems but should be recognized as a management tool under circumstances where it has not provided desirable results until otherwise found unsuitable. In most of the biocontrol efforts, substantial pest suppression has been achieved. According to DeBach and Rosen (1991), upto 1988, natural enemies were introduced against 416 insect pest species and permanent control was obtained against 39.4% species (184) and of these 18% were completely controlled, 17.8% were substantially controlled and in 3.6% cases partial control was achieved with a reduction in pesticide use by about 50%. Taking into consideration the successful exploitation of these natural enemies in other countries, a total of 384 introductions were successful by then. These statistics reveal wide applicability of biological control, success of which would depend upon the target crop to be covered, pest to be controlled and natural enemy to be used.

### *Crop factors*

Feasibility of biological control depends upon injury tolerance level, value and type of the crop. Crops which can tolerate injury to a greater extent are more vulnerable to biological control. Likewise, for low value crops biological control is economically viable. For high value or cash crops, one can't take risk by sole dependence on biological control as it is slow acting and even a small pest damage to the crop may be undesirable. For such crops, natural enemies should be properly manipulated so that they are not prohibititively expensive and biological control may be reckoned as component of integrated pest management. Nature or type of the crop, whether annual or perennial, is also another consideration. Perennial ecosystem (fruit crops, forest-trees, etc.) provides stable and relatively undisturbed environment congenial for biological control. Biological control is more difficult with annual or short duration crops where frequent cropping practices, crop rotations and competition with weeds, demand application of weedicides, often create hinderance for natural enemy activity and induce inherent instability in the agroecosystem. Even introducing plant diversity may result in better biological control by providing more favourable conditions to the natural enemies, e.g., planting sorghum with cotton reduces pest damage to cotton by enhanced natural enemy activity.

### *Pest factors*

Feasibility of biocontrol also depends upon the type and number of pests to be tackled on the crop. Indirect and sessile pests like scale insects, whiteflies, mealybugs and aphids which do not directly damage the marketable produce, are more vulnerable than the direct pests. Even direct pests which inflict injury to the fruit or maketable produce sometimes can be successfully controlled, e.g., control of the olive scale, *Parlatoria oleae* (Colvee), by *Aphytis maculicornis* (Masi). Low economic injury level of direct pests makes them less likely but not impossible target for biological control. Although introduced pests are likely condidates for biological control by imported natural enemy, yet native pest suppression has also been successful by exotic natural enemy. Amenability of indigenous pests to biological control may be because of presence of less effective natural enemies in the ecosystem or because of suppressed activity of an effective natural enemy due to detrimental human activity. Thus by introducing a desirable natural enemy and adopting compatible farm practices even indigenous pest becomes a good subject for biological contol. Success also depends upon the number of pests and level of control to be achieved on a crop. Fewer the pests better are chances of success. Earlier concept that tropical pests are more susceptible than the pests in temperate region does not now hold true as success has been obtained to suppress pests even in temperate regions. Hence, each pest problem has to be analysed on its own merits.

### *Natural enemies*

Availability of effective natural enemy which has kept pest suppressed in an environment elsewhere is the key factor for feasibility of any biological control programme. A numerically rare natural enemy encountered in a regulated pest population may be more effective as even at a low level it regulates the pest population in absence of competitors. A common or abundant natural enemy may not be able to do so as they are abundant at abundant prey level (DeBach and Rosen,

1991). Inspite of possession of all desirable qualities, a natural enemy may not be able to express its effectiveness because of control strategy adopted for suppression of other pest(s) for which effective natural enemy is not available. Under such circumstances, feasibility would depend upon availability of selective pesticide or pesticide resistant strain of the natural enemy.

## NATURAL ENEMIES

Perhaps every living being is having one or other natural enemy which is a naturally occurring organism that causes mortality or debility of the one with which it is associated. These are parasites, predators and pathogens. Enemies of insects are either entomophagous (insectivorous) feeding on the whole body or part thereof (parasites and predators) or entomogenous which grow on or in the body of insects (microorganisms). Usually entomogenous organisms are pathogenic and cause disease-like condition in the host and hence, more accurately referred to as entomopathogens. All these natural enemies have been used in biological control programmes but parasites have been of most importance followed by predators.

### Entomophages

Both parasites and predators feed on other animals and their victim is known as host and prey, respectively. Though both these terms are often used interchangeably, in this chapter the term prey has been used with natural enemy.

A parasite is an organism living and feeding on or in a larger animal which when numerous, may irritate and debilitate host occasionally causing its death (e.g., tape worms), may complete life cycle on a single host (e.g., lice) or be parasitic during part of their life cycle (e.g., mosquitoes, fleas) or have complicated life cycle requiring several host species (e.g., tape worms). Askew (1971) used the term protelean parasite for an insect species in which only immature stages are parasitic and the term parasitoid, according to him, is an insect parasite of an invertebrate host often an arthropod, which nearly destroys its host. Although considered as a parasite, parasitoid is a special kind of predator, parasitic only in immature stages, often of the size of the host, kills it and requires only one host (prey) for its development into a free living adult (e.g., braconid wasps). Predator is a free living animal often larger and stronger than its victim which spends a limited period of time with an individual of one or more species of prey for feeding on it and requires more than one prey to reach maturity.

### *Parasitism and biological relationship*

Parasitism is a qualitative term referring to a kind of symbiosis in which one party (the parasite) lives at the expense of the other (the host) contributing nothing to the relationship and frequently destroying the host in the process. But parasitization is a quantitative term referring to the proportion of a host population attacked by parasites (Sweetman,1958). Simple to complex biological relationships exist between parasites and hosts and that forms the basis of classification of parasitism. On the basis of position of feeding on or in the host, the parasitism may be ectoparasitism or endoparasitism. Ectoparasites may glue the egg on host body anywhere or at a specific site, while endoparasites usually deposit egg in the haemocoele (body cavity) or embed it in specific organ like ganglion, muscles, intestine, etc. Based on adaptation or necessasity, it may be obligate (exclusive parasitic existence) or facultative (non-parasitic but can adapt to parasitic or a free life) or incidental (parasite establishing on a host with which it is not ordinarily associated). Based on number of host species attacked, parasitism may be monophagous (specific), oligophagous (associated with a few or closely related species) or polyphagous (many species). On the basis on number of parasites developing on or in the host, it may be solitary or gregarious. Usually one parasite deposits one or more eggs or larvae at one time in the host. If more, all are capable of completing development in the host. Some polyembryonic species present a special case of obligatory gregarious development wherein from a single egg large number of parasitoids develop. Based on stage of the host attacked, it may be egg, larval, pupal or adult parasitism expressed as E, L, P or A parasite, respectively, if they complete development in the stage in which parasitism occurs, or as egg-larval ($E_L$) or larval-pupal ($L_P$) parasite if parasitism occurs in egg or larval stage and the parasitoid adult emerges from the host larva or pupa, respectively.

Complex parasitism arises when more than one species of natural enemies act as parasite on one prey and on basis of sequence of attack, parasitism may be hyperparasitism or primary parasitism depending upon whether the attacked host is a parasite or not and thus if parasite attacks another parasite, it becomes hyperparasite. Hyperparasitism may be secondary (if attacked host is primary parasite), tertiary (attacked host is secondary parasite) or quaternary (attacked host is tertiary parasite). The secondary parasitism may be direct if secondary parasite attacks primary parasite within or without its non-parasitic host or indirect if it attacks the non-parasitic host (i.e., secondary host) even regardless of presence or absence of its primary parasite but is capable of development if the latter becomes available. Although hyperparasitism helps in maintaining and regulating the population of primary parasites, yet it is not desirable from biocontrol view point as it may adversely affect the beneficial performance of the primary parasite. Autoparasitism or adelphoparasitism is a special type of hyperparasitism in which the female develops as a primary parasite but the male behaves as a secondary parasite of the female of its own species (e.g., *Encarsia* and *Coccophagus*). Based on intensity of attack and the parasite-parasite relationship, the parasitism is known as simple parasitism when it results from a single attack upon a host, the parasite laying one or more eggs at a time on/in it; superparasitism if the single host is exploited by more parasites of the same species than can normally develop upon it; or multiple parasitism (multiparasitism) when individual host is attacked by parasites of different species. Thus intraspecific or interspecific competition hampers the development of parasitites. Superparasitism may result in the formation of small or malformed adults or death of competitors. Multiparasitism often results in survival of intrinsically superior parasite while others die. Cleptoparasitism is a special type of multiparasitism in which the adult parasite, incapable of paralizing and parasitizing its host, utilizes already paralyzed host by other species after killing the latter. These three types of parasitism are undesirable for efficient implementation of biological control programmes.

***Biological characters of entomophagous agents***

Parasites may parasitize the host by using ovipositor which is a powerful tool of Hymenoptera to place the egg at appropriate place on or in the body. Hosts residing in protected site or habitat e.g., in leaf mines, tissue galleries, burrows, pupa, cocoon or under waxy armature (scale insects) are usually parasitized by ectoparasites and the eggs are glued by mucilaginous accessory gland secretion or embedded by stalk or pedicel in the integument. Such parasites often paralyse their host temporarily or permanently by injecting venom through ovipositor in their host and are known as idiobionts if the paralysis is permanent or the host is killed, and koinobionts if after paralysis the host continues to function. Paralysis makes the host docile and facilitates the parasite to feed on body without being dislodged by the otherwise active host. On the other hand, endoparasites lodging their eggs inside the body usually do not paralyse the host but have to have special adaptation to get oxygen from host body either through entire and thin integument (cutaneous respiration) or through posterior caudal vesicle (e.g., some braconids) or retaining connection with their own egg-stalk protruding through the body of the host (e.g., encyrtids). Some parasites lay their eggs away from the host randomly on foliage or aerial parts of plants. Such minute eggs may be thick shelled known as microtype eggs which measure 0.02–0.4 mm. These have to be ingested along with foliage by the host in order to gain access into intestine where they hatch and larvae penetrate into body cavity for further developement. Eggs of some parasites hatch into active apodous long-lived fusiform larva (planidium) e.g., some tachinids with macrotype elongate eggs of 0.4–0.9 mm size and some Hymenoptera. These larvae actively search for a suitable host and are prone to high risk which is compensated by high fecundity of the species. Such parasites usually undergo hypermetamorphosis and second instar onwards larvae are different from the first instar.

During larval development most of the species moult 3–5 times and the host body contents are almost utilized leaving behind the exoskeleton as a collapsed structure in ectoparasites. Usually by the end of larval development, integument of the host parasitized by endoparasite(s) becomes smooth and hardened, maintains the typical shape of the body and transforms into a mummified body (e.g., aphids, scale insects). Others which do not mummify their host, may escape through the host cuticle and pupate in cocoon outside the body of the host (e.g., braconids).

Pupae formed in protected sites, e.g., inside host, mines etc., are not encased in cocoons. But external feeders or those emerging from the host as full-fed larva either enter the soil for pupation or form cocoon outside. Dipteran pupae either are free (in most Barchycera and Nematocera) or are in puparium (hard and tanned integument of last instar larva) in cyclorrhaphan Diptera.

Adults of most parasites are free living. They meet their food requirement by feeding on nectar, water, fruit juices, honey dew excreted by aphids and other homopteran insects, even haemolymph oozing out from the puncture drilled during oviposition or intentionally made by stabbing ovipositor in host body. Some drill the prey body and its cover, e.g., scaly armature, cocoon of pupae or puparium and use ovipositor as tempelate to make feeding tube, a fine bore or capillary, supported by secretions from alkaline glands associated with the reproductive system and coagulated host blood or body fluids. This tube is used to suck the haemolymph of the prey. Excessive feeding may mutilate and kill the prey. Thus many parasites (e.g., *Aphytis, Nasonia*) do predatory feeding also. If the food is insufficient or the parasite is unable to find host, oosorption (absorption of the egg from the ovariole) may take place. It is a conservative phenomenon of reproductive output for economy of parasitism and is correlated with high searching capacity and long life of the female.

Parasites may by bisexual exhibiting a balanced sex ratio of 1 : 1 (some Hymenoptera and mostly Diptera) but in those having obligate or facultative parthenogenesis the sex ratio is variable. In most parasitic Hymenoptera, facultative parthenogenesis is observed. Laying of unfertilized eggs by a mated female is determined by many intrinsic and extrinsic factors and may pose problem in arrhenotokous species (male developing from unfertilized eggs). Some species may be uniparental (e.g., some *Aphytis, Trichogramma*) and these are thelytokous (only female progeny develops from unfertilized eggs). Parasites may have one generation in a year corresponding to one generation of the host (univoltine) or many generations (multivoltine), each lasting for 10 days to a month's duration in summer and longer in winter.

As far as predators are concerned, many characteristics of parasites are applicable to them but unlike parasites, the sex ratio is even in predator species and parthenogenesis is rare. It is the searching behaviour of larva as well as adult which forms the basis for predator-prey population dynamics. Often predators have longer premating period than parasites which may vary from 4–6 days to several months. Predators devour their prey by biting and chewing type of mouth parts (e.g., beetles and dragon flies) or by piercing and sucking type of mouth parts (e.g., bugs, larvae of lace wings, flies and mites) and for capturing the prey, some have specially modified organs (legs, mouthparts, etc.). Adult and larval food may be same, as in coccinellids, or different as in predacious Diptera and Neuroptera where nectar, honey dew and pollen are important sources of energy.

***Major groups of entomophagous insects***

According to Sweetman (1958), five orders of insects which contain known species of parasitic insects are Strepsiptera (commonly known as stylops or stylopids, all parasitic often on species of non-economical importance), Coleoptera and Lepidoptera (with some species of practical use in both), Hymenoptera and Diptera. Only Hymenoptera followed by Diptera are of practical use as biocontrol agents. At least 15% of the known species of insects which exceed 820,000 are parasitic (Askew, 1971). However, predatism by insects on insects is much more general than parasitism and in addition to 11 orders which include predatory species (but not parasitic on insects), four of the five orders containing insect-parasites also have predacious forms. With the exception of Hymenoptera, predatism is much more common than parasitism, in these entomophages. Predacious insects of all orders are important in natural pest suppression but those which have found importance in biological control come from Coleoptera, Neuroptera, Hymenoptera, Diptera and Hemiptera (Heteroptera). Other invertebrate predators with promise in biological control of crop pests include mites (Acarina) and spiders (Araneida). Amongst vertebrates, frogs (Amphibia), birds (Aves), moles and shrews (Insectivora) and some mice (Rodentia) play a significant role in natural suppression of many insect pests.

***Parasitic insects***

(a) *Hymenoptera*: Amongst Hymenoptera, most important parasites occur in superfamilies Ichneumonoidea (Ichneumonidae and Braconidae), Chalcidoidea and Proctotrupoidea. In the first two, the last abdominal sternite is longitudinally divided, ovipositor arises from anterior to tip of the abdomen and is furnished with a pair of narrow exserted sheaths reaching to its tip, trochanters 1- or 2-segmented. However, in the Proctotrupoidea, the last abdominal sternite of the female is not divided longitudinally, the ovipositor arises from the tip of the abdomen and without a pair of sheaths, trochanter one segmented, and pronotum extending laterally back to tegulae. Ichneumonoidea are usually large parasitic wasps (though some are small) associated with larvae and pupae of host. They have pterostigma (black patch on costal margin of forewing) and well developed venation extending to the latter, antennae long, filiform with more than 16 segments, and trochanter two segmented. On the other hand, Chalcidoidea are small, their body bare or sparsely haired and with metallic sheen, pterostigma absent, venation of forewings much reduced, antennae geniculate and pronotum not reaching tegulae.

*Chalcidoidea:* Chalcidoids are most versatile parasites, their size ranging from <0.25 mm to over 12 mm and these attack species in nearly all orders with a preference for Coleoptera, Diptera, Homoptera and Lepidoptera. Although it includes over 30 families, yet the better known and exploited families are Aphelinidae, Chalcididae, Encyrtidae, Eulophidae, Eurytomidae, Mymaridae, Pteromalidae and Trichogrammatidae.

Most of Aphelinidae are parasitic on Homoptera (scale insects, mealybugs, aphids, whiteflies, leaf hoppers, etc.) and some important genera are the ectoparasitic *Aphytis*, and the endoparasitic *Aphelinus, Coccophagus* and *Encarsia*. *Aphytis* has played a major role in classical biological control programmes against armoured scale insects and some to mention are on citrus with *A. holoxanthus* DeBach against the Florida red scale, *Chrysomphalus aonidum* (L.) in Israel, *A. melinus* DeBach against the California red scale, *Aonidiella aurantii* (Maskell) in California, *A. lepidosaphes* Compere against the purple scale, *Lepidosaphes beckii* (Newman) in California, *A. lingnanensis* Compere against the citrus snow scale, *Unaspis citri* Comstock in Florida, *A. yanonensis* DeBach & Rosen against the arrowheaded scale, *U. yanonensis* (Kuwane) in Japan and southern Europe, and *A. paramsculicornis* DeBach & Rosen against the olive scale, *Parlatoria oleae* (Colvee). on olive, deciduous trees and ornamentals in California. Many naturally occurring species regulate scale populations to low levels. Introduction of *Aphytis* spp. and *Encarsia perniciosi* Tower in India in 1958–60 against the San Jose scale, *Quadraspidiotus perniciosus* Comstock, considerably reduced the pest population in Himachal Pradesh, Uttar Pradesh and Kashmir. The citrus black fly, *Aleurocanthus woglumi* Ashby has been successfully suppressed in many countries by introduction of *Eretmocerus serius* Silv. from Malaysia and in Mexico by *Encarsia opulenta* Silv., *E. clypealis* Silv. and *E. smithi* Silv. from India and Pakistan. The control of the woolly apple aphid, *Eriosoma lanigerum* (Hausmann) in India and other countries by introduction of *Aphelinus mali* (Hald.) is a well known example.

Chalcididae are primarily solitary and internal parasites of Lepidoptera, Coleoptera and Diptera, some occur as hyperparasites on the latter and Ichneumonids. These are relatively large with shiny and highly sculptured body and greatly enlarged denticulate hind coxae. *Brachymeria* is the best known genus and *B. nephantidis* Gahan is an important pupal parasite of the coconut caterpillar, *Opisina arenosella* Walker in India.

Encyrtidae contains many highly beneficial insects crowned with successful biological control of many homopteran and lepidopteran pests. Its important genera are *Ooencyrtus* and *Anastatus* (both egg parasites), *Copidosoma* and *Litomastrix* ($E_L$ parasites), *Anagyrus, Encyrtus, Adelencyrtus* and *Comperiella*. Outstanding example of successful use of an encyrtid is the control of Rhodes grass mealybug, *Antonina graminis* (Mask.), in the USA in 1959–60 by importing *Neodusmetia sangwani* Subba Rao from India. This species also provided very good control of this pest in Bermuda and northern Brazil. Other successes include control of *Aonidiella citrina* (Coq.) in the USA by *Comperiella bifasciata* How., and *Ceroplastes rubens* Maskell in Japan by *Anicetus beneficus* Ishii & Yasumatsu. *Anicetus dodonia* Ferr. is an important parasite of *Ceroplastes* in India. After mid eighties, *Leptomastrix dactylopii* Howard was introduced in India from Brazil against the citrus mealybug, *Planococcus citri* (Risso) which established in Karnataka and spreaded to adjoining states.

Trichogrammatidae encompasses minute egg parasites primarily of Lepidoptera but also attack eggs of Coleoptera, Neuroptera, Diptera, Homoptera, Orthoptera and Hymenoptera. These are minute (<1 mm) having three segmented tarsi, short broad and apically rounded forewings studded with pubescence in rows, and with marginal and stigmal veins united into a strongly arched curve. Some important genera in this family are *Trichogramma, Trichogrammatoidea, Lathromeris, Megaphragma* (on thrips), *Oligosita* (on Homoptera), *Ufens* (on Lepidoptera) and *Doirania*. *Trichogramma* is the most frequently exploited genus in biological control programmes because it can be easily and economically multiplied in laboratory. Another important genus of practical use is *Trichogrammatoidea*. *Trichogramma* is being used in over 50 countries over 32 million ha per year mainly against *Ostrinia* spp., followed by *Helicoverpa*, pyralids, tortricids, noctuids, oleuthrids and pierids in decreasing order. Although more than 145 species exhibiting much interspecific variations are known, yet decreasing order of use for most important species is *T. evanescens* Westwood, *T. dendrolimi* Matsumara, *T. pretiosum* Riley, *T. Brassicae* Bezdenko and *T. nubilale* (Smith, 1996). Small scale multiplication for local periodic releases is being carried out throughout the world including India, but on larger scale commercial production is confined to the USA, Canada, Mexico, Germany, France, the Netherlands, Switzerland, China and former USSR. More than half the countries are using Angoumois grain moth, *Sitotroga cerealella* Olivier as factitious host for multiplication of *Trichogramma* while other small-egg laying species are the Mediterranean meal moth, *Anagastra kuehniella* Zeller (in France and Canada) and rice moth, *Corcyra cephalonica* (Stainton). The larger eggs of silkworm and *Ostrinia nubilalis* (Hubner) are being used in China. Tricho-cards having ultraviolet irradiated eggs glued on it are exposed to *Trichogramma* females for a day @ 1 : 10 (parasite female : host egg ratio) or for life long survival of the parasite @ 1 : 30 ratio. First release should be aimed at just before host-oviposition and unparasitized factitious host eggs should be exposed when target pest is low, so that when pest egg laying picks up, already parasite females are sufficient in number to cause above 80% parasitisation. The interval between two releases is variable but averages 5–7 days. Total releases may on an average vary from 200,000–600,000 wasps/ha in agricultural crops like corn, cotton and tomato and several millions/ha in arboreal situations depending upon leaf surface area and disappearance rate of the parasite. The cost-benefit ratio may vary from 1 : 8 to 1 : 25 (Li, 1994). In India, use of *Trichogramma* has been recommended on sugarcane against *Chilo* spp. (*T. chilonis* Ishii @ 50,000/ha, 8 releases at 10-day interval on 30 or 60-day old crop for shoot or other borers), rice against *Scirpophaga incertulas* (Walker) (*T. japonicum* Ashmead and *T. chilonis* @ 50,000/ha, 2 releases at 30 and 37-day after transplantation) and cotton bollworms (*T. chilonis* @ 150,000/ha, 6 releases at 7-day interval starting from 40 days after plantation) (Singh, 1997).

Mymaridae are also minute yellow or black egg parasites but with four tarsomeres, long narrow forewings with very long marginal fringe. Mymarids have been used successfully in South Africa where the eucalyptus weevil, *Gonipterus scutellatus* Gyll., was controlled by *Patasson nitens* (Gir.) from Australia, and in Hawaii where sugarcane leafhopper, *Perkinsiella saccharicida* Kirk. was partially controlled by using three species, viz., *Anagyrus frequens* Perk., *Paranagyrus optabilis* Perk. from Australia and *P. perforator* Perk. from Fiji (Sankaran and Nagarkatti, 1982).

Eulophidae contains many parasitic genera which usually are external parasites of larvae of Diptera, Lepidoptera and Hymenoptera but a few have been successful in biological control, e.g., *Pediobius parvulus* Ferr. from Java gave outstanding control of the coconut leaf-mining hispa, *Promecotheca reichei* Baly in Fiji. Amongst *P. foveolatus* (larval parasite) and *Tetrastichus ovulorum* Ferr. (egg patasite) of *Epilachna* spp. in India, the former provided good parasitism of the Mexican bean beetle, *Epilachna varivestis* Mulsant, in the USA in late sixties and currently the pest is successfully managed by annual inoculative releases (Singh, 1997).

*Ichneumonoidea:* Braconids are entirely beneficial species acting as external or internal and solitary or gregarious parasites of Lepidoptera (Braconinae and Microgastrinae as larval and Cheloninae as egg-larval parasite) and Coleoptera (Braconinae, Cosmophorinae and Doryctinae) and also Diptera including fruit flies (Opiinae). Earlier, braconids parasitizing aphids were placed in sub-family Aphidiinae but now it has been elevated to the family level as Aphidiidae. Important braconid genera in India are *Apanteles, Chelonus, Cotesia, Bracon, Stenobracon, Rogas, Opius, Microptilis, Macrocentrus, Aphidius* and *Lysiphlebus*. *Cotesia flavipes* Cam., a good larval parasite of graminaceous stem borers in India, provided satisfactory suppression of the sugarcane stem borer, *Diatraea saccharalis* (Fabr.) in Barbados within three years of its single introduction made in 1967. *C. Subandinus* Mues., a promising parasite of potato tuber moth, *Phthorimaea operculella* Zeller in Argentina, provided high levels of parasitism in Australia and Zambia and has also established in India, New Zealand, Tanzania and Cyprus. *Aphidius smithi* Sharma & Subba Rao, introduced from India to the USA in 1958, gave considerable control of the pea aphid, *Acyrthosiphon pisum* (Harris). Considerable control of *Bactrocera (Dacus) dorsalis* (Hendel) and *B. cucurbitae* (Coq.) has been obtained in Hawaii by introduction of *Biosteres arisanus* (Sanan) (= *Opius oophilus* Fullaway), *Opius fletcheri* Silv. (both from India), *B. tryoni* (Cam.) (from Australia) and *O. fullawayi* Silv. (from South Africa).

Ichneumonidae is a very large family making about 20% of all parasitic insects and majority are primary parasites mainly of Lepidoptera followed by Coleoptera and Hymenoptera (sawflies and wood boring Siricidae). Some genera of common occurrence in India are *Campoletis, Rhyssa, Diadegma, Eriborus, Xanthopimpla, Charops* and *Isotima*. Most of these are solitary endo- or ecto-parasites. Many species play important role in natural pest suppression, e.g., *Isotima javensis* Rohw. is a well known parasite of sugarcane top borer and *Campoletis chlorideae* Uchida of *Helicoverpa armigera* (Hubner).

*Proctotrupoidea (Serphoidea):* This superfamily includes two important families Platygastridae and Scelionidae which are primary internal parasites; the former of cecidomyiid larvae and homopterous nymphs and the latter as egg parasites of Lepidoptera, Hemiptera and Orthoptera. *Amitus* (viz., *A. Spiniferus* (Brethes) and *A. hesperidum* Silv.) is an economically important genus of Platygasteridae which has provided successful biological control of whiteflies, *Aleurocanthus spiniferus* Quaint in Guam and *A. woglumi* Ashby in Mexico. *Amitus aleurolobi* Mani occurs on sugarcane aleyrodid, *Aleurolobus barodensis* (Maskell) in India. *Allotropa* is an important parasite of mealybugs. Amongst scelionids, *Telenomus* is a well known genus in India as an important parasite of sugarcane stem borers. *T. proditor* Nixon was introduced form Papua New Guinea against *Achaea janata* (Linn.) in 1963 and has given 60–80% parasitism. Another introduction was of *T.*

*alecto* (Crawford) in India in 1966 against sugarcane borer eggs which has established in some pockets of Uttar Pradesh. Another genus *Trissolcus* is an important parasite of bug eggs.

(b) *Diptera*: Several better known families of Diptera include some of the valuable and promising species of parasitic insects like Tachinidae, Sarcophagidae, Cryptochaetidae, Conopidae, Pipunculidae, Pyrgotidae, Nemestrinidae, Cecidomyiidae and Bombyliidae. However, species of importance in biological control of insects hail from the first three families, the most promising being Tachinidae.

Tachinidae resemble house flies but are robust, have bare arista on antennae, bristly legs, abdomen covered with marginal, lateral and discal hair, and veins on wings often do not reach the posterior margin. These are common endoparasites of larvae of Lepidoptera (Exoristinae and Tachininae), larvae and adults of Coleoptera (Exoristinae and Dexiinae), and nymphs and adults of pentatomid, pyrrhocorid and coreid Hemiptera (Gymnosomatinae). Their larvae are broadly cylindrical and tapering a little anteriorly, have indistinct body segmentation and possess dark conspicuous and sclerotized posterior spiracles. Adult flies may be ovo-viviparous (laying membranous, microtype, macrotype or pedicillate eggs) and eggs hatch in short duration after oviposition on foliage, in soil or on/in host larvae. Larviparous females may do external or internal host-larviposition. The offspring per female may vary from 50 (*Rondania*) to at least 13,000 (*Echinomyodes*). Pupation usually takes place outside the host body in puparium. *Bessa remota* (Aldrich) introduced from Malaysia into Fiji in 1925 to control the coconut moth *Levuana iridescens* Bethune-Baker, provided highly satisfactory control by 1929. *Lixophaga diatraeae* (Townsend) was successful in St. Kitts against sugarcane borer and this alongwith *Cotesia flavipes* Cameron brought about economic control of sugarcane borer in Barbados. *Eucelatoria bryoni* Sabrosky, originating from southern USA, was introduced in India in 1979 against *H. armigera*. It was multiplied, released and recovered from this pest on tomato, cotton, pigeonpea, lablab and tobacco in South India (Karnataka, Andhra Pradesh and Kerala). *Sturmiopsis inferens* Townsend is an indigenous parasite of sugarcane stem borers which can be reared on *Galleria mellonella* L. or pink stem borer *Sesamia inferens* (Walker) (Singh, 1994a). Its releases are recommended @ 125 gravid females/ha after 35–45 days of plantation. *Stomatomyia (Spoggosia) bezziana* Baranoff, is a common parasite of the coconut leaf eating caterpillar, *Opisina arenosella* Walk., in Sri Lanka, which was introduced into India during 1962–1964 but its releases did not provide satisfactory results (Singh, 1994b).

Sarcophagidae (fleshflies and blowflies) look like tachinids but arista is hairy or plumose upto its middle and bare distally, and large bristles are usually present on distal part of the abdomen. Some are ovo-viviparous while others are larviparous. Although many are parasitic on some Orthoptera (especially Acrididae), larvae of Lepidoptera, some Hymenoptera (including bees) and adult Coleoptera, but their effectiveness in pest suppression is yet to be proved. Others are predacious, scavengers or parasitic on mammals. Cryptochaetidae (flies looking like Agromyzidae) are endoparasitic on some coccid scales. *Cryptochaetum iceryae* (Willistron) is a well known species which was first introduced into California from Australia in 1888 against the cottony cushion scale on citrus and later into many other countries.

***Predatory arthropods***

As mentioned earlier, predators come from almost all orders of insects and mites but only Coleoptera, Neuroptera, Hemiptera, Diptera and phytoseiid mites contain economically important predatory creatures.

(a) *Coleoptera*: About 40 of the approximately 150 families of Coleoptera exhibit entomophagous traits (Schlinger and Doutt, 1964) but the principal predacious families are: Coccinellidae (ladybird beetles), Carabidae (ground beetles), Staphylinidae (rove beetles), Cicindelidae (tiger beetles), Lampyridae (glow worms and fire flies), Cantharidae (soldier beetles), Meloidae (blister beetles), Histeridae, and aquatic beetles of Dytiscidae (diving beetles), Gyrinidae (whirligigs) and Hydrophilidae families and the first two are of significance in biological control.

Coccinellid beetles, the ladybirds or ladybeetles, have been widely used in practical biological control of soft-bodied insects like aphids, scale insects, coccids and mites. Both adults and larvae are predacious. These are hemispherical and have brightly coloured elytra in various shades of red, yellow, brown and black, which are often spotted. Head is small and attached to notched prothorax, antennae short and clavate, and tarsi are three segmented. Larvae usually are dark-coloured and bear brightly coloured spots on dorsum. Some of successful biocontrol programmes have been with two Australian coccinellid beetles, viz., vedalia *Rodolia cardinalis* (Mulsant) against cottony cushion scale in California, India and other countries and *Cryptolaemus montrouzieri* Mulsant against mealy bugs in many countries, and West Indian *Cryptognatha nodiceps* Marshall against coconut scale *Aspidiotus destructor* Sign. in Fiji. In 1988, *Curinus coeruleus* Mulsant of Mexican origin was introduced in Bangalore for the control of *Hetropsylla cubana* Crawford on subabul trees and it has established in South Indian states of Karnataka, Andhra Pradesh and Tamil Nadu as well as in Maharashtra. Its beetles were released @ 1000–5000/ha area. The most common species in India is the seven spotted coccinellid, *Coccinella septempunctata* L., which is of migratory nature and is abundant during spring and early summers on a variety of aphids in forests and on field crops in the hilly regions. *Chilocorus nigrita* (Fab.) and *C. bijugus* Mulsant are potential diaspine scale predators especially on sugarcane and pome and stone fruit trees, respectively. These can be multiplied on diaspine scales *Hemiberlesia lataniae* (Signoret), *Aonidiella aurantii* (Maskell) and *Quadraspidiotus perniciosus* (Comstock) reared on pumpkins and are released @ 10–15 beetles/tree or 1,500 beetles/ha spread over 100 release spots. Other potent ladybirds feeding on scale insects are *Sticholotis madagassa* Weise, *S. marginalis* Kapur, *Pharoscymnus horni* Weise and *P. flexibilis* (Mulsant), while those feeding on aphids are *Hippodamia variegata* (Goeze), *Cheilomenes sexmaculata* (Fab.), *Ballia eucharis* Mulsant, *Coelophora* spp., *Oenopia* spp., *Scymnus* spp., and *Platynaspis* spp. Coccinellid beetles play a significant role in the natural suppression of scale and aphid pests and hence need to be conserved in the ecosystem.

Most of Carabidae are predacious on soil inhabiting soft bodied insects like lepidopterous larvae, white grubs, grasshoppers, maggots, snails and other small creatures. Mostly ground beetles are black but some are brilliantly coloured (yellowish, reddish, metallic blue, golden, bronze or green) and measure 2 to over 25 mm in size. In most species, adults are long lived with broad elytra, narrow pronotum and still narrower head, the latter bearing long 11 segmented antennae between the eyes and mandibles, and long selender legs for fast movement. Their larvae are active, elongate and taper towards posterior end where spine-like projections are present. They have sharp projecting mandibles for ingesting the liquid food. The life cycle of carabids may vary from months to several years. Both larvae and adults are predacious and live almost in similar environment under stones or bark of trees, in rotton wood and such situations. *Calosoma sycophanta* (Linn.) was introduced from Europe into North America for the control of the gypsy moth, *Lymantria dispar* L., and brown tail moth, *Nygmia phaeorrhoea* (Donovan). Several carabids (e.g., *Calosoma himalayanum* Gestro, *Calathus* sp., *Broscus* sp., etc.) occur in the Himalayan region feeding on larvae mainly *Lymantria obfuscata* Wlk. *Chaenia pictus* feeds on *Chilo partellus* Swinhoe larvae. Carabids feeding on *Opisina arenosella* Walker in South India are *Parena nigrolineata* (Chaudoir), *Calleida splendidula* (Fabricius) and other species.

(b) *Neuroptera*: Most of the neuropterans are predatory. Adults are weak fliers having long antennae and lace like wings carrying profuse venation, which often get attracted to the light source at night. Adults may or may not be carnivorous but their elongate larvae are predacious, possessing long curved mandibles projecting forward from the head in forceps like manner for grasping, piercing and sucking the body contents of its prey.

Green lacewings and brown lacewings constitute two important groups under Chrysopidae and Hemerobiidae families, respectively, which feed upon small soft bodied agricultural pests like aphids, scale insects, mealy bugs, whiteflies, jassids, psyllids, mites, young larvae and a variety of eggs. Chrysopids are small to medium sized insects, yellow to green in colour bearing filiform antennae usually longer than the body, and widely separated large golden yellow eyes. Eggs are laid singly or in groups borne on long stalk. Larvae have spiny cuticle and short stout legs. Many species have a tendency to carry the dead host on their body by these spines as trash to camouflage and are thus difficult to locate. Pupation occurs in almost spherical and tightly woven cocoon. Hemerobiidae are small brown, golden or metallic coloured and their antennae are elongate and filiform. Their larvae are smooth with short curved mandibles and maxillae, have comparatively longer body and make rapid movement. The females lay oval to elongate eggs on foliage which are without stalk and even cocoons are oval or elongate and loosely woven. Lacewings are quite effective in natural biological control.

In India, 65 species belonging to 21 genera have been recorded from various crop ecosystems and amongst these, *Chrysoperla carnea* (Stephens), *Apertochrysa crassinervis* (Esben Peterson), *Mallada boninensis* (Okamoto) and *M. astur* Banks are the most common, the last two being trash carriers (Singh, 1994b). The first two have been used in cotton ecosystem. *C. carnea* at present is extensively being used all over the country since its mass multiplication procedure has been standardized. Its egg, larval and pupal periods are 3–4, 6–8 and 5–7 days, respectively. The female lives for 30–35 days and lays 600–800 eggs at 27°C and 70% relative humidity. Field releases are made @ 50,000–100,000 first-instar larvae/ha, 4–6 larvae/plant, or 10–20 larvae/fruit plant, depending upon the pest situation, one to two releases are recommended. The larvae are released on the plants along with saw dust or dropped from the corrugated paper strips (Singh, 1994b). Releases are recommended on cotton, tobacco, sunflower, groundnut, fruits and vegetable crops (Singh, 1997).

(c) *Hemiptera*: Hemiptera is a vast assemblage of diverse and specialized minute to large insects possessing piercing and sucking mouth parts. Members of some families of sub-order Heteroptera (now-a-days recognized as order) are of predacious nature most of which inject into their victim the digestive enzymes contained in saliva for paralyzing and pre-oral digestion before ingesting the body contents. Their fore wings are hemelytra type or reduced while hind wings are membranous. Anthocoridae, Miridae, Reduviidae, Lygaeidae and Pentatomidae are important families, adults and nymphs of which are of predacious nature.

Anthocorids are small elongate to oval and flattened bugs occurring as predators of thrips, aphids, leafhoppers, psyllids, spider mites, eggs and young larvae of many crop pests; their ocelli are absent, rostrum and tarsi 3-segmented and there is no vein on the membrane of hemelytra. *Orius* spp. have preference for thrips while *Anthocoris* (e.g., *A. minki* Dhorn) prefers to feed on aphids. In the north west Himalayas, these bugs occur as good predators in orchards and forest ecosystems (Rao and Ghani, 1972; Gupta and Bhalla, 1993; Gupta *et al*., 1988). *Cardiastethus pygmaeus* Poppins is a key predator of coconut leaf caterpillar (Singh, 1994a).

Though largely phytophagous, some genera of Miridae (lacebugs) e.g., *Tytthus (Cyrtorhinus), Deraeocoris* and *Stethoconus* are predacious. These are also small to medium sized delicate bugs without ocelli but rostrum is 4-segmented and tarsi are usually 3-segmented. Importation and successful use of Australian *Tytthus mundulus* (Breddin) into Hawaii for the control of sugarcane leafhopper, *Perkinsiella saccharicida* Kirkaldy in early twenties is an outstanding example. *Tytthus lividipennis* (Reuter) and *T. chinensis* (Stal) are important predators of eggs and early nymphal instars of leaf- and plant-hoppers, and eggs of rice hispa and gall midges which prevent pest flare up. Multiplication procedure for the former has been standardized. Field release @ 70 nymphs per square metre in brown planthopper infested area is recommended and in endemic areas, 1 : 4 predator-pest ratio needs to be maintained. In Kerala, *Stethoconus praefectus* (Distant) is recommended for release at 1 : 20 predator-prey ratio where the tingid, *Stephanitis typicus* (Distant), is a pest of coconut palm, jackfruit and banana (Singh, 1994b). *Deraeocoris indicus* Ballard is a common predator of *Myzus persicae* (Sulzer) and *Aphis gossypii* Glover in India (Sankaran and Naggarkatti, 1982).

Reduviidae (assassin bugs) is the largest family of large predacious land Heteroptera and 129 species have been identified as predators of more than 140 insect pests of agriculture and forestry. Being polyphagous, these may not be useful as natural enemies of specific pests but are valuable in situations where a variety of pests occurs as they kill many more than their actual need (Ambrose, 1996). Reduviids have geniculate antennae with many (upto 40 intercalary) segments, strong and short 3-segmented rostrum, 3-segmented tarsi and hemelytra with strong veins. In India, *Platymeris laevicollis* Dist. of Zanzibar origin, was introduced by CIBC in 1968 to control the rhinoceros beetle on coconut and reduced the beetle population in Kerala in seventies. Laboratory reared bugs released in large numbers in Kerala, Lakshadweep and Karnataka provided some control of this pest but there was no permanent establishment or subsequent build up of the predator population. Its important genera in India are *Rhinocoris, Sycanus, Acanthaspis, Brassivola, Coranus* and *Endochus* (Ambrose, 1996).

Though Lygaeidae and Pentatomidae contain primarily phytophagous bugs, *Geocoris* of the former (predator of leafhoppers and mites) and *Prillus* and *Picromerus* of the latter (Asopinae, predacious on caterpillars and chrysomelid beetles) are good predators.

(d) *Diptera*: Predatory dipterans are many more than the parasitic ones and about 37 families are partially or completely predatory on insects, important being Cecidomyiidae, Bombyliidae, Asilidae, Syrphidae, Chamaemyiidae, Anthomyiidae, Calliphoridae and Sarcophagidae (Sweetman, 1958; DeBach and Rosen, 1991). Adults of Asilidae (robberflies) prey upon flying insects and those of Anthomyiidae feed on flies, while larvae of many species of Syrphidae, Cecidomyiidae and Chamaemyiidae are predacious on soft bodied insects.

Entomophagous syrphids (hoverflies, sunflies or flowerflies) are brightly coloured flies with stripes, bands or spots of yellow on blue, black or metallic background, making them resemble to wasps. They are 3–25 mm long having peculiar spurious vein on wing membrane passing through radio-medial cross vein. Their elongate green or tan larvae are apodous (leg less) and narrower towards anterior extremity with pointed mandibles used to suck body fluids of their prey. Majority of them prey on aphids and a few feed on coccids, leafhoppers, whiteflies, thrips and lepidopteran larvae. Important predatory genera are *Metasyrphus*, *Episyrphus*, *Scaeva*, *Eupeodes*, *Sphaerophoria*, *Melanostoma*, etc. These predatory species lay eggs singly amongst or in immediate vicinity of gregarious prey. Immediately after hatching, sluggish maggots begin feeding on the prey and to complete development one may destroy several hundred prey. They pupate in puparium on foliage or in soil. Their adults are pollen and nectar feeders and hence have importance in pollination of crops.

Chamaemyiidae are small (a few mm in size) grey to blackish flies, the larvae of which are predatory mainly on aphids and also on some coccids and mealybugs. Like syrphids, they also lay minute eggs singly in or near prey colonies and their small yellowish white maggots feed on prey. It has several generations in a year. Its important genera are *Leucopis*, *Chamaemyia* and *Paraleucopis*, the former being the best known. Cecidomyiidae (midges) is a large family of minute delicate flies clothed with hair which possess long antennae bearing whorls of hair. A number of predatory species hail from *Aphidoletes, Lestodiplosis, Arthrocnodax, Microdiplosis, Diadiplosis* and *Feltiella*. The life cycle, on an average, is completed in about three weeks. Egg laying, predation and life cycle is like that of syrphids and chamaemyiids but larva pupates in a loosely woven cocoon.

(e) *Acarina*: Many species of mites are predacious on a number of small insects but those of Phytoseiidae family have been used in the biological control of phytophagous mites. Some species of *Amblyseius*, *Phytoseiulus*, *Metaseiulus* and *Typhlodromus* prey upon red spider mites and eriophyid mites and are effective in their natural and applied biological control. Predacious mites are usually distinguishable by their larger size, often longer legs and active movements. Many *Amblyseius* species can be reared on castor pollen grains but *Phytoseiulus* and several other predatory mites require tetranychid mites for their production. *A. tetranychivovus* (Gupta) has been successfully multiplied in India whereas *P. persimilis* Athias-Henriot was imported from United Kingdom in 1984 and since then is being maintained on *Tetranychus urticae* Koch. Its releases are recommended at 10–60 mites/plant. In green houses, these are released @ 100/m$^2$ against the red spider mite. *Amblyseius* spp. are effective against thrips as well. Another species used in glass houses is *P. riegeli*. Hemisarcoptidae is another important family, species of which are predators of scale insects.

Although predators exist in great variety and exhibit much diversity, yet only a few species have been exploited for mass production mainly because of availability of limited prey species which can be economically raised in large numbers in the laboratory. At the Project Directorate of

Biological Control of the Indian Council of Agricultural Research, Bangalore, it has been possible to rear *Chrysoperla carnea* (Stephens) larvae on encapsulated semi-synthetic diet. A diet encapsulation machine has recently been installed which will facilitate large scale multiplication of chrysopids. Such efforts need to be made to commercially utilize the biocontrol agents.

**Entomopathogens**

As compared with entomophages, lesser number of entomopathogens (several thousands) have been described which inflict varying degree of mortality in the pest populations exposed to them. While entomophages are active searchers for their prey, pathogens are not so, rather their impact becomes evidently expressed among dense pest populations. The most common diseases of insects are caused by bacteria, viruses, fungi, entomopathogenic nematodes and protozoa, which may actively or passively find route to internal milieu of the host after crossing the barriers posed by the integument. Usually, bacteria, viruses and protozoa enter the host body cavity via mouth and alimentary canal while fungi and nematodes gain access through natural openings or penetration through the cuticle. Important references pertaining to insect pathogens and their use in biological control include the classic publications of Steinhaus (1946, 1949, 1963) and DeBach (1964) followed by those of Burges and Hussey (1971), Burges (1981), Cantwell (1974), Weiser (1977), Allen *et al.* (1978), Deacon (1983), Fuxa and Tanada (1987), Battu *et al.* (1993), Tanada and Kaya (1993) and Maddox (1994).

***Bacteria***

The bacteria (Class Schizomycetes, Order Eubacteriales) causing diseases in insects broadly belong to two categories, the non-spore formers and spore formers. The former group contains such diverse families as Enterobacteriaceae (*Aerobacter*, *Enterobacter*, *Proteus* and *Serratia*), Streptococcaceae (*Streptococcus*), Micrococcaceae (*Micrococcus*) and Pseudomonadaceae (*Pseudomonas*). Spore formers have been placed under only one family, Bacillaceae (*Bacillus*). Most of the non-spore forming bacteria are potential pathogens which may remain in the alimentary canal as relatively harmless creatures but prove highly pathogenic once they find entry into the body cavity (haemocoele). Many stress factors, like poor food quality and biotic factors, temperature extremes, etc., may facilitate their entry into the body cavity. Although non-spore forming bacteria play a vital role in natural disease outbreak of pest populations, yet it is the spore formers which have been largely exploited in suppression of pests. Since spore is a resistant form, its formation enables them to survive for longer period outside the body of the host. Among spore formers, viz, *Bacillus thuringiensis* Berliner, *B. popilliae* Dutky and *B. Sphaericus* Meyer and Neide, the first two have been used in microbial control while commercial products of the third one are yet to be developed.

*Bacillus thuringiensis* Berliner: This bacterium, which has found world-wide use in pest control and first recorded as the cause of "sotto" disease of silkworm in Japan in 1901, was described by Berliner in 1915, again isolated by Mattes in 1927 in Germany and was given the name *B. thuringiensis*. The first commercial preparation ('Sporiene' in France) was available by mid forties but the first product released for field use in the USA was in 1958 after the problem of variability in effectiveness of the product was overcome. Since then it has been marketed in different countries under different trade names.

*B. thuringiensis* (*Bt*) is a common soil inhabiting weak pathogen which is aerobic, facultative Gram-positive, rod shaped bacterium (3–5 μ x 1–1.2 μ), sometimes motile by lateral flagella. It grows well at 15–40°C and 30°C is the ideal temperature for commercial production on soybean and molasses based substrates. It sporulates readily upon aging and produces a crystalline body or crystal outside the spore wall (exosporium) both being surrounded by cell wall of the vegetative cell. Thus from each bacterial cell, a spore and a crystal (parasporal body) are produced after lysis which can be separated by centrifugation from cultural media to be further formulated as commercial products. There are many subspecies (strains, serotypes, serovars, varieties or pathovars) of *Bt* and on the basis of most commonly used method of subspecies identification (based on the flageller or H-antigens, also known as H-serotyping), 34 serovars (27 antigen groups and 7 subgroups) have been distinguished (deBarjac and Frachon, 1990). Only serotype 2 (*finitimus*) forms crystal inside the exosporium. *Bt* is very closely related to the other commonly found soil inhabiting spore former *B. cereus* Frankland and Frankland but the latter does not produce the crystal along with the spore. Diversity prevailing in the species is reflected from the fact that Martin and Travers (1989) were able to get about 1000 isolates from soil samples obtained from five continents and that Martin (1994) was able to recover >20,000 bacterial isolates from soil in one summer of 1991 and >40% of these were identified as *Bt*.

*B. thuringiensis* produces many toxins which may be contained in crystal in the bacterial cell as endotoxin (known as δ-endotoxin), or get diffused through the cell wall into the culture medium as exotoxins (α, β and γ). Usually spores and crystals are required for toxicity to some insects but most of susceptible species need ingestion of crystals alone. The crystal contains δ-endotoxin which is a high molecular weight protein (25–230 kilodalton or kd). In most of the cases, this protein exists in three size designations, viz. 125–138, 65–75 and 25–28 kd which are encoded by at least 18 genes, *cry I* and *cry IV*; *cry II, cry III* and *cry IV*; and the *cyt* genes, respectively. Their products, the toxins (designated as Cry I to Cry IV, and Cyt) contain about 1155–1207, 633, 644–659, 643–675 and 248 amino acids and are toxic to Lepidoptera, Diptera and Lepidoptera, Coleoptera, Diptera, and Diptera, respectively (Hofte and Whiteley, 1989; Gill *et al*., 1992). The Lepidoptera specific crystals are bigger and bipyramidal while those toxic to both Lepidoptera and Diptera are small and cuboidal, and those toxic to Coleoptera are square and flat. Mostly crystals contain more than one protein (polypeptides) which are held together by complex interaction. These are heat labile, insoluble in water and organic solvents but soluble in alkaline reducing conditons (pH ≥ 8.9). Susceptible insect gut pH lies between 10.2 and 10.5 and is ideal for dissolution of the crystal. The digestive enzymes (proteases) present in the gut along with pH activate the toxin (protoxin) by removing C-terminal amino acids. The polypeptide so obtained in case of 130–140 kd protoxin after limited proteolysis is reduced to 60–70 kd activated toxin which is less stable than the protoxin. The activated toxin has three structural regions, viz., N-terminal toxic domain, conserved C-terminal region and variable region between the two. The latter two appear more critical in toxin selectivity determined by binding of toxin to its receptors on mid gut cells. The Cry toxin polypetides bind with specific plasma membrane receptors (a glycosylated protein) on apical membrane of brush border of columnar cells in the mid gut epithelium, while the Cyt toxin initially binds with unsaturated phospholipids of the plasma membrane. This binding more likely results in formation of non-selective pore (0.6–1 nm) at the binding site which leads to osmotic imbalance due to influx of water, cations and other smaller molecules. Consequently gut pH drops below 9 and it is an ideal pH for germination of spore (which otherwise is prevented above 9 pH). The vegetative cells then invade the tissue and cause a lethal septicaemia. Heimple and Angus (1959) furnished the first convincing account on the mode of action of Bt and classified the susceptible insects into three types, *viz.*, type I (e.g., *Mamestra brassicae* (Linn.)) which are extermely sensitive to the toxin, gut paralysis and death occurring in 1–7 hours following an increase in blood pH; type II (most of the larvae) in which gut paralysis occurs but without rise in pH of haemolymph and death occurs in 2–7 days; and type III larvae (e.g., *Anagastra kuhniella* (Zeller)) toxin is not lethal but death results from septicaemia after germination of spores in the mid gut. Ultimately cell contents of enteric epithelium are released into the gut. Thus sometime after ingestion of crystals, mid gut paralysis occurs, consequently the larva stops feeding though it may die after some days. Subsequent symptoms are vomiting, diarrhoea, sluggishness, and finally paralysis and death.

The other toxins (the exotoxins) which are of less importance as compared with δ-endotoxin are produced by some strains of *B. thuringiensis*. These are: α-exotoxin which is a phospholypase-C, a water soluble enzyme, to which sawfly larvae are sensitive; β-exotoxin, commonly known as thuringiensin (also called flyfactor, thermostable exotoxin, etc.) is an ATP-analogue of low molecular weight (701 d) which is thermostable, water soluble and causes DNA-dependent RNA-polymerase inhibition; and γ-exotoxin which is an enzyme (or a group of enzymes) that clears egg-yolk agar and attacks phospholipids *in vitro*. Out of these exotoxins, only thuringiensin has broad spectrum of insecticidal activity against some Coleoptera, Diptera, Hymenoptera, Lepidoptera and Orthoptera, besides being toxic to vertebrates. It has been of some interest in the erstwhile USSR (Soviet block) but is a prohibited product by the US Environmental Protection Agency because of high mammalian toxicity. Although it can be easily removed by washing, yet the strains not producing it have been selected and are being used commercially.

Host specificity of the *Bt* is because of number and type of genes in a particular isolate hence there are large number of pathotypes, e.g., *B. t. israelensis* active against larvae of mosquitoes and blackflies, *B. t. tenebrionis* active against Coleoptera, and *B. t. kurstaki* active against larvae of Lepidoptera. Several varieties are now commercially available which have different spectra of activity against Lepidoptera and selected strains having least effect on beneficial insects like honeybee and silkworm. The efficiency of the bacterium is affected by the variety selected, method of production and manner of formulation. Susceptibility may differ within a single variety as it is readily inactivated by exposure to ultraviolet irradiation (within 10 minutes) but less quickly by the sunlight (more than an hour). Temperature of 18–20°C is most favourable for *Bt* application. Larvae of early instars are more susceptible. Hence, application of the formulation in the evening hours would be desirable. Thorough coverage of the crop is important. Adjuvants like sticker, spreader and wetting agents also affect the effectiveness. By genetic engineering processes, it has been possible to make tailor-made strains with increased persistence, virulence and increased or modified host range. Now transgenic plants with *Bt* toxin genes are available which cause high larval mortality and reduce pest damage on crops like tobacco, tomato, cotton, potato, etc. However, environmentalists are raising voice against their use fearing possible side effects. Furthermore, recently some species like *Plutella xylostella* (Linn.) and *Plodia interpunctella* (Hubner) have been reported to develop resistance to *Bt*, therefore, use of *Bt* in combination with other chemicals would enable a more judicious use of both, which would also delay the onset of insect resistance (Gill *et al.*, 1992).

*Bacillus popilliae* Dutky: These bacteria cause milky disease in grubs of scarabaeid beetles. The disease is called so because of milky white appearance of the grubs due to refractile spores produced in the haemolymph. *B. popilliae* is Gram-negative, non-motile rod shaped (1.3–5.2 μ x 0.5–0.8 μ) fastidous organism which grows only in rich medium and sporulates readily only in diseased host. No satisfactory method has so far been found for its mass production *in vitro*. It is only being mass produced *in vivo* and therefore, cost of production is high and a time consuming process. The larvae are injected with precise quantity of spores in aqueous suspension, transferred individually to a rearing chamber containing soil and grass seeds, incubated at 30°C for 10–18 days, diseased grubs are rinsed and refrigerated. Finally, these are crushed and mixed with inert carrier and to make a product having 100 million viable spores per gram of the finished product. At least two bacterial types are associated with two types of diseases. Grubs infected with type A disease, caused by *B. popilliae* var *popilliae*, are pure white, while type B disease infected grubs show transition from white to brown over winter and bacterium associated with this is *B. p. lentimorbus*. The former is crystalliferous and the latter is non-crystalliferous bacterium.

The bacterium enters the grub body orally by ingestion of spores. After germination, these penetrate the alimentary canal, enter in the haemocoele and multiply rapidly in blood and at last when grub body becomes depleted of food material, it sporulates. Since the death occurs after several days to 2–3 weeks of ingestion of spores, the role of crystal in causing toxicity is remote. Further, both crystalliferous and non-crystalliferous varieties are toxic to grubs, the younger larvae being more sensitive to the former and older more so to the latter. Despite slow rate of action and ineffectiveness to adult beetle, which is also a crop damaging stage, it is rated as a good microbial pesticide having narrow host range, high pathogenicity, long persistence (for years in the soil) and safety to vertebrates. Once inoculated in the soil, it would remain effective for a number of years. It is marketed as "Doom" and "Japidemic" by Fairfax Biological Laboratory Inc. and "Milky spore" by Reuter Laboratory Inc. in the USA. This was used in Gujarat in late seventies against *Holotrichia* spp and some reduction in grub population (20–25%) was obtained. However, a new indigenous strain identified as *B. p.* var *holotrichiae*, was isolated in Gujarat (Vyas *et al.*, 1986). Development of semi-synthetic diet facilitated production of large number of diseased grubs (Vyas *et al.*, 1991). In field tests, it provided 70% grub mortality (Vyas *et al.*, 1991a). Predatory birds like cattle egret, house sparrow, Indian mynah, etc. help in dispersal of the pathogen (Vyas *et al.*, 1988).

***Viruses***

Like any living organism, insects are attacked by a variety of viruses which are obligate intracellular submicroscopic pathogenic entities composed of the fundamental viral nucleic acid embedded in a protein matrix which in turn is encased in a limiting membrane (virion or virus particle without limiting membrane is nucleocapsid). These are characterized on the basis of shape and symmetry of particles; type, molecular weight and strandedness of the nucleic acid; site of replication of the virus; serological reaction, etc.

At least 1270 insect virus associations have been described of which 71% were found in Lepidoptera, 14% in Diptera, 7% in Hymenoptera, 5% in Coleoptera, and remaining 3% in Orthoptera, Isoptera, Hemiptera and Neuroptera (Entwistle and Evans, 1985). In 1982, International Committee on Taxonomy of Viruses, proposed upto 13 families or subfamilies for insect viruses, most of which infect not only insects but also vertebrates and even plants. These viruses may be of occluded (virions enclosed in protein body or inclusion/occlusion body, IB or OB), or non-occluded types. Most of the viruses (above 75%) found in insects are of occluded type, perhaps because of ease of detection. Major families to which occluded viruses belong are Baculoviridae (baculoviruses including nuclear polyhedrosis and granulosis viruses, NPV and GV), Reoviridae (cytoplasmic polyhedrosis virus, CPV) and Poxviridae (entomopox viruses, EPV). Since viruses of the latter two families have potential to attack vertebrates as well, these have not been exploited for microbial control of insect pests. Amongst non-occluded viruses associated with insects, important groups are iridescent viruses (Iridoviridae), densoviruses (Parvoviridae), polydnaviruses, ascoviruses, small RNA viruses (Picornaviridae), Rhabdoviridae, caliciviruses and nodamura virus, all of these are known to be associated with vertebrates. Thus, baculoviruses have been isolated only from arthropods (primarily from insects) and are unusual among invertebrate viruses in having no known related vertebrate viruses. One subgroup of Baculoviridae (subgroup C) is peculiar being of non-occluded type and because of its specificity to grubs of rhinoceros beetle on coconut palm, it has been used in its control, as described later.

Since occluded viruses are enclosed in proteinaceous occlusion bodies which impart protection to individual virions, this is an important feature for consideration for microbial control. No such OBs have ever been seen in mammalian cells and hence these viruses present no real environmental hazards. NPV and GV belong to subgroup A and subgroup B of Baculoviridae, respectively, which are rod shaped (approx. 250 nm x 50 nm) and enveloped virions with a circular, double stranded DNA. But NPVs are occluded in polyhedral inclusion body (0.5–15 μ in diameter with actual mean size of 0.8–2 μ) while GVs are occluded in cigar shaped oval capsules often 0.3–0.5 μ in size. Several to many virions may be contained in an IB of NPV while usually one (rarely 2) virion is present in an IB of the GV. Further, NPV replicates in the nucleus but GV may multiply in nucleus or the cytoplasm surrounding it. CPV on the other hand are icosahedral to isometric (50–70 nm) particles lacking an envelope and are enclosed in polyhedral inclusion bodies 0.5–15 μ (usually 1–7 μ) in size, resembling those of the NPV but the polyhedra are relatively unstable and virions near surface can be removed easily. CPV replicates in cytoplasm rather than in the nucleus of the host cell and contains double stranded RNA as nucleic acid. Viruses used in the microbial control and their field application are described below.

***Baculoviruses***: NPVs are of two morphological types, *viz.*, single nucleocapsid (SNPV) type with only one nucleocapsid per occluded virion and multiple nucleocapsid (MNPV) type in which variable number of nucleocapsids per virion is present in occluded body (OB), shape of which is usually governed by the virus. It contains the protein, polyhedrin. Most NPVs have been isolated from larvae of Lepidoptera (88%), Hymenoptera (6%) and Diptera (5%). Whatever the case may be, these find entry in the insect body by ingestion of polyhedra. Highly alkaline mid gut pH (9.5–11.5) brings about dissolution of OBs within a few minutes thereby releasing virions from it. These penetrate through the peritrophic membrane and reach microvilli of columnar epithelial cells of mid gut. The virion envelope fuses with the plasma membrane of apical portion of the microvillus and only nucleocapsid enters the cell (leaving behind the envelope) within 4 hours of infection. The nucleocapsids move rapidly to enter the nucleus (as it happens with NPV) via nuclear pores and release the DNA in it or may uncoat at nuclear pore before passing into the nucleus (as observed in GV). Within several hours (8–12 hours for MNPV of *Autographa californica* (Speyer)), the nucleus enlarges, virogenic stroma appears at virus replication sites, nucleocapsids are synthesized and get embedded in protein matrix. These do not develop into PIBs in mid gut of Lepidoptera but are budded off along with plasma membrane from the base of the infected cells into the body cavity to further spread the infection to sensitive tissue (fat body, hypodermis, tracheal cells, muscular and nerve sheath, ganglia and pericardial cells) where usually PIBs are produced after further multiplication of the virus. In Lepidoptera, infection is polyorganotropic (affecting many organs) but in Diptera (mosquitoes) and Hymenoptera (sawflies), infection is strongly mono-organotropic, affecting only mid gut cells. In advanced stage of infection, lepidopteran larva becomes sluggish, its cuticle gets discoloured and turns darker at death. Infected larvae have a tendency to climb to a higher point on the host plant and often die in hanging posture remaining attached to the substratum by crochets of the pseudopods. The cuticle often ruptures to release polyhedra which further contaminate the food plant. The death may take place in 4 days to 3 weeks depending upon various factors associated with the virus, the host and the environment. During post-infection period the infected larva may continue to feed and cause damage to the crop. To minimize the damage, it is desirable to treat the crop at appropriate time when early larval instars of the pest are available in the field. From one host larva, variable number of PIBs (6–43 x $10^9$ polyhedra/g of larva) can be obtained and for making standardized dose application of NPV, one larval equivalent (LE) is rated equal to 6 x $10^9$ PIBs, and the dose per hectare varies from 50 to 1000 LE (usually 100–250 LE). The NPV of *Helicoverpa* spp. (Elcar), *Lymantria dispar* (Linn.) (Gyp-check), *Orygia pseudosugata* (McDunnough) (Biocontrol-1), *Neodiprion sertifer* (Geoffroy) (Neo-check), *N. lecontei* (Fitch) (Lecont-virus) and *Mamestra brassicae* (Linn.) (Mamestrin) are produced commercially *in vivo* in some countries, *viz.*, USA, Canada, Britain and Finland. However in India, the NPV studied are of *H. armigera, Spodoptera litura* (Fab.), *Spilosoma obliqua* Walker, *Mythimna separata* (Walker) and *Amsacta moorei* (Butler) and small scale multiplication is being done by State Agricultural Universities, Indian Council of Agricultural Research Institutes, some private biocontrol laboratories and non-governmental organizations. The production of these viral preparations can be triggered if adopted as a cottage industry.

GV (subgroup B) are exclusively known to be associated with Lepidoptera and about 150 viruses are known in this subgroup, which are identified on the basis of morphology and the associated host species. Some of the known GVs are from *Agrotis segetum* (Denis & Schiffermuller), *Cydia pomonella* (Linn.), *Phthorimaea operculella* (Zeller), *Plutella xylostella*, *Pieris brassicae* (Linn.), *P. rapae*, *H. armigera*, *Spodoptera frugiperda* (J.E. Smith), *Chilo infuscatellus* Snellen, *Euproctis chrysorrhaea* (Linn.), etc. These are more host specific than the NPVs, being cross-infective at genus level or to closely related families. Their mode of action is similar to that of NPV. Since the size of IBs of GV is smaller than that of NPV, yield per gram body weight of the larva is more. One LE of GV is considered equivalent to 4 x $10^{10}$ IBs.

Non-occluded *Oryctes* virus (subgroup C), as the name implies, does not form IBs and nucleocapsid is always enveloped singly within the virion. Damage by the coconut rhinoceros beetle, *Oryctes rhinoceros* (Linn.), has been significantly reduced in Fiji, Western Samoa, Mauritius, Papua New Guinea and India by using this virus (DeBach and Rosen, 1991; Singh, 1994a). It is lethal to larvae and also multiplies in the gut epithelium of the beetles which spread the disease by defaecation. The best method to spread the disease is to release the infected beetles which are obtained by immersing these in viral suspension for 2–3 minutes and then made to crawl in contaminated saw dust for 24 hours (Deacon, 1983).

*Cytoplasmic polyhedrosis viruses (Reoviridae)*: These have been reported from about 200 insects mainly Lepidoptera. CPVs infect the cytoplasm of the mid gut epithelium of larvae. The infection is not always lethal but it may slow down the larval growth, consequently infected larvae appear small bodied with a large head. The adults formed from infected larvae have reduced longevity and fecundity. CPVs are less host specific and less pathogenic than baculoviruses and therefore, are less appropriate as microbial control agent. Apparently, the only one to be registered as a control agent was a CPV of the pine caterpillar, *Dendrolimus spectabilis* in Japan in 1974, though others are doubtlessly exploited indirectly as natural mortality agents (Deacon, 1983).

*Other viruses*: Insect viruses other than above mentioned families, act as naturally occurring mortality factors and have little potential to be developed as microbial insecticide. However, myxoma virus which is a member of pox-group (characterized by 290 nm x 390 nm virions with double stranded linear DNA) and does not pose threat to man and domestic animals, has been successfully used to control rabbits in the fifties in New South Wales (Australia) where mosquitoes acted as vectors and in France where rabbit flea acted as a vector. It brought about 99% control in Australia. Now this disease is occurring mainly as an annual mosquito-borne epizootics in spring and summer and has a fatality rate of 30% or less. Symptoms of myxomatosis appear in 5–7 days and death occurs 2–4 days later (Deacon, 1983).

*Field applicability*: Insect viruses considered for microbial control exhibit wide variations in host specificity ranging from highly species specific to a broad host range. SNPV of the spruce sawfly, *Gilpinia hercyniae* (Hartig), is restricted to a single species while that of *Helicoverpa zea* (Boddie) is infective to this genus only. GVs are also highly specific. On the other hand, MNPV of the alfalfa looper, *Autographa californica* (Speyer), is capable of infecting 34 species belonging to 29 genera and 9 families. Likewise, MNPV of *Anticarsia gemmatalis* Hubner has a broad host range having cross infectivity to 9 species of 6 genera. Why is it so, is not known. NPVs with broad host range have an advantage over the monospecific ones as greater number of hosts is available for its continued activity. These viruses are pathogenic to larvae but the adults infected or contaminated with some of the viruses do play an important role in their transmission by transovum (external or surface contamination of egg) rather than by transovarial transmission (incorporation of the virus inside the egg). In Lepidoptera, contamination of the adult is of external nature while in sawflies having gut infection of the NPV in adult stage, true tansovum transmission occurs. Infectivity of the virus may vary intrinsically (decrease with increasing larval age) as well as at population level (proportion of susceptible individuals in a population). Once the population has acquired the primary inoculum after contamination with the viral spray application, it is the proportion of infective larvae which brings about the secondary spread of the inoculum through disintegration of the cadaver and contamination of the foliage and soil. The greater host density results in density-dependent mortality. Further spread of the disease by predators (bird and insect predators) and parasites has also been reported. Viruses persist longer in soil and soil acts as their repository. Their spread is brought about by abiotic factors like wind and rain as well as irrigation water. Like *B. thuringiensis*, viruses are quite sensitive to ultaviolet irradiation and hence viral preparation lose activity under direct exposure to sunlight (e.g., *Spodoptera litura* NPV inactivated in a day on exposure to sunlight) and most of these do not persist significantly for more than two days (99% loss in 2–20 days on foliage depending upon the type of virus, e.g., *Trichoplusia ni* (Hubner) NPV degrades slower than *Helicoverpa* NPV, and nature of the foliage, e.g., on cotton there is rapid inactivation of *H. armigera* NPV). To minimize photoinactivation, NPV preparations are recommended to be sprayed in the evening. Further, addition of UV-reflectants (like aluminium oxide or aluminium powder and Tinopal) and UV-absorbants (buffalo black, carbon, egg albumin, molasses, milk, soy hydrolysate, brewer's yeast, brilliant yellow stain, etc. (Entwistle and Evans, 1985) may further enhance the efficacy of the NPV formulations. Further, crude preparation is more UV-tolerant than the purified one and persists longer. Addition of sticker and wetting agents which prevents formation of aggregates may bring about better dispersal of the formulation. Thus adjuvants in formulation are likely to affect the efficacy of the viral insecticide.

*Fungi*: Over 750 species of fungi are associated with insects, some of them which cause serious diseases and prove lethal offer potential in pest management. These infect broader range of insects than do other microorganisms, some having broader host range (e.g., *Beauveria* and *Metarhizium*) and some are specific (some species of *Entomophthora*) to their hosts. Fungi infect insects by attacking host cuticle and hence these are suitable biocontrol agents for insect pests with sucking type of mouth parts which cannot ingest other pathogens infecting through gut wall (e.g., bacteria, viruses and protozoa). Fungi are also particularly important for the control of Coleoptera because viral and bacterial diseases are rare among them (Hajek and St. Leger, 1994). Fungi also play a significant role in disease epizootics (when unusually larger proportion of population is infected) as natural mortality factor.

*Groups of entomopathogenic fungi*: Insect pathogenic fungi (Division Eumycota) hail from almost all subdivisions, *viz*., Mastigomycotina, Zygomycotina, Ascomycotina, Basidiomycotina and Deuteromycotina. The important genera are *Coelomyces* (Class Chytridiomycetes), and *Lagenidium* (Class Oomycetes) from Mastigomycotina, both being aquatic fungi and specific to mosquitoes; *Entomophthora, Entomophaga, Erynia, Zoophthora, Massospora* (specific to Cicadidae), *Neozygotes, Conidiobolus,* and *Mucor* (all from Class Zygomycetes) from Zygomycotina; *Cordyceps* of Class Pyrenomycetes from Ascomycotina; and *Aschersonia* (Class Coelomycetes, Order Sphaeropsidales) which parasitizes scale insects and whiteflies; and *Beauveria, Metarhizium, Verticillium, Hirsutella* (infecting mites and leafhoppers)*, Paecilomyces, Nomuraea, Fusarium* and *Aspergillus* from Class Hyphomycetes, Order Moniliales, belong to Deuteromycotina. Out of this vast assemblage of species only 10 have been or are presently being developed as mycoinsecticides and all of these are from Deutromycotina (primarily first four genera), the main reason being their saprophytic nature and low nutritional demand which favours easy cultivation on standard mycological media and production of large number of asexual spores of infective nature in culture media at temperature range of 20–25°C (do not grow at all at 37°C, so do not pose serious hazards to mammals except spores of some allergy causing species). These are also known as imperfect fungi which seldom, if ever, produce sexual stages and are distinguished by morphological and developmental features of spore bearing hyphae, the conidiophores and shape and size of the spores, the conidia.

Another important group of fungi which frequently causes epizootics in insects is from Order Entomophthorales, Class Zygomtycetes, which are characterized by the production of forcibly discharged and asexually produced conida from insect cadaver (except *Massospora*) and thick walled non-flagellate resting spores (zygospores) produced inside the host body. Usually these fungi are fastidious to grow on culture media. More than 200 species of Entomophthoraceae in 5 or 6 genera are known to parasitize arthropods, frequently aphids and Diptera (Ferron, 1978; 1985). Natural fungal infection may occur at low level in insect population (10–20% infected individuals), known as endemic mycosis representing pre-epizootic phase. Epizootic phase of infection occurs under favourable conditions resulting in high population decline. Such natural population collapses have rarely been experienced with fungi imperfecti and frequent epizootics have been observed with Entomophthorales (Ferron, 1978). Thus these pathogens act in a delayed density-dependent manner.

*Mode of infection*: A nice account on mode of infection by entomogenous fungi has been given by Ferron (1978) and Hajek and St. Leger (1994). Cuticle of the insect poses a barrier for fungal infection. High humidity is the first requirement for germination of infective fungal spores (propagules). Usually before germination of the spore, it swells and secretes adhesive mucus to facilitate its adhesion on cuticular surface. The germ tube emerging from the spore may penetrate the epicuticle directly or forms an appressorium (an organ of holdfast) from which infection pegs arise to penetrate the outer epicuticle which primarily is a mechanical barrier. The inner epicuticle (made of lipoprotein stabilized by quinones) is probably breached by enzymes produced by the pathogen. The penetrating peg may behave as a straight penetrating hypha or by lateral extension may form a penetrant plate loosening the cuticle and facilitating spread of cuticle degrading fungal enzymes (proteases, lipases, chitinase, etc.) in procuticle. The degree of thickness and sclerotization of the procuticle also pose resistance for fungal activity. Under such circumstances infection may take place through intersegmental (arthodial) membrane or spiracles. From penetrant plate may arise vertically growing penetrant hypha or laterally growing irregularly walled secondary hyphal body that later on grows down, penetrates through epidermal layer of cells and reaches haemocoel. So long the hyphae are in cuticle, infection can be lost at the moulting time. Once the fungus has entered the body cavity it has to encounter the cellular defense of the host which it does by producing enzymes and metabolites of low molecular weight, the toxins. Some of the fungal toxins have been characterized and these may be cyclic depsipeptides like destruxins (*Beauveria, Entomophthora* and *Metarhizium*) which paralyse the host cells, beauvericin (*B. bassiana* (Balsamo) Vuillemin and *Verticillium lecanii* (Zimmerman) Viegas), bassianolide (*B. bassiana*) and antibacterial cordycepin (*Cordyceps*) or linear peptides like cytochalacins (*M. anisopliae* (Metchnikoff) Sorokin) which paralyse host cells, leucinostatins (*Paecilomyces*) and efrapeptins (*Tolypocladium*) which exhibit antimicrobial activity. Pigments like biochrome (bassianin and tenellin) and dibenzo-quinones (oosporin) are responsible for the colour change in the insect body. By forming yeast-like budding cells or hyphal bodies, known as blastospores, the fungus gets readily dispersed in the body cavity. Later, these germinate to produce new hyphae.

The toxins bring about progressive degeneration of the host tissue primarily affecting the structural integrity of the cellular membranes. Lethally infected insect exhibits behavioural change such as reduction in feeding as the first apparent change in infected host, change in egg laying behaviour (e.g., carrot fly infected with *Entomophthora* lays eggs away from the host), sluggishness and lack of co-ordination associated with perturbed electric activity of nervous circuit, disturbance in metamorphosis, increase in oxygen consumption and movement to elevated locations just before death. However, suboptimal inoculum may disturb the fecundity and diapause of surviving adults and reduce the resistance to abiotic factors. With the death of the insect, parasitic phase of the fungus ends and it starts growing saprophytically to fill the body with hyphae and the cadaver becomes mummified. Depending upon type of fungus, it may internally form resting spores (chlamydospores, sporanges, etc.) or when high humidity exists, the hyphae penetrating through the cuticle form conidiophores which produce conidia (spores) copiously.

*Application of fungal preparations*: Some fungi which have been exploited for short-term season-long control are used by adding relatively large amount of inoculum in the agroecosystem (e.g., $2.5 \times 10^{13}$ conidia/ha for foliage feeders or as much as $10^{16}$–$10^{17}$ spores/ha for soil inhabiting insects like white grubs). There exists positive correlation between number of infective spores and mortality by mycosis (Agarwal, 1990). High spore mortality is caused by high temperature, desiccation and solar radiation, and fungistatic activity of soil micro-organisms, that is why application of higher doses is desirable. Since fungi generally infect all developmental stages of insects, these can be applied against any appropriate stage when environmental conditions are favourable. However, susceptibility of different stages to many diseases may vary considerably. Development of infection by a virulent strain is governed by abiotic factors, quantity of infective spores and physiological condition of the host. Temperature (5–35°C, optimum being 20–30°C) and humidity (usually high, above 75%) favour germination of spores and create congenial conditions for epizootic mycosis. However, for spore production some species need darkness and some (e.g., *Entomophthora muscae*) produce high conidia even at low humidity (50%). Although higher mortality rate has been associated with the more number of infective spores, it is virulence of the pathogen which is of greater significance. Success depends upon selection of an optimal strain. Even 10 spores of low-temperature tolerant strain of *M. anisopliae* cause 70% mortality of a pasture scarab whereas $10^5$ spores of a heterologous strain of the same fungus may be needed to obtain same mortality (Hajek and St. Leger, 1994). As the physiological stresses enhance the infection process and yield higher mortality, susceptibility of the population can be increased artificially by using sublethal doses of insecticides or by using fungal preparation in combination with other mortality factor, e.g., some microbes (Agarwal, 1990; Ferron, 1978). Creating conditions favourable for fungal infection also result in better efficacy and epizootic development. Applying fungal formulation in the evening after irrigating the fields properly increases humidity in the microclimate and favours disease development. Further, improved formulations of the mycoinsecticides having some adjuvant of antidesiccating nature for increasing the moisture level may prove more effective. Some situations like control of soil pests in rainy season and in mushroom houses and green houses where temperature and humidity can easily be manipulated, offer attractive environment for use of fungal pathogens. *V. lecanii* which has a wide host range and has potential for mite and thrips control in greenhouses, can be recommended (Rombach and Gillespie, 1988). Fungi registered for commercialisation are *B. bassiana* (as Boverin; Naturalis L) against spittle bugs, mosquitoes, boll weevils, rhinoceros beetles, etc., *M. anisopliae* (as Bio-1020; Metaquino) for the control of coleopterous pests, *V. lecanii* (as Vertalec; Mycotal) for greenhouse aphids and whiteflies and *Hirsutella thompsonii* (as Mycar) against citrus rust mites. Others which hold promise for commercialisation are *Nomuraea rileyi* (Farlow) against lepidopterous larvae and *Aschersonia* against whiteflies and scale insects. Developing fungicide resistant strains of mycoinsecticide is drawing attention and such strains would find better place in the IPM. Use of pheromone traps dosed with fungal preparation has been suggested to spread *Zoophthora radicans* in crucifer crops via infected males (Hajek and St. Leger, 1994). Some fungal pathogens, though not on large scale, have been used for the control of crop pests in India, *viz.*, *M. anisopliae* on rhinoceros beetle, *V. lecanii* on coffee green scale, *Coccus viridis* (Green), *Beauveria* sp. on tobacco caterpillar, *Spodoptera litura* (Fab.), *Fusarium oxysporum* on rice brown leafhopper, *Nilaparvata lugens* (Stal), *N. riley* on diamondback moth, *Plutella xylostella* (L.), etc.

***Nematodes***

Insect-nematode associations that are beneficial to mankind are at present being exploited for pest suppression with much interest. Although nematodes as insect parasite have been known since the 17th century yet insect nematology found its sound roots only in 1929 after Glasner and colleagues carried out outstanding research and subsequently used *Steinernema (Neoaplectana) glaseri* (Steiner) for the control of grubs of the Japanese beetle, *Popillia japonica* Newman. Since then, insect nematology is developing at a variable pace. Insect parasitic (entomophilic) nematodes primarily are endoparasitic (Aphelenchida being ectoparasitic) living inside host body either in blood cavity (haemocoel) or in the alimentary canal and feeding upon tissue and body organs or in the alimentary canal. They fall in four groups (Orders) *viz.*, Rhabditida, Tylenchida, Aphelenchida and Mermithida. Except Rhabditida, all possess a strong pointed and cuticularized stylet in the buccal cavity and thus two groups can easily be distinguished. According to Poinar (1975, 1979), 27 families or higher categories of nematodes are entomogenous and most important families that have found use in biological control are Mermithidae (*Reesimermis*), Steinernematidae (*Steinernema*), Heterorhabditidae (*Heterorhabditis*) and Neotylenchidae (*Deladenus*).

Entomogenous nematodes are associated with about 19 insect orders for phoresis (*Parasitorhabditis*, *Aphelenchoides*, etc.), facultative (*Steinernema*, *Heterorhabditis*, *Deladenus*, etc.) and obligate (mermithids) parasitism. The facultative parasites are capable to reproduce and develop under free living conditions while obligate parasites are often host specific. Infection by entomogenous nematodes may be passive (accidental ingestion of egg and entry through natural openings) or by active (penetration though the body wall). Infection may result in reduced activity, retarded growth, lowered fecundity and sterility. Death occurs due to mechanical destruction of host tissues or secondary pathogenic infection during exit of nematodes, or bacterial infection when nematode is entomopathogenic and behaves as a vector.

***Biology***

(a) *Mermithids:* The mermithids are obligate parasites of interest for insect suppression. *Heteromermis truncata* is a widely studied parasite of lepidopterans which sometimes kills more than 90 per cent of the larval population. Infective larvae may survive in soil for several months.

While encountering a host larva, it penetrates the integument by its stylet to reach the haemocoel, absorbs food through its smooth skin and rapidly grows to reach a full length of about 30 cm in a month's period, remaining characteristically coiled in the host body cavity. Later, it leaves the host as preadult, enters soil upto 30 cm depth and remains coiled for 2 weeks to months to undergo two moults. Copulation occurs in soil as tight cluster. Egg laying starts 20–40 days after mating and a female has a capacity to lay 1000–2000 eggs of about 180–200 μ size in 8 months. Eggs hatch in 3 months as II-stage infective larvae.

In nature mermithids play a vital role in the suppression of larval populations of lepidopteran pests but obligate parasitism and complex and long life cycle limit their mass multiplication for use as biocontrol agent.

(b) *Entomopathogenic nematodes:* All members of Order Rhabditida act as vector for bacteria and cause disease-like symptoms (septicaemia) and hence are entomopathogenic. Nematodes of its two important families, Steinernematidae (with two genera, *Steinernema* and *Neosteinernema*) and Heterorhabditidae (with single genus *Heterorhabditis*) have found use in biocontrol of insect pests. Anterior part of the alimentary canal of infective juveniles (dauerlarvae) is modified as bacterial chamber containing symbiotic bacteria (*Xenorhabdus* in *Steinernema* and *Photorhabdus* in *Heterorhabditis*). These are Gram-negative, non-spore former, anaerobic and facultative rod shaped bacteria belonging to family Enterobacteriaceae. Several species of these bacteria (at least 5) have been described.

The infective juvenile enters the host through natural openings (spiracles, mouth and anus) or by penetrating through the cuticle (*Heterorhabditis* can do so by using sharp hook above stoma) and in haemocoele it releases bacteria which multiply at very fast rate and cause septicaemia. Thus nematode acts as a vector for entry of the bacterium. However, the nematode itself is also capable of producing toxin to kill its host. The infective juvenile becomes a feeding III stage larva and feeds on bacteria and other metabolites produced by septicaemia. The host may die in 1–3 days. Nematodes moult through IV-stage to become males and females of first generation. After mating females of the first generation lay eggs in host body which hatch to pass through four moults to become second generation males and females. These after mating lay eggs and their late II-stage larvae cease feeding, attain bacterial pellet in bacterial chamber and finally become III-stage infective juveniles ensheathed in II-stage larval cuticle. These double sheathed larvae leave the host cadaver to attack the new host as infective juveniles. In *Heterorhabditis*, first generation adults are hermaphrodites while males, females and hermaphrodite progeny is produced in second generation. Generally this whole process takes 7–10 days at 25°C in *Galleria mellonella* (Linn.) as its host and from one host larva, 0.1 million to over 1 million infective juveniles may be produced. The survival of these ensheathed larvae in the soil depends upon biotic (natural enemies) and abiotic (temperature, humidity, soil type and moisture, etc.) factors and surviuval varies from weeks to months. Currently, 16 species of *Steinernema* and 7 species of *Heterorhabditis* are known to exist. For detailed biology see Wouts (1984, 1991) and Nguyen and Smart (1994).

(c) *Neotylenchid,* Deladenus *spp.*: *Deladenus* belongs to neotylenchid group of Tylenchida which parasitizes the wood wasp, *Sirex noctiliol* (Fab.), a serious pest of pine trees (mainly *Pinus radiata* D. Don) in Europe, Asia, Australia and New Zealand where it has killed thousands of hectare of trees. The nematode *D. siricidicola* Bedding carries two symbiotic fungi of *Amylostereum* spp. and it can be mass cultured on fungus. Thus it has mycelophagous as well as entomophagous relationship.

The infective adult female nematode penetrates the wood wasp larva upon coming in contact with it. Its reproductive system expands in host pupa and lays thousands of eggs inside the host. The juveniles hatching from eggs infect reproductive system of the pupa of wood wasp thereby sterilizing the female. The infected eggs and symbiotic fungi are deposited by the affected wood wasp female into the trees. The juveniles escape the eggs and feed on fungi for further development as mycetophagous adults. This nematode has provided substantial control of wood wasp in Australia (Coppel and Mertins, 1977).

***Field efficacy and use of entomopathogenic nematodes***

The pest species which harbour in soil or live in cryptic or protected habitat are vulnerable to entomopathogenic nematodes like cutworms, white grubs, flea beetles, root worms, mole crickets, webworms, carpenterworms, armyworms, plume moth, crown borers, leafminers, stem borers, sciarid flies, etc. Entomopathogenic nematodes have advantage over chemical control as they actively find their host by foraging strategy like hunters or wait for their prey as ambushers. They have a narrow host range, thus do not cause indiscriminate mortality. Further, they perpetuate by infecting fresh host and remain effective for long period (weeks to months) in favourable environment. However, being sensitive to desiccation and solar radiation they are not suitable for foliar application.

For successful application, soil should be moist (neither too dry nor near saturation). Adequate temperature range for infectivity is 18–32°C. To avoid solar radiation, application should be made during early morning hours or in the evening. These nematodes need be applied in high concentration (above 2.5 billion/ha or 250,000/m$^2$) and spray volume needed depends upon soil factors (250–5000 l/ha). Nematodes can be applied with conventional equipment to the soil by broadcasting at planting time, injection into borer galleries, as baits, through drip irrigation or by overhead sprinkler system. Difference in field efficacy between *Steinernema* and *Heterorhabditis* perhaps is due to their vertical distribution, the former being more effective near the soil surface (e.g. cutworms, mole crickets, etc.) while the latter with a greater tendency to move down has superior host seeking capability (e.g., white grubs, root weevils, etc.).

Steinernematids and hetrorhabditids can be reared *in vivo* (in insect hosts) and for it rice moth larvae or greater wax moth larvae may be used. By *in vitro* methods, they can be mass multiplied on solid medium using beef or chicken offal or pork liver as substrate, or on liquid medium in fermentation chambers, the latter being specially suited for large scale commercial production. The commercially available formulations in the USA, UK, Switzerland, Germany, Canada and other countries are various strains of *S. carpocapsae* Weiser (the most studied nematode, and its DD-136 strain widely used), *S. feltiae (=bibionis)* Filipjev, *S. reobravis, S. scapterisce, H. bacteriophora (=heliothidis)* Poinar and *H. megidis*. Thus out of 23 known species, 6 have been commercialized (Smart, 1995). In India, some biocontrol laboratories are producing *S. feltiae* and *H. bacteriophora* for field use on small scale.

## APPROACHES OF BIOLOGICAL CONTROL

Biological control is achieved through one of three approaches alone or in combination, *viz.* classical biological control, augmentation and conservation. The classical biological control, which is on parity with introduction, involves introducing an effective exotic natural enemy and establishing it in new environment. After introducing, management of natural enemy is of significance to harness maximum benefits by employing techniques of conservation which permit natural enemies to express their full effectiveness. Sometimes, due to some inimical factors (weather extremes, pesticidal application, asynchronization of the release with vulnerable stage of the pest, etc.), population collapse of introduced natural enemy may occur. Then one has to rely upon augmentation invloving releases of laboratory reared natural enemies to supplement their already existing populations for obtaining desired level of control. Though more emphasis has been on classical biological control, augmentaiton and conservation constitute important aspects of biological control for management of indigenous natural enemy and both these techniques have yet to be fully exploited for different pest situations.

### Classical biological control

Classical biological control is based on the discovery, importation of exotic natural enemies against established exotic pests (though indigenous pest may also be the likely target for it), and finally their long-term establishment in the new environment. Often insect pests have been defying quarantine boundries laid down for their further spread. The new environment may be more favourable and offer poor resistance to population build up of the introduced insect because of prevalence of ineffective indigenous enemy complex. According to Hoy (1994), feasibility of classical biological control against alien pests rests with three assumptions, *viz.* (1) the pest is regulated by natural enemies, (2) there is no effective natural enemy complex in new environment and (3) introduced natural enemy may be more effective if is without hyperparasites. If in such environment an effective additional mortality factor in the form of a natural enemy is introduced from ecologically similar area where such a pest is being maintained at non-economic status, it is anticipated to exercise effective suppression of the new pest. This forms the basis of classical biological control. Classical biological control requires several steps for its successful implementation (Coppel and Mertins, 1977; Pschorn-Walcher, 1977; Dent, 1991) such as:

#### *Identification of the pest, area of its origin and distribution*

Exact identification of the pest is the first and foremost step since every pest has its own natural enemy complex. Once the pest is identified as exotic, access to literature would be undertaken to get maximum biological details about the pest, its origin, geographical distribution, ecological requirements and existing natural enemies. Significance of identifying the area of origin or natural home or home land of the pest lies in the fact that in most of the successful biological control programmes, potential agents have been introduced from the home land of the pest where well balanced natural enemy complex is established over long time association between pests and their natural enemies (an old association between pest and natural enemy). Failure to identify correctly the centre of origin may lead to introduction of less effective natural enemy. Literature on distribution of the pests, areas where the pest is regulated by natural enemies and areas having target host-plants and climatic conditions almost similar to the country/place where the introduction is intended, provide useful wealth of information for probable natural enemy import.

*Foreign exploration and selection of suitable natural enemy*

Generally, foreign exploration is desirable in order to get thorough check of the natural enemy complex and to ascertain identity of the pest in endemic area (DeBach, 1974). Only under natural conditions, tritrophic interaction (plant-pest-natural enemy) can be truely understood. Collection of natural enemies should be at least for one complete season of the pest activity and from wide geographical area especially from undisturbed environment so as to sample complete natural enemy complex from all life stages of the pest. Two schools of thought exist, the first advocating selection of the best species and, the other favouring multiple species introduction (Coppel and Mertins, 1977). Supporters of the first opinion favour introduction of most promising natural enemy to avoid potential adverse competition. It can be selected by sequential releases, evaluating each introduction for pest regulation until the best species is found. However, no best set of criteria is available for evaluating a natural enemy. Further, it is time consuming and an expensive procedure. In the case of multiple species introduction, every potential natural enemy is introduced in the new habitat with the hope that the best one would become established. The concept of multiple species introduction, the so called empirical approach (Ehler,1990) is supported by the ecological principles of competitive displacement and coexistenece. Simultaneous or sequential multiple importation of diverse natural enemy species is the only practical method of obtaining the best species or a combination of species. According to Gauses law, different species occupying identical ecological niches (ecological homologues) can not coexist indefinitely in the same habitat and better species utilizing same stage of the host would competitively displace the other one. But species which attack different life stages of the pest are expected to excercise complementary action (sequence theory of Howard and Fiske, 1911). Those favouring multiple species introduction, suggest for introducing the natural enemy from both, the ecologically favourable centre of distribution (where genetic diversity is maximum bacause of outbreeding and lethal gene frequency is high due to heterozygosity) and ecological margins of a species (more inbred homozygous populations with lower genetic load) so that amongst the greater genetically diverse populations of the species, the best suited genotype is sorted out for the new situation through natural selection process. In case, a genetically superior strain or species is already identified for a pest, it would be advisable to introduce it rather than going for multiple species introduction.

Natural enemy which efficiently suppresses the closely related pest species in home land of the pest or elsewhere can also be an option for classical biological control. In general, increasing species diversity of natural enemies per given prey species is inversely correlated with regulation of a prey population at low levels. In other words, we should seek to reduce the number of natural enemy species established against a given prey species by introducing highly specific, more effective enemy capable of replacing them and producing better results (DeBach and Rosen, 1991). According to Hokkanen and Pimental (1984), such agents which constitute new pest-natural enemy associations will be more effective and exert control for longer than an old association because co-evolution of the natural enemy and its prey leads to increased resistance in prey thereby reducing its effectiveness. They claimed that with the use of new associations, it would be possible to increase the chances of success from about 7 : 1 to about 4 : 1. However, Waage and Greathead (1988) found no significant difference for the cases involving old and new associations.

### *Quarantine and multiplication*

The natural enemy being introduced should be subjected to rigorous quarantine procedures so that the unwanted contaminating species as well as hyperparasites could be eliminated. For it, the natural enemy has to be reared either on its own host or on a factitious one for one or more generations under controlled and confined conditions in the country of its origin and/or quarantine section of the importing country. When the sample size is small, it can be efficiently handled and scrutinized. But such samples have restricted genetic variability, loss of adaptability would be phenomenal and the success of establishment would be threatened (Dent, 1991). The small limited stock selected from the natural enemies received in the quarantine section has to be multiplied to sufficient numbers for further release and establishment in the field. For this purpose extensive rearing facilities and equipment would be needed. Intended target pest is the preferred laboratory host but if it is unavailable during certain period, a factitious host may be used. Artificial diet for rapid multiplicaiton of effective natural enemies and/or their prey has its own significance and advantage provided these are cost effective. This would cut short the procedures involved for multiplication of plants, their pests and finally natural enemies. Laboratory reared insects will need to be acclimatized before release if the rearing conditions are not similar to the actual environment. Now-a-days more research is being conducted in this direction and many species can be easily mass multiplied for releases.

### *Releases and establishment of the natural enemy*

Once the safety and suitability of natural enemy has been established and it has been multiplied to a required number, it would be better to attempt the earliest field release for early establishment. Such areas will act as loci of local source of a well adapted material for further releases at other places. Before making releases, first step is to ensure the presence of the appropriate stage of the target pest in sufficient numbers at different release sites. A large number of factors affect the establishment and effective control, viz. climate; prevailing natural enemy complex of the target pest; habitat, topography and intensity of agriculture; availability of alternate hosts and food (nectar and pollen) for the adults; proper site selection and number of sites; adequate number of natural enemies to be released and frequency of releases; and proper timing of the release especially for univoltine species. According to Coppel and Mertins (1977) releases are probably best made in either early morning or late evening hours when weather conditions are favourable. Initially, releases should be restricted to a few sites meeting requirements of the natural enemy and colonization be done in cages to minimize the dispersal and isolate it from competitors. Once established, the cages are removed.

Beirne (1975) critically analysed the success and failure of 159 species introduced in Canada and concluded that successful establishment is affected by the number of individuals released, greater the number more are chances of success, and if numbers are below some minimum (about 5000), the probability of success is low (about 10%). While 40% introductions were successful with releases of 5000–31200 individuals, success was 78% with releases of over 31200 individuals. Further, greater the number of individuals per release (minimum about 800), and greater the number of releases (70% of species released more than 20 times), greater is the possibility of establishment. In stable and discrete environment (e.g. orchards), even with smallest releases good success was obtained but in stable and diverse environment (e.g., forests), success was lesser with larger numbers; lowest rate of success was in the unstable monoculture of annual agricultural crops. An effective exotic natural enemy usually becomes established within a year (DeBach, 1974); if establishment is not evident within 3 years of release, then the hope of success is little (Clausen, 1951) and after 3 years, it would be better to attempt introduction of other species or different strain of the unsuccessful species (Coppel and Mertins, 1977).

Effective classical biological control depends upon establishement of one or more natural enemies in the new environment and the success estimates of establishment have been 16–34% (Hoy, 1994). According to DeBach and Rosen (1991), chances of obtaining a degree of control that would not need insecticidal control included more than one-third of all target pest species (35.6%). This strategy remains an effective option for the control of exotic pests since it offers somewhat permanent control, has few associated risks and provides cost effective control. Data as presented by Greathead and Waage (1983) on biological control from some projects of the International Institute for Biological Control in tropical areas reveal the cost-benefit ratio varying from 5 (for the control of sugarcane borer in Brazil) to 346.5 (for the control fo coconut leafminer in Sri Lanka) and overall cost-benefit ratio is estimated to be 1 : 30 (DeBach and Rosen, 1991).

## Augmentation

Augmentation refers to direct efforts and activities to increase populations or beneficial effects of natural enemies of both native and exotic pests. Thus augmentation deals with manipulation of an established natural enemy by enhancing its efficiency. It may be achieved through (1) periodic colonization of mass reared or naturally collected natural enemies, (2) genetic improvement of their attributes and (3) chemical manipulation of their behaviour. It should be given the lowest priority in biological control research endeavours and should be resorted to if importation or conservation strategies fail because with possible exception of genetically improved strains, augmentation techniques are more expensive to develop and apply (DeBach and Rosen, 1991). Augmentation efforts should be restricted to natural enemies which have been demonstrated to be effective in prey regulation.

### *Periodic releases*

Depending upon the number of natural enemies released, interval between releases and frequency of releases, the periodic releases may by inoculative or inundative. Inoculative release, also known as accretive release, referes to release of relatively small number of individuals at an appropriate time so that with increase in pest density, the beneficial organisms also increase naturally and their progeny affect the control for subsequent generations. Usually such releases are preferred at an early season of the crop when pest population starts building up and are, therefore, also known as seasonal inoculative releases. For example, release of *Aphelinus mali* (Haldeman) in apple orchards with warming up of the weather after winter @ 1000/ha against the woolly apple aphid, release of *Cryptolaemus montrouzieri* @ 10–20 beetles/tree against citrus mealy bugs in citrus orchards at the beginning of summer, and releases of predatory mites on almond and apple fruit trees to reduce population growth of phytophagous mites. However, sometimes in certain areas more small scale releases are required to be made in a year to supplement natural population and offset negative effects of weather, e.g. releases of *Aphytis melinus* DeBach against the California red scale in citrus orchards (Rosen, 1985; DeBach and Rosen, 1991).

In inundative releases, biotic agents are released in excessive quantity than are actually effective so as to get immediate control by the released agent rather than by its progeny. In such cases, repetition of treatment is required and thus such releases are more or less equivalent to application of insecticides. Inundative releases are most economical against the pests having one (univoltine species) or a few discrete generations in a year. Further, annual row-crops against which classical bioligical control is not much promising, agumentation by inundative releases espcially in green house system holds promise (Nordlund, 1997). Often insects that have been used on large scale for this strategy are egg parasite *Trichogramma* spp. against many lepidopteran pests and a general predator *Chrysoperla* spp. (*C. carnea*, *C. rufilabris* and *C. sinica*) against homopteran and lepidopteran pests and mites. Insect pathogens (e.g., *B. thuringiensis*, baculoviruses, fungi like *Metarhizium*, *Beauveria* and *Aschersonia*) are especially suitable for inundative release.

Augmentation is extensively being employed in glass houses for the control of spider mites, leafminers, thrips, aphids and whiteflies. By 1985, about 8000 ha area under glasshouses was covered by augmentative releases over 23 countries; in the Netherlands whiteflies on tomato, spidermites on cucumbers and thrips on sweet peppers are controlled by *Encarsia formosa* Gahan, *Phytoseiulus persimilis* (Athias-Henriot) and *Amblyseius cucumeris*, respectively (van Lenteren, 1986; van Lenteren and Woets, 1988). King *et al.* (1985) listed 28 natural enemies that have been evaluated in augmentative releases for agricultural pest control in the USA and out of these only two have been extensively used in inundative releases. Biocontrol attempts with *Trichogramma* sometimes have resulted into utter failure indicating that some strains may be specific to a pest or habitat. Otherwise also, there exist very minute differences in different species which often can be overlooked during the identification. Hence, it is imperative to select desirable strain of *Trichogramma* to obtain desired level of success. In India so far 79 species of exotic predators and parasites have been obtained, out of which 53 have been successfully multiplied in the laboratory, 34 shipped to different centres and 21 have been established in the field (Singh, 1994a).

Augmentation is a preventive strategy, not a curative one, hence it is justifiable when there is high probability of the problem to occur (Dent, 1991). Success of release programme is mainly dependent upon readily available higher quality natural enemy that can easily be mass-produced in commercial insectary in a cost effective way. Unless cheap and simple methods are known, augmentation is likely to remain confined to a few species. Continuous breeding under artificial conditions may result in inbreeding depression and quality deterioration in terms of reduced fitness, undesirable male biased sex-ratio, altered host preference because of use of factitious host or artificial media, improper field acclimatization and adaptability of natural enemy produced commercially under controlled conditions. Hence, mass reared unit should have a broad based and genetically diverse stock of natural enemies. For quality maintenance, it is imperative to collect biotype/species from the pest on target crop and induct the field collected material in the laboratory stock for reducing the risk of inbreeding. Usually natural enemies are reared on their prey and thus three trophic levels have to be maintained. Development of artificial diets for host and/or natural enemy would improve rearing efficiency. However, it is quite difficult to develop artificial diet for endoparasitoids. Beside production aspect, storage, packaging and transportation of natural enemy is also important. If proper storage conditions are known, continuous production can be taken up and availability in sufficient numbers at the need of the hour does not become a constraint. Actual cost of augmentation which depends upon number of natural enemies to be released for effective control, should also be economically feasible.

*Genetic improvement*

An effective natural enemy lacking in some desired attributes can be improved by genetic manipulations through artificial (laboratory) selection, hybridization (use of heterosis) and recombinant DNA techniques. The desired traits that have been considered in laboratory selection of some natural enemies are host finding, host preference, host synchrony, climatic tolerance, increased fecundity, insecticide resistance, non-diapause and thelytoky. However, demonstration of field effectiveness of such improved strain is practically difficult. Some constraints to initiate genetic improvement are (1) identification of factor(s) limiting efficacy of the natural enemy, (2) prevalence of genetic variability, (3) documented field efficacy of genetically improved natural enemy and (4) justified cost-benefit ratio. Nevertheless, genetic improvement of at least the predatory mites through laboratory selection has been made for traits like fecundity (*Phytoseiulus persimilis* Athias-Henriot), pesticide resistance (*Amblyseius*, *Metaseiulus*, *Phytoseiulus* and *Typhlodromus*), non-diapause (*M. Occidentalis* (Nesbitt) and temperature tolerance (*P. persimilis*) (Hoy,1988). Some other parasites selected for resistance to insecticides are *Aphytis melinus* DeBach (parasite of California red scale, against carbaryl), *Trioxys pallidus* (Haliday) (the walnut aphid parasite, against azinphos methyl) and *T. chilonis* (against endosulfan) (Hoy, 1988; Singh, 1997). If such strains are able to establish and survive, only then their effectiveness can be documented provided that it replaces the other. How long such improved strains are capable to retain superiority for the selected trait depends upon the genetic inheritance of such character, which is often polygenic and tends to diminish with segregation of genes in the nature. At least, the character being goverened by the single dominant gene is stable and it gradually spreads in the field population. Also, genetical improvement can be stable in thelytokous species.

The potential of using recombinant DNA technology through genetic engineering in creating improved individuals having desirable traits is presently a well known fact (Beckendorf and Hoy, 1985), which has recently been demonstrated by cloning sheep and calves. In the Indian Institute of Sciences, Bangalore, work is in progress to induct genes resistant to Hepatitis virus E and B, genes for producing human growth hormone and for getting more and better quality silk, into the mulberry silkworm. This technology comprises identification of desirable genes for improved character, cloning and inserting these into the genome of natural enemies so that it becomes a counterpart of its own genome and expresses itself in appropriate tissue at right time, and finally be transmitted to the progeny. A day is not far off when such a transformed natural enemy strain would be available for field releases. Through this technique it would also be possible to broaden the range of activity of much specific microbial insecticide like *B. thuringiensis* and baculoviruses. Genetic engineering may also provide solution to the problems of storage and persistence of microbial insecticide like fungi.

***Chemical manipulation of natural enemies***

As mentioned earlier, semiochemicals (pheromones and allelochemicals) are the behaviour modifying chemicals which determine prey finding and its acceptance by the natural enemy. These may be manipulated for the advantage of natural enemies for their increased activity. If such chemicals are identified and synthesized, these can be used to attract natural enemy on target crop, retain them in the area of release and get better parasitization/predation, e.g., long-chain hydrocarbon tricosane present in scales of *H. armigera* acts as kairomone to *Trichogramma* (Ananthakrishnan, 1992). Proper and timely application of such chemicals is also important so as to prevent over-stimulation of the natural enemy to remain confined to treated areas and spare the patchily distributed prey. Host-marking pheromones may also be used to improve egg distribution and reduce superparasitism and hyperparasitism (Rosen, 1985).

## REFERENCES


Agarwal, G.P. 1990. Entomogenous fungi in India and management of insect pests. *Indian Phytopathology*, **43**(2): 131–142.

Allen, G.E., Ignoffo, C.M. and Jaques, R.P. 1978. *Microbial Control of Insect Pests: Future Strategies in Pest Management Systems*. NSF-USDA-University of Florida, Gainsville, FL.

Ambrose, D.P. 1996. Assassin bugs (Insecta—Heteroptera—Reduviidae) in biological control: success and strategies, a review. *In:. Biological Control of Insect Pests, an Indian scenario*. D.P. Ambrose (ed). Adeline Publishers, Tirunelveli, Tamil Nadu, pp. 262–284.

Ananthakrishnan TN. 1992. Chemical ecology in biological control. *In: Emerging Trends in Biological Control of Phytophagous Insects*. T.N. Ananthakrishnan (ed). Oxford & IBH Publishing Co. Pvt. Ltd., New Delhi, pp. 59–67.

Askew, R.R. 1971. *Parasitic Insects*. Heinemann Educational Books Ltd., London.

deBarjac, H. and Frachon, E. 1990. Classification of *Bacillus thuringiensis* strains. *Entomophaga* **35**(2): 223–240.

Battu, G.S.; Ramakrishnan, N. and Dhaliwal, G.S. 1993. Microbial pesticides in developing countries: Current status and future trends. *In: Pesticides: Their Ecological Impact in Developing Countries*. G.S. Dhaliwal and B. Singh (eds). Commonwealth Publishers, New Delhi, pp. 417–468.

Beckendorf, S.K. and Hoy, M.A. 1985. Genetic improvement of arthropod natural enemies through selection, hybridization or genetic engineering techniques. *In: Biological Control in Agricultural IPM Systems*. M.A. Hoy and D.C. Herzog (eds.). Academic Press, Orlando, FL. pp. 167–187.

Beirne, B.P. 1975. Biological control attempts by introductions against pest insects in the field in Canada. *Canadian Entomologist*, **107**: 225–236.

Burgess, H.D. (ed). 1981. *Microbial Control of Insect Pests, Mites and Plant Diseases, 1970–1980*. Academic Press, London.

Burgess, H.D. and Hussey, N.W. 1971. *Microbial Control of Insects and Mites*. Academic Press, London and New York.

Cantwell, G.E. (ed). 1974. *Insect Diseases*, 2 vols. Marcel Dekker, New York.

Clausen, C.P. 1951. The time factor in biological control. *Journal of Economic Entomology*, **44**: 1–9.

Carson, R.L. 1962. *Silent Spring*. Houghton Mifflin, Boston.

Coppel, H.C. and Mertins, J.W. 1977. *Biological Insect Pest Suppression*. Springer-Verlag, Berlin, Heidelberg, New York.

Deacon, J.W. 1983. *Microbial Control of Plant Pests and Diseases*. Van Nostrand Reinhold (UK) Company Ltd., Wokingham, UK.

DeBach, P. (ed). 1964. *Biological Control of Insect Pests and Weeds.* Chapman and Hall, London.

DeBach, P. 1974. *Biological Control by Natural Enemies*. Cambridge University Press, London, New York.

DeBach, P. and Rosen, D. 1991. *Biological Control by Natural Enemies* (2nd edition). Cambridge University Press, Cambridge.

Dent, D. 1991. *Insect Pest Management*. CAB International, Wallingford.

Ehler, L.E. 1990. Introduction strategies in biological control of insects. *In: Critical Issues in Biological Control*. M. Mackauer, L.E. Ehler and J. Roland (eds). Intercept, Andover, UK, pp. 111–134.

Entwistle, P.F. and Evans, H.F. 1985. Viral control. *In: Comprehensive Insect Physiology, Biochemistry and Pharmacology*,. G.A. Kerkut and L.I. Gilbert (eds). vol. 12. Pergamon Press, Oxford, pp. 347–412.

Ferron P. 1978. Biological control of insect pests by entomogenous fungi. *Annual Review of Entomology,* **23**: 409–442.

Ferron, P. 1985. Fungal control. *In: Comprehensive Insect Physiology, Biochemistry and Pharmacology*. G.A. Kerkut and L.I. Gilbert (eds). vol.**12**. Pergamon Press, Oxford, pp. 313–346.

Fuxa, J.R. and Tanada, Y. (eds). 1987. *Epizootiology of Insect Diseases*. John Wiley and Sons, New York.

Gill, S.S., Cowles, E.A. and Pietrantonio, P.V. 1992. The mode of action of *Bacillus thuringiensis* endotoxins. *Annual Review of Entomology,* **37**: 615–636.

Greathead, D.J. and Waage, J.K. 1983. Opportunities for biological control of agricultural pests in developing countries. *World Bank Technical Paper* No. 11.

Gupta, P.R. and Bhalla, O.P. 1993. Conservation of biotic agents in apple orchards of Himachal Pradesh. *Journal of Insect Science,* **6**: 204–206.

Gupta, P.R. and Chander, R. 1992. Role of coccinellids in suppression of woolly apple aphid. *In: Emerging Trends in Temperate Fruits Production in India.* K.L. Chadha, D.K. Uppal, R.N. Pal, R.P. Awasthi and S.A. Ananda (eds). National Horticulture Board, Technical Publication No. 1, Gurgaon, pp. 263–267.

Gupta, P.R.; Thakur, J.R. and Dogra, G.S. 1988. Management of peach leaf curl aphid, *Brachycaudus helichrysi* (Kalt.), infesting stone fruits in Himachal Pradesh. *Journal of Tree Science,* **7**: 26–30.

Hagen, K.S. and Franz, J.M. 1973. A history of biological control. *In: History of Entomology.* R.F. Smith, T.E. Mittler and R.N. Smith (eds). Annual Reviews Inc., Polo Alto, California. pp. 433–476.

Hajek, A.E. and St. Leger, R.J. 1994. Interactions between fungal pathogens and insect hosts. *Annual Review of Entomology,* **39**: 293–322.

Heimpel, A.M. and Angus, T.A. 1959. The site of action of crystalliferous bacteria in Lepidoptera larvae. *Journal of Insect Pathology,* **1**: 152–170.

Hofte, H. and Whiteley, H.R. 1989. Insecticidal crystal proteins of *Bacillus thuringiensis. Microbiology Review,* **53**: 242–255.

Hokkanen, H. and Pimentel, D. 1984. New approach for selecting biological control agents. *Canadian Entomologist* **116**: 1109–1121.

Horen, D.J. 1988. *Ecological Approach to Pest Management.* Elsevier Publishers, London.

Howard, L.O. and Fiske, W.F. 1911. The importation into the United States of the parasites of the gypsy moth and brown-tail moth. *United States Department of Agriculture Bur. Entomological Bulletin,* **91**: 1–312.

Hoy, M.A. 1988. Biological control of arthropod pests: traditional and emerging technologies. *Americal Journal of Alternative Agriculture,* **3** (2): 63–68.

Hoy, M.A. 1994. Parasitoids and predators in management of arthropod pests. *In: Introduction to Insect Pest Management.* R.L. Metcalf and W.H. Luckmann (eds). IIIrd edition. John Wiley and Sons Inc., New York. pp. 129–198.

Huffaker, C.B. and Messenger, P.S. (eds). 1976. *Theory and Practice of Biological Control.* Academic Press, New York.

King, E.G.; Hopper, K.R. and Powell, J.E. 1985. Analysis of systems for biological control of crop arthropod pests in the U.S. by augmentation of predators and parasites. *In: Biological Control in Agricultural IPM systems.* M.A. Hoy and D.C. Herzog (eds). Academic Press, Orlando, FL, pp. 201–227.

van Lenteren, J.C. 1986. Parasitoids in the greenhouse: successes with seasonal inoculative release systems. *In: Insect Parasitoids*. J.K. Waage and Greathead, D.J. (eds). Academic Press, London, pp. 341–374.

van Lenteren, J.C. and Woets, J. 1988. Biological and integrated pest control in greenhouses. *Annual Review of Entomology,* **33**: 239–269.

Li, L.Y. 1994. Worldwide use of *Trichogramma* for biological control on different crops: a survey. *In: Biological Control with Egg Parasitoids*. E. Wajnberg and S.A. Hassan (eds). CAB International, Oxon, pp. 37–51.

MacArthur, R.H. and Wilson, E.O. 1967. *The Theory of Island Biogeography.* Princeton University Press, Princeton.

Maddox, J.V. 1994. Insect pathogens as biological control agents. *In: Introduction to Insect Pest Management*. R.L. Metcalf and W.H. Luckmann (eds). JohnWiley & Sons, Inc., New York. pp. 199–244.

Martin, P.A.W. 1994. An iconoclastic view of *Bacillus thuringiensis* ecology. *American Entomologist* (Summer), pp. 85–90.

Martin, P.A.W. and Travers, R.S. 1989. Worldwide abundance and distribution of *Bacillus thuringiensis* isolates. *Applied Environmental Microbiology,* **55**: 2437–2442.

Nguyen, K.B. and Smart, G.C. Jr. 1994. *Neosteinernema longicurvicauda* n. gen., n. sp. (Rhabditida: Steinernematidae), a parasite of the termite *Reticulitermes flavipes* (Koller). *Journal of Nematology,* **26**: 162–174.

Nordlund, D.A. 1997. Biological control by augmentation of parasitic and predaceous arthropods—an important component of integrated pest management. *In: IPM System in Agriculture.* R.K. Upadhyay, Mukerji, K.G. and R.L. Rajak (eds). Vol. **2**, *Biocontrol in Emerging Biotechnology*. Aditya Books (Pvt.) Ltd., New Delhi, pp. 1–24.

Poinar, G.O. 1975. *Entomogenous Nematodes.* E J Brill, Leiden.

Poinar, G.O. 1979. *Nematodes for Biological Control of Insects.* CRC Press, Boca Raton.

Pschorn-Walcher, H. 1977. Biological control of forest insects. *Annual Review of Entomology,* **22**: 1–22.

Rao, V.P. and Ghani, M.A. (eds). 1972. *Studies on predators of* Adelges *spp. in the Himalayas*. Misc. Pub. No.3, Commonwealth Agricultural Bureaux, Slough, England.

Rombach, M.C. and Gillespie, A.T. 1988. Entomogenous Hyphomycetes for insect and mite control on greenhouse crops. *Biocontrol News and Information,* **9** (1): 7–18.

Rosen, D. 1985. Biological control. *In: Comprehensive Insect Physiology, Biochemistry and Pharmacology*. G.A. Kerkut and L.I. Gilbert (eds). vol. **12**. Pergamon Press, Oxford. pp. 413–464.

Sankaran, T. and Nagarkatti, S. 1982. Insects as biological control agents. *In: Agricultural Entomology*. P.D. Srivastava, Jotwani, M.G., R.A. Agarwal, S.R. Wadhi, R.K. Bhanotar and R.K. Bhatnagar (eds). vol. 1. All India Scientific Writers' Society, Paschim Vihar, New Delhi. pp. 54–74.

Schlinger, E.I. and Doutt, R.L. 1964. Systematics in relation to biological control. *In: Biological Control of Insect Pests and Weeds.* P. DeBach (ed). Chapman and Hall, London, pp. 247–280.

Singh, S.P. 1994a. *Fifteen Years of AICRP on Biological Control.* Technical Bulletin No. 8, ICAR, Project Directorate of Biological Control, Bangalore.

Singh SP. 1994b. *Technology for Production of Natural Enemies.* Technical Bulletin No. 4, ICAR, Project Directorate of Biological Control, Bangalore.

Singh, S.P. 1997. Biological control of pests in India. *In: IPM System in Agriculture*, vol. 2 *Biocontrol in Emerging Biotechnology*. R.K. Upadhyay, K.G. Mukerji and R.L. Rajak (eds). Aditya Books (Pvt.) Ltd., New Delhi. pp. 216–327.

Smart, G.C. Jr. 1995. Entomopathogenic nematodes for the biological control of insects. *Journal of Nematology,* **27** (4 S): 529–534.

Smith, S.M. 1996. Biological control with *Trichogramma*: advances, successes, and potential of their use. *Annual Review of Entomology,* **41**: 375–406.

Steinhaus, E.A. 1946. *Insect Microbiology*. Comstock Publishing Co., Ithaca.

Steinhaus, E.A. 1949. *Principles of Insect Pathology*.McGraw Hill, New York.

Steinhaus, E.A. 1956. Microbial control—the emergence of an idea. *Hilgardia,* **26**: 107–160.

Steinhaus, E.A. 1963. *Insect Pathology, an Advanced Treatise*, 2 vols., Academic Press, New York.

Strong, D.R., Lawton, J.H. and Southwood, R. 1984. *Insects on Plants.* Blackwell, Oxford.

Sweetman, H.L. 1958. *The Principles of Biological Control.* Wm. C. Brown, Dubuque, Iowa.

Tanada, Y. and Kaya, H.K. 1993. *Insect Pathology*. Academic Press, New York.

Vet, L.E.M. and Dicke, M. 1992. Ecology of infochemical use by natural enemies in a tritrophic context. *Annual Review of Entomology*, **37**: 141–172.

Vyas, H.G.; Joshi, D.P.; Patel, K.C.; Yadav, D.N. and Dodia J.F. 1986. Natural incidence of milky disease of white grubs in Gujarat. *Indian Journal of Agricultural Sciences,* **56**: 213–214.

Vyas, R.V.; Parasharya, B.M. and Yadav, D.N. 1988. Dispersal of milky disease organism *Bacillus popilliae* var. *holotrichiae* of white grub (*Holotrichia consanguinea*) through birds. *Indian Journal of Agricultural Sciences,* **58**: 243–244.

Vyas, R.V.; Yadav, D.N. and Patel, R.J. 1991. Studies on mass production of *Bacillus popilliae* var. *holotrichiae*, a promising pathogen of white grub in Gujarat. *Gujarat Agriculture University Research Journal,* **17**: 30–37.

Vyas, R.V.; Yadav, D.N. and Patel, R.J. 1991a. Efficacy of *Bacillus popilliae* var. *holotrichiae* against white grub (*Holotrichia consanguinea*). *Indian Journal of Agricultural Sciences,* **61**: 80.

Waage, J.K. and Greathead, D.J. 1988. Biological control: challenges and opportunities. *Philosophical Transactions of the Royal Society of London,* B **318**: 111–128.

Weiser, J. 1977. *An Atlas of Insect Diseases.* Academic Press, Prague.

Wouts, W.M. 1984. Nematode parasites of Lepidoptera. *In: Plant and Insect Nematodes*. W.R. Nickle (ed). Marcel Dekker, Inc., New York, pp. 655–696.

Wouts, W.M. 1991. *Steinernema (Neoaplectana)* and *Heterorhabditis* species. *In: Manual of Agricultural Nematology*. W.R. Nickle (ed). Marcel Dekker, Inc., New York pp. 855–897.

*Pest Management in Horticulture Crops*
L.R. Verma, A.K. Verma and D.C. Gautam (eds)


*Chapter 4*

# Cultural, Mechanical and Regulatory Methods in Pest Management

*N.P. Kashyap and P.K. Mehta*

## INTRODUCTION

THE INCREASING USE OF insecticides has many well documented adverse ecological consequences. The publication of 'Silent Spring' in 1962 by Rachel Carson aroused public concern about excessive and indiscriminate use of pesticides. This realization led to the concept of integrated pest management as an environmentally safe alternative to toxic chemicals. Presently, integrated pest management is accepted as a control strategy in which a variety of methods are combined for long term pest control. This integration of different approaches reduces the frequency and severity of pest problems with minimum pesticide application in the field situations and thus has revived the interest in cultural, mechanical and regulatory methods of insect control. In the context of pest management, no implication is made that either of these methods should function as effective control practices. Most frequently, each practice is applicable to one or a few insect species of a large pest complex. Assumption is made that cultural, mechanical and regulatory strategies in combination with other compatible direct control practices will help in checking the insect pest menace in the field.

Historically, cultural and mechanical practices were the farmers first choices for preventing crop losses. However, farmers were probably not aware that the yield increase resulting from these practices was also due to decrease in the damage caused by pests. The introduction of agronomically improved varieties that were highly susceptible to pest damage further aggravated the insect pest problem. Another important component of integrated pest management which hitherto remained neglected is the legislative method through imposition of plant quarantine measures. This is a regulatory programme of managing insect pests associated with the movement of plants and plant materials amongst countries, and also within a country, amongst different ecological zones. Integrated pest management envisages plant quarantine as the first and foremost method of pest management options in addition to being environmentally safe and economical.

Non-chemical methods of insect control were mostly developed empirically before the synthetic pesticide era and in consequence have rarely been evaluated through such standards that would meet present day criteria. Environmental and economic benefits could, however, have a significant intrinsic value, not only from increased diversity within the agro-ecosystem but also from a reduction in insecticide use. So in order to strengthen the concept of integrated pest management, some of the cultural, mechanical and regulatory methods have been modified to enhance crop productivity. Some of these practices are discussed in this chapter in relation to insect pests of various horticultural crops.

## CULTURAL CONTROL

Cultural control of insect pests is through the manipulation of the environment in such way so as to render it unfavourable to the pest. Historically, these were farmers' most important tools in preventing crop losses. Before the discovery of modern plant protection measures, the farmers evolved many cultural practices largely through trial and error experiences so as to minimise the damage caused by various insect pests (Smith *et al.*, 1976).

Any change to normal cultural practices, made for whatever reasons, may make survival conditions less favourable for insect pest depending on their individual biological requirements. Modification of cultural practices requires a great deal of ecological information about the pests if it has to be effective in pest control programme.

In the past, when a few control measures were available, cultural practices were the only efficient tools to minimise pest problems but with widespread use of insecticides, the possibilities of using cultural control were largely neglected. However, due to total reliance on insecticides for the control of pests, a number of problems have arisen. There is now considerable revival of interest in the potential of cultural control procedures as they represent an important component in the development of a coherent, holistic approach to pest management.

The objectives of the cultural control are:

i. to reduce the initial colonization of the pest,
ii. to minimize reproduction and survival rate and
iii. to increase dispersal if colonization is achieved.

Some of the important cultural methods used in the management of insect pests are discussed below.

### Crop rotation

Crop rotation involves sequential planting of botanically unrelated crop plants in normally a three to five years cycle. Corp rotation normally reduces and delays pest attack rather than giving complete control. Though control may be significant within a given field, it is a less effective restrain over an area as a whole. The practice was developed for the purpose of maintaining soil fertility and soil structure and benefits of suppressing pests have been incidental to a large extent. Generally, crop rotation is most effective against pests that have a narrow host range and limited mobility, yet it is effective in reducing many soil pests which multiply successfully under grasses. Various crop midges are weak fliers and are affected by crop rotation.

Rotations have been found to be effective against the colorado potato beetle, *Leptinotarsa decemlineata* (Say), by delaying the time of infestation (Wright, 1984). In potato–wheat rotation, the oviposition and first appearance of beetle were delayed when compared with an unrotated potato field. This delay was attributed to physical and environmental barriers that slow down emigration from wheat by overwintering adults (Lashomb and Ng, 1984). In Europe, for the control of *wireworms, Agrotis* spp. *Chafers, Melolontha melolontha and Amphenallon solatitialis,* the most typical rotations involve grasses, legumes and root crops.

### Time of sowing

The manipulation of the time of sowing helps minimizing pest damage by producing asynchrony between host plant and the pest so that pest emergence fails to coincide with a critical crop growth stage. The time of sowing alters the period of crop development to allow young plants to become established to a tolerant stage before the attack occurs and thus reduces the susceptible period of the crop before a pest becomes abundant.

Sugarbeet plants sown early on well cultivated and fertilized soil grow vigorously and become increasingly tolerant to aphid colonization and virus infection by beet fly, *Pegomyia beatae* (Jepson and Green, 1983) and Swede midge, *coptarinia nastwrtii* (Bardner *et al.*, 1971).

In Punjab, early and late sown pea crops are damaged to a lesses extent by the leaf miner, *Chromatomyia horticola* (Goureau) than those sown on normal sowing dates (Kumar *et al.*, 1991). On the contrary, in the mid-hill region of Himachal Pradesh, late sown crop suffers low incidence of the leaf miner (Sood and Mehta, 1993). Pea crop sown early is more likely to be attacked by weevil, *Sitona lineatus* (Dore *et al.*, 1991) and stem fly (Bali and Quershi, 1997).

Early transplantation of brinjal crop suffers higher incidence of the brinjal shoot and fruit borer, *Leucinodes orbonalis swinhoe* and jassid, *Amarasca bigutulla bigutulla* (Chaudary and Kashyap, 1992). The incidence of the green peach aphid, *Myzus persicae* (Sulzer) was reported to be less on early sown cumin (Gupta and Yadav, 1989) as compared to late sown crop. Soy beans sown between 10–30 June suffered less incidence of bean fly, *Ophiomyia phaseolin* and resulted in higher yield (Negasi and Abate, 1986).

No infestation of potato tuber moth, *Phthorimaea operculella Zeller* was observed in the crop planted early which resulted in higher yields (Ahmadabai and Hessan, 1991). In Egypt, early planted onions were found to be less susceptible to *Thrips tabaci* and *Eumerus amochus* (Hyder and Sherif, 1990).

### Irrigation

Irrigation is a common practice in many crops and can be manipulated for pest control. Under conditions of moisture stress, irrigation helps to minimize the effects of pest injury as the plants grow vigorously and thus compensate to some extent the loss of tissues due to pest feeding. However, irrigation can also be harmful to some insects like aphids which are easily washed off from the plants by overhead irrigation. Soil insects may be killed by the pressure of swelling soil particles in saturated soils.

Additionally, ample water availability causes physiological changes in the plants. Some sucking insects such as aphids and thrips survive poorly on well irrigated plants and benefit from periodic wilting of plants. Plooding of fields is recommended for reducing the attack of cutworms, army worms, termites, white grubs, etc. Frequent overhead irrigation in potato can control the potato tuber moth effectively.

In the erstwhile USSR, on the basis of long term observations, it was found that irrigation reduces the damage caused by grape *phylloxera* (*Viteus vitifoliae* Fitch), winter moth, cutworm (*Agrotis segetum* Schiff), the dust beetle (*Opatrum sabulosum*) and the black beet weevil (*Psalidicum maxillosum*). Resistance to pest attack was increased by irrigation (Minoranskii, 1985). The percent damage by cut worm, *A. segetum* to carrot was 6, 33 and 48 and red beet 0, 4 and 2, respectively, with 3, 1 and no irrigation (Kobro, 1991).

The physical effects of irrigation can be both beneficial and detrimental to pest population. Water droplets can cause dislodging, drowning and suffocation or may drive off pests and kill soil inhabiting species. For example, in pea, the incidence of *Chrotogonus trachypterus, Ophiomyia phaseoli*, *Caliothrips indicus* and *Cofena spectra* decreased with irrigation water but at the same time the incidence of leaf miner, *C. horticola* increases (Parsad *et al.*,1987b). The application of straw mulch and trickle irrigation in potato in USA is reported to reduce the larval population of the Colorado potato beetle *L. decemlineata* in the first generation. This also reduces the proportion of stems having more than 50% defoliation as compared to unirrigated and unmulched plots (Stoner, 1993).

## Intercropping

One of the features of modern agriculture is an everincreasing tendency towards the use of monocultures. These have allowed agriculture to become intensified and more easily manageable. However, the crops grown in a intercropping system will improve soil fertility and availability of alternative sources of nutrients as well as reducing the incidence of insect pest. Intercropping seems to have three main effects on insects which result in lowest pest numbers (van Emden, 1989).

Intercropping may reduce pest damage by attracting the pest to a less valuable crop or the one where the pest is less serious for some reason, e.g. intercropping of potato with chillies (capsicum), onions and peas reduces the infestation of potato tuber moth attack compared to potato grown alone (Lal, 1991). Intercropping of pea with non-host plants like cereals (barley) and oilseed (linseed) brings about considerable suppression of most of the insect-pests in early maturing variety (Arkel) through changed canopy and resultant changes in microclimate. The pest problems further accentuates when chickpea is sown as an intercrop (Prasad *et al.*, 1987 a). According to Gold *et al.,* (1989), in Columbia (USA), intercropping of casava with cowpea (*Vigna unguiculata*) resulted in lower number of aleyroloids, *Aleurotrachelus socialis* and *Trialeurodes variabilis* than monoculture casava and these differences persisted upto 6 months after cowpea harvest.

The host plant finding behaviour of insects may be disrupted by the close juxtaposition of two important species. Several crop pests such as cabbage butterfly, *Pieris brassicae* (L) and cabbage aphid *Brevicoryne brassicae* (L) are influenced by the crop background in their colonization behaviour as inter-cropping removes the contrast between seedlings and bare soil in the same way as dense planting.

Mustard intercropped with cabbage has been reported to be successful for the management of major lepidopterous pets on cabbage. Planting of 15 rows of cabbage followed by a row of mustard reduced the incidence of the diamondback moth, *Plutella xylostella* (L) and the cabbage webwebber, *Crocidolomia binotalis Zeller* (Srinivasan and Moorthy, 1991a; Khan *et al.*, 1991), Intercropping of tomato with cabbage caused maximum reduction of P. *xylostella* and C. *Binotalis* larvae on cabbage. This reduction in incidence can be attributed to possible release of volatile substances from tomato plants which may inhibit oviposition by incoming moths (Srinivasan and Moorthy, 1993). Cabbage intercropped with carrot was also reported to have lower number of *P. xylostella* on cabbage than in monoculture (Varela and Guharay, 1988). Fewer alates of *B. brassicae* were encountered on cabbage intercropped with dill (*Anethum gravesleps*) than when grown alone (Kenny and Chapman, (1988). In Netherlands, infestations of cabbage by *Mamestra brassicae*, *Delia radicum* and *Thrips tabaci* were reduced by intercropping with white clover or subterranean clover (Theunissen *et al.*,1992).

Intercropping may also increase the impact of natural enemies because one of the intercropped plant may provide a honey or nectar source to attract adults of natural enemies for feeding and at the same time may provide shelter and humid conditions near the ground level to the predators living on the ground. According to Weich and Wnak (1991), intercropping of cabbage with French beans and white clover resulted in increase in beneficial insects by 32.4% whereas populations of *B. brassicae* and *Myzus persicae* (Sulzer) were reduced in both the cropping system.

**Sanitation**

Eradication of harmful weed hosts, timely destruction of crop residues, removal of scrub or shelter in which insects harbour, involve huge works solely for the control of insect pests, unlike other cultural control methods which relate to the removal of crop residues from fields. This is one of the basic method to eliminate pest over wintering sites and reduce the spread of infestation. For example, the heaps of creepers of cucurbits give protection to red pumpkin beetles during winter. So, maintenance of cleanliness in the field, helps in checking the pest infestation.

In USA, autumn burning or careful removal of dormant asparagus fein with or without subsequent spring tillage inhibited overwintering of aphid, *Brachycolus asparagi*. (Halfhill *et al.*,1985; Coceano, 1989). Destruction of weed hosts near hop plantation has been advised for the management of black caterpillar, *Vanessa cashmirensis* (Sood and Kashyap, 1997). Burning of weeds at the edge of onion fields resulted in reduced damage by stem borer, *Hydracia mongoliensis* (Iinuma and Saito, 1988).

Some pests, being inside the plant tissue, produce prominent symptoms on their hosts. However, these can be prevented from further multiplication and spread by removing and destroying them. For fruit fly control in cucurbits, guava, peach, and that of brinjal shoot and fruit borer, *Lucinodes orbonalis* should be frequently examined and infested plant parts should be collected and destroyed by burying deep in the soil. Fruit fly infested material should be buried at least 60 cm deep so that fly emergence could be checked (Mann.,1990). In Cuba, *Apate monachus* attacks citrus plants in nursery and its eradication is achieved by burning the infested plants (Martinez and Tamajo, 1988). Burning of dried pomegranate fruits resulted in killing of overwintering stages of insects and provide protection to trees (Koshkarova and Zhigarevich, 1988). Heaps of pruned branches of grapes should be destroyed so that they do not attract the ovipositing females of *Amphicerus bimaculatus* (Campolin *et al*., 1989). Keeping the mango orchards clean and avoiding overcrowding of trees helps in checking mango hoppers attack (Batra and Khangura, 1990). Good orchard sanitation, removal of weeds and destruction of ants protecting the harmful sap sucking insects have been found useful in reducing pest infestation in citrus orchards (Raheja and Tewari. 1996).

Collection and destruction of fallen plum and apricot fruits and debris harbouring cocooned larvae of plum fruit moth, *Cydia* funebrana Trisk reduced the inoculum of pest for the ensuing year (Bhalla and Gupta, 1993).

Clean cultivation in mango orchards in north eastern regions of India helps in reducing fruit infestation by mango pulp weevil, *Sternochaetus frigidus* from 85 to 63 percent (Bhattacharya and Khound, 1995). Clean cultivation and sanitation prevent the occurrence of flea beetle, *Nodostoma subcastetum*, *N. viridipennis* and *Colaspis hypochlora* in banana (Tondon, 1993).

## Pruning

Pruning of dead branches, scrapping of unhealthy stems, patching-up wounds with clay and tar and removal of bark keep away the insects which hibernate in such places. Mild pruning of peach trees and burning of pruned branches in Ludhiana resulted in reduction of the carry over of flat headed peach tree borer insect, *Sphenoptera dadkhani.* When the branches were left in field, there was only 35.5% reduction in infestation level. In a heavily pruned orchard this reduction was 61.1 percent when the pruned branches were burnt and 43.2% when left in the orchard (Chaudhary *et al.*, 1993).

Pruning of dry wood from citrus trees during December-January helps in checking citrus decline due to citrus psylla, *Diaphorina citri* Kuwayama. The pest causes drying up of the adjacent branches because certain toxins are injected into the tree while feeding (Batra and Khangura, 1990). Pruning of infested branches of apple also results in reduction of *Paropta paradoxa* (Lepidoptera: Cossidae) infestation by 7.7 -10.3% in Egypt (Kinawy *et al.*, 1991). Removal of thick bark and injured shoots of peach and apple resulted in lesser incidence of oriental fruit moth, *Grapholita molesta* in Northern China (Zhao *et al.*, 1989).

Prunning of current vegetative growth which usually carries eggs during December has been suggested as an important strategy for the effective management of peach leaf curling aphid, *Brachycaudus helichrysi* Kaltenbach by Gupta *et al.* (1988). Proper training and pruning of stone fruit trees to avoid direct exposure of the stem to sun and covering the bark with craft paper or muslin cloth provide some protection against flat headed stem borer, (Bhalla and Gupta, 1993).

In Rajasthan, pruning of the branches of ber (*Zizyphus mauritiana* Lark.) reduced the intensity of attack of gall forming mites, *Laryacarus transitans* (Ewing.) and *Eriophycas cernuus* (Sharma, 1992). Singh *et al.* (1975) recommended pruning of gall bearing shoots in mango as an effective tool to control mango shoot gall psyllid, *Apsylla cistellata.*

Grape vines severely infested and damaged by the grapes vine bud mite, *Colomerus vitis* should be pruned back to 6 buds for one season, encouraging the healthier buds at nodes 5 and 6 to produce the next seasons growth while reducing the yield loss by *C. vitis* from 56 to 35% (Denhill, 1991). Pruning of affected parts have been found to be very effective in controlling coccids attack on tropical fruits (Tandon, 1993). Removal of infested mulberry branches during winter reduced the population of *Baris deplanta* by 70.6–83.3 per cent (Zen *et al.*, 1990).

### Trap crops

Smaller plantings made earlier than the main planting may effectively divert the insect attack from the crop at risk by using more attractive food source. For the system to succeed the pest must have a narrow host range and the trap crop selected should not be different from the main crop. Trap crops must obviously be attractive to the target pest and tolerant to heavy attack.

The major benefit of trap cropping is that insecticides are seldom required to be used on main crop and this also enhances the natural control of pests e.g. when maize is used as a trap crop in cucurbits, *Bactocera cucurbitae* Cog., spraying only maize will protect cucurbit crop from fruit fly attack (Anonymous, 1988).

Use of African tall marigold (cv. Golden Age) as a trap crop for the management of the tomato fruit borer, *Helicoverpa armigera* (Hubner) was exploited in Banglore (Srinivasan and Moorthy, 1993) as was mustard exploited for major cabbage pests, *P. xylostella*, *C. binotalis*, *H. armigera*, *B. brassicae* and *L. erysimi* (Srinivasan and Moorthy, 1991a, b). Mustard (*Bassica hirta*) sown one week after broccoli transplanting showed no reduction of broccoli yield and reduced the number of *B. brassicae* by increased predation by syrphid larvae (Kloen and Altieri, 1990).

In American samoa planting of *Coloeus blumei* as a trap crop in or around *Colocasia esculentum* plantation controlled *Spodoptera litura* (Fatuesi *et al.*, 1991. The infestation of fruit fly, *B. cucurbitae* on melon was reduced in Pakistan by using squash (*Cucurbita pepo* var. *melopepa*) as a trap crop.

### Mulching

Mulches have their origin in early agriculture where covering soil and plants with organic or inorganic matter was used to form a protection barrier against frost or warm soil to improve plant growth. Mulches, both living or non-living, have also been used as means of controlling insect pests as they may encourage natural enemies by providing shelter and a favourable micro-environment. More recently the development of mulches has taken a very significant step forward with the use of transparent polyethylene film for solar heating of soil.

In citrus, the incidence of mites, *Phyllocoptruta oleivora, Brevipalpus phoenicis* and laprosis virus was reduced with green cover of *Ageratum conyzoids* and *Eupatorium paniciflorum* (Gravena *et al.*, 1993). In Nigeria, mulching with grass in tomato was found to be more effective than maize or sorghum stalks for the control of tomato fruit borer, *H.armigera* (Uvah *et al.*,1991).

Mulching of cauliflower plots in New Zealand resulted in lesser colonization of *B. brassicae* in mulched plots as compared to control plots (Lovei and Bycroft, 1992). Straw mulching reduced the larval incidence of *L. decemlineata* on potato during first generation (Stoner, 1993).

In Hungry, transparent blue or black plastic foil mulches were used between rows of cucumber grown for seed. The incidence of viral inoculum in the plants was reduced by 70% in plots mulched with transparent foil and by 77% with blue plastic foil as compared with non-mulched plots. The use of the black plastic mulch did not reduce the virus incidence (Basky, 1984).

Populations of chrysomelid, *Phyllotreta cruciferae* and aphid, *B. brassicae* were much lower on cabbage grown inter planted with living mulches (*Agrotis palutri*, *Fertuca rubra* and *Trifolium repens*) than on cabbage in bare ground (Andow *et al*., 1986). In Bangalore, use of nylon net for raising tomato nursery followed by using aluminium surfaced polythylene mulch drastically reduced the incidence of TLCV and white fly population (Singh, 1997).

### Tillage

The type and timing of tillage can markedly influence the soil environment and affects the survival of insect-pests or their natural enemies. Many insects live or hibe in suitable temperature and humidity conditions relatively near the soil surface. These conditions can be disturbed by ploughing. Insect pests are affected either indirectly or by the creation of inhospitable conditions or by exposing the insects to their natural enemies or directly by physical damage inflicted during the actual tillage process (Stinner and House, 1990). Many of these insects will be eaten by birds, other pupae and eggs may be buried by ploughing to a depth from which they fail to reach the soil surface after emergence. Deep ploughing for instance can be applied as an indirect measure to bury or expose the pupae or diapausing stages of insect pests such as *Helicoverpa* and *Spodoptera* species (El Amin and Ahmed, 1991). Deep ploughing was also found effective in managing cut worms population in vegetables in Tehran (Iran) (Nikkhoo and Moiihi, 1993).

Raking up and hoeing of soil around melon plants, mango and other fruit trees serve to destroy pupae of fruit flies. For the control of fruit flies hoeing with disc against pupating larvae is effective (Mann, 1989). When the soil in a mango orchard in UP (India) was raked periodically with various combinations of timing of irrigation, raking 4 times (May, June, August, October) afforded best control of egg hatching of mango mealy bug, *Drosicha mangiferae* (Chandra *et al*.,1990). Digging or ploughing repeatedly around the trees during summer kill the eggs of mango mealy bug, *Drosicha mangiferae* by exposing them to sun heat, ants, birds etc. (Batra and Khangura, 1990). Tandon (1993) recommended summer ploughing to control mealy bug attack in Jack fruit since eggs are exposed to natural enemies and high temperature.

Deep ploughing has been recommended as a method of management of defoliating and fruit eating beetles in temperate fruits by Bhalla and Gupta (1993) as it exposes the white grubs to predators and adverse climatic conditions.

### Fertilizers

Excessive use of the organic nitrogenous fertilizers generally created congenial conditions for many insect-pests. On the other hand, application of potash and sometimes phosphorus also either singly or in combination with nitrogen results in lower incidence of many insect-pests. High nitrogen and low potash and phosphorus application in brinjal resulted in higher infestation of *L. orbonalis* and *A. bigutulla bigutulla* at Ludhiana (Singh and Brar, 1990). For effective management of citrus black/white flies excessive application of nitrogen should be avoided (Tandon, 1993). Similarly, a reduction in the incidence of a number of insect-pests including *Empoasca kerri*, *Autographa nigrisigna* and *S. litura* on cowpea, aphids and thrips in chilies following increased application of K, either singly or in combination with P, has been reported (Dhaliwal and Arora, 1993).

### Harvesting practices

Harvesting practices involving crop maturity, time of harvest or cutting practices can be of immense importance in reducing the damage caused by a number of insect pests. Early harvest can often reduce losses from pest damage and in some cases optimum planting time to avoid pest attack may also influence optimum harvest time. Early harvest means the removal of crop from the field, before it has reached maximum yield i.e. has fully ripened or dried. In other words, some loss in yield or additional expense to prepare the crop for storage or market may have to be sacrificed to obtain an acceptable quality of product free of insect damage.

For example, early harvest of potato has been useful against the potato tuber moth, *P. operculella*, because longer the tubers are left in the field, greater is the damage. Timely harvesting also applies to brassica attacked by the cabbage looper, *Trichoplusia ni* and the small white butterfly, *Pieris rapae* (Coaker, 1987).

In Italy, early harvesting of olive is recommended to reduce damage by *Bactocera oleae* (Delrio, 1995). Eagle and Barnes (1983) suggested that in California (USA), almonds should be harvested early, at 95–100% hull split of nuts on the tree, to minimise the infestation of navel organge worm, *Amyelius trapsitella*. The most effective control methods for cocoa pod borer, *Conopomorpha cramerella* are frequent harvesting of ripe pods and husks to prevent pupation in field (Mumford and Ho, 1988).

## MECHANICAL CONTROL

Mechanical control involves the use of manual operations and machanical devices. These methods are quite efficient under some situations, but may require repeated use. It is advantageous over other methods since it require home labour or low cost equipment and also does not pose residue problem or health hazards. The disadvantage of this method is that they cannot be practised on commercial scale and destroy the pest when adequate damage by pest has already been done. Pest control by this method is based on mechanical destruction and exclusion.

### Hand picking

Hand picking of egg masses, inactive adults, large sized larvae in orchards, small farm and kitchen garden is sometimes practicable when their population is low during initial stages of infestation. Collection and killing of pre pupae and pupal cocoons in apple orchards during April-June forms a part of intensive strategic pest suppression programme against *Cydia pomonella* (Bhalla and Gupta, 1993).

Similarly infestation of tent hairy caterpillar, *Malacosoma indica* can be checked to some extent by collection and destruction of conspicuous egg clusters and mechanical destruction of silken ducts. Tandon (1993) reported handpicking of egg masses, larvae and their subsequent destruction to be a effective insect pest management component in case of bag worms, *Kaphene cuprae, Acanthopsyche* minima, slug caterpillars *Belippa laleana*, *Contheyla returda*, leaf roller, *Erionota thrax* in banana; lemon butterfly, *Papilio demoleus*, *P. helemus*, on citrus; slug caterpillars, *L. lepide* on citrus, fig, mango and pomegranate.

In view of high risk to biotic agents by insecticides, it is advisable to collect egg masses and larvae of Indian gypsy moth, *Lymantria obfucata* in apple orchards which get trapped in gunny burlaps tied on the trunk near the ground level (Bhalla and Gupta, 1993).

Some other insect pests have also been reported to be well manipulated by collection and destruction of egg masses and larvae like cabbage butterfly, *P. brassicae* and cabbage stem borer, *Helulla undalis* on cole crops (Panwar, 1995), hop caterpillar, *Vanessa cashmirensa* on hops (Sood and Kashyap, 1997), hairy caterpillars, tobacco caterpillar and brinjal hadda on different vegetable crops(Singh, 1984).

**Exclusion by screens and barriers**

This technique involves the use of devices which prevent insects from having assess to their food materials. A number of sticky and slippery bands at the basal end of tree trunk are utilized for preventing mealy bug nymphs from climbing trees. Banding with coaltar and grease (2 : 1) was effective for two weeks. Singh (1984) recommended alkathene paper band (200 gauge, 30 cm) combined with grease to plus the space between the trunk and band on lower end. The alkathene bands were found to be most effective in preventing the mango mealy bug nymphs reaching the tree canopies (Chandra *et al.*,1991). A combination of alkathene and grease was found effective by Tandon and Lal (1982).

In Queensland, application of sticky bands on the trunks of custard apple trees reduced the number of ants, *Rheidole megaciphala* in the trees and resulted in lower level of infestation of mealy bug (*Planococus citri*), (Murray, 1982).

The presence of a close fittings vinyl wrap on young (45 cm stem dia.) flowering dogwood was found to have a significant effect on the incidence of dogwood borer (*Synanthodus scitula*) attack (Owen *et al.*,1991) . Use of tylon coated fabric strips as barriers around concord grape trunks significantly reduced the bud injury by cutworms (Wright and Cone, 1983).

Allegro (1992) obtained effective results in controlling principal xylophagous insects of poplar using barriers made of Alfoglan D60 applied to the basal meter of the stem. Sticky barrier bands made of duct tape and tangle foot reduced the late instar larvae population densities of gypsy moth in oak (Thorpe *et al.*,1993).

Non woven materials made from strands of heat fused polypropylene can be used to protect plants from insects and viruses (Gregorie, 1994). In Florida (USA) the yield of squash increased dramatically with the use of covers (super banded polythylene) and these excluded almost all insects. (Webb and Linda, 1992). Millar and Isman (1988) also used super banded polyester row cover in cauliflower for the protection against anthomyiid, *Delia radicum*.

Against some insect pets such as *D. radicum* on radish and chinese cabbage, *Psila rosae* on carrot and *M. brassicae* on chinese cabbage, nets are quite effective (Haselr and Konrad, 1988). McKinlay (1990) also reported that light weight fabric cover over cabbage field significantly reduced damage by lepidopterous larvae (*P. brassicae, M. brassicae, P. xylostella*) as well as that caused by larvae of *D. radicum*.

Fruits wrapped in paper bags or cloth or straw remain protected to the extent of 95 per cent against fruit flies and other insect pest. In India, bagging of pomegranate fruits with polythene or muslin cloth bags reduced the incidence of fruit borer *(Virachola isocrate*) (Shukla and Prasad, 1983). In California (USA), covering of apple fruits with brown paper bags resulted in fewer fruits being damaged by codling moth, *Cydia pomonella* and improved the fruit quality (Bentley and Viveros, 1992). Similarly, melon fruit fly, *Bactrocera cucurbitae* Coq. infestation could be checked to a greater extent when bitter gourd fruits (2–3 cm) were covered with paper bags (Fang and Chang, 1987; Fang 1989).

## REGULATORY CONTROL

One of the methods of crop protection is to exclude the pests from entering the area in which the host plants are growing to minimise the chance of attack due to lack of exposure. This method of exclusion of the pests is enforced through certain legal measures commonly known as quarantine or regulatory measures. Plant quarantine is an endeavour, by legislative measures and through the policing of these measures to exclude from a defined geographical area such as countries, states or region, pests or diseases of plants exotic to that particular geographical area and assessed to be potentially serious to agricultural, horticultural or forestry industry (Roher, 1991). The term quarantine has a traditional rather than factual background. It simply means 'forty' i.e. a long enough period which allows for the latent cases of infestations to develop and be detected before any person/plant or plant material is allowed to land (Wadhi, 1982). These regulatory measures aims at preventing the introduction and spread of foreign pests and diseases into the country and through legally enforced procedures plant parasites are intercepted at the points of entry into geographical and political subdivisions.

In earlier days, there were no restrictions on the transportation of plants and animals from one country to another. The regulatory measures and acts are of comparatively recent origin. The adoption of such steps by different countries of the world has arisen out of the fact that extensive damages have been caused not by indigenous pests, but by exotic ones which have been introduced alongwith plants, plant parts or seeds in the normal channel of trade or individual interest. Instances may be cited of the introduction of grape phylloxera, *Phylloxera vitifoliae* from USA to France which caused destruction of French vineyards and Mexican boll weevil (*Anthonomus grandis*) whose original home was in Mexico or Central America, got entered in USA during the year, 1892

In India, the San Jose scale, *Quadraspidiotus perniciosus* is a pest of apple introduced about 77 years ago now causing concern to apple growers of Himachal Pradesh, Jammu & Kashmir. Potato tuber moth, *Phthorimaeae operculella* which entered from Italy in 1900 is an established field and store-house pest of potato all over the country. Woolly apple aphid, *Eriosoma lanigerum* an introduced serious pest of apple, introduction of golden nematode of potato, *Heterodera rostochiensis* into India, are examples showing that many destructive pests have entered into this country and have established themselves causing extensive damage. This clearly indicates that the process of pest dissemination still continues, thus justifying the need for effective plant quarantine.

In order to restrict the introduction of exotic pests into our country. The movement of plant/plant materials is regulated by prescribing the relevant rules/regulations. The first quarantine act came into operation in USA during 1905, followed by Australia in 1909. The first plant quarantine legislation in the world was enacted in Netherlands in 1977 prohibiting the import of coffee plants and seeds from Ceylon (Kumar *et al.*,1996). In India the earliest record of plant quarantine work dates back to 1907, when the Govt. of India recognised the danger of introducing serious pests with the importation of cotton seed and in consequence ordered the fumigation of all such seeds at the port of entry (Anonymous, 1910). However, it was only in 1914 that the Govt. of India passed the 'Destructive Insect and Pests Act' (ii of 1914) (DIP) which is commonly referred to as pests act. Plant quarantine regulations are needed to:

i. prevent the introduction of plant pests/diseases not already known to occur/widespread in the country.
ii. control pests and diseases already introduced in the country by restricting their further spread and by endeavouring to eradicate them and
iii. provide facilities and services for the import/export of plants/plant materials.

The plants/plant materials are imported into the country as per regulations enshrined under the plants, fruits and seeds (regulation of import into India) Order, 1989, (PFS Order) issued under Destructive Insects and Pests Act, 1914 to prevent the introduction of exotic pests/diseases/weeds into the country. Plant quarantine comprises two aspects, namely domestic quarantine and foreign quarantine.

**Domestic quarantine**

Inspite of best effort a new pest may gain entry into the country. So domestic quarantine aims at stopping further spread of exotic pests/diseases already introduced into the country by regulating the inter state movement of plants/plant materials. Some of the important existing domestic quarantine measures already in force in the country to prevent pests are given below (Kumar *et al.*,1996; Wadhi, 1982):

i. To prevent the spread of codling moth, *Carpocapsa pomenella*, movement of apple and walnut plants including the fruits are prohibited from Ladakh and Kargil districts of Jammu & Kashmir.
ii. To prevent the spread of San Jose scale, *Quadraspidiotus prenicicosus,* transport of host materials is prohibited unless accompanied with certificate of free of the pest from the Director of Agriculture.
iii. To prevent the dispersal of fluted scale, *Icerya purchasi*, the movement of materials from Karnataka, Kerala and Tamil Nadu (infected area) to rest of the country is prohibited.
iv. To prevent the spread of potato cyst nematode, *Globodera rostochiensis* and *G. pallida*, movement of potato from Tamil Nadu and UT is prohibited except for table potato accompanied with special permit issued by plant protection advisor to Govt. of India.

**Foreign quarantine**

The foreign quarantine deals with the movement of plants/plant materials between the countries in order to reduce the risk of importation of exotic pests and diseases into the country. In India, the goal of foreign quarantine is realized by observance of PFS order, 1989 (Anonymous, 1989).

The foreign quarantine has helped to keep out from our country such dangerous pests as Colorado potato beetle, *Leptinotarsa decmlineata*, the mediteranean fruit fly, *Ceratatus capitata* and cotton boll weevil, *Anthonomus grandis.*

**Plant quarantine procedures**

For enforcing plant quarantine regulations, a number of procedures are adopted. Some of the more important and generally applied procedures are:

*Embargo*: Total ban on imports is applied when there is a serious danger of introducing a devastating pest or disease.

*Unrestricted entry*: This procedure is for propagation purposes, i.e. meant only for consumption as food, medicine or in industry the import is allowed as free entry subject to occasional checks.

*Inspection*: This is indespensable procedure in plant quarantine despite its limitations. It serves three useful purposes:

i. to verify if the consignment contains prohibited/unrestricted materials or the contents are permissible under specified conditions etc.

ii. to verify if the consignment has been shipped in compliance with the regulations and if it is accompanied with proper documents and

iii. to determine if the consignment is free from pests and diseases i.e. to intercept plants of unknown or questionable economic importance that are not present or widely distributed with in the country.

*Detention:* This procedure deals with the import for quarantine observations. This is particularly useful for propagating materials which are suspected to deep inside the tissue and hence their presence cannot be brought out at the time of inspection. The detained material is held under quarantine isolation.

*Treatment:* The import of plants and plant materials is often subject to fumigation as an additional measure of safety. This treatment may be given before export or immediately on import.

*Compulsory Eradication:* This function relates to the isolation of the introduced pests and their elimination through rigidly applied control measures.

Through the implementation of quarantine regulations the management of insect pests becomes easy as the threat of introduction of new pests is considerably reduced.

Lal *et al.* (1988) reported the interception of two dead specimens of cotton boll worm, *Anthonomus grandis* in postal consignments of cotton seed received in India from USA. The bostrichid, *Prostephanus truncatus* has been intercepted three times in India. Two of these interceptions were from consignments of maize seeds from Mexico and other of rice from Phillipines (Verma *et al.*,1988). Twenty four species of insects, 7 of which were of plant quarantine importance, were intercepted in logs imported to India through Madras from far East in 1990. The important species were *Cerobatus sexsulcatus, Heterobostryelus qequalis*, *Opisthenoxys ochraceus*, *Plattypus curtus* and *Xyloterechus bravicornis* (each from Malaysia) and *Xyloborus percforanis* from Malaysia, Papua, new Guinea and Vietnam) (Krishnasawmy *et al.*,1991), Ghodeswar *et al.*,1992) also reported about interception of 40 species of Coleoptera on logs being imported into Karnataka.

Quarantine intrerceptions of *Frankliniella bondri*, *F. intosa*, *F. lilovora*, *F. occidentalis* and *F. schultzeri* in Netherlands were reported by Vierbergen (1992 b). Mango stone weevil, *Steronchetus mangiferae* (F.) is a blemish on Indian mango export because of countries like USA have put an import embargo depriving the country of much needed foreign exchange. Recently, mango stones of cv. Bappakai, Kurukan, starch and EC 95862 meant for export to Oman were found to have hidden infestation (2 -44%) of stone weevil (Kapoor *et al.*,1997). Significant quarantine interceptions at the Lincoln Plant Protection Centre in Newzealand were made which include the vilanid, *Abasverus advena* in ginger root from Australia, the pyralid, *Ephestia kuehniellia* in Turkish figs, larvae of *Dacus xanthodes* in melons from Tanga, were found (Anonymous, 1986). Satoh *et al.* (1986) found that a total of 1013 individuals of *D. dorsalis* from mango, guava and 1 of *D. ratifrons* from litchi were found in the material imported from Japan.

During check inspections of imported plant materials by the plant health and seeds inspectrate of the UK Ministry of Agriculture, Fisheries and Food, a number of insect pests were intercepted, but the insects of particular interest were *L. decemlineata*, *L. trifoli*, *H. armigera*, *P. operculella* (Seymour *et al.*,1985). Browne (1985), presented data on 20 species of scolytids and platypodids intercepted on imported timber as pests in Japan and described further 8 new species. American moth, *Parmyelois transitella* was found in various lots of walnuts imported into Italy from California in 1987 (Trematerra, 1988). Several adults of *Daphorina citri*, a pest of citrus were received that had been intercepted at an airport in France in fruit originating from Handrus (Buckhardt & Martinez, 1989).

Gonazalez and Barra (1985) reported that during last one decade the presence of crickets, *Microgrylus pallipes* has resulted in the rejection at USA ports of consignments of pea, apple and melon exported from Chile. In Chile, the pomecous fruits are sent for quarantine inspection for infestation with San Jose scale, *Q. perniciciosus*, before export to Europe (Lucke,1990). Hot water quarantine treatment is given to florida mangoes to destroy all the instars of carribean fruit fly, *Anastrapha suspensa* (Sharp *et al.*, 1989). In Australia control of arthropods on cut flowers, plants and parts of plants imported into the country is achieved by quarantine fumigation or insecticidal dips (Turner, 1985).

During past decade, India has witnessed the inadvertment introduction of at least three devastating pests the coffee berry borer, *Hypothenemus hampei*, the serpentine leaf miner, *Liriomyza trifoli* and the spiralling white fly, *Aleryrodicus dispersus*. There has also been intra country spread of the banana psudostam weevil, *Odiopurus longicollis* from North Eastern States to Kerala, Tamil Nadu and Karnataka (Ranjith, 1997).

There are 26 plant quarantine and fumigation stations located at 10 international airports, 9 seaports and 7 land frontiers. But for the sake of safety, the import is allowed at present only through plant quarantine and fumigation stations based at Amritsar, Bombay, Calcutta, Madras and New Delhi.

The enhancement of quarantine facilities at major ports and airports of exit/entry is urgently required as an effective implementation of plant quarantine which will help in strengthening the pest management programme in India.

## REFERENCES

Ahmadabadi, A. N. and Hessan, A. R. 1991. Effect of time of planting and harvesting on yield and infestations level of potato tuber moth in Aula Variety. *Seed and Plant* 7 (1): 7.

Allegro, G. 1992. Mechanical control of the principal xylophagous insects of poplar using trunk barrier. *Informatore Agrari* **46** (11): 91–95.

Andow, D, A.; Nicholson, A. G.; Wien, H. C. and Wilson, H. R. 1986. Insect population on cabbage grown with living mulches. *Environmental Entomology,* **15** (2): 293–299.

Anonymous. 1910. *Correspondance regarding measures to prevent the introduction of insect pests and fungoid diseases into India through importation of foreign plants.* Government Printing, Calcutta. pp. 49.

Anonymous. 1986. Lincoln plant protection centre. *Weta,* **8** (2): 34–35.

Anonymous. 1988. *Package of practices for fruit crops.* Punjab Agricultural University, Ludhiana.

Anonymous. 1989. *Plants, fruits and seeds (Regulation of import into India) order.* 1989. Gazette of India No.**706**, Part II Section 3 -sub Section (ii).

Bali, G. and Querishi, Q. G. 1997. Effect of date of sowing on stem fly infestation in pea. *First National Symposium on Pest Management in Horticultural Crops: Environmental Implications and Thrusts* Oct 15–17 Banglore. pp. 44.

Bardner, H. M.; Edwards, C. A.; Arnolds, M. K. and Regerson J P. 1971. The symptoms of attack by swede midge (*Contarinia nastwrtii*) and effects of yield of swede. *Entomologia Experimentalis et Applicata,* **14**: 223–233.

Basky, Z. 1984. Effect of reflective mulches on virus incidence in seed cucumbers. *Protection Ecology,* **6** (1): 57–61.

Batra, R. C. and Khangura, J. S. 1990. Bioecology and management of key pests of fruit trees. *In: Proceedings Summer Institute on Key Insect Pests of India Their Bioecology with Special Reference to Integrated Pest Management*. Punjab Agricultural University, Ludhiana. pp. 258–272.

Bentley, W. J. and Viveros, M. 1992. Brown-bagging Granny Smith apples on trees stops codling moth damage. *California Agriculture,* **46**(4): 30–32.

Bhalla, O. P. and Gupta, P. R. 1993. Insect pests of temperate fruits. *In Advances in Horticulture* Vol.3—*Fruit Crops*: Part 3. K.L. Chadha and O.P. Pareek (eds). Malhotra Publishing House, New Delhi. pp. 1557–1589.

Bhattarcharya, B. and Khound, J. N. 1995. Integrated management of mango pulp weevil, *Sternochetus frigidus* (Fbr.) (Coleoptera: Curculionidae). *National Symposium on Integrated Pest Management in Agriculture* December 29–30, 1995, Nagpur. p5.

Browne, F. G. 1985. Bark beetles and ambrosia beetles intercepted at Japanese ports with descriptions of new species. *Kontyu,* **53**(1): 190–198.

Buckhardt, D. and Martinez, M. 1989. Note on the presence in Honduras of an important enemy of citrus, *Diaphorina citri* Kwiayama (Homoptera: Psyllidae). *Bulletin de la Societe Entomologique de France,* **94**(1–2): 65–66.

Campoline, M.; Corazza, F. and Lunghini, D. 1989. Damage by *Amphicerus bimaculatus* to grapes in Lazio. *Informatore Agrario,* **45** (5): 93–98.

Chandra,, A.; Bhatti, D. P. S. and Singh, K. M. 1990. Note on the effect of soil raking and irrigation on survival and hatching of eggs of mango mealy bug, *Drosicha mangiferae. Current Agriculture,* **13** (1 & 2): 103–104.

Chandra, A.; Bhatti, D. P. S. and Singh, K. M. 1991. Mechanical control of mango mealy bug by tree trunk banding. *Indian Journal of Entomology,* **53** (1): 76–79.

Chaudhary, O. P. and Kashyap, R. K. 1992. Effect of some management practices on the incidence of insect pests and yield of egg plant (*Solanum melongena* L.) in India. *Tropical Pest Management,* **38** (4): 416–419.

Chaudhary, O. P.; Ratual, H. S. and Sandhu G S. 1993. Management of flat headed borer, *Saphenoptera dadkhani* Obenberger on peach in India. International Journal of Pest Management, **39** (1): 75–79.

Coaker, T. H. 1987. Cultural methods: The Crop. *In: Integrated pest Management*. A.J. Brown, T. H. Coaker and P.C. Jepson (eds). Academic Press, New York. pp. 69–88

Coceano, P. G. 1989. A new danger for Italian asparagus growers. *Informatore Agrario,* **45** (24): 73–74.

Delrio, B. 1995. Integrated control of olive pests. *Informatore Fitopatolegico,* **45**(12): 9–15.

Denhill, G. B. 1991. A pruning technique for saving vine yards severely infested by the grape vine bud mite *Colomerus vitis* (Pangenstecher) (Eriophyidae). *Crop Protection*, **10**(4): 310–314.

Dhaliwal, G. S. and Arora, R.1993. Changing status of Inasect -pests and their management strategies. In *Changing Scenario of Indian Agriculture.* K.S.Gill, G.S. Dhaliwal and B.S. Hansara (eds). Commonwealth Publishers, New Delhi. pp. 98–145.

Dore, T.; Barrier, C. and Carriov, Y. 1991. Management of proteinaceous peas and damage by weevils in Seine-et-Marna. *Comptes Rendus de Academia d Agriculture de France,* **77**(8): 137–146.

Eagle, C. E. and Barnes, M. M. 1983. Cultural control of naval organgeworm in almond orchards. *California Agriculture,* **37**(9/10): 19.

El Amin, E. T. M. and Ahmed, M. A. 1991. Strategies for integrated cotton pest control. in Sudan. I. Cultural and legislative measures. *Insect Science and Application,* **12** (5/6): 547–552.

Fang, M. N. 1989. Studies on using different bagging materials for controlling melon flies on bitter gourd and Sponge gourd. *Bulletin Taichung District Agriculture Station,* **25** 3–12.

Fang, M. N. and Chang, C. P. 1987. Population changes, damage of melon fly in the bitter gourd garden and control with paper bag covering method. *Plant Protection Bulletin (Taiwan),* **29** (1): 45–51.

Fatuesi, S.; Tauililli, T,. F. and Vargo, A. 1991. Cultural methods of pest control on taro (*Colocasia esculentum* Schoot.) in American Samoa. *Micropesica,* **3**: 123–127.

Ghodeswar, K. S.; Minakarishnakutty. C.; Tippana, G. M.; Varkey, K. J.; Sivagnanam, V.; Arunchalaam, A. S.; Thankam, K. and Raghu, S. 1992. The interceptions of insect pests and fungi on imported wooden logs through the ports of New Mangalore and Karwar (Karnataka) from 1989–1991. *Plant Protection Bulletin,* **44**(3): 1–5.

Gold, C. S.; Altieri, M. A. and Bellotti, A. C. 1989. The effects of intercropping and mixed varieties of predators and parasitoids of casava whiteflies (Hemiptera: Aleyrodidae) in Columbia. *Bulletin of Entomological Research,* **79** (1): 115–121.

Gonazalez, R. H. and Barra, M. A. 1985. Crickets of quarantine importance. Revista Fruticola, **5**(3): 99–100.

Gravena, S.; Coletii, A. and Yamamato, P. T. 1993. Influence of green cover with *Ageratum conyzoids* and *Eupatorium pauciflorum* on predatory and phytopagous mites in citrus. *Bulletin OLLB/SROP*, **16** (7): 104/114.

Gregoire, P. 1994. Non-woven materials and protection from insects and viruses. *Review of Agricultural Entomology,* **82**(1): 621.

Gupta, B. M. and Yadav, C. P. S. 1989. Incidence of aphid, *Myzus persicae* (Sulz.) on cumin in relation to sowing date variation. *Indian Journal of Entomology,* **51** (1): 60–63.

Gupta, P. R.; Thakur, J. R. and Dogra, G. S. 1988. Management of peach leaf curl aphid, *Brachycaudus helichrysi* (Kalt) infesting stone fruits in Himachal Pradesh. *Journal of Tree Science,* **7**: 26–30.

Halfhill, J. E.; Gefre, J. A. and Tamaki, G. 1985. Cultural practices inhibiting overwintering survival of *Brachyoclus asparagi* (Homoptera: Aphididae). *Journal of Economic Entomology,* **77**(4): 954–956.

Haselr, A. and Konard, P. 1988. Pest management with nets on vegetable crops. *Pflanzenschutzberichte,* **49**(1): 1–8.

Hyder, M. F. and Sherif, L. S. 1990. Ecological aspects and developing method of onion pest control. *Bulletin Entomological Society of Egypt,* **16**: 119–126.

Iinuma, A. and Saito, O. 1988. Additional notes on damages by the onion stem borer, *Hydracia mongolensis* Urbahn and its control. *Annual Report Society of Plant Protection North Japan,* **39**: 233–234.

Jepson, P.C. and Green, R.E. 1983. Prospectus for improving control strategies for sugar beet pests in England. *Advances in Applied Biology*. T.H. Coaker (ed.). Academic Press, London. **7**: 175–242

Kapoor, M. L.; Verma, B.R. and Bhalla, S. 1997. Interception of mango stone weevil, *Sternochaetus mangiferae* (F.) (Coleptera: Curculio nidae)- A quarantine view point. *In: First National Symposium in Pest Management in Horticultural Crops: Environmental Implications and Thrusts*. October 15–17, Bangalore. pp. 9.

Kenny, G.J. and Chapman, R.B. 1988. Effects of an intercrop on the insect pests, yield and quality of cabbage. *Newzland Journal of Experimental Agriculture,* **16** (1): 67–72.

Khan, H.K.; Nagraj, G. N. and Reddy, S. V. S. 1991. Integrated pest management demonstrations in cabbage. *Plant Protection Bulletin,* **43** (3–4): 11–12.

Khan, L.; Inayatullah, C. and Manzoor-ul-Haq. 1992. Control of melon fruit fly, *Dacus cucurbitae* Coq. (Diptera: Trypetidae) on melon in Pakistan. *Tropical Pest Management,* **38** (3): 261–264.

Kinawy, M. M.; Tadros, A. W. and Abdallah, F. F. 1991. Effect of horticultural, mechanical and chemical treatments on the reduction of *Paropta paradoxa* H-

Schaeff (Lepidoptera: Cossidae) infestation in apple orchards. *Bulletin Faculty of Agriculture University of Cario,* **42** (1): 129–137.

Kloen, H. and Altieri, M. A. 1990. Effect of mustard (*Brassica hirta*) as a non crop plant on competition and insect pests in broccoli (*Brassica oleracea*) *Crop Protection,* **9** (2) : 90–96.

Kobro, S. 1991. Irrigation against cut worm. *Gartheryrket,* **81**(21): 22–23.

Koshkarova, D.D. and Zhigarevich, G.P. 1988. Agrotechnics of the protection of pomegrante against pests and diseases. *Subtropicheskie Kul'tury,* **1**: 128–131.

Krisahnasamy, N.; Muthaiyan, M.C.; Murali, R.; Kumarasamy, M. and Ahmed, M. 1991. Note on coleoptera pests intercepted during 1990 at Madras port, India. *Quaterly News. Asia and Pacific Plant Protection Comm.* **34** (3–4): 1–4.

Kumar, D.; Mavi, G.S. and Singh J. 1991. Impact of agrotechnical practices on the incidence of pea leaf miner, *Chromatomyia horticola* (Goureau) on peas. *J ournal of Insect Science,* **4** (2): 180–181.

Kumar, R.N.; Mukerji, K.G. and Upadhyay, R.K. 1996. Role of plant quarantine in integrated pest management, *In: IPIM System in Agriculture*. R.K. Upadhyay: K.G. Mukerji and Rajak R L (eds). Vol. 1 *Principles and perspectives*. Aditya Books Pvt. Ltd. New Delhi. pp. 414–429.

Lal, B.; Verma, B.R. and Wadhi, S.R. 1988. Quarantine risk in cotton seed import by post. *Indian Journal of Plant Protectio,*. **16** (2): 241–43.

Lal, L. 1991. Effect of intercropping on the incidence of potato tuber moth, *Phthorimaea operaculella* (Zeller). *Agriculture Ecosystem and Environemnt,* **36**(3–4): 185–190.

Lashomb, J. H. and Ng, Y. S. 1984. Colonization by colorado potato beetles, *Leptinotarsa decemlineat* (Say) (Coleoptera: Chrysomelidae), in rotated and non-rotated potato fields. *Environmental Entomology,* **13**: 1352–6.

Lovei, G. L. and Bycroft, B. 1992. The effect of mulch on the incidence of insect pests, natural enemies and plant damage in organically grown cauliflowers. *Proceedings 45h Newzealand Plant Protection Conference*. 11–13 August, 1992: 62–63.

Lucke. E. 1990. Phytosanitary pre-inspections of pomaceos fruits for San Jose scale (*Quadraspidiotus perniciosus* Comstock) in Chile. *Gesunde Pflahzen,* **42** (3): 97–102.

Mann, G. S. 1989. The response of guava fruit fly (*Dacus dorsalis* Hendel) larvae in different tillage practices at Ludhiana. *76th Session of the Science Congress*. Jan. 7–12, 1989, Madurai, Tamilnadu Absst. No. 278.

Mann, G. S. 1990. Bioecology and management of fruit flies. *In: Proceedings Summer Institute on Key Insect Pests of India and Their Bioecology with Special Reference to Integrated Pest Management*. PAU, Ludhiana. pp. 236–257

Martinez, D. and Tamajo, D. I. 1988. *Apate monachus* attacking citrus plants in the nursery. *Centro Agricola,* **15** (3): 92–93.

McKinlay, R. G. 1990. Pest management in vegetable brassicas without insecticides. *Monograph BCPC,* **45**: 205–208.

Millar, K. V. and Isman, N. B. 1988. The effect of a superbonded polyster row cover cauliflower yield loss caused by insects. *Canadian Entomologist,* **120** (1): 45–47.

Minoranskii, V. A. 1985. An effective method of controlling crop pests. *Zaschita Rastenii,* **2**: 21–22.

Mumford, J. D. and Ho S H. 1988. Control of cocoa pod borer (*Conopomorpha cramerella*). *Cocoa Grower's Bulletin,* **40**: 19–29.

Murray, D. A. H. 1982. Effects of sticky banding of custard apple tree trunks on ants and citrus mealy bug, *Planococcus citri* in Southeast Queenland. *Journal of Agriculture and Animal Science,* **39**(2): 141–146.

Negasi, F. and Abate, T. 1986. Studies on the control of bean fly *Ophiomyia phaseoli* (Tryon) (Diptera: Agromyzidae). *Ethopian Journal of Agricultural Sciences,* **8** (1): 47–49.

Nikkhoo, F. and Moiini, M. 1993. Study of cut worm biology in the vegetable growing areas around Tehran and determination of its control measures. *Applied Entomology and Phytopathology,* **58** (1–2): 1–3.

Owen, N. P.; Sadof, C. S. and Raupp, M. J. 1991. The effect of plastic tree wrap on borer incidence in dog-wood. *Journal of Arboriculture,* 17 (2): 29–31.

Pawar, V. P. S. 1995. *Agricultural insect pests of crops and their control*. Kalyani Publishers, Ludhiana pp. 286.

Parsad, D.; Singh, K. M.; Katiyar R N and Singh R N. 1987a. Impact of Intercropping on the plant growth, pest incidence and crop yield of pea, *Pisum sativum* L. *Indian Journal of Entomology,* **49**(2): 153–172.

Parsad, D.; Singh, K. M.; Katiyar, R. N. and Singh, R. N. 1987 b. Influence of irrigation on the crop growth, pest incidence and crop yield of pea, *Pisum sativum* L. *Indian Journal of Entomology,* **49**(3): 330–344.

Raheja, A. K. and Tiwari, G. C. 1996. *Integrated pest management in 50 years of Crop Science Research in India*. R.S. Pardoa and K.L. Chadha (eds). Indian Council of Agricultural Research, New Delhi. pp. 55–71.

Ranjith, A. M.; Rao, D. S. and Thomas, J. 1997. How effective is the quarantine? *In: First National Symposium on Pest Management in horticultural crops: Environmental Implications and Thrusts* Oct. 15–17, 1997, Bangalore. pp. 10.

Rohwer, G. G. 1991. Regulatory plant pest management. *In: Handbook of pest management in Agriculture*. D. Pimental (ed). CRC Press, USA. pp. 285–310

Roush, R. T.; How, C. W.; Ferro, D. N. and Tingey W M. 1990. Insecticide resistance in the colorado potato beetle: Influence of crop rotation and insecticide use. *Journal of Economic Entomology,* **83**(2): 315–319.

Satoh, I. Y.; Satoh, M. and Ohki, S. A. 1986. Study on the frequency of finding of fruit flies infesting the fruit imported as an baggage. *Research Bulletin of the Plant Protection Service Japan*.

Seymour, P R.; Roberts, H. and Kilby, L. J. 1985. Insects and other invertebrates found in plant material imported into England and Wales. *Reference Book Ministry of Agriculture, Fisheries and Food* UK No.442/83 iii pp. 68.

Sharma, A. 1992. Pruning provides management of gall mites in ber. *In: National Symposium Recent Advances in Integrated pest Management*, October 12–15, 1992, PAU, Ludhiana. p 26.

Sharp, J. L.; Ouye, M. T.; Hart, W.; Ingle, S.; Hallman, G.; Gould, W. and Chew, V. 1989. Immersion of Florida mangoes in hot water as a quartantine treatment for carribean fruit fly (Diptera: Tephritidae). *Journal of Economic Entomology,* **82** (1): 186–88.

Shukla, R.P. and Prasad, V. G. 1983. Comparative efficacy of various treatments for controlling pomegranate fruit borer, *Virachola isocrates* (Fabr.). *Entomon,* **8**(4): 381–383.

Singh, D. and Brar, K. S. 1990. Bioecology and management of insect pests of vegetable crops. *In: Proceedings Summer Institute on Key Insect Pests of India and Their Bioecology with special reference to Integrated Pest Mangement*, PAU, Ludhiana. pp. 274–289.

Singh, G.; Kumar, A. and Everett, T. R. 1975. Biological observations and control of *Apsylla cistellata* Buckton. *Indian Journal of Entomology,* **31** (1): 46–50.

Singh, H. 1984. *House hold and kitchen garden pests*. Kalyani Publishers, Ludhiana.

Singh, S. J. 1997. Effect of coloured polythylene mulching, coarse net and neem oil in the management of tomato leaf curl virus. *First National Symposium on Pest Management in Horticultural Crops*: *Environmental Implications and Thrusts* Oct 15–17, 1997, Bangalore. pp. 55.

Smith, R. F.; Apple, J. L. and Bottrell, D. G. 1976. *The origin of integrated pest management concepts for agricultural crops, in integrated pest management*. J.L. Apple and R.F. Smith (eds). Plenum Press, New York. pp. 1–16

Sood, A. K. and Kashyap, N. P. 1997. Bioecology and management of *Vanessa cashmirensis* Kollar on Hops in dry temperate regions of Himachal Pradesh. *First National Symposium on Pest Management in Horticultural Crops: Environmental Implications and Thursts* October 15–17, Bangalore. p67.

Sood, P. and Mehta, P. K. 1993. Influence of date of sowing on incidence of pea leaf miner, *Chromatomyia horticola* (Goureau) on pea. *New Agriculturist,* **4**(1): 92–93.

Srinivasan, K. and Moorthy, P. N. K. 1991a. Indian Mustard as a trap crop for management of major lepidopterous on cabbage. *Tropical Pest Management,* **37** (1): 26–32.

Srinivasan K and Moorthy P N K. 1991b. Mustard plants trap major cabbage pests. *Indian Farming,* **40** (11): 11–12.

Srinivasan, K. and Moorthy, P. N. K.1993. Changing scenario of pest problems in Solanaceous, Malvaceous and leguminous vegetables and their management. *In: pests and pest management in India—The changing scenario*. H.C. Sharma and M. Veerabhadra Rao, M (eds.). Plant Protection Association of India, Hyderabad. pp. 103–115,

Stinner, B. R. and House, G. J. 1990. Arthropods and other invertebrates in conservation-tillage agriculture. *Annual Review of Entomology,* **35**: 299–318.

Stoner, K. A.. 1993. Effects of straw and leaf mulches and trickle irrigation on the abundance of colorado potato beetles (Coleoptera: Chrysomelidae) on potato in connecticut. *Journal of Entomological Science,* **28**(4): 393–403.

Tandon, P. L. 1993. Insect and mite pests of tropical fruits. *In Advances in Horticulture.* K.L. Chadha and S.P. Pareek (eds). *Vol. 3—Fruit Crops*. Malhotra Publishing House, New Delhi. Part 3 pp. 1527–1555

Tandon, P. L. and Lal, B. 1982. Effectiveness of different banding materials for preventing migration of *Drosicha mangiferae*. *Indian Journal of Entomology,* **43** (2): 188–191.

Theunissen, J.; Bouij, C. J. H.; Schelling, G. and Noor, L. J. 1992. Intercropping white cabbage with clover. *Bulletin OLLB/SROP,* **15** (4): 104–114.

Thorpe, K. W.; Webb, R. E.; Ridgway, R. L.; Venables, L. and Tatman, K. M. 1993. Sticky barriers bands affects density of gypsy moth (Lep: Lymantriidae) and damage in oak canopies. *Journal of Economic Entomology,* **86** (5): 1497–1501.

Trematerra, P.1988. *Paramyelois transitells*(Walker) american moth found in Walnuts Imported from California. *Informator Fitopatologico,* **38** (12) 51–55.

Turner, J. W. 1985. Disinfestation of imported flowers and plants. *Proceedings of the 4th Australian Applied Entomological Research Conference Adelaide* 24–28 Sept.,1984.

Uvah, I. I.; Ajayi, O. and Olarewayi, J. D. 1991. Evaluation of insecticides and mulching materials for control of tomato fruit worm, *Heliothis armigera* Hubner at Kadaua Nigeria. *Nigerian Journal of Entomology,* **10** (1–2): 35–40.

van Emden, H. F. 1989. *Pest Control* (2nd Edn.) Edward Arnold, London pp. 117.

Varela, O. G. and Guharay, F. 1988. The use of multiple cropping (Cabbage-carrot) as a component of integrated pest management of cabbage defoliators. *Revista Nicaraguense de Entomologia,* **4**: 49–56.

Verma, B. R.; Lal, B. and Wadhi S R. 1988. A quarantine note on *Prostephamus truncatus* (Horn)—a pest that needs to be watched. *Current Science,* **57** (15): 863–864.

Vierbergen, B. 1992. Interceptions of species of genus *Frankliniella* in the Netherlands. *Proceedings Section Experimental and Applied Entomology of Netherlands Entomological Society,* **3**: 175–180.

Wadhi S R. 1982. Plant quarantine in India. *In: Agricultural Entomology.* P.D. Srivastava, M.G. Jotwani, R.A. Agarwal, S.R. Wadhi, R.K Bhanotar and R.K Bhatnagar (eds). All India Scientific Workers Society, New Delhi. pp. 158–171.

Webb, S. E. and Linda, S. B. 1992. Evaluation of spunbonded polyethylene row covers as a method of exclusing insects and viruses affecting full grown squash in Flordia. *Journal of Economic Entomology,* **85** (6): 2344–52.

Wiech, K. and Wnuk, A. 1991. The effect of intercropping the cabbage with white clover and frenchbean on the occurrence of some pests and beneficial insects. *Folia Horticulture,* **3**(1): 39–45.

Wrigt, R. J. 1984. Evaluation of crop rotation for control of colorado potato beetles (Coleoptera: Chrysomelidae) in commercial potato fields on long island. *Journal of Economic Entomology,* **77**: 1254–9.

Zen, M. E.; Wu, H. P.; Xu, J. S.; Fang, L L.; Li. M. and Yao, Y. G. 1990, Study on the life history and control of *Baris deplanta. Zheiiang Nongye Kexue,* **4**: 187–190.

Zhao, Z. R.; Wang, Y. G. and Yan G Y. 1989. A preliminary report on the oriental fruit moth in north Jiangsu. *Insect Knowledge,* **26** (1): 17–19.

*Pest Management in Horticulture Crops*
L.R. Verma, A.K. Verma and D.C. Gautam (eds)


## *Chapter 5*

# Insecticides in Pest Management

*A. Nath*

## INTRODUCTION

PEST MANAGEMENT IS THE integration of several measures of keeping the pest population down to a reasonable low level and the chemical control is its important component. An organism that is of nuisance value to man is called a pest and the chemical used to suppress its population is called a pesticide. This chapter gives a brief account of the classification, limitations, advantages, methods of application, equipment used, precautions and laws regulating the use of pesticides in India. Most of the information contained in this chapter pertains to the use of pesticides in horticulture.

## CLASSIFICATION OF PESTICIDES

Pesticides may be organic or inorganic. Considering place of action, insecticides can be stomach poisons, contact poisons, systemic poisons, fumigants, repellents, and attractants. Pesticides mainly consist of insecticides, fungicides, rodenticides, herbicides and fumigants. Throughout the text, the broader term pesticide is used in place of insecticide because the underlying principles are not only applicable to insects but also to mites, nematodes, rodents, etc.

Insecticides are the most powerful tools in pest management during emergency situations when the pest populations are about to cross the economic injury levels. No other method of pest control can be as effective under such situations, and we have to depend upon insecticides for pest control needs. Four main advantages of the use of pesticides in pest management programmes are namely:

- It is the only practical method of pest control when pest population is close to economic threshold. It is an emergency intervention when all other pest control methods have failed.
- Pesticides have rapid action in reducing pest population. Contrary to bio-control agents that take long time for the suppression of pest population, insecticides rapidly kill pest

population. The effect of pesticides is so dramatic that sometimes owing to this property they tend to be over used.

- Pesticides have wide range of activities. There are large numbers of chemicals from which one can choose an insecticide according to ones need. There are chemicals with long persistence and those with short residual action. Highly selective chemicals and those with range of activity against several pests are available. Chemicals with contact, fumigant and systemic action (those that enter into plant system, symplast and are toxic to sap sucking insects) are available. Soil insecticides are chemicals that remain active in soil for a long time. There are chemicals that are active against pest organisms other than insects like nematodes (nematicides), mites (acaricides), rodents (rodenticides) and weeds (weedicides).
- Cost/benefit ratio is favourable with pesticides. Pesticides are relatively inexpensive depending upon the number of steps involved in their synthesis. The low cost of DDT and HCH is due to their single step synthesis in contrast to synthetic pyrethroids, methoprene or Diflubenzuron involving multi-step synthesis in their manufacturing process. All future pesticides are expected to be isolates from microorganisms and are thus likely to be relatively inexpensive.

## LIMITATIONS IN THE USE OF PESTICIDES

Most of the horticulture crops such as fruits and vegetables have a characteristic eye appeal. No one likes to eat an apple with red spots of scale insect or black marks of sooty blotch or fly speck. The farmers, therefore, tend to use more than necessary amounts of pesticides. This excessive use of pesticide creates several problems as follows:

### Resistance to pesticides

Resistance is the heritable ability acquired by a strain to tolerate doses of a toxicant that would prove lethal to the majority of individuals, in a normal population of the same species. The first report of development of resistance to an insecticide was in 1914 in the San Jose Scale, *Quadraspidiotus perniciosus* (Comstock) against lime sulphur. Since then a large number of pests have acquired resistance to one or another pesticide so much so that resistance today is the biggest limitation in the use of pesticides. Pesticide resistance is generally due to segregation of pre-existing genes responsible for resistance. It is due to frequent applications of a chemical or related chemicals over large area that wipes off the susceptible individuals. The surviving residual population can tolerate doses of pesticides lethal to normal population of the pest. There are certain genes in a pest population that amplify resistance (produce multiple copies of resistance gene) and in such a population the build up of resistance are faster than in normal course.

There are several types of genes responsible for resistance, like dominant, semi-dominant, recessive or semi-recessive. Similarly the resistance to a pesticide may be due to a single gene or multiple genes. In the later case the development of resistance is much faster. Cross-resistance is the development of resistance to a chemical in a pest that could lead to resistance to related chemicals without being exposed to them. A DDT resistant insect can also be cross resistant to lindane or dieldrin. On the contrary, an insect resistant to a chemical if is resistant to chemicals unrelated to each other then that would be a case of multiple resistance. Pest resistant to DDT if resistant to synthetic pyrethroids without being exposed to them is a case of multiple resistance. The European red mite, *Panonychus ulmi* (Koch) and citrus red mite, *P. citri* (Mc Gregor) have become resistant to almost the entire range of acaricides in Europe. Unfortunately, these two mite species are assuming serious proportions in India too.

Management of pesticide resistance can be through reducing the selection pressure of pesticides and by encouraging natural control agents, cultural methods for controlling pesticide resistant populations. Metcalf (1980) suggested following methods of pesticide management:

- Reduce the frequency of pesticide applications, by preferring chemicals with short persistence so that pesticide screening is of one stage of the pest i.e., either of egg, larva, pupa or adult. Preference should be for the use of ovicides as eggs have poorly developed detoxification system. One should apply pesticide only when the natural enemies are unable to control the pest population. It is particularly important with mite pests, as predatory mites alone are able to control phytophagous mites.
- Early detection of pesticide resistance is important. Detection of resistance can either be by determination of median lethal doses or by assaying activities of enzymes like glutathione transferases, mixed function oxygenases or esterase in pest species. One can locate the amplification genes in a resistant pest population by using suitable DNA probes.
- Avoid use of mixtures of insecticides; instead mix insecticides with synergists for the control of resistant insects.
- Choice of the chemical should have genetic basis taking into account cross or multiple resistances in a pest population. For example, if resistance to organophosphorous pesticide is leading to resistance to carbamate then they should not be used alternatively instead, a treatment with organochlorine or synthetic pyrethroids insecticide should be followed.
- Neem seed kernel extract is very effective against pesticide resistant strains of insects particularly against diamondback moth, *Plutella xylostella*..

**Flare up of secondary pest population**

Highly persistent insecticides with range of activity cause upset in natural balance resulting in increase in the pest population that was kept under control by natural enemies. DDT when first used against citrus pests in the United States in 1947 increased the population of cottony cushion scale, *Icerya purchasi* Maskell. In Himachal Pradesh, India, there has been tremendous increase in population of relatively unknown European red mite, *P. ulmi* on apple following heavy applications of pesticides.

The management of resurgent secondary pest population can be by observing realistic population thresholds, proper timing of spray, spot application of insecticides, reducing application rates and by using pest-specific chemicals.

**Adverse effect on non-target species**

An estimate shows that less than 1% of the sprayed chemical reaches the target pest while rest of it reaches non-target substrates and may cause pollution of the environment. The total number of target pests is less than 1000 whereas pesticides are detrimental to about 200,000 biological species (Pimental, 1971).

***Natural enemies:*** Adoption of spray schedules without considering the presence of pests results in formation of biological deserts (Vanden Bosch and Stern, 1962). These deserts are devoid of parasites and predators, and due to this the target pest population goes up and/or results in resurgence of secondary pests. In either case, there are requirements for more chemical treatments and eventually insecticides become useless. Generally it takes 2 to 4 years to re-establish the natural enemy complex in biological deserts (Smith, 1970).

***Honeybees and other natural pollinators:*** Pollinators not only increase crop yields but also maintain the biodiversity in the ecosystems. Honey bees in particular, are extremely susceptible to crop protection chemicals because they spend early life in hives almost free of toxic chemicals.

In insecticide management programme, insecticide applications should not be done on crops in bloom. However, in an emergency situation, select the less harmful chemical, formulation and time of application. The following precautions, if taken, will minimise pesticide hazard to bees:

- Avoid sprays on a crop in blossom when pollinators are working.
- Liquid formulations are less toxic than wettable powder formulations.
- Ground applications are less hazardous compared with aerial applications.
- Apply insecticide in the evening while bees are not foraging. Early morning sprays can be toxic to bees.
- Methyl parathion and carbaryl are always toxic and, therefore, in no case their application should be done in areas where bees are working.
- It is safe to apply malathion, endosulfan and fenvalerate on blossom when bees are not working.

***Wildlife:*** Some insecticides are highly toxic to non-target vertebrates and non-vertebrates. The most susceptible animal is fish that dies even when endosulfan or toxaphene residues are at parts per billion levels. One of the uses of endosulfan in India is for killing unwanted fish in ponds. Like most other animals, fish accumulates DDT residues in fat. While insecticides are not only toxic to fish but they also kill stone flies, May flies and midge larvae, which are the staple food of fish. DDT causes thinning of egg shell in predatory birds like bald eagle. Insecticide management in such situations requires use of chemicals less toxic to wildlife at minimum dose. Bait treatments and spot treatments are less harmful to wildlife than the normal sprays.

***Phytotoxicity:*** Carbaryl, if sprayed on apple fruits at early stages of development, causes fruit drop. Diazinon causes splitting of bark on apple trees. Recently, fenitrothion on Golden Delicious variety of apple caused phytotoxicity on leaves. Lindane is toxic to cucurbits and quinalphos to crucifers. Phytotoxicity is some times due to mixing emulsifiable concentrates of different pesticides and in that case the concentration of surface-active agents sometimes go beyond the tolerance level of the crop. One should not mix more than two pesticides, i.e., a fungicide and an insecticide and in no case two insecticides should be mixed. Mixing of incompatible pesticides can reduce the effectiveness of both and cause phytotoxicity. Phytotoxicity can also result from poor quality solvent used in an emulsifiable concentrate. Pesticides can adversely affect pollination by inhibiting pollen germination, pollen tube formation, fruit bud formation and fruit setting (Verma and Jindal, 1997).

## PROBLEMS RELATING TO PESTICIDE RESIDUES

In the past, pest control revolved around use of persistent chemicals on fruit crops so that single application gave lasting protection against pests. Generally, less than 1% pesticide falls on the pest and the rest drifts off into the water and soil. Fish and birds pick up the persistent pesticides and biomagnify them 103—107 folds particularly at the upper end of the food chain. It is now becoming increasingly apparent that one can not avoid taking in low levels of pesticides through food, water and air. With proper management of pesticides the risk involved with these residues can be reduced to negligible level. A level of 1 : 40,000 or less is an acceptable risk. Dermal exposure of pesticides is important in fruit orchards to those engaged in different operations like training and pruning harvesting, etc. Such exposures can be lethal. One should not enter an orchard before 48 h following highly toxic chemical and 24 h after spraying with other chemicals.

The amount of chemical one takes over the entire lifetime, without appreciable harm to the health is termed as acceptable daily intake (ADI). Its precise definition is the daily exposure level of a toxicant during the entire lifetime of a human being that appears to be without appreciable risk based on all the facts known at that time. The unit of its expression is mg/kg/day. Long-term feeding experiments determine the maximum level of a toxicant that causes no effects (no-effect level, NOEL). ADI can be calculated by dividing NOEL by a safety factor of more than ten (commonly 100) depending upon differences in sensitivity in individuals and in species and reliability of data. The legal limit of a pesticide permitted on a crop is called maximum residue limit (MRL). As defined by *Codex Alimentarius Commission*, MRL is the maximum concentration of pesticide residues legally permitted in or on a commodity and expressed as mg/kg. The estimation of MRL is dependent on maximum residues that occur following good agricultural practices that envisage application of pesticide at the lowest concentration, frequency and the right stage of the crop growth for good control of the pest. Calculation of MRL involves analysis of samples from supervised field trials following good agricultural practices. Pesticide application is at a little higher dose as a safety factor; to safeguard the interests of the farmers and confirmation of results obtained by analysing samples from farmer's field that uses chemicals according to the recommended practices. MRL is always lower than the safe level and, therefore, chemicals on a commodity above MRL do not necessarily constitute health risk. However, MRLs provide useful guidelines for knowing whether the producers have followed good agricultural practices and indicate if there was inadvertent food contamination due to an accident.

There are countries that have no MRLs instead they make sure the adoption of good agricultural practices. However, there are wide variations in MRLs in different countries. The food importing countries feel concerned about the safety of the imported food. On the other hand food-exporting countries consider MRLs non-tariff barrier to the movement of food. The rejections of Indian food export consignments are worth Rs 1000 crore ($ 300 million) every year because of the presence of excess insecticide residues.

Several studies on dissipation of pesticides on different crops reveal that the initial deposits dissipate at the fastest rate mainly due to volatilisation losses. Action of rain and dew, photochemical reactions too dissipate residues. Enzymatic destruction in the tissues of plants and animals, dilution's due to the growth of tissues and through post harvest treatments and processing are some of the ways for destruction of pesticide residues. Pesticide dissipation follows first order kinetics and from these curves, one can compute safe pre-harvest waiting periods (time to reach MRL) following last applications.

Pesticide residues present can deteriorate quality of fruits and vegetables for processing. Traces of fungicides prevent fermentation in grapes while cereals become unsuitable for making bread. Potatoes become tainted due to the presence of traces of lindane.

**Acute toxicity hazard**

Acute toxicity of a chemical is the toxicity following single exposure either oral or dermal. It is important for those involved in manufacture, formulation, transportation and spray operations. Some of the present day insecticides that can cause acute toxicity are aldicarb, carbofuran, phorate, methyl parathion, etc. These insecticides have oral LD50 (Median lethal doses kill half the population) of less than 15 mg/kg body weight. The classification of insecticides on the basis of toxicity is as given in the Table below:

**Table:** Toxicity classes of the available pesticides

| *Toxicity category* | *Extremely toxic* | *Highly toxic* | *Moderately toxic* | *Slightly toxic* |
|---|---|---|---|---|
| Acute LD50 (mg/kg weight of test animals) oral | 1–50 | 51–500 | 501–5000 | >5000 |
| Acute LD50 (mg/kg weight of test animals) dermal | 1–200 | 201–2000 | 2001–20000 | >20000 |
| Colour of the band on label | Bright red | Bright yellow | Bright blue | Bright green |
| Symbol and signal word | Skull & cross bones and the word "POISON" in red. | Word "POISON" in red. | Word "DANGER" | Word "CAUTION" |
| Warning statement | Keep out of the reach of children. If swallowed or if symptoms of poisoning occur call physician immediately. | Keep out of reach of children | Keep out of reach of children | |

In pest management programs only safe pesticides should be used. One should wear gloves, protective clothing, and masks and should maintain strict hygiene while handling pesticides. Field workers should observe re-entry period into the field/orchard following pesticide application. It is 24 h for normal pesticides and 48 h for highly toxic pesticides. One should dispose off unused pesticides properly and should store pesticides out of reach of children. Handling of pesticide should be by adult males only and not females as the later are more susceptible to pesticides.

### Pesticides banned for use in India from April 1997

Pesticides that have been banned in India are Aldrin, calcium cyanide, chlordane, copper acetoarsenite, dibromochloropropane, endrin, ethyl mercury chloride, heptachlor, menazon, nicotine sulphate, nitrofen, paraquate dimethyl sulphate, pentachloro nitrobenzene, pentachlorophenol, phenyl mercury acetate, sodium methane arsenate, tetradifon and toxaphene.

### Pesticides restricted for use in India from April 1997

Pesticides that have been restricted for use in India are Aluminium phosphide, BHC (banned for use on vegetables, fruits, oilseed crops and preservation of food grains), captafol (use permitted for seed dressing only), carbaryl (not permitted for spray on crops in flowering stage; b carbaryl should be less than 0.05%), chlorbenzilate (permitted for use in bee keeping), DDT (use permitted only in exceptional cases), dieldrin (restricted to use for locust control only), ethylene dibromide, lindane (smoke generating formulation for indoor use prohibited), methyl bromide, methyl parathion (permitted for use on crops where honey bees are not working), phenyl mercury acetate and sodium cyanide.

### Pesticides refused registration in 1997

Pesticides that have been refused registration in 1997 are Ammonium sulfamate, azinphos ethyl, azinphos methyl, binapacryl, calcium arsenate, carbophenothion, chinomethioate (Morestan), dicrotophos, EPN, fentin acetate, fentin hydroxide, lead arsenate, leptophos (Phosvel), mephosfolan, mevinphos (Phosdrin), 2 4 5-T, thiodemeton, disulfoton and vamidothion.

### Pest management rating of some of the insecticides in use in India

Pest management rating of some of the insecticides in use in India (Metcalf, 1980) is as follows:

| *Pesticide* | *Overall rating* |
|---|---|
| Bacillus thuringiensis | 3.0 |
| Methoprene | 4.3 |
| Trichlorfon | 4.3 |
| Malathion | 5.3 |
| Carbaryl | 6.0 |
| Diflubenzuron | 7.0 |
| Dicofol | 7.3 |
| Oxydemeton methyl | 7.7 |
| Fenvalerate, permethrin | 7.7 |

| | |
|---|---|
| Phosphamidon | 9.0 |
| Chlorpyrifos | 9.7 |
| Endosulfan | 9.7 |
| Methyl parathion | 9.7 |
| Lindane | 10.0 |
| Phorate | 11.7 |
| Carbofuran | 12.0 |
| Aldicarb | 12.3 |

This rating takes into account the toxicity to man, domestic animals, birds, bees and fish besides persistence in the environment. Pesticides with rating of 3.0–7.0 (including neem) are for general use in pest management; from 8.0–10.0 require skilful supervision for use, and >11.0 are for restricted use only.

## SAFE USE OF PESTICIDES

Pesticides in pest management require need-based application rather than blindly following spray schedules and should be used at doses that give reasonable control of the pest. Complete control of the pest is neither feasible, nor desirable in pest management. Use of pesticides in pest management requires pesticide intervention at the weak points in the life of the pest. It is an emergency intervention when all other pest control efforts fail to control the pest or as preventive treatment at the lowest dose with minimum damage to the environment. Its application should coincide with the most vulnerable stage of the pest This requires detailed knowledge of the pest and of the insecticide. Insecticide chosen for use in pest management should be target specific and its application should coincide with the most vulnerable stage of the pest. The method of application should not harm the non-target species. One can improve the selectivity of a chemical by considering the behaviour of the pest species.

### Target specific pesticides

There was a time when the typical doses of insecticides were 20–30 kg/ha as against present day chemicals like deltamethrin that gives good control of lepidoptrous pests at 10–20 g/ha. This development has considerably reduced the environmental load of pesticides. The expectations are that new pesticides of 21st century will provide good control of field/orchard pest at 1–2 g/ha.

***Pesticide selectivity***: Pesticide selectivity is the inherent quality of pesticide and is different from selective use of pesticides. Pesticide selectivity can be enhanced by proper timing, formulation and by different methods of application.

There are insecticides that mimic the insect hormone. These hormones appear at certain stage of the development of pest species particularly during moulting. These hormone mimics when sprayed at a time when they are not present in the insect and thus cause disruption in the life processes. Since these hormones are specific to a group of insects, they do not constitute risk to non-target species. The most notable insecticide that acts as juvenile hormone mimic is methoprene. It is most efficient when applied to late instar larvae before pupation.

Chitin is present in animals belonging to Phylum Arthropoda and in some fungi. Chemicals inhibiting chitin synthesis are specific against species belonging to Arthropoda. Diflubenzuron is a commercial insecticide belonging to this class and is the most toxic against immature insects. Acaricides like dicofol are effective against mites and are not toxic to insects or other forms of life

Biological insecticides, such as *Bacillus thuringiensis* and viruses are specific and control a number of insect pests. Viruses although efficient in pest control, yet are slow in their action and therefore cannot prevent crop damage.

***Stage Specificity:*** All pests have certain stages in their life cycle that is more vulnerable. Control of over-wintering red mite and San Jose scale on apple during winter and that of peach leaf-curling aphids by spraying insecticide during early spring are some of the examples. These pests are difficult to control during summer as during this period, pesticide applications are harmful to their natural enemies. Borer pests are very difficult to control as they enter into plant tissues and the vulnerable stages are adult, egg or early stages only.

**Method of application**

Foliar application and broadcasting of insecticide granules are the most inefficient methods of pest control. Doses suggested by pesticide vendors are generally much higher than those required for the control of pests. Sometimes doses lower than those suggested give good control of the pest and at the same time are less toxic to natural enemies of the pest. Less persistent pesticides are more target specific than more persistent ones. Dichlorvos, if sprayed in the evening is less harmful to bees than more persistent chlorpyrifos or carbaryl. It is not always necessary to treat the entire crop for good control. One can treat only the infested patches, border rows or every second row or the trap crop for pest control. In this way most of the natural enemies escape mortality on the main crop.

Systemic insecticide sprayed for the control of aphids, thrips or mites enters into the plant system and is safe to parasites, predators and pollinators. Granules of phorate, carbofuran, aldicarb applied at the time of planting as side dressing are safe. They also enter into the plant system and their high concentrations are present in growing tips or new leaves where most of the pests feed.

Seed treatment is the most efficient use of insecticides for the control of pests. Lindane at 15–25 g/h as seed treatment in vegetables gives good control of wire-worms and equivalent control with foliar application requires 2.5–5 kg/h. The disadvantage of using seed treatment is that sometimes seed treatment with insecticide reduce germination and treated seeds are lethal if eaten by birds.

Pesticide selectivity can also be achieved by using granules in soil. The application does not require any special equipment, there is no problem of drift and the treatment gives protection over a long time without harming the natural enemies.

A substantial reduction in pesticide doses can be achieved by considering the behaviour of the pest. Gravid melon fly, *Dacus cucurbitae* Coquilett enters the tomato field to oviposit and leaves at dusk to rest in the adjacent vegetation. The control strategy should be to treat adjacent vegetation during its attack. It will not only save pesticide but also considerably reduce the pesticide residues in tomato fruits (Nishida and Bess 1950; Nishida 1954). Spray of dichlorvos mixed with *gur* gives good control of fruit flies on several fruits and vegetables. This treatment becomes difficult because infestation of fruit flies is generally in rainy season and flies prefer to lay eggs in ripe fruits. In such situation dichlorvos with gur in bait stations would provide significant control of fruit flies in orchards.

In Himachal Pradesh, good control of apple fruit moth, *Archips pomivora* in field can be achieved through spray applications synchronising with peak moth emergence. The adult moth emergence can be monitored by using light or pheromone traps.

## PESTICIDE APPLICATION EQUIPMENT

It consists of different types of sprayers, dusters, mist machines and aircraft's. The broad categories of application equipment are manually operated or power operated. Application of powder formulations is carried out through dusting equipment and those of liquid formulations through sprayers. Motorised knapsack mist blower can apply both dust and liquid formulations. The choice of spray equipment depends upon the crop, availability of water and labour, nature of pest and insecticide, area of treatment, etc. The groups of spraying and dusting equipment are as under:

### Dusters

***Hand operated dusters***

*Rotary type:* The blower sucks the dust from the hopper and discharges it out with speed.

*Plunger type:* The pump generates air blast and passes it into the dust chamber causing the dust to blow out through the discharge outlet

*Bellows type:* The air blast produced by expansion and contraction of bellows passes through the dust chamber and discharges out the dust.

***Power operated dusters***

*Rotary type:* The air is sucked by a blower and discharged out through the blower outlet.

*Air jet type*: the air jet from the blower enters into the hopper that agitates the dust and whirls it out, simultaneously the part of the air jet blows it out.

### Sprayers

***Manually operated***

- Foot sprayers
- Rocker sprayers
- Pressure retaining knapsack sprayers (Battery sprayers)
- Compression sprayers (Pneumatic knapsack sprayers)
- Knapsack sprayers
- Stirrup pumps

- Nursery sprayers

*Power sprayers*

The main parts of the sprayer are pump barrel, pressure chamber and discharge line. The discharge line consists of:

- Required length of delivery hose;
- Spray lance consisting of a) Cut off-valve (b) extension rod (c) gooseneck (d) Spray nozzle spray boom spray rig spray gun
- Nozzles

It is very important to select right type of nozzle for proper use of pesticides The nozzles have discharge rate from 250 ml to 2500 ml per minute. Nozzles are of following types:

- Cone spray type
- Flat fan spray type
- Flood jet type

Swath width is the space covered effectively by the pattern of a nozzle. While spraying, hold the nozzle at a specific distance from the sprayed surface (generally 45 cm). Longer the distance from the surface the broader will be the swath width.

For good control of pests and diseases, the selection of nozzle with a specific discharge rate, use the correct dose of technical material and correct volume of spray are important. The sprayed area depends on the row width as well as the height of the plants.

Several types of equipment are available for spraying and dusting various pesticides. Selection of proper equipment is important for getting maximum advantage out of it. It is, therefore, important to consider the following factors while selecting pesticide application machinery.

*Recommendation:* The most important factor to consider is the right pesticide at the right dose (active ingredient/h) that would kill the maximum proportion of pest organism with minimum adverse effects.

*Dose:* The minimum quantity of the pesticide in ready to use form (L/kg) required to spray or dust to achieve maximum control of pests. The spraying should be such that it covers uniformly leaf or plant surface area for its maximum effectiveness. For small volumes, use low volume or ultra low volume sprayer as it sprays fine droplets. The manually operated equipment provides large droplet size.

*Droplet size:* This is also an important factor as it directly influences the effectiveness of the pesticide. The properties of the spray fluid, type of nozzle, operating pressure, velocity of spray fluid emitted and the air movement determine the droplet size. Optimum droplet size is 20 to 30 microns. For systemic pesticides droplet size is not very important.

*Water:* In areas of water scarcity, preference should be given to dusting over spraying.

*Urgency:* Under pest epidemic, use of power operated sprayer is most suitable so that maximum area is covered in minimum time.

*Labour:* Where labour is costly and scarce, use of power operated pest control equipment is suitable.

*Crop size:* Treatment of field and vegetable crops can be done by manual sprayers/dusters whereas, orchard crops will need power operated equipment's that can produce enough pressure to ensure proper coverage of big trees.

*Function:* The main function of a sprayer is to atomise the spray fluid (solution, emulsion or suspension) into small droplets and ejection of the same with some pressure for uniform distribution. Another function is to regulate the pesticide dose to avoid excessive application that might prove harmful or wasteful.

The mechanism of atomisation of a liquid in a sprayer is as follows:

i) forcing a liquid through the nozzle by hydraulic pressure as in foot sprayer and ordinary power sprayer or by air pressure as in compressed air sprayer.

ii) by the use of high velocity air stream into which flows liquid or coarsely atomised liquid as in motorised knapsack sprayer,

iii) by centrifugal force, wherein the liquid is fed under low pressure to the centre of a high speed rotating device and is atomised by centrifugal force as it leaves the periphery.

**Types of sprayers**

According to the source of motive power, the available spraying machine can be classified as (a) manually operated and (b) power operated. Manually operated sprayers are particularly useful use in small orchards and fields, nursery, etc. They also find use in hilly regions where it is not possible to use power sprayers, as they require high speed of operation not always possible in hilly locations.

***Hydraulic sprayers***

For discharging the spray, the manually operated conventional sprayers depend on hydraulic pressure developed by the direct action of a pump on the spray fluid. This pressure forces the liquid through the nozzle that atomises it into droplets of various sizes and reproduces a spray pattern characteristic of the nozzle. A variety of high pressure, large volume hydraulic pumps is employed for spraying pesticides. Such equipment usually required 500–1000 litres of spraying fluid per hectare of field crop. Using various types of nozzles and lances, one can regulate the quantity of spray material per unit area and the height to which the spray reaches. In hydraulic sprayers, the droplet size varies between 300–400 microns. The application of fluid is done by means of a hand lance fitted with one or more nozzles. Such hand spraying at high pressure produces complete coverage. These sprayers have special advantage where spraying is discontinuous as in spraying trees, certain crops and in spraying in households and public health programs. These types of sprayer include syringes, Bucket pump (stirrup pump), knapsack sprayer, rocking sprayer, foot sprayer, and wheel-barrow sprayer.

### *Compression sprayers*

These sprayers are also known as pneumatic sprayers because they employ air pressure for forcing the liquid through the nozzle for atomisation. A part of the container is kept empty so that adequate air pressure can develop over the spray fluid in the tank. These sprayers do not have provision for agitation of the spray fluid. As such one should not spray pesticide material requiring agitation, like wettable powders, with these sprayers. Compressed air sprayers are common in use in gardens and small farms. They are also the mainstays of most public health projects involving insect control the world over because of their simple design, operation and relatively low cost.

### *Power operated sprayers*

Mechanically operated equipment powered by gasoline engines for the application of the spray fluid is currently replacing the manually operated type as such equipment's cost less and are fast. Both hydraulic and compressed air sprayers can be used as power operated sprayers when fitted with engines. They have a high capacity for spray fluid and higher rate of discharge of spray fluid. These motorised sprayers may be mounted on stretcher, wheelbarrow or tractor.

### *Low pressure, low volume sprayers*

They include mist blowers, fog generating machines and ULV applicators. With these sprayers one can apply high concentration and low volume of the insecticide. Here atomisation is by the injection of the liquid into high air velocity. The sprayers have various names such as blower type applicators or air-blast sprayers or spray blowers, wet dusters, air blowers, vapour sprayers and mist blowers. Atomisation increases by increasing the flow volume of air. In these sprayers, it is the air that carries the particles to the target surface. Volume of chemical used per hectare is low and the number of droplets deposited per unit area is large. This method is cheaper and more efficient if adopted judiciously. These sprayers have smaller tank and save time, as refilling is less frequent and are particularly suitable for area where water is scarce or transportation of large quantities of water is not possible. These include spray blowers, mist blowers and fogging machines. The main difference is on the size and number of droplets required per unit area.

### *Ground ULV applications*

A number of motorised knapsack sprayers adopted for ULV application is available. The efficiency of such applicators depends on the droplet size, density volume of air, etc. The droplet size varies from 30–150 microns.

### *Aerial application*

LV and ULV application of insecticides are now common for the control of some of the important pests giving excellent control. Aircraft fitted with boom and nozzles for ULV or LV sprayings serve the purpose. One should use fixed wing planes and helicopters equipped with spraying applicators for large-scale aerial applications. It's a very good device to control the epidemics of pest in short time.

The usefulness of any insecticide depends upon its proper application influenced by the properties of insecticide, the nature of pest or pest complex and the place where application is being made.

## INSECTICIDE APPLICATION STRATEGY

### Timing of application

Application of insecticides on prophylactic or on a fixed calendar schedule irrespective of the occurrence or level of pest infestation is generally undesirable since fewer applications are often sufficient if timed accurately. Besides being economical, this will reduce selection pressure for development of resistance also. However, application on schedules find favour since forward planning is easier and the user knows what exactly the requirement of chemical and time are.

### Application sites and placement

The importance of restricting insecticides to the actual target is fundamental to good pest management. It depends on the behaviour and habitat of the pest. Among the three species of mango hoppers, two species prefer to rest on leaves and shoots while the third species is mostly resting on the lower side of horizontal branches and on the stem. So it is important to spray the leaves, shoots and panicles but also the lower side of branches and stem. Likewise, many pests have wide host range, mostly weeds existing in the vicinity of crops and fruit trees. Foe example, mango mealy bug first instar nymphs, after hatching, emerge out of soil and climb up the mango tree and also they heavily infest the weed, *Clerodendron infortunatum* in the orchard. The weed should also be sprayed to prevent the spread of the pest. Similarly, one should avoid spraying at the time when crops are in full bloom in orchards. This will help in conservation of natural enemies and pollinators.

### Spray coverage and distribution

The biological efficacy of an insecticide depends upon three factors, the mean level of deposit (dosage), distribution and coverage of plant surface.

Both, nature of pest and mode of action of insecticide determine the requirements for degree of coverage needed in good pest control. Pests that are sessile like scales require a thorough coverage of spray so that pesticide may reach to them while in flying insects (hopper, flies, etc.) even a partially covered crop will be sufficient. There is no need to cover the complete crop as these pests meet the pesticide because of their mobility. Similarly contact insecticides need thorough coverage while in the case of systemic insecticides there is no need of complete coverage because toxicant is absorbed in the plant tissues and is eventually distributed in the plant system.

### Volume of spray fluid required

The volume of spray fluid required for each application depends on many factors such as the size of the crop, type of pest or disease and type of machine being used. The following table gives the volume of spray fluid required on different crops while using different equipment.

Table showing volume rate of different crops (litre/ha).

| *Name of equipment* | *Field Crops* | *Trees and bushes* |
|---|---|---|
| High volume | > 600 | >1000 |
| Medium volume | 200–600 | 500–1000 |
| Low volume | 50–200 | 200–500 |
| Very low volume | 5–50 | 5–200 |
| Ultra low volume | <5 | <50 |

It depends on the droplet size and the density per square unit. After ascertaining these two, calculate volume of spray.

$$n = 60/l \times (100)3\ Q/d \text{ (Theoretical for flat surfaces)}$$

Where n = number of droplets/sq. cm; d = droplet diameter; Q = volume of spray required (litre/hectare).

## PROBLEMS RELATED WITH PESTICIDE APPLICATION

**Washing, evaporation and drifting**

Temperature, moisture, wind velocity, wind direction and rains are very important factors to be taken into account while making application.

Wind speed is an important factor that influences the deposit of the droplets. When wind speed is high the droplets tend to drift rather than settle down and thus the very purpose of spray application is lost. Concentrated spray applications should be made only when wind velocity is less than 8 km/hr. The drift can be minimised by reducing distance between nozzle and the target, increasing droplet size by reducing pressure and by using suitable formulation. Dusts tend to drift more than sprays.

**Mixing and handling of pesticide concentrate**

Pesticide preparation should be made in open space or well-ventilated room. One should always use a funnel for pouring concentrated pesticide so as to avoid spillage. Operator should wear protective clothing, eye protector, gloves etc. Safe disposal of containers is important. These should be incinerated, preferably decontaminated, broken before burying deep in soil. Excess of spray fluid should be disposed off properly. Care should be taken not to wet the clothes gloves with pesticides. Operator should not chew, drink and smoke during the operation. At the end of spraying the operator should change clothes and should wash thoroughly. In case of any discomfort during spraying, the operator should immediately stop spraying. In case of poisoning, medical help should be sought.

**Care and maintenance of equipment**

Mechanically efficient and trouble free machines pay good dividends in pest control work. The following measures will keep the machines in good working condition.

- Clean the machine properly after use and store it in a dry place.
- Drain the sprayer tank and flush it with clean water. Wash the pump, nozzle, strainers, etc. with kerosene oil.
- Keep the machines in working order by following a regular maintenance programme.
- Regularly lubricate washers, valves, agitator shaft, bearings, plungers and packing and other parts likely to go dry when not in use.
- Always stock spare parts, washers, nozzles, etc. for replacement.
- Inspect pump valve set plunger cups and packing at least once in a year to check leakage from these parts.
- Keep gun and nozzle parts clean to avoid waste of spray material.
- Test pressure regulator and pressure gauge occasionally for efficient functioning of the sprayer.

## INSECTICIDE ACT

The Insecticide Act, 1968 governs the registration and regulation of pesticides in India. The objective of this act is to regulate the import, manufacture, sale, transport, distribution and use of pesticides to prevent risk to human beings or animals. This act came into effect with the enforcement of the insecticides rules, 1971. The rules of the Act were amended in 1993. The implementation of the Insecticide Act is the responsibility of both central and state governments. The Centre government looks after the policy making and registration certificates for import and manufacture of pesticides, while the state governments issue licenses for manufacture and sale of pesticides in the state.

The responsibility for monitoring the quality of pesticides and their proper usage lies with the governments. To meet these objectives, states notify Appellate Authority, Licensing Officer, Insecticide Inspectors, Insecticide Analyst, etc.

The central government has constituted the Central Insecticide Board to advise the Central and State Governments on technical matters concerning the Act. Centre has a Registration Committee for the registration of pesticides, besides other functions assigned to them by the Act and the Rules.

## CONCLUSION

Pesticides certainly have an important role to play in pest management till we have a better alternative provided we use them judiciously. Presently the subsidy given on pesticides is same irrespective of their environmental implications. To encourage the use of safer pesticides higher subsidy should be given to safer chemicals compared to more toxic chemicals. Bio-pesticides need to be encouraged over synthetic chemicals. There is a need to constantly monitor the pest and natural enemy populations for proper management of pesticides. Next generation pesticides are going to be microbial isolates, unlike present day synthetic chemicals and will be greener it is to be seen whether insects will continue to develop resistance to them or not. Susceptibility of pests to pesticides is a non-renewable genetic resource and there is a need to conserve it by proper management of pesticides alongwith pest management as the cost of development of newer pesticides are going to be very high.

## REFERENCES

Bindra, O. S. and Singh, H. 1971. *Pesticide application equipment.* Oxford and IBH Publishing Co. New Delhi, pp. 307.

Lal, O. P. 1996. *Recent Advances in Indian Entomology.* APC Publication Pvt. Ltd. New Delhi. pp. 392

Metcalf, R. L. 1980. Changing role of insecticides in crop protection. *Ann. Rev. Entomol.* **25**: 219–256

Nishida, T. 1954. Further studies on the treatment of border vegetation for melon fly control. *J. Econ. Entomol.* 47: 226–229

Nishida, T. and Bess, H. A. 1950. Applied ecology in melon fly control. *J. Econ. Entomol.* **43**: 877–883

Pimental, D. 1971. *Ecological Effects of Pesticides on Non-Target Species.* Executive Office of the President. Office of Science and Technology. US Government Printing Office, Washington, D.C.

Smith, R. F. 1970. Pesticides: Their Use and Limitations in Pest Management. *In*: *Concept of Pest Management.* R. L. Rabb and F. E. Guthrie (eds). North Carolina State University, Raleigh, North Carolina. pp. 103–113.

vanden Borch, R and Stern, V. M. 1962. The integration of chemical and biological control in arthropod pests. *Ann. Rev. Entomol.* 7: 367–386.

Verma, L. R. and Jindal, K. K. 1997. *Fruit Crops Pollination.* Kalyani Publishers, New Delhi. pp. 363–372.

*Pest Management in Horticulture Crops*
L.R. Verma, A.K. Verma and D.C. Gautam (eds)


## *Chapter 6*

# Insect Growth Regulators: Current Status and Scope in the Management of Fruit Crop Pests

*Gursharan Singh and Darshan Singh*

## INTRODUCTION

The annual consumption of pesticide in India has increased tremendously, from meagre 4000 metric tons in 1960–61 to 90,000 to 1,00,000 metric tons in 1992–93. Although the average annual consumption of pesticides is the lowest in India, yet they have produced a number of ecological and economical imbalances and the country shares the global concern about their undesirable side effects. Research is being pursued all over to develop viable alternatives with emphasis on minimal use of synthetic pesticides. Development of biorational pesticides including insect growth regulators (IGRs) is one such potential possibility. A large number of insect growth regulators have been developed which either mimic the action of various insect hormones or inhibit the synthesis of chitin in the new cuticle during moulting of insects. These compounds being biodegradable are environmental friendly and have unique modes of action, specificity and low toxicity to non target organisms. IGRs include compounds such as analogues of neuropeptides, juvenile hormones, moulting hormones and their antagonists and chitin synthetase inhibitors. In this chapter, discussion is centred around the potential uses of IGRs in the management of insect pests of fruit crops.

## NEUROPEPTIDES (NPs)

Different types of neurohormones are secreted from various parts of the central nervous system of insects. A pentapeptide, proctolin, was the first neuropeptide identified (Starratt and Brown, 1975) and since then as many as 40 different neuropeptides have been characterised in insects (Holman *et al.*, 1990). These NPs are oligopeptidic in nature and are known to a series of physiological processes of insects. The potential of their exogenous application in insect pest control is limited, mainly because of their inability to penetrate the insect cuticle and being proteinaceous, these are liable to enzymatic degradation in gut after their ingestion. The most promising use lies in insertion of genes responsible for production of such neurohormones into microbes and plant species. Inspite of these suggestive uses till date, no neurohormone has been successfully employed in insect pest control.

## MOULTING HORMONE AGONIST AND ANTAGONISTS (MHAs)

Prothoracic glands of insects secrete a steroidal hormone called moulting hormone. This hormone, called alfa-ecdysone gets hydroxylated into 20-hydroxyecdysone (beta-ecdysone) and other ecdysteroids (20, 26-dihydroxyecdysone) which are responsible for various processes of moulting, growth and maturation of insects (Watkinson and Clarke, 1973). A large variety of phyto-ecdysteroids have also been reported from several plant species especially those belonging to Pteridophyta, Gymnospermae and Angiospermae (Singh and Bhathal, 1994). Synthetic analogues of moulting hormones can be exogenously applied to enhance the internal titer which cannot be metabolized or excreted. This hormonal imbalance will result in moulting promotion into adult midget. A moulting inhibitor RH-5992 (3,5-dimethylbenzoic acid 1-(1.1-dimethylethyl)-2-(4-ethylbenzoyl) hydrazide) was evaluated under field conditions in Europe (Heller *et al.*, 1992). When used against the codling moth, *Cydia pomonella* (L.) on apple in France, the fruit infestation was drastically reduced in treated trees and the extent of marketable fruits increased to 86.2 per cent from 11.7 per cent in untreated trees. At this dose excellent control of the Oriental fruit moth, *Cydia molesta* (Busck) on peach was achieved in Italy. When applied @ 96 g a.i./ha at the time of egg hatching in grape moth, *Lobesia borrana* Den. and Schiff., on grapes in France, it resulted in 97 per cent reduction in larvae per fruit cluster. Similar results were also obtained in Germany for the control of *Eupoicilia ambiguella* Hubner on grapes.

At present, the use of MHAs on insect pest control is limited. The steroidal analogues are costly to synthesize and are poor in penetrating the wax layer of insect cuticle because of their lipophilic nature. Moreover, their steroidal structure is also viewed skeptically because of their lipophilic regulatory role in crustaceans and proximity to steroid hormones in higher animals.

Moulting hormone antagonists like azosterols and non-steroidal amines inhibit the conversion of 29 carbon atom phytosterols into 27 carbon atom cholestrol, a precursor for synthesis of ecdysone in insects (Walker and Svoboda, 1973). The first group of compounds to possess anti-ecdysone activity was the derivatives of 2,4-dehydroxy-6-keto-cholestane (Hora *et al.*, 1966) some of which are structurally related to cheilanthones isolated from plants. Some phytochemicals like ajugalactone found in *Ajuga decumbens* Thumb. (Koreeda *et al.*, 1970), ajugarins from *A. remota* Benth (Kubo *et al.*, 1976), plumbagin from roots of African shrub, *Plumbago capensis* Tunb. (Joshi *et al.*, 1988 and Joshi and Sehnal, 1989) and azadirachtin from neem, *Azadirachtra indica* A. Juss. (Gracia *et al.*, 1990; Sieber and Rembold, 1990) are known to possess antiecdysone type activities. At present the use of such chemicals in insect pest control is subject to srutiny mainly because of high cost of azosterols and their probable interference with steroid hormone regulation in higher animals (Saxena, 1983).

## JUVENILE HORMONE AGONISTS AND ANTAGONISTS

Juvenile hormone is responsible for maintenance of juvenile characters in insects and inhibits their metamorphosis. Exogenous application of juvenile hormone upsets their development and results in formation of different types of interstage intermediates and supernumerary instars. At present five different juvenile hormones (JH-0, JH-I, JH-II, JH-III and 4-methyl JH-I also called ISO-JH-O) have been isolated and characterized (Richards, 1981; Bergot *et al.*, 1981).

Natural juvenile hormones are rapidly metabolized by inherent esterases of insects and are also unstable in sunlight. Based on JH-model, stable JH analogues (JHAs) were developed by diminishing the ester cleavage on introduction of double bond and substitution of isopropyl for methyl ester. Epoxide hydration was eliminated by replacement of epoxide of tertiary methoxyl. In doing so, the first commercial JHA, methoprene, was evolved by Zoecon Corporation of USA in 1973. It was followed by synthesis of other analogues like hydroprene and kinoprene. These analogues too were unstable to UV-radiation, hence lacked full potential for their commercial use. Thereafter, UV-stable analogues like epofenonane were developed (Schmid *et al.*, 1978) followed by fenoxycarb and pyriproxyfen. In these compounds, isoprene groups were substituted by linked phenoxy groups giving increased stability and substantial improvement in biological activity (Menn *et al.*, 1989). The JHAs have been tested on large scale trials for the control of various important pests of agriculture, forestry and public health importance. Methoprene was the first JHA to be registered in USA in 1973 for commercial use in mosquito control, while in UK it was registered in 1980 for the control of pharaoh's ants. At present, several formulations of methoprene have been cleared for commercial use in the control of mosquitoes, manure breeding flies, cigarette beetles, pharaoh's ants, fungus gnat and fleas. Another JHA, kinoprene (Enstar) is being recommended for the control of greenhouse homopterous insects. Several field trials with more recent JHA like fenoxycarb, pyriproxyfen and epofenonane have been undertaken. Fenoxycarb is registered in Switzerland for the control of several fruit pests such as grape moths, *Lobesia botrana* Denn. and *Shiff., Eupoecilia ambiguella* Hubner (Charmillot *et al.*, 1987), plum moth, *Grapholita funebrana* Tr. (Charmillot and Bloesch, 1987), codling moth, *Cydia pomonella* (L.) (Charmillot *et al.*, 1989) and summer fruit tortrix, *Adoxophyes orana* F.V.R. (Chormillot and Brunner, 1989). Its formulation, insegar, has also been approved for use against *E. ambiguella, L. botrana* and other Lepidoptera infesting grapevines, apple, peach and olive in Italy (Menke, 1988; Oberhofer, 1988).

Among the lepidopterous insects infesting apples, *C. pomonella* is a very serious quarantine pest which needs to be controlled through quarantine treatments to check its spread to countries where it is not present. Juvenile hormone analogues have been found to be very useful in quarantine treatments because of their specificity and low mammalian toxicity. Application of one such analogue, methoprene, reduced fecundity and adult survival of the moth (Brown and Brown, 1982). Methoprene and fenoxycarb are also known to affect diapause (Brown, 1985) and development of the moth (Gelbic and Sehnal, 1973; Charmillot, 1989). Early sprays of fenoxycarb also controlled *C. pomonella* in Western Galilee (Ben-Avraham and Yizhar, 1987). Another JHA, pyriproxyfen had ovicidal effects on freshly laid (1-day old) eggs of *C. pomonella* and *C. molesta* (Yokoyama and Miller, 1991). Exposure of adults of both the moth species, reduced the egg viability but did not affect fecundity. Further, exposure of both sexes of *C. molesta* was more effective in reducing egg viability (94 % mortality) than exposure of either sex alone.

Some analogues of JHs have also been evaluated against the leafrollers infesting apple. Epofenonane has proved to be a very promising control agent of the major leaf-roller pest, *Adoxophyes orana* (Fisch V. Roesl.) in apple orchards in the Netherlands (Jong *et al.*, 1981).

Similarly, fenoxycarb is also known to disturb metamorphosis of various leafrollers, viz. *A. orana, Archips podana* (Sc.) and *Pandemis heparana* (Den. and Schiff.) damaging apple trees. NKI-35120, a derivative of fenoxycarb was very effective when sprayed on apple trees in the field trials in Hungary (Varjas, 1992). Results revealed that fenoxycarb matched with broad spectrum insecticides as regards its activity against the leafrollers, particularly *A. orana* (Monnet, 1986). Application of fenoxycarb on lime, also gave satisfactory control of the citrus leaf miner, *Phyllocnistis citrella* St. (Pena and Duncan, 1994). Larval population was considerably reduced during a week after the spray.

Fenoxycarb have also been tried for the control of various hemipterous pests. When used on pear against *Psylla pyri* Schmidb., it resulted in the formation of nymphal-adult intermediates and also delayed hatching when applied on the psyllid before ovipostion or on newly deposited eggs (Larguier and Rivenez, 1990). Results of field trials with fenoxycarb (Insegar 25%) have shown very encouraging trends. Its application @ 7.5 and 12.5 g/hl against the black scale, *Saissetia oleae* (Olivier) with that of Pomex (carbaryl 49%) at 57 g/hl in a mandarin orchard in Italy revealed that fenoxycarb compared very well with carbaryl and methidathion and moreover it fits better in pest management programme due to its low toxicity to man and domestic animals (D Anna and Leonardi, 1990). Results of these findings were also endorsed by Lampson and Morse (1992) who observed significant reduction in populations of *S. oleae* upon treatment with fenoxycarb and methoprene.

The Florida wax scale, *Ceroplastis floridensis* (Comst.) is an important pest of citrus in Egypt. Its development was arrested upon treatment with fenoxycarb and R-20458 (Eisa *et al.*, 1991). There was no adult emergence at the end of first generation. Total inhibition of egg hatching was observed upon treatment with another JHA, dofenapyn Pyriproxyfen @ 0.05 to 0.1 per cent resulted in satisfactory control of the California red scale. *Aonidiella aurantii* (Maskell), chaff scale. *Parlatoria oleae* Colvee, and cottony cushion scale, *Icerya purchasi* Maskell in citrus orchards in Israel (Gokkes *et al.*, 1989; Peleg, 1989).

Pyriproxyfen prevented oviposition or nymph production in the green peach aphid, *Myzus persicae* (Sulzer), when 1-day old nymphs were dipped in 50 ppm concentration (Hatakoshi *et al.*, 1991). Also, treatment with 5 ppm pyriproxyfen or 5 or 50 ppm methoprene reduced oviposition as compared with untreated nymphs. Two sprays of 0.05 per cent pyriproxyfen at 45 days interval were found to be effective for the control of citrus whitefly, *Dialeurodes citri* (Ashmead) on citrus in Israel (Gokkes *et al.*, 1990).

Juvenile hormone antagonists also hold promise as fourth generation pesticides. Two chromenes, viz., 7-methoxy-2, 2-dimethylchromene (Precocene-1) and 6, 7-dimethoxy 2,2-dimethylchromene (Precocene-II) were the JH-antagonists isolated from common bedding plant, *Ageratum houstonianum* Mill. (Bowers, 1976; Bowers *et al.*, 1976). On exogenous application these resulted in precocious metamorphosis of milk weed bug, *Oncopeltus fasciatus*. Other chromenes like phenolic chromene and a hydroxyethyl chromene were also isolated from *Ageratum conyzoides* L. (Fagoonee and Umrit, 1980; Vyas and Mulchandani, 1984). Based upon these natural models, several derivatives with JH-antagonistic activity like fluoromevalonate (Quistad *et al.*, 1981), compactin (Monger *et al.*, 1982), ETB (Edwards *et al.*, 1983; Staal, 1977) and a variety of imidazoles (Kuwano *et al.*, 1983) have been synthesized. These chemicals at present are viewed as excellent research tools with promise as future pest control agents.

## CHITIN SYNTHETASE INHIBITORS (CSIs)

The enzyme, chitin synthetase, is crucial in the final step of polymerization of acetylglucosamine units to form long chain nitrogenous polysaccharide, chitin, in the cuticle of insects (Candy and Kilby, 1962). Chitin is one of the chief components of insect cuticle and provides strength to the exoskeleton. Thus, those chemicals which inhibit the activity of this enzyme have great potential as potent IGRs in the management of insect pests.

A benzoyl phenyl urea analogue as CSI was first discovered in 1970 by Philips-Duphar Co. which was a combination of a herbicide dichlobenil with urea diuron. On treatment symptoms appear during moulting and ecdysis of old skin is disrupted (Retnakaran *et al.*, 1985). The severity of disruption, depends upon the dose and insect species. In case of complete disruption, the insect dies in pharate condition. But in some cases of incomplete ecdysis, the old cuticle is shed off from the insect body but for the old head capsule which remains attched to the mandibular region of the new head capsule. Moulting disruption can also occur during larval-pupal moult in which larval pupal intermediates as envisaged in JHA treatment, are formed. In later stages of treatment, the adult emergence is interfered which results in inhibition of flight and reproduction in adults. The adult treatment results in laying of unviable eggs. Ovicidal effects have also been deciphered on direct treatment of eggs. Though CSIs are known to produce their effects mainly on ingestion but recently 'Novaluron' has been found to act both through ingestion and contact (Ishaaya and Horowitz, 1998). Limitations in the use of CSIs include the delayed mortality and activity only a certain times during development, apparently during chitin synthesis.

### Lepidoptera

Several commercial formulations of CSIs have been evaluated against different species of moths, leafminers and leafrollers. Amongst moths, the codling moth, *Cydia pomonella* (L.) and the Oriental fruit moth, *Cydia molesta* (Busck) were mainly the target insects. Treatment of apple leaves with 0.04 per cent suspension of diflubenzuron resulted in total sterility of both sexes of *C. pomonella* (Velcheva, 1983). These observations were also confirmed by Kholchenkov (1983). Incorporation of the CSI into artificial diet also proved lethal to the first and second instar larvae of the moth; the LC50 being 4.82 and 8.1 ppm, respectively (Elliott and Anderson, 1982). The CSI proved as effective as azinphosmethyl at the same concentration (Anderson and Elliott, 1982). Another CSI, triflumuron proved even more effective against *C. pomonella* than similar treatment with diflubenzuron. A satisfactory control of *C. molesta* on peach crop was also provided by diflubenzuron and hexaflumuron (Vermeulen and Parra, 1982; Kyaparissoudas, 1989). Also teflubenzuron and triflumuron proved more effective against *C. molesta* on apple and pear than azinphosmethyl (Pollini and Bariselli, 1993).

Diofenolan, a new CSI, applied @ 15 g a.i./100 litre performed equal or better than commercial standard insecticides against *C. pomonella* (Streibert *et al.*, 1994). The first application was done at the beginning of the oviposition followed by six applications until the end of moth flight. The CSI, at rates ranging from 10–30 g a.i./100 litre, also proved effective against some other important lepidotperous pests, e.g. *Cryptophlebia leucotreta, Phyllocnistis citrella Stainton* and *C. molesta*. Another CSI, hexaflumuron provided satisfactory control of *C. molesta* and peach twig borer, *Anarsia lineatella* on peach (Kyaparissoudas, 1989). Canson *et al.* (1992) have given details of physico-chemical and toxicological properties of the hexaflumuron and its formulated product,

Consul, alongwith its activity against pests of apple and pear including *C. pomonella, Leucoptera scitella* Zell. and *Lithocalletis blancardella* F. Teflubenzuron and triflumuron proved more effective against *C. molesta* on apple and pear than azinphos methyl (Pollini and Bariselli, 1993). A pre-blossom spray of triflumuron was found to be significantly better than diflubenzuron against *C. pomonella* (Glen *et al.*, 1982). Complete *Scitella* Zell. mortality of eggs of leafminer *Leucoptera* and *L. blancardella* F. on apple was achieved upon treatment of leaves with 0.1 per cent concentration of diflubenzuron before oviposition (Injac, 1981). Reduction of larval population of *L. malifoliella* with application of diflubenzuron (0.05%) at the start of oviposition, was also reported by Dulic and Injac (1982).

Kiwi fruit, an important crop of New Zealand, is infested by leafroller species like *Ctenopseutis obliquana* (Walker), *C. herana*, *Planotortrix octo*, *P. excessana* (Walker) and *Epiphyas postvittana* (Walker) (Follas *et al.*, 1994). A new CSI, Lufenuron, was evaluated against these species at 5 a.i./100 litre at 28-day interval gave significant reduction in fruit damage in this crop as well as on apple. Following two sprays of diflubenzuron on walnut plantations in China, the damage by a fruit borer, *Atrijuglans hetaohei* Yang could be reduced from 100 per cent to 5 per cent (Zhang *et al.*, 1992).

### Hemiptera

Several species of hemipterous insect pests of fruit crops including the scale insects and mealy bugs when subjected to treatment by some CSIs have given promising results, e.g. the California red scale, *Aonidiella aurantii* (Maskell), a major pest of citrus, has developed resistance to many conventional insecticides. Recently, experiments with buprofezin (Applaud 50 WP) resulted in the registration of this compound with 'Citrex' (an oil with a narrow distillation range) for the control of this scale on citrus (Vermeulen *et al.*, 1990). This combination is used on citrus in the early part of the season to prevent movement of the scale onto the fruit before late summer. Later, Grout and Richards (1991) determined the efficacy of buprofezin plus oil against *A. aurantii* at two different phenological times, viz., crawler emergence and midway between two peaks of crawler emergence (inter-emergence). The first treatment was significantly more effective than inter-emergence treatment only in dualcohort populations during winter when the inter cohort periods were longest. In populations with three cohorts, application timings did not influence treatment efficacy much, probably because the time between crawler emergence and inter emergence was shorter and within the effective residual period for buprofezin plus oil. Earlier, Yarom *et al.* (1988) conducted experiments study relating to the effectiveness of buprofezin against *A. aurantii* and the Mediterranean black scale, *Saissetia oleae* (Olivier). LC50 for the first and second instar of *A. aurantii* was 0.127 and 0.135 mg a.i./litre and for *S. oleae* 0.008 and 0.031 mg a.i./litre, respectively. Buprofezin had a moderate effect on male pupae of *A. aurantii* with an LC50 of 93 mg a.i./litre. At 100 mg a.i./litre, buprofezin suppressed embryogenesis of both the species, resulting in significant reduction in *A. aurantii* crawlers and in strong suppression of egg viability in *S. oleae*. Buprofezin was also found to be quite persistent, the RLT50 values (residual lethal time by which potency is reduced to 50% of the original value) following treatments at 125 mg/litre under indoor and outdoor conditions were 87.9 and 33.5 days, respectively.

Diofenolan at the rate of 20 g a.i./100 litre water per hectare, applied at the time of first and second crawler movement provided excellent control of *A. aurantii* under high population pressure in South Africa (Streibert *et al.*, 1994). Not only the California red scale but the control of whole citrus scale complex (Diaspididae) was obtained in Spain at the above concentration (Streibert *et al.*, 1994). Diofenolan also provided satisfactory control of the soft scale, *Ceroplastes floridensis* (Coccidae) at the rate of 5 to 10 g a.i./100 litre. Previously Eisa *et al.* (1991) observed inhibition of development of *C. floridensis* upon treatment with two CSIs, triflumuron and teflubenzuron. Teflubenzuron also appreciately reduced the population of *S. oleae* (Lampson and Morse, 1992).

Diofenolan also exhibited excellent control of some species of scales infesting apple and other pome fruits. At much lower doses than insecticides, the CSI proved effective against the San Jose scale, *Quadraspidiotus perniciosus* (Comstock) and other armoured scales like *Q. ostreaeformis* (Curtis) and *Lepidosaphes ulmi* (L.) (Streibert *et al.*, 1994). The compound was also effective against the white peach scale, *Pseudaulacaspis pentagona* at the rate of 15 g a.i./100 litre.

Three species of the armoured scale insects, viz., *Hemiberlesia rapax* (Comstock), *H. lataniae* (Signoret) and *Aspidiotus nerii* Bouch. are serious pests of Kiwi fruit in New Zealand. Buprofezin has been found to be highly effective against the armoured scales and hence registered for pre-flowering use (Tomkins *et al.*, 1994). The compound at 12.5 g a.i./100 litre lowers the fruit infestation to 5.8 per cent from 37.4 per cent in unsprayed vines.

Avocado crop in South Africa is seriously infested by heart shaped scale, *Protopulvinaria pyriformis* (Cockerell) The scale feeds on leaves and produces excessive quantities of honeydew which promotes the growth of sooty mould fungi and consequently the leaves and fruits become stained. In field trials, CGA-211446, a new CSI, was found to be more effective in suppressing the larval populations of the scale in comparison to the registered product, buprofezin (Steyn *et al.*, 1993).

Mealybugs on citrus could be effectively controlled with the two CSIs, buprofezin and diofenolan. Two sprays of diofenolan @ 20 g a.i./100 litre against *Planococcus citri* performed as good as chlopyriphos (Streibert *et al.*, 1994). *P. citri* also experienced 91 to 100 per cent nymphal mortality upon application of buprofezin (25 WP) (Mendal *et al.*, 1991). Also more than 80 per cent inhibition of egg hatch was observed. Buprofezin also gave satisfactory control of the citrus woolly whitefly, *Aleurothrixus floccosus* Mask. (Viaene, 1992).

### Coleoptera

Only a very limited work seems to have been done relating to the evaluation of CSIs against the coleopteran insects infesting fruit crops. Diflubenzuron against the sugarcane root and stalk borer weevil, *Diaprepes abbreviatus* L., which is an important pest of citrus in West Indies and has also established on citrus in the USA, when applied in citrus groves by air, reduced the egg hatch (Schroeder *et al.*, 1976). A spray oil, FC 435–36 (Simanton and Trammell, 1966) recommended for the control of citrus scale was also found to have an adverse effect on physical attachment of weevil eggs (Schroėder *et al.*, 1977). It was concluded that better control of *D. abbreviatus* could be achieved by a combination of the CSI plus oil (Schroeder and Sutton, 1978). No detectable residues of diflubenzuron, 4-chlorophenyl urea, 4-chloroaniline, was found in the finished pulp, fruit juice, processed liquor, molasses, prewash or afterwash water and emulsion water fractions (Schroeder *et al.*, 1980). Treatment with another CSI, triflumuron reduced the hatching rate to 1 and 2 per cent in weevil eggs caged on treated trees (Beavers and Schroeder, 1981). Flufenoxuron evaluated against *Capnodis tenebrionis* (L.) infesting apricot was found to have poor ovicidial effect (Malagon *et al.*, 1991).

## EFFECT OF INSECT GROWTH REGULATORS ON NON-TARGET ORGANISMS

Insect growth regulators certainly have an edge over the conventional insecticides being selective in action and possessing low toxicity to non-target organisms including parasitoids, predators and pollinators, etc. Diflubenzuron did not show any considerable adverse effect on *Aphelinus mali* (Hald.), a parasitoid of the woolly apple aphid, *Eriosoma lanigerum* (Hausm.) (Ravensberg, 1981). The CSI was also found to be non-toxic to the predacious mites, *Typhlodromus occidentalis* Nesbitt and *Zetzellia mali* (Ewing) and this treatment did not increase populations of phytophagous mites, *Panonychus ulmi* (Koch) and *Aculus* spp. (Anderson and Elliott, 1982). The parasitism of eggs of *Cydia pomonella* (L.) by *Trichogramma minutum* Riley and *T. pretiosum* Riley remained unaffected on application of diflubenzuron @ 10.06 kg a.i./ha and Alsystin 480 F (triflumuron) @ 1.00 and 0.41 kg a.i./ha) (Hagley and Laing, 1989). Buprofezin was found to be non-toxic to the predator, *Amblyseius ehari* Amitai and Swirski, of *Panonychus citri* (McG.) on citrus in western Japan (Kashio, 1983).

An excellent example of selectivity is presented by diofenolan, an IGR which has shown promise against lepidopterous pests and scale insects in pome fruits and citrus in Spain (Sechser *et al.*, 1994). Populations of *Cales noacki* and other hymenopterous parasitoids, *Chrysoperla carnea* (Stephens), *Conwentzia* spp., *Scymnus* spp. and spiders survived in the diofenolan treated plots at the same levels as in untreated plots. On citrus trees in Egypt, diofenolan caused only a small reduction of *Aphytis* spp. larvae parasitic on the citrus purple scale. Similarly, when used for the control of Codling moth, diofenolan showed no effect on the predatory mite *Typhlodromus pyri* in an Italian apple orchard.

## CONCLUSIONS

Relative safety of insect growth regulators to non-target organisms including human beings and other mammals, their selectivity and easy biodegradability make them eco-friendly and better alternatives as compared to the conventional insecticides, particularly in edible commodities which are generally consumed as raw. Of all the IGRs evaluated against insectpests of various fruit crops, the juvenile hormone analogues (JHAs) and chitin synthesis inhibitors (CSIs) hold promise for field level management.

Amongst JHAs, fenoxycarb, epofenonane, pyriproxyfen have shown great potential for the control of apple and citrus pests though several other UV-stable analogues have also been registered and are commercially available. The JHAs are quite selective and safe; toxicity of Insegar (containing 25% fenoxycarb) to man and domestic animals is low and is also harmless to bees. Antijuvenile hormone analogues (AJHAs), although seem interesting and useful research tools, yet none have shown sufficient promise as viable insect control agents; their limited specificity for hemimetabolous insects being the major hinderance in the commercial exploitation.

Many CSIs have been tried at laboratory and field level trials and several are commercially available for the management of insect pests of fruit crops like diflubenzuron, triflumuron, teflubenzuron, hexaflumuron, flufenoxuron, lufenuron and buprofezin. The potential exhibited by diflubenzuron and buprofezin, particularly against the pests of apple and citrus had been very encouraging. Scale insects and some lepidopterous pests seem to be quite amenable to diofenolan whose low mammalian toxicity and selectivity to beneficial insects hold the potential to make it an excellent insect control agent in pest management programmes of fruit crops.

Neuropeptides, though master regulators of various physiological processes in insects, are unlikely to be used as direct insect control agents because of their unstability and biodegradability. Moulting hormone agonists and antagonists too do not hold much promise in the management of insect pests because of their costly synthesis involving complex structures.

## REFERENCES


Anderson, D.W. and Elliott, R.H. 1982. Efficacy of diflubenzuron against the codling moth, *Laspeyresia pomonella* (Lepidoptera: Olethreutidae), and impact on orchard mites. *Canadian Entomologist,* **114**(8): 733–737.

Beavers, J.B. and Schroeder, W.J. 1981. Effect of BAY SIR 8514 in *Diaprepes abbreviatus* in citrus on Florida. *Florida Entomologist,* **64**(3): 361–362.

Ben-Avraham, E. and Yizhar, Y. 1987. Trials for the integrated control of apple pests in Western Calilee, *Hassadeh,* **67**(5): 918–924.

Bergot, B.J.; Baker, F.C.; Cerf, D.C.; Jamieson, G. and Schooky, D.A. 1981. Qualitative and quantitative aspects of juvenile hormone titres in developing embryos of several insect species; discovery of a new JH-like substance extracted from eggs of *Manduca sexta*. *In: Juvenile Hormone Biochemistry.* G.E. Pratt and G.T. Brooks (eds). Elsevier/North Holland Biomedical Press, Amsterdam. pp. 33–45.

Bowers, W.S., 1976. Discovery of insect antiallalotropins. *In: The Juvenile Hormones.* L.I. Gilbert (ed). Plenum Press, New York. pp.398–408.

Bowers, W.S.; Ohta, T.; Cleere, J.S. and Marsella, P.A. 1976. Discovery of antijuvenile hormones in plants. *Science,* **193**(4253): 542–547.

Brown, J.J. 1985. Influence of methoprene, low temperature and starvation on the incidence of diapause in the codling moth (Lepidoptera, Tortricidae: Olethreutinae). *Annals of the Entomological Society of America,* **78**: 316–321.

Brown, M.R. and Brown, J.J. 1982. Effect of methoprene on the fecundity and fertility of the codling moth, *Cydia pomonella. Annals of the Entomological Society of America,* **75**(3): 257–260.

Candy, D..J. and Kilby, B.A. 1962. Studies on chitin synthesis in the desert locust. *Journal of Experimental Biology,* **39**: 129–140.

Canson, B.De.; Faucon, M. and Decanson, B. 1992. Hexaflumuron, Insect growth regulator on pear and apple trees. *Phytoma*, **444**: 83–85.

Charmillot, P.J. 1989. Insect growth regulators (IGR), mimics of juvenile hormone as morphogenetical and ovicidal means of control against orchard torticids. *Entomologia Experimental et Applicata,* **51**: 59–69.

Charmillot, P.J.; Baillod, M.; Bloesch, B.; Guignard, E.; Antonin, P.; Frischknecht, M.L.; Hoehn, H. and Schmid, A. 1987. An insect growth regulator used for its ovicidal action in the control of the grape moths, *Lobesia botrana* Den. & Schiff. and *Eupoecilia ambiguella* Hb. *Revnu suisse de Viticulture. d'Arboriculture et. d. Horticulture,* **19**(3): 183–191.

Charmillot, P.J. and Bloesch, B. 1987. An insect growth regulator (IGR) approved for its ovicidal action against the plum moth, *Grapholita funebrana Tr. Revue Sussie de Viticulture, d' Arboriculture et.d. Horticulture,* **19**(2): 87–92.

Charmillot, P.J.; Bloesch, B. and Benz, M. 1989. Control of the codling moth *Cydia pomonella* L. with fenoxycarb and teflubenzuron.*Revue Suisse de Viticulture. d.Arboriculture et.d. Horticulture,* **21**(3): 187–193.

Charmillot, P.J. and Brunner, J.F. 1989. Summer fruit tortrix. *Adoxophyes orana*: Life cycle, warning system and control. *Entomologia Hellenica,* **8**: 17–26.

D'Anna, R. and Leonardi, M. 1990. Preliminary tests in the control of *Saissetia* on citrus trees using fenoxycarb. *Informatore-Agrario,* **47**(16): 89–90.

Dulic, K. and Injac, M. 1982. Regulation of populations of the apple leaf miner (*Leucoptera scitella* Zell.) by the method of integrated protection. *Zastita Bilja,,* **33**(3): 333–342.

Edwards, J.P.; Bergot, B.J. and Staal, G.B. 1983. Effects of three compounds with antijuvenile hormone activity and a juvenile hormone analogue on endogenous juvenile hormone levels in Tobacco hornworm, *Manduca sexta. Journal of Insect Physiology* **29**: 83–89.

Eisa, A.A.; El-Fatah, M.A.; El-Nabawi, A. and El-Dash, A.A. 1991. Inhibitory effects of some insect growth regulators on developmental stages, fecundity and fertility of the Florida wax scale, *Ceroplastis floridensis. Phytoparasitica* **19**(1): 49–55.

Elliott, R.H. and Anderson, D.W. 1982. Factors influencing the activity of diflubenzuron against the codling moth *Laspeyresia pomonella* (Lepidoptera: Olethreutidae). *Canadian Entomologist,* **114**(3): 259–268.

Fagoonee, I. and Umrit, G. 1980. Biology of *Dysdercus flavidus* Sign. and its control by *Ageratum conyzoides. Revue agri. Sucr. lle Maurice,* **59**: 122–128.

Follas, G.; Welsh, R.; Bland, S.; Patterson, T.M. and Popay, A..J. 1994. Lufenuron-an insect growth inhibitor for the control of leaf-rollers in apples and kiwi fruit. *Proceedings of 47th New Zealand Plant protection Conference, New Zealand,* 9–11 August, 1994. pp.341–345.

Garcia, E.S., Luz, N. Azambuja, P. and Rembold, H. 1990. Azadirachtin depresses the release of prothoraciotropic hormone in *Rhodnius prolixus* larvae-Evidence from head transplantations. *Journal of Insect Physiology,* **36**: 679–682.

Gelbic, I. and Sehnal, F. 1973. Effects of juvenile hormone mimics on the codling moth, *Cydia pomonella* (L.) (Lep, Olethreutidae). *Bulletin of Entomological Research,* **63**: 7–16.

Glen, D.M.; Milson, N.F.; Wiltshire, C.W. and Ruby, M.P. 1982. SIR 8514 compared with diflubenzuron for control of orchard moths and apple sucker. *Annals of Applied Biology,* **100**: 6–7.

Gokkes, M.; Eshel, G. and Tadmor, U. 1989. field trials for the control of chaff scale and Florida wax scale in citrus orchards with Tiger (Pyriproxyfen). *Hassadeh,* **69**(11): 2019.

Gokkes, M.; Tadmor, U.; Carmon, D. and Shneider, R. 1990. An uncorroborated field study for the control of citrus white fly, *Dialeurodes citri* (Ashmead). *Alon Hanotea,* **44**(8): 759–761.

Grout, T.G. and Richards, G.I. 1991. Effect of buprofezin applications at different phenological times on California red scale (Homoptera: Diaspididae). *Journal of Economic Entomology,* **84**(6): 1802–1805.

Hagley, E. A.C. and Laing, J.E. 1989. Effect of pesticides on parasitism of artificially distributed eggs of the codling moth,, *Cydia pomonella* (Lepidoptera: Tortricidae) by *Trichogramma* sp. (Hymenoptera: Trichogrammatidae). *Proceedings of the Entomological Society of Ontario,* **120**: 25–33.

Hatakoshi, M.; Shono, Y.; Yamamoto, H. and Hirano, M. 1991. Effects of the juvenile hormone analog pyriproxyfen on *Myzus persicae* and *Unaspls yanonensls. Applied Entomology and Zoology,* **26**(3): 412–414.

Heller, J.J. Mattioda, H., Klein, E. and Sagenmuller, A. 1992. Field evaluation of RH 5992 on lepidopterous pests in Europe. *Proceedings of the Brighton Crop Protection Conference, Pests and Diseases*, 1992, Brighton, Nov. 23–26, 1992, British Crop Protection Council. Farnham, U.K. pp. 59–65.

Holman, G.M., Nachman, R.J. and Wright, M.S. 1990. Insect neuropeptides. *Annual Review of Entomology,* **35**: 201–218.

Hora, J., Labler, L., Kasal, A., Cerny, V., Sorm, F. and Slama, K. 1966. Moulting deficiencies produced by some sterol derivatives in an insect (*Pyrrhocoris apterus* L.). *Steroids,* **8**: 887–914.

Injac, M. 1981. Results of laboratory experiments on the ovicidal effects of diflubenzuron on eggs of various ages of the apple leaf miners *Leucopteru scitellu* Zell. and *Lithocolletis blancardella* F. Zastita Bilja, **32**(3): 241–249.

Ishaaya, I. and Horowitz, A.R. 1998. Insecticides with Novel Modes of Action: an Overview. *In: I. Ishaaya and D. Degheela (eds). Insecticides with novel modes of action: Mechanisms and Application*, Narosa Publishing House. pp. 1–24.

Jong, D.J.de., Minks, A.K. and De Jong, D.J. 1981. Studies on *Adoxophyes orana,* the major leafroller pest in apple orchards in the Netherlands. *Mitteihungen der Schweizerischen Entomologischen Gesell Schaft,* **5493**: 205–214.

Joshi, N.K. and Sehnal, F. 1989. Inhibition of ecdysteroid production by plumbagin in *Dysdercus cingulatus. Journal of Insect Physiology,* **35**: 737–741.

Joshi, N.K., Lathika, K.M., Banerji, A. and Chadha, M.S. 1988. Effects of plumbagin on growth and development of red cotton bug, *Dysdercus cingulatus* Fab. *Proceedingsof the Indian National Science Academy* B, **54**: 43–46.

Kashio, J. 1983. Toxicity of certain insecticides and acaricides toward *Amblyseius eharai* Amitai and Swirski (Acarina: Phytosceidae). *Bulletin of the Fruit Tree Research Station, Japan-D-Kuchino-Tsu,* **5**: 83–92.

Kholchenkov, V.A. 1983. The effectiveness of Dimilin against the codling moth and phytophagous mites on apple orchards. *Khimiya-V-Sel'Skom-Khozyaistve,* **2**: 23–25.

Koreeda, M., Nakanishi, K. and Goto, M. 1970. Ajugalactone, an insect moulting inhibitor as tested by the chilo dipping method. *Journal of American Chemical Society,* **92**(25): 7512–7513.

Kubo, I., Lee, Y.W., Balogh-Nair, V., Nakanishi, K. and Chapya, A. 1976. Structure of Ajugarins. *Journal of Chemical Society,* **22**: 949–950.

Kuwano, E., Takeya, T. and Eto, M. 1983. Terpenoid imidazoles: new antijuvenile hormones. *Agric. Biol. Chem.* **47**: 921–923.

Kyaparissoudas, D.S. 1989. Simultaneous control of *Cydia molesta* and *Anarsia lineatella* in peach orchards of northern Greece by combining mating disruption and insecticide treatments. *Entoimologia-Hellenica,* **7**: 13–16.

Larguier, M. and Rivenez, M. 1990. The types of secondary action of fenoxycarb on the pear psyllid. *Phytoma,* **420**: 22–24.

Lampson, L.J. and Morse, J.G. 1992. Impact of insect growth regulator on black scale, *Saissetia oleae* (Olivier) (Homoptera: Coccidae) and inter-tree dispersal. *Journal of Agricultural Entomology,* 9(3): 199–210.

Malagon, J., Garrido, A., Busto, T.del and Del-Busto, T. 1991. Note on the effect of flufenoxuron on oviposition by *Capnodis tenebrionis* (L.) (Col.; Bupestidae). *Investigation Agraria Production-Υ-Protection-Vegetables,* **6**(1): 129–132.

Mendel, Z., Blumberg, D. and Ishaaya, I. 1991. Effect of buprofezin on *Icerya purchasi and Planococcus citri. Phytoparasitica,* **19**(2): 103–112.

Menke, F. 1988. What is known of Insegar in Viticulture. *Obstbau Weinbau,* **25**(6): 89.

Menn, J.J.; Raini, A.K. and Edwards, J.P. 1989. Juvenoids and neuropeptides as insect control agents: Retrospect and prospect *In: Progress and Prespects in Insect Control. Monograph* No. 43. British Crop Protection Council Brighton, U.K. pp. 89–105.

Monger, D.J.; Lim, W.A.; Kezdy, F.J. and Law, J.H. 1982. Compactin inhibits insect HMG Coa reductase and juvenile hormone biosynthesis. *Biochem. Biphys. Res. Commun.* **105**: 1374–1380.

Monnet, Y. 1986. Against fruit peel tortricids: A substance that interferes with insect metamorphosis. *Phytoma,* **374**: 28–30.

Oberhofer, H. 1988. Insegar, a new growth regulator. *Obstbau Weinbau,* **25**(6): 183–186.

Peleg, B.A. 1989. Evaluation of the insect growth regulator pyriproxyfen (Tiger) as a control agent for the California red scale and the cottony cushion scale. *Alon Hanotea,* **43**(6): 681–686.

Pena, J.E. and Duncan, R. 1994. Control of the citrus leaf- miner in South Florida. *Proceedings of the Florida State Horticultural Society,* **106**: 47–51.

Pollini, A. and Bariselli, M. 1993. *Cydia molesta*: Pest on the increase and defence of pome fruits. *Informatore-Agrario, 49*(14): 19–21.

Quistad, G.B.; Cerf, D.C.; Schooley, D.A. and Staal, G.B. 1981. Fluoromevalonate acts as an inhibitor of insect juvenile hormone biosynthesis. *Nature,* **289**: 176–177.

Ravensberg, W.J. 1981. The natural enemies of the woolly apple aphid. *Eriosoma lanigerum* (Hausm.) (Homoptera: Aphididae), and their suceptibility to diflubenuron. *Mendedelingen-van-de-Faculteit-Landbouw wetenschappen-Rij Ksunvyersiteil-Gent.* **46**(2): 437–441.

Retnakaran, A.; Granett, J. and Ennis, T. 1985. Insect growth regulators. *In: Comprehensive Insect Physiology, Biochemistry and Pharmacology.* C.A. Kerkut and L.I. Gilbert (eds). Vol. **12**. Pergamon Press, Oxford. pp. 529–601.

Richards, G. 1981. Insect hormones in development. *Biological Review,* **56**(4): 501–549.

Saxena, R.C. 1983. Naturally occurring pesticides and their potential. *In: Chemistry and World Food Supplies: The New Frontiers.* L.W. Shemilt (ed). Pergamon Press, Oxford. pp. 143–161.

Schmid, A.; Van der Molen, P.J.; Jucker, W.; Baggiolini, M. and Antonin, P. 1978. The use of insect growth regualtor analogues of the juvenile hormone against summer fruit tortrix moth, *Adoxophyes orana* and other pests. *Entomologia Experimentalis at Applicata,* **24**: 64–82.

Schroeder, W.J., Beavers, J.B., Sutton, R.A. and Selhime, A.G. 1976. Ovicidal effect of Thompson-Hayward TH-6040 in *Diaprepes abbreviatus* on citrus in Florida. *Journal of Economic Entomology* **69**: 780–782.

Schroeder, W.J.and Sutton, R.A. 1978. Diaprepes abbreviatus: supression of reproductive potential on citrus with an insect growth regulator plus spray oil. *Journal of Economic Entomology,* **71**: 69–70.

Schroeder, W.J.; Sutton, R.A. and Beavers, J.B. 1980. *Diaprepes abbreviatus*: fate of diflubenzuron and effect on non-target pests and beneficial species after application to citrus for weevil control. *Journal of Economic Entomology,* **73**(5): 637–638.

Schroeder, W.J.; Sutton, R.A. and Selhime, A.G. 1977. Spray oil effects on *Diaprepes abbreviatus* on citrus in Florida. *Journal of Economic Entomology,* **70**: 623–624.

Sechser, B.; Reber, B. and Wesiak, H. 1994. Selectivity of diofenolan (CGA 59205) and its potential for integrated scale control.*Proceedings of the Brighton Crop Protection Conference-Pests and Diseases.* pp. 1193–1198.

Seiber, K.P. and Rembold, H.. 1983. The effects of azadirachtin on the endocrine control of moulting in *Locusta migratoria*. *Journal of Insect Physiology,* **29**: 523–527.

Simanton, W.A. and Trammell, K. 1966. Recommended specifications for citrus spray oils in Florida. *Proceedings of the Florida State Floricultural Society,* **79**: 26–30.

Singh, D. and Bhathal, S.S. 1994. Role of insect growth regulators in integrated pest management. *Journal of Insect Science,* **7**: 1–9.

Staal, G.B. 1977. Insect control with insect growth regulators based on insect hormones. *In: Natural Products and the Protection of Plants.* G. Marini-Bettolo (ed). Elsevier, Amsterdam. pp. 353–377.

Starratt, A.N. and Brown, B.E. 1975. Structure of the pentapeptide proctolin a proposed neurotransmitter in insects. *Life Science,* **17**: 1253–1256.

Steyn, W.P.; Toit, W. J du.; Villiers, E.A. de.; Du Toit W..J. and De-Vukkuersm, E.A. 1993. Effect of insect growth regulator, CGA 211446, on the third instar of the heart shaped scale on avocados. *Yearbook-South African Avocado Growers Association,* **16**: 116–117.

Streibert, H.P., Frischknecht, M.L. and Karrer, F. 1994. Diofenolan a new insect growth regulator for the control of scale insects and important lepidopterous pests in deciduous fruit and citrus. *Proceedings of the Brighton Crop Protection Conference Pests and Diseases*. pp. 23–30.

Tomkins, A.R., Follas, G., Thomson, C., Wilson, D.J. and Popay, A.J. 1994. A new insect growth regulator, CGA 59205, with promising activity against scale insects. *Proceedings of the 47th New Zealand Plant Protection Conference New Zealand.* pp. 337–340.

Varjas, L. 1992. Use of juvenoid, a derivative of fenoxycarb in the control of lepidopterous pests in orchards. *Acta Phytopathologica et Entomologica Hungarica,* **27**(1–4): 643–648.

Velcheva, N. 1983. Chemical sterilisation of the codling moth *Laspeyresia pomonella* L. (Lepidoptera: Tortricidae), *Gradinarka-i-Lozaroska-Nauka,* **20**(1): 27–35.

Vermeulen, J.B. and Parra, E. 1982. Effectiveness of insecticides for the control of *Grapholitha molesta on peach Plaguicidas-Υ-hormones*: 7–8.

Vermeulen, J.B.; Sureet, S.; Krause, M.; Hollings, N. and Nel, A. 1990. *A guide to the use of pesticides and fungicides in the Republic of South Africa.* 34th edition, Department of Agricultural Development, Pretoria, South Africa.

Viaene, N. 1992. Application of buprofezin to control citrus woolly whitefly (*Aleurothrixus floccosus* Mask) in a citrus nursery in Zaire. *Fruits Paris,* **47**(2): 303–308.

Vyas, A.V. and Mulchandani, N.B. 1984. New chromenes from *Ageratum conyzoides. In: A.S.U. Choughley and M.R. Heble (ed.). Progress Report. Bhabha Atomic Research Centre, Trombay. Bio Organic Division.* pp. 7.

Walker, W.F. and Svoboda, J.A. 1973. Suppression of Mexican bean beetle development by foliar applications of azasterols and reversal by cholesterol. *Entomologia Experimentalis Applicata,* **16**: 442–426.

Watkinson, I.A. and Clarke, B.S. 1973. The insect moulting hormone system as a possible target site for insecticidal action. PANS, **19**(4): 488–506.

Yarom, I.; Blumberg, D. and Ishaaya, I. 1988. Effects of buprofezin on California red scale (Diaspididae)and Mediterranean black scale (Coccidae). *Journal of Economic Entomology,* **81**(6): 1581–1585.

Yokoyama, V.Y. and Miller, G.T. 1991. Potential of pyriproxyfen as a quarantine treatment for codling moth and Oriental fruit moth (Lepidoptera: Tortricidae). *Journal of Economic Entomology*, **84**(3): 842–947.

Zhang, W.Z.; Sun, X.; Gao, Z.L.; Wang, H.J. and Fu, Y.Z. 1992. Control of *Atrijuglans hetaohei* Yang (Lep: Heliodinidae) with diflubenzuron. *Chinese Journal of Biological Control,* **8**(3): 129–130.

*Pest Management in Horticulture Crops*
L.R. Verma, A.K. Verma and D.C. Gautam (eds)


## *Chapter 7*

# Role of Semiochemicals in the Management of Insect Pests of Fruit Crops

*Gursharan Singh and Darshan Singh*

## INTRODUCTION

CHEMICALS ARE KNOWN to be used by organisms to communicate with other organisms and with their surrounding environment. Among insects, the phenomenon of chemical communication has fascinated man since centuries but it was the French naturalist, Henry Fabre (1879) who first documented his observations scientifically. He observed that placing a single female emperor moth, *Saturnia pavonia* in a screen cage on his window ledge, attracted 150 marked male emperor moths released several kilometers away during a period of eight evenings.

Chemicals used for communication between organisms are known as semiochemicals (Law and Regnier, 1971). The semiochemicals are subdivided into two major groups, pheromones and allelochemicals, depending upon the interactions are intraspecific or interspecific, respectively. Karlson and Luscher (1959) coined the term pheromone to define chemicals emitted by an individual which cause a change in behaviour of another individual of the same species. Pheromones are subdivided according to their function, the main types being sex pheromones which bring together opposite sexes for mating; aggregation pheromones which attract both sexes and generally serve to capitalise on a food source as well as bring together opposite sexes for mating; alarm pheromones which alert other individuals to some source of danger; trail pheromones which mark a trait laid by pioneering individuals to some source of food or refuge, and social pheromones which govern interactions within organised societies (Justum and Gordon, 1989).

At present, four types of allelochemicals are recognised, i.e. allomones, kairomones, synomones and apneumones (Norlund, 1981). An allomone is a substance produced or acquired by an organism that when it contacts an individual of another species in the natural context, evokes in the receiver a behavioural or physiological response that is adptively favourable to the emitter (Brown, 1968; Norlund and Lewis, 1976). In contrast, kairomone evokes in the receiver a behavioral or physiological response that is adaptively favourable to receiver (Brown *et al.*, 1970; Norlund and Lewis, 1976). In the complementary terminology an allomone can be described as a repellent and a kairomone as an attractant which repel and attract an organism, respectively. Similarly a synomone evokes in the receiver a behavioural or physiological response that is adaptively favourable to both emitter and the receiver (Norlund and Lewis, 1976). There are some interesting chemical mediated interactions in which the chemical(s) mediating the interactions between individuals of different species originate from a non-living source. These chemicals cannot be covered under allomones, kairomones, or synomones. Instead, Norlund and Lewis (1976) proposed another term, apneumone for such chemicals evoking a behavioural or physiological reaction that is adaptively favourable to the receiving organism but deterimental to an organism of another species that may be found in or on the non-living material.

## USES OF SEMIOCHEMICALS

The credentials of semiochemicals are impeccable in comparison with the conventional chemical insecticides which apart from environmental pollution/hazards have created serious problems such as residues in food and feed, development of resistance and resurgence in target insects and elimination of natural enemies. Insect semiochemicals, being natural products, do not themselves kill insects, let alone other animals and plants. Because of their high biological activity, only small amounts are required to produce an effect. They are generally remarkably specific for the target pest, unlike many insecticides which are brood-spectrum, killing not only the target pest but other insects and higher organisms as well.

Among all classes of semiochemicals, pheromones and kairomones (attractants) have been appreciably exploited in the context of pest management. Sex and aggregation pheromones have been investigated most intensively in terms of their chemistry, behaviour activity and application to manipulation of insect behaviour. Their applications include use of pheromone-baited traps for control by mass-trapping, combining with insecticides in a 'lure and kill' approach and to disrupt matings of the target species using sex pheromones (Justum and Gordon, 1989; Smith, 1992; McVeigh *et al.*, 1993; Krishnian and Divakar, 1995).

### Semiochemical traps for monitoring

Traps baited with pheromones or other semiochemicals have significant advantage over other insect sampling techniques such as using light or suction. These traps are simple and cheap to construct and maintain, require no power source and attract only the target pest unlike other non-specific devices where the catch has to be sorted out by trained entomologists. Traps baited with sex and aggregation pheromones can be used to monitor the target pests species. A wide range of trap designs is available (Muirhead-Thomson, 1991) most involving some form of adhesive surface, water of dry funnel as the traping medium. The pheromone is formulated in a slow-release dispenser which typically contains 1 mg or less of pheromone in the case of sex pheromones, although more with the aggregation pheromones and remains active for several weeks in the field. These traps can be used as sensitive devices for simply detecting the presence of the pests and in principle, they can be calibrated to provide a quantitative measure of the pest density which it may be possible to relate to prospective levels of pest damage (Wall, 1989). These days pheromone traps are extensively used for monitoring insect pests. Even in 1988, lures for more than 230 species were commercially available (Inscoe *et al.*, 1990).

### Semiochemical traps for mass trapping and lure-and-kill

Traps baited with aggregation pheromones have been very successful for mass trapping of coleopterous insect pests. However, traps baited with synthetic female sex pheromones could not be very effective since even a single male can generally fertilise many female insects. Semiochemicals attracting female insects stand much better chances for control of insect pests by trapping. For example, the sex pheromone produced by male coffee white stem borer, *Xylotrechus quadripes* Chev. has recently been identified at The Natural Resource Institute, U.K (Hall, 1998) and traps baited with the synthetic pheromone have been shown to attract female weevils in India.

The more successful use of mass trapping has been with lure-and-kill technique baited with semiochemicals in which the target insect is attracted to a source of insecticide rather than to a mechanical trap. The advantage over mass trapping is that many more attractive sources can be employed while the advantage over mating disruption method is that less pheromone is used. A great success has been achieved in the control of fruit flies through this technique (Cunningham *et al.*, 1990).

### Semiochemical for mating disruption

Although mass trapping with sex pheromones has generally proved unsuccessful, yet the alternative approach of permeating the atmosphere with synthetic sex pheromones to disrupt communication between the sexes and hence reduce mating frequency as originally proposed by Shorey and co-workers (1972) has been much more effective. The exact mechanism by which this works is not clear and possibly differs according to the type of formulation and even with the particular species being targeted, but the effect probably involves a combination of both false trail-following and sensory fatigue (Sanders, 1997). Availability of controlled release formulations is a pre-requisite for this technique to be effective. More fundamental difficulties inherent in the mating-disruption technique involve its specificity for the target pest which is highly desirable for an integrated pest management strategy and the need to treat relatively large or isolated areas (Hall, 1995).

## MANAGEMENT OF INSECT PESTS OF FRUIT CROPS WITH SEMIOCHEMICALS

Cotton pink bollworm, *Pectinophora gossypiella* is the insect where a large number of extensive field trials have been conducted in USA, Egypt, India, Pakistan, China, Peru and Greece. Operations on other major crops include orchard and vineyard pests where there is a strong need to minimise conventional insecticides because these are high-value food crops. Of all the semiochemicals, the ones maximally exploited for the management and/or monitoring the insect pests of fruit crops include the sex pheromones and the kairomones described under the following heads.

### Codling moth

Codling moth, *Cydia pomonella* (Linn.) is a major pest of pome fruits in most temperate regions of the world (Barnes, 1991). The pest can be controlled with conventional insecticides and microbial pesticides, but the strict regulatory measures adopted by the developed countries have resulted in the cancellation of registration of many insecticides. Further, resistance to guthion, one of the most commonly used insecticides and cross-resistance to several other classes of pesticides have also been documented (Varela *et al.*, 1993). Pheromone-baited mating disruption of the moth has emerged as an alternative method to suppress its population. The major biologically active component of codling moth pheromone is the conjugated diene alcohol, E8, E10-dedcatienol (Roelofs *et al.*, 1972), commonly known as Codlemone. Rubber or plastic dispensers containing the pheromone were hung in the orchard just before the flight period of the moth (Charmillot, 1990). The rubber tubing dispenses used were small tubes, 2 mm in internal diameter and 4 mm in external diameter. These dispensers were filled with undiluted pheromone and sealed at both ends by a metal clip. Hercon dispensers have also been tested; cut into pieces of 6.45 $cm^2$ and each unit contained 40 mg of Codlemone. With both kinds of dispensers, the borders of the orchards were treated by placing one dispenser per tree, 4 to 5 metre apart. In case of both the dispensers, the catches of males in the pheromone traps located in the middle of the trial orchard were always almost completely inhibited compared with the control traps placed in the same region. However, a high reduction in catches is not always a gurantee of successful control (Moffitt and Westigard, 1984). Conversely, the inhibition of catches may be considerably weak when some traps are placed at particular places, such as top of the tree, border of the orchard etc., but this does not mean that the mating disruption technique has failed (Mani, 1986).

Madsen and Carty (1979) have also achieved good control of codling moth in an isolatd orchard in British Columbia. Mass trapping reduced the number of eggs deposited resulting in low population of larvae.

During 1979–94, a large number of mating disruption trials were carried out against *C. pomonella* on apples over an area of 220 ha in Switzerland (Mani *et al.*, 1996). Very good results (< 1% attack on fruits) were obtained in 60 per cent cases, good results (1–2% attack) in 15 per cent cases and insufficient control (>2% attack) in 25 per cent orchards. Reasons for inadequate control proved to be a large emerging population or insufficient evaporation rate of the pheromone. Failure was also as a result of unsuitable orchard conditions (too small, low uniformity of plant stands, irregular ground plant, insufficient isolation). Moth catches and mating rates of females exposed in the field indicated that pheromone concentrations at the borders of the orchards and especially on the tops of the trees, did not always reach levels sufficient to completely prevent communication between males and females. Pheromone-mediated mating disruption of the moth was also tried in commercial apple and pear orchards in British Columbia in 1991 and 1992 using Isomate-C dispensers (Judd *et al.*, 1996). A single treatment of 1000 dispensers/ha released the pheromone (Codlemone) at calculated rates of 14.9, 15.2m 16.6 and 17.5 g/ha from Ist May to 30$^{th}$ September in Kelowna, Summerland, Cawston and Oliver, respectively. Damage in the pheromone treated orchards ranged from 0 to 1.52 per cent against 43.5 per cent in the untreated control orchards, thus making Isomate-C commercially viable in British Columbia. As an exercise in resistance management, an Isomate-C plus mating disruption programme with supplementary insecticide intervention, was implemented in a block of apple orchards in South Africa (Barnes *et al.*, 1997). By the end of the third season, insecticide applications were reduced by an average of 32 per cent and fruit infestation limited to an average of 0.3 per cent. Another formulation (Sirene) containing 0.16 per cent Codlemone and 6.0 per cent permethrin was shown to have a different mode of action from mating distruption and to be effective in controlling *C. pomonella* in orchards (Charmillot *et al.*, 1997).

In pear orchards in the USA field trials were conducted to compare the ability of Codlemone and isomers of Codlemone to disrupt the pheromonal communication of *C. pomonella* (McDonough *et al.*, 1996). When the test disruptant was an equilibrium mixture of Codlemone and its isomers (61% Codlemone + 39% isomers), the percentage communication disruption was 86.8 compared to 68.7 for Codlemone. When the disruptant was (8E, 10Z)-8, 10-dodecadien-l-ol (94%, E, Z, 3% E, E), the percentage disruption was 86.4 compared to 62.7 for Codlemone. The results showed that the previously reported superior disruptant potency (relative to Codlemone) of compositions containing Codlemone with a high percentage of isomers was not a result of the proximity of the dispensers to the traps.

A comparison of cost of using mating disruption versus traditional insecticide applications for controlling *C. pomenella* in Washington apple orchards revealed that sucessful mating disruption prevented the moth from oviposition and consequently there were no larvae to damage fruits (Williamson *et al.*, 1996). On an average, mating disruption was found to be more expensive by $ 188.22/ha. Mating disruption was, however, more economical under low pest population pressure. A risk analysis or price/cost sensitivity analysis confirmed that changes in pheromone prices, cover spray use and cullage could alter the outcome of the economic analysis to where pheromone control would be economically desirable control method.

Exposure of E8, E10-dodecadienol, the major component of codling moth pheromone, to sunlight and air resulted in extensive degradation, primarily to non-volatile oligomers and polymers (Miller, 1995). Contents and surface, residues of new and field-aged pheromone dispensers were analysed; the tacky film on the surface of field aged dispensers was primarily non-volatile decomposition products and the pheromone remaining in the reservoir of field-aged dispensers degraded significantly. Carbon black, which physically blocks light and 2 phenylenediamine antioxidants were effective stabilizers that hindered pheromone degradation.

Experimentation with mating disruption technique to control the codling moth has paved the way for its practical field application. The fact that this control strategy is totally innocuous to useful fauna and the environment, makes it fit in the integrated pest management programmes. However, on consideration of mode of action of pheromone and dispersal activity of codling moth, Charmillot (1990) listed few topological and ecological restrictions for successful use of mating disruption technique. These are:

- The mating disruption technique against this moth must be exclusively reserved for isolated orchards of not less than 3 hectare.
- Isolation from the external infestation sources of the moth must be at least 100 meter. When the mating disruption technique is applied only on part of a set of juxtaposed orchards, a buffer zone of 40 to 50 meter must be kept, where mating disruption technique and classical insecticide treatments should be used concurrently.
- The borders of the orchards must be protected by small dispensers placed at a maximum distance of 4 to 5 meter from one another, but inside the orchard bigger ones can be placed with more space between them.

- The mating disruption technique should not be used when the borders are irregular or when the fields are too elongated, because the proportion of borders where matings are more likely to occur becomes too important in comparison with other area.
- Difficulties may also appear in crops with heterogeneous crown sizes or with trees of more than 5 meter in height. Experience has shown that on the top of the canopy the attractant is rapidly removed by wind.
- The initial populations must not have more than 2 or 3 diapausing larvae per tree so that the tolerance threshold is not exceeded during the first year of application of the mating disruption technique, because insect density decreases gradually in successive generations.
- If the half-emission time of the attractant is sufficiently long, the dispensers can be put in place sometime before the beginning of the flight period of the codling moth to avoid all matings at that time. The second application should be made 40 days after the first.
- Finally, it should be noted that placing of dispensers, which takes from half an hour to an hour per hectare according to the size of orchard, should not discourage anyone from using the mating disruption technique.

**Oriental fruit moth**

The Oriental fruit moth, *Graphalita molesta* (Busck) is a world famous pest of deciduous fruits resulting in huge economic loss to peaches and nectarines (Rice and Kirsch, 1990). It also attacks apple, pear, plum, apricot, cherry and almond. Availability of synthetic sex pheromones for *G. molesta* has provided an opportunity to use these semiochemicals as alternative to conventional insecticides. Identification of major component of *G. molesta* female pheromone, (Z)-8 dodecenyl acetate (Z8012: AC), was made by Roelofs *et al.* (1969) and mating disruption trials were conducted. In Australia, where *G. molesta* is a key pest of stone-fruits, a pheromonal control was achieved through mating disruption technique for the first time in 1985. In all trials, the catches of males in pheromone traps and the proportion of mated females taken at food lures (brown sugar solution plus terpenyl acetate) were significantly lower in the treated orchards than in controls (Vicker, 1990). Similarly, level of fruit and shoot damage was lower in pheromone-treated area than in control. Thus, it concluded that pheromones released from suitably placed dispensers, at rates exceeding 6 mg/ha/hr, were as effective as the standard insecticide schedule in controlling *G. molesta* (Rothschild, 1975). Control of *G. molesta* by mating disruption over a large peach growing area was attempted by Vickers *et al.*, 1985. The treatment reduced the population size; as indicated by the pheromone trap and feeding lure captures, level of infestation and the mating success of wild females. The mating status of wild females is considered to be the most meaningful criterion of mating disruption (Rothschild, 1981). Rice and Kirsch (1990) conducted mating disruption trials of the moth with pheromone dispensers, each containing a 75 mg blend of (Z)-8-dodecenyl acetate (93%), (E)-8-dodecenyl acetate (6%) and (Z)-8-dodecenol (1%). Dispensers were placed in the upper third of the tree canopy (peach and nectarine) at heights of 2.5 to 4.0 metre and at densities of 1000 dispensers per hectare. The results revealed that the pheromone treatment invariably achieved control levels equal to or better than those measured in the control blocks.

An area wise mating disruption programme with Isomate-M was initiated for the control of *G. molesta* during 1991–92 in 1200 hectare of peaches and nectarines in the Tulbagh valley of South Africa. The project was an outstanding success. By the end of the season, only 69 adults of *G. molesta* were caught in pheromone traps, shoot strikes were reduced to isolated occurrences and no infested fruit was recorded from the treated area (Barnes *et al.*, 1997).

### Tephritid fruit flies

Male lures and food baits have been extensively tried the worldover for monitoring or controlling tephritid fruit flies, viz., the Oriental fruit fly, *Bactrocera* (= Dacus) *dorsalis* Hendel, Mediterranean fruit fly, *Ceratitis capitata* (Wiedemann) and melon fruit fly, *B. cucurbitae* (Coquillett). Interestingly, a powerful male lure was available for each of the three species; methyl eugenol for the Oriental fruit fly (Howlett, 1915); Trimedlure for Mediterranean fruit fly (Beroza *et al.*, 1961) and Cuelure for the melon fruit fly (Alexander *et al.*, 1962). The first and most efficient male lure, methyl eugenol was discovered serendipitously and identified by Howlett in 1912 in India (Howlett, 1915). This discovery lay unexploited for decades until Loren Steiner rediscovered its efficiency on *Bactrocera* (=Dacus) *dorsalis* (Steiner, 1951).

Although pheromones have not yet proved to have practical applications in the management of tephritid fruit flies, the so-called male lures have great utility in programmes against certain species from Paleotropics and Oceania (Cunningham *et al.*, 1990). The male lures are used in large scale dtection-trapping arrays in the continental United States where there is a continuous influx of accidental introductions of these species through movement of infested fruit past agricultural quarantine barriers.

Methyl eugenol was first used in a male eradication programme for *D. dorsalis* by Steiner and coworkers on the Island of Rota (north of Guam) in the Marianas (Steiner *et al.*, 1965). Infact, they absorbed a mixture of methyl eugenol and insecticide (naled) onto cane-fibre squares and either threw them out of an aeroplane in uninhabited areas or placed them in trees in village areas. This method, either alone or in combination with another technique such as sterile male technique, subsequently has been used a number of times to control or eradicate populations of *D. dorsalis* in the Marianas (Steiner *et al.*, 1970), the United States (Cunningham *et al.*, 1990) and Japan (Koyama *et al.*, 1984; Ushio *et al.*, 1982). It was also used to eradicate an infestation of *Dacus zonatus* (Saunders) in California in 1987 (Cunningham *et al.*, 1990).

In 1987, 15000 traps baited with methyl eugenol, 33000 with Trimedlure and 7000 with Cuelure were deployed in the populated area of California (Cunningham *et al.*, 1990). Traps baited with trimedlure for the Mediterranean fruit fly were deployed at the rate of five traps per square mile (2/km$^2$) in Los Angles basin throughout the year. Oriental fruit fly traps with methyl eugenol were deployed at the lower rate of two traps per square mile (0.8/km$^2$), both because the lure is relatively more powerful and because the fly itself is more mobile and therefore more likely to encounter a trap. The lower trapping rate reflected the confidence in the male annihilation eradication system.

An easily applied eradication system for use in urban California was recommended against introductions of *B. Dorsalis* (Chambers *et al.*, 1974). The programme consisted of application of methyl eugenol: insecticide: thickening agent in the ratio of 7 : 1 : 2 giving the foumulaton a custard-like consistency and was applied as 5 ml spots to telephone poles and roadside trees, 180 cm or more above the ground. These spots are applied at the rate of 600 per square mile (230/km$^2$) throughout the treatment zone. A treatment zone was a minimum of square miles (23 km$^2$) i.e. 1.5 mile (2.4 km) radius. The treatments were applied every 2 weeks for hypothetical generations past the last fly-find date such that usually six applications were given over a 3-month period.

Probably because of the potency of methyl eugenol, little efforts have been made to find an equivalent or an improved lure. Beroza and Green (1963) tabulated a large number of chemicals that were evaluated as attractants for a number of insects including *B. dorsalis* in olfactometer tests. Recently, DeMilo and coworkers (1994a) found that 8 out of 44 selected analogues of vertarole showed promise as attractants for male *B. dorsalis* in week-long field tests when compared with a standard methyl eugenol lure. 4-propyl and 4(and 5)-allyl-2-ethoxy-1-methoxybenzene were the more effective of the eight promising lurs showing high level of initial and persistent attraction and could be used as potential alternatives to methyl eugenol.

Commercially synthesized Trimedlure, tert-butyl-4 (and5)-chloro-2-methylcyclohexane-1-carboxylate (TML) is a mixture of eight isomeric trisubstituted cyclohexanes. The mixture is made up of four trans (=90–95%) and four cis (=5–10%) isomers (Leonhardt *et al.*, 1982). Although TML has been used extensively for years to detect and monitor population outbreaks of *C. capitata*, its decreased field persistence in hot weather and tendency to form crystals during storage in cool weather and on cotton-wick dispensers (Leonhardt *et al.*, 1987) are characteristics that decrease its ease of deployment and cost effectiveness. Hence there is an urgent need to define further the importance of individual substitutes of TML and also to identify more potent, stable and persistent attractants. Accordingly a wide variety of analogues reflecting changes at the 1–2-,4-and 5-positions were synthesized for evaluation against *C. capitata* in field plot tests. DeMilo *et al.* (1994b) evaluated 68 structural variants of TML as attractant for *C. capitata* to determine the importance of substituents. The 1-tert-butyl carboxylate and 4 (and 5)-chloro substituents in TML can be replaced selectively with no significant reduction in attractiveness. However, the 2-methyl substituent appears to be very important for high levels of attraction. Some halogen analogues were found to be as persistent or more persistent than TML at high levels of attractiveness. Ethyl 4(and 5)-bromo-trans-2-methylcyclohexane-1-carboxylate were very persistent lures, continuing to catch moderate number of *C. capitata* through day 11 and 15, respectively.

*Bactrocera latifrons* (Hendel) is native to South and Southeast Asia and was first detected in Honolulu in 1983. It is currently an established pest posing serious quarantine risk in Hawaii and contiguous United States (Flath *et al.*, 1994). Several thousand traps baited with male lures are deployed routinely in California and several southern states to detect the colonizing populations of accidentally introduced tephritid (Cunningham *et al.*, 1990). The use of male lures has been effective in detecting, estimating density and delimiting distribution of *B. dorsalis, B. cucurbitae* and *C. capitata*. However, it is not known whether *B. latifrons* exhibits the same parapheromone response to a single attractive chemical compound which is so common in males of many tephritid fruit flies. Studies conducted by Flath and coworkers (1994) led to the discovery of an attractant, α-ionol which attracted wild males of *B. latifrons* more than females. The males showed more attraction to α–ionol than hydrolyzed protein bait.

Food baits are also useful for monitoring or controlling tephritids. The most important of these are the protein hydrolysates of corn, soybean or yeast (Foster and Harris, 1997). Microbial fermentation of these baits produces volatile chemicals that attract a range of tephritids. Food baits have the advantage over male lures as they attract females as well as males (Chambers, 1978). Protein hydrolasate baited toxic baits have been used to eradicate *C. capitata* in the mainland United States during the later part of this century (Steiner *et al.*, 1961; Chambers, 1978; Jackson and Lee, 1985).

Besides *C. capitata*, the Mexician fruit fly, *Anastrepha ludens* (Loew) and other *Anastrepha* spp. are the most damaging tephritid pests in Latin America (Enkerlin *et al.*, 1989). Traps baited with protein solutions were used to detect the presence of *C.. capitata* and *Anastrepha* fruit flies. Steiner (1955) found the acid hydrolysate (NuLure) and yeast hydrolysates to be equally atractive to *C. capitata* when tested on an equal-solids basis. Addition of sodium borate (borax) to NuLure or other acid hydrolysates increased the bait pH and attraction of *Bactrocera tryoni* (Froggatt) (Bateman and Morton, 1981). Heath *et al.* (1994) conducted field trials to determine the attractiveness of NuLure as compared with a standard bait of pelletized torula yeast plus sodium borate for *C. capitata*, *A. ludens* and other *Anastrepha* spp. and found that addition of 10 per cent borax to 10 per cent NuLure solution increased the bait pH which corresponded directly with increase in number of females (*C. capitata*) trapped. With NuLure at 8.6 pH, significantly higher number of females were trapped than with traps baited with torula yeast plus borax pellets.

McPhail traps, belt shaped invaginated glass traps with a water reservoir (Newell, 1936) baited with torula yeast (TY) and borax pellets (4 : 5, TY : borax) are currently used for monitoring *C. capitata* and *Anastrepha* spp. throughout citrus growing regions of the world. McPhail traps baited with liquid baits are cumbersome and have numerous disadvantages (Heath *et al.*, 1995). Because of the difficulties associated with use of either liquid baited McPhail traps or sticky-coated traps, a dry plastic insect trap for monitoring populations of tephritid fruit flies was developed by Heath and coworkers (1995). The plastic trap consists of acetate film which is painted to provide a visual cue and contains a toxicant panel that also provides a visual cue and a feedingstimulant to flies that have entered the trap. Moreover, unlike McPhail traps which catch large number of miscellaneous Diptera, the plastic traps were highly specific and catch few non-target flies.

**Insect pests of grapes**

The European grape berry moth, *Eupoecilia ambiguella* Hubner, and the grapevine moth, *Lobesia botrana* Den. And Schiff., are the important pests on grape plants in Europe (Descoins, 1990). Both species are harmful not only because the larvae directly attack the grapes but also because they promote growth of a pathogenic fungus, *Botrylis cinerea*. Accurate determination of the flight periods of the adults and estimation of population densities are a prerequisite for the management of these pests. In northeastern Canada *Endopiza viteana* (Clemens) is the most important grape berry moth whose larvae feed exclusively upon the flowers and fruits of both wild and cultivated grapes (Roberts and Simpson, 1982). Another insect, the western grapeleaf skeletonizer, *Harrisina brillians* Barnes and McDunnough, is a serious pest of California grape vineyards (Curtis *et al.*, 1989).

The sex pheromone of *E. ambiguella* was identified in Switzerland by Arn *et al.* (1976) of which is-o-dodecenyl acetate (Z9-C12: AC) was the major component, essential for luring males. Catches were markedly increased by adding dodecyl acetate (C12: Ac), a minor component of the secretion. Sex pheromones containing 1000 μg of Z9-C12: Ac and 5000 μg of C12: Ac impregnated on rubber septa were used commercially in Europe. In addition to dodecyl acetate, Arn *et al.* (1986) discovered another saturated acetate, octadecyl acetate (C18: Ac), which acted as a synergist to the two component blend (Z9-C12: Ac/C12: Ac) and the male catches in traps turned out to be excellent (between 96.4 to 100 per cent) in the vineyards (Descoins, 1990). It was observed that efficiency was higher when the population density was low.

Field trials conducted in the USA resulted in successful control of grape berry moth *E. viteana* with a commercial pheromone product, the Shin-Etsupheromonedispenser (Dennehy *et al.*, 1990). The pheromone component used was cis-9-dodecenyl acetate (Z9–12: Ac).

Roelofs and coworkers (1973) suggested that (E, Z)-7–9-dodecadienyl acetate (E7-Z9-C12: Ac) might be a sex attractant for the males of the grapevine moth. Late the structure was confirmed by the biological activity of the synthesized compound and identification of this chemical in the female glands (Buser *et al.*, 1984). Commercial sex attractants are made of 1000 or 100 μg of E7, Z-9-C12: Ac(95 to 98% steric purity) impregnated on rubber septa, sometimes with anti-UV preparations or antioxidants. Normally dispensers are placed at the beginning of the second generation (June-July) because it has been demonstrated that the disruption technique is useless for the first generation (April) when the vines have no leaves (Descoins, 1990). A drastic decrease in the number of males caught in the traps within the treated area, was observed. Reduction of damage varied from 30–90 per cent depending on the initial population density.

The need for monitoring populations of the western grapeleaf skeletonizer *Harrisina brillians*, led to the identification of a sex pheromone from female abdomen washes and volatiles (Curtis *et al.*, 1989). Although four compounds were identified, only 2-butyl-(Z)-7-tetradecenoate was shown to be active behaviourally (Myerson *et al.*, 1982; Soderstrom *et al.*, 1985). Peak catches of male *H.brillians* were obtained at a dose of 100 μg per septum (estimated release rate of 8.6 ng/hr) (Curtis *et al.*, 1989). In a comparison of 500 μg dosage of the pheromone and unmated female *H. brillians* moths in tarps, female-baited traps attracted 2.5 times as many males as did pheromone-baited traps. It seems that additional pheromone components are produced by calling females of *H. brillians.*

### *Carpophilus* beetles and other pests

Nitidulid beetles of the genus *Carpophilus* are serious pests of ripening stone-fruits in southern Australia, particularly apricot, peach and nectarine (Hely *et al.*, 1982). *Carpophilus* spp., primarily dried fruit beetles (*C. hemipterus* (L.), *C. mutilatus* Erichson and C. *davidsoni* Dobson) (James *et al.*, 1995) cause damage by entering the fruits after resulting in complete destruction of the crop. Control with conventional insecticides is often unsatisfactory and short-lived because of difficulties in targeting the beetles and immigration from surrounding untreated areas. Synthetic aggregation pheromones have shown considerable potential for the management of *Carpophilus* spp. though monitoring and mass-trapping of the populations. Male-produced aggregation pheromones have been identified and synthesized for a number of *Carpophilus* spp. including *C. hemipterus* (Bartelt *et al.*, 1992), *C. mutilatus* (Bartelt *et al.*, 1993) and *C. davidsoni* (Bartelt and James, 1994). In order to evaluate the potential of mass trapping as a management option for *Carpophilus* spp. Peng and Williams (1991) compared 9 trap designs for nitidulids and found 5 funnel type traps very efficient in trapping the dusky sap beetle, *Carpophilus lugubris* Murray. In a series of experiments conducted in stone-fruit orchards in souther Australia, water-based funnel type traps, baited with synthetic aggregation pheromone and fermenting bread dough, trapped 3- to 7-fold as many *Carpophilus* beetles (James *et al.*, 1996). The efficacy of dry funnel traps (but not pipe traps) appeared to be improved by using water filled collecting bottles. Water-based funnel traps showed considerable potential for the management of *Carpophilus* spp. through mass-trapping.

Nitidulid beetle area also known to inflict serous damage to the developing crop of date *Phoenix dactylifera* L. in southern California (Warner *et al.*, 1990). Male produced aggregation pheromones to which both sexes of *C. mutilatus, C. hemipterus* and *C. freemani* Dobson responded were synergised by food volatiles such as those from fermenting bread dough or decomposing fruits (Bartelt *et al.*, 1992, 1994). Combinations of aggregation pheromones for these species were evaluated as trap baits in a date garden in southern California (Bartelt *et al.*, 1993). Each species was strongly attracted to its own pheromone and species specificity was high. Generally, responses to cospecific pheromones were not inhibited by the presence of other pheromones. The effectiveness of the pheromone combinations would simplify the use of the pheromones for mass trapping and one multispecies type lure could replace a series of specific lures for individual species with no or slight loss of efficacy for any target species.

Peach tree borer, *Synanthedon exitiosa* (Say) and lesser peach tree borer, *S. pictipes* (Grote and Robinson) are important sesiid species damaging cherry, almond, plum and peach (Snow, 1990). Larvae of *S. exitiosa* attack at the base of the tree or the roots at or below the ground surface, whereas *S. pictipes* confines its attack to the above ground parts of the tree. Tumlinson and coworkers (1974) discovered (Z, Z)-3, 13-octa-decadienyl acetate (ZZA) and (E, Z)-3–13-octa-decadienyl acetate (EZA) as attractants for both the species McLaughlin *et al.* (1976) were first to report that sexual communication of these borers could be disrupted by permeation of the atmosphere with their respective pheromones. They also reported that the two isomers seemed equally effective against both the species. Disruption was greatest when the pheromone was evaporated from the top of peach tree. Trials were conducted in the orchards using Conrel hollow-fibre dispensers, Hercon vinyl-laminated dispensers or Shin-Etsu rope dispensers that were prepared commercially to provide the isomeric mixture and dosage rate and it was found that one dispenser per tree of EZA gave season-long protection against *S. pictipes* and also suppressed *S. exitiosa* population during early spring through mid-summer, and one application of ZZA was required in August for *S. exitiosa*, control with pheromones was as good or usually better than with insecticides and the pheromone was cost-competitive with insecticides.

The peach twig borer, *Anarsia lineatella* Zeller is one of the key pests of almond and other stone-fruits in many parts of the world (Summers *et al.*, 1959). Sex pheromone baited traps are widely used as a convenient tool for monitoring the efficacy of control methods, the size of population and flight periods of male moths so that further control measures can be instituted as required (Rice and Jones, 1988). However, sex pheromone based monitoring of the borer has not been without problems. For instance, the initial identification of the pheromone blend was hindered by the presence of trace amounts of inhibitory compounds in the synthetic pheromone blends developed for field trials, rendering the blends only weakly attractive to moths (Roelofs *et al.*, 1975). The poor performance of some commercial pheromone lures resulted in the identification of two potentially inhibitory compounds, Et-decenyl acetate and E7-decenyl acetate (Miller and Rice, 1992). These compounds when mixed with the standard peach twig borer pheromone blend (E5-decenyl alcohol and acetatel: 4) drastically reduced the attractiveness of the pheromone blend. Recently, Miller and Rice (1996) identified 5-decynyl acetate as an antagonist of the pheromone of the borer. Addition of 5-decynyl acetate to an otherwise attractive pheromone blend resulted in virtually complete shut down of trap catches. Identification of pheromone antagonists highlighted the importance of stringent quality control in pheromone manufacture and the need for field testing of batches of pheromones, particularly those used for manufacturing monitoring lures before their release for sale.

## CONCLUSIONS

Considering various problems resulting from the use of conventional insecticides, employment of semiochemicals for the management of insect pests comes as a great relief since these highly biologically active chemicals are not-toxic to plants and animals and thus environmentally acceptable. Fruits being high value cash crops which generally are consumed raw, demand the minimum or no use of insecticides. Pheromones (sex and aggregation) and attractants (kairomones) are two main categories of the semiochemicals, currently employed to manage the insect pests of fruit crops.

Synthetic sex pheromones are being used to monitor the population and the control of insect pests through mass trapping, lure and kill, and disruption of their mating. There are several examples of the satisfactory control of insect pests of orchards and vineyards. Codling moth and Oriental fruit moth could be managed through the use of their synthetic sex pheromones which are commercially available. Successful mating disruption of *C. pomonella* prevented the moth from oviposition and consequently there was no damage. Generally mating disruption is found to be more expensive under heavy infestation but economical under low population levels (Williamson *et al.*, 1996). Though pheromones could not prove their worth in the management of tephritid fruit flies, male lures and food baits have been found to be of great utility for monitoring and controlling these flies. An effective male lure is available for each of the three species of important fruit flies, i.e. methyl eugenol for the Oriental fruit fly, *B. dorsalis,* for Mediterranean fruit fly, *C. capitata* and cue lure for the melon fruit fly, *B. cucurbitae*. Successful control of European grape berry moth, *E. ambiguella* and grape vine moth *L. botrana* could be achieved by employing their synthetic sex pheromone traps. Synthetic aggregation pheromones have also shown considerable potential for the management of *Carpophilus* beetles through monitoring and mass trapping of the populations.

Despite the obvious advantages and considerable success in the management of insect pests, the semiochemicals could not be adopted universally since these could not live upto early expectations. This, however, does not mean that semiochemicals lacked potential but it certainly underlines the need to understand the factors that determine field efficacy of these chemicals. For instance, various factors such as trap shape, placement, weather and population ecology will influence the catch of the target species. Pertinent information relating to the effective range of pheromone traps, relationship between trap catches and economic threshold levels (ETLs) of the pest; the strength and positioning of sources for mating disruption, mass trapping, etc. is not available for all the target insect pests. Pheromone traps have been used to detect insects at remarkably low population densities, but numbers caught often do not correlate well with density of the insects in the field (Jayaraj *et al.*, 1995).

One limitation of using mass trapping technique is, that trap only traps male moths whereas it is the female moth which lays eggs and produces larvae responsible for causing the damage to crops. Moreover, there is need to treat relatively large and isolated areas to ensure success with mating disruption technique. Also, the mechanism of mating disruption is poorly understood (Carde, 1990) and effectiveness of the technique depends upon a range of insect responses to the same stimulus and the interaction of this stimulus with other stimuli and environmental conditions (Pickett, 1991). Mating disruption is economical only under low levels of population and non-economical under heavy infestation.

Presently, it is unlikely to have large-scale production and use of semiochemicals till a market is established which seems to be a difficult task due to uncertainities over registration procedures apart from several other fundamental difficulties.

## REFERENCES

Alexander, B.H.; Beroza, M.; Oda, T.A.; Miyashita, D.H. and Mitchell, W.C. 1962. The development of male melo1 fly attrctants. *Journal of Agricultural and Food Chemistry*, **10**: 270–76.

Arn, H.; Rauscher, S.; Buser, H.R. and Guerin, P. 1986. Sex pheromone of *Eupoecilia ambiguella*: Female analysis and male response to ternary blend. *Journal of Chemical Ecology*, **12**: 1417–29.

Arn, H.; Rauscher, S.; Buser, H.R. and Roelofs, W.L. 1976. Sex pheromone of *Eupoelcia ambiguella*: cis-9-dodecenvl acetate as a major component. *Z. Naturforsch*, **31C**: 499–503.

Barnes, M.M. 1991. Tortricids in stone and pome fruits. *In: Tortricid pests. their biology natural enemies and control.* L.P.S. van der Geest and H. Evenhuis (eds.). Elsegier, Amsterdam. pp. 313–327.

Barnes, B.N.; Blomefield, T.L.; Witzgall, P. and Arn, H. 1997. Coading growers towards mating disruption: the South African experience with *Grapholita molesta* and *Cydia pomonella* (Lepidoptera, Tortricidae*). Bulletin OILB, SROP*, **20**: 45–56.

Bartelt, R.J.; Veter, R.S.; Carlson, D.G. and Baker, TC. 1993. Male-produced aggregation pheromone of *Carpophilus nutilatus* (Coleoptera: Nitidulidae). *Journal of Chemical Ecology*, **19**: 107–18.

Bartlet, R.J.; Vetter, R.S.; Carlson, D.G. and Baker, T.C. 1994. Responses to aggregation pheromones of five *Carpophilus* species (Coleoptera: Nitidulidae) in a California date garden. *Environmental Entomology*, **23**: 1534–43.

Bartelt, R.J.; Weisleder, D.; Dowd, P.F. and Plattener, R.D. 1992. Male-specific tetraene and triene hydrocarbons of *Carpophilus hemipterus:* structure and pheromonal activity. *Journal of Chemical Ecology*, **18**: 379–402.

Bateman, M.A. and Morton, T.C. 1981. The importance of ammonia in proteinaceous attractants for fruit flies (family: Tephritridae). *Australian Journal of Agricultural Research*, **32**: 883–903

Beroza, M. and Green, N. 1963. *Materials tested as insect attractants*. USDA-ARS Agriculture, No. 239.

Beroza, M.; Green, N.;, Gertler, S.I.; Steiner, L.F. and Miyashita, D.H. 1961. New attactants for the Mediterranean fruit fly. *Journal of Agricultural and Food Chemistry*, **9**: 361–365.

Brown, W.L. Jr. 1968. An hypothesis concenrning the function of the metapleural glands in ants. *American Naturalist*, **102**: 188–91.

Brown, W.L.; Eisner, T. and Whittarker, R.H. 1970. Allomones and kariomones: transspecific chemical messengers. *Bio Science*, **20**: 21–22.

Buser, H.R.; Rauseher, S. and Arn, H. 1984. Sex pheromone of *Lebesta botrama*: E7, Z9-dodecadienyl acetate in the female grape vine moth. *Z. Natur. Forsch.*, **29**: 731–748.

Carde, R.T. 1990. Principles of mating disruption. *In: Behaviour-modifying Chemicals for Iinsect Pest Management*. R.L. Ridgway, R.M. Silverstein and M.N. Inscoe (eds). Marcel Dekker, Inc., New York. pp.47–71.

Chambers, D.L. 1978. Attractants for fruit fly survey and control. *In: H.H. Shorey and J.J. Mekelvey (Eds.). Chemical Control of Insect Behaviour: Theory and Applications*. Wiley and Sons, New York. pp. 327–344.

Chamebrs, D.L.; Cunninghama, R.T.; Lichty, R.W. and Thrailkill, R.B. 1974. Pest control by attractants: a case stdy demonstrating economy, specificity and environmental acceptability. *Bio Science*, **24**: 150–152.

Charmillot, P.J. 1990. Mating disruption technique to control codling moth in Western Switzerland. *In: Behaviour-modifying chemicals for insect pest managemnet*. R.L. Ridgway, R.M. Silverstein and M.N. Inscoe (eds.). Marcel Dekker, New York. pp. 165–182.

Charmillot, P.J.; Hofer, D.Q.; Witzgall, P. and Arn, H. 1977. Control of codling moth, *Cydia pomonella* L. by an attract and kill formulation. *Bulletin OILB SROP*. **20**: 139–40.

Cunningham, R.T.; Kobayashi, R.M. and Miyashita, D.H. 1990. The male lures of tephritid fruit flies. *In: Behaviour Modifying Chemicals for Insect Management*. R.L. Ridgway, R.M. Silverstein and M.N. Inscoe (eds). Marcel Dekker Inc., New York. pp. 255–267.

Curtis, C.E.; Landolt, P.J.; Heath, R.R. and Murphy, R. 1989. Attraction of western grapeleaf steletonizer males (Lepidoptera: Zygaenidae) to S-(+)-2butyl-(2)-7- tetradecenoate. *Journal of Economic Entomology*, **82**: 454–57.

DeMilo, A.B.; Cunningham, R.T. and McGovern, T.P. 1994a. Structural variants of methyl eugenol and their attractiveness to oriental fruit fly (Diptera: Tephritidea). *Journal of Economic Entomology*, **87**: 957–964.

DeMilo, A.B.; Cunningham, R.T. and McGovern, T.P. 1994b. Trimedlure: Effects of structural modifications on its attractiveness to Mediterranean fruit fly males (Diptera: Tephritidae). *Journal of Economic Entomology*, **87**: 1494–1501.

Dennehy, T.J.; Roelofs, W.L.; Taschenberg, E.F. and Taft, T.N. 1990. Mating disruption for control of grape berry month in New York vineyards. *In: Behaviour Modifying Chemicals for Insect Management*. R.L. Ridgway, R.M. Silverstein and M.N. Inscoe (eds.). Msarcel Dekker Inc., New York. pp. 233–240.

Descoins, C. 1990. Grape berry moth and grape vine moth in Europe. *In: Behaviour modifying chemicals of insect management*. R.L. Ridgway; R.M. Silverstein and M.N. Inscoe (eds.). Marcel Dekker Inc., New York. pp. 213–222.

Enkerlin, D.L.; Gareia, R. and Lopez, F. 1989. Mexico, Central and South America. *In: World Crop pests* A.S. Robinson and G. Hooper, Elsevier, Amsterdum (eds.). *Vol. 3A. Fruit flies, their biology, natural enemies and control.* Amsterdum. pp. 83–90.

Fabre, J.H. 1879. Sorvenirs Entomologiques Paris, Delagrave.

Flath, R.A.; Cunningham, R.T.; Liquido, N.J. and McGovern, T.P. 1994. Alpha-ionol as attractant for trapping *Bactrocera latifrons* (Diptera: Tephritidae). *Journal of Economic Entomology*, **87**: 1470–76.

Foster, S.P. and Harris, H.P. 1997. Behavioural manipulation methods for insect pest managemnet. *Annual Review of Entomology*, **42**: 123–146.

Hall, D.R. 1995. Current status of the application of pheromones in insect pest management. *In: Processdings of the National Workship on Use of Insect Pheromones in Integrated Pest Management*. K, Krishaniah and B.J. Divakar.(eds). Plant Protection Association of India, Hyderabad, India. pp. 5–13.

Hall, D.R. 1998. A rose by any other name? An inaugural lecture delivered at the University of Greenwish. pp. 42.

Heath, R.R.; Epsky, N.D.; Bloem, S.; Bloem, K.; Acajabon, F.; Guzman, A. and Chambers, D. 1994. pH effect on the atractivenss of a corn hydrolysate to the Mediterranean fruit fly and several *Anastrepha* species (Diptera: Tephritidae). *Journal of Economic Entomology*, **87**: 1008–1013.

Heath, R.R.; Epsky, N.D.; Duzman, A.; Bueben, B.D.; Manukian, A. and Meyer, W.L. 1995. Development of a dry insect trap with food-based synthetic attractnat for the Mediterranean and Mexican fruit files (Diptera: Tephritidae). *Journal Economic Entomology*, **88**: 1307–1315.

Hely, P.C.; Pasfied, G. and Gallately, J.G. 1982. Insect pests of fruit and vegetables in NSW, Inkata, Melbourne.

Howlett, F.M. 1915. Chemical reaction of fruit flies. *Bulletin of Entomological Research*, **6**: 297–325.

Inscoe, M.N.; Leonhardt, B.A. and Ridgway, R.L. 1990. Commercial availability of insect pheromones and other attractants. *In: Behaviour-modiftying chemcials of insect management*. R.L. Ridgway, R.M. Silverstein and M.N. Inscoe (eds.). Marcel Dekker Inc., New York. pp. 631–715.

Jackson, D.S. and Lee, B.G. 1985. Medfly in California. 1980–1982. *Bulletin of Entomological Society of America*, **31**: 29–37.

James, D.G.; Faulder, R.J. and Bartelt, R.J. 1995. Fauna and seasonal abundance of *Carpophilus* spp. (Coleoptera : nitidulidae) in four stone fruit growing regions of south-eastern Australia as determined by pheromone trapping. *Journal of Australina Entomological Society*, **34**: 327–333.

Jamesm, D.G.; Batelt, R.J. and Moore, C.J. 1996. Trap design affect on capturer of *Carpophilus* spp. (Coleoptera: Nitidulidae) using synthetic aggregation pheromones and co-atytratants. *Journal of Economic Entomology*, **89**: 648–653.

Jayaraj, S.; Sathian, N. and Muthuswami, M. 1995. Insect sex pheromones: innovative tools in IPM. *In: Proceedings of the National Workshop on use of insect pheromones in Integrated Pest Management*. K. Krishaniah and B.J. Divakar (eds). Plant Protection Association Association of India, HYderabad, India. pp. 5–13.

Judd, G.J.R.; Gardiner, M.G. and Thomson, D,.R. 1996. Commercial trials of pheromone-mediated mating disruption with Isomate-CR to control codling moth in British Columbia apple and pear orchards. *Journal of Entomological Society of British Columbia*, **93**: 23–24.

Justum, A.R. and Gordon, R.F.S. 1989. Pheromonoes: Importance to insects and role in pest management. *In: Insect Pheromones in Plant Protection*. A.R. Justum and R.F.S. Gordon (eds.). John Wiley, U.K. pp. 1–13.

Karlson, P. and Luscher, M. 1959. Pheromones, a new term for a class of biologically active substances. Nature (London), **183**: 155–176.

Koyama, J.; Teruya, T. and Tanaka, K. 1984. Eradication of the Oriental fruit fly (Diptera: Tephritidae) from Okivawa islands by a male annihilation method. *Journal of Economic Entomology*, **77**: 468–472.

Krishniah, K. and Divakar, B.J. 1995. Pheromones in integrated pest management. *Proceedings of the National Workshop on Use of Insect Pheromones in Integrated Pest Management.* Plant Protection Association of India, Hyderabad, India. pp. 100.

Law, J.H. and Regnier, F.E. 1971. Pheromones. *Annual Review of Biochemistry,* **40**: 533–548.

Leonhardt, B.A.; MeGovern, T.P. and Plimmer, J.R. 1982. Capillary go analysis of trimedlure, the attractant for the "medfly" *Journal of High Resohution Chromatography Communication*, **5**: 430–431.

Leonhardt, B.A.; Cunningham, R.T.; Rice, R.E.; Harte, E.M. and MeGovern, T.P. 1987. Performance of controlled relases formation of trimedlure to attract the Mediterranean fruit fly, *Ceratitis capitata. Entomologia Experimentalis et Applicala*, **94**: 45–51.

Madsen, H.F. and Carty, B.E. 1979. Codling moth (Lepidoptera: Olethreutide) suppression by male removal with sex pheromone traps in three british Columbia Orchards. *Canadian Entomologist*, **111**: 627–630.

Mani, E. 1986. Field trials to control codling moth by mating disruption. 1979–85. *In: Seventh Symposium on Integrated Plant Protection in Orchards*. E. Dickler, L.J.M. Blomers and A.K. Minks (eds.). IOBC/WPRS. pp. 166–169.

Mani, E.; Schwaller, F. and Horn, H. 1996. Control of codling moth with the mating distruption technique in German-speaking Switzerland. *Obst. und. Weinban*, **132**: 718–722.

McDonough, L.M.; Chapman, P.S.; Weissling, T.J. and Smithhisler, C.L. 1996. Efficacy of nonpheromone communication disruptants of codling moth (*Cydia pomonella*): effect of pheromone isomers and of distance between calling females and dispensers. *Journal of Chemcial Ecology*, **22**: 415–423.

McLaughlin, J.R.; Doolittle, R.E.; Gentry, C.R.; Mitehell, E.R. and Tumlinson, J.H. 1976. Response to pheromone traps and disruption of pheromone communication in lesser peach tree borer and peach tree borer. *Journal of Chemical Ecology*, **2**: 73–81.

McVeigh, L.J.; Hall, D.R. and Beevor, P.S. (eds.) 1993. Pheromone technology in Europe and Developing Countries. *Proceedings of Working Group Meeting, "Use of Pheromones and other Semiochemical in Integrated Control*. IIOBC/WPRS Bulletin, **16**(10): pp. 372.

Millar, J.G. and Rice, R.E. 1996. 5-decyn-l-yl acetate: powerful anatagonist of peach twig borer (Lepidoptera: Gelechiidae) sex pheromone. *Journal of Economical Entomology*, **89**: 131–133.

Millar, J.G. 1995. Degradation and stabilizing of E8, E10- dodecadienol, the major component of the sex pheromone of the codling moth (Lepidoptera: Tortricidae). *Journal of Economic Entomology*, **88**: 1425–1432.

Millar, J. G. and Rice, R.E. 1992. Re-examination of the female sex pheromone of the peach twig borer field screening of minor constitutents of pheromone gland extracts and of pheromone analogues. *Journal of Economic Entomology*, **85**: 1709–1716.

Moffitt, H.R. and Westigrd, P.H. 1984. Suppression of the codling moth (Lepidoptera: Tortricidae) population on pears in southern Oregon through mating disruption with sex pheromone. *Journal of Economic Entomology*, **77**: 1513–1519.

Muirhead-Thomson, R.C. 1991. Traps responses of flying insects. Academic Press, London, pp. 287.

Myerson, J.; Hadden, W.F. and Soderstrom, E.L. 1982. See-butyl (Z)-7-tetradecenoate: A novel sex pheromone component from the western grapeleaf skeletonizer.; *Harrisina brillians. Tetrahodron Letters*, **23**: 2757–2760.

Newell, W. 1936. Progress report on the Key West (Florida) fruit fly eradication project. *Journal of Economic Entomology*, **29**: 116–120.

Norlund, D.A. 1981. Semiochemicals: A review of the terminology. In: D.A. Norlund, R.L. Jones and W.J. Lewis (eds.). *Semiochemicals their role in Pest Control*. John Wiley and Sons, New York. pp.13–28.

Norlund, D.A. and Lews, W.J. 1976. Terminology of chemcial releasing stimuli in intraspecific and interspecific interactions. *Journal of Economic Entomology*, **2**: 211–220.

Peng, C. and Williams, R.N. 1991. Effect of trap design, trap height and habitat on the capture of sap beetles (Coleotpera: Nitidulidae) using whole wheat bread dough. *Journal of Economic Entomology*, **84**: 1515–1519.

Pickett, J.A. 1991. Pheromones: will their promise in insect pest control ever be achieved? *Bulletin of Entomological Reseach*, **81**: 229–232.

Rice, R.E. and Jones, R.A. 1988. Timing post-bloom sprays for peach twig borer (Lepidoptera: Gelechiidae) and San Jose Scale (Homoptera: Diaspididae). *Journal of Economic Entomology*, **81**: 293–299.

Rice, R.C. and Kirscll, P. 1990. Mating disruption of Oriental fruit moth in the United States. *In: Behaviour-midifying chemcials for Insect Management*. Ridgway, R.M. Silverstein and M.N. Inscoe (ed.). Marcel, Dekker Inc. New York. pp. 193–211.

Roberts, W.P. and Simpson, C.M. 1982. Pest management programme for grapes series: Monitoring and predicting spray dates for the grape berry moth on the Niagara Peninsula. *OMAF Agdox*, 212/632.

Roelofs, W.L.; Bartelr, R.J.; Hill, A.S.; Carde, R.T. and Water, L.H. 1972. Codling moth sex attractant-field trials with geometrical isomers. *Journal of Economic Entomology*, **65**: 1276.

Roelofs, W.L.; Comean, A. and Selle, R. 1969. Sex pheromone of the oriental fruit moth. *Nature,* **224**: 723.

Roelofs, W. L.; Kochansky, J.; Anthion, E.; Anthion, R.E. and Carde, R.J. 1975. Sex pheromone of the peach twig borer, *(Anarsia lineatella)*. *Environmental Entomology*, **4**: 580–582.

Roelofs, W.L.; Kochansky, J.; Carden, R.; Arn, H.and Rauscher, S. 1973. Sex attractant of teh grape vine moth. *Lobesia botrana. Mill Schweiz, Entomol. Ges.*, **46**: 71–73.

Rothschild, G.H.L. 1975. Control of Oriental fruit moth *Cydia molesta* (Busck) (Lepidoptera: Tortricidae) with synthetic female pheromone. *Bullein of Entomological Research*, **65**: 473–490.

Rothschild, G.H.L. 1981. Mating disruption of lepidopterous pest: Current status and future prospectus. *In: Management of Insect Pests with Semio-chemicals*. E.R. Mitchell (ed.). Plenum Press, New York. pp. 207–228.

Sanders, C.J. 1997. Mechanism of mating disruption in moths. *In: Inserct Pheromone Research: New Directions*. R.T. Carde and A.K. Minks (eds.).Champman and Hall, New York. pp. 333–346.

Shorey, H.H.; Kaae, R.S.; Gaston, D.K. and McLaughlin, J.R. 1972. Sex pheromone of Lepidopera: XXX. Disruptioj of sex pheromone communication in *Trichoplusiani* as a possible means of conrol. *Environmental Entomology*, **1**: 641–645.

Smith, A. 1992. Semiochemicals in Crop Protection. *PJB Publishers C/O AGROW Reports, 18/20 Hill Rise, Richmond, Surrey TW 10–6UA*, pp. 160.

Snow, J.W. 1990. Peach tree borer and lesser peach tree borer control in the United States. *In: Behaviour-modifying chemcials for Insect Management*. R.L. Ridgway; R.M. Silverstnein, and M.N. Inscoe (eds.). Marcel Dekker Inc., New York. pp. 241–253

Soderstrom, E.L., Brandl, D.G., Myerson, J., Buttery, R.G. and Mackey, 1985. Sex phermone for attracting western grapeleaf skeletonizer (Lepidoptera: Zygaenidae). *Journal of Economic Entomology*, **78**: 799–801.

Steiner, L.F. 1951. Lures for *Dacus dorsalis.* Proceedings of the Hawaiian Entomological Society, **14**: pp. 204.

Steiner, L.F. 1955. Bait sprays for furit fly control. *Agricultural Chemi*stry, **10**: 32–34, 113–114.

Steiner, L.F.; Hart, W.G.; Harris, E.J.; Cunningham, R.T.; Ohinata, K. and Kamakahi. 1970. Eradication of the oreintal fruit fly from the Mariana islands by the methods of male annihilation and sterile insect release. *Journal of Economic Entomology*, **63**: 131–135.

Steiner, L.F.; Mitchell, W.C.; Harris, E.J.; Kozuma, T.T. and Fujimoto, M.S. 1965. Oriental fruit fly eradication by male annihilation. *Journal of Economic Entomology*, **58**: 961–964.

Steiner, L.F.; Rohwer, G.G.; ayers, E.L. and Christenson, L.D. 1961. The role of attractants in the recent Mediteranean fruit fly eradication programme in Florida. *Journal of Economic Entomology*, **54**: 30–35.

Summers, F.M.; Donaldson, D. and Togashi, S. 1959. Control of peach twig borer on almonds and peaches in California. *Journal of Economic Entomology*, **52**: 637–639.

Tumlinson, J.H.; Yonce, C.E.; Doolittle, R.E.; Heath, R.R.; Gentry, C.R. and Mitchell, E.R. 1974. Sex perhomones and reproductive isolation of the lesser peach tree borer and the peach tree borer. *Science*, **185**: 614–616.

Ushio, S.K.; Yoshioka, K.; Masuku, K. and Waki, K. 1982. Eradication of the oriental fruit fly from Amami islands by male annihilation (Diptera: Tephritidae). *Japan Journal of Applied Enomology and Zoology*, **26**: 1–9.

Varela, L.G.; Welter, S.C.; Jones, V.P.; Burunner, J.F. and Riedl, H. 1993. Monitoring and characterization of resistance in codling moth (Lepidoptera: Tortricidae) in four western states. *Journal of Economic Entomology*, **86**: 1–10.

Vickers, R.A. 1990. Coriental fruit moth in Australia and Canada. *In: Behaviour-modifying chemcials for Insect Management*. R.L. Ridgway, R.M. Silverstein and M.N. Inscoe (eds.). Marcvel Dekker Inc., New York. pp. 182–192.

Vickers, R., Rothschild, G.H.L. and Jones, E.L. 1985. Control of Oriental fruit moth. *Cydia molesta* (Busck) (Lepidoptera: Tortricidae) at a disrict level by mating disruption with synthetic female pheromone. *Bulletin of Entomological Research,* **75**: 625–634.

Wall, C. 1989. Monitoring and spray timings. *In: Insect Pheromones in Plant Protection*. A.R. Justum and R.F.S. Gordon (eds.). John Willey, U.K. pp. 39–66.

Warner, R.L.; Barne, M.M. and Laird, E.F. 1990. Reduction of insect infestation and fungal infection by cultural practice in date gardens. *Environmental Entomology*, **19**: 1618–1623.

Williamson, E.F.; Folwell, R.J.; Knight, A. and Howell, J.F. 1996. Economics of employing pheromones for mating disruption of the codling moth, *Carpocapsa pomonella. Crop Protection*, **15**: 473–477.

*Pest Management in Horticulture Crops*
L.R. Verma, A.K. Verma and D.C. Gautam (eds)


## *Chapter 8*

# Indigenous Technical Knowledge for Pest Management

*L.R. Verma*

## INTRODUCTION

AS MAN BECAME CIVILIZED, he started cultivation, gradually of monocultures which led to the necessity for storage of harvested food grains. This induced many insects to abandon their original feeding habits, habitats and turned to feeding upon the plentiful and easily available food produced by man and thus became pest. Deforestation by human population and other needs forced the forest dwelling insects to migrate to other areas in quest for food and spread of the human population offered further conditions for the pest to multiply and flourish from place to place. The final impetus to the increase in pest population was given by the use of insecticides that eliminated not only their natural competitors, but also their natural enemies (parasites and predators) that were earlier keeping them below the economic threshold (Srivastava, 1996).

Out of all the animals present in this universe, 70 percent are the insects. It is estimated that insects destroy annually on an average 15 percent of crop produced in India. In order to check the destruction caused by insect pests, initially people used their indigenous technical knowledge (ITK) only.

In recent years, the large scale application of pesticides, primarily insecticides, has made cultivation of some cultivated crops and fruits difficult due to the harm brought to the bio-environment by large scale destruction of natural bio-enemies causing pest resurgence, development of pesticide resistance and consequent secondary pest outbreaks.

Insect based problems are the major bottlenecks in realizing higher yields. The pest attacks of crops at the reproductive stage cause direct loss. The management practices employed to contain these rely mostly on insecticides. This has led to the problem of residues, resurgence and resistance to insecticides. Our efforts to combat these pests by the indiscriminate use of pesticides has created several environmental hazards and these necessitated the re-orientation of our strategies to pests and disease management in a eco-friendly manner.

Perhaps the only solution to this problem at hand lies in the adoption of eco-friendly approaches which are not destructive to natural enemies but gradually remove sizeable proportions of pest populations and tend to keep their populations in check. Physical (devices and procedures used to change physical environment of pest populations), and mechanical (mitigating pest populations by cultural practices) methods of pest control are the oldest of all such insect control methods. These are rooted in simple practices that man, as a farmer, has learnt from his long and close association with pests. These aid him in reducing pest populations to low levels. These include both direct and indirect measures which may be preventive or corrective in nature but are essentially slow acting, often eco-friendly, cost effective compatible with other methods of pest control. These characteristics make them amenable to blend better with integrated pest management practices (IPM) even though they do not bring about an immediate or drastic reduction in pest populations. Even the modern concept of pest control does not emphasize the outright eradication of pests but focuses on maintaining their populations at levels which do not cause economic losses.

Therefore, there is a need for developing system with greater reliance on non-pesticidal management and eco-friendly pesticides of botanical and biological origin. Traditional pest management practices with indigenous technical knowledge of farmers are reported to be effective in managing pests and diseases to a large extent.

Indigenous technology knowledge is an eco-friendly and cost-effective for controlling the insect pest. The rural people have good knowledge for these methods because there are traditional and their use does not require any specific training. On the other hand the use of chemical pesticides leads to

- increased environmental pollution,
- very high cost of chemical pesticides,
- damage in soil texture and several others environmental problems,
- adverse effect on human health and insects develop resistance to the pesticide.

Indigenous knowledge is the inter-generational wisdom of local inhabitants to perform their livelihood operations in a most eco-friendly manner. Since this knowledge is transferred orally from one generation to the next, it is dynamic in dissemination and scientific in indigenous experimentation, receiving constant stimuli from outside. However, indigenous not only stands for ingrained intrinsic knowledge, but is also amenable to modifications based on latest technical know how by local inhabitants through native means to suit their daily requirements. Therefore, to discard any indigenous knowledge on connotations of superstition, conservatism, primitivism etc. by modern science would only result into a failure of the developmental networks. Evidences have shown that the developmental projects which overtly rejected already acquired knowledge of the local inhabitants, have failed to achieve their targets.

## MAIN ATTRIBUTES OF THE INDIGENOUS TECHNICAL KNOWLEDGE IN RELATION TO PEST MANAGEMENT

- Appropriate, viable, cost effective and sustainable
- Developed over time and continue to develop

- Based on human experiences on mass scale
- Dynamic and changing
- Tested in most cases over centuries use
- Adaptable to local culture and environment
- Weightage on minimizing risk than maximizing profit
- Free from indiscriminate application
- All ITK not equally useful
- Operate with in a given socio-economic and spatial setup
- Ignored and marginalized by outsiders
- Knowledge-practice congruence
- Traditional inheritance
- Biological synergy.

## DIVERSITY OF INDIGENOUS KNOWLEDGE AND PEST MANAGEMENT

Indigenous knowledge systems are:

- Adaptable skills of local people usually derived from many years of experience that have been often communicated through "oral traditions" and learned through family members over generations (Thrupp, 1989).
- Time tested physical, mechanical, biological and eco-friendly chemical control methods, which paved the way for mitigating the pest population (Venkatratram, 1990).
- Strategies and techniques developed by local people to cope with the changes in the socio-cultural and environmental conditions.
- Practices that are accumulated by farmers due to constant experimentation and innovation.
- Trial-and-error problem solving approaches by groups of people with an objective to meet the challenges they face in their local environments (Rolig and Engel, 1992).
- Decision-making skills of local people that draw upon the resources they have at hand.
- Farmers can be excellent conservators of natural resources. They have rich knowledge concerning different methods of pest managements. Though years of practical adopted experience, indigenous knowledge widely developed by rural communities. Their knowledge generation is based on a sophisticated undertaking of environment and culture. Some of the traditional technologies of pest management are relevant even today on some of the commercial farming systems. Indigenous knowledge based organizations still serve as a forum for identifying problems, dealing with them through local level experimentation and exchange of information with other societies. In the last few decades, there has been divergence on the traditional system with focus on highly toxic chemical methods of pest management leading to environmental degradations and decline of social cultural value (Davis, 1996).

## INDIGENOUS CULTURAL PRACTICES TO PROTECT CROPS FROM INSECT PESTS

Historically, cultural and technical practices were the farmer's most important methods of preventing crop losses. Prior to the emergence of crop protection sciences, even before the broad outlines of the biology of pests were understood, the farmers evolved many cultural practices largely through trial and error experiments to minimize the damage caused by insect pests (Smith *et al.*, 1976).

The prophylactic practices included field and plant sanitation, good seeds, proper cultivation, winter management etc. The cultural methods consisted of deep ploughing, flooding, weeding, trap crops etc. On the other hand, cultural practices like closer plant spacing, sowing time, cultivation, harvesting procedure etc. developed as a part of the modern agricultural technology for exploiting the full genetic yield potential of hybrid varieties have simultaneously proved favourable to most of the insect pests (Ayyer, 1938).

Traditional cultural control of crop pests includes all the crop production and management techniques which are required and hence utilized by the farmers to maximize their crop productivity and/or farm income. It include decisions on crop varieties to be grown, time and manner of planting, tillage, field and crop sanitation, application of fertilizers and irrigation, harvesting time and procedure, and even off season operations.

The purpose of traditional cultural control practices is to make the environment less favourable for the pest and/or more favourable for its natural enemies (Dhaliwal and Arora, 1993). Some insect problems can be overcome through the use of indigenous practices, cultural practices, or at least through slight modification of field operations (Chapman and Carter, 1982). Following are the indigenous cultural practices performed by farmers to protect crops:

### Ploughing, hoeing and basin preparation

Cultural practices like ploughing, hoeing and basin preparation influence directly the survival of soil inhabiting insect pests and other arthropods and nematodes. Insect are most vulnerable when in the pupal stage and most insect pests pupate in the soil which furnishes a protective habitat. Birds like the king crow, the myna, the starling, etc. pick up the exposed pupae following there cultural operations. Some insects live in the upper layer of the soil. Then eggs are exposed during soil preparation and subsequently desiccate. Many insects like cutworms, grubs of the root borer and white grubs, which feed on the root system of plants are also exposed to the vagaries of the elements during basin preparation and hoeing. Deep ploughing carried out in winter helps in reducing the over wintering population of several pest.

The time and type of tillage influence the soil environment and affect the survival of insect pests or their natural enemies, e.g. fall ploughing is often helpful in reducing the over wintering population of *Helicoverpa* sp. Deep ploughing is also helpful in minimizing infestation of armyworm in cereal crops. Raking up and hoeing of the soil around fruit trees serve to destroy pupae fruit flies. (Verma, 1998)

### Hand picking of pests

Hand picking of pests and their destruction is another time tested method of pest control. Right from picking lice from human hair, clothes and even animals to the manual separation of pests from stored grain. This method can prove effective in curtailing pest population on some crops. Insects which lay eggs in conspicuous and easily illuminated places, e.g. tobacco caterpillar, gypsymoth, hairly caterpillar, sugarcane top borer, epilachna beetle, etc. can be easily eliminated. Early instar larvae of such insects often feed in congregations and later disperse to the entire plant/tree and in the field, making control difficult.

Hand picking demands alertness, patience and keen observation. The collected pests are destroyed by immersing them in kerosinized water or by deep burying. Nocturnal insects responding positively to light, e.g. defoliating beetles, moths of Bihar hairy caterpillar, tomato fruit borer, tobacco caterpillar, and cerambycid beetles etc. are collected, using light source or by trapping them in a light-trap and are subsequently destroyed.

However, it is a labour intensive and time consuming process. Knowledge of egg laying behavior, location of eggs and larvae on infested crop is essential. Hence, only a well trained person would prove effective.

**Planting time**

The manipulation of planting time helps to minimize pest damage by producing asynchrony between host plant and the pest or synchronizing insect pests with their natural enemies or crop production with available alternative host plants of the pests or by production followed by destruction of crop residue before the insect can enter diapause. For example, Hessian fly (*Magetiola destructor*) is controlled by planting dates over a delayed time, so that most of the overwintering adults emerge and die consequently (Teetes, 1993).

**Seed rate of spacing**

Use of high seed rate is recommended in those crops where removal of infested plants is helpful in minimizing the incidence of insect pest (Dhaliwal and Arora, 1993).

**Plant diversity**

Deterrence of colonization is probably one of most promising means of controlling insect pests through intra-field diversity, because only a little additional diversity in the crop field may leave a profound effect on colonization by insects. The diversity of crop system can be increased by inter-cropping, trap-cropping, presence of weeds in the crops which are also suitable host plant for a particular pest as they may reduce feeding damage to the main crop by diverting the pest (Cromarttrie, 1993).

**Crop rotation**

Part of the value of crop rotations is their ability to prevent the build up of insect pests. Generally, crop rotation is most effective against pests that have a narrow host range and limited range of dispersal. Crop rotation includes the use of cereals and pulses to minimize the damage caused by insect pests. Crop rotation is most effective against pests that have a narrow range of dispersal capacity. Changing to a different crop by rotation helps to break the cycle, changing the selection pressure for a given species. Crop rotation also permits the use of different herbicides (Shenk, 1994).

### Nutrient management

Irrigation, manuring and fertilization are practices that, in general, promote rapid growth and shorten the time of susceptible stages available for attack, providing the crop with greater tolerance and the opportunity for insect damage. However, such practices may also change the physiology of host plants, making them lush and succulent and thus encouraging pest survival and multiplication.

The organic manures release nitrogen in small quantities over a longer period and hence their application does not generally lead to pest outbreaks. Green manures and bio-fertilizers based on *Rhizobium, Azolla, Azospirillum and Azotobacter* can serve as important components of organic manurring. Integrated nutrient management must, therefore, form an integrated part of IPM programme (Jayaraj *et al*., 1994).

### Pruning of fruit trees

Pruning of fruit trees is an important practice performed during the dormant season to create a proper frame and provide symmetry to the tree and ensure proper and balanced growth in the ensuing season. Set procedures of pruning are followed by using secateurs, knives, pruning, saw, etc. Some inconspicuous pests which hibernate as egg/nymphs (e.g. some aphids, leaf hoppers, etc.) in limited numbers are automatically removed along with pruned branches/twigs thereby mitigating the severity of their attack in the next season. Further, pruning of old, damaged and weak portion of the tree encourages new growth which is healthier and stronger.

Thus, pruning aids pest control as it releases overwintering population, improves general health of the tree and helps in maintaining a balanced foliage distribution with proper aeration and sunlight penetration. Pruned plants usually have lesser pest infestation and do not easily succumb to pest attacks.

However, since pruning is both an art and skill it demands that the pruner possesses basic knowledge of overwintering behaviour and sites of pest infestation to deliver maximum benefit from this mechanical method of pest control.

### Water management

The amount of moisture in the soil has a profound effect on the survival of many insect pests. The quality and quantity of nutrients available in the host plants are also greatly affected due to change in the soil moisture level (Dhaliwal and Arora, 1993).

In situation, where crops are not solely rainfed, the management of water may be used as a means of pest control. Drip irrigation will favour a crop rather than a weed and will provide the crop with a competitive advantage. Flooding is also an effective practice for the control of a number of weed and of course is usefully employed in this way with vast areas for pest management. However, in general, water supplies are managed primarily to promote healthy and vigorous plant growth, with pest control usually only considered of secondary importance.

## POTENTIAL OF INDIGENOUS TECHNICAL KNOWLEDGE IN PEST MANAGEMENT

The increasing public awareness about the contamination of environment by toxic chemical residues both in developed and developing countries and public perception about the use of eco-friendly methods for the control of pests of agricultural and public health importance has necessitated the search and development of non-chemical method.

The potential of several secondary chemical compounds in plants to disrupt specific physiological mechanisms involved in nutrition, reproduction, metamorphosis and behaviour of insects could be exploited in the control of insect pests as an eco-friendly alternative to the conventional pesticide use. In this context, use of extracts of such plants like neem (*Azadirachta indica* A. Tuss) is now gaining momentum since several active insecticidal neem compounds, including azadirachtin, are feeding inhibitors and growth disruptors for most insect orders (Schmutterer, 1990).

- Feeding deterrence of neem compounds is explained by behavioral antecedent effect due to perception by the peripheral chemorecptors in insects (Shoonhoven, 1981; Simmonds and Blaney 1984; Mordue *et al.,* 1985; Dorn and Trumm, 1993). Such growth inhibitory effect of neem ranging from delay in moulting with the production of deformed insects, to the complete inhibition of growth at higher doses, have been reported in various insect species (Schmuttener, 1990).
- During last five decade, more than 2000 plant species belonging to different families and genera, especially higher plants (Arnason *et al.*, 1989) have been reported to contain biologically active principles with multifacial effects against various insects (Grainge and Ahmed, 1998).
- Most of the study till date have been restricted to agricultural pests and no such studies have been conducted against forest insect pests in India except against *Atteva fabriciella* (Ahmed *et al.,* 1991) and *Closlera cupreata* (Ahmed *et al.*,1996).

The antifeedant effects of *Acorus calaumus* (Sweetflage) rhizome extract on several insect species are evident from the literature (Subramanium 1994; Dixit *et al.*, 1956; Mukerjee and Govind, 1959; Paul *et al.*, 1965; Yadav, 1971; Jacob and Sheila, 1994). The antifeedant and repellent action of *Acorus calamus* and other plant extract against several insect species were well reported by earlier workers (Saxena and Srivastava, 1992; Teotia and Tiwari, 1977; Edeg and Davis, 1980 and Tripathi and Singh, 1994).

## EVIDENCE OF USING INDIGENOUS BIO-INSECTICIDE FOR CROP PROTECTION

### Cattle urine

There are a number of instances which show that cattle urine especially cow urine has been used us an insecticide for different kinds of insect pests of cultivated crops:

- Control of termites by preparing a spray solution by mixing *Calotropis sp*. and *Leptadenia pyrotechnica* extracts in cattle urine (Balusinh, 1992).
- Control of castor semilooper by use of goat urine and neem leaves can be controlled by soaking 2 kg of neem leaves in 2–3 litre of goat urine. This mixture is then distilled. About 500 ml of the distillate is diluted with 15 liters of water and sprayed over crops (Darbar, 2000).

- Pests and diseases are controlled in cotton and *Barleria* sp. by pulverizing leaves of Agava, Vettikeltai (*Strycchnos nuxvomica*) and soaking in cattle urine. The resultant solution is diluted in water in 1 : 8 and sprayed against pests and diseases of cotton and *Barleria* sp. (Thyammal, 1994).
- 5 kg. of *Lantana camara* leaves are soaked in five liters of cow urine for 3 days. Then it is diluted in 90 liters of water and sprayed against pest (Sundra Raman, 2000).

### Use of wood ash on and around vegetables crops

Fruits wood is a major source of energy in the hills, consequently, wood ash is available in plenty. It is used to wash and clean utensils as well as clothes. It is a common practice to sprinkle wood ash on vegetable crops, especially growing in kitchen garden and to spread it around plants to ward off pests and to enhance nutrient status of the soil. To achieve this a thick layer of ash is spread on the soil around plants and it is also sprinkled on foliage to protect it against a variety of pests. This is because it is a source of phosphorus for plants and it also acts as a physical poison usually causing abrasion of epicuticular waxes and thus exposing pests to death through desiccation. It also interferes in the chemical signals emanating from the host plants thus obstructing the initial host location by pests. The treated foliage further becomes unpalatable for foliage feeders like cutworms, caterpillars, grasshoppers, etc. But since ash provides only temporary protection against pests, insecticides which have quick knock down effect have replaced the use of wood ash today.

### Kerosene oil for killing borers

Shoot and stem borers bore and riddle the twigs, branches and even the main trunk of fruit trees and push out faecal matter and frass through an exit hole. Farmers have been using kerosene oil to kill these tissue borers using a flexible metallic wire which is inserted through the hole made by the borer into the gallery to clean it. Then a small bung made of cloth soaked in kerosene oil is inserted into the hole and finally, it is plugged using a paste of cowdung and clay. The insertion of the metallic wire into the gallery causes physical injury to the larvae which are destroyed in the process. The oil vapour emanating gradually from the cloth bung fill the closed gallery, suffocating the pests and ultimately the larvae die. The death of the borer is indicated by unopened plugged hole. The drawback of this practice is that kerosene vapours act slowly and this treatment is not sure and definite method to annihilate the borer. Farmers now prefer to use quick acting insecticides for assured results..

### Bajagun weed for reducing rat menace

Common *cowitch, Mucuna prurita*, vernacularly known as bajagun, is a leguminous herbaceous annual plant found throughout the plains of India. It also grows in the low lying areas of Himachal Pradesh. Its longitudinally ribbed turgid pods are covered with dense pale brown or grey stinging bristles (trichomas) which upon coming in contact with the skin cause discomfort and itching attributed to presence of a histamine liberating proteinase, mucunain. The effect persists for 3–5 minutes. These hair also contain serotonin which causes pain besides itching. It was an old practice to powder leaves and pods of the plant and scatter it over main ways and in the live burrows of the rats to drive them away from cultivated fields to reduce rat damage to the crop. The easy availability of various effective rodenticides from the market have resulted in this practice gradually dying out.

**Use of walnut and sweetflag leaves against pests in stored grains**

In rural areas of Himachal Pradesh, it is an old practice to use walnut leaves and leaves of pond weed, commonly known as sweetflag, *Acorus calamus* as a protection for both grain and clothes against insect damage. To achieve protection, a layer of leaves of walnut is spread over grain stored in gunny bags. Likewise, shade dried leaves of sweetflag are powdered and put over grain stored in gunny bags to protect it from damage due to stored grain pests.

Walnut leaves are astringents and the aqueous extract of fresh leaves possesses bactericidal action while mature leaves contain 9 to11 percent tannin. Tannins are known to act as feeding deterrents. Sweetflag leaves and rhizomes have many chemical ingredients including an essential oil, the oil of *Calamus*, which primarily contains *asarone*. Depending upon quantity of *asarone (cis as well as trans)* its effect on insects may be attractant, antifeedant, repellent, antigonadal or insecticidal. However, since the active principal is present only in meager amounts in the leaves, this treatment may not be able to afford protection for long periods.

**Cow dung and clay mixture**

The practice of applying a thin paste made of cow dung, clay and cow urine, as a disinfectant, on the floor of mud (kutcha) houses in rural areas is an age old practice. Similarly, the application of a paste, of good consistency, to treat wounds and injured limbs of fruit trees has been in vogue, in villages, since long. Such a paste is also applied to pruned twigs/limbs.

A fine slurry is prepared by through mixing of the clay, cow dung and cow urine in a container. The paste is then applied manually using bare hands or a locally made brush. Cow dung and cow urine possess complex degrading substances and may possess antibacterial properties. Addition of clay results in better adhesion of other constituents to the treated surface. Sealing of wounds/cuts prevents access of pathogens to the otherwise exposed surface. To some extent, the already present pathogen is also dealt by using the applied paste. Pruned ends of twigs and cuts are also favorable spots for the settlement and establishment of woolly apple aphid on apple trees. Covering such sites with the cowdung paste hastens healing and prevents aphid settlement. The paste is also used for the treatment of cankered limbs of the trees. The entire affected region is removed by a disinfected knife and is then covered with freshly prepared cow dung paste. As is evident, the application of the paste is a laborious procedure and proper sealing of the injured region is required. Sometimes, it does not prove to be an effective method for the treatment of canker.

- Traditional pest control practices are generally prophylactic in nature and are frequently the first line of defense against crop pests.
- These tactics are environmentally safe, as they do not leave significant disruptive effects on the non-target organisms and environmental quality.
- These are usually compatible with other components of IPM like resistant varieties, biocontrol agents and even insecticides.

- The chances of development of resistance to a bio-insecticide are considered remote.

In spite, of these advantages, traditional pest management practices are not widely used in modern agricultural scenario, due to a number of limitations like—

- ITK practices alone are not sufficient to bring insect pest population below economic injury level, so these have to be supplemented by other IPM strategies.
- A ITK practice directed against the pest may well increase the chance of survival and multiplication of other pests.
- Many of the ITK practices are effective only when undertaken on an area wide basis.

## ITK MOST THREATENED SOURCE

Despite wide adaptability and great relevance the indigenous technology bank for pest control is fast depleting due to:

- inherent offensive strategies of the modern technology on the belief basis of native know how,
- lack of scientific evidences in support of a traditional knowledge ignoring the fact that traditional knowledge is mostly based upon the meta-physical ingredients which cannot be measured under present limits of the modern science,
- lack of documentation,
- lack of interest in younger generation due to silence of modern education systems on ITK and
- preference for hi-tech chemical products.

It is obvious that a mere symbolic presence of local traditions and their superficial understanding check the serious pest management problems.

## CONSEQUENCES OF DISREGARDING INDIGENOUS KNOWLEDGE SYSTEMS

Undermining farmer's confidence in their traditional knowledge leads to become increasingly dependent on outside expertise. Small-scale and marginal mountain farmers are often portrayed as backward, obstinately conservative, resistant to change, lacking innovative ability and even lazy (IFAP, 1990). The International Federation of Agricultural Producers (IFAP) enumerated certain reasons for such a perceptions.

- Lack of understanding of traditional agriculture, which further leads to a communication gap between promoters and practioners giving rise to myths.
- The accomplishments of farmers often are not recognized, because they are not recorded in writing or made known.
- Poor involvement of farmers and their organizations in integrating, consolidation and disseminating what is already known.

## FUTURE STRATEGIES FOR ITK STRENGTHENING

In a nutshell, horticultural pest management is possible through complementary integration of traditional and modern systems which would be cost effective, equitable and sustainable. However, sincere efforts would have to be undertaken by development planners, policy makers and researchers on the following broad aspects;

- Scientific documentation of ITK for pest management.
- Refinement oriented scientific interventions.
- Participation of farmers.
- Complementary blending of modern and traditional technologies.

## CONCLUSIONS

Undoubtedly, it is clear that tried and time tested traditional pest management technologies with wide-adaptations are often of more value than modern technologies designed to yield quick results with serious ecological consequences.

The indigenous knowledge of pest management was systematically pursued through incorporation of natural processes, such as cultural, mechanical, physical, biological and non-toxic chemicals etc. These methods have minimum harmful effects on the eco-systems because of their objectivity to sustain the biological interactions matching with the demand of production agriculture (Verma, 1998).

There is an urgent need for both identification, rediscovery and documentation of sustainable traditional methods which are sensitive to biotic stress and to an inequitable agrarian socio-economic structure.

## REFERENCES

Ahmed, Mukhtar, Dayal, R. and Borai, P. 1996. Evaluation of insecticidal properties of *Dalbergia stipulacea* against poplar defoliator, *Clostera cupreata* (Lepidoptera: Notodontidae). *Ann. For.* **4**: 47–51.

Ahmed, Mukhtar; Gupta, B.K. and Bhandari, R.S. 1991. Efficacy of some plant extract against *Ailanthus webworm, Atteva fabriciella. Indian Journal of Forestry,* **14**: 5–7.

Arnason, J.T., Philogene, B.J.R. and Moramd, P. 1989. *Insecticides of plant origin*. ACS Symp. Ser., 387. American Chemical Society, Washington D.C. pp. 213

Ayyar, T.V.R. 1938. *Handbook of Economic Entomology for South India*. Government Press, Madras, India.

Balusinh, P.G. 1992. Leaves of *Calotropis gigantia*. *Honey Bee*, **3**: 19.

Chapman, S.R. and Carter, L.P. 1982. *Crop production, principles and practices*. Surjeet Publications, Delhi, India. pp. 164–193.

Cromarttrie, Jr, W.J. 1993. The environmental control of insects using crop diversity. *In: CRC Handbook of Pest Management in Agriculture*. D. Pimental (ed.).Vol. I. CBS Publishers and Distributors, New Delhi, India. pp. 183–216.

Darbar, A. B. 2000. Control of Castor Semilooper. *Honey bee*, **9**: 19.

Davis, M. 1996–1997. Indigenous people and intellectual property rights, Research paper no. 20, *Information and Research Services*, Department of the Parliamentary Library, Canberra, 30 June, 1997.

Dhaliwal, G.S. and Arora, R. 1993. Changing status of insect pests and their management strategies. *In: Changing Scenario of Indian Agriculture.* K.S. Gill; G.S. Dhaliwal and B.S. Hansra (eds). Common wealth Publishers, New Delhi, India, pp. 1–55.

Dixit, R.S.; Petri, S.L. and Rangnathan, S.K. 1956. Evaluation of insecticidal activity of solvent extract and stem principal of rhizome of *Acorus calamus* against house fly, *Musca domestica*, Mosquito *C. fatigens* and carpet beetle, *A. vorax. Jour. Sci. Indust. Res.* **15**: 16–22.

Dorn, A. and Trumm, P. (1993). Effect of azadirachtin a regulation of midgut peristalsis in *Locusta migratoria*; *Abstracts World Neem Conf.* Bangalore, India. pp. 44.

Edig, S.H. and Davis, C.R.F. 1980. Repellency of rape seed extracts against adults of *Tribolium eastneum* and *Tribolium confusion* (coleoptra: Tenigrionidae). *Canadian J. Ent.* **112**: 971–974.

Grainge, M. and Ahmed, S. 1988. *Handbook of plants with pest control properties.* A Wiley Interscience Publication, John Wiley and Sons, New York. pp. 470.

IFAP 1990. *Sustainable farming and role of farmer's organization.* The Netherlands International Federation of Agricultural Process, Wageningen.

Jacob, Sosamma and Sheila M.K. 1994. Studies on antifeedent of some plant products against the leaf caterpillar, *S. docillis* Butle., on brinjal and wooly bear, *Pericallia ricinii* Fabr. on castor. *Indian J. Ent,* **56**: 276–278.

Jayaraj, S.; Ananthakrishan, T.N. and Veeresh, G.K. 1994. *Biological Pest Control in India: Progress and Perspectives.* RGICS Project No. 2, Rajiv Gandhi Institute for Contemporary Studies, New Delhi, India.

Mordue, A.J.; Cotte, P.K. and Evan, K.A. 1985. Azadirachtin: its effects on gut mobility, growth and moulting in locusta. *Physiological Entomology*, **10**: 431–437.

Mukerjee, T.D. and Govind, R.R. 1959. A plant insecticide *Acorus Calamus* Linn. *Indian J. Ent.*: 194–205.

Paul, C.F.; Agarwal, P.N. and Ausat, A. 1965. Toxicity of solvent extract of *Acorus calamus* to be some grain pests and termites. *Indian J. Ent.* **27**: 114–117.

Rolig, N. and Engel, P. 1992. The development of concept of agricultural knowledge information system (AKIS): Implication for extension. *In:. Agricultural Extension, World Wide International Evolution and Forces for Change.* W..M. Rivera and D.G. Gustatson (eds). Elsevier Science Publishing Company, New York. pp. 125–138.

Saxena, B.P. and Srivastava, J.B. 1972. Effect of *Acorus calamus* Linn. oil vapours on *Dysderus Cingulatus* Fabr. *Indian J. Expt. Biol.* **10**: 391–393.

Schmutterer, H. 1990. Properties and potential of natural pesticides from the neem tree, *Azadirachta indica. Annu. Rev. Entomol.* **35**: 271–297.

Schoonhoven, L.M. 1981 Perception of azadirachtin by some lepidopterous larvae. *Proc. Ist. Int. Neem Conf.* Rottach-egern. pp. 105–108.

Shenk, M.D. 1994. *Cultural Practices for Weed Management, in weed Management for Developing Countries.* R. Labrada; J.C. Caseley and C. Parker (eds). FAO Plant Production. and Protection. pp. 163–170.

Simmonds, M.S.J. and Blaney, W.M. 1984. Some neurophysiological effect of Azadirachtin on lepidoperous larvae and their feeding response. *Proc. 2nd Int. Neem Conf.* Rauischholzhausen, 1983. pp. 163–180.

Smith, R.F.; Apple, J.L. and Bottrell, D.G. 1976. The origin of integrated pest management concepts for agricultural crops. *In: Integrated Pest Management.* J.L. Apple and R.F. Smith (eds). Plenum Press, New York, USA.. pp. 1–16.

Srivastava, K.P. 1996. *A Book of Applied Entomology*, Vol. II. Kalyani Publishers, Ludhiana, India.

Subramanium, T.V. 1994. Sweetflage, *Acorus calamus*, A potential source of valuable insecticides. *Jour. Bombay Nat. Hist. Soc.* **48**: 338–341.

Sundara Raman, S.R. 2000. Casuarina against crop diseases. *Honey bee*, **11**: 16.

Teetes, G.T. 1993. The environmental control of insects using planting times and plants spacing. *In: CRC Handbook of Pest Managemet in Agriculture.* D. Pimental (ed.). Vol. 1. CBS. Publishers and Distributors, New Delhi. pp. 169–182.

Teotia, T.P.S. and Tiwari, G.C. 1977. Insecticidal properties of drupes of Dharek, *M. azadirach* and rhizome of sweetflage, *A. calamus* against adult of *S. cereaeella* Oliv. *Indian J. Ent.* **39**: 222–227.

Thyammal, 1994. Pests and diseases of cotton and *Barleria* sp. *Honey bee*, **5**: 15.

Tripathi, A.K. and Singh, D. 1994. Screening of natural products for insect antifeedant activity: (1) Plant extract. *Indian J. Ent.* **56**: 129–133.

Thrupp, l. 1989. Legitimizing local knowledge sensitized packages or environment for third war people. *In: Indigenous Knowledge System: Implications for agriculture and international development studies in technology and social change.* D.M. warren; L.J. Kerneer and S.O. Titiola, (eds)., Ames, Iowa State, University, US. pp. 138–153.

Venkatratram, L. 1990. Traditional approach, farmer's wisdom for sustainable agriculture. *Kissan World* (Feb. Issue). pp. 20–22.

Verma, L.R. 1998. *A glimpse of Indigenous Technology Knowledge for Watersheds Management in Upper North-West Himalayas of India.* Participatory Watershed Management Training in Asia (PWMTA) Programme GCP/RAS/IGI/NET, FAO(UN), Kathmandu, Nepal, P. iv+97.

Yadav, R.L. 1971. Use of essence of *Acorus calamus* Linn. as an insecticide against the pulse beelte, *Bruchus chinensis. Z. Angew. Ent.* **68**: 289–294.

*Pest Management in Horticulture Crops*
L.R. Verma, A.K. Verma and D.C. Gautam (eds)


*Chapter 9*

# Genetic Engineering Techniques for Insect Pest Management

*D.C. Gautam*

PROTECTION OF CROPS from insect pests has been an important concern of agriculturists and horticulturists. Insect pests cause great damage to crops of horticultural and agricultural importance as they have short life cycles coupled with great capacity for multiplication. The high genetic plasticity of the insect pests as is shown by the widespread development of insecticide resistance and their genetic diversity have complicated the problem of pest management. Large quantities of pesticides are used throughout the world to control insect pests and excessive use of these chemical pesticides has become detrimental to maintain ecological stability. A number of alternatives to these chemical pesticides are now available. And so more emphasis is being laid on Integrated Pest Management (IPM) which is based on the principle of not to exterminate plant pests but to establish a system of co-existence. In fact, IPM is a flexible, ecological based pest control strategy that uses a combination of techniques applied at specific times, aimed at specific crops and pests.

Genetic engineering and other recent technologies can effectively help in our war against pests. Exploitation of natural biological diversity and use of recombinant DNA technology hold great promise for insect pest control. Now all these technologies are put to use to devise ways to control the insect pests in such a way that there are minimum adverse effects on environment and useful organisms. Development of biopesticides, which are environment friendly, has been a great step in this direction as their use is getting wider acceptance world over.

In this article, attempt has been made to consolidate the vast information available on biopesticides and to synthesize a brief informatory material covering almost all the aspects of this most rapidly expanding area of research and application.

## BIOPESTICIDES

Biological pesticides or biopesticides are defined as a group of products that are distinguishable from conventional chemical pesticides by their nontoxic mode of action, their specificity to target pests and most importantly by the fact that they are, or can be, produced by living organisms (Matten *et al.* 1993). The US Environmental Protection Agency has officially defined a biopesticide as belonging to one of three groups: microbial pesticides, biochemical pesticides and transgenic plant pesticides.

- Microbial pesticides include microorganisms (bacteria, viruses, fungi) and their products, usually proteins as active ingredient which may be naturally occurring or may be altered via manipulation of genetic material. Mode of action of microbial pesticides is through production of toxin or pathogenicity.
- Biochemical pesticides contain semiochemicals, pheromones, insect growth regulators, repellents and enzymes as active pesticidal ingredient. Biochemical pesticides are of natural origin and act through nontoxic mode of action. Semiochemicals or pheromones which are emitted by a plant or animal affect the behavior of similar or other species. Hormones are synthesized in one part of organism and translocated to another where they have controlling or regulating activity. Natural plant growth regulators inhibit, modify or stimulate plant growth. Repellents result in mating disruption. Enzymes or protein molecules control gene expression or catalyze reactions.
- Transgenic plant pesticides are plants genetically altered via introduction of genetic material for the purpose of imparting or increasing the production of a pesticide.The active pesticidal ingredient could be considered the pesticidal substance produced from, or modified as a result of introduced genetic material.

Biopesticides are now becoming an important component of intergrated pest management programmes. These are receiving great attention as these are considered to be the potential agents to decrease the use of chemical pesticides, a step towards safer food and cleaner environment. Biopesticides are not simply the replacement for hazardous substances but, in fact, these act as supplements to other methods of pest management (Hall and Menn, 1999).

The major constituent of biopesticides is bioinsecticides comprising of microbials (Bt) and natural products. Bts are the most important group of biopesticides produced by fermentation. These do not produce discrete chemicals but yield insecticidal crystal proteins. These can be manipulated and targeted against specific insect pests (Carlton, 1992, 1993). Natural and recombinant baculoviruses, singly or jointly with other control agents are already in use against insect pests. These have received much attention because of their small genome and simple molecular organization facilitating genetic manipulations (Bonning and Hammock, 1996). Mycoinsecticides (entomopathogenic fungi) are ecofriendly and potential insect control agents (Feng *et el.* 1994). Production of engineered plants (transgenic crops) is the outcome of recent emerging technologies.

Fermentation-derived insecticides are also the recently introduced group of insect control products. Biofungicides are very effective agents for the control of fungal diseases (Papavizas, 1985). Bioherbicides are the most suited for the control of undesirable vegetation or weeds. Pheromones are very effective in causing mating distruptions in insect pests and thus are potential insect control agents.

All these components of biopesticides require proper formulations and delivery systems for efficient action. These crop protection agents are more specific, environment friendly and effective alternatives or supplements to the conventional approaches of insect pest control.

## MICROBIAL PESTICIDES

The most important characteristic of microbial pesticides is their ability to replicate in the host insect pest. Successful infection results either replication as in case of viral and fungal pathogens or toxemia (possibly with replication) in case of bacterial pathogens. The ability to replicate make these biopesticides quite distinct from chemical pesticides. Microbial pesticides produce large quantities of infectious units as the primary inoculum leads to the production of secondary inoculum which is very high than the initial applied dose.

### Bacillus thuringiensis

*Bacillus thuringiensis* has been developed as a control agent for a number of insect pests. Bt is not only the most successful commercial microbial pesticide with world wide application for protection of crops and environment, but also it has started replacing conventional insecticides in several areas of application. *Bacillus thuringiensis* subspecies ***kurstaki*** (Btk) is now widely used against lepidopteran pests of a number of crops (Entwistle *et al.* 1993).

*Bacillus thuringiensis* (Bt) is a Gram-positive soil bacterium and can be readily isolated from insects, soil and stored product dusts. These crystalline spore forming bacteria, though small in number, are commonly found in most samples of natural field materials. Many strains of Bt have been isolated from insects but epizooitics (widespread, rapidly spreading infections) are extremely rare. Occassionally, Bt has been found to be enzooitics (ever present, but not at outbreak proportions) in insects in the field.

#### *Life cycle of Bt*

Life cycle of Bt consists of two distinct stages: vegetative cell division and sporulation. The production of δ-endotoxin takes place during sporulation. The spores exhibit no metabolic activity and are able to remain viable under extreme environmental conditions such as dessication and extreme low availability of nutrients to which the vegetative cells are more sensitive. Bt persists in the soil mainly as spores. In favourable conditions, spores germinate and undergo subsequent cell division.

#### *Bt genes and delta-endotoxins*

Bt is a major microbial insecticide and a source of genes encoding several proteins toxic to insects (Cannon, 1996). Main toxins of Bt are delta-endotoxins or insecticidal crystal proteins (ICPs), a group of structurally related proteins that are present as crystalline inclusion bodies in sporulated cultures. Each endotoxin is a product of a single gene. This has enabled genes to be inserted into plants and express as toxin production in the plant tissue. A feature common to many Bt subspecies is the presence of complex array of plasmids. The plasmids are of two kinds: small with no known function and large plasmids sharing homologous DNA sequences. The main function assigned to these is the production of entomocidal toxins. Presence of one or several toxin genes on same or different replicons including very large plasmids is now established feature of δ-endotoxin genes. Several plasmid genes encoding δ-endotoxins are part of composite structures which include several mobile genetic elements. There are 96 genes reported to code for δ-endotoxins described so far (Estruch *et al.* 1997).

Bt produces crystalline inclusions during sporulation (Hofte and Whiteley, 1989). These inclusions consist of proteins with insecticidal properties and thus are very effective against many insect pests. Crystal proteins are categorised into four major classes: Lepidoptera specific (Cry I), Lepidoptera and Diplera specific (Cry II), Coleoptera specific (Cry III) and Diptera specific (Cry IV), although there are exceptions. The Cry proteins of Bt, also referred to as delta-endotoxins, comprise a diverse group of insecticidal agents which are very effective. In fact crystal proteins are protoxins that are converted into smaller toxic peptides in the midgut of insect pests. These toxins cause osmotic imbalance by creating pores in the cell membranes of mid gut epithelium of susceptible insect pests.

In addition to endotoxins, other proteins effective against certain insect pests are Vip insecticidal proteins produced by Bt during vegetative growth of bacteria. Vip proteins are insecticidal proteins different from δ-endotoxins and are effective against a wide range of lepidopteran insect pests and thus this makes these potential candidates for use in conjuction with δ-endotoxins.

### *Bt production*

Bt product is produced by fermentation process; the broth contains a mixture of proteins, bacterial metabolites, spores and growth media components. Two types of methods are used for producing Bt products: i. production by surface cultures on semisolid media; ii. Production by liquid media in submerged cultures. With semisolid media, nutrients are contained in a coarse porous matrix with a large surface area. Oxygen transfer, humidity and temperature are maintained either by placing media on shallow trays, or keeping them in aerated drums. But semisolid media are difficult to sterilize. So generally, submerged cultures are preferred. Spores and crystals are readily produced by aerated liquid fermentation. These are easily harvested and when properly formulated, have a long shelf life.

Though sporulated cultures may be used directly in pest control, but Bt preparations are normally further processed to make their physical properties more suitable to field applications. Bt based products are relatively inert and can be readily incorporated into the user friendly formulations (Bryant, 1994). Concentrated Bt preparations are either wettable powders, wettable granules (also called dry flowables), or liquid, flowable suspension, concentrates which must mix with water to give homogenous suspension. Granular formulations are also applied directly while concentrates are diluted with water and sprayed on to plants.

### *Mode of action*

Bt has a distinct yet complex mode of action. Once Bt protoxin is ingested, it is first proteolysed to an activated toxin protein. The toxin acts directly on the membrane of mid gut epithelial cells, causing gradual increase in membrane permeability, followed by swelling and lysis of cells. The pores are formed in the membrane. The intoxication is associated with immediate feeding inhibition, leakage of ions from gut to hemolymph resulting in ionic imbalance in hemolymph. In some species, this is sufficient to cause death while in others, changes in gut and hemolymph conditions permit vegetative propagation of Bt within the insect resulting in septicemia which contributes to or causes death.

### *Target insect pests*

Bt strains generally have a narrow spectrum of insecticide activity as compared to conventional insecticides and exhibit great toxicity towards only one order of insect species. Even within one group, sensitivity differences are found in species. For example, tobacco budworm *Heliothis virescens* and diamondback moth *Plutella xylostella* are controlled effectively with Bt based (HD1) product easily while fall armyworms (*Spodoptera* spp) are difficult to control. So multispecies pest complexes continue posing problems. But if the target pests are fewer in number, Bt based products are very advantageous. Most of the Bt bioinsecticide products are produced using naturally occurring strains of Bt. Twardus (1989) reported that in USA derivatives of Bt strain HD1 subspecies *kurstaki* have been found very effective against gypsy moth *Lymantria dispar.* Industrial products based on Bt are now well recognised weapons for pest control.

### *Bt efficacy*

Bt efficacy is also affected by the timing of spray applications, spray coverage and larval feeding behavior. Spores and crystal proteins are inactivated by sunlight (Morris, 1983). But encapsulation of crystal proteins within the host cell improves their persistence. Synergistic toxicity toward insect pests is exhibited by certain combinations of Cry proteins (Lee *et al.* 1996) Even the presence of spores can also synergize the activity of Cry proteins against insect pests. Kronstad *et al.* (1983) reported that *cry* genes are almost localized on large plasmids. Some of these can be transferred from one Bt strain to another (Gonazalez *et al.* 1982) and through these natural processes of plasmid transfer, trans-conjugants are produced which have enhanced insecticidal properties.

### *Genetic manipulations*

Recombinant DNA technologies have made possible the production of improved strains. The first insecticidal protein encoding gene from Bt was cloned, sequenced and expressed in *Escherichia coli* in 1981 and this paved the way for using these insecticidal proteins in transgenic plants (Schnepf and Whiteley, 1981). Recombinant plasmids can be transferred from donor Bt to recipient Bt through conjugation and transduction. With the use of electroporation, now the cloned *cry* genes can be introduced into Bt through a variety of *Escherichia coli*–Bt shuttle vectors which employ plasmid replicons derived from other Gram-positive bacteria or from native Bt plasmids (Arantes and Lereclus, 1991).

Genetic manipulations of Bt has resulted in having a number of bioinsecticide products based on trans-conjugant strains of Bt (genetically modified Bt strains) or recombinant Bt strains. Even transgenic microorganisms may act as toxin delivery systems. δ-endotoxin genes have been cloned and expressed in a wide variety of microorganisms from baculoviruses to cyanobacteria to improve their delivery and efficacy to insect pests. Further improvements in efficacy of Bt products in horticulture and agriculture will make these strong alternatives to other strategies. And that will certainly widen their role in integrated pest management programmes.

**Baculoviruses**

Insect pest control through baculoviruses involves use of a virus as insecticide. These viruses are quite specific and typically do not kill predatory and parasitic insects. Viral multiplication in the target insect pests increases the amount of the virus in the system further and thus provides an important tool against insect pests. These are very useful in controlling those pests which produce many generations in a season. Most baculoviruses have a narrow spectrum of activity and less residual activity.

***Natural and recombinant baculoviruses***

Baculoviruses are natural as well as recombinant. Natural baculoviruses of Family *Baculoviridae* are Nucleopolyhedroviruses commonly called nuclear polyhedrosis viruses (NPVs) and Granuloviruses commonly called granulosis viruses (GV). Most commonly used microbial insecticides are NPVs. GVs are used against those insect pests for which no NPVs are effective. NPVs are reported to occur most commonly in lepidopteran insects but they are found in almost all the insect orders. These infect almost all the tissues. Some NPVs have narrow host range while others have broad host range infecting species from different genera. GVs infect only lepidopteran pests. Baculoviruses are very useful where a single lepidopteran species is a major pest for most of the growing season of a crop.

***Mode of action***

Baculoviruses (NPVs) produce resting structures called polyhedra containing one or more infective virions. Polyhedra yield is high in natural and low in recombinant baculoviruses. When the insect larvae feed in the contaminated foliage, the polyhedra dissolves in mid gut releasing virions which then invade mid gut epithelial cells. Replication of virus results in increase in number of pathogen which invades helmocel and colonizes remainder of the insect. The life cycle is completed when the insect dies and release large quantities of polyhedra onto the leaf surface.

***Formulations***

Affected larvae, when start dying, can be harvested, grounded up and processed into a product for application. Liquid and dust formulations of viruses have been developed and tested with or without additives which generally improve the efficacy of baculoviruses.

***Target insect pests***

Cunningham and Entwistle (1981) reported that European spruce sawfly *Gilpinia hercyniae* is effectively controlled by NPV in North America. The infected larvae become virus factories for distributing the viruses in the environment. The virus can be disseminated both vertically from adults to larvae as well as horizontally from one infected individual to other. Gelernter *et al.* (1986) made a study on effectiveness of nuclear polyhedrosis virus and chemical insecticide for the control of *Spodoptera exigua* on head lettuce and found it very effective when used in appropriate amounts. NPV is most effective against beet armyworm, one of the most important pests of vegetable crops throughout the world which has developed resistance to most of the chemical insecticides. GV is very effective against western grape leaf skeltonizer *Harrisina brillions* (Stern and Fedirici, 1990). Since larvae of this pest are gregarious, infection of even a single larva results in infecting the whole population which is destroyed within a week.

### *Genetic manipulations*

The techniques of genetic engineering are now widely used to enhance insecticides properties of NPVs while maintaining their pest specific characterization. More virulent forms are produced through genetic alterations either by insertion of genes that express an insecticidal protein, and for this, virus is utilized as a vector for carrying and expressing gene or by deletion of genes that prolong host survival.

Insertion of juvenile hormone esterase (JHE) gene into nucleopolyhedrovirsus AcNPV isolated from alfalfa looper *Autographa californica* reduced the survival time of *Trichoplusia ni* (Hubner) (Hammock, *et al.* 1990, Bonning and Hammock 1996). Similarly, O'Reilly and Miller (1991) reported that deletion in EGT gene of AcNP results in early death of infected fall armyworm, *Spodoptera frugiperda*.

The technology of isolating genes that encode insecticidally active proteins and inserting them into baculovirus is rapidly evolving. Baculoviruses have host specificity and environment compatibility. Because of these, attributes they are readily integrated into IPM programmes as they do not adversely affect the beneficial insects and other non-target organisms.

### Transgenic bacteria and viruses as Bt toxin (Micro-mediated delivery systems)

For effective delivery, living recombinant microorganism are produced. Bt toxin genes are inserted into microorganism that are associated with target insect habitats. These transformed organisms colonize and continue to synthesize sufficient quantities of toxin to prevent insect damage.

For example, Bt genes can be inserted into bacteria that colonize leaf surface and roots externally such as *Pseudomonas cepacia* (Stock *et al.* 1990) and *Pseudomonas fluorescens* (Obukowicz *et al.* 1986) or internally such as *Rhizobium leguminosarum* (Skot *et al.* 1990). But concerns have been raised about release of such recombinant microorganism as horizontal gene transfer may occur to other bacterial species. DNA developments paved the way for expressing certain genes in bacteria that are killed to provide a non-living, cellular delivery system. Improved toxin delivery is attained by killing the cells during stationary growth in a manner that stabilized them to enhance field persistence of the toxin.

Bt toxin genes can also be inserted into insect baculovirus (NPV) to accelerate the speed of kill as well as to have a virus-mediated toxin delivery system (Martens *et al.* 1990, Merryweather *et al.* 1990).

### Mycoinsecticides (Entomopathogenic fungi)

Entomopathogenic fungi are commonly used for controlling insect pests (Ferron, 1978). These are environment friendly and can be readily isolated from soil. The most common species are *Metarhizium anisopliae, Beauveria bassiana* and *Verticillium lecanii.* Of these, *Beauveria bassiana* has a broad potential as a viable insect control agent.

Many species of entomopathogenic fungi can be produced in submerged cultures. Hyphomycetes typically produce single celled hyphal bodies called blastospores. Liquid fermentation technology is the developing area of mycoinsecticide research because conidia of many pathogens can not grow on solid substrates (Romabach *et al.* 1988). Now mass culture systems are developed in which conidia are produced at a liquid-air or solid-air interface (Goettel, 1984). After profuse sporulation when the culture matures, the spores are dried and harvested. A number of hypomycetes species can be induced to produce true conidia which are thick walled cells and persist in the environment under a broad range of conditions.

*Metarhizium anisopliae* is very effective against black wine beetle *Otiorhynchus sulcatus* (Reinecke *et al.* 1990). *Beauveria bassiana* is effective against European cornborer *Ostrinia nubilalis* and Colorado beetle *Leptinotarsa decemlineata* and *Verticillium lecanii* is effective for the control of aphids and thrips (Feng *et al.* 1994). The use of oil has been shown to enhance the efficacy of mycoinsecticides as it enables fungal pathogen to remain active under condition of low humidity (Batemann *et al.* 2000).

*Beauveria bassiana* produces toxin called beauvericin. Infection of *Beauveria bassiana* begins when a conidiospore lands on a host and germination tube penetrates the larval integument. Hyphal bodies circulate throughout the hemolymph and larvae get filled up with mycelium. Conidiospores are produced and they break through the skin and completely cover the larvae.

## BIOCHEMICAL PESTICIDES

Insect pest management through behavior controlling chemicals, referred to as semiochemicals, is emerging fast. It is based on the chemical messages that organism use to communicate.

### Plant volatiles

Plants produce signals in the form of volatile substances as a response to chemicals in saliva of insect pests. When attacked by pests, plants release such semiochemicals to attracting predators and parasitoids-insects that lay eggs inside insect larvae which are then devoured from within. This plant-predator communication raises the hope of providing nontoxic form of pest control (Whitfield, 2001).

The production of plant volatiles was first noticed by Dicke and Sabelis (1998). Later it has been observed that volictin, a compound found in insect oral secretions of plant and herbivore origin may interact with octadecanoid pathway and results in increased production of volatile compounds in corn plants that are attractive to natural enemies of pests (Pare *et al.* 1998).

These chemicals can be used to spray onto crops. Plants can be induced to produce such substances using jasonate (a plant hormone) as spraying of jasonate on tomato plants resulted in two fold increase in parasitism of *Spodoptera exigua* caterpillars by *Hyposoter exiguae* wasps (Thaler, 1999).

Tobacco plants have been reported to have different blends of volatile compounds by day and night (De Moraes *et al.*, 2001). At night, tobacco plants that are being attacked by caterpillars of *Heliothis virescens* emit a specific blend of volatile chemicals which repel the larvae. During the day, the plants emit blends that attract parasitic or predatory insects. Both these responses appear to benefit the plant as well as insect species involved.

Producing plant varieties which respond more strongly to the attack by the pests could be a very effective strategy for protection of crops. If the volatile compounds are analysed and genes producing them are identified then genetically engineered plants could be produced that emit blends of defensive volatiles at appropriate time. That would reduce the use of environmentally detrimental pesticides significantly (Ryan, 2001).

**Pheromones**

Sex pheromones are chemical messages between individuals of same species which facilitate mating (Mayer and McLaughlin, 1991). Pheromones have great potential for control of insect pests as release of large amounts of synthetic pheromones could interfere with mate location (Gaston *et al.* 1967). Disruption of mating is generally achieved by manipulating certain aspects of sexual behavior of insect pests (Carde and Minks, 1995)

Pheromones are target specific, exploit certain behavior of the pest and result in reduced use of chemical insecticides. Pheromones may be extracted from the insect pest or synthesized in the laboratory. They are then used in minute amounts in pest control strategies. Insect pests vary in their susceptibility to pheromone mediated control. Some pests are very sensitive while others are not. For controlling major pests of crops, mating disruption technology must be used with other management practices. Resistance to pheromones has not been reported and these are non-toxic to natural enemies.

Pheromones have been successfully used to control pink bollworm *Pectinophora gossypiella* and codling moth *Cydia pomonella*. Miller *et al.* (1990) demonstrated that pink boll worm males approach and contact the pheromone laden dispensers. Incorporation of small amount of chemical insecticide in the stickers used to adhere the pheromone dispensers to crop foliage, proved very effective as males coming in contact with stickers were killed resulting in decrease in population. Moffitt and Westigard (1984) used sex pheromone for mating disruption in codling moth, which causes considerable damage to pome fruits and observed suppression of the pest population.

**Neem botanical**

Neem functions not only as an insect growth regulator but it also acts as a behavior modifying substance also. It has poor contact action so it has to be ingested by the insect pests. In certain insect pests, it acts as a deterrent to feeding and oviposition (Schmutterer, 1995) but does not have adverse effect on natural enemies and pollinators. The desirable attributes of neem are: efficacy at low concentration, broad spectrum of action, minimal non-target toxicity and no persistence in environment.

The neem tree *Azadirachta indica* (A. Juss) is a tree of tropical regions. Azadirachtin is found in varying degrees in all parts of plant but highest concentration is in seeds. Neem seeds contain oil which is removed before extraction of azadirachtins. Neem oil itself is a crop protectant. Two important azadirachtins found in neem are azadirachtin '*aza* A' and '*aza* B'. Since the active component is very susceptible to photodegradation, so the formations have to be planned in such a way so as to protect the active ingredients.

Azadirachtin is a liminoid with great insecticidal activity. Though it affects insects in different ways but its main modes of action are to act as insect growth regulator, feeding and oviposition deterrant. Azadirachtin is structurally similar to natural insect hormone ecdysone. That's why it interferes with and stops the moulting cycle of insect. Immature insects die during development resulting in decrease in population of insect pests. Because of these properties and being environmentally safe, it is considered a very important agent for insect control

Neem is active against a number of pests. Lowery and Isman (1994) reported that neem is very effective against aphids but the efficiency is influenced by the systemic movement of the azadirachtin in different crops. These workers further reported that neem has minimal effect on natural enemies of aphids. Naumann *et al.* (1994) observed that neem does not affect the pollinators including honey bees indicating that it is quite compatible for integrated pest management.

### Other botanicals

Other botanical (insecticides) in use are given below:

| *Botanical* | *Source* | |
|---|---|---|
| Nicotine | *Nicotiana tabacum* | (Solanaceae) |
| Pyrethrum | *Chrysanthemum cineraraefolium* | (Asteraceae) |
| Quassia | *Quassia amara* | (Simaroubaceae) |
| Rotenine | *Derris elliptica* | (Leguminosae) |
| Ryania | *Ryania speciosa* | (Flacourtiaceae) |
| Sabadilla | *Schoenocaulon officianale* | (Liliaceae) |

These botanicals are broad spectrum insecticides consisting of complex mixture of active ingredients and can have more than one mode of action in insect pests.

## TRANSGENIC PLANT PESTICIDES

Different constraints of controlling insect pests through chemical pesticides led to the new developments in transformation technology enabling the stable introduction of foreign genes into the many important crop species. In contrast to animal cell, plant cells are totipotent. Each living cell is capable of regenerating into an entire plant identical to one from which cell was obtained.

The ability to insert foreign genes into plant cells and tissues and to regenerate viable, stable and fertile novel plants, has been a revolutionary development which can be called gene revolution.

Recombinant DNA technology has made it possible to employ special enzymes to cut and splice sections of DNA. The specific genes can be removed from one species, modified or tailored as necessary and inserted into another species. Engineering of insecticidal genes from *Bacillus thuringiensis* into crop plants has been carried on large scale resulting in growing a large number of transgenic crops.

Transgenic plants express toxins from *Bacillus* thuringiensis. Plant breeders have been improving cultivars of crop plants and now this potential of improvement of crop plants is greatly enhanced by tissue culture and molecular biology techniques. The most recent approach involves cloning of the desirable gene from bacteria and express its function in the plant. So this systemic toxin reaches to burrowing as well as surface feeding insect pests. So it is possible now to transform any crop with a tailored gene to produce desirable chemical products (Estruch *et al.* 1997). Many species of monocots herbaceous and woody dicots have been transformed (Gasser and Fraley, 1989) with useful traits such as insect, herbicide tolerance and resistance to virus diseases. Transgenic corn, *Zea mays* L. and cotton, *Gossypium hirsutum* L. are now widely grown.

The first reported case of δ-endotoxin gene expressed in plants for insect pest control occurred in 1987 (Barton *et al.* 1987, Vaeck *et al.*, 1987). Tobacco plants, *Nicotiana tabacum* were developed that produced enough endotoxin to kill tobacco hornworm *Manduca sexta* L. placed on transformed leaves. Estruch *et al.* (1997) highlighted that the genetically engineered plants are better equipped to resist the attack of economically important insect pests.

### *Methods of plant transformation*

The transformation of plants can be achieved in different ways: The most widely used methods of transforming plants are: Agrobacterium mediated transfer of DNA (Hinchee *et al.* 1988), bombardment of cells with DNA coated particles (Klein *et al.* 1987) and electroporation (Rhodes *et al.* 1988, Toriyama *et al.* 1988).

### *Agrobacterium mediated transfer of DNA*

This approach exploits the tumour or gall production ability of *Agrobacterium tumefaciens* in dicotyledonous plants. Tumour inducing (Ti) plasmids in virulent strains of this bacterium which are responsible for DNA transfer and appearance of disease symptoms, contain two sets of sequences. DNA which is transferred to plant (T-DNA) and virulent (vir) genes which are not transferred to plants. T-DNA regions are flanked by border sequences. For transformation, Ti genes-disarmed bacterium is used which can infect the plant cell but does not induce tumour. The desired gene is inserted between flanked border sequences of Ti plasmid where the tumour gene is normally located. The bacterium then inserts the desired gene into the plant chromosome like a normal infection of normal plant tissue in nature.

### *Microprojectile bombardment method*

The second method uses high velocity microprojectiles for delivering nucleic acids in living cells. Cell suspensions, immature embryos or ovaries in inbred lines can also be subjected to microprojectile bombardment to produce regenerated plants. In this technique, a gene gun is used to inject DNA coated particles into the plant cell. The injected DNA is then incorporated into its own system by the plant cell. This single cell is then developed into a complete plant through tissue culture techniques. Even multiple copies of the bioengineered plant can be produced. When the plants reach a certain size, they can be transferred to a field and grown like any regular crop.

### *Electroporation*

This method involves the electrporation of protoplast in a DNA solution. High voltage is applied across protoplast-DNA suspension which causes small holes to form temporarily in the protoplast membrane through which DNA can pass. This method has been successfully applied for the production of transgenic maize and rice.

As all major crops of horticultural and agricultural importance have to be protected from insect pests, and genes from Bt are useful for developing insect resistant cultivars for certain specific pests, the technology of producing transgenic resistant cultivars is becoming popular.

*Important transgenic crops resistant to different insect pests*

The most important transgenic crops produced are:

- Transgenic maize (*Zea mays* L.) has *cry1A(b)* gene from Bt and is resistant to southwestern corn borer *Diatraea grandiosella*, fall armyworm *Spodoptera frugiperda* (Williams *et al.* 1997). Transgenic avidin maize is reported to be resistant to stored insect pests (Kramer *et al.* 2000).
- Transgenic rice (*Oryza sativa* L.) has modified *cry1A(b)* gene from Bt and is resistant to rice leafhopper *Cnaphalocrosis medinalils,* striped stem borer *Chilo suppressalis* (Fuzimoto *et al.* 1996).
- Transgenic soybean (*Glycine max* L.) has *cry1A(c)* gene from Bt and is resistant to soybean looper *Pseudolusia includens*, velvet bean caterpillar *Anticaassia gemmatalis* Hubner, corn earwom and tobacco budworm (Hinchee *et al.* 1988).
- Transgenic cotton *(Gossypium hirsutum* L has delta endotoxin or modified endotoxin genes from Bt and is resistant to cabbage looper *Trichoplusia ni* Hubner, beet armyworm *Spodoptera exigua* Hubner (Perlak *et al.* 1990, 1991).
- Transgenic potato (*Solanum tuberosum* L.) has modified version of δ-endotoxin gene *cry3A* from Bt and is resistant to Colorado beetle *Leptinotarsa decemlineata* Say (Perlak *et al.* 1993) and *cry5* gene from Bt provides resistant to potato tuber moth *Phthorimaea operculella* Zeller (Douches *et al.* 1996).
- Transgenic tomato (*Lycopersicon esculentum* L.) has δ-endotoxin gene from Bt that provides resistant to bollworm, tobacco budworm and tobacco liornworm (Fischoff *et al.* 1987).
- Transgenic eggplant (*Solanum melongena* L. has *cry3A* gene from Bt and is resistant to Colorado potato beetle *Leptinotarsa decemlineata* Say (Jelenkovic *et al.* 1996).
- Transgenic *Popular* spp. have δ-endotoxin gene from Bt and are resistant to gypsy moth *Lymantria dispar* and forest tent caterpillar *Malacosoma disstria* Hubner (Strauss *et al.* 1991).

These are the examples of some of the most important transgenic crops. The US Environmental Protection Agency regulates transgenic pesticidal crops. Keeping in view the constant exposure of the insect pests to transgenic plant pesticides, long term insect resistant management (IRM) strategies are essential. In this direction, EPA has imposed two structured refuge requirements for Bt cotton (Matten *et al.* 2000).

## OTHER BIOPESTICIDES FOR INSECT PEST CONTROL

The other biopesticides which are now widely used for insect pest control are the spinosyns and entomopathogenic nematodes.

### *The spinosyns (Actinomycetes natural products)*

A variety of natural products are now put to use to control insect pests. The latest addition is the class of fermentation derived insect control agents called spinosyns. While searching for new microorganisms with insecticidal potential, soil samples from were analyzed. Extracts from fermentation broth in such fermentation screening programme were found to have insecticidal activity. The microorganism involved was identified as *Saccharopolyspora spinosa* (Mertz and Yao, 1990) which is a soil Actinomycete and produced a family of new, unique macrolides (containing macrocyclic lactone) called spinosyns (Sparks *et al.* 1995). Due to their complex nature these could be obtained only through the process of fermentation. More than 20 spinosyns have been isolated and identified. Wild strain of *Saccharopolyspora spinosa* produces spinosyns A-J and spinosyns A and D are most effective against lepidopteran larvae.

Spinosyn A is most active against larvae of rice stem borer *Chilo suppressalis*, loopers *Trichoplusia ni*, American bollworm *Heliothis armigera*, the cotton bollworm (*Helicoverpa zea*, tobacco budworm *Heliothis virescens* and armyworms *Spodoptera exigua, Spodoptera littoralis, Spodoptera frugiperda* (Sparks *et al.* 1995; Thompson *et al.* 1995). In addition to these, they are also active on some homopteran species such as aster loophopper *Macrosteles severini*.

Spinosyns act on central nervous system and increase the spontaneous activity resulting in contraction of involuntary muscles causing tremors. These do not persist in environment, are least toxic to beneficial insects and no cross resistance develops in insect pests. These attributes make these insect control agents unique.

### *Entomopathogenic nematodes*

Entomopathogenic nematodes are used in inundative biological control and are very effective against those insect pests which reside in soil or cryptic habitats. Prophylactic application to soil surrounding seeds and seedling are very useful. The two genera of entomopathogenic nematodes which have emerged effective biological control agents are *Steinernema* and *Heterorhabditis* (Burnell and Stock, 2000). The infective juveniles (IJs) invade the insect host (Grewal *et al.* 1997).

*Steinernema feltiae* is very effective against mushroom sciarid *Lycoriella auripila* (Richardson, 1987), *L. solani* (Tomalak, 1994), *L. mali* (Grewal *et al.* 1993; Nickle and Cantelo, 1991) fungus gnat *Bradysia coprophila* (Harris *et al.* 1995). Similarly *Steinernema* sp is effective against corn earworm *Helicoverpa zea* (Boddie) and fall armyworm *Spodoptera frugiperda* (Raulston *et al.* 1992). *Heterorhabditis* sp is very effective against black wine weevil *Otiorhynchus sulcatus* (Simon, 1981).

Nematodes formulations are developed by conserving IJ nematodes either by physical immobilization (restricting their movement with limited stored energy reserves) or by physiological immobilization (reducing their oxygen comnsumption). Nematodes are successfully formulated as flowable gels or gel forming polyacramide. IJs are also encased in water-dispersible-granules. These formulations are advantageous as they extend nematode storage stability for longer periods as well as enhance nematode tolerance to temperature extremes.

Successful multiplication and pathgenecity of these nematodes is attributed to symbiotic associations with specific bacteria. Nematodes acts as vectors to transport bacteria into the host in which they proliferate and bacteria create conditions necessary for survival and reproduction with in host (Burnell and Stock, 2000). Entomopathogenic nematodes can be mass processed in the insect or on artificial medium. Insects, in fact, serve as a bioreactor for mass production of these nematodes.

## BIOPESTICIDES FOR CONTROL OF FUNGAL PATHOGENS AND WEEDS

Biopesticides for the control of fungal pathogens and weeds are biofungicides and mycoherbicides.

### *Biofungicides*

Biofungicides are used for the control of fungal pathogens of plants which may be soil borne pathogens, foliar diseases or post-harvest rots during storage. Two biocontrol agents act as biofungicides, the bacterium *Burkholderia cepacia* and fungus *Gliocladium virens.* At present, liquid and dry formulations are used for delivering these biological control agents.

*Burkholderia cepacia*, a Gram-negative bacterium, generally found associated with rhizosphere and roots of corn, has broad spectrum antifungal activity against a range of fungal pathogens (Hebbar *et al.* 1992)

*Gliocladium virens* is a common soil saprophyte that produces antibiotic metabolites. The most common metabolite produced is gliotoxin which has antifungal activity. It is very effective against *Pythium ultimum.* (Roberts and Lumsden,1990).

Most biofungicides require an intact living cell for function, Intact cell dependent biofungicides require delicate and sophisticated approaches in the design of commercial formulations. This provides stability during shelf life.

### *Mycoherbicides*

Fungal pathogens are frequently used to control unwanted plant species. A number of strategies are used for the effective control of weeds. Generally, the pathogens are isolated from the regions where weed and pathogen species are indigenous and released in those areas where weed has become problematic. The pathogens such as rust and fungi are capable of self-dissemination and this makes these useful for controlling noxious weeds (Wapshere *et al.* 1989). In other approach, indigenous microbial biocontrol agents are greatly supplemented utilizing native pathogens occurring on weeds on infested areas (Auld and Morin, 1995). For effective control, the information on the biology of target weed and that of pathogen is important.

## CONCLUSION

Future management strategies will rely on the use of combination of different strategies. As ever increasing concern for deployment of environmentally acceptable insect pest control strategies is gaining momentum, there is great potential for the long-term development and growth of pesticides. With the emerging technologies, biopesticides could be made more specific and effective by improving formulations, refining application techniques and enhancing microbe and crop mediated delivery systems. The scope of the biopesticides in integrated pest management can be increased by exploiting the diversity of the wild type isolates allied with creation of recombinant strains of microbes and high quality cultivars of transgenic plants. This in turn will also encourage the growth and diversity of the beneficial organisms and thus will enhance plant defenses.

Genetic manipulations can afford solutions which conventional methods can not. Genetically engineered resistant varieties of plants will not only increase the productivity but also enhance the quality.Utilization of the principles of natural plant defense mechanisms as reported recently may reveal very interesting arrays of new products. With all these components available, it is now a reality to protect the crops through genetic engineering and other recent technologies which are environment friendly.

## REFERENCES


Arantes, O. and Lereclus, D. 1991. Construction of cloning vectors for *Bacillus thuringiensis. Gene,* **108**: 115–119.

Auld, B. A. and Morin, L. 1995. Constraints in the development of bioherbicides. *Weed Technol.* **9**: 638–652.

Barton, K.A.; Whiteley, H.R. and Ning-Sun, Y. 1987. *Bacillus thuringiensis* delta endotoxin expressed in transgenic *Nicotiana tobacum* provides resistant ot lepidopteran insects. *Plant Physiol.* 85: 1103–1109.

Bateman, R. P.; Alves, R. T.; Cross, J. V.(ed.); Gilbert, A. J.(ed.); Glass, C. R.(ed.) Taylor, W. A.(ed.); Walklate, P. J.(ed.) and Western, N. M. 2000. Delivery systems for mycoinsecticides using oil based formulations. Pesticide Applications, University of Surrey, guildford, UK, 17–18 January 2000. *Aspects of Applied Biology,* **57**: 163–170.

Bonning, B. C. and Hammock, B. D. 1996. Development of recombinant baculovirus for insect control. *Ann. Rev. Entomol.* **41**: 191–210.

Bryant, J. E. 1994. Commercial production and formulation of *Bacillus thuringiensis*. *Agric. Ecosys. Environ.* **49**: 31–35.

Burnell, A. M. and Stock, S. P. 2000. *Heterorhabditis, Steinernema* and their bacterial symbionts-lethal pathogens of insects. *Nematology,* **2**(1): 31–42.

Cannon, R. J. C. 1996. *Bacillus thuringiensis* use in agriculture: a molecular perspective. *Biol. Rev.* **71**: 561–636.

Carde, R. T. and Minks, A.K. 1995. Control of moth pests by mating disruptuion: successes and constraints. *Ann. Rev. Entomol*. **40**: 559–585.

Carlton, B.C. 1992. Development of improved bioinsecticide based *on Bacillus thuringiensis. In*: *Pest Control with Enhanced Environmental Safety.* S. O.Duke, J. J. Menn and J. R. Plimmer (eds.). (ACS Symposium Series, Vol.524, no 18, American Chemical Society, Washington DC, pp. 258–266.

Carlton, B. C. 1993. Genetics of Bt insecticidal crystal proteins and strategies for the construction of improved strains. *In*: *Proceedings of the XVIII Beltsville Symposium.* R. D. Lumsden and J. L. Vaughn (eds.). (ARC USDA, Beltsvelle, MD, pp. 326–337.

Cunningham, J. C. and Entwistle, P. F. 1981. Control of sawflies by baculovirus. *In*: *Microbial Control of Pest and Plant Diseases.* H. D. Burges (ed). Academic, London. pp. 379–407.

De Moraes, M. C.; Mescher, M. C. and Tumilinson, J.H. 2001 Caterpillar-induced nocturnal plant volatiles repel conspecific females. *Nature,* **410**: 577–580.

Dicke, M. and Sabelis, M. W. 1998. How plants obtain predatory mites as body guards. *Netherland J. Zool.* **38**: 148–165.

Douched, D. S.; Liswidowati, A.; Hadipermadi, W. P.; Hudy, P.; Westedt, A. and Grafus, E. 1996. Progress in development and evaluation of Bt transgenic potatoes with resistance to potato tuber moth (*Phthorimaea operculella*). *Abstract of the 80th Annual Meeting of the Potato Association of America, Annual Meeting, Idaho Fall, ID*. August 11–15, 1996. *Am. Potato J.* **73**: 8.

Entwistle, P. F.; Cory, J. S.; Bailey, M. J. and Higgs, S. (eds.). 1993. *Bacillus thuringiensis, An Environmental Biopesticide: Theory and Practice.* John Wiley and Sons, New York.

Estruch, J. J.; Carozzi, N. B.; Desai, N.; Duck, N. B.; Warren, G. W. and Koziel, M. G. 1997. Transgenic plants: An energing approach to pest control. *Nature Biotechnol.* **15**: 137–141.

Feng, M. G.; Poprowski, T. J. and Hjachatourians, G. G. 1994. Production, formulation and application of entomopathogenic fungus *Beauveria bassiana* for insect control: current status. *Biocontrol Science Technol.* **4**: 3- 34.

Ferro, P. 1978. Biological control of insect pests by entonogenous fungi. *Ann. Rev Entomol.* **23**: 409–422.

Fischoff, D.A.; Bowdish, K.S.; Perlak, F.J.; Marrone P.G.; McCormicke, S.M.; Niedermeyer, J. G. *et al.* 1987. Insect tolerant transgenic tomato plants. *Bio/Technology*, **5**: 808–813.

Fuzimoto, H.; Itoh, K.; Yamanoto, M.; Kyozuka J. and Shimamoto, K. 1996. Insect resistant rice generated by introduction of a modified delta-endotoxin gene of *Bacillus thuringrensis*. *Bio/Technology*, **11**: 1151–1155.

Gasser, C. and Fraley, R. T. 1989. Genetically engineered plants for crop improvement. *Science* **244**: 1293–1299.

Gaston, L. K.; Shorary, H. H. and Sario S. A.. 1967. Insect population control by use of sex pheromone to inhabit orientation between sexes. *Nature,* **213**: 1155.

Gelernter, W. D.; Toscano, N. C.; Kido, K and Federici, B. A. 1986. Comprasion of nuclear polyhedrosis virus and chemical insecticide for control of beet armyworm *Spodoptera exigua* (Lepidoptera: Noctuidae) on head lettuce. *J. Econ. Entomol.* **79**: 714–717.

Goettel, M. S. 1984. A simple method for mass culturing entomopathogenic Hyphomycete fungi. *J. Microbiol. Methods*, **3**: 15–20.

Gonzalez, J. M. Jr.; Brown, G. J. and Carlton, B. C. 1982. Transfer of *Bacillus thuringiensis* plasmid coding for δ-endotoxin among strains of *B. thuringiess and B. cereus*. *Proc. Natl. Acad. Sci. USA*, **79**: 6951–6955.

Grewal, P. S.; Matsuura, M. and Converse, V. 1997. Mechanism of specificity of association between the nematode *Steinernema scapterisci* and its symbiotic bacterium. *Parasitology*, **114**: 483–488.

Grewal, P. S.; Tomalak, M.; Keil, C.B.O. and Gaugler, R. 1993. Evaluation of genetically selected strains of *Steinernema feltiel* against the mushroom sciarid *Lycoriella mali*: *Ann.Biol.* **123**: 659–702.

Hall, F.R. and Menn, J. J. 1999 (eds.) *Biopesticides: Use and Delivery*. Humana Press, Totowa, New Jersey.

Hammock, B. D.; Bonning, B. C.; Posse, R. D.; Henzlik, T. N. and Maeda, S. 1990. Expression and effects of juvenile hormone esterase in a baculovirus vector. *Nature*, **344**: 458–460.

Harris, M. A.; Oetting, R. D., and Gardener, W. A.. 1995. Use of entomopathogenic nematodes and a new monitoring technique for the control of fungus gnats, *Bradysia corpophila* (Diptera: Sciaridae) in Floriculture. *Biol. Control*. **5**: 412–418.

Hebbar, K. P.; Atkinson, D.; Tucker, W and Dart, P. J. 1992. Suppression of *Fusarium moniliforme* by corn root associated *Pseusomonas cepacia*. *Soil Biol. Biochem*. **24**: 1009–1020.

Hinchee, M. A.; W., Conner-Ward, V. D.; Newll, C. A.; McDonnell, R. E.; Sato, S. J.; Gasser, C. S. *et al.* 1988. Production of transgenic soybean plants using Agrobacterium mediated DNA transfer. *Biotechnology,* **6**: 915–922.

Hofte, H. and Whiteley, H. R. 1989. Insecticidal crystal proteins of *Bacillus thuringiensis*. *Microbiol. Rev*. 242–255.

Jelenkovik, G.; Billings, S.; Chen, Q.; Lashomb, J. and Ghidu, G. 1996. Engineering transgenic egg plant (*Solanum melogena* L.) resistant to Colorado potato beetles *(Leptinotarsa decemlineata* Say*)*. *Hort. Science*, **31**: 572.

Klein, T.M.; Wolf, E.D.; Wu, R. and Stanford, J. E. 1987. High velocity microprojectiles for delivering nucleic acids into living cells. *Nature,* **327**: 70–73.

Kramer, K. S.; Morgan, T. D.; Thorne, J. E.; Dowell, F. E.; Baily, M. and Howard, J. A. 2000. Transgenic avidin maize is reisstant to storage insect pests. *Nature Biotechnology*, **18**(6): 670–674.

Kronstad, J. W.; Schnepf, H. E. and Whiteley, H. R.1983. Diversity of locations for *Bacillus thuringiensis* crystal protein genes. *J. Bacteriol*. **514**: 419–428.

Lee, M.K.; Curtiss, A.; Alcantara, E. and Dean, D. H. 1996. Synergistic effect of the *Bacillus thuringiensis* toxins *Cry*IAa and *Cry*IAc on gypsy moth, *Lymantria dispar*. *Appl. Environ. Microbiol*. **62**: 583–586.

Lowry, D. T. and Isman, M. B. 1994. Insect growth regulation effects of neem extracts and azadirachtins on aphids. *Entomol. Exp. Appl.* **72**: 77–84.

Marten, J. W. M.; Honee, G.; Zuidema, D.; Vanlent, J. W. M.; Visser, B. and Vlak, J. 1990. Insecticidal activity of a bacterial crystal protein expressed by a recombinant baculovirus in insect cells. *Appl. Env. Microbiol*. **56**: 2764–2770.

Matten, C.; Schneider, W.; Slutsky, B. and Milewski, E. 1993. Biological pesticides and US Environmental Protection Agency. *In*: *Advanced Engineered Pesticides*. L. Kim (ed.). Marcel Dekker, New York, p324.

Matten, S. R.; Dugger, P. (ed.) and Richter, D. 2000. EPA regulation of transgenic pesticidal crops and insect resistance managementment for Bt cotton. *Proceeding Beltwide Cotton conferences, San Antinio, USA*, 4–8 January, 2000, vol. 1: 71–76.

Mayer, M. S. and McLaughlin, J. R. (eds.). 1991. Handbook of Insect Pheromones and Attractants, CRC, Boca Raton, FL.

Merryweather, A. T.; Weyer, U.; Harris, M. P. G.; Hirst, M.; Booth, T. and Possee, R. D. 1990. Construction of genetically engineered baculovirus insecticides containing the Bacillus thuringiensis sub sp kurstaki HD73 delta endotoxin. *J. Gen. Virol*., **71**: 1535–1544.

Mertz F. P. and Yao, R. C.1990. *Saccharopolyspora spinosa* sp nov. isolated from soil collected in sugar mill rum still. *Int. J. Syst. Bacteriol* **40**: 34–39.

Miller, E.; Staten, R.T.; Nowell, C. and Guard, J. !990. Pink bollworm (Lepidoptera: Gelechiidae) point source density and its relationship to efficacy in attracticide formulations of gossyplure. J. *Econ. Entomol*. **83**: 1321–1325.

Moffitt, H.R. and Westingard, P.H. 1984. Suppression of the codling moth (Lepidoptera: Torticidae) population on pear in southern Oregon through mating disruption with sex pheromone. *J. Econ. Entomol.* **77**: 1531–1519.

Morris, O. N. 1983. Protection of *Bacillus thuringiensis* from inactivation by sunlight. *Can. Entomol.* **115**: 1215–1227.

Naumann, K.; Currie, R.W. and Isman, M.B. 1994. Evaluation of the repellent effects of neem indecticide on foraging honey bees and other pollinators. *Can. Entomol.* **126**: 225–230.

Nickle, W. R. and Cantelo, W. W. 1991. Control of mushroom-infesting fly, *Lycoriella mali* with *Steinernema feltiae*. *J. Nematol*. **23**: 145–147.

Obukowicz, M. G.; Perlak, F. J.; Kusano-Kretzmer, K.; Meyer, E. J.; Bolten, S. L. and Watrud, L. S. 1986. *Tn5*-mediated integration of the delta-endotoxin gene from *Bacillus thuringiensis* into chromosomes of root colonising Pseudomonads. *J. Bacteriol*. **168**: 982–989.

O'Reilly, D. R. and Miller, L. K.1991. Improvement of a baculovirus pesticide by deletion of the *egt* gene. *Bio/Technology* **9**: 1086–1089.

Papavizas, G. C. 1985. *Trichoderma* and *Gliocladium*: biology, ecology and potential for biocontrol. *Ann. Rev. Phytopathol*. **23**: 23–54.

Pare, P. W.; Alborn, H. T. and Tumlinson, J. H. 1998. Concerted biosynthesis of an insect elicitor of plant volatiles. *Proc. Natl. Acad. Sci. USA*. **95**: 13971–13975.

Perlak, F. J.; Deaton, W. R.; Amstrong, T. A.; Fuchs, R. L.; Sims, S. R.; Greenplate, J. T. and Fischoff D.A. 1990. Insect resistant cotton plants. *Bio/Technology*, **8**: 939–942.

Perlak, F. J.; Fuchs, R. L.; Dean, D. A.; McPherson, S. L. and Fischoff, D. A. 1991. Modification of the coding sequence enhances plant expression of insect control protein genes. *Proc. Natl. Acad. Sci.* **88**: 3324–3328.

Perlak, F. J.; Stone, T. B.; Muskopf, Y. M.; Peterson, L. J.; Parker, G. B.; McPherson, S. A. *et al.* 1993. Genetically improved potatoes: protection from damage by Colorado potato beetles. *Plant Mol. Biol.* **22**: 313–321.

Raulston, J. R.; Pair, S. D.; Loreza, J.; Cabanillas, H. E. 1992. Prepupal and pupal parasitism of *Helicoverpa zea* and *Spodoptera frugiperda* (Lepidoptera: Noctuidae) by *Steinernema* sp in cornfields in the lower Rio Grand Valley. *J. Econ. Entomol.* **85**: 1666–1670.

Reinicke, P.; Andersch, W.; Stenzel, K and Hartwig, J. 1990. BI0 1020, a new microbial insecticide for use in horticultural crops. *In*: *Proceedings of the Brighton Crop Protection Conference, Pests and Diseases,* Vol **1**, pp. 49–54.

Richardson, P. N. 1987. Susceptibality of mushroom pests to insect-parasitic nematodes *Steinernema feltiae* and *Heterorhabditis heliothidis*. *Ann. Appl. Biol.* **111**: 433–438.

Rhodes, C. A.; Pierce, D. A.; Mettler, I. J.; Mascarenhas, D. and Detmer, J. J. 1988. Genetically transformed maize plants from protoplasts. *Science,* **240**: 204–207.

Roberts, D. P. and Lumsden, R..D. 1990. Effect of extracellular metabolites from *Gliocladium virens* on germination of sporangia and mycelial growth of *Pythium ultimum*. *Phytopathology,* **80**: 461–465.

Romaback, M. C.; Aguda, R. M. and Roberts, D. W. 1988. Production of *Beauveria brassiana* (Deuteromycotina, Hyphomycetes) in different liquid media and subsequent conidiation of dry mycelium. *Entomophaga,* **33**: 315–324.

Ryan, C. A. 2001. Night moves of pregnant moths. *Nature*, **410**: 530–531.

Schmutterer, H. (ed.) 1995. *The Neem Tree*. VCH, Weinheim, Germany.

Schnepf, H.E. and Whiteley, H.R. 1981. Cloning and expression of the *Bacillus thuringiensis* crystal protein gene in *Escherichia coli*. *Proc. Natl. Acad. Sci. USA*. **78**: 2893–2897.

Simon, W. R. 1981.Biological control of *Otiorhynchus sulcatus* with heterorhabditid nematodes in green house. *Neth. J. Plant Patholo*. **87**: 149–153.

Skot, L.; Harrison, S. P.; Nath, A.; Mytton, L. R. and Clifford, B. C. 1990. Expression of insecticidal activity in *Rhizobium* containing the delta-endotoxin gene cloned from *Bacillus thuringiensis* subsp. *tenebrionis*. *Plant Soil*, **127**: 285–295.

Sparks, T. C.; Thompson, G. D.; Larson, L. L.; Kirst, H. A.; Jantz, O. K. and Worden, T. V. 1995. Biological characteristics of the spinosyns: a new class of naturally derived insect control agents. *In: Proceedings of the 1995 Beltwide Cotton Production Conference, National Cotton Council, Memphis, TN* pp. 903–907.

Stern, V. M. and Fedruci, B. A. 1990. Biological control of the western grape leaf skeltonizer with granulosis virus. *CA Agriculture*, **44**: 21–22.

Stock, C. A.; Mcloughlin, T. J.; Klein, J. A. and Adang, M. J. 1990. Expression of Bacillus thuringiensis crystal proteins gene in *Pseudomonas cepacia* 526. *Can. J. Microbiol*. **36**: 879–884.

Strauss, S. H.; Howe, G.T.; Goldfarb, B.; Neale, D. B. and Kinlaw, E.S. 1991. Prospects for genetic engineering of insect resistance in forest trees. *Forest biotechnology Forest Eco. Manage.* **43**: 181–209.

Thaler, J.S. 1999. Jasmonate-inducible plant defences cause increased parasitism of herbivores. *Nature*, **399**: 686–688.

Thompson, G. D.; Busacca, J. D.; Jantz, O. K.; Larson, L. L. and Sparks, T. C. 1995.; Spinosyns: an overview of new natural management systems. *In: Proceedings of the 1995 Beltwide Cotton Production Conference.* National Cotton Council, Memphis, TN, 1039–1041.

Tomalak, M. 1994. Selective breeding of *Steinernema feltiae* (Filipjev) (Nematoda Steinernematidae) for improved efficacy in control of a mushroom fly *Lycoriella solani* Winnertz. (Diptera: Sciaridad. *Biocontrol Sci. Technol.* **4**: 187–198.

Toriyama, K.; Akimoto, Y.; Uchimiya, H. and Hinata, K. 1988. Transgenic rice plants after direct gene transfer into protoplasts. *Biotechnology*, **6**: 1072–1074.

Twardus, D. 1989. USDA Forest service gypsy moth aerial suppression/eradication projects-1989. *Gypsy Moth News* **20**: 2.

Vaeck, M.; Reynaerts, A.; Hofte, H.; Jansens, S.; De Beukeleer, M..; *et al* 1987. Transgenic plants protected from insect attack. *Nature*, **328**: 33–37.

Wapshere, A. J.; Delfosse, E. S. and Cullen, J. M. 1989. Recent developments in the control of weeds. *Crop Protection*, **8**: 227–250.

Whitfield, J. 2000. Making crops cry for help. *Nature,* **410**: 736–737.

Williams, W. P.; Sager, J. B.; Hanten, J. A.; Davis, F. M. and Buckley, P.M. 1997. Evaluation of transgenic corn for resistance to fall armyworm and southwestern corn borer. *Crop Sci*. **37**: 957–962.

# Part II

# Horticultural Pest Management: Practices

*Pest Management in Horticulture Crops*
L.R. Verma, A.K. Verma and D.C. Gautam (eds)


*Chapter 10*

# Insect and Mite Pests of Apple and Pear

*J.P. Sharma and J.R. Thakur*

## INTRODUCTION

IN INDIA, APPLE (*Malus sylvestris* (Millers) and pear (*Pyrus* species) are cultivated from Kashmir to North-eastern hills at an altitude ranging between 1500 to 3000 m above mean sea level. Apple is the most important amongst the temperate fruits. In Himachal Pradesh alone, it occupies an area of 85,631 hectares that is about 41 per cent of the total area under fruits and 64.1 per cent of the area under temperate fruits (Singh, 2001). Pear comes next in order of importance among the temperate fruits. There are over 500 insects feeding on apple throughout the world (Singerland and Crosby, 1930). Though in India, more than 200 species have been reported feeding on apple trees (Bhalla and Pawar, 1977) yet, only a few occur in numbers enough to warrant control measures. In pear also, as many as 70 insect species have been recorded in India (Butani, 1979). But there is hardly any pest exclusive to pear as all the pests recorded on pear are also found on other fruit trees like apple, plum and other stone fruits. Many insect pests of these two crops are polyphagous and damage both the crops. The host–wise distribution of the pests of common occurrence to both apple and pear is given below:

**Apple**: San Jose scale, woolly apple aphid, apple root borer, apple stem borer, Quetta borer, apple leaf folders, tent hairy caterpillar, plum hairy caterpillar, Indian gypsy moth, *Zeuzera* shoot borer, defoliating beetles, codling moth and apple fruit moth.

**Pear:** Apple root borer, plum hairy caterpillar, bark eating caterpillar, *Zeuzera* shoot borer, codling moth and leaf and flower damaging beetles.

Several workers have reviewed the information on insect pests of apple and other temperate fruits in India (Pruthi and Batra, 1960; Butani, 1979; Bhalla and Gupta, 1993; Sharma and Thakur, 1996). In this chapter, an attempt has been made to present a critical analysis of information on bionomics and management of economically important pests of apple and pear.

Since most of the pests of these fruits are common to both the crops, therefore, to avoid repetition, these have been described under five categories i.e. sap suckers, defoliators, frugivores, borers and phytophagous mites.

## SAP SUCKING PESTS

San Jose scale and woolly aphid are the two key pests in India. Other scale and aphid species are of minor importance. San Jose scale attacks both apple and pear, whereas woolly apple aphid is a pest of apple and crab apple only.

### San Jose scale (*Quadraspidiotus perniciosus* (Comstock)) (Diaspididae)

This pest is believed to be a native of China but was noticed for the first time by the fruit growers at San Jose in California in 1873. It was introduced in India (Kashmir) from France in 1906 and has now spread all over the temperate areas at 1000–2200 m altitude infesting large number of plants. Bhalla and Gupta (1993) and Sharma and Thakur (1996), have reviewed the work done on this pest in India. Pruthi and Rao (1951) have listed 51 plant species of 15 families as host of this scale, 25 species of Rosaceae being the preferred ones. It is one of the serious pests of apple and pear in India. The pest has unsystematic distribution on the tree. Nymphs and adults cause damage by sucking cell sap from aerial parts. A typical halo like discoloration appears around the feeding site, particularly on the young twigs and fruits, within 24h of settling of the crawlers. Growth of the cambium is impaired and the associated xylem and phloem cells differ in number, size, and arrangement. The attack on fruits is more towards calyx end. Young, heavily infested twigs may die within a few months, and undetected infestation may kill young trees within three years. Bud burst is delayed in infested trees in spring, and leaves are unevenly and sparsely distributed. Fruit infestation directly reduces market value, and even a single infested fruit on a tree is taken as an economic threshold for initiation of control measures in USA (Whalon and Croft, 1984).

In India, the bionomics of this pest has been studied by various workers (Sahai and Joshi, 1965; Dinabandhoo and Bhalla, 1980; Masoodi and Trali, 1987). Mode of reproduction is sexual ovo-vivipary. The female has eggs, 2 nymphal instars, each of 13–28 and 8–9 days duration, respectively. The male passes through 4 instars. The duration of first to fourth instar is 12, 6, 2 and 4–6 days, respectively, the third and fourth instars being pre-pupal and pupal. This pest overwinters as first-instar black cap scale (stage before first moult) in Kashmir and other colder regions, but in Shimla and Kullu valley, all stages have been noticed to overwinter with an abundance of first-instar nymphs during December- January, and second-instar nymphs in February-March. In Kullu valley, the crawler emergence has been recorded as early as in second week of February due to heavy encrustation of scales on the tree. After resuming their activity in February- March, the scales mature within four weeks. The males (winged) fertilize the females (wingless) and die. The fecundity of the female varies between 300–400 eggs. The eggs mature into tiny nymphs in the ovisac of the female in about a month. At $27^0$ C and 50 per cent relative humidity, the scale takes 51–60 days for completion of life cycle with a nymphal duration of 25–30 days for both sexes and adult female survives for 26–30 days (Sahai and Joshi, 1965). The number of generations is mainly governed by temperature and to some extent by humidity and rainfall and varies from 3 (higher altitude areas) to 6 (lower valley areas like Kullu). In valley areas where the weather is comparatively warmer the number of generations is more. Under laboratory conditions at Solan in Himachal Pradesh as many as 8 generations have been raised at $28+1^0$ C temperature. The spread of this pest takes place through the infested nursery stock, fruits, wind, cloths, and tools.

*Management*

***Application of miscible oils and insecticides:*** In earlier reports, use of 8% diesel oil emulsion has been reported to be effective (Pruthi and Rao, 1951), but later on, application of 3% miscible oils in winter was found effective against this pest (Sharma and Bhalla, 1962, 1965). Further experimentation revealed that late dormant application of superior miscible oils (like Hindustan Petroleum Spray oil E) at 2% gave better control than mid- or early- dormant application (Gupta *et al.*, 1982). In lower valley areas, where heavy infestation results in encrustation on the tree the crawler emergence is early, therefore the dormant spray alone is insufficient to provide effective control. In such areas, an additional application of summer spray with some effective insecticides or release of biotic agents is essential. Among the insecticides tested, chlorpyriphos, fenitrothion, phosphamidon, methyl-demeton, and dimethoate have been found effective (Dinabandhoo and Bhalla, 1975; Sud *et al.*, 1975; Thakur and Hameed, 1980a, b; Bhardwaj, 1988; Khajuria and Sharma, 1992). Insecticide sprays should be synchronized with crawler emergence in last week of April to last week of May. Application of miscible oils in late dormancy has no adverse effects on the natural enemies and pollinators. However, thorough coverage of apple trees with the spray, use of only recommended good quality miscible oils at the right stage, proper concentration, and application on a clear sunny day are essential to get the desired efficacy. Since first-instar crawlers are the most susceptible, so proper surveillance to know the crawler emergence in April-May is essential.

***Scale free nursery plants:*** Spread of San Jose scale is mainly through infested nursery stock. So, use of scale-free nursery plants is of prime importance. The infested nursery plants can be made free from infestation by the use of 3% miscible oils. Application of 0.75 to 1 g a.i. aldicarb or carbofuran per plant has also given effective control of San Jose scale, woolly aphids, and nematodes in nursery plants.

***Biological control:*** Bhalla and Gupta (1993) and Singh (1993) have reviewed the work done on bio-control of this pest in India. Coccinellids (*Chilocorus bijugus* Mulsant, *C. rubidus* Hope, *Pharoscymnus flexibilis* (Mulsant), and *Sticholotis marginallis* (Kapur)), and parasitoids (*Encarsia perniciosi* (Tower), *Aphytis diaspidis* Howard, and *Aphytus (proclia* group)) have been mass multiplied and released in various localities of Himachal Pradesh, Jammu and Kashmir, and Uttar Pradesh. *C. bijugus* and *P. flexibilis* among predators and *E. perniciosi* and *Aphytis (proclia* group) among parasitoids have shown promising results. Mass multiplication and repeated release of these biocontrol agents can substitute summer application of insecticides. *C. bijugus* can be multiplied in the laboratory on San Jose scale raised on pumpkins at 27 $^0$ C and 55 per cent relative humidity in 33.5 days with a biotic potential of 92 eggs, longevity of 87 days and host consumption of 35.5/grub and 67/adult/day (Jalali and Singh, 1989). In Thanedhar area of Himachal Pradesh *C. bijugus* has given significant control of this pest within two years when it was released at the rate of 10–20 beetles/tree (Rawat *et al.*, 1988). Repeated releases of *E. perniciosi* and *Aphytis* (proclia group) at 1.16 and 2.81lakh adults have been found very effective in Jammu and Kashmir (Masoodi and Trali, 1987; Masoodi *et al*., 1989).

**Woolly apple aphid *Eriosoma lanigerum* (Hausmann)**

Woolly apple aphid is one of the most destructive pests of apple in the world. This pest is native of America and is cosmopolitan in distribution except for the hotter part of the tropics. In India, it is an exclusive pest of apple and crab apple. It was first recorded at Conoor (Tamil Nadu) in 1989 and Kumaon and Shimla hills in 1909 (Lal and Singh, 1947). Now it is widespread in all the apple growing areas (Rao, 1960; Sachan and Gangwar, 1980; Thakur and Dogra, 1980). The biology of this pest has been studied by various workers (Rahman and Khan, 1941a; Fotedar and Kapur,1943; Lal and Singh, 1947; Gautam and Verma, 1982, 1983a b; Thakur, 1984). Thakur and Dogra (1980) have reviewed the research work done on this pest in India. Bhalla and Gupta (1993) have discussed in detail the information on bionomics and management of this pest in India.

It infests the trees throughout the year, damaging both the root and shoot systems. First-instar nymphs keep on moving from aerial to root- zone, and vice- versa depending upon the climate and population build up. To establish on mature wood, they require a cut or injury, but can readily establish on young twigs and nursery plants. The adults and nymphs suck plant sap, and as a result, phloem cells, cambium, and xylem around the feeding site undergo excessive cell division and elongation resulting into formation of galls or knots. These galls interfere with nutrient uptake of the tree and splitting of these galls also acts as an entrance for a variety of pathogens. The vitality of the infested tree is significantly reduced resulting in poor yields.

This pest reproduces by parthenogenetic viviparty. The duration of development is longer (32–51days) in winter with poorer progeny per female (20–34 nymphs), while in summer, the duration is shorter (11–13days) but the fecundity per female is higher (30–116 nymphs). The growth and development is mainly temperature dependent. The reproduction ceases during winter (below $3^0$C). The pest over-winters as egg or young nymph on the roots of the host tree. The young nymphs, without a cover of woolly mass, also survive in cracks and crevices of the bark on the aerial parts of the plants. With the warming up of weather in spring, the pest multiplies rapidly. If unchecked by the biotic agents and human intervention, the population goes on increasing until the rainy season when it declines but again flares up in the autumn. The duration of development is about a month during July to November when the alates are produced but their significance in the spread of infestation needs further investigation. Usually, viviparous parthenogenetic alates lay progeny of a few apterous male and female nymphs (5–8 nymphs/alate) and oviparous apterous female lays an egg on the crown of the tree (Fotedar and Kapur, 1943; Gautam and Verma, 1982). Fotedar and Kapur(1943) reported that alates produced in May-July in small numbers in Kashmir reproduced asexual progeny which resembled that of apterous forms and hence such alates spread the infestation. The modes of dispersal of this pest are infested nursery, trees touching each other, air currents and chance dispersal by insects, birds, animals and man.

***Management***

This pest can be effectively controlled by the integrated use of chemicals, biocontrol agents, resistant rootstocks and plantation of pest free nursery plants. Since, infested nursery plants are the main source of spread of this aphid, therefore application of granules of aldicarb, carbofuran, disulfoton or phorate at 0.5g a. i. per plant in situ is effective and recommended to get healthy nursery pants. Before plantation, it is better to treat the whole plant with chlorpyrifos(0.04%) or fenitrothion (0.05%) and discard all the nursery plants with galls. Plants grafted on resistant rootstocks should be preferred for plantation. Among rootstocks, Merton Stocks; 778,779,789, and 793 are tolerant. Rootstocks of M 21, M 25, and MM series, *Malus baccata* L. *var himaliaca* and grafted plants of variety Fanny are also resistant (Thakur and Dogra, 1980; Rai and Tripathi, 1984).

The aphid remains covered with flocculant waxy mass that creates a barrier for penetration of the insecticides. However, if thorough application of insecticides is made with high volume sprayer, aerial population of the aphid can be checked either by contact or systemic insecticides like phosalone, endosulfan, methyl parathion, fenitrothion, chlorpyrifos, phoxim, DDVP, vamidothion, phosphamidon, menazon, dimethoate, formothion, etc. (Thakur and Dogra, 1980). Since most of these insecticides are injurious to biotic agents, therefore, comparatively less harmful ones to the endoparasitoid, *Aphelinus mali* (Hald) are: vamidothion (0.05%), followed by phosphamidon (0.03%) (Hameed *et al.*, 1974), and phosalone(0.05%) (Gaffar *et al.*, 1989). When natural enemy activity is low, it would be desirable to treat selectively heavily infested trees instead of whole orchard to safeguard the bioenvironment. The best strategy, however, would be drenching the infested tree basins with any of the above mentioned insecticides against the root infestations followed by banding of the infested trees with persistent granular insecticides such as aldicarb or phorate at 2.5–3 g a. i. per tree in April, and again in October- November to check migration of the aphid (Chandler and Dogra, 1977).

Bhalla and Gupta (1993) have reviewed the work done in India on the biocontrol of this pest. Releases of endoparasitoid, *A. mali* in summer and rainy season is recommended where insecticide use is minimum, and if at all required on some trees, spot application of insecticide is given. Natural population of coccinellids, syrphids, and chrysopids exercises a good control of this aphid on aerial parts, if present in good numbers (Thakur *et al.,* 1988; Rawat *et al*., 1988). Activity of natural enemies in autumn is surpassed by multiplication rate and fecundity of the aphid. Therefore, natural population of biotic agents has to be supplemented by their mass releases. For this, the endoparasitoid, *A. mali* was introduced in Kullu valley in the thirties which, in short time, gave 98 per cent control (Rahman and Khan, 1941b). Later on this parasite was also released in Shimla hills (H.P.), Chaubatia (U.P.), Coimbatore (Tamil Naidu), Shillong (Meghalaya), and Kinnaur, where it has established and manifested 5–60 per cent control (Rawat *et al*., 1989). This parasitoid is giving satisfactory control (>90%) of woolly apple aphid in lower Kullu valley areas (below 1500 m a.s.l.), while at higher elevations like Kinnaur, it is less effective.

**Other sap-sucking pests**

In addition to San Jose scale and woolly apple aphid, two species of mealy bugs, sixteen of scales and three of aphids have been recorded on apple trees whereas, one species of mealy bug, ten of scales, four of aphids, two of white flies and one of psylla have been recorded feeding on pear trees in India (Butani, 1979). However, all these species are of minor importance and are observed more damaging on hosts other than apple and pear. A few of them have been described below.

Green apple aphid, *Aphis pomi* (de Geer) *(mali* Fabricious) is a minor pest of apple but has also been recorded on pear. Its infestation is more commonly found in apple nurseries and on the young and tender leaves of apple trees in the orchards. Nymphs and adults suck the cell sap from the leaves, twigs, branches, blossoms, and young fruits. As a result of desapping, the affected leaves curl up, blossoms are shed and the young fruits drop prematurely while the quality of ripening fruits is greatly impaired. Normally the natural enemies like coccinellids, syrphids, and chrysopids take care of its sporadic population in the orchards. However, it can be easily controlled by an application of any systemic insecticide like, dimethoate, methyl demeton or quinalphos.

Plum scale, *Eulecanium* spp. (Coccidae: Homoptera) is primarily a pest on stone fruits, but also attack apple. Miscible oils sprayed against San Jose scale also control other scale insects.

Pear psylla, *Psylla pyricola* (Foerster) is a minor pest of pear found all over the hilly regions of India. Nymphs and adults feed inside the central shoot causing twisting of leaves and stunted growth. Eggs are laid on flower buds during March-April and two generations are completed on fruit trees during March to May; thereafter the adults migrate to other hosts. It has been easily controlled by the use of any of the systemic insecticides mentioned above against the scales or aphids.

## FOLIAGE FEEDERS

Various species of leaf eating caterpillars (Lepidoptera), beetles & weevils (Coleoptera) have been reported damaging apple and pear in India. Most of these defoliators are polyphagous and attack a number of other temperate fruits also. In apple, out of 84 species of Lepidoptera recorded feeding on leaves, only *Archips termias* Meyrick (Tortricidae), *Caloptelia zachrysa* Meyrida, *C. roscipenella* Hubner (Gracillaridae), *Porthesia scintillance* (Walker) (Notodontidae) and Indian gypsy moth, *Lymantria obfuscata* Walker (Lymantridae) have been reported to cause moderate to heavy damage in some pockets of Himachal Pradesh (Chander and Dogra, 1983, Sharma and Khajuria, 1994)). In Kullu valley, Indian gypsy moth has emerged as a major pest in various localities. Tent hairy caterpillar, *Malacosoma indica* Wilk (Lasiocampidae), a slug caterpillar, *Agalope hyalina* (Walker) and apple giant caterpillar, *Actias selene* Hubner are sporadic pests while all other lepidopteran species are of minor importance in apple. Among coleopteran defoliators, various species of defoliating and fruit eating beetles have been recorded as sporadic pests. In pear, all the lepidopteran defoliators are of minor significance. The larvae of Indian gypsy moth do not feed on the leaves of pear whereas, *Archips* spp., *M. indica, A. selene* and few other species have been reported as minor pests. A number of polyphagous chafer beetles have been recorded as sporadic pests on the leaves of pear. Among these the most damaging beetle is *Brahmina crinicollis* Burmeister. The description of important defoliators is given below.

### Leaf rollers/folders

Apple leaf roller, *A. termias* (Tortricidae) is the most important among the eight species of leaf rollers infesting apple. The larvae attack the foliage and also cause injury by scrapping fruit skin causing up to 41% fruit loss during storage (Verma and Chander, 1992). Verma (1985) studied in detail the bionomics and control of this pest. It overwinters in larval stage in fallen leaves or cracks and crevices. Small buff coloured adults emerge in May. Three generations are completed in a year. The maximum damage to leaves is done in June and to fruits in late maturing cultivars. Eggs are laid on dorsal surface of the leaves and on depressed fruit surface. Black headed green coloured larvae hatch out of the eggs in 8–16 days and feed on young buds, blossom, and terminal leaves after binding them with silken threads, which act as hiding sites for feeding. The larvae moult six times and become full-fed in 20–37 days. The adults emerge in 1–3 weeks after pupation.

The males of this moth are attracted to oblique-banded leaf roller pheromone (OBLR), which have been used for monitoring the moths in May (Bhardwaj, 1983, 1987). Carbaryl (0.1%), malathion (0.05%) or endosulfan (0.05%) sprayed 3 weeks prior to harvest, when moth activity is maximum, gives satisfactory control of this pest in storage.

In Kullu valley, two species of apple leaf folders, *Caloptelia zachrysa* Meyrida and *C. roscipenella* Hubner (Gracillaridae: Lepidoptera) have been found to damage over 90% terminal leaves of new growth in young and old apple trees during June- August. The above mentioned insecticides are effective against these species also.

**Indian gypsy moth *Lymantria obfuscata* walker**

Earlier it was a pest of deciduous forest trees (oaks, elder, willows, poplars, false acacia etc.) and some ornamentals in north-west Himalayas (Beeson, 1941), but now it is recorded infesting various temperate fruits except pear. Its incidence was higher in Kashmir (Masoodi *et al.*, 1986) and Uttar Pradesh (Gupta, 1976) than in Himachal Pradesh (Bhalla and Powar, 1977). But, for the last over 8 years, this pest has been appearing in moderate to severe form on fruit trees particularly on apple in Himachal Pradesh especially in Kullu valley where this pest has been recorded on various cultivars of apple, plum, apricot, cherry and persimmon (Sharma and Khajuria, 1994). Besides fruits, the outbreak of this pest has also been observed on various agro-forestry trees. Plantation of trees such as poplars, *Eucalyptus, Ulnus*, oak, etc., in the vicinity of the orchards has acted as a source of infestation to apple and other fruit trees. Bhalla and Gupta (1993) have reviewed the work done on this pest in India. This species was often confused with *L. dispar*, the gypsy moth widely occurring in Asia, Europe, Canada and U.S.A. owing to the identical serological reaction of nuclear polyhedrosis virus. However, it has serological differences in different stages of development of adults and is different in morphology of larva, pupa and adult.

The pest is univoltine and polyphagous. Complete defoliation and fruit damage by the larvae have been recorded in badly infested apple trees. Smaller larvae damage both leaves as well as fruits. In fruits, larvae scrap the skin and make holes. The badly damaged fruits fall off, while those which recover become deformed with scars on the skin. The gregarious nocturnal larvae feed voraciously the foliage with low tannin content (1%) (Masoodi, 1985). Larval development and survival had negative correlation with the tannin contents in leaves. Repeated defoliation for 7 years has been reported to cause 9–36 per cent mortality of the trees. In Kullu valley at Seobagh, repeated complete defoliation of an 8 year's old apple tree between April to June has resulted in killing of that tree within one season. The larvae eat up the entire leaf except the hard veins.

The pest diapauses in egg stage from June- July to March-April. Larvae emerge out of the eggs in March–April, coinciding with the appearance of leaves, feed during night, and take shelter during the day in groups on the bark under stones, weeds, and other hiding sites. The larvae become full fed in 65–100 days and pupate at their hiding places in June. Adults (2 males: 1 female) emerge out of the dark reddish brown pupae in June- July. The male is active flier while the female has weak and small wings in comparison to the body and survives for 4–7 days and 11–31 days, respectively. Mating takes place within a day of emergence and lasts for 1–2.5 hours. A single female lays 150–400 spherical light grayish-brown eggs in 1–3 clusters in June-July on bark, in crevices, and holes of walls under the stones, logs, etc. Egg clusters are conspicuous as these are covered with a layer of yellowish brown scales. In eggs the embryonic development continues for 23–32 days before the pest diapauses as pharate larva in the egg.

*Management*

Biological control agents play a crucial role in suppression of this pest. As many as 47 parasitoids and predators have been recorded on this pest in Kashmir, apart from a nuclear polyhedrosis virus (NPV), which causes 30- 40% larval mortality in Jammu and Kashmir, and over 90% mortality in Kullu valley. The outbreak of NPV in the larvae is directly correlated with the frequency of rains during May-June. Phenthoate (0.03%), DDVP (0.05%), endosulfan (0.05%), fenvalerate (0.03), methyl parathion (0.1%)and chlorpyrifos (0.04%) have been found to be effective against the larvae.

Keeping in view the behaviour of the larvae to take shelter in groups during day time, an effective and economic control of this pest on apple in Kullu valley has been achieved by:

a) Practising mechanical control by locating larvae during day time, collecting the caterpillars by using a broom and dipping them in kerosenised water; b) laying hiding traps of grass or rags of gunny bags around tree trunks in the evening with the first appearance of caterpillars in spring, search for the resting caterpillars next morning/day and kill mechanically, repeat this operation till trap catch is stable at zero for orchard and agro forestry trees, c) giving spot spray treatment with methyl parathion (0.1%) or chlorpyrifos (0.04%) or lindane (0.05%) where mechanical killing is not practicable and d) mechanical destruction of egg masses in winter. If the attack is observed at the green tip stage, the insecticides like chlorpyrifos (0.02%) sprayed at pink bud stage against the thrips also give effective control of this pest.

**Tent hairy caterpillar *Malacosoma indica* (Wlk.)**

It is a univoltine, polyphagous and sporadic pest of a number of deciduous forest trees and temperate fruits including apple and pear. Apple is its preferred host. Larvae are voracious defoliators. After 2–3 days of hatching, they move gregariously to safer places and spin triangular or quadrangular tents of silk at forking of twigs. Larvae rest in the tent (hibernaculum) during day and come out of the resting places during night to feed on leaves, leaving behind only the midribs and portions of hard veins. The size of the tent expands with an increase in size of the larvae (Joshi and Agarwal, 1979). Eggs are about one mm long, shining and elongated thimble shaped. Full-grown caterpillars are 40 –45 mm long, blackish-brown in colour; pupae are 18–21 mm long, smooth and dark-brown in colour. Adults are light brown to light red in colour with a wing expanse of 29 to 32mm and 35 to 37 mm in males and females, respectively. Eggs are laid round the branches of trees as broad bands or rings of 200 to 400 eggs during May- June. Eggs are well protected with a coating of dark brown glue like substance. Incubation period is about 9 to 10 months. Eggs hatch between mid-March to mid-May. Larval period is 39 to 68 days and pupal period is 11 to 12 days. Moths emerge in May- June.

Mechanical collection and destruction of the conspicuous egg clusters at the time of pruning and destruction of silken tents during the daytime with the help of a pole carrying rags soaked in kerosene checks the infestation. The larvae may be collected and dipped in kerosenized water. Spraying of diazinon (0.03%), carbaryl (0.05%) or endosulfan (0.05%) is effective at the time of outbreaks.

The slug caterpillar, *A. hyalina* is a minor pest of apple. Its outbreak was observed only in 1976 (Bhalla *et al.*, 1976). Caterpillar feed on sprouting spurs and bark, and later shift to foliage and tender fruits. It can be effectively controlled by spraying phosalone (0.05%), carbaryl (0.1%) or malathion (0.02) (Chander, 1978).

Apple giant caterpillar, *Actias selene* Hubner, Apple hawk moth, *Langia zeuzeroids* Moore are other minor pests of apple and pear causing damage to isolated neglected trees.

**Defoliating and fruit eating beetles**

Defoliating beetles are polyphagous and damage a number of temperate fruits and forest trees. Several species of Scarabaeidae, Cetoniidae, Curculionidae and Chrysomelidae feed on the terminal growth of the fruit trees (Butani, 1979; Bhalla and Gupta, 1993). Majority of defoliating beetles belong to Scarabaeidae; sub family Melolonthinae. In India, about 70 species of these beetles have been recorded; 57 in Uttar Pradesh, and about 36 in Himachal Pradesh (Gupta *et al.*, 1977; Bhalla and Pawar, 1977; Sharma and Khajuria, 1994). Adults of most of the species are defoliators, while a few damage fruits and flowers. Larvae, known as white grubs, feed on the root systems of variety of plants and inflict more damage on annual crops, pasture grasses, vegetable crops and nursery of fruit trees.

The maximum emergence of adults from soil occurs during June- August whereas, the total period of emergence extends from March to September depending upon the species, habitat and climatic conditions. The adults emerge at dusk, congregate on the preferred host, mate for 5–15 minutes, continue feeding during night and at dawn enter the soil for hiding and oviposition. The damage and abundance of the species varies from locality to locality. Most of the species complete their life cycle in one year. Eggs are pearly white and laid singly in the soil at 3–8 cm depth and hatch within 3 weeks. First instar larva feed on the organic matter while the subsequent 2 instars feed on live roots. The larval stage lasts for 80–300 days. Before pupation, larvae go deeper and pupate in earthen cells within which the transformed adults remain for few weeks to several months until first rain in summer (Veeresh, 1988).

Some important species on apple, pear and other deciduous fruits are: *Holotrichia* spp., *Melolontha* spp., *Mimela* spp., *Adoretus* spp., and *Anomala* spp. In Kullu valley, *Anomala polita* Baln., *Brahmina crinicollis* Burmeister, *B. coriacea* Hope, *Holotrichia longipennis*, and *Adoratus* spp., are major defoliators (Sharma and Khajuria, 1994). Cetoniids are flower beetles, which are diurnal, and feed on flowers, pollen. nectar, and ripening fruits. Important species belong to *Clinteria, Mimela, Oxycetonia*, and *Proctelia*. Veeresh (1988) reviewed the information on white grubs as pests of fruits and forest trees in India.

In Kullu valley, *B. crinicollis* is the most damaging beetle on flowers and leaves of fruit trees. Pear is its most preferred host followed by apple and other fruit trees. In pear, cv. Red Bartlett is preferred where up to 100 per cent flower damage has been observed in some localities (Sharma and Khajuria, 1999). The maximum emergence of this species is recorded in April.

*Management*

Installation of light traps helps in monitoring the density and diversity of beetles to undertake control measures. Collection and destruction of beetles after dusk by shaking non-bearing fruit trees and other preferred forest trees can give effective control, if undertaken as a campaign on a community basis. Spraying of orchard/trees, if severe defoliation appears all of a sudden, with carbaryl (0.1%) or Methyl- parathion (0.05%) need to be undertaken. It is better to spray preferred forest trees near the vicinity of the orchards as and when the beetles appear.

In endemic areas, control of white grubs should be undertaken by drenching the crop with chlorpyrifos 1.25–1.5 kg a.i./ha before egg laying. Application of granular insecticide (e.g. carbofuran 1–2 kg a.i./ha) would be economic only in case of cash crops. Where available, entomopathogenic nematodes (*Steinernema feltie* and *Heterorhabditis bacteriophora*) can also be applied to the soil along with irrigation water or drenched into the root system of the plant (Gupta, 2001).

## FRUGIVOROUS INSECTS

Insect pests that cause direct damage by affecting fruits are either fruit borers (apple fruit moth and codling moth) or fruit scrappers and external feeders (fruit eating beetles, caterpillars, and thrips that infest blossom as well as young fruits).

### Codling moth *Cydia pomonella* (L.)

Codling moth is a native pest of Eurasia but has now been introduced into North America and most of the world wherever apples are grown. In India, this pest is confined to Leh and Kargil districts of Ladakh in Jammu and Kashmir, where its incidence ranges from 3 to 96% (Pawar *et al.*, 1981). So far it has not entered other parts of Kashmir and Himachal Pradesh. In Ladakh, it is suspected to have come from North-West Frontier Provinces. It was recorded in mid thirties in Baluchistan and Afghanistan. Besides apple, it attacks apricot, walnut, pear, and quince. The larvae make holes from side or from bottom end of the fruit and bore to the core, and even at seeds. Biology and control of this pest have been studied by various workers, and reviewed by Bhalla and Gupta (1993). From the over wintering population, there is a protracted emergence of the adults leading to overlapping generations. In Ladakh, Pawar *et al.* (1981) noticed adult emergence during June- July, and second generation adults started emerging in lesser numbers from third week of August resulting in overlapping populations of larvae in early September. Full-fed larvae leave the fruit and spin silken cocoons under loose bark or in crevices, on stumps, brickbats, clods, and under stones where they over winter as such or in their pre-pupal stage.

*Management*

(i) The spread of this pest needs to be prevented by strict domestic quarantine, which has already been imposed in Jammu and Kashmir since 1977 by prohibiting movement of apple and walnut fruits from Leh and Kargil. Nevertheless, intensive pest suppression is required through the following ways (Pawar *et al.,* 1984).

(ii) Mechanical collection and killing of the pre-pupae in cocoons from April to June

(iii) Deep burying of fallen fruits during August

(iv) Weekly releases of egg parasitoids *Trichogramma* embryophagum and *T. caoeciae pallidum* at 2000 adults per tree, which has resulted in 29 per cent parasitism in June.

(v) Mass trapping of the males with Codling moth lure (E, E-10, 12-DODECADIEN-1–01) traps at 1 trap/10 trees. In addition to this, two sprays of DDVP or phosphamidon (0.04%) (One in second half of the June and another two to three weeks later) depending upon the moth catch in the pheromone traps. If the catch is fewer than 5 moths/week/trap, spray is not required.

**Apple fruit moth *Argyresthia conjugella* Zeller**

Apple fruit moth is a major pest of apple in Europe, Asia and North America (Belosels'kaya, 1963). The main host of this pest is the sorb tree, *Sorbus aucupria* L, a member of family Rosaceae (Ahlberg, 1927), but it has also been reported feeding on cherry, plum and pear (In India it has been reported only on apple. It was perhaps due to parallel pestilence behaviour, the apple fruit moth was initially misreported from India as codling moth, *C. pomonella* (Sharma and Bhalla, 1964) and later as tortricid, *Eucosma* sp. It was only in 1984 that this pest was properly identified as apple fruit moth, *A. conjugella,* a pest of apple in Kinnaur District of Himachal Pradesh (Khajuria *et al*., 1986). Later on it was also reported from the Lahaul Valley of H.P. Recently, the incidence of this pest has also been noticed by the growers in apple orchards located at higher reaches of Kullu and Chamba (Pangi and Bharmaur areas) districts of the state. Therefore, in India, at present this pest is confined to only dry temperate regions of Himachal Pradesh located at an altitude of 2445 to 3000 m (Sharma *et al.,* 1988a). This pest has specific ecological requirements and is confined to specific areas of dry temperate zone having annual isotherms very close to ±0 to 10°C.

The average incidence of this pest on various apple cultivars may reach 70 per cent. Short statured apple cultivars with compact and luxuriant canopy favour moth activity. Golden delicious is the most preferred cultivar followed by Red Gold, Royal Delicious, Red Delicious and crab apple. The average number of larvae developed per fruit varies from 2.23 (Royal Delicious) to 4.52 (Red Gold). The number of seeds per fruit in self-incompatible cultivars had positive correlation with the number of larvae developed but this relationship did not hold in self-fruitful cultivars (Khajuria *et al.,* 1987).

The pest is univoltine and overwinters in pupal stage from September to June. Adults appear in first week of June, whereas infestation on fruits occurs 17 days after adult emergence, and continues up to third week of August. Neonate larvae enter the fruit from the sides, but the calyx end is preferred. At the site of entrance, a crystalline white deposit is formed which is visible from a distance. First visible symptom of this crystalline white deposit is appropriate time for initiation of control measures. The larva makes an ingoing gallery and reaches the core of the fruit. It enters the seed. Feeds on the perisperm for 2–4 days before it moults. Subsequent three instars feed on the cotyledons, reduce it to a lump of brown frass and occupy whole of the seed. After feeding for

18–21 days, the larva comes out of the fruit by making an exit gallery and a hole of 1mm diameter on the fruit surface. In this way, the fruits are rendered unfit for human consumption. The larval emergence from the infested fruits occurs during mid-July to September. After descending to the ground, with the help of silken thread, the larva webbed a two-layered silken cocoon, either under tree basin or in bunds and pupates within 3 to 4 days. The pupae are normally found under stones, clods, and below the fallen leaves in shady and moist places (Khajuria *et al.*, 1989).

*Management*

Fenthion (0.04%), fenitrothion (0.05%), methyl- parathion (0.05%) or deltamethrin (0.03%) sprayed twice at fortnightly interval during June- July starting with first visible symptom of attack (i.e. appearance of first entry hole of neonate larva on the fruit as indicated by a white crystalline globule on the fruit) gave effective control of this pest (Sharma *et al.,* 1988 b). The natural mortality of the larvae and pupae has been recorded up to 59.6 per cent due to predatory ant, *Myrmica* species (20–49% in larvae), an unidentified ichneumonid parasitoid (1.7%in larvae), unidentified fungal pathogen (up to 40%in pupae), larvae pathogens (12.8% in larvae within the fruit), and due to sun burn/exhaustion (22.5–32.5%) which indicated that only 40.3 per cent individuals developed successfully into adults (Khajuria *et al.,* 1992).

**Fruit flies *Dacus* spp.**

Fruit flies have been observed as minor pests of apple and pear in India. They are more serious on other fruits such as peach, apricot, plum, etc. *Dacus dorsalis* Hendel*, D. zonatus* (Saunders), and *D. ciliatus* (Low) have been recorded on apple and *D. dorsalis* on pear during the rainy season (Butani, 1979).

The localized attack of fruit flies can be controlled by sanitary and other cultural methods such as burying of dropped fruits, prompt disposal of unmarketable fruits, and attracting the flies to poisoned bait traps (Malathion 0.1%+protein hydrolysate +sugar to make a slurry). Methyl – eugenol can be used as an attractant for *D. dorsalis*.

**Blossom thrips**

Thrips damage the plant parts by their asymmetrical rasping and sucking mouthparts. These are the pests with sporadic outbreaks at higher altitudes than in valley areas. The important blossom infesting species of thrips are: *Frankliniella dampfi* Priesner, *Taeniothrips rhopaloantennalis* Shunisster, *Thrips flavus* Schrank, *T. florum* Schmutz and *Haplothrips tenuipennis* Bagnall (Thakur *et al.*, 1973). They are associated with blossom of almost all crops including apple and pear. Damage inflicted to the blossom would depend on the population of thrips. Mishra *et al.,* (1987) suggested that population up to 15 thrips per flower did not affect fruit set in apple.

In apple, pear and other fruit crops, *T. flavus* is of most common occurrence. Its infestation has been observed on 78 plant species belonging to 26 families (Veer, 1985). The adults and nymphs lacerate floral parts and delicate and unfolding leaves of vegetative buds resulting in the development of brown spots. The lesions on androecium and gynoecium deteriorate fruit quality and reduce fruit set and yield. The set fruits may drop. Heavily infested flowers bear a sick and faded appearance with indication of early senescence. Adults of this pest appear with the onset of spring.

The eggs are inserted into the host plant tissue. Hatching takes place within 4–10 days. Three nymphal instars are elongate, white or yellowish in colour. Full-fed nymphs fall to the ground and pupate in soil. The development period is 9–21 days and the generations are many depending upon the climatic conditions. In fruit trees, after fruit set, there is general decline in population of thrips. However, some population can be seen breeding on terminal leaves.

Coccinellids, anthocorids (*Orius* sp.), and lygaeids (*Goecoris* sp.) feed on thrips (Bhalla and Gupta, 1993). Chemical control is detrimental to predators and pollinators during flowering. Therefore, spraying of insecticides is never undertaken during flowering. However, if population is too high, spraying of fenitrothion (0.05%) or chlorpyrifos (0.04%) can be undertaken at pink bud stage when predator activity is very low.

### Fruit eating caterpillars and beetles

Various species of leaf rollers, lymantrids, zygaenids, lycaenids and defoliating beetles feed on fruits of apple and pear. Major species have already been described above under defoliators.

## BORERS

### Root borer *Dorysthenes hugelii* Redtenbacker

It is a pest of fruit and forest trees in temperate region of India up to an elevation of 2150 m. Its incidence has also been reported from apple growing areas of Nepal (personal communication). In fruits, apple is its most preferred host. In Himachal Pradesh, apart from apple, it infests apricot, peach, cherry, pear and walnut, besides forest trees. The bioecology and control of this pest have been studied in Kumaun hills of Uttar Pradesh and mid-hills of Himachal Pradesh (Singh, 1941; Verma and Chander, 1988; Sharma and Khajuria, 1997; Sharma *et al*, 1999). The apple orchards planted on sandy-to-sandy loam soils are the worst affected where its infestation goes up to 100 per cent. The grubs bore or girdle the roots. Affected roots get severed, and branches wither and die. Such trees loose upright posture, become shaky, stunted and weak with reduced or no linear shoot growth, leaves are small, pale green and often show wilting at noon, trees tilt or get easily uprooted with wind and usually die within 3 years. Infested trees become susceptible to the attack of secondary pests like stem borers and shot hole borers and diseases like canker. Such apple trees on seedling rootstock also show symptoms of abnormal splitting of bark showing 2.5–15 cm long oval longitudinal cuts on the bark of a branch (when the root of that particular side is infested) or branches (when more number of roots are infested) or all over the trunk, and branches (when almost all the roots are eaten away). The number of cuts varies depending upon the extent of infestation. The bark at these cuts hardens and gets detached from the hard wood, thus giving an appearance of oval cuts. Infested trees with less than 5 such cuts appear healthy and produce normal crop. This is the most appropriate time to apply control measures.

It completes life cycle in about 4 years. Chestnut red adults start emerging with the onset of pre- monsoon showers in the second fortnight of June. Emergence continues for 4–5 weeks. Adults get attracted to the source of light at night, mate, lay up to 300 elongated (3 x 1 mm) eggs, singly or in clusters of 2–4 in soil within 1–10 cm depth. The eggs are glued to the hard surface preferably the roots of the trees. The maximum eggs are laid within 1.25m radius around the tree trunk. Eggs hatch in about 3 weeks. As soon as the neonate larva encounters the roots, it bores inside and starts feeding. Full- fed larvae measures 7- 10 cm after 3.5 years of feeding. It pupates in earthen cells at 30 to 60 cm depth after leaving the roots.

*Management*

Earlier, Verma and Chander (1988) had suggested the application of aldrin dust at 150–200 g per tree basin in July followed by drenching the tree basins with aldrin (0.01%) or methyl parathion (0.05%) in winter and again in March with either of these two insecticides or with chlorpyrifos (0.05%) for the control of this pest. Presently the control strategy consisting of a) collection and destruction of grubs by exposing the roots of infested trees (marked earlier on the basis of symptoms of attack) and while preparing the tree basins, b) drenching of tree basins up to 1.25m radius (since the grubs remain confined to 1.25m around the tree trunks and up to 30 cm depth) with chlorpyrifos (0.06%) or triazophos (0.12%) after exposing the infested roots in winter between November to March, c) dusting of tree basins with lindane (1.3% dust) or methyl parathion (2%dust) @ 300 g/tree basin during mid-June to mid- August against the eggs, neonate larvae and adults, d) installation of light trap in the orchard during June- July for collection and destruction of light attracted adults, e)complete removal of decaying roots of dead trees from the orchard, f) supplementing the lost root system with nurse grafts and g) protection of infested trees from the secondary pests (like stem borers) and pathogens (like canker) has been recommended against this pest (Sharma and Khajuria, 1997).

**Stem and shoot borers**

Apple Stem Borer, *Apriona cinerea* Chevrolat is adestructive pest of apple, peach, fig, and a number of forest trees (Butani, 1979; Singh and Prasad, 1985). The larvae bore into branches, stem and tree trunk just under the bark, and sometimes feed on heart wood. The larva cuts 8–9 circular holes on the bark of each branch at a distance of 10 – 15 cm and tunnels through, and ejects frass. The trunk is hollowed inside at several places. Affected trees break with moderate wind. Adults are ashy grey to greyish yellow long horned beetles (26–50mm long) that appear in July-August, feed on bark and leaves and girdle young shoots. The female lays an egg in a chamber excavated in 2–3 cm thick branches. The egg is covered anteriorly by the flap of the bark. Incubation period is 5–8 days and the larva bores down often making a peripheral tunnel and cuts circular holes at some distance for ejecting faecal pellets. In the hills, it continues feeding until November and resumes activity in March. Meawhile, it keeps on moving down from shoots to branches and then to the stem. There may be more than 6 grubs in trunk region. In the winter of second year, full-fed creamy-white elongate (6–7 cm long) larva excavates a pupal chamber covered with coarse wood fibres and over winters in it. In the following spring, it pupates and immature adult is formed in May-June. The adult emergence starts in July-August and continues till September.

Cherry stem borer, *Aeolesthes holosericea* Fab. is a main pest of cherry and apple, but also infests other fruits. This pest has been described in detail under pests of cherry.

Quetta Borer, *A. sarta* Solsky, is a major pest of walnut in Kashmir but it also damages apple and other temperate fruits and agro forestry trees.It has been described in detail under the pests of walnut.

Other cerambycids *viz*., fig borer, *Betocera rufomaculata* (de Geer) and *B. horsfieldi* Hop, which are pests of tropical fruits, but also damage stems of apple and walnut in Kullu valley and Shimla hills (Pruthi and Batra, 1960). Shoot borer, *Linda nigrosculata* (Fairm.) is another important pest of apple in north-east hill region. The yellowish larva bores upward and cuts holes in the twig to throw out frass pellets. Drying of twigs is pronounced in October to April due to larval damage (Sachan and Gangwar, 1980).

**Shot hole borers**

These borers are sporadic pests and attack weak, declining, decaying, and poorly managed trees. About a dozen species have been reported from India. Bhalla and Sharma (1963) recorded *Scolytoplatypus raja* Blandaf on apple trees in Himachal Pradesh. Small cylindrical beetles with deflexed head initiate damage. They burrow in the bark making small perforations, exuding dark fluid. They tunnel into sap- wood and hardwood of the trunk and branches, and make pin holes in which eggs are laid. Grubs make small vertical parallel tunnels in the tree trunk and branches, thus obstructing the normal flow of the sap. Severe infestation results in wilting and yellowing of the foliage followed by death of the tree.

**Lepidopterous shoot borers *Zeuzera* spp.**

Larvae of leopard moth, *Zeuzera* spp. are polyphagous and bore into shoots and stem of Cherry, apple and other fruit and forest trees. Information on its biology has been given under pests of cherry.

***Management of stem and shoot borers***

Most of the borers are pests of poorly managed orchards. Whenever a plant gets weakened due to poor nutrition, rocky strata, disease or any other abnormality, it becomes susceptible to borers, particularly shot hole borers. So, growers must follow the sound horticultural practices and grow healthy trees. Clean orchard practices are essential. All the dead or dying branches should be cut and burnt before leaf fall.

Pruning and destruction of newly attacked branches and killing the older larvae by inserting a flexible-wire into the live tunnel are effective. If attack is severe, satisfactory control can be achieved by inserting solid fumigant (0.5 g paradichlorobenzene or 0.2 g aluminium phosphide) or a cotton wick soaked either in liquid fumigant (petroleum, PDCB in kerosene oil, mixture of chloroform and creosate in 2 : 1 ratio or a mixture containing naphthalene flakes, creosate and carbon disulfide in 5 : 2 : 1 ratio) or an insecticidal emulsion (0.1% DDVP, methyl-parathion, etc.) in larval gallery already cleaned by iron spoke and plugging the hole with plaster. The holes, which get re-opened, should be treated again.

Proper training and pruning of trees to avoid direct exposer of the stem to sun keep the borers at bay, particularly shot hole borers. Spraying of shot hole borer infested trees with methyl- parathion (0.1%), fenitrothion (0.1%) or DDVP (0.1%) is more effective than swabbing with these insecticides. Maintaining tree vigour is better than chemical treatment.

## PHYTOPHAGOUS MITES

Phytophagous mites have now been recognized as primary pest of fruit crops including apple throughout the world. In Himachal Pradesh, until 1991, phytophagous spider mites were of minor importance. Since then, these mites have been appearing in epidemic form all over the state. Over reliance on pesticides has been mainly attributed towards their outbreak During a short span of 8 years, these mites have now attained the status of major pest in apple growing regions of the country particularly in Jammu Kashmir and Himachal Pradesh. Presently 4 species of mites have been observed damaging apple trees. These include European red mite, *Panonychus ulmi* (Koch), two-spotted mite *Tetranychus urticae* Koch, apple rust mite, *Aculus sp*. and Brown mite, *Bryobia sp*. So far only *P. ulmi* and *T. urticae* have been recorded causing significant damage to apple trees in Jammu & Kashmir, Himachal Pradesh and Uttar Pradesh. Presently *P.ulmi* and *T. urticae* occur in 70 : 30 ratio in apple orchards (Awasthi et al., 1999). The work on bioecology and management of these two major species of apple mites have been mainly conducted in Kullu valley and Shimla district of Himachal Pradesh (Bhardwaj, 2001)

**European red mite, *Panonychus ulmi* (Koch)**

It is a pest of deciduous fruit trees throughout the world. In India, it was reported for the first time in 1974 from Northwestern Himalayan region of Jammu & Kashmir, Himachal Pradesh and Uttar Pradesh (Prasad, 1974). Later on in 1992–93 it was observed seriously attacking some apple orchards in Kullu valley as well as in Thanedhar areas of Shimla district in Himachal Pradesh (Kumar and Bhalla, 1993; Khajuria and Sharma, 1996). Surveys conducted during the past 5–6 years in apple growing regions of Himachal pradesh have indicated its continuous spread from barely negligible in 1992 to 74% in 2000 (Bhardwaj and Bhardwaj, 2000). The pest has now established as a major pest in almost all the apple growing areas of Himachal Pradesh and Jammu & Kashmir.

It is a polyphagous species and cause damage to apple pear, peach, apricot, plum, walnut, jack fruit, rose and *Hibiscus* (Prasad, 1974). In Himachal Pradesh it is primarily a pest of apple, but pear trees within an infested apple orchard are also attacked by this mite. The adult is red in colour and small in size (0.4mm long). The colour of immature stages and adult vary considerably from bright red to pale yellowish green. The damage is caused by the young ones and the adults by sucking the sap from the leaves and tender parts. The mites also inflict direct physical injury by making at least 20 incisions into the leaf tissue per minute which cause extensive mechanical damage. Initially the mite attack appears in the form of a minute speckling of the leaves. Further damage renders the leaves loose their bright green colour which become dull green and finally brownish-green or bronzed. Consequently, the photosynthetic activity is adversely affected, size of the leaves is reduced and transpiration rate is abnormally increased. The affected leaves become brittle and fall off prematurely only after physical disturbance such as torrential rain, storm etc. The white exuviae of mites remain attached to the underside of leaves imparting a look as if attacked by powdery mildew. Continuous mite injury leads to undersized and poorly coloured fruits, production of fewer buds, and reduction of return bloom and fruit load in the following season.

The mite passes through four stages, namely, egg, larvae, protonymph and deutonymph before becoming an adult. The winter is passed in egg stage. Winter eggs are red, spherical and laid singly from August to November on spurs, around the buds, young shoots, cut ends, cracks and crevices and sometimes on stock end and calyx end of the fruits. Usually a female lays 20–50 eggs (0.15mm dia.). The eggs start hatching from March-end coinciding with pink bud stage and continue up to first week of May. The larva has six legs and moves to the underside of leaves for feeding. The mite becomes adults after three moults. There is a resting stage preceding each moult. The immature stages from egg to adult are completed in about 14 days and it takes one month to complete a generation. Hot and humid weather is highly favourable for their multiplication. The spring/summer eggs, laid underside of the leaves preferably near the midrib and veins, are globular and somewhat flattened in shape and transluscent when freshly laid but turn reddish to dark-orange before hatching. The incubation period varies from 4.5 to 14 days. The total development period is dependent on weather conditions. It varies from 10 to 22.5 days in females and 8–19.5 days in males. The mite has 5–8 generations a year.

**Two-spotted mite *Tetranychus urticae* (Koch)**

It is highly polyphagous and most widespread pest recorded feeding on more than 150 species of economic importance including many fruit trees all over the world. In Himachal Pradesh it has been found attacking apple trees in many apple-growing belts. The adult female is oval in shape and small in size (0.8mm). The mites produced in summer are light in colour varying from pale-yellow to greenish with two dark-red spots on the dorsal side. The hibernating mites produced during autumn are uniformly pink with no spots on their body. The immature and adult stages suck plant sap from the underside of the leaves as a result small whitish spots appear on the upper leaf surface. With the increase in infestation, the leaves show characteristic speckles which later turn bronzed. It spins a fine silken web near the feeding site and later extend it to cover the adjoining leaves and branches. In India the work on its life cycle on apple has been carried out at Kullu (Khajuria and Sharma,1999) and at Shimla hills (Bhardwaj and Bhardwaj, 2000).

It overwinters as orange coloured oval shaped females (0.8mm size) mostly under the loose bark and also in the ground cover in debris, fallen leaves, cracks of soil surface etc. The females on the ground start developing colonies on broad-leaved weeds from the beginning of March. The females migrate to the apple trees in the second week of May when the weeds start drying. However the over wintering females on apple trees start feeding on apple foliage in early April. The females on apple lay spherical trnsluscent eggs singly on lower surface of leaves mostly on the tips of the leaf hairs and knit webs. These eggs turn white opaque before hatching. The eggs hatch in 5–14 days depending upon the temperature. The young larvae are oval, dirty white in colour and start feeding under the leaves near the webs. It moults to protonymph and deutonymph stages before becoming adult. The life span of female and male is 30–35 days and 15–20 days, respectively. The development is highly temperature dependent. The life cycle is completed in 6 days at $35^{0}$C, 12 days at 20°C and 55 days at 10°C.

***Management of mites***

The mite pests are almost completely regulated by natural enemies. In Himachal Pradesh, six different species of predators viz., black coccinellid [*Stethorus punctum* (Le Conte)], mirid [*Chrysopa* spp.], anthocorid [*Orius insidiosus* (Say)], black hunter thrips [*Leptothrips mali* (Fitch)], spiders and the predatory mites [*Amblyseius fallacies* A-H, *Zetzellia mali* Ewing and *Typhlodromus* spp?] have been recorded feeding on mites (Khajuria and Sharma, 2001). Indiscriminate use of pesticides kill the predatory complex and help establish mites as major pests. Fungicides, particularly dithiocarbamates and mancozeb, have been reported toxic to natural enemies. Therefore it is imperative to conserve these natural enemies, augment their activity and restrict the use of pesticides toxic to these predators.

Miscible oils used against San Jose scale are also effective against the over wintering eggs of *P. ulmi* when applied during late dormancy (Bhardwaj and Kumar, 1994). Miticides like Dicofol (0.05%), dinocap (0.024%), fenzaquine (0.005%), fenpyroximate (0.0025%), profenofos (0.05%), propargite (0.0285%) and clofentezine (0.02%) have been found effective against the summer population of mites (Khajuria and Sharma, 1994; Khajuria and Sharma, 1999). Summer spray oil formulations *viz.,* Orchex-796 (1.5%) and IPOL (2%) have been found effective in checking hatchability (95–100%) of winter eggs of *P. ulmi* when applied nearly at pink bud stage and a month later. Summer spray oil formulations, *viz*., Orchex-796, D-C-Tron Plus, and IPOL orchard spray oil, each at 1 % concentration during May June and endosulfan (0.05%) have been found effective in checking the summer populations of both the mite species (Bhardwaj, 2001). For winter eggs, miticides should be applied only when 10% spurs are found infested with winter eggs. Application of miticides during active season is recommended only if 4–5 mites per leaf are observed per 100 leaves drawn randomly from the orchard (Bhardwaj, 2001).

Thus an integrated approach involving a) the conservation and augmentation of natural enemies by restricting the use chemicals detrimental to the predators, b) application of oils against the winter eggs of *P. ulmi*, c) need based application of miticides and/or summer spray oils against the summer population of mites and d) destruction of the pruned wood before the egg hatch and keeping the tree basins free from the weeds, should be followed for the control of these pests.

## REFERENCES

Ahlberg, O. 1927. (*Argyresthia conjugella* Zell., a report on investigations in 1927–26). *Meddelande fran Centralanstalten for forsoksyasendet pa jorbruksomradet* No **324**, pp. 127.

Awasthi, R.P.; Verma, H.S., Sharma, R.D., Bhardwaj, S.P. and Bhardwaj, S.V. 1999. *Decline in productivity of apple in Himachal Pradesh: Causes and Remedial strategies.* Bulletin Directorate of Extension Education, YSP UHF, Solan, 18pp.

Beeson, C.F.C. 1941. *The Ecology and Control of the Forest Insects of India and the Neighbouring Countries.* F.R.I., Dehra Dun.

Belosel's Kaya, Z.G. 1963. The mountain-ash moth, *Argyresthia conjugella* Zell. (Lep. Yponomeutidae) as pest of the fruits of apples and mountain-ash (In Russian). *Entomologicheskoe Obzrenie* **4**, 720–729.

Bhalla, O. P. and Gupta, P.R. 1993. Insect pests of temperate fruits. *In: Advances in Horticulture: Fruit Crops.* K.L. Chadha and O.P Pareek (eds). *Volume* **3** Malhotra Publishing House, New Delhi, India. pp. 1557–1589.

Bhalla, O. P. and Pawar, A. D. 1977. *A Survey Study of Insect and Non-insect Pests of Economic Importance in Himachal Pradesh.* Bulletin Department of Entomology and Zoology, College of Agriculture, Solan, Himachal Pradesh, India.

Bhalla, O. P. and Sharma, P.L.1963. A new record of shot hole borers *Scolytoplatypus raja* Blandat (Scolytidae: Coleoptera) of apple in HP. *Current Science,* **32**: 86.

Bhalla, O. P.; Dogra, G. S. and Thakur, J. R, 1976. Outbreak of *Agalope hyalina* Kollar (Lepidoptera: Zygaenidae) on apple in Himachal Pradesh. *Indian Journal of Entomology,* **38**: 377–379.

Bhardwaj, S. P. 1983. Performance of pheromone traps and dispensers in apple orchards. *In: Proceedings of 10th International congress of Plant Protection.* British Crop Protection Council, Brighton, U.K.

Bhardwaj, S. P. 1987. Investigations on response of lepidopteran sex pheromones of temperate fruit pests in Himachal Pradesh, India. *Agriculture, Ecosystem and Environment*, **19**: 87–91.

Bhardwaj, S.P. 1988. Effect of summer application of insecticides on San Jose Scale (*Q. perniciosus*) in orchards of apple (*Malus pumila*). *Indian Journal of Agricultural Sciences,* **58**: 79–80.

Bhardwaj, S.P. 2001. Red spider mite-a major threat to apple production. *In: Productivity of Temperate Fruits.* K.K.. Jindal and.D.R. Gautam (eds). YSP UHF, Nauni, Solan. pp. 409–416.

Bhardwaj, S.P. and Bhardwaj, S. 2000. European red mite—an emerging problem of apple. *In: Horticulture Technology Vision 2000 and Beyond.* V.K. Sharma and. K. C. Azad (eds). *Volume 2 Protection and Preservation*. Deep and Deep Publishers Pvt. Ltd. New Delhi. pp. 428–432.

Bhardwaj, S.P. and Kumar, R. 1994. Effects of miscible oil on infestation of the Europian red mite on apple. *In: National Symposium on Emerging Trends in Pest Management*. Indian Society of Pest Management and Economic Zoology, Solan, Himachal Pradesh, India.

Butani, D.K. 1979. *Insects and fruits.* Periodical Expert Book Agency, New Delhi, India.

Chander, R. 1978. Control of *Zygaena* moth of apple in Himachal Pradesh. *Haryana Journal of Horticulture Science,* 7: 120–122.

Chander, R. and Dogra, G. S. 1977. A technique of granular application for woolly apple aphid control. *Entomologists Newsletter,* 7(9/10): 42.

Chander, R. and Dogra, G.S. 1983. Lepidopterous insects infesting apple in Himachal Pradesh. *In: Proceedings of Workshop on High Altitide Entomology and Wild Life Ecology*. Zoological Survey of India, Calcutta, West Bengal, India.

Dinabandhoo, C.L. and Bhalla, O.P. 1975. Comparative studies on dormant and pre-bloom sprayings for the control of San Jose scale. *Indian Journal of Agricultural Sciences,* **45**: 60–62.

Dinbandhoo, C.L. and Bhalla, O.P. 1980. Bionomics of San Jose scale, *Quadraspidiotus periciosus* Comstock (Hemiptera: Diaspididae) infesting temperate fruits. *Journal of Entomological Research,* **4**: 63–67.

Fotedar, M.R. and Kapoor, A.P. 1943. first record of seasonal forms and oviparous reproduction of woolly aphid, *Eriosoma, lanigerum* (Hausm.) from Kashmir, India. *Current Science*, **12**: 84–85.

Gaffer, S. A., Fazli, M.A. and Lone, A. H. 1989. Selectivity of some common insecticides to woolly apple aphid parasite, *Aphelinus mali* (Hald.). *Indian Journal of Plant Protection,* **17**: 127–130.

Gautam, D.C. and Verma, L.R. 1982. Occurrence of sexuparae and sexual morphs of woolly apple aphid (*Eriosoma lanigerum* Hausmann) in Shimla hills, India. *Current Science,* **51**: 481–483.

Gautam, D.C. and Verma, L.R. 1983a. Life history of sexuparae and sexual morphs of woolly apple aphid *(Eriosoma lanigerum* Hausmann). *Proceedings of Indian Academy of Sciences (Animal Science),* **92**: 247–251.

Gautam, D.C. and Verma, L.R. 1983b. Seasonal biology and reproductive behaviour of woolly apple aphid (*Eriosom lanigerum* Hausmann*). Indian Journal of Horticulture,* **40**: 119–123.

Gupta, B.P.1976. Life history and control of Indian gypsy moth, *Lymantria obfuscata* Wlk. Research report on some stone and nut fruits. *All India Co-ordinated Fruit Improvement Project, Ranikhet.*

Gupta, B.P., Joshi, N. K. and Joshi, L.D. 1977. Some important defoliating and fruit eating beetles of temperate fruit trees. A survey and control. *Progressive Horticulture,* **8** (4): 53–60.

Gupta, P.R. 2001. Integrated pest management approach for major insect pests of temperate fruits. *In: Productivity of Temperate Fruits*. K.K. Jindal and D. R. Gautam (eds). Dr Y.S.Parmar University of Horticulture and Forestry, Nauni, Solan, India. pp. 397–408.

Gupta P.R.; Bhardwaj S.P. and Bhalla O.P. 1982. Control of San Jose Scale, *Quadraspidiotus pernicious* (Comstock) with miscible oil and their combination with Fenitrothion. *Journal of Entomological Research,* **6**: 140–54.

Hameed S.F.; Adlakha, R.L. and Sud, V.K. 1974. Control of *Eriosoma lanigerum* Hausm.by systematic insecticides. *Indian Journal of Agricultural Sciences,* **44**: 301–303.

Jalai, S.K. and singh. S.P. 1989. Biotic potential of three coccinellid predators on various diaspine hosts. *Journal of Biological Control,* **3**: 20–23.

Joshi K.C. and Agarwal S.B. 1979. Tent caterpillar and its control. *Indian Horticulturist,* **24**(1): 11–13.

Khajuria, D.R. and Sharma, J.P. 1992. Efficacy of some insecticides for the control of San Jose scale *Quadraspidiotus perniciosus* (Comstock) in Kullu valley. *In: NHB Technical Communications.* University of Horticulture and Forestry, Solan, Himachal Pradesh, India. pp. 277–280.

Khajuria, D.R. and Sharma, J.P. 1994. Toxicity of various pesticides against the European Red mite, *Panonychus ulmi*, Koch. *In: National Symposium on Emerging Trend in Pest Management.* Indian Society of Pest Management and Economic Zoology, Solan, Himachal Pradesh, India. (abstract).

Khajuria, D.R. and Sharma, J.P. 1996. Outbreak of phytophagous mites on apple in Kullu valley of Himachal Pradesh. *Indian Journal of Plant Protection,* **24**: 134–138.

Khajuria, D.R. and Sharma, J.P. 1999. Biology of European red mite and two spotted red mite on apple under Kullu valley conditions. *Annual Report, 1999–2000.* Regional Horticultural Research Station, Bajaura, Kullu. pp. 39–40.

Khajuria, D.R. and Sharma, J.P. 2001. Natural enemies of phytophagous mites and their role in integrated mite management in apple ecosystem. *In: Proceedings of Symposium on Biocontrol Based Pest Management for Quality Crop Protection in Current Mellenium,* (July 18–19,2001), PAU, Ludhiana, India. Extended Abstract, pp. 182–183.

Khajuria, D.R.; Sharma, J.P. and Dogra, G.S. 1986. Incidence of apple fruit moth (*Argryesthia conjugella* Zell.) in Himachal Pradesh, India. *Tropical Pest Management,* **32**: 350.

Khajuria, D.R.; Sharma, J.P. and Dogra, G.S. 1987. Studies on varietal preference of apple fruit moth, *Argyresthia conjugella* Zeller (Yponomeutidae: Lepidoptera). *Indian Journal of Horticulture,* **44**: 289–93.

Khajuria, D.R., Sharma, J.P. and Dogra, G.S. 1989. Seasonal incidence of apple fruit moth in relation to temperature and humidity in Himachal Pradesh. *Indian Journal of Ecology,* **16**(1): 56–59.

Khajuria, D.R., Sharma, J.P. and Dogra, G.S. 1992. Role of natural mortality factors in suppression of apple moth, *Argyresthia conujugella* Zeller. *Indian Journal of Plant Protection,* **20**: 92–94.

Kumar, R. and Bhalla, O.P. 1993. An epidemic outbreak of phytophagous mites on apple in Kullu valley of Himachal Pradesh. *Current Science,* **64**(10): 709.

Lal, K.B. and Singh, R.N. 1947. Seasonal history and field ecology of the woolly aphids in the Kumaun Hills. *Indian journal of Agricultural Sciences,* **17**: 211–218.

Masoodi, M.A. 1985. Growth response of *Lymantria obfuscata* Walker in relation to tannin content in different host foliage. *Indian Journal of Entomology,* **47**: 422–426.

Masoodi, M.A. and Trali, A.R. 1987. Seasonal history and biological control of San Jose scale *Q. perniciosus* (Comstock) (Diaspididae: Homoptera) on apple in Kashmir. *Journal of Biological Control,* **1**: 3–6.

Masoodi, M.A.; Trali, A.R.; Bhat, A.M.; Tikku, R.K. and Nehru, R.K. 1986. Incidence of parasites of *Lymantria obfuscata* (Lymantridae: Lepidoptera) in Kashmir. *Entomophaga,* **31**: 401–404.

Masoodi, M.A.; Trali, A.R.; Bhat, A.M.; Tikku, R.K. and Nehru, R.K. 1989. Establishment of *Aphytis sp. proclia* of San jose scale in Kashmir. *Indian Journal of plant Protection,* **17**: 71–73.

Mishra, R.C.; Verma, A.K. and Gupta, P.R. 1987. Insecticides for the control of apple blossom thrips. *Indian Journal of Horticulture,* **44**: 115–20.

Pawar, A.D., Tuhan. N.C. Balasubramanian, S. and Parri, M. 1981. Distribution, damage and biology of codling moth, *Cydia pomonella* (L.) *Indian Journal of Plant Protection,* **9**: 111–114.

Pawar, A.D., Tuhan, N.C., and Parry, M. 1984. Management of codling moth, *Cydia pomonella* (L.) (Lepidoptera: Olethreutidae) at Khalsi (Ladakh), Jammu and Kashmir (India). *Pesticides,* **18**(10): 39–42.

Prasad, V. 1974. *A catalogue of mites of India*. Indira Acarology Publishing House, Ludhiana (Punjab). pp. 320.

Pruthi, H.S. and Batra, H.N. 1960. *Important fruit pests of North-West India*. Bulletin No. 80. *Indian Council of Agricultural Research, New Delhi, India*

Pruthi, H.S. and Rao, V.P. 1951. *San Jose scale in India*. Bulletin No. 71. Indian Council of Agricultural Research, New Delhi, India.

Rahman, K.A. and Khan, A.W. 1941a. Biology and control of woolly aphids *Eriosoma lanigerum* Hausmann (Aphididae, Rhynchota) in the Punjab. *Indian Journal of Agricultural Sciences,* **11**: 119–29.

Rahman, K.A. and Khan, A.W. 1941b. Observations on *Aphelinus mali* hald. in the Punjab. *Indian* Journal *of Agricultural Sciences,* **11**: 265–278.

Rai, K.M. and Tripathi, G.M. 1984. Resistance of various rootstocks to the attack of apple wooly aphid, *Eriosoma lanigerum* (Hausmann). *Progress Horticulture,* **16**: 359–362.

Rao, V.P. 1960. Record of woolly apple aphid in Sikkim. *FAO Plant Protection Bulletin,* **10** (1–4): 3–7.

Rawat, U.S.; Pawar, A.D. and Singh, R. 1989. Establishment of *Aphelinus mali* (Hald.) in cold and dry zones of Kinnaur district of Himachal Pradesh. *Journal of Insect Science,* **2**: 109–13.

Rawat, U. S., Thalur, J.N. and Pawar, A.D. 1988. Role of natural enemies in integrated control of San Jose scale and woolly aphid in Himachal Pradesh. *In: Proceedings of National Symposium on Integrated Pest Control: Progress and Perspectives.* Association for advancement of Entomology, Trivandrum, Kerala, India. pp. 182–187.

Sachan, J.N. and Gangwar, S.K. 1980. Insect pests of apple in Meghalaya. *Bulletin of Entomology,* **21**: 113–21.

Sahai, B. and Joshi, L.d. 1965. Bionomics and biological control of San Jose scale(*Q. perniciosus* Comstock) in U.P. *Punjab Horticultural Journal,* **5**: 37–43.

Sharma, J.P. and Khajuria, D.R. 1994. Survey and monitoring of defoliating beetles and other defoliators of temperate fruit trees. *In: National Symposium On Emerging Trends in Pest Management (June 28–30,1994)*. Indian Society of Pest Management and Economic Zoology, Solan, Himachal Pradesh, India (abstract).

Sharma, J.P. and Khajuria, D.R. 1997. Some observations on the bioecology and management of apple root borer, *Dorysthenes hugelii* Rdt. in Himachal Pradesh. *In: National Seminar on Plant Protection towards sustainability* (Dec. 22–24, 19997). Plant Protection Association of India. NPPTI, Rajendranagar, Hyderabad. Abstract pp. 13–14.

Sharma, J.P. and Khajuria, D. R. !999. Studies on defoliating beetles and other defoliators of fruits and forest trees. *Annual report, 1999–2000*. Regional Horticultural Research Station, Bajaura, Kullu. pp. 37–39.

Sharma, J.P.; Khajuria, D. R. and Dogra, G.S 1988a. Studies on apple fruit moth, *Argyresthia conujugella* Zell.(Yponomeutidae: Lepidoptera) identification, distribution and extent of damage in India. *Tropical Pest Management,* **34**: 189–92.

Sharma, J.P.; Khajuria, D.R. and Dogra, G.S. 1988b. Chemical control of apple fruit moth,*Argyresthia conujugella* (Lepidoptera: Yponomeutidae). *Indian Journal of* Agricultural *Sciences,* **58**: 378–381.

Sharma, J.P.; Khajuria, D.R. and Thakur, J.R. 1999. Studies on the population distribution of apple root borer, *D. hugelii* Rdt. in apple tree basins. *In: National Seminar on Himalayan Horticulture Horti-Vision 2020* (Dec. 2–4th 1999). Deptt. of Horticulture, HP. Shimla. Abstract p27.

Sharma J.P. and Thakur, J.R. 1996. Integrated management of insect pests of apple. *Plant Protection and Environment.* D.V.R Reddy, H.C. Sharma, T. B. Gour and B. J. Divakar (eds). Plant Protection Association of India. pp. 223–238.

Sharma, P.L. and Bhalla, O.P. 1962. Preliminary studies on the control of San Jose scale, *Q. perniciosus* Comstock with miscible oils as compared to diesel oil emulsion. *Himachal Horticulture,* **3**: 235–237.

Sharma, P.L. and Bhalla, O.P. 1964. Record of Codling moth in Himachal Pradesh, India. *Indian Journal of Entomology,* **26**: 129–30.

Sharma, P.L. and Bhalla, O.P. 1965. Studies on the San Jose scale, *Q. perniciosus* (Comstock). *Indian Journal of* Entomology, **27**: 323–330.

Singerland, M.V. and Crosby, C.R. 1930. *Manual of Fruit Insects*. McMillan, New York, Usa. pp. 503.

Singh, J. 2001. Developmental approaches for optimizing productivity of temperate fruits. *In*: K.K. Jindal and D. R. Gautam. *Productivity of Temperate Fruits* University of Horticulture and Forestry, Nauni, Solan, India. pp.21–30.

Singh, P. and Prasad, G. 1985. Poplar stem borer, *Apriona cinerea* Chevrolat (Coleoptera: Cerambycidae), its biology and control. *Indian Forester,* **111**: 925–939.

Singh, R.N. 1941. The life history, biology and ecology of apple root borer, *Lophostenus hugelii* Redtenbach in Kumaon. *Indian Journal of Agricultural Sciences,* **11**: 925–939.

Singh, S.P. 1993. Biological control of insect pests. *In:* Chada, K.L. and Pareek, O.P. *Advances in Horticulture: Fruit Crops*- Volume 3 Malhotra Publishing House, New Delhi, India. pp. 1592–1614.

Sud, V.K.; Hameed, S.F. and Sharma, L.D. 1975. The insecticidal control of San Jose scale on apple. *Journal of Horticulture Science,* **50**: 165–68.

Thakur, A.K. and Hameed, S.F. 1980a. Biological performance of some organophosphorus insecticides against *Quadraspidiotus perniciosus* Comstock on apple. *Proceedings, Indian Academy of Sciences(B),* **89**: 587–601.

Thakur, A.K. and Hameed, S.F. 1980b. Toxicity of some organophosphorus insecticides to the crawlers of San Jose scale, *Quadraspidiotus perniciosus* Comstock. *Pesticides,* **14**(2): 10–12.

Thakur, J.N.; Pawar, A.D. and Rawat, U.S. 1988. Observations on the correlation between population density of apple wooly aphid and its natural enemies and their effectiveness in Kullu valley (HP). *Plant Protection Bulletin,* **40** (2): 13–15.

Thakur, J.R. 1984. *Studies on the biology and control of woolly aphid Eriosoma lanigerum (Hausmann) with special reference to alates and root population.* Ph.D. thesis. College of Agriculture, Solan.

Thakur, J.R. and Dogra, G.S. 1980. Woolly apple aphid, *Eriosoma lanigerum,* research in India. *Tropical Pest Management,* **26**: 8–12.

Thakur, J.R.; Bhalla, O.P. and Sharma, P.L. 1973. Survey of Thysanoptera in Himachal Pradesh. *Indian Journal of Entomology,* **35**: 345–346.

Veer, V. 1985. Observations on the host preference and biology of *Thrips flavus* Schrank (Thysanoptera: Thripidae) from Dehra Dun (India). *Ann. Entomon*. **3**: 39–47.

Veeresh, G.K. 1988. White grubs as pests of forest and fruit trees in India and their management. *In: Tree Protection.* V.K. Gupta and N.K. Sharma (eds.). ISTS, Solan, Himachal Pradesh, India.

Verma, K.L. and Chander, R. 1988. Root borer, *Dorysthenes hugelii* Redtenbatcher (Cerambycidae: Coleoptera) and its management in apple orchards in Himachal Pradesh: Abstract in *National Symposium on Tree Protection.* ISTS, Solan (H.P.)

Verma, T.D. 1985. *Studies on Tortricids Infesting Apple with Special reference to Biology and control of Archips pomivora Meyrick.* Ph.D. Thesis. Himachal Pradesh Krishi Vishwa Vidyalaya, Solan, Himachal Pradesh, India.

Verma, T.D. and Chander, R. 1992. Behaviour and seasonal history of apple leaf roller, *Archips pomivora* Meyrick (Lepidoptera: Totricidae). *Journal of Tree Science,* **11**(1): 62–66.

Whalon, M.E. and Croft, B.A. 1984. Apple IPM implimentation in North-America. *Annual Review of Entomology,* **29**: 435–470.

*Pest Management in Horticulture Crops*
L.R. Verma, A.K. Verma and D.C. Gautam (eds)


## *Chapter 11*

# Aphids as Pests of Horticultural Crops

***D.C. Gautam***

## INTRODUCTION

APHIDS ARE A subject of great interest as these have attracted the attention of man because of the unusual phenomena such as cyclic parthenogenesis, polymorphism, vivipary and telescoping of generations occurring in their life cycle (Dixon, 1985). These tiny sap sucking insects have great economic importance as not only they cause great damage to crops of agricultural and horticultural importance by sucking the plant sap but transmit large number of virus diseases also. These problems are further aggravated because of aphids' great capacity to reproduce in order to exploit the host plants fully.

At present, there are about 4,000 species of aphids described so far. Of these, around 250 species feed on horticultural and agricultural crops throughout the world (Blackman and Eastop, 1984). Not all have pest status as only about 50 aphid species are considered as key pests.

## HETEROECY

The aphid species that are heteroecious occur on two taxonomically unrelated host plant species with host alternation while the aphid species that are autoecious live on one (monophagous) or few related host plants. The polyphagous species belong to subfamily Aphidinae. Though temperate regions are rich in aphid fauna but they are abundant in tropics also where they infest all kinds of groups of plants (Dixon, 1987). In many regions, there occur many introduced species along with the native ones.

In heteroecious aphids, there is migration from a primary host to one or several species of secondary host. The aphid life cycle is under genetic control but it is modified by environmental factors. There is tendency among the pest aphids particularly those on agricultural crops, to be polyphagous i.e. to have a wider host range than related economically unimportant species. Generally most of the pest aphids such as *Cavariella aegopodii* on Umbelliferae and *Sitobion avenae* on Gramineae. restrict their feeding to species within plant family as summer generations. But these are also able to colonize a wide range of reserve hosts including economically important plants. Polyphagous species have the ability to exploit reserve hosts in many different families (Blackman and Eastop, 1984) and in many cases wild plants are most often used as hosts (Eastop, 1981).

The significant feature of aphids has been their evolutionary adaptation as it has been primarily directed towards synchronization of their life cycles with seasonal cycles of their hosts. Primary hosts of heteroecious species are conifers while secondary hosts are dicotyledons or monocotyledons. This shows that the aphids were primarily connected to conifers (primary hosts), then with dicotyledons and secondarily with monocotyledons (secondary hosts).

The question arises what is the advantage of heteroecy? It actually enables an aphid species to live on two hosts in succession. Host alternating aphids are generally holocyclic and abundant in temperate regions and they produce eggs through sexual generation for overwintering especially survival in severe winters when anholocyclic forms are unable to survive. These species have abundant food supply for rapid reproduction. But the evolving process does not stop here as is evident from many species reverting to monoecy due to either selection pressure or favourable conditions. Aphids adapted to one food plant may become anholocyclic and this may lead to speciation process. Some aphid species have the ability to complete their life cycle successfully on either the primary or secondary host thus accelerating the evolutionary process that proceeds towards elimination of sexual generation.

Commonly each aphid species has only one species of primary host plant and when there are two or more species they are closely related. But there are exceptions also. In temperate climates, most species of aphids have a single, annual sexual generation resulting in the production of cold resistant, over wintering eggs. In the tropics, reproduction may be entirely parthenogenetic (found throughout the year) on plants in suitable condition. In dry tropics, aphids are seldom found during hottest season. When temperatures increase suddenly, aphids may be killed on plants and each autumn, the area is repopulated. Often only a small part of the gene pool of a species in its region of origin is represented in the introduced population (the founders effect). The introduced genotypes are frequently anholocyclic i.e. able to survive the winter without a sexual phase.

In all aphids, parthenogenetic generations exploit actively growing plants while the sexual generation produces over wintering eggs. These features being common in almost all the aphids, indicate that evolution of aphids as a group has been primarily in relation to temperate climate with severe cold winters. However, variations also occur in different geographic regions. Sexual pressure on secondary hosts may favour the return of sexual forms to primary host.

## HOLOCYCLY

Holocycly in aphids is an adaptation to temperate conditions as fertilized egg is the stage of over wintering. Parthenogenesis in aphids is of considerable importance as it evolved as an adaptation. Anholocycly is a subsequent development and combining thelytoky with viviparity was a way for rapid multiplication of aphids. Under suitable conditions generally the aphids resort to continuous thelytokous reproduction. Long photoperiods, high temperatures and adequate food supply all help aphids to multiply thelytokously (Blackman, 1980).

Cosmopolitan aphid species are anholocyclic. The populations generally originate from a single parthenogenetic female but later on give rise to new biotypes through mutation which can be selected out rapidly by the new environment. Anholocycly seems to have evolved simultaenously with the adaptation to one of the hosts. Apomictic parthenogenesis evolved from automictic parthenogenesis. Cyclic parthenogenesis in aphids and their heteroecious life cycles are the major evolutionary developments which have enabled aphids to exploit their food plants fully especially agricultural crops.

## NATURE OF DAMAGE

Aphid pests cause damage directly by sucking the sap from the phloem. They also cause damage indirectly by producing honeydew that results in faster aging of the leaf and also due to development of sooty moulds, there is reduced photosynthesis. Moreover, great damage is caused through transmission of virus diseases of plants. Damage from the virus diseases can be related to the size of the vector population. Those aphid pests which induce galls may lead to secondary infections by fungus especially through wounds created by aphid pests. Many aphid species form galls which may be true or pseudogalls. In true galls, the plant tissues grow around and completely enclose the aphid and its colony. Pseudogalls are those which do not enclose the aphids such as on apple young leaves infested by rosy leaf curling aphid which first develop colouration and pocket galls on leaves of red currant *Ribes rubrum* L. which are caused by the currant aphid.

Some species of aphids develop the highly sophisticated mechanisms for living. The galls of *Fordini* act as small incubators. Many successful generations are produced in these galls under optimal conditions. On one hand, the aphid colonies are protected from the external environment while on the other hand the aphids get abundant food supply. The galls offer the best possible conditions for producing numerous copies of these genotypes that have survived selection. This also raises the probability of these aphids to be represented in the natural populations in the subsequent year (Wool, 1980).

## VIRUS TRANSMISSION

Aphids are the most significant group of vectors transmitting viruses of economically important agricultural crops. Plant viruses are luteoviruses. These are small isomeric and phloem restricted on their plant host. All luteoviruses are actually serologically related (Martin and D'Arcy, 1990). These infect and replicate only in the sieve elements and the parenchyma cells of the phloem located in the vascular bundles of host plants (Waterhouse *et al.* 1987). Aphids while probing the plant tissues, penetrate the vascular bundles to feed the phloem. This they achieve by utilizing their tissue piercing stylets (Evert *et al.* 1973). The time period between acquisition and inoculation is called the latent period. Aphids' ability to retain infectivity after leaving the host plant is called persistence. The time needed to initiate infection is called inoculum. Acquisition time is the time of feeding on host plant.

Viruses may be transmitted through non-persistent or persistent mode of transmission. The viruses when acquired and transmitted within a few minutes are called stylet borne or non-persistent as they are eliminated from the vectors immediately (Kennedy *et al.* 1962). But many aphid vectors transmit persistent viruses as they need many hours to be acquired and even more to be inoculated into the plants. Luteoviruses are transmitted in a persistent manner by aphid. Vectors, generally, can retain their viruses for life.

Steps in persistence transmission involve: retention of viruses for days/weeks, acquisition from and transmission to long periods and delay after acquisition of viruses. In many cases aphid transmission is the only way of transmission such as is known for barley yellow dwarf virus (BYDV) which is transmitted by more than 20 species of aphids (A'Brook and Dewar, 1980). This transmission is in a persistence manner. Aphids such as *Myzus persicae* move frequently while colonizing a crop and transmit semi- persistent viruses after feeding as little as 15 minutes and those viruses can be transmitted upto two days to a succession of plants.

## MODE OF ACQUIRING VIRUSES

Aphids have evolved as ideal vectors of plant luteoviruses because of their tissue specificity of viruses and mode of feeding. Penetrating deep into the phloem sieve elements by stylets tips, the aphids initiate feeding process by contracting the muscles of the sucking pump. Decrease of pressure in the oesophagous results in drawing phloem sap into the fore gut. This facilitates the transfer of viruses in the phloem to the alimentary canal of the vectors. Chitin lining, in the fore gut, does not allow penetration of viruses through it. So the virus passes through anterior mid gut into the hind gut. Much of virus particles may pass to honey dew (Gildew, 1990).

Most aphids that transmit viruses in temperate regions are host-alternating species. Primary hosts on which these aphids normally over-winter do not usually harbour viruses of agricultural crops. Because of changes in the life cycle of many species, the return migration to primary host does not occur. That means, they overwinter on herbaceous crops, which are infected with viruses. The regions where winged forms arrive early are likely to be affected severely because over wintering on herbaceous plants renders aphids infected with viruses.

## WHY DO APHIDS TRANSMIT VIRUSES?

Aphids are certainly benefited from the transmission of plant viruses. Rather viruses and aphid vectors are benefited mutually. When the plants are infected with viruses, there occurs increase in concentration of free amino acids. This change, in fact, improves the food quality resulting faster development in aphids and production of more progeny. Aphids are not affected by the infected plant or by carrying viruses. So transmission of viruses by aphid vectors is a very profitable proposition. But what do the viruses gain? Aphids are the mean of dispersal of viruses which in return improve the quality of plant sap as food for the aphid vectors.

## APHID SPECIES AS VECTORS

Of the aphid species, 228 species have been reported to transmit virus diseases of plants (Eastop, 1977). Aphidinae has adapted itself to heteroecious angiosperms while other subfamilies are associated with trees. The viruses of the herbs have been intensively studied than those of the trees.

Important aphid vectors listed in Eastop (1983) are as:

- The pea aphid *Acyrthosiphon pisum* is a vector of more than 30 viruses including non-persistently transmitted viruses of beans, peas, beet, clover, cucurbitae and crucifers and persistently transmitted viruses of peas and beans.

- The American large and red raspberry aphid *Amphorophora agathonica* is a vector of several non-persistently transmitted viruses including raspberry (black) necrosis and *rubus* net viruses.
- The large raspberry aphid *Amphorophora idaei* is the only vector of raspberry (black) necrosis virus in Europe.
- The green citrus aphid, (spiraea aphid) *Aphid citricola* (also known as *A. spiraecola*) is a vector of several non-persistently transmitted viruses including bean common mosaic, beet mosaic, or at least one strain of citrus tristeza potato virus Y and tobacco etch viruses. It also occurs on apple (where it is confused with *A. pomi*), pear, and some other Compositae.
- The groundnut (peanut) aphid also called cowpea or black legume aphid *Aphis craccivora* transmits about 20 viruses non-persistently, including broad bean mosaic, beet mosaic, cabbage black ring spot, cowpea mosaic, cucumber mosaic, alfalfa mosaic, pea mosaic viruses. *A. craccivora* is wide spread in warm temperate, subtropical and tropical regions. It lives mainly on Leguminosae.
- Black bean aphid *Aphis fabae* transmits about 35 viruses. Those transmited non-persistently include potato virus Y and viruses causing mosaics of beet, cucumber, peas and soybean. The persistently transmitted viruses are beet yellow net and potato leaf roll.
- The melon and cotton aphid *Aphis gossypii* is a vector of 50 non-persistently transmitted viruses of banana, phaseolus beans, beet, cucurbits, potato, radish, soybean, strawberry, tobacco and tomato.
- The buckthorn-potato aphid *Aphis nasturtii* is recorded as a vector of nine non-persistently transmitted viruses including beet mosaic, cabbage black ringspot and potato viruses A, Y, and aucula mosaic, and persistently transmitted potato leaf roll virus.
- The glass house-potato aphid *Aulacorthum solanii* transmits about 40 viruses including both persistently and non-persistently transmitted viruses of beet and potato. It is polyphagous.
- The cabbage aphid *Brevicoryne brassicae* transmits about 15 viruses non-persistently. It also transmits cauliflower mosaic virus in a persistent manner.
- The willow carrot aphid *Cavariella aegopodii* transmits seven viruses non-persistently and many others persistently.
- The strawberry aphid *Chaetosiphon fragaefolii* transmits four strawberry viruses persistently and two non-persistently.
- The currant-sowthistle aphid *Hyperomyzus lactucae* transmits 10 viruses non-persistently and lettuce necrotic yellow virus persistently.
- The American turnip aphid *Lipaphis erysimi* transmits 10 viruses non-persistently.
- The potato aphid *Macrosiphum euphorbiae* transmits more than 50 viruses including non-persistently transmitted viruses of beans, beets, crucifers, clovers, chrysanthemums, tulips, and persistently transmitted viruses of beet, pea, tomato, potato etc.
- The rose grain aphid *Metopolophium dirhodum* transmits barley yellow dwarf virus persistently.
- The peach-potato aphid *Myzus persicae* which occurs on diverse herbaceous plants transmits about 120 plant viruses. Many are non-persistently transmitted but about a dozen including potato leaf roll, pea leaf roll are persistently transmitted.

- The banana aphid *Pentalonia nigronervosa* transmits abaca bunchy top and banana bunchy top viruses persistently and banana mosaic virus non-persistently.
- The cereal leaf aphid *Rhopalosiphum maidis* transmits seven viruses non-persistently as well as persistently
- The bird-cherry aphid *Rhopalosiphum padi* transmits non-persistently four viruses of dicotyledons and three of monocotyledons which it does not colonize. It also transmits persistently barley yellow dwarf and maize leaf fleck viruses.
- The green bug *Schizaphis graminum* transmits barley yellow dwarf and millet red leaf viruses persistently and sugarcane mosaic virus non-persistently.
- The grain aphid *Sitobion avenae* transmits barley yellow dwarf and radish yellow viruses persistently and bean yellow mosaic and pea mosaic viruses non-persistently.
- The grain aphid *Sitobion miscanthi* transmits barley yellow dwarf and millet red leaf viruses persistently.
- The black citrus aphid *Toxoptera citricola* transmits citrus tristeza, citrus vein enation, abaca mosaic and pea mosaic viruses.

Virus particles are bound to mouthparts when an aphid probes an infected area and if released into the cells of an uninfected plant in a subsequent feed, can cause infection. A more subtle mechanism is involved when virus particles are taken up by the aphid gut, circulate in the haemolymph, reach the salivary gland and then pass into a new plant with the injected saliva.

## CLONES

Aphid populations generally consist of a number of clones. Aphid developing from fertilized egg in spring reproduces to produce clones. A number of clones develop from fertilized eggs which are subjected to screening by the environmental changes. Those which are fully adapted survive while others non-adaptive are eliminated. So only a few genotypes are left during autumn which contribute to the gene pool of the following year. What leads to this? Primarily the tough competition for food resource between clones results in selecting only the fittest genotypes. The interesting thing is that each genotype is represented not by one individual but by all the members of a clone which may be in thousands (Blackman, 1981).

## GENETIC VARIATIONS IN APHID POPULATIONS

Aphids exploit fully the advantages of parthenogenetic reproduction and at the same time produce sexual generation in autumn. Thus a diversity of genotypes are produced through sexual reproduction which spend winter as eggs. These genotypes encountering environmental conditions are subjected to various selection pressures and thus are eliminated. Remaining genotypes build up large populations of genetically identical individuals. Some of these produce sexual generation and the egg stages are the source of future generations of aphids. The sexual generation has dispersive phase before it and resting stage afterwards (Blackman, 1981, 1985).

The new genotypes in the eggs encounter entirely new set of conditions on hatching, such as new season, hosts/succession of hosts and vagaries of weather. But the genetic recombinations provide the adaptation to the new environmental pressures of new season. In fact, the genetic recombinations, which occur only in sexual generation preserve the integrity of species gene pool. In aphids, the genetic diversity is maximum in egg stage and minimum before sexual reproduction. Actually this kind of life cycle has consequences for genetic variability in aphid populations. The reproductive potential of genotypes is determined by the intrapopulational and ecological forms. Minor differences in the beginning lead to major differences in frequency of clones. This lowers the genetic variability. Reduction in genetic variability may even be caused by founder effect during search for new host. This raises the point of occurrence of genetic variability which is generally through mutation, anticyclic selection on different hosts as well as in different seasons.

In those species, which have holocyclic populations raise the genotype variability by recombination (Dixon 1985, 1987; Wohrmann and Tomuik, 1988) and this compensates the effects of drift and selection during the parthenogenetic phase.

Occurrence variable enzyme loci can also be explained as a result of random drift (founder effect and bottleneck effects) and lack of sexual reproduction (Wohrmann and Tomuik, 1988). Asexual reproductions have a low genetic heterogeneity and strong local and temporal unstable population structure, while populations with cyclic parthenogenesis have a high degree of genetic heterogeneity. Genetic systems of organisms are evolved as adaptation to survival and thus are finely tuned to their needs.

## APHID BIOTYPES

Many species of aphids develop biotypes which have the ability to break plant resistance to aphids (PRA). Aphid biotypes (Smith, 1989) are the phenotypic traits consisting of many genotypes (Putreka and Peters, 1989). The biotype variation is important to developing PRA programmes. This very fact has generated much interest in those involved in management programs. There occurs a variation in the biotypes as they have differential ability to damage plant resistance. These genetic variants in aphid populations warrant evaluation of various plant cultivars in order to select those cultivars that have highest degree of multibiotype resistance.

## HOW THE EVOLUTION OF BIOTYPES OCCURS?

The biotype evolution occurs through induction, muation and selection (Mayr, 1982). Exposure to unfavourable environmental conditions compels a species to adaptation which can be achieved through gene amplification due to duplication or translocation that changes gene expression due to position effects. The phenotype plasticity can be imparted through induction of other gene regulatory mechanisms.

Mutations result in change in gene or chromosomal complement producing phenotypes that are different to wild type. Mutations are of recessive nature. The life cycle of the aphids plays a key role in expression of mutations. Changes may be in the form of gene duplication and translocation.

Gene duplication may be responsible for increase in level of esterase-4-activity associated with organo-phosphate resistance in *Myzus persicae* (Blackman and Devonshire (1978). In aphids, translocation (1, 3) has been reported in *Myzus persicae* aphids (Blackman *et al.* 1978).

Translocations are actually chromosome structural mutations that involve the relocation of a chromosome segment to a non-homologous chromosome.

Determining the genetic structure of a population is essential for understanding adaptations of aphids. Different selection pressures operate (Blackman, 1985). So the genotype composition of aphid population differs from season to season such as spring/autumn. Population structure will be generated by host availability as well as the fitness of each genotype on the hosts.

Crossing of two biotypes of *Amphorophora idaei* yielded progeny that comprised of four biotypes when evaluated on two resistant cultivars of host plants (Briggs, 1965). Thus these strains result from segregation of two gene loci. But the strains of raspberry aphids that colonize most raspberry varieties are least common. Similarly crossing of three biotypes of *Schizaphis graminum* produced progeny that yielded 25 biotypes when evaluated on 8 resistant and susceptible cultivars. These 25 biotypes were the product of independent assortment of these genes during meiosis.

## VARIANTS

Development of new variants such as in case of alfalfa aphid, enable them to colonize previously resistant varieties of alfalfa through mutation. The mutational events occur rarely (Blackman, 1981). Once the mutation occurs, it multiplies so rapidly in the parthenogenetic generations because an aphid with an advantageous mutation passes this trait to all its progeny or descendants. As the reproductive potential of parthenogenetic aphid is very high, the new mutant takes over the whole population.

## INSECTICIDE RESISTANCE

Aphid pests have developed insecticide resistance and thus made the management very difficult. Many pest species have undergone life cycle variations and changes at the chromosomal or genic levels have possibly been involved in the world wide build up of high level of insecticide resistance to most of the organophosphates (Blackman, 1980).

For example, in *Myzus persicae*, exposure to the insecticides applied for controlling this aphid pest, there was development of low resistance in those aphids with normal karyotype. This development is due to changes in esterase regulatory system of selective gene duplication. However, in those which developed high resistance are translocation heterozygote, and this resistance was attained by preserving multigene copy regions on the chromosome involved in the translocation or by modifying esterase regulatory system.

## APHID FORECASTING

For management of aphid pests, forecasting is very important for making rational control decisions. Crops can be protected from aphid pests, if forecasting is given and proper appropriate action is taken. The forecasting can be made on the basis of information on the aphid populations in previous years as well as on biology of the pest aphids. Much depends upon environmental conditions.

Previous years' data on these aspects are very useful for aphid forecasts. Aphid infestation can be monitored on the plants and this can provide database for forecasting (Way *et al*. 1977). The infestation can be compared with those of the previous years. For this, information especially on outbreak and non-outbreak years is useful in predicting the general risks in a year (Tatchell, 1982a, b). This has been very well established in some aphids where spring migrants can be related to an established economic threshold.

Aphid populations are affected by temperature. In favourable temperature conditions, the aphid populations survive and build up easily while extreme temperatures retard and kill the aphids. Temperature even influences the life cycle of aphids. In severe winter conditions, the holocyclic populations survive as egg stage but the anhocyclic populations are affected the most.

Temperature records can help in forecasting. Mild winters favour anholocyclic aphids. If weather is warm, anholocyclic aphids will survive winters resulting in early build up of aphid colonies in comparison to holocyclic populations. Migration, after mild winters occur earlier so this results in rapid spread of virus diseases.

For management of those aphids that cause great damage and have developed insecticide resistance, it is important to apply the control at appropriate time. So it is very important for agriculture practices to know how pest aphid populations are structured in order to get optimized forecasting systems.

## MONITORING

Infestation and colonization are generally done by the alate aphids. Generally, these pests respond predominantly to visual cues i.e. flying actively and landing. Many a time, colour affects the landing behaviour of aphids (Gibson and Rice, 1989). That is why the red cultivars of cabbage and brown cultivars of lettuce are having least infestation while green cultivars are affected the most. Weather is a major disturbing factor in population dynamics of aphids.

How can the aphids be detected? For the successful monitoring of aphid population, it is very important to know which species are infesting the crops. Aphid infestation can be detected by cast skins on the leaves, curling of the leaves, distortion of the new growth, honeydew, sooty molds on leaves and yellow spots on upper surface of leaves.

## MANAGEMENT OF APHID PESTS

The populations of many pest aphids are generally kept in check by natural controls such as environment (high winds, heavy rains and high temperature) and natural enemies. A number of approaches are used for the control of aphid pests.

### Chemical control

Chemical method is widely used for the control of aphid pests but it depends largely on the appropriate timing of application. Many aphid pests have developed resistance to insecticides especially organophosphates, carbamates and pyrethroid insecticides. Moreover, these insecticides may cause adverse effects on the natural enemies.

Most commonly used insecticides reported in literature for the control of aphid pests are endosulphan, phosphamidon, dimethoate, carbaryl, etc. There is extensive literature available on the chemical control of different aphid species.

**Neem botanical**

Azadirachtin, a tetranortriterpenoid isolated from seeds of neem tree *Azadirachta indica* A Juss (Meliaceae) is a potent antifeedant affecting a wide range of aphid pests.

**Cultural control**

Increasing environmental awareness has resulted in the introduction of new crop management strategies. Some aphid species can be effectively managed through cultural techniques. For example, populations and damage of green bug can be reduced in winter wheat and grain sorghum by adopting the practice of reduced tillage which curtails soil disturbance but at the same time results in increase in surface residue.

The crop residue and the undisturbed soil, as a result of reduced or no tillage, in fact, provides reflective surfaces that influence flight behaviour and host finding in *Schizaphis graminum* (Rondani) (Burton and Burd, 1990). Light waves of no tillage surface repel alighting aphids by masking the attractiveness of host plants. As the influx of the migrating aphids is reduced, so the build up of aphid population is affected.

'No tillage' system when combined with other feasible cropping practices can effectively bring down the aphid populations, this means the reduced use of chemical pesticides. In addition to conserving soil moisture, it is inexpensive and environmentally sound.

**Control through water management**

Significance of water management for pest suppression has been greatly recognized. Strategically, aimed irrigation of alfalfa fields may be instrumental in promoting fungal epizooitics among aphid pests therein (Hall and Dunn, 1957). The effectiveness of entomogenous fungi can be enhanced in alfalfa fields against alfalfa aphid *T. maculata* by extra periods of irrigation. The fields which are well drained and able to withstand additional water retain a blanket of humidity which facilitate more rapid and complete spread of the fungi throughout the aphid population.

**Control through microbial organisms**

Aphid lethal paralysis virus (ALPV) has great potential as a biological control agent for cereal aphids (von Wechmar *et al.* 1990). The other aphid virus which affects aphids is *Rhopalosiphum padi* virus (RhPV).

ALPV is reported to be very effective against *Diuraphis noxia, Rhopalosiphum padi* and *Sitobion avenae* aphid populations on small grain. The decrease in population of these aphids corresponds with increase in aphids testing positive for ALPV. Under natural field populations, dense populations of *Rhopalosiphum maidis* on wheat disappeared entirely within a week after infection of ALPV (Laubscher and von Wechmar, 1990).

ALPV is very effective against *Diuraphis noxia* as aphid deaths occur before ALPV replicates to detectable levels in the affected aphid. But variation in this aspect is observed in aphid species.

The *Rhopalosiphum padi* is a better replicating host while *D. noxia* is a sensitive host. ALPV contents of *Rhopalosiphum padi* are highly contagious and can cause deaths of 5 to 8 aphids as well their progeny. That's why *D. noxia* populations never build up so easily in those regions where *R. maidis* occur in large numbers.

In natural conditions, virus infected aphids stick and disintegrate on wheat stem and leaves. This results in providing the contagion for further progeny of aphids which are already colonizing the same plant (von Wechmar *et al.* 1990).

## Fungal pathogens of aphids

Fungi are the significant microbial pathogens of insect pests and have been reported to affect aphids. The dense populations of aphids provide an easy target for fungal pathogens which facilitate dispersal, thus promoting the spread of epizooitics.

Fungal entomopathogens always attack their hosts by directly penetrating the exoskeleton and so do not require ingestion with subsequent infection through the gut wall. Fungal pathogens which affect aphid populations have narrow host specificities and thus do not affect non-target organisms (Goettel *et al.* 1989).

Infective unit of most fungi is a conidium. The vegetative growth of most insect pathogens is some sort of rapidly proliferating hyphal bodies, short rod like or subglobose cells instead of usual long hyphal filaments. Hyphal bodies proliferate rapidly and circulate readily in the blood to all parts of infected host. Fungus fills the whole body and the death of the host occurs because of the toxins produced by the fungus or by filling the host body cavity which results in destruction of tissues.

For effective control, the formulations are more important. A number of techniques are available to mass produce fungal cell types using rich array of inexpensive components for cultural media (Soper and Ward, 1981; Samson *et al.* 1988). While preparing the fungal formulations, the biology of the fungus and that of the host aphid alongwith details of environment are important.

Many fungi have a natural capacity to dessicate and on rehydration start normal development. Most of entomopathogenic fungi have the characteristics of marcescence (dried mycelial preparation). There are many examples of the control of aphid pests through fungi. A few important examples are as follows:

- *Pandora neoaphidis* is a very effective pathogen of pea aphids. Only the Austrian biotype of *Acyrthosiphon pisum* has been found resistant to some strains of fungal pathogens (Milner, 1985).
- Some of the European aphids have been reported to develop natural resistance to *Conidiobolus obscurus* (Papierok and Wilding, 1981).
- Inundative mycoinsecticide treatment strategies are effective for control of aphids. Fungus based preparations are used to cause high mortality rapidly.
- Conidial product of *Verticillium lecanii* is effective against aphids (Hall, 1981). But the effectiveness is affected by temperature and humidity. Temperature between 15 to 25$^0$C and humidity above 85% favour infection.

- *Beauveria bassiana* is used against some aphids (Feng *et al*. 1990). Russian wheat aphid seems to be highly susceptible to nearly all of the most common aphid pathogenic fungi.
- Fungal entomogenous offer great advantages over pesticides making them very attractive alternatives in future. Once introduced successfully, these can survive from year to year and protect expanding areas as they disperse.
- *Conidiobolus obscurus* can infect many aphid genera while *C. coronatus* and *C. thromboides* have wide range of hosts in insects including aphids.
- *Neozygites fresenii* infects aphids of genus *Aphis*.

Most common aphidicilous hyphomycetes, *Beauveria bassiana* and *Verticillium lecanii* have broad host specificities.

### *Fungal endophytes*

Aphids may be affected by the presence of fungal endophytes. Fungal endophytes have been associated with high levels of resistance to insect herbivores. The greenbug *Schizaphis graminum* (Rondani) is affected by fungal endophytes.

Presence of fungal endophytes in plant tissues enhance plant resistance to Russian Wheat Aphid and the fungal endophytes infest large numbers of *Festuca* sp. Kindler *et al*. (1990) stated that screening of germ plasm for genetic resistance to RWA through fungal endophytes should be done. This should also be applied to all those aphid pests which can not be easily controlled.

## Plant resistance

Plant resistance offers tremendous economic potential and environmentally sound means for managing aphids. Host plant resistance to aphids is simply inherited and is mostly dominant.

### *Aphid resistant cultivars*

Development of aphid resistant cultivars provides a more attractive aphid management option. Many crop plants can be tested for aphid resistance. One of the earlier examples of aphid resistance in crop plants is in peas to pea aphid *Acyrthosiphon pisum* (Harris) (Searle, 1933). Now aphid resistance has been developed in large numbers of crop cutivars and hybrids.

Webster (1990) listed many such cultivars which are resistant to aphids:

- Many cultivars of alfalfa are resistant to alfalfa aphid *A. kondai* Shinji, pea aphid *A. pisum* (Harris) and the spotted alfalfa aphid *Therioaphis maculata* (Buckton).
- Cultivars of barley, wheat, oat and sorghum are resistant against greenbug *Schizaphis graminum* (Rondani).
- Cultivars of cabbage are resistant against cabbage aphid *Brevicoryne brassicae* L.
- Cultivars of corn are resistant against corn leaf aphid *Rhopalosiphum maidis (*Fitch).
- Cultivars of lettuce are resistant against green peach aphid *Myzus persicae* (Sulzer).
- Cultivars of tobacco are resistant against tobacco aphid *Myzus nicotiana* Blackman.
- Cultivars of wheat are resistant to greenbug *Schizaphis graminum.*

In order to reduce the use of chemical pesticides, resistant cultivars are developed. Alfalfa resistant cultivars proved to be very successful against spotted alfalfa aphid in USA (Painter, 1968). Similarly, resistant hybrids were developed to use against a new green bug biotype C in USA (Weibel *et al.*, 1972). But the most important crops are alfalfa and rye. Alfalfa crop is important from dairy industry point of view as it is widely used for this purpose. Its germplasm is now available with resistance to a number of pest species such as pea aphid and the spotted alfalfa aphid (Sorenson *et al.* 1990).

The earlier record of aphid resisitant cultivars is of apple cultivar 'Winter Majetin' to woolly apple aphid, *Eriosoma lanigerum* Hausmann (Lindley, 1831). In late 1800s, the grape *Phylloxera vitifoliae* (Fitch) devastated the grape crop in France causing huge losses to wine industry in France (Painter, 1951). This was followed by introduction of resistant American vines which yielded significant results. There are many instances such as spotted alfalfa aphid which after a period of time developed ways of breaking down the plant resistance.

Host plant resistance has been an effective strategy for controlling 4 known biotypes of large raspberry aphid *Amphorophora idaei* Borner and the viruses they transmit. In red raspberry, *Rubus idaeus* L., resistance to the aphid is confered by different combinations of 14 genes. This resistance to *A. idaei* is caused by antixenosis (i.e. reduced settling of alate and apterae) and antibiosis (i.e. reduced growth rate of nymphs) in each of these genetic systems studied. Most important components in this include interference with initial settling responses of *A. idaei* during host recognition and acceptance (Birch *et al.* 1990).

Birch *et al.* (1990) further reported that the plant volatiles are involved in short range host recognition and acceptance by *A. idaei.* The dense layers of long hairs on the lower surface may trap the released volatiles in a concentrated boundary layer.

Partial host resistance in Gramineae can make a substantial contribution to reducing the damage caused by cereal aphids, reducing the use of chemical pesticides. This is conferred by hydroxanic acids (HX) (Thackray *et al.* 1990) occurring in cereals. In particular, the compound 2,4-benzoxazin-3-one (DIMBOA) have been found to be involved in the resistance of cereals to aphids. HX may even be important in preventing BYDV transmission.

Plant resistance to aphids is generally governed by single genes which can be inserted into the plants easily. However, in some cases many genes are involved in conferring resistance such as in corn leaf aphid, resistance is controlled by several genes. In wheat, the aphid resistance is conditioned by a single dominant gene. The green bug resistant wheat germ plasm is the outcome of interspecific hybridization conferring aphid resistance (Sebesta and Wood, 1978).

There is great scope in developing aphid resistant cultivars as new information is coming from the use of biotechnology to fingerprint biotypes. New strategies can be used in deploying resistant genes.

The tobacco aphid *Myzus nicotiana* Blackman is an important insect pest of commercial tobacco *Nicotiana tabacum* L. In this aphid species, morphs changed from green to red and the insecticide resistance is associated with primarily red morphs which develop some special characteristics such as fast development, high reproductive potential, ability to tolerate high temperatures. Since this aphid pest is difficult to control through chemical pesticides, cultural and biological approaches for management of these pests are required. Development of aphid resistant cultivars is required as this would be one of the most attractive strategies for this aphid pest control. The aphid resistant germplasm has to be identified (Jackson and Silsson, 1990).

There is a gene-for-gene (virulent gene-resistance gene) relationship between biotypes of aphids and the plant cultivars as is evident from the biotypes of *Therioaphis maculata* and resistant alfalfa clones that show a gene-for-gene (virulent gene-resistance gene) interaction. It is further supported by the fact that duplicated genes produce same gene product (Puterka and Peters, 1989). When resistance is recessively inherited in the plant, aphid virulence is dominantly inherited or it can be other way round. Such gene-for gene interactions are common even in host parasite relationships (Christ *et al.* 1987). This has to be kept in view while deciding for insect or pathogen resistance.

Foremost among the factors affecting the response of the aphid pests to their resistant hosts is the development of biotypes. The biological strains of the pest species which are indistinguishable morphologically, exhibit diverse reactions to genetically stable variety of hosts.

In *Acyrthosiphon pisum* (Harris), resistance is apparently overcome by greater vigour and large size, in successful biotypes (Harrington, 1945). Lock and key theory (Painter, 1951) or gene for gene relationship (Flor, 1956) has been illustrated in spotted alfalfa aphid *Therioaphis maculata* (Buckton) on alfalfa. Nielson and Don (1974) reported that the response of each of the recognized biotypes of *T. maculata* varies in its virulence towards clones of 'Hayden' alfalfa. A particular clone may be susceptible to biotype A but resistant to biotype H and other biotypes because only A carries that gene which can overcome the gene for resistance in the plant. Other clones may be susceptible to H but not to A. Briggs (1965) reported similar observations for *Amphoraphora rubei* (Kaltenbach) on raspberry.

Combination of several genetic factors determines the resistance in each variety/cultivar as well as also the development of biotypes. The gene for gene relationship arises through selection in as much the same manner as insecticide resistance (Painter, 1951).

### *Effect of temperature on resistant cultivars*

Temperature affects the performance of the resistant cultivars. Effect of temperature has been documented by Dahms and Painter (1940), Albrecht and Chamberlain (1941) who reported that at higher temperature, aphid resistant strains of alfalfa perform better perhaps due to increased growth. Similar findings were also reported in green bug *Schizaphis graminum* (Rondani) on resistant sorghum (Wood and Starks, 1972). But contrary to this, higher temperature may have opposite effect on aphid resistance in wheat (National Academy of Sciences, 1969). Selective pruning has been useful in aphid control on street trees in California (Olkowski, 1974).

### Control through predators and parasites

Damage to agricultural crops can be reduced utilizing the natural enemies of aphids. Predators and parasites are now reared by commercial producers as natural enemies. Aphids build up large populations parthenogenetically. So generally the predators and parasites which reproduce sexually can not compete with aphid populations. Rapid multiplication of aphids may cause great damage to host plants. More over if the aphids disperse widely over the habitat, they are unlikely to be eliminated by natural enemies such as predators and parasites.

Aphid pests of deciduous fruits such as *Aphis pomi* De Geer (apple aphid), *Eriosoma lanigerum* Hausmann (woolly apple aphid), *Myzus persicae* (Sulzer) and *Dysaphis plantaginea* (Passerini) (rosy apple aphid) can be effectively controlled through predators and parasites.

Biological insect pest suppression was and still is the most promising areas in which extensive and intensive research is required. Ladybird beetle received much attention and acclaim in biological control. Kirby and Spence (1867) recommended and used them for aphid control. These workers described various predatory wasps, lace wings larvae, syrphids and lady beetles for aphid control. Syrphid larvae feed only on aphid species, and *Laricobius erichsonii* Rosenhauer (Coleoptera: Derodontidae) preys on bark aphids (Adelgids and Pineus) (Franz, 1958).

*Trioxys pallidus* (Haliday) was imported from southern France to California to control walnut aphid, *Chromaphis juglandicola*, an exotic monophagous pest of walnuts (vanden Bosch *et al.* 1964). This species is a highly host specific endoparasitoid of developing aphids (vanden Bosch *et al.* 1962, 1970). This braconid lacked genetic make up to survive in extreme summer heat and low humidity. New strains resulted in effective biological suppression. *Chrysopa carnea* Stephens (Neuroptera: Chrysopidae) is one of the most important chrysopids in biological pest suppression. It feeds on honeydew, nectar and pollen. The larvae of *C. carnea* feed on garden pests and with their sucking mouth parts, rapidly empty the fluid contents of aphids.

***Predators***

Coccinellids are the most important group of aphid predators (Frazer, 1988). Other groups of predators are Anthocoridae, Carabidae, Cecidomyiidae, Neuroptera and Syrphidae.

Ant attended aphids are generally protected from predators and parasites as they remove these enemies from aphid colonies. But still many predators have managed to prey effectively even in ant attended colonies. For example, *Coccinella divarricata* Olivier (lady bird beetle) habitually feed on aphid colonies attended by *Formica rufa* L., a large voracious ant.

When the predator encounters a prey, it starts intensive search which is generally area concentrated (Carter and Dixon, 1986). *Aphis pomi* De Geer is effectively controlled by a predator *Aphidoletes ophidimyza* (Rondani) (Prokopy and Croft, 1994).

From control point of view, volatile or contact cues are very important for searching prey. In many cases, honeydew acts as a searching, feeding and oviposition stimulant in predators and parasites (Hagen, 1986). Natural enemies are reported to use even volatile compounds emanating from plants (Dicke and Sabelis, 1988) to locate host habitats. The synomones act as attractant to many predators. Hagen (1986) reported that chrysopid predator *Chrysoperla carnea* Stephens seems to use plant produced volatile synomones to locate potential prey habitats.

***Parasites***

Aphid parasites are generally oligophagous, attacking several species of aphids. The life cycle of parasites are closely linked to that of their hosts (Stary, 1988).

The hymenopterous parasites insert their eggs into the body of the aphids. The parasite larva develops within the host, finally killing it, sticks the skin of the dead aphid to the surface of the leaf and spins cocoon around it. This cocoon together with the skin of the dead aphid is called a mummy. In each parasitized aphid only one parasite develops to maturity.

Parasites are very specific. For example, woolly apple aphid is effectively controlled by aphelinid *Aphelinus mali* Haldeman. Both these, pests and its parasite, are imported from USA to different countries. Spotted alfalfa aphid, *Therioaphis trifolii* (Buckton) is effectively controlled by *Trioxys complanatus* Quilis. Coevolved conspicuous relationship with aphids has enabled aphid parasites to exercise effective control. Parasites use aphids as obligatory hosts (Carver, 1989). Even in *Myzus persicae* satisfactory control has been obtained by means of parasite *Aphidius matricariae* Haliday (van Lenteren *et al.* 1980).

Mustard oil produced by cultivars of cruciferous crop plants attracts *Diaertiella rapae*, a parasite of cabbage aphid *(Brevicoryne brassicae)* (van Embden, 1986).

### Parasitoids

Aphids are attacked by many parasitoid species so they have received much attention as biological control agents. The parasitized aphids are consumed from inside and their skin remains appear as mummies from which adult parasitoids emerge such as *Monoctonia pistaciaecolai. Pachyneuron? Leucopiscida* Mani attacks the aphids and also parasitizes primary parasitoid *M. pistaciaecola* inside *Smynthurodes betae.*

The walnut aphid, *Chromaphis juglandicola* (Kaltenbach) is effectively controlled by parasitoid *Trioxys pallidus* Haliday (Schlinger *et al.* 1960). This parasitoid has also been reported to be very effective against filbert aphid, *Myzocallis coryli* (vanden Bosch *et al.* 1970).

### Parasitoides of gall aphids

Galls protect the aphids from parasitoids and predators. Free living aphids are attacked by many parasitoid species, but gall forming aphids have few recorded parasitoides (Stary, 1976).

*Smynthurodes betae* Westw (Homoptera, Aphididae) forms two types of galls on its primary host *Pistacia atlantica* (Burnstein and Wool, 1990). This aphid species is effectively controlled by parasitoids. A small oval gall is formed on the mid rib of a new unfolding leaflet. The descendants form spindle shaped galls.

As compared to numerous species of parasitoids known from free living aphids, gall forming species, especially true galls, are attacked by few. Two species of parasitoides attacking true gall forming aphids are: *Monoctonia pistaciaecola* Stary and *Areopraon lebelleyi* Mani. *Areopraon lebelleyi* enters galls of *Schizoneura ulmi* after they open to release aphids (Stary, 1976). All the instances show that perhaps no aphid parasitoid evolved the ability to oviposit through the gall walls and get the aphids inside.

Aphids have evolved a large number of defense mechanisms such as body shake, walk, dropping (Klingauf, 1967), uptake of toxic allelochemicals compounds (Wink and Witte, 1991), sterile soldier castes (Aoki, 1978, Ito, 1989) and alarm pheromone. Underground populations of woolly apple aphid are inaccessible to predators/parasites. The variety of the defense mechanisms has evolved in close association with predators and parasitoids as well as in response to attack by their natural enemies.

Aphid control by natural enemies is affected by many things. Parasites tend to remain on same host while aphids move from winter to summer hosts (Stary, 1988). Aphids have a very high rate of multiplication as compared to natural enemies. Aphids infest hosts easily but there is late arrival of natural enemies.

Effectiveness of control can be obtained by continuing host plant resistance and biological control as well as by attracting natural enemies through semiochemicals. This can also be achieved by developing cultivars that have architecture such as hairy/vein pattern etc.

Principles of genetics could be applied to mass production of parasitoids. Through these techniques their reproductive potential, vigour and longevity could be increased. Similarly their adaptation to climate or habitat into which they are released is also enhanced.

## REFERENCES

A'Brook, J. and Dewar, A. M. 1980. Barley yellow dwarf virus infectivity of alate aphid vectors in West Wales. *Ann. Appl. Biol.* **96:** 51–58.

Albrecht, H. R. and Chamberlain, T. R. 1941. Instability of resistance to aphids in some strains of alfalfa. *J Econ. Entomol.* **34**: 551–556.

Aoki, S. 1978. Two pemphigids with first instar larvae attacking predatory intruders (Homoptera, Aphidoidea). *New Entomology*, **27**: 7–12.

Birch, A. N. E.; Jones, A. T.; Woodford, J. A. T. and Jones, D.R. 1990. Behavioural responses of the virus vector aphid *Amphorophora idaei*, to resistance to red raspberry. *Proceedings: Aphid plant interactions: Populations to Molecules.* Stillwater, Oklahoma USA. p265.

Blackman, R. L. 1980. Chromosomes and Parthenogenesis in Aphids. *In: Insect Cytogenetics.* R. L. Blackman, G. M. Hewitt and M. Ashburner (eds). Blackwell Publ., London. pp. 133–148.

Blackman, R. L. 1981. Species, Sex and Parthenogenesis in Aphids. *In: Evolving Biosphere.* P. L. Forey (ed.). Cambridge University Press. pp. 75–85.

Blackman, R. L. 1985. Aphid Cytology and Genetics. *In: Evolution and Biosystematics of Aphids. Proc. Int. Aphid Symp*. Jablona, Poland. pp. 171–236.

Blackman, R. L.; Takada, H and Kawasaki, K. 1978. Chromosomal rearrangement involved in insecticide resistance of *Myzus persicae*. *Nature*, **271**: 450–452.

Blackman, R. L. and Devonshire, A. L. 1978. Further Studies on the genetics of carboxylesterase regulatory system involved in resistance to organophosphorus insecticides in *Myzus persicae* (Sulzer). *Pesticide Sci.* **9**: 517–521.

Blackman, R. L. and Eastop, V. F. 1984. *Aphids on the World's Crops: An Identification Guide.* Wiley, Chichester. pp. 466.

Briggs, J. B. 1965. The distribution, abundance and genetic relationships of four strains of the *Rubus* aphid (*Amphorophora rubi* (Kalt.) in relation to raspberry breeding. *J. Hort. Sci.* **40:** 109–117.

Burnstein, M. and W D. 1990. Host trees, Gall forming aphids and Parasitoides: *Smynthurodes betae* Westw. on *Pistacia Atlantica* 1. Ecology of the Gall forming aphid. *Proceedings: Aphid Plant Interactions: Populations to Molecules.* Stillwater, Oklahoma USA. p 255.

Burton, R. L. and Burd, J. D. Cultural control of greenbug. *Proceedings: Aphid Plant Interactions: Populations to Molecules.* Stillwater, Oklahoma USA. p 262.

Carter, M. C. and Dixon, A. F. G. 1986. Assessment of habitat quality by coccinelid larvae. *In: Ecology of Aphidophaga.* I. Hodek (ed). Academia Prague and Dr. W. Junk, Dordrecht, 89–97.

Carver, M. 1989 Biological Control of Aphids. *In: Aphids: Their Biology, Natural Enemies and Control.* A. K. Minks and P. Harrewijn (eds). *World crop Pests* **2B** Elsevier, Amsterdam. pp. 141–166.

Christ, B. J.; Person, C. O. and Pope, D. D. 1987. The genetic determination of variation of parthenogeneticity. *In: Populations of Plant Pathogens: Their Dynamics and Genetics*. M.S. Wolfe and C.E. Caten (eds) Blackwell Sci. Publ. Oxford. pp. 7–19.

Dahms, R. G. and Painter, R. H. 1940. Rate of reproduction of pea aphid on different alfalfa plants. *J. Econ. Entomol.* **33**; 482–485.

Dicke, M. and Sabelis, M. W. 1988. Infochemicals in Tritrophic Interactions. Ponsen en looijen, Wageningen. pp. 235.

Dixon, A. F. G. 1985. *Aphid Ecology.* Blackie: Glasgo and London.

Dixon, A. F. G. 1987. Aphid reproductive tactics. *In: Population Structure, Genetics and Taxonomy of Aphids and Thysanoptera.* J. Holman, J. Pellikan, A. F. G. Dixon and L. Weismann (eds). The Hague: SPB Academic Publishing. pp. 3–18.

Eastop, V. F. 1977. Worldwide Importance of Aphids as Virus Vectors. *In: Aphids as Virus Vectors.* K.F. Harris and K. Maramorosch (eds.). Academic press, New york. pp. 3–62.

Eastop, V. F. 1981. The wild hosts of aphid pests. *In: Pests, Pathogens and Vegetation*. J.M. Thresh (ed). Pitman London. pp. 285–298.

Eastop, V. F. 1983. The biology of the principal aphid virus vectors. *In: Plant Viruses Epidemiology*. R. T. Plumb and J.M. Thresh (eds). Blackwell Scientific Publications, Oxford.

Evert, R. F.; Escherich, W.; Eichhorn, S. E. and Limbach, S. T. 1973. Observation on penetration of barley leaves by the aphid *Rhopalosiphum maidis* (Fitch.) *Protoplasma* **77**: 95–110.

Feng, M. -G.; Johnson, J. B. and Kish, L. P. 1990. Virulence of *Verticillium lecanii* and an aphid-derived isolate of *Beauveria bassiana bassiana* (Fungi: Hypomycetes) for six species of cereal infesting aphids (Homoptera: Aphididae). *Environ. Entomol*. **19:** 815–820.

Flor, H. H. 1956. The complementary genic system in flax and flax rust. *Adv. Genet*. **8**: 29–54.

Franz, J. 1958. Studies on *Laricobius erichsonii* Rosenh. (Coleoptera: Derodontidae) a predator on Chermesids. *Entomophaga* **3**: 109–196.

Frazer, B. D. 1988. Predators. *In: Aphids: Their Biology, Natural Enemies and Control.* A. K. Minks and P. Harrewijn (eds). *World Crop Pests* **2B** Elsevier, Amsterdam, 217–230.

Gibson, R. W. and Rice, A. D. 1989. Modifying Aphid Behaviour. *In Aphids: Their Biology, Natural Enemies and Control*. A. K. Minks and P. Harrewijn (eds). *World crop Pests* **2C** Elsevier, Amsterdam, 209–224.

Gildew, F. E. 1990. Barley Yellow dwarf Virus transport through aphids. *Proceedings Aphid-Plant Interactions: Populations to Molecules*. Stillwater Oklahoma, USA. pp. 165–177.

Goettel, M. S.; Poprawski, J. J.; Vanderberg, J. D.; Li, Z. and Roberts, D. W. 1989. The safety of fungal biocontrol agents to invertebrates. *In: Safety of Microbial Insecticides.* M. Laird; L. A. Lacey and E. W. Davidson (eds). pp. 209–231. Boca Raton, CRC. Press.

Hagen, K. S. 1986. Ecosystem Analysis: Plant Cultivars (HPR), *Entomophagous* species and Food Supplements. *In: Interaction of Plant Resistance and Parasitoides and Predators of Insects.* D. J. Boetheland and R. D. Eikenbary (eds). Wiley, New York, pp. 157–197.

Hall, I. M. and Dunn, P. H. 1957. Fungi on spotted alfalfa aphid. *Calif. Agr*. **11**: 5. 14.

Hall, R. A. 1981. The fungus *Verticillium lecanii* as a microbial insecticide against aphids and scales. *In: Microbial Control of Pests and Plant Diseases (1970–1980)*. H. D. Burges (ed). London Academic Press. pp. 483–498.

Harrington, C. D. 1945. Biological races of the pea aphid. *J. Econ. Entomol*. **38**: 12–22.

Ito, Y. 1989. The evolutionary biology of sterile soldiers in aphids. *Trends in Evolution and Ecology,* **4**: 69–73.

Jackson, D.M. and Silsson, V. A. 1990. Mechanism of resistance in *Nicotiana tabacum* to the potato aphid. *Proceedings: Aphid plant interactions: Populations to Molecules.* Stillwater, Oklahoma USA. p 281.

Kennedy, J. S.; Day, M. E. and Eastop, V. F. 1962. *A Conspectus of aphids as vectors of plant Viruses.* Commonwealth Inst. Entomol. pp. 114.

Kindler, S. D.; Bren, J. P. and Springer, T. L. 1990. Endophytes in *Fescue* species and associated enhancement of Russian wheat aphid resistance. *Proceedings: Aphid Plant Interactions: Populations to Molecules.* Stillwater, Oklahoma USA. p258.

Kirby, W. and Spence, W. 1867. *An Introduction to Entomology*. 7th edition Longmans, Green.

Klingauf, F. 1967. Alewehr-unt-Meidereak tionen von Blattlausen (Aphididae) bei Bedrohung durch rauber and Parasiten. *Zeitschrift fur Angewandte Entomologie*, **59**: 266–317.

Laubscher, J. M. and von Wechmar, M.B. 1990. The effect of Aphid lethal Paralysis Virus infection on Natural populations of aphids: *Rhopalosiphum padi, Diuraphis noxia* and *Sitobion avenae. Proceedings: Aphid Plant Interactions: Populations to Molecules.* Stillwater, Oklahoma USA. p245.

Lindley, G. 1831. *A Guide to the Orchard and Kitchen Garden*. London. Longman, Rees, Orme, Brown and Green.

Martin, R. R. and D'Arcy, C. J. 1990. Relationship among luteoviruses based on nucleic acid hybridization and serological studies. *Intervirology,* **31**: 23–30

Mayr, E. 1982. *The Growth of Biological Thought: Diversity, Evolution and Inheritance.* University Press, Cambridge, Harvard Mass. pp. 974.

Milner, R. J. 1985. Distribution in time and space of resistance to pathogenic fungus *Erynia neoaphidis* in the pea aphid, *Acyrthosiphon pisum. Entomol Exp.Appl.* **37**: 235–240.

National academy of Sciences 1969. *Insect Pest Management and Control.* Pub. 1695. Washington, D.C.

Nielson, M.W. and Don, H. 1974. Interaction between biotypes of the spotted alfalfa aphid and resistance in alfalfa. *J.Econ. Entomol*. **67**: 368–370.

Olkowski, W. 1974. A model ecosystem management program. Proc. Tall Timbers Conf. Ecol. Animal Contr. *Habitat Management*, **5**: 103–117.

Painter, R. S. 1951. *Insect Resistance in Crop Plants*. New York: Macmillon.

Painter, R.H. 1968. *Crops that Resist Insects Provide a Way to Increase World Food Supply.* Kansas State Agric. Exp. Stn. Bull. No. **520.**

Papierok, B. and Wilding N. 1981 Etude du comportement de plusiers souches de Conidiobolus obscurus (Zygomycates: Entomophthoraceae) vis-à-vis des pucerons *Acyrthosiphon pisum* et *Sitobion avenae* (Homoptera: Aphididae). *Entomophaga*, **26:** 241–249.

Prokopy, R. J. and Croft, B. A. 1994. Apple Pest Management. *In: Introduction to Pest Management.* R. L. Metcalf and W. H. Luckmann (eds). John Wiley & Sons, New York.

Puterka, G. J. and Peters, D. C. 1989. Inheritance of greenbug virulence to Gb2 and Gb3 resistance genes in wheat. *Genome,* **32:** 109–114.

Samson, R. A.; Evans, H. C. and lagte, J. P. 1988. *Atlas of Entomopathogenic Fungi*. Berlin. Springer-Verlag. pp. 187.

Schlinger, E. I.; Hagen, K. S. and van den Bosch, R. 1960. Imported French parasite of walnut aphid established in California. *California agriculture,* **14**: 3–4.

Searle, E. M. 1933. Resistance of peas to aphids. Wis. Agric Exp. Stn. Bull. 425: 94–95.

Sebesta, E. E. and Wood, E. A. Jr. 1978. Transfer of greenbug resistance from rye to wheat with X-rays.. *Agron Abstr.* pp. 61–62.

Smith, C. M. 1989. Plant resistance to insects, a fundamental approach. John wiley & Sons, New york. pp. 286.

Soper, R. S. and Ward, M. G. 1981. Production, Formulation and Application of Fungi for Insect Control. *In: Biological Control of Crop Production.* B.L. Papavizas (ed). Montclair, Allenheld, Osmun and Totowa. pp. 161–180.

Sorenson, E. L.; Stutville, D. L. and Horber, E. K. 1990. Registration of KS153BA3P4 alfalfa germplasm. *Crop sci*. **30:** 750.

Stary, P. 1976. *Aphid Parasites (Hymenoptera) of Mediterranean Area*. Czechoslovak Acad. of Sciences and W Junk. The Hague.

Stary, P. 1988. Parasites: Aphidoidea. *In: Aphids: Their Biology, Natural Enemies and Control.* A. K. Minks and P. Harrewijn (eds). *World crop Pests* **2B** Elsevier, Amsterdam, 171–184.

Tatchell, G. M. 1982a. Aphid migration and forecasting as an aid to decision making. *Proceedings 1982. British Crop Protection Council Symposium. Decision Making in the Practice of Crop Protection.* BCPC: Monograph No. **25**: 99–112.

Tatchell, G. M. 1982b. The interpretation of suction trap samples and aphid forecasting in Britain. *In: Euraphid Brussels 1982.* J. Bernard (ed). Commission des Communautes Europeenes, Brussels. pp. 63–68.

Thackray, D. J.; Wratten, S. D.; Edwards, P. J. and Niemeyer, H. M. 1990. Resistance to aphid Sitobion avenae in gramineae in relation to Hydramic acid levels. *Proceedings: Aphid plant interactions: Populations to Molecules*. Stillwater, Oklahoma USA. pp. 268.

van Embden, H. F. 1986. The potential for managing indigenous natural enemies of aphids on field crops. *Phil. Trans. R. Soc. Lond.* **B 318:** 183–201.

van Lenteren, J. C.; Ramakers, P. M. J.; and Woets, J. 1980. Integarted control of pests in greenhouses. *In: Integrated Control of Insect pests in The Netherlands.* A. K. Minks and P. Gruys (eds). Pudoc, Wageningen. pp. 109–118.

vanden Bosch, R; Schlinger, E. L. and Hagan, K. S. 1962. Initial field observations in California on *Trioxys pallidus* (Haliday), a recently introduced parasite of walnut aphid. *J. Econ. Entomol.* **55**: 857–862.

vanden Bosch, R; Schlinger, E. L.; Hall, J. C.; Puttler, B. 1964. Studies on succession, distribution and phenology of imported parasites of *Therioaphis trifolii* (Monell) in southern California. *Ecology,* **45:** 602–621.

vanden Bosch, R.; Frazer, B. D.; Davis, C. S.; Messanger, P. S.; Hom, R. 1970. *Trioxys pallidus* an effective new walnut aphid parasite from Iran. *Calif. Agric.* **24(11)**: 8–10

von Wechmar, M. B.; Laubscher, J. and Williamson, C. 1990. The pathogenesis of aphid lethal paralysis virus in *Diuraphis noxia* and other smallgrain aphids. *Proceedings: Aphid plant interactions: Populations to Molecules.* Stillwater, Oklahoma USA. p246.

Waterhouse, P. M.; Gildew, F. E. and Johnstone, G. R. 1987. *The Luteovirus Group Association of Applied Biologists Description of Plant Viruses.* No. **339**. p9.

Way, M. J.; Cammell, M. E.; Alford, D. V.; Gould, H. J.; Graham, C. W.; Lane, A.; Light, W. I. St G.; Rayner, J. M.; Heathcote, G. D.; Fletcher, K. E. and Seal, K. 1977. The use of forecasting in chemical control of blackbean aphid *Aphis fabae* Scop., on spring sown field beans, *Vicia fabae. Plant Pathology*, **26**: 1–7.

Webster, J. A. 1990. Yellow sugarcane aphid (Homoptera: Aphididae): detection and mechanism of resistance among Ethiopian sorghum lines. *J. Econ. Entomol.* **83**: 1053–1057.

Weibel, D. E.; Starks, K. S.; Wood, E. A. and Morrison, R. D. 1972. *Sorghum* cultivars and progenies rated for resistance to greenbug. *Crop sci.* **12**: 334–336.

Wink, M and Witte, L. 1991. Quinolizidine alkaloides in *Macrosiphum albifrons* and *Aphis genistae* (Homoptera, Aphididae). *Entomologia Generalis*, **15**: 237–254.

Wohrmann, K. W and Tomuik, J. 1988. Life cycle strategies and genotypic variability in populations of aphids. *J. Genet.* **67**: 43–52.

Woods, E. A. and Starks, K. S. 1972. Effect of temperature and host plant interaction on the biology of three biotypes of the green bug. Environ. *Entomol*. **1**: 230–234.

Wool, D. 1980. Galls and Genetic Variations in the Fordini. (Aphididae, Homoptera). *Bull. Soc. Bot. Fr. 127. Actual bot.* **1**: 197–198.

*Pest Management in Horticulture Crops*
L.R. Verma, A.K. Verma and D.C. Gautam (eds)


*Chapter 12*

# Insect Pests of Peach, Plum and Apricot

*J.R. Thakur and P.R. Gupta*

## INTRODUCTION

PEACH, PRUNES, PLUM and apricot occupy a distinct place amongst cultivated fruits because of their excellent dessert quality and wide spread use in canning and processing industry. In India these are grown in wide range of climatic conditions, from sub-tropical to temperate regions. The major commercially cultivating provinces are Himachal Pradesh, Jammu and Kashmir, sub-mountain region of Uttar Pradesh, Punjab, Haryana, North-East hill region and other hilly areas in South India. Since peach requires the warmest climate of all the temperate fruits, its cultivation is generally restricted to mid- and lower-hills. However, in recent years, the development of low chilling cultivars belonging to Honey and Florda groups has made it possible to grow peach in the sub-tropical plains of Punjab, Haryana, Rajasthan, Western UP and Jammu and the cultivation is expanding rapidly as its production per unit area is among the highest in stone fruits.

Plum has immense adaptability to climatic conditions and is grown throughout temperate zone as well as sub-tropical regions where about 200 chilling hours are completed. Apricot requires warm temperate climates and is thus cultivated in mid hill regions and dry temperate zone. Although stone fruits are grown in the wider climatic zones yet the quality of fruits in the hills are far superior to those produced in the plains.

Stone fruit trees are attacked by a number of root, foliage, stem and fruit feeding insects from the very beginning of their growth till fruit maturity. Most of these pests are common to peach, plum and apricot and occur almost in all the regions. In general, peach attracts more pests in comparison to plum and apricot. Butani (1979) listed 80 species of insect-pests of peach, 60 of plum and 30 of apricot in the country but a few of these are of major importance.

## SAP THRIPTERS

Aphids and scales are the major sapsuckers of peach, plum and apricot in India. The other groups of minor importance are white flies, hoppers and bugs.

**Peach leafcurling aphid *Brachycaudus helichrysi* (Kalt.)**

*The peach leafcurling aphid* is the key pest of peach. Besides, the other aphid species recorded on peach are *Lachnus persicae* (Cholodkovsky) in Kullu (Pruthi and Batra, 1960) and Shimla (Sharma and Bhalla, 1964), *Myzus persicae* (Sulzer) in Kullu and Shimla (Pruthi and Batra, 1960). On plum, aphids recorded are *Aphis gossypii* Glover in Solan, *A. pomi* De Geer, in Kullu and Shimla (Bhalla and Verma, 1988), *L. persicae* in Kullu (Pruthi and Batra, 1960), *Rhopalosiphum maidis* (Fitch.) in Solan (Bhalla and Verma, 1988) while on apricot – *L. persicae* (Cholodkovsky) is recorded in Kullu (Pruthi and Batra, 1960) in Himachal Pradesh.

Peach leaf curling aphid is distributed from the eastern Himalayan states of Arunachal, Nagaland, Manipur and hilly areas of Assam to the Western Himalayan states of Uttar Pradesh (hilly region), Himachal Pradesh and Jammu & Kashmir upto an elevation of 7000 feet. The pest was first recorded from Tarujabba, a place in Pakistan, and later from Afghanistan (Ghulamullah, 1941). It is comparatively more serious on the low chilling cultivars grown in the plains of Punjab, Haryana, Rajasthan, western Uttar Pradesh and Jammu. It is a regular and serious pest of plum and peach in Israel, Japan, Australia, Egypt, Argentina, North pacific region of America and almost all stone fruit growing countries of Europe.

The aphid damages floral and vegetative buds of its primary hosts (Prunus sp.) by sucking sap due to which the unfolding leaves curl up tightly, remain smaller, get distorted and later turn pale and drop. Heavy infestation can curl almost every leaf on the tree, which causes reduction of fruit size as well as production of irregular shoot and twig growth. Numerous aphids inside the flowers suck the vital sap and render them sickly and dull. Consequently, poor fruit set occurs. There may be premature flower and fruit drop and subnormal development of left over fruits (Bhalla and Gupta, 1993).

*B. helichrysi* is extremely polyphagous pest infesting about 175 plant species belonging to 115 genera of 49 families (Ghosh and Verma, 1988). In Indian sub-continent it is serious pest on peach and nectarines and hence is popularly called as peach leaf-curling aphid. However, in Europe and American countries its main hosts are plum, prunes and damson and called as leaf curling plum aphid. The other hosts in order of preference are almond, plum, behmi and apricot. All hosts belonging to genus Prunus are primary hosts. The principal summer hosts reported are Golden rod, *Erigeron canadensis, Trifolium pretanse* and Cineraria sp. at higher altitude in Uttar Pradesh and Himachal Pradesh, and at lower altitude Ageratum *conyzoides*, *Chrysanthemum* spp. and sunflower (Lal and Siddiqi, 1952; Verma, 1984).

The biology of this aphid has been extensively studied by in India (Lal and Siddiqi, 1952; Batra, 1953; Sharma *et al*. 1968; Verma, 1984; Gupta and Thakur, 1993. The aphid remains on its primary hosts from autumn to beginning of summer (October to April-May) and on secondary hosts during summer (May-September). Depending on altitude vis-a-vis temperature, the autumn migrants gynoparae fly back on its primary hosts and settle on both sides of leaves (preferably on dorsal side) during October-November and deposit a progeny of oviparae. Low chilling cultivars, viz., Flordasun, Shan-e-Punjab, Sunhaven, Sunred Nectrin and Sunred Gold harbour comparatively more populations of gynoparae because of late shedding of leaves in the mid hills (Thakur *et al*., 1991). Newly laid oviparae nymphs are green but the colour of the adults depends on food i.e. nymphs feeding on leaves are green while those feeding on bark are brown. Arrival of the alate males synchronizes with the maturation of oviparous gynoparae and females are more numerous than males. The gynoparae posses a road pyriform abdomen with a brown genital plate anterior to the cauda while the males have an elongated and tapering abdomen with a black ring-like structure, the genital armature at its tip. These characters make the sexes easily distinguishable (Gupta and Thakur, 1993).

Apterous oviparae move on the twigs after attaining sexual maturity and settle on or behind the buds. These are fecundated by alate males on peach foliage or buds. The successful mating lasts for 2 to 5 minutes. The eggs are laid usually singly on the scales or any other rough place/crevice on one-year growth during November to January. The number of eggs per tree is usually low and sporadic in distribution. Freshly laid eggs (0.51 x 0.28 mm) are green but turn black after 2–3 days. The eggs hatch in 20 to 28 days but those laid during December hatch in 45–50 days. The aphid over winters in nymphal stage. The neonate fundatrices (stem mothers) make random movements to settle in bud axils. The development of fundatrix and laying of progeny is mainly dependant on state of the host. In mid hills, stem mothers commence larviposition on low chilling cultivars with the break of dormancy from mid January onwards whereas in late flowering cultivars, the process is delayed upto March. The brown coloured fundatrix lays about 50 young ones in its adult life span of about a fortnight. Young one matures in 10 days. The number of generations of fundatrigenae varies from 3 to 8. Migrants (alates) appear in the colonies during April-May depending on cultivar, altitude and crowding, and shift to secondary hosts.

Application of systemic insecticides like methyl demeton 0.025 per cent and dimethoate 0.03 per cent at pink bud stage was suggested by Sharma *et al.* (1968). They also recommended a post-bloom spray, 10 days after petal fall or a month after the pre-bloom spray, in mid-and lower-hills. Later, Gupta *et al.* (1988) did not favour the second spray because of low population intensity and intense biotic agent activity.

Since yellow leaves during October-November attracted alate gynoparae and males, early shedding of leaves or application of insecticides is recommended (Thakur *et al.*, 1991). Regular pruning of current year's growth during December to reduce egg population, pre-bloom application of either systemic or contact insecticides with high volume sprayer and avoiding post-bloom spray to encourage biotic agents like coccinellids, syrphids, anthocorid bugs, etc was recommended by Gupta *et al.* (1988). They also found that some peach cultivars like Stark Early Giant, Early White Giant and Flavour Crest escaped infestation and ranked them as moderately susceptible (10–30 per cent leaf whorl infestation) against this pest.

## Peach mealy aphid *Hyalopterus pruni* (Geoffroy)(=*arundinis* Fabricius)

The peach mealy aphid, *Hyalopterus pruni* (Geoffroy) has been reported from Punjab, Jammu & Kashmir, and Maharashtra on peach, apricot and plum. Nymphs and adults congregate on the ventral surface of the leaves and suck sap. Aphids also secret honeydew which falls on the surrounding leaves and appear as if sprayed with oil. In the case of severe infestation, the affected leaves get twisted, the fruit size is reduced considerably and the fruits drop prematurely. This species has also been reported as a pest of plum, peach, cherry and apricot in Iraq, Pakistan, Italy, and USA and on almond in Tunisia (Butani, 1979, Jerraya, 1997).

The aphids breed parthenogenetically throughout the year except during winter months when sexual forms appear. Eggs are laid on peach twigs around December-January and the aphid overwinters in egg stage. The eggs hatch in the following spring giving rise to apterous viviparous females. As in the case of peach leaf-curling aphid, alates appear during May which fly to the alternate hosts, i.e. *Arundo donax* and *Phragmites karka* (Pruthi and Batra, 1960). During autumn, the alates return to the peach trees and lay the sexuales. Similar life cycle was reported in Iraq where the alternate host is *Pharagmites communis* (Mohamad and AlMallah, 1990).

In Kashmir, the pest is managed by its natural enemies like *Coccinella undecimpunctata*, *Episyrphus balteatus* and *Sphaerophora indiana* during May-June (Bhagat and Masoodi, 1988). Chemical control. suggested for peach leaf curling aphid are also effective against this species.

### Peach black aphid *Pterochlorus persicae* (Cholodkovsky)

The peach black aphid *Pterochlorus persicae* (Cholodkovsky), also called peach stem aphid, is occasionaly a serious pest of peach, apricot, cherry, and plum. It is widely distributed in India, Pakistan, Afghanistan and the Mediterranean region. In Italy the most common host is peach, followed by almond and other Prunus spp. but occasionaly found on orange, apple, pear and quince (Butani, 1979; Ciampolini *et al.*, 1997). The large sized, dark brown aphid, resembling in colour to the bark of the each tree, congregates on stem, branches and 1–2 year growth and sucks sap from the bark. Continuous draining off of cell sap by large number of aphids devitalizes the tree, adversely affecting fruit set and fruit size. These aphids also secrete copious quantity of honeydew, which attracts ants, honeybees, wasps and a number of flies. Sooty mould developing on the honeydew turns the surrounding ground and the vegetation black. In North-Western Frontier Province, the pest is found throughout the year but in Punjab the aphid appears in the beginning of February, gradually the population increases to its maximum during May-June and continues to be high until the first monsoon shower in July (Bindra and Bakhetia, 1968). In Egypt the aphid has 18 generations in a year with two peaks of alate oviparous females during April and October (Darwish *et al.*, 1989).

The aphid can be effectively controlled by spraying fenitrothion 0.05 per cent, chlorpyriphos 0.03 per cent, oxydemeton methyl 0.025 per cent or dimethoate 0.03 per cent. Natural enemies exert low or sporadic activity against this pest.

### Peach green aphid *Myzus persicae* (Sulzer)

*Myzus persicae* (Sulzer) is a polyphagous pest infesting large number of crops and wild plants. It is distributed throughout the world causing more harm as a vector of various virus diseases. In India it also infests peach, plum, almond, apple, apricot, cherry, citrus and pear. In western countries, peach is the main host of this species but in India peach leaf curling aphid *B. helichrysi* is the dominating species on peach which does not allow any other aphid species to invade the tree. However, its sexual life cycle on peach has been studied by Verma and Chauhan (1993) in the plains of Western Uttar Pradesh. However, this aphid is not a serious pest of peach, plum and apricot inIndia. Chemical control measures suggested for peach leaf-curling aphid are also effective against this species.

### Plum scale *Eulecanium tiliae* (L.) and *E. coryli*

The unarmoured scale *Eulecanium tiliae* (L.) infests plum and apple in Himachal Pradesh (Mishra and Bhalla, 1974; Sharma and Dogra, 1982), *Rosa damascena* in Turkey, horse chestnut, apple, cherry, hawthorn and many shrubs in British Columbia, and quince, elm, apricot, pear, walnut and rose in the former Soviet Union. The other species *E. coryli* (L.), infests plum, quince, apricot, almond and cherry in Kashmir (Zaka-ur-Rab, 1982), plum and peach in Uttar Pradesh (Gupta and Rai, 1971, apple in Quebec, American elm in Canada, grapevine, shrubs, ornamental and wild trees in Niagara Peninsula of Ontario. In India these are sporadic and serious pests of stone fruits especially plum in hilly areas of Uttar Pradesh, Himachal Pradesh and Kashmir. Nymphs and adults suck sap from the leaves and one year old twigs and in case of heavy infestation the leaves and fruits appear dull on the tree.

The life cycle of *E. tiliae* was studied by Sharma and Dogra (1982) in Himachal Pradesh and that of *E. coryli* by Gupta and Rai (1971) in Uttar Pradesh. Whereas, *E. tiliae* was reported to reproduce parthenogenetically, *E.coryli* was found to reproduce sexually. Alate males appear during March-April, fertilize the female within 20–48 hours and die. The eggs of *E. coryli* are bigger (0.413 × 248 mm) having incubation period of 50–95 days whereas *E.tiliae* eggs measure 0.363 x 0.203 mm and have shorter incubation period of 30 days. Both the species have high fecundity (200–1500 eggs/female). The eggs remain protected underneath the dome shaped body of the female measuring 3.3 mm in length, 2.5 mm in width and 3.1 mm in height. Adult females secrete honeydew during March-April which attracts large number of flies, wasps, honey bees, ants and other insects. The eggs are laid throughout April and the female die during the process. The eggs hatch during May-July and the crawlers leave the dome shaped body of the mother through the anal slit. They move on to the adjoining leaves and settle on both surfaces preferably on the underside (ventral), along mid rib and other veins. The nymphs enter into second instar in August and migrate to 1 to 2 year old twigs before leaf fall (September-October). The scales overwinter in second instar and adult females are formed in March. There is only one generation in a year.

The sporadic nature of the pest is due to poor dispersal, single generation in a year and high parasitization. Mishra and Bhalla (1974) recorded 73 per cent parasitization by the aphelinid, *Coccophagus* sp. nr. *ishii* in Himachal Pradesh. Sharma and Dogra (1986) recorded *Coccophagus* sp. and *Blastothrix sericea* (Encyrtidae) parasitizing second instar nymphs and adult females to the extent of 7.2 and 40.7 per cent respectively. The fungus *Rhinocladiella* sp. Caused 13.6 per cent mortality of the second instars resulting in total mortality of 89.3 per cent on leaves and 65.8 per cent on twigs.

The scales can be effectively managed with dormant spray oils applied during December-January. However, post bloom spray during May with monocrotophos 0.04 per cent, oxydemeton methyl 0.025 per cent, quinalphos 0.04 per cent or fenitrothion 0.05 per cent is also effective (Sharma and Dogra, 1986). Gupta and Rai (1971) found dimethoate 0.03 per cent, phosphamidon 0.03 per cent and diazinon 0.03 per cent highly effective when sprayed during June. However, the maturity period of most of the plum cultivars coincides with spray during June, it is advisable to postpone the insecticide application to post harvest stage.

### White peach scale, *Pseudaulacaspis pentagona* (Targioni)

The white peach scale, *Pseudaulacaspis pentagona* (Targioni) is a serious pest of peach trees in India, Ceylon, Japan, Australia, Trinidad, Jamaica, England and southern Europe, China and Turkey, (Pruthi and Batra, 1960, Erkilic and Uygun, 1997). In India it has been reported from Assam, Bengal and Uttar Pradesh as a pest of peach, almond, apple, cherry and plum (Singh,

1964) while in other countries it is known to infest mulberry, papaya, fig, walnut and kiwi (Butani, 1974; Pedata et al., 1995). Nymphs and adult females suck the cell sap from stem and tender shoots and occasionally from leaf veins. The devitalized trees have comparatively less growth and poor quality fruits.

The female lays 25–34 eggs, which hatch in 8–12 days. The males have shorter nymphal period 8–10 days as compared with females, which take 3 weeks or more to mature. The life cycle in the case of females is thus completed in 8–9 weeks during summer and 21–22 weeks during winter. In India, the insect completes four generations in a year but in Italy three complete and one partial generations are known (Gardonna and Viggiani, 1988).

Pesticides recommended for the control of plum scale are also effective against *P. pentagona*. Mineral oils prove more effective against the pest and safer to the natural enemy complex (Gardonna and Viggiani, 1988).

### San Jose scale *Quadraspidiotus perniciosus* (Comstock)

The San Jose scale, *Quadraspidiotus perniciosus* (Comstock), a well-known polyphagous pest, also infests peach, plum and apricot throughout the temperate and sub-tropical regions. The young plants of stone fruits are more vulnerable to the attack. The pest can be controlled with dormant spray oils. and other pesticides. For details, refer to pests of apple.

## FRUGIVOROUS PESTS

The larvae of these pests feed or bore into the fruits, cause direct damage to the produce and hence are pests of economic significance. Fruit moth, fruit flies and codling moth are the major pests whereas fruit eating beetles and thrips are of minor importance.

### Plum fruit moth *Cydia* sp.

Tortricid moths (*Cydia* spp.) damage fruits of plum and apricot fruits in Himachal Pradesh. On apricot the pest has been identified as *Cydia* sp. (? *funebrana*) (Tre) (Sharma and Gupta, 1986). It causes 20–100 per cent fruit infestation in apricot at places situated between 1500 to 2100 m altitude in high rainfall areas and upto 3000 m amsl. in dry-temperate zone. In plum, 35–65 per cent fruits are infested in some localities. The larvae feed on the flesh of the fruit by making galleries of various shapes and sizes. The attack is conspicuous around the stone in apricot and is characterized by the presence of brownish excretal pallets of the caterpillar. A number of larvae, depending upon stage of the fruit development may be present in each fruit.

Moths are seen on wings at dusk during April end and May. They usually lay eggs singly on the fruit surface. The larvae enter the fruit on hatching, become fully fed in when fully fed in 21–24 days and leave the fruit for spinning the cocoon on or near the tree. The severely infested fruits may drop. There are two generations in a year and the second generation larvae are more destructive. These become full-fed by August end, enter into diapause and larvae pupate in April.

Clean cultivation, collection and destruction of fallen fruits and debris harbouring cocooned larvae, reduce the pest population in the ensuing season. Orchards sprayed with dormant spray oil (2 per cent) in mixture with fenitrothion 0.05 per cent do not suffer damage (Bhalla and Gupta, 1993). Orchards with high infestation history should be sprayed with fenitrothion 0.05 per cent, quinalphos 0.05 per cent, malathion 0.1 per cent or carbaryl 0.1 per cent in the first week of May followed by second spray in first fortnight of June. However, near ripening fruits should not be sprayed. Synthetic sex attractant "Denatail" tested against *C. Funebrana* in the former USSR was reported to successfully disorientate males (Sazonove and Sunducova, 1987).

**Fruit flies *Bactrocera* sp.**

The fruit flies are cosmopolitan pests that infest a large variety of fruits. Their attack reduces the yield and affects quality. Several species belonging to this genus and sub- family Dacinae, viz., *Bactrocera (=Dacus) dorsalis* (Hendel), *B. zonatus* (Saunders), *B. Cucurbitae* Coq., *B. correcls* (Bezzi), *B. Duplicatus* (Bezzi), *B. maculipennis* Dolesechall) and *B. ciliatus* (Loew) have been recorded on various temperate fruits of which *B. dorsalis* and *B. zonatus* are minor to moderate pests of apple, peach and pear in some pockets (Bhalla and Gupta, 1993). The former is widely distributed in the oriental region and infests all fleshy fruits. The latter, however, infests peach, plum and apricot in mid-hill regions. The activity of *B. dorsalis* was reported by Narayanan and Batra (1960) on guava during March, on loquat, apricot and plum during April-May, on peach and fig during June, on mango during June to August. After August the activity shifts again to guava, apple, citrus and ripe banana. In the mid-hills of Himachal Pradesh, the pest is active from April to November (Gupta *et al.* 1990) in orchards of apricot, plum and peach coinciding with the ripening of each fruit. However, according to Mann (1996), the activity of *B. dorsalis* in the plains of Punjab remains throughout the year, being low during winter. Temperature below 20 oC is considered to be detrimental to the fly (Liu, 1983).

Adult females lay about 200 eggs in small batches of 2–10 in its life span of 1 to 4 months inside the ripening fruits by making punctures with their ovipositors. On hatching the maggots feed on the pulp and fruit becomes soft, ferments and drops. The attack is more serious on late maturing varieties of peach. The eggs are elongate, shiny white, slightly curved and narrow at both ends and often hatch within 2 days. Yellowish maggots feed on the fruit pulp for 1 to 4 weeks. Full-grown third instar maggots measuring 8–10 mm, come out of the infested fruits and jump to suitable places for pupation. Pupation occurs in the puparium and the pupal period may vary from a week in summer to 5 months or more in winter. In the temperate region, the pest hibernates as adult from October to March/April (Singh, 1964).

Monitoring of activity of fruit flies can be carried out successfully with the help of methyl euginol baiting traps. Since eggs and maggots remain inside the fruits, they cannot be made targets for control. However, populations in the subsequent generations can be reduced by adopting sanitation methods such as collection and deep burying of infested fruits and prompt disposal of unmarketable fruits. Pupae in the soil can be destroyed by ploughing and digging the tree basins. Adult flies can be managed by spraying the crop with fenthion 0.05 per cent, fenitrothion 0.05 per cent (Sharma *et al.*, 1973; Shukla *et al.*, 1984) or by adopting bait spray containing malathion (0.1 per cent), protein hydrolysate (0.3 per cent) and sugar (0.3 per cent) at resting sites of the fruit flies (Narayanan and Batra, 1960) or attracting the flies to poison trap baits.

**Codling moth *Cydia pomonella* (L.)**

The codling moth, *Cydia pomonella* (L.) is a pest of apple confined to Leh and Kargil districts of Ladakh in Jammu & Kashmir (Pawar *et al.*, 1981). It also attacks apricot, walnut, pear and quince. It is suspected to have come from North-West Frontier provinces where it was recorded in mid-thirties (Bhalla-and Gupta, 1993). Strict inter and intra-state quarantine measures should be adopted to stop its further spread in adjoining areas. For details refer to chapter on apple.

## XYLOPHAGOUS PESTS

Several species of insects bore into the bark and wood of stone fruit trees. These wood borers are larval stages of beetles or moths. Burrowing may weaken the tree or may kill it depending upon the severety of infestation. The most important xylophagous borers that attack peach, plum and apricot trees are the flat-headed tree borer, bark eating caterpillar and shot hole borer. Stem borers, root borers, twig borers and shoot borers are other minor pests of stone fruits.

**Flat-headed peach tree borer *Sphenoptera lafertei* Thomson and *S. dadkhani* Obenberger**

The buprestid borers are among the worst enemies of fruit and other trees in temperate and sub-tropical regions of the world. *Sphenoptera lafertei* Thomson is a secondary invader as it has been recorded reproducing only in weakened stone fruit trees. The pest is widely distributed in northwest India and in south west Afghanistan (Ahmed, 1940), Uttar Pradesh, around Delhi, Haryana and Punjab. Obenberger (1955) identified an other species (*S.dadkhani* sp.n.) from specimens collected in Pakistan, and later reported from Punjab (Chaudhary *et al.*, 1993).

Adults feed on the foliage but do not cause serious damage. Grubs tunnel under the bark, in the inner phloem and the shallow, broad and irregular burrows thus formed are packed tightly behind them with fine frass and excreta. As a consequence of continuous feeding, sap wood (phloem and cambium) becomes powdery and the bark over it dries and splits. Adults prefer older, weak, injured, under stress, sun burnt parts of the stem and branches for egg laying and thus cause serious damage to such trees. As a result of attack, fruit setting is adversely affected. Heavily infested branches become weak, are unable to bear the weight of the developing fruits and may even break on slight disturbance during harvesting. The dead bark and exposed hardwood of the stem and branches also become vulnerable to termite attack. Among the stone fruits, apricot trees are the least affect (Sharma *et al.*, 1990). At Hisar, Lakra *et al.* (1980) reported *S. lafertei* damage on peach trees to the extent of 45 per cent.

Beeson (1919) reported *S. lafertei* as a pest of deodar tree *Cedrus deodara*. Presently, the pest shows great preference for cherry, peach plum and almond trees (Sharma *et al.*, 1990). Besides, the grubs also bore into stems and branches of loquat, pear and apple trees.

The life history of the beetle was extensively studied on peach in the plains of Haryana and Punjab by Batra and Renjhen (1950), Lakra *et al.*, (1980) and Chaudhary *et al.* (1996). The sturdy, flattened, elongate with shining metallic copper luster body measuring 7–9 mm. The adults start emerging in early March till May by making oval exit holes. The adult's longevity is 24–33 days. The mating process lasts for 2 hours. The females lay whitish eggs which later turn dark coloured in cracks and crevices on the sun burnt parts of stem and branches. The grubs bore through the bark and start tunneling under the bark. The gum oozes out of the entry point. The incubation period is of 20 days during spring and the grub stage lasts for 2 months in summer and 6 months in winter. The pupal stage occupies 10 days. The shortest duration of development from egg to adult emergence is about 3 months. The winter is passed in the grub stage. The legless pale yellowish grub has the largest prothoracic segment and well developed mandibles and a club shaped body. While feeding under the bark grubs usually turn half of the abdomen on one side. The full-grown larva bores sub-cylindrical chamber 4–6 mm long, 2–4 mm broad and 10 mm deep into the hard wood where it pupates after 1–2 days of prepupation. The peak period of activity is from April to October. The three overlapping generations are completed between April-June, July to September and October to March respectively. According to Chaudhary *et al.* (1996), three distinct generations are completed in 87, 83 and 210 to 208 days, respectively.

For the management of these borers, maintain tree vigour by following sound horticultural practices and planting stone fruit orchards at proper sites. Trees, planted on south facing slopes in hilly areas remain under stress and hence prone to borer attack. Healthy trees are attacked to a lesser extent by buprestid and excessive gum secretion from healthy trees kills the larva. Proper training and pruning of the plants provide good canopy over the stem and save it from direct exposure to sun. It also helps to keep the plant straight. Trees tilting towards north get more sunburn and heavy attack of borer.

Providing cover over the south facing parts of the stem and branches by wrapping dry grass, craft paper or gunny bags during winter to protect these from direct exposure to sun. The cover, provided during active season, acts as barriers for oviposition and saves the tree from borer attack. Burning of dead or heavily infested branches destroys the borer population to a greater extent (Chaudhary *et al.*, 1993).

The beetle is vulnerable to insecticidal sprays only in the egg, first instar grub and adult stage. Sharma *et al.* (1963) suggested 2–4 swabbings with insecticides, the treatments synchronizing with laying of brood. Spraying with lindane 500 ml a.i./ha during April, June and September (Chaudhary *et al.*, 1993), methyl parathion 0.05% (Sharma and Chander, 1992) or chlorpyriphos 0.05% synchronizing with egg laying provides complete control of the pest.

**Bark eating caterpillar, *Indarbela quadrinotata* Walker**

The bark-eating caterpillar is a polyphagous pest widely distributed in India infesting fruit, forest and ornamental trees. Verma and Khurana (1974) found the species feeding on aonla, pear, *jamun,* pomegranate, *ber*, guava, plum, orange, mango, litchi and peach. In Himachal Pradesh, the pest has been recorded damaging mango, loquat, pomegranate, citrus, Grewia optiva Drummond, *Bauhinia variegata* Linn. and plum (Bhalla and Pawar, 1977; Verma, 1985). Peach and pear trees are also attacked at lower elevations. Neglected and old trees suffer the more than vigorously growing trees. The larvae feed on the bark and make galleries over it with loose webbings containing chewed material and excreta pallets. As a consequence of injury, movement of the sap is checked and the vigour and fruiting capacity of the tree are adversely affected.

Eggs are laid in clusters in crevices of the bark during June. After 2–3 days of hatching the caterpillars bore into the stem or branches, usually at forking or dried cut wounds and tunnel downward. The larva becomes full-fed by December but continue feeding till March-April. Pupation takes place inside the tunnel in March-April. There is only one generation in a year. The incubation, larval and pupal periods vary from 8–10 days, 10 to 11 months and 12 to 25 days, respectively (Butani, 1979).

The typical behaviour of the caterpillar like nocturnal habit, shallow hiding hole, feeding under protection of web and single generation in a year make the pest less vulnerable to chemical control. However, clean orchard practices, removal of webs and killing the caterpillars by inserting a wire in the hole, plugging the holes with cotton soaked in fumigant like carbon bisulphide, chloroform, benzene, petrol, etc. have been recommended against this pest by various workers. Srivastava (1972) found kerosene as an effective, easy to handle, cheap and commonly available fumigant. A practice of injecting insecticide with a syringe into the feeding hole and plugging is also in vogue. The most effective insecticides include dichlorvos 0.025%, fenitrothion 0.05% and methyl parathion 0.05%. Placement of phorate granules (Thimet 10 G) @ 1g per hole in plum trees is also effective. However, spot treatments is feasible only in small trees like peach but not in grown up trees.

**Peach twig borer *Anarsia lineatella* Zeller**

The peach twig borer, *Anarsia lineatella* Zeller, is a serious pest of peach, nectarine, almond and apricot in North Africa, North America, Europe, Iraq, South China, and the Asia Minor. In India, Fletcher (1932) reported the caterpillars damaging twigs of peach, plum, apricot and almond in Kashmir. The pest also occurs in Pakistan. Just after hatching, the caterpillars bore into the twigs and later also damage the young fruits. Janjua (1942) studied the biology of the pest in Baluchistan. Small greyish moths (1.2–1.4 cm wing span) lay about 170 eggs singly or in small groups of 2–3, on shoots, twigs, fruits or leaves. Eggs are oval, 0.4 to 0.5 mm long and white in colour. The egg, larval, prepupal and pupal period ranges from 6–11, 18–35, 3–9 and 6–14 days respectively and the life cycle is completed in 40–70 days. There are three generations in a year and overwintering takes place in larval stage. The third instar larva spuns a silken web in the bark of the tree and overwinters under it. The overwintering larva resumes feeding by the end of March and bores into the new growth. Small galleries are made on twigs or terminal buds and the pre-pupation takes place on the bark in cocoons.

The gelechiid is successfully managed in peach, nectarine and almond orchards by monitoring with sex pheromones and sprayed with microbial pesticides based on *Bacillus thuringiensis* Berliner in California. Mating disruption with sex pheromones combined with post-bloom sprays with *B. thuringiensis* are highly effective (Quarles, 1997). The pest can also be controlled by spraying carbaryl 0.1 per cent or endosulfan/fenitrothion 0.05 per cent.

**Shot hole borers *Scolytoplatypus pusullis* Eggers and *Scolytus nitidus* Schadl.**

Scolytid beetles, commonly called bark beetles or shot hole borers, are secondary invaders to the weakened deciduous and conifer trees in temperate region. These beetles prefer trees drying from neglect, lack of moisture, diseases or injuries caused by animals or other insects. *Scolytoplatypus pusullis* Eggers and *Scolytus nitidus* Schadl. have been reported to infest peach, plum and apricot in Himachal Pradesh (Bhalla and Pawar, 1977). Small dark brown and black cylindrical beetles (2.5 mm long) about half-broad to their body length, tunnel into the sapwood and hardwood of the trunk and branches of the tree and lay eggs inside. Grubs make small parallel galleries leading out from a shorter central gallery and pupate at the end of it. Adults come out by eating small holes in the bark; hence the common name: shot hole borer. Numerous galleries obstruct the normal flow of the sap and holes perforate the infested branches. Heavy infestation results in wilting and yellowing of the foliage followed by death of the tree.

To control shot hole borer, destroy dying or injured trees in which it breeds. Remove infested branches and broken or weak ones that might become infested. Keep the trees vigorous by cultivating, fertilizing and watering them and by controlling insects and diseases that reduce their vitality. Chemical control methods suggested for flat-headed borer are also effective against this pest.

### Stem borers

The other species of the stem borer, viz., apple stem borer, *Apriona cinerea* Chevrolat, stem borer Aeolesthes *holosericea* Fab. and Quetta borer, *Aeolesthes sarta* Solsky, are minor pests of peach, plum and apricot and can be controlled by the methods already suggested. For details refer to chapter on apple pests.

### Apple root borer *Dorysthenes hugelii* Redtenbacher

Apple root borer a major pest of apple is occasionally found damaging peach, plum and apricot roots in the hilly areas of North India. The grubs feed on the main root and the trees become shaky and weak. For details refer to chapter on apple pests.

### Shoot borer *Zeuzera* spp.

Larvae of the leopard moth, *Zeuzera* spp., tunnel the one-year-growth of plum occasionally but the attack is generally low. For details refer to chapter on apple pests.

### Termites *Odontotermes* sp. and *Microtermes mycophagus* Desneux

Termites or white ants are polyphagous pests and generally attack dead or dying fruit trees. Its attack on peach, plum and apricot trees occur in subtropical, mid hills, and valley areas upto an altitude of 1500 metre amsl. The damage is generally restricted to the trees having sun burnt dead bark/wood on trunk and branches infested with flat-headed borer. The species recorded damaging apricot trees in the mid hills of Himachal Pradesh are *Heterotermes indicola* Wasmann and *Odontotermes* sp. whereas in the plains *Microtermes mycophagus* Desneux is a common pest of peach trees (Butani, 1974).

In subtropical areas the pest remains active throughout the year but the major damage is caused during August to December. In mid hills and valley areas also the activity is restricted upto November. In such places winged sexual forms appear during March-April and after mating lose their wings and re-enter the soil to start new colonies. This social insect has various forms in the colony but the workers are more destructive and cause heavy losses to the fruit trees. The queen is comparatively bigger and lays one egg in every 2–3 seconds during her entire life span of 7 to 10 years (Butani, 1979).

The sun burnt, borer attacked areas on stem and branches should be cleaned and treated with kerosene oil/cresote oil. If mud galleries appear on tree trunk or branches remove the same and treat the area with kerosene oil/Cresote oil or chlorpyriphos 0.04 per cent. The tree basin should also be drenched with insecticide.

## DEFOLIATOR PESTS

Various species of beetles and caterpillars defoliate the stone fruit trees. Majority of these are polyphagous and sporadic in nature. Veeresh (1988) have reviewed the work done on white grubs as pests of fruits and forest trees in India.

### Defoliating and fruit eating beetles

The terminal growth of these stone trees are damaged by several species of beetles belonging to Scarabaedae, Cetoniidae, Curculionidae and Chrysomelidae (Butani, 1979; Bhalla and Gupta, 1993). As many as 70 species of these beetles have been recorded in India. Of these, about 57 have been reported from U.P. and 36 from H.P. (Gupta *et al.*, 1977; Bhalla and Pawar, 1977). Adults of these polyphagous species are mostly defoliators while a few damage fruits and flowers.

Cockchafers or May June beetles (Scarabaeidae) inflict more injury as adults and grubs than other beetles. Their larvae feed on the fine roots. Adults emergence occurs after the first rain in summer from March to September depending upon species, habitat and climatic conditions. However, maximum emergence takes place during monsoon (July-August) at dusk. On emergence, the beetles tend to congregate on the preferred host, immediately mate for 5–15 minutes, continue feeding during night and at dawn and enter into the soil for hiding and oviposition. Most of the species complete their life cycle in a year. Pearly oblong white eggs are laid singly in the soil at 3–8 cm depth which hatches within 3 weeks. The first instar feeds on the organic matter and the subsequent 2 instars feed on live roots. The larval stage lasts for 80–300 days. Pupation takes place deep into the soil in earthen cells and pupal period lasts for a few weeks to several months (Veeresh, 1988).

In Kullu valley, *Maladera insanabilis* (Bronske) has been recorded as the predominant species on plum, peach and apricot. In addition, Anomala *polita* Blanch., *Brhamina crinicollis* Burmeister, *B. coriacea Hope*, *Holotrichia longipennis* Blanch. and *Adoretus* sp. were also found defoliating the terminal growth of these fruit trees.

Flower beetles (Cetoniidae) inflict injury to flowers and ripening stone fruits. They are diurnal and have brightly coloured flattened bodies. Important species belong to *Clinteria, Mimela, Oxycetonia* and *Protaetia*. Their larvae feed on decaying wood or live roots. Sporadic blue chrysomelids appear on peach during June-July, which defoliate the trees besides damaging the fruits.

The density, diversity and period of activity of Scarabaeid beetles can be monitored effectively by installing light traps in fruit orchards but it is not an effective measure for their control (Tripathi and Gupta, 1985). The adults are difficult to control because of their mobility, yet collection and destruction of beetles after dusk by shaking non-bearing fruit trees and other host trees can give effective control, if followed as a campaign. Spraying with carbaryl 0.2 per cent, methyl parathion 0.05 per cent or monocrotophos 0.05 per cent has been found effective (Bhalla and Gupta, 1993).

### Defoliating caterpillars

Larvae of over 80 species feed on the foliage of temperate fruits (Chander and Dogra, 1983) but only a few are regular pests. In stone fruits, the Indian gypsy moth, *Lymantria obfuscata* Walker., tent hairy caterpillar, *Malacosoma indica* Walker and apple leaf-roller, *Archips pomivora* Meyrick are important sporadic pests.

*Lymantria obfuscata* Walker, the Indian gypsy moth was earlier recorded as a pest of deciduous forest trees (Oaks, elders, willows, poplars, phalsa, acasia, etc.) and some ornamentals in North-West Himalayas but now it has been observed infesting apple, plum, apricot and some other fruits trees. The pest causes more damage to apple than stone fruits. It is a univoltine species, the caterpillars of which appear in March-April and remain active upto June-July. The adults appear in June-July. Male moths are active fliers while the females have weak and small wings. Males and females survive for 4–7 and 11–31 days, respectively. The female lays 150–400 spherical, light greyish brown eggs in clusters on the bark of trees, cracks and crevices in walls, stones, logs, etc. The eggs are covered by a layer of yellowish brown anal-tuft of the female. The diapausing eggs hatch in March and larvae become full fed in 65–100 days. Pupation period is 9–21 days.

Biotic agents play a crucial role in the suppression of *L. obfuscate* populations. Rishi and Shah (1985) reported 47 parasitoids and predators associated with this species in Kashmir. They also reported 30–40 per cent larval mortality due to nuclear polyhedrosis virus (NPV). Masoodi *et al.* (1987) recommended the use of endosulfan (0.05%) and fenvalerate (0.03%) for its control but keeping in view the high risk to biotic agents, it is advisable to collect egg masses and larvae trapped in gunny burlap tied on the trunk near the ground level and destroy them.

## REFERENCES

Ahmed, T. 1940. A survey of insect fauna of Afghanistan. I. General feature of the country and its fauna. *Indian Journal of Entomology,* 2(2): 159–176.

Batra, H.N. 1953. Aphids infesting peach and their control. *Indian Journal of Entomology,* **15**: 45–51.

Batra, H.N. and Renjhen P.L. 1950. Preliminary observations on the biology of the buprestid (*Sphenoptera lafertei* Thomson) of certain fruit trees in Northwest India. *Indian Journal of Entomology,* **12**(2): 159–165.

Beeson, C. F. C. 1919. *The Ecology and Control of the Forest Insects of India and the Neighbouring Country.* FRI Dehradun.

Bhagat, K. C. and Masoodi, M. A. 1988. Natural enemies of mealy plum aphid, *Hyalopterus arundinis* Fab. (Aphididae: Homoptera) in Kashmir. *Journal of Advanced Zoology,* **9**(2): 145–147.

Bhalla, O. P. and Gupta, P. R. 1993. Insect pests of temperate fruits. *In: Advances in Horticulture.* K. L. Chadha and O. P. Pareek (eds). Vol. 3. *Fruit crops* Part 3. Malhotra Publishing House, New Delhi. pp. 1567–1589.

Bhalla, O. P. and Pawar, A. D. 1977. *A Survey Study of Insect and non-Insect Pests of Economic Importance in Himachal Pradesh.* Bulletin, Department of Entomology and Zoology, College of Agriculture, Solan. pp. 80.

Bhalla, O. P. and Verma, A. K. 1988. Annotated list of host plants and associated aphids in Himachal Pradesh. *In: Trends in Tree Science.* P. K. Khosla and R. N. Sehgal (eds). Indian Society of Tree Scientists, Solan. pp. 333–350.

Bindra, O. S. and Bakhetia, D. R. C. 1968. The chemical control of the peach stem aphid, *Pterochlorus parsicae* (Cholodkovsky). *PAU Journal of Research,* **8**(1): 48–51.

Butani, D. K. 1974. Insect pests of fruit crops and their control. 9. Peach. *Pesticides,* **8**(9): 25–31.

Butani, D. K. 1979. *Insects and Fruits.* Periodical Expert Book Agency, New Delhi pp. 415.

Chander, R. and Dogra, G. S. 1983. Lepidopterous insects infesting apple in Himachal Pradesh. *In: Proceedings of Workshop on High Altitude Entomology and Wild Life Ecology.* Zoological Survey of India, Calcutta, West Bengal, India.

Ciampolini, M.; Farnesi, I. and Capella, A. 1997. The spread of the woolly aphid of *Prunus* (*Pterochlosides persicae*). *Informator Agrario,* **53**(8): 105–107.

Chaudhary, O. P.; Rataul, H. S. and Sandhu, G. S. 1993. Management of the flat headed borer, *Sphenoptera dadkhani* Obenberger on peach in India. *International Journal of Pest Management*, **39**(1): 75–79.

Chaudhary, O. P.; Rataul, H. S. and Sandhu, G. S. 1996. Biology of flat-headed borer, *Sphenoptera dadkhani* Obenberger (Buprestidae: Coleoptera) on peach. *Journal of Insect Science*, **9**(1): 48–51.

Darwish, E. T. E.; Attia, M. B. and Kolaib, M. O. 1989. Biology and seasonal activity of giant brown bark aphid *Pterochlorus persicae* (Cholodk.) on peach trees in Egypt. *Journal of Applied Entomology*, **107**(5): 530–533.

Erkilic, L. B. and Uygun, N. 1997. Studies on the effects of some pesticides on white peach scale, *Pseudaulacaspis pentagona* (Targ.) (Homoptera: Diaspidae) and its side effects on two common scale insect predators. *Crop Protection,* **16**(1): 69–73.

Flecher, B. T. 1932. Life histories of Indian Microlepidoptera (II series) Alucitidae (Pterophoridae), Torticina and Gelechidae, *Science Monograph* 2. Imperial Council of Agricultural Research, New Delhi.

Gardonna, A. P. and Viggiani, G. 1988. Observations on the white peach scale (*Pseudaulacaspis pentagona* (Targ. Tozz.). *Annali della Facolta di Scienze Agrarie della universita degli studi di Napoli,* **22**: 1–10.

Ghosh, L.K. and Verma, KD. 1988. A new host record of *Brachycaudus helichrysi* (Kaltenbach) in India (Homoptera: Aphididae). *Journal of Aphidology,* **2**: 66–68.

Ghullamullah. 1941. Aphidae and some other Rhynchota from Afganistan. *Indian Journal of Entomology*, **3**: 225–313.

Gupta, B. P.; Joshi, N. K. and Joshi, L. D. 1977. Some important defoliating and fruit-eating beetles of temperate fruit trees-a survey and control. *Progressive Horticulture*, **8**(4): 53–60.

Gupta, B. P. and Rai, K. M. 1971. Biology and control of *Eulecanium (Lecanium) Coryli* (Linn.) (Holeoptera: Coccidae). *Progressive Horticulture*, **2**(4): 21–30.

Gupta, D.; Verma, A. K. and Bhalla, O. P. 1990. Population studies on fruit flies infesting fruit crops. *Indian Journal of Agricultural Sciences,* **60**(7): 471–474.

Gupta, P. R. and Thakur, J. R. 1993. Sexual generation and overwintering of the peach leaf-curling aphid, *Brachycaudus helichrysi* (Kalt.) in Himachal Pradesh, India. *Annals of Applied Biology*, **122: 215–221.**

Gupta, P. R.; Thakur, J.R. and Dogra, G.S. 1988. Management of peach leaf curl aphid, *Brachycaudus helichrysi* (Kalt.) infesting stone fruits in Himachal Pradesh. *Journal of Tree Sciences,* 7: 26–30.

Janjua, N. A. 1942. On the biology of *Anarsia lineatella* Zeller in Baluchistan. *Indian Journal Entomology* **4**: 137–44.

Jerraya, A. 1997. On the population dynamics of *Hyalopterus pruni* Geoffroy (Homoptera: Aphididae) in Tunisia. *Journal of Applied Entomology*, **121**(7): 373–382.

Lakra, R. K.; Gupta, D. S. and Bhanot, J. P. 1980. Nature and extent of damage and varietal susceptibility to peach borer, *Sphenoptera lafertei* Thomson. *Indian Journal of Entomology*, **42**(2): 233–239.

Lal, K. B. and Siddiqi, Z. A. 1952. Biology of the peach leaf-curling aphid in the Kumaun Hills. *Indian Journal of Entomology*, **14**: 191–196.

Liu, Y. C. 1983. Population studies on the oriental fruit fly, *Dacus dorsalis* Hendel in Central Taiwan. *Fruit flies of Economic Importance.* R. Fide Cavalloro (ed). CEC/IOBC Symposium Athens 1982.Rotterdam Balkema). pp. 62–65

Mann, G. S. 1996. Seasonal incidence and build up of *Bactrocera dorsalis* Hendel on mango in Punjab. *Journal of Insect Science*, **9**(2): 129–132.

Masoodi, M. A.; Kaul, V. K., Bhagat, K. C. and Singh, D.1987. Control of *Lymantria obfuscata* Walker on walnut in Kashmir. *Indian Journal of Plant Protection,* **15**: 15–17.

Mishra, R. C. and Bhalla, O. P. 1974. Occurrence of Eulecanium sp.? tiliae (L.) on plum and apple in Himachal Pradesh. *Current Science,* **43**: 115–120.

Mohamad, M. A. and Al Mallah, N. M. 1990. Preliminary field observations on the ecology and biology of apricot leaf aphid, *Hyalopterus pruni* (G) (Homoptera: Aphididae) in Mosual region Iraq. *Arab Journal of Plant Protection*, **8**(1): 1–5.

Narayanan, E. S. and Batra, H. N. 1960. *Fruit Flies and their Control*. Indian Council of Agricultural Research, New Delhi.

Obenberger, J. 1955. A new species of *Sphenoptera* damaging the peach trees in Pakistan. *Acta Entomology Museum Natural Prague*, **29**(429): 89–91.

Pawar, A. D.; Tuhan, N. C.,;Blasubramanain, S and Parry, M. 1981. Distribution, damage and biology of codling moth, *Cydia pomonella* (L.). *Indian Journal of Plant Protection,* **9**: 111–14.

Pedata, P. A.; Hunter, M. S., Godfray, H. C. J and Viggiani, G. 1995. The population dynamics of the white peach scale and its parasitoids in a mulberry orchard in Campania, Italy. *Bulletin of Entomological Research*, **85**(4): 531–539.

Pruthi, H. S. and Batra, H. N. 1960. Important Fruit Pests of North-West India. *Bulletin 80*, Indian Council of Agricultural Research, New Delhi.

Quarles, W. 1997. Mating disruption for the peach twig borer. *IPM-Practitioner,* **19**(5–6): 1–8.

Rishi, N. D. and Shah, K. A. 1985. Survey and biocecological studies on the natural enemy complex of Indian gypsy moth, *Lymantria obfuscata* Walker (Lepidoptera: Lymantriidae). *Journal of Entomology Research*, **9**: 82–93.

Sazonov, A. P. and Sundukova, N. E. 1987. Synthetic pheromones against the plum fruit moth. *Zashchita—Rastenii – Moskva*, **6**: 48–49.

Sharma, H. K. and Chander, R. 1992. Effectiveness of different methods for the control of flat headed peach tree borer *Sphenoptera lafertei* Thomson on stone fruits. *In*: *Emerging Trends in Temperate Fruit Production in India*. K. L. Chadha, D. K. Uppal, R.N. Pal, R. P. Awasthi and S. A. Ananda (eds). NHB Technical Communications 1, Gurgaon. pp. 281–283

Sharma, H. K.; Chander, R. and Thakur, J. R. 1990. Incidence of flat-headed peach tree borer (*Sphenoptera lafertei*)(Coleoptera: Buprestidae) on stone fruits in Himachal Pradesh. *Indian Journal of Agricultural Sciences,* **60**(10): 705–707.

Sharma, J. P. and Dogra, G. S. 1982. Biology and seasonal history of plum scale, *Eulecanium* sp.? *tiliae* (L.) in Himachal Pradesh. *Bulletin Entomology*, **23**: 74–84.

Sharma, J. P. and Dogra, G. S. 1986. Studies on the control of the plum scale *Eulecanium* sp.? *tiliae* (L.) (Homoptera: Coccidae) through chemical and natural enemies. *Indian Journal of Entomology*, **48**(3): 258–263.

Sharma, P. L.; Attri, B. S., Shandil, R. L. and Bhalla, O. P. 1968. Biology and control of peach leaf curl aphid, *Brachycaudus (Anuraphis) helichrysi* Kalt. *Indian Journal of Entomology*, **30**: 377–379.

Sharma, P. L. and Bhalla, O. P. 1964. A survey study of insect pests of economic importance in Himachal Pradesh. *Journal of Entomology*, **26**: 318–31.

Sharma, P. L.; David, A. C. and Bhalla, O. P.1963. Flat headed borer, *Sphenoptera lafertei* Obb. and its control in Himachal Pradesh. *Himachal Horticulture*, **4**(1): 1–5.

Sharma, P. L., Srivastava, S. and Dhaliwal, H. S. 1973. Chemical control of the peach fruit flies. *Pesticides,* 7(10): 20–21.

Sharma, S. K and Gupta, P. R. 1986. Occurrence of *Cydia* sp.? *funebrana* (Treitschke) as apricot fruit borer—A new record from India. *Journal of Bombay Natural History Society*, **83**: 677–678.

Shukla, R. P.; Prasad, V. G. and Tandon, P. L. 1984. Effectiveness of different insecticides against oriental fruit fly *Dacus dorsalis* Hendel. *Indian Journal of Horticulture*, **41**: 307–309.

Singh, C. 1964. Temperate fruit pests. *In*: *Entomology in India*. Entomological Society of India, New Delhi.

Srivastava, O. S. 1972. Chemical control of bark eating caterpillar, *Indarbela quadrinotata* (Walker) in guava trees. *Indian Journal of Agricultural Sciences*, **42**: 847–848.

Thakur, J. R.; Chauhan, Usha and Dogra, G. S. 1991. Autumn population and settling behaviour of peach leaf curl aphid, *Brachyacudus helichrysis* (Kalt.) on each peach varieties in Himachal Pradesh. *Journal of Aphidology*, **5**(1&2): 77–85.

Tripathi, G. M. and Gupta, B. P. 1985. Studies on collection of beetles on light trap at Chaubattia in UP. *Hill Progressive Horticulture*, **17**(4): 347–353.

Verma, A. N. and Khurana, A. D. 1974. Further new host record of *Indarbela* sp. (Lepidoptera: Metarbelidae). *HAU Journal of Research*, **4**(3): 253–254.

Verma, K. L. 1984. *Studies on the biology, natural enemies and chemical control of the peach leaf curl aphid Brachycaudus helichrysi (Kaltenbach)*. Ph.D. Thesis, HPKVV, Solan.

Verma, K. D. and Chauhan, R. S. 1993. Life cycle of potato vector, *Myzus persicae* Sulzer in India. *Current Science*, **65**(6): 488–489.

Verma, T. D. 1985. Incidence and chemical control of bark eating caterpillar, *Indarbela quadrinotata* Walker on plum trees. *Indian Journal of Agricultural Sciences*, **55**(2): 131–132.

Veeresh, G. K. 1988. White grubs as pest of forest and fruit trees in India and their management. *In: Tree Protection*. V. K. Gupta and N. K. Sharma (eds.). Indian Society of Tree Scientists, Solan.

Zaka-ur-Rab, M. 1982. Predation of the plum scale, *Eulecanium coryli* (L.)(Homoptera: Coccidae) by *Ballia bayaderae* (Coleoptera: Coccinellidae) in Kashmir. *Journal Bombay Natural History Society*, 79: 436.

*Pest Management in Horticulture Crops*
L.R. Verma, A.K. Verma and D.C. Gautam (eds)


*Chapter 13*

# Pests of Almond, Cherry and Walnut

*J.P Sharma and D.R. Khajuria*

## INTRODUCTION

ALMOND, CHERRY AND WALNUT are important fruits of the temperate region of the world. In India, these fruits are mainly grown in the Himalayas. Almond and walnut are highly nutritious nut fruits that fetch exorbitant price in the market. Both these fruits have tremendous export potential in case international standards are maintained. Besides, walnut is also known for its valuable timber. Its plantation is widespread in the Himalayas between 1200–2150 m altitude. Almond is not only grown for its dried nuts but also for marketing as green almonds. In India, almond cultivation is found at an altitude of 900 -1500 m and in river valley areas of dry temperate zone. Cherry is a high value cash crop. It is delicious, delicate and highly attractive fruit. Sweet cherry is mostly taken as dessert and sour cherry is used for canning or cooking. The cultivation of cherry is confined to a typical temperate climate mostly above 1300 m altitude.

These three fruit crops are subjected to attack by several species of insect and non-insect pests which reduce the yield and quality of these fruits. The orchardist has to fight a constant battle to maintain the production capacity of these fruit trees and protect their crop from the ravages of pests. The trees are attacked from nursery stage to fruiting stage and from root to fruit. A larger number of pests of these three tree fruits are polyphagous and of frequent occurrence. Some of these have restricted host range whereas few are specific.

## ALMOND

Cultivated almond, *Prunus dulcis* (Miller) D.A. Webb. (Syn. *Prunus amygdalus* Batsch) is native to Central Asian mountain areas (India, Iran and Pakistan) from where it spreaded as far as the Mediterranean basin. Presently, almond is cultivated mainly in countries between the latitudes 36°E and 45°N, although under certain mild climatic conditions, cultivation may extend further north. It is also cultivated in South Africa, South America, Australia and India. The only limiting factor in almond cultivation is the spring frost, particularly during full bloom. Of the insect-pests found on almond trees, a few species of beetles and weevils and occasionally some aphids cause severe damage whereas scale insects, stem borers, leaf eating caterpillars and fruit borers are of minor importance. Besides, dried nuts are attacked and spoiled by various storage pests namely, *Cadra cautella* (Walker), *Sitotroga cerealella* (Olivier) and also by *Laemophloeus* spp., *Oryzaephilus mercator*, *O. surinamensis* (Linnaeus) and *Tribolium castaneum* Herbust (Butani, 1979).

## SAP SUCKING PESTS

### Aphids

Several species of aphids have been recorded to damage almonds all over the world. These include *Anuraphis spp.*, *Brachycaudus helichrysi* (Kaltenback), *B. (Acaudus) persicae* Pass, *B. (Thuleaphis) amygdalinus* (Schout), *Hyalopterus pruni* Geaffr and two other *Hyalopterus* species, *Myzus persicae* (Sulzer) and *Pterochloroides persicae* Chalodkovsky (Butani, 1979; Mustafa and Hamdan, 1989; Rolli, 1974; Triggiani, 1973). All these species are polyphagous pests and have a wide range of host plants. Nymphs and adults suck sap from leaves, petioles, blossoms and fruits. Infestation is higher during spring (February—April) when there is fresh flush of leaves. Infested leaves become curled, pale and fall down. Development of fruits is arrested and many may drop down prematurely. Besides, there is exudation of honeydew on which sooty mould develops, which further hinders the growth and fruiting capacity of the tree.

Peach shoot spotted aphid *Pterochloroides persicae,* also called the giant brown bark aphid or black aphid is comparatively more common and harmful to almond in South West Europe and Central Asia. It is brownish with black spots and white patches. It attacks older branches forming dense colonies and a lot of sooty mould (El-Tariki and El-Sharif, 1987).

Peach leafcurl aphid *Brachycaudus helichrysi,* also known as the green plum and weed aphid is a key pest of peach in India but it also causes curling of leaves in some cultivars of almond. It is a problem until June when it migrates to herbaceous plants and returns to almond in autumn when the amphigonic females lay eggs which overwinter. The control schedule is the same as recommended for peach.

The green peach aphid *Myzus persicae* is a common pest of peach trees and has also been recorded on almond. The yellowish green aphid is pear shaped, with fairly long antennae and a pair of cornicles at the end of the body. Toxins emitted by aphids while feeding cause curling, distortion, occasional foliage discoloration and premature leaf drop. Feeding aphids also excrete copious amounts of honeydew, upon which sooty fungus grows that causes smutting of leaves and fruits.

*Hyalopterus amygdali and H. Pruni* appear in early spring covering the leaves with dense colonies until May when the winged forms migrate to herbaceous secondary hosts. The aphids reduce plant growth and fruit set in the current year and also in the following year.

Management

The peach leafcurl aphid, *B. helichrysi* can be controlled by application of systemic insecticides at pink bud stage. *M. persicae,* was controlled in Tunisia with two sprays of mevinphos @ 2.5 g ai., dimethoate @ 1.5 g ai, oxydemeton methyl @ 1.25 g ai. Or phosphamidon @ 1 gai. per tree in the beginning of March and repeated in early April (Rolli, 1974).

## Scale insects

The San Jose scale, *Quadraspidiotus perniciosus* (Comstock), unarmoured scales, *Eulecanium tiliae* (Linnaeus) and *E. coryli* (L.), cottony cushion scale, *Icerya purchasi* Maskell, ulmi comma scale, *Mytilococcus ulmi*, almond dark grey scale, *Pelomphala inopinata*, olive parlatoria, *Parlatoria oleae*, white peach scales, *Pseudaulacaspis pentagona* (Targioni) and *Chrysomphalus ficus* Ashmead, have been recorded on almond from different parts of the world (Bhalla and Gupta, 1993; Behdad and Barooti, 1978; Butani, 1979). The scale insects suck sap from leaves and tender shoots, mature ones migrate to stems and branches and occasionally attack fruits as well. They are polyphagous pests. *Q. perniciosus* is the only pest of importance as it often causes serious damage to nursery plants and young trees. For the control of scale insects spray miscible oils (as mentioned in apple) during January-February or post-harvest spray of systemic organophosphate insecticides like methyl demeton (0.025%) or dimethoate (0.03%) as recommended on plum.

# DEFOLIATORS

## Defoliating beetles

As many as 70 species of leaf eating beetles belonging to several families including Scarabaeidae, Cetonidae, Curculionidae and Chrysomelidae have been recorded to cause damage to pome and stone fruits in India (Bhalla and Pawar, 1977; Gupta *et al.*, 1977; Sharma *et al.*, 1969). Almost all these are polyphagous species. Almond is also an important host of leaf eating beetles especially the almond beetle, *Mimastra cyanura* (Hope), ber beetle, *Adoretus nitidus* Arrow, and chafer beetle, *Holotrichia consanguinea* Blanchard. Of these, *M. cyanura* is the most common pest reported from the northwestern states of Himachal Pradesh, Uttar Pradesh and Jammu and Kashmir (Butani, 1979). The grubs live in soil and feed on roots or decaying organic matter. The adults start appearing around May and the maximum activity is reported during July- August when the swarms of adults may be seen defoliating the trees, whereas the apricot weevil, *Emperorrhinus defoliator* Marshal, apple weevil, *Myllocerus dicolor* Boheman, grey weevil, *M. maculosus* Desberdes and plum weevil, *Amblyrrhinus poricollis* are of minor importance.

## Almond weevil (*Myllocerus lactivirens* Marshal)

The almond weevil, *M. laetivirens* is the most serious pest of almond and pear being its preferred hosts. It has also been recorded on apple, apricot, ber, citrus, falsa, loquat, mango, peach, pear, plum, pomegranate, etc. Adult weevils congregate on the ventral surface of leaves, nibble irregular holes and gradually eat away the entire leaf leaving only the mid rib. The weevils first consume the tender leaves and then move on to older ones.

The eggs are broadly oval, about 0.6 mm long, creamy yellow, smooth, transparent and shiny. They are fixed on the leaves by means of a sticky substance secreted by the females. The newly hatched grubs are creamy white, stout and finely setose; full grown grubs are about 4 mm long, stout with large conspicuous spiracles, no legs but short erect setose and oval armature well developed. Adult weevils are 3–4 mm long and pale metallic green in colour. They are poor fliers but active feeders both during day and night. These weevils start emerging in May, their population increases with onset of rains and becomes maximum around August followed by a decline in

October. Oviposition begins by the end of July and continues till the beginning of September. Eggs are laid in batches of 40 to 50 in the soil. The eggs are difficult to locate in the soil due to their small size and cryptic colouration. Incubation period is 4 to 5 days. On hatching, the grubs burrow deeper in the soil (200 to 300 mm) and feed on roots. The full grown grubs come up to pupate in the upper 25 mm of soil. The larval period lasts for 10 months and pupal period is of 4 to 5 days. Overwintering occurs in pupal stage in the soil.

Installation of light trap provides information on the appearance and abundance of defoliating beetles but is not an effective method of control (Tripathi and Gupta, 1985). Collection and destruction of beetles after dusk by shaking non-bearing trees can give effective control if done as a campaign at the right time in an area (Veeresh, 1988; Yadav *et al.*, 1977). Insecticides effective against the weevils are carbaryl 0.1 per cent or methyl parathion 0.05 per cent.

**Defoliating caterpillars**

Tent caterpillar, *Malacosoma indica* Walker is a major pest of apple, but also attacks almond. The caterpillars damage the newly opened leaves and hinder the growth and development of the tree. The infestation can be checked by collection and destruction of shining and elongated thimble shaped egg masses during pruning. The silken tents of these caterpillars should be collected and destroyed by burning.

*Trabala vishnou* (Leavere) is a sporadic pest of castor but also feeds on the foliage of almond, guava, *jamun* and pomegranate. The caterpillars feed during night and hide during daytime under the leaves or in cracks and crevices of the stems. Full grown caterpillars are 60 to 68 mm long, pinkish brown in colour with tufts of dark hair arising from blue spots. Cocoons are tent shaped, tough and papery interwoven with larval hair. Moths are pale green with a large reddish brown patch and a dark spot on fore wings. Wing expanse of moths is 48 to 55 mm. There are four generations in a year (Beeson, 1941).

## STEM AND SHOOT BORERS

Flat-headed peach tree borer, *Sphenoptera lafertei* Thompson (Coleoptera: Buprestidae) attacks trunks and branches of almond tree exposed to sun light. Almond is perhaps the least preferred host among the stone fruits (Sharma *et al.*, 1990). Under normal conditions, its infestation in almond does not cause any marked adverse effect on growth parameters (Sharma and Gautam, 1994). The pest can be controlled by using physical barrier of grass or gunny bags as described under cherry.

Quetta borer, *Aeolesthes sarta* Solsky (Coleoptera: Cerambicidae) is a major pest of apple, apricot, peach, fig, etc. But also damages almond trees in India, Pakistan, Afghanistan and China. Its grubs bore inside the trunk and main branches, making horizontal tunnels. For details refer to stone fruits.

Shot hole borers, *Scolytus* spp. (Coleoptera: Scolytidae particularly *Scolytus amygdali* Guer, attack weak, declining, decaying and poorly managed trees. It is a minor pest of almond.

Peach twig borer, *Anarsia lineatella* Zeller (Lepidoptera: Gelechidae) occurs almost worldwide and attack almond in addition to peach, plum and apricot. It causes damage by feeding on new leaves, shoots and developing nuts. It also encourages damage by *Amyelois transitella* (Wlk.) whose ovipositing females are attracted to almond fruits with hulls prematurely split by *Anarsia*

(Reil *et al.*, 1981). The caterpillars generally bore into buds, flowers, shoots and fruits. The overwintering larvae of the first generation attack shoots making them to wither while the second generation larvae attack fruitlets which drop off or ripen prematurely. Gummy exudation at the entrance hole can be seen on infested fruits.

The males of *A. lineatella* can be monitored with the help of sex pheromone traps (1 trap per 0.84 acre placed atleast 192 feet apart) which are rechanged fortnightly (Youngman and Barnes, 1986). At high population densities, non sticky IPS traps are more effective in capturing males than the sticky Pherocon IC Traps, whereas at low population densities, Pherocon IC traps are better (Kehat *et al.*, 1994). However, pheromone trap catches have not been found to be good indicators of female moth activity (Barnett and Hendricks, 1992). In California the timing of diazinon spray at the rate of 2 lb a.i./acre for the control of this pest has been recommended at 400 day—degrees after the appearance of males in pheromone traps (Rice and Jones, 1988). The ULV application of *Bacillus thuringiensis* Berliner by aircraft has also shown promise as an alternative to conventional insecticides in California (Roltsch *et al.*, 1995).

Fruit Borer, *Heliothis armigera* Hubner (Lepidoptera: Noctuidae) has been observed on almond, boring the tender fruits of varieties like Mercede and IXL in April (Butani, 1979). Spraying with 0.05 per cent endosulfan is effective for its control.

## CHERRY

There are two types of cherries—sweet cherry, *Prunus avium* Linnaeus and sour cherry, *P. cerasus* Linnaeus. Cherry has its origin to region between Caspian and Black Seas. Being a favourite food of many birds, the seeds were scattered over various parts of the continents of Eastern Hemisphere. At present it is grown in all the temperate countries of the world.

There are over 60 species of pests that inflict direct or indirect damage to cherry in various parts of the world. About 40 have been reported as pests of cherry from India (Butani, 1979). Only a few species are serious pests of cherry while most others species are polyphagous and cause only minor injury to cherry.

## STEM AND ROOT BORERS

### Flat-headed stem borer, *Sphenoptera lafertei* Thomson

The flat-headed stem borer is the most serious pest of cherry in India and the single important factor responsible for dwindling of cherry plantations in Himachal Pradesh. The pest is restricted to India, Afghanistan and Pakistan. In Himachal Pradesh, it prefers cherry followed by peach, plum, apricot and almond. Whereas in the plains, peach is the primary host of this pest. Apart from stone fruits, it has also been recorded on apple and various woodland and shade trees (Sharma *et al.*, 1963; Singh, 1964).

The most conspicuous symptom of its attack on cherry is the oozing out of gum globules from the damaged portions of the plants. Small, oval exit holes (1 cm × 0.6 cm) of beetles are also found at the site of attack. The leaves of infested trees gradually turn pale and growth is arrested. The pest attacks trees of all ages but the infestation is more severe on weak, injured and sun-burned trees with main stem and branches exposed to the sunlight. Occasionally plants are also infested at the graft union. The newly emerged larvae feed on the fresh bark by making minute, irregular galleries and later bore into the sapwood (phloem and cambium). The larvae often girdle the branches and stem, killing them outrightly. The extent of damage in various cultivars ranges between 51 to 100 per cent with an average of 74.47 per cent in Kullu, while on other stone fruits infestation varies from 24.5 (almond) to 53.4 per cent (peach) in Himachal Pradesh (Sharma *et al.*, 1990).

The emergence of beetles from overwintering larvae/pre-pupae starts from the first week of March. The beetles are coppery black in colour. The male and female beetles measure on an average 7.5 and 9.4 mm in length and 2.3 and 3.1 mm in width, respectively. The newly emerged beetles can be seen basking in the sun on cherry leaves in the morning. The beetles are not very active fliers. They feed on foliage, mate and lay white, spherical eggs singly or in batches of 2–3 in crevices all over the tree trunk and main branches exposed to the sun. The colour of the eggs changes to black before hatching. The incubation period on cherry varies from 12 to 20 days and averages to around 17 days. The female lays upto 140 eggs in its life span of 14 to 20 days. The neonate white larva with a small yellow head and greatly enlarged prothorasic segment bores into the bark. A lump of resinous gum oozes out at the entry hole. As the larva grows, it penetrates deeper into the sapwood. There are four instars. The typically flat-headed full grown larva averages 21.4 mm in length and 3.41 mm in width at the prothorax, eats up a small pupation chamber (4–6 × 2–4 mm, 8–10 mm deep) in the hardwood and pupates in it. The larval duration varies from two and a half months in summer to over 6 months in winter. Prepupal and pupal periods last for 1–2 and 8–10 days, respectively.

This pest can be effectively controlled by covering the trunk and exposed areas of main branches with 1 cm layer of dried grass or gunny bags from February end to October every year. Before applying this physical barrier to prevent egg laying by the females, it is essential that the dead bark is removed and trunk and other infested portions sprayed or swabbed with 0.2 per cent methyl parathion. In the absence of physical barrier, this insecticidal treatment has to be repeated 4 times in a year during April, May, June and September. Proper training and pruning of the tree to avoid direct exposure of the stem and main branches to sun is important to keep the pest at bay. This can be done by keeping enough branches on south and south-western side of the tree.

**Cherry stem borer *Aeolesthes holosericea* Fab.**

The cherry stem borer is reported from India, Srilanka, Bangladesh, Myamar, Malaysia and Thailand. Apart from cherry it also infests apple, guava, apricot, crab apple, mulberry, peach, plum, walnut, etc.

The newly emerged grubs bore inside the bark, feed on the inner layer by making zigzag galleries and later bore further inside to feed on the outer layer of sapwood. The feeding grubs throw out the fibrous frass and faecal matter which is a clear indication of their presence in live galleries. As a result of feeding, the vitality and productivity of the tree is greatly impaired. Such weak trees become vulnerable to the attack of secondary pests like shot hole borers and die out, within 2–3 years. The infestation is generally seen in neglected orchards.

The adults appear between April to July and remain active till the end of October. The female lays upto 92 elliptical (2.5 mm long) eggs in cracks and crevices of the bark. The eggs hatch in 7–12 days. The grubs first feed on inner layer of the bark and later on sapwood. The grubs become full-fed in 27–32 months and enter the hardwood for pupation. Pupation takes place either in

October-November or March-April. Pre pupal and pupal periods last for 3–150 and 40–100 days, respectively. The beetles emerging from pupae formed in October remain within the stem throughout winter and spring while those emerging from pupae formed in April rest only for 6 weeks. Thus a single life cycle is completed in 32 to 36 months (Rahman and Khan, 1942).

The pest can be controlled by pruning and destruction of newly attacked branches and killing the older larvae by inserting flexible wire into the live tunnels. In case of severe attack, satisfactory control can be achieved by inserting solid fumigant (0.5 g paradichlorobenzene (PDCB), 0.2 g aluminium phosphide) or a cotton wick soaked either in liquid fumigant (petroleum, PDCB, kerosene oil, mixture of chloroform and creosote in 2 : 1 ratio, a mixture containing nephthalene flakes, creosote and carbon disulphide in 5 : 2 : 1 ratio) or an insecticide emulsion (0.1% DDVP, methyl parathion) in larval gallery already cleaned by iron spoke and plugging the hole with mud plaster (Bhalla and Gupta, 1993). The live tunnels remaining after the treatment should be treated again till the grub is finally killed.

Another stem borer that infests cherry is the Quetta borer, *Aeolesthes sarta* Solsky. This pest is a minor pest of cherry and major pest of apple, peach, fig, etc.

**Leopard moth *Zeuzera multistrigata* Moore**

The larvae of leopard moth are polyphagous pests of many tree fruits including cherry and apple on which the attack is severe while walnut, pear, plum, quince, forest trees, ornamental woody trees and shrubs are also attacked to a lesser extent.

The early instar larvae of this pest feed on the cambium and causes gummosis on trunks, branches and twigs of cherry, while the later instars penetrate into the hardwood producing 20–25 cm long tunnels from March-April until the following year. Foliage of the infested tree turns yellow or brown and drops off prematurely. White moths with blue/black markings on the wings appear in June-July and survive for 10–23 days. The female lays about 600 eggs singly or in groups of 3–8 in cracks and crevices of the tree which hatch in 8–9 days. The larval stage lasts for 20–22 months. The full grown larva measures 4–6 cm, pupates just below the bark near the opening of the tunnel during April-May. The pupal period lasts for 4–8 weeks. The life cycle is completed in 20–26 months (Bhardwaj,1982). Control measures suggested for the cherry stem borer (*A. holosercea*) are also effective against *Zeuzera* species.

**Root Borer *Dorysthenes hugelii* Rdt.**

Occasionally found feeding on roots of cherry trees, this species is a major pest of apple and minor pest of several other tree fruits. The infested trees become weak and often get uprooted. One to two larvae are enough to kill a 4-year old cherry tree. The pest can be controlled by drenching with chlorpyrifos (0.04%).

## SAP SUCKING PESTS

### Aphids

Several species of aphids, namely, the apricot leaf aphid, *Hyalopterus pruni* G., black cherry aphid, *Myzus cerasi* (F.), peach black aphid, *Pterochlorus persicae* Cholodkovsky, and green peach aphid,

*Myzus persicae* (Sulzer) have been reported damaging cherry from various parts of the world including India, North America and Europe (Butani, 1979; Hogmire, 1995; Mohmad and Almallh, 1990). In addition, *Brachycaudus helichrysi Kaltenbach* has been reported from cherry orchards in Ukrain where it is reported to be a virus vector (Kuprii and Damrose, 1986). In India, *M. cerasi* is the most common recurring pest which requires frequent suppression.

The black cherry aphid is a European species that causes serious injury to sweet cherries and occasionally to tart cherries. Although, the aphid can survive on wild cherries, yet it prefers commercial cherry varieties. it also favours sweet cherry varieties over tart ones. 'Napolean' 'Black Tartarian', 'Schmidt' and 'Widsor' are the most susceptible varieties, while 'Dikeman' and 'Yellow Spanish' are not seriously injured. Tart cherry varieties such as 'Montmorency', 'Early Richmond' and 'English Morello' are attacked less frequently and the injury is slight. Alternate summer hosts of this aphid include water cress, pepper grass, and other members of the mustard family. The adults and nymphs are shiny black in colour and can be easiy identified as there is no other black colour species of aphid attacking cherry. Adults measure 3 mm in length and exist in both winged and wingless forms. Nymphs are smaller in size but identical in appearance to the wingless adults. The shiny black, oval eggs measure less than 1.5 mm in length.

The aphid overwinters in egg stage on the bark of small branches.The eggs begin to hatch as soon as cherry buds break in April and young nymphs move into the buds to feed on the foliage. Within 3 to 4 weeks, the wingless females, called stem mothers, rapidly establish colonies parthenogenetically on the new growth. Two to three generations are produced on cherry by early July when most of the aphids (winged females) move to alternate hosts (mainly of mustard family) for the summer, but a few remain on cherry during the entire season. In September—October winged males and females return to cherry trees, mate and lay overwintering eggs.

Feeding by adults and nymphs causes curling and stunting of leaves and growth distortions in the stem. Heavy infestations may killl young trees, and reduce the quantity and quality of crop and return bloom on mature trees. Honeydew from aphids also causes the growth of black sooty fungus.

Trees should be scouted in early spring to detect the presence of stem mothers on actively growing shoots. It is important to determine aphid severity before the leaves curl. While no thresholds have been established for mature trees, young trees cannot tolerate even low numbers of aphids. Delayed application of dormant oil and insecticides to control other cherry pests usually controls the black cherry aphid. Sometime the aphid's natural enemies (syrphid flies, lace wings and lady bird beetles) are abundant enough to keep the population suppressed. Earwigs (*Forficula spp.*) were effectively used for the control of aphids on wild cherry in Germany (Schonbeck, 1988) spraying with dimethoate, methyl demeton or malathion @ 1 ml per litre water before or at bud swell stage is suggested to control the pest. In case of severe attack, the spray should be repeated after petal fall (Hogmire, 1995).

### Coccids

San Jose scale *Quadraspidiotus perniciosus* (Comstock), soft scale, *Eulecanium tiliae* (L.), white peach scale, *Pseudaulacaspis pentagona* (Targioni), and mango mealy bug, *Drosicha mangiferae* (Green) have been recorded infesting cherry trees from India and (Behdad and Barooti, 1978; Butani, 1979; Kozar and Konstantinova, 1981; Stimmel, 1982). These insects suck the sap from leaves and tender shoots and in case of severe infestation, growth and fruiting capacity of the affected tree is impaired. However, all these polyphagous species are of minor importance to cherry.

## FOLIAGE FEEDERS

Cherry leaves are damaged by species of leaf rollers, defoliating caterpillars, beetles and weevils (Bhalla and Pawar, 1977; Butani, 1979; Chander and Dogra, 1983). Larvae of over 80 species feed on the foliage of temperate fruits including cherry but only a few are regular pests in some pockets.

### Leaf folders/rollers

Among the species of leafrollers found on deciduous tree fruits, *Archips termias* (Meyrick) causes minor damage to cherry. Caterpillars roll the leaves and feed within, the damaged leaves get skeletonized. In the case of severe infestation, leaves and fruits are webbed together and larvae feed inside on the soft tissues and ripening fruits. Mechanical collection and destruction of rolled leaves and webbed bunches of flowers and fruits have been suggested to prevent the spread of the pests. Spraying of phosphamidon (0.03 %) or endosulfan, fenitrothion (0.05%) is effective to control these pests.

### Leaf defoliators

The foliage of cherry is damaged occasionally by apple giant caterpillars, *Actias selene* Hubner; tent caterpillar, *Malacosoma indica* (Walker), Indian gypsy moth, *Lymantria obfuscata* Walker and other *Lymantria spp.* which are otherwise major pests of apple. Besides, a number of beetles and weevils also feed and defoliate cherry plantations. These include, the cherry black beetle, *Stalugosoma albella* Palls; cherry green beetle, *Mimela passerini* Arrow; cherry gold luster beetle, *Anomala flavipes* Arrow; chafer beetles, *Adoretus bimarginatus* Ohaus, *Holotrichia longipennis* Blanchard, *Hoplosternus furicaudus* (Ancy), *Serica clypeata* Brenske, *S. muculosa* Brenske, *S. marginella* Hope; flea beetles, *Altica caerulescens* Baly, *Pachnephorus impressus* Rosenburg; apricot weevil, *Emperorrhinus defoliator* Marshall; and cherry weevil, *Myllocerus lefroyi* Marshall. None of these species is a major pest on cherry and as such no control measures are generally warranted. However, in case of severe attack a spray of 0.1 per cent carbaryl or 0.05 per cent endosulfan is enough to control these pests.

### Blossom beetles

Two species of the blossom beetles, *Protaetia impavida* Janson and *P. neglecta* Hope are important pests of cherry and other tree fruits. The former feeds on cherry blossoms and fruits while the latter feeds on leaves, flowers and fruits. Apart from cherry, these beetles also feed on apple, peach, pear, plum, etc., in the entire Himalayan range in India and Nepal.

Adults of *P. impavida* are stout, rather convex, shining blue black beetles having elytra closely set with large crescentic punctures and minute white markings. Beetles of *P. neglecta* are slightly bigger, 20 to 22 mm long, stout, convex, bronze coloured with elytra strongly and rugosely punctured and having intricate pattern of grey lines. Both the species are minor pests and often no control measures are required for these beetles.

## WALNUT

Walnut is an important temperate fruit found growing in vast regions right from the Carpathian mountains in the eastern Europe, across Turkey, Iraq, Iran, Russia and Afghanistan to the north western Himalayas. It is believed to have originated in Iran and the areas surrounding it from where the returning armies of Alexander the Great brought it to Europe. It is now grown extensively in south eastern Europe, Iran, Himalayan region and China. There are 18 more species of walnut (genus *Juglans*) which are grown extensively in almost all the temperate countries of the world. In India, walnut is mainly grown in Jammu and Kashmir, Himachal Pradesh and hills of Uttar Pradesh.

Several species of insects damage various parts of walnut tree. As many as 30 species of pests have been reported from India (Butani, 1979). Of these, the walnut aphid, walnut weevil, bark boring beetles and leaf eating beetles cause severe damage. Additionally minor pests like San Jose scale, leopard moth, leaf eating caterpillars, chafer beetles, apple root borer, etc., are also found feeding on this species. The red flour beetle, fig moth, khapra beetle and saw-toothed grain beetle are important pests of stored walnut in India.

### Walnut green aphid *Chromaphis juglandicola* Kaltenbach

It is a common and sporadically severe species of aphids found on walnut in India and elsewhere in the world. The nymphs and adults suck sap from underside the leaves and tender shoots between April and September. The peak period of activity of the aphid is late rainy season. In severe infestation, the affected parts become pale, curled up and disfigured. The leaves fall, kernels shrivel and sooty mould develops on honeydew. The growth of the tree is adversely affected and plants get stunted. Overwintering occurs in egg stage.

The aphid can be effectively controlled by spraying systemic insecticides like dimethoate (0.03%), methyl demeton (0.025%) or phosphamidon (0.03%) (Butani, 1979). In the USA, a parasitic wasp, *Trioxys sp.* has been found to be an effective biocontrol agent of this pest (Hoy *et al.*, 1991). Pawar and Parry (1989) reported a predator (*Calvia punctata*) on this aphid.

Apart from this species, the apple aphid, *Aphis pomi* (de Geer), dusky veined walnut aphid, *Callaphis juglandis* Kalt., and *Panaphis juglandis* (Georgi) have also been reported infesting walnut trees. Bio-ecology of *C. juglandis* was studied in Kashmir, India and a syrphid, *Metasyrphus corollae* (Fabr.) was reported to be an effective biocontrol agent of this aphid (Bhagat *et al.*, 1994; Masoodi *et al.*, 1989).

San Jose scale *Quadraspidiotus perniciosus* (Comstock) (Diaspididae: Homoptera) is becoming an important pest of walnut in mid hill regions of the Himalayan range. The pest has a number of generations each of 5–6 weeks duration, during active period of development. The pest overwinters in nymphal stage from mid October to mid March.

## DEFOLIATORS

### Leaf eating beetles

Walnut blue beetle, *Monolepta erythrocephala* (Baly) is a serious pest of walnut in Himachal Pradesh and Uttar Pradesh. Apart from walnut it infests apple, pear, grapevine, etc. The beetles feed on leaf epidermis leaving behind a network of veins. The beetles are active from April to September and cause maximum damage during June to August. Adult beetles are 4–5 mm long having reddish brown head, black body and bluish elytra which are broader at the base than at prothorax. These beetles can be controlled by spraying endosulfan (0.05%).

The other species of beetles namely, *Altica caerulescens* Baly, *Cerogria nepalensis* Hope, *Dactylispa lohita* Maulik, *Hispa daina* Chapman, *Galerucida rutilans* Hope, *Platypria hystrix* Fabricious, *Phyllotreta chotanica* Duvier and *Triclina* spp. have also been reported feeding on walnut leaves in India. All these species occur sporadically and rarely cause economic loss (Butani, 1979).

Chafer Beetles namely,*Adoretus versutus* Harold, *Anomala rugosa* Arrow, *Brahmina coriacea* (Hope), *Holotrichia longipennis* Blanchard and *Mimela pusilla* Hope cause sporadic damage to leaves of walnut trees during June-July. These polyphagous beetles are nocturnal in habit. They feed, mate and lay eggs during night. The leaf eating beetles can be controlled by spraying methyl parathion 0.05% or carbaryl 0.1% on the affected trees.

### Leaf eating caterpillars

Indian Gypsy moth, *Lymantria obfuscata* Wlk. (Lymantriidae: Lepidoptera) has only one generation a year and overwinters in egg stage. Eggs hatch in March and emerging caterpillars feed on buds and leaves. The caterpillars attack the foliage during night and hide under rocks, stones or clods during the day. The pest causes partial to complete defoliation.It can be effectively controlled by spraying endosulfan (0.05%) or fenvalerate (0.03%) (Masoodi *et al.*, 1987).

Walnut blue, *Chaetoprocta sp.* causes severe defoliation of wild walnut plantations in Kinnaur, Himachal Pradesh. *Actias selene* (Hubner), *Belippa laleana* Moore, *Oxyambulyx sericeipennis* Butler have also been reported defoliating walnut trees from Assam. Besides, caterpillars of *Malacosoma indica* Walker and wild silkworm namely, *Antheracea roylei* Moore, *Odina wodier* and *Saturnia simla* Westwood have also been recorded feeding on walnut foliage from Himalayan range. All these are minor pests and can be controlled by spraying endosulfan (0.05%).

## BORERS

### Stem and shoot borers

Long horned beetles, *Batocera horsfieldi* Hope is a serious cerambycid borer of walnut in Kumaon hills, Uttar Pradesh, Himachal Pradesh and Darjeeling (West Bengal). The grubs feed on inner side of the bark making zig zag tunnels. Later, the grubs bore down to the surface of sapwood and go upto the centre of heartwood. Usually one or two tunnels are found near the base of the tree but as many as 16 tunnels have been observed upto 6 metre length of the tree trunk (Beeson, 1941). Adults feed on the bark of young twigs but the damage caused is negligible. Attacked trees are neither killed nor their vigour is affected much but the timber is damaged severely.

The oval, brown colour eggs are laid singly on the bark of lower one metre portion of the trunk. Full grown grubs are 90–150 mm long, pale yellow in colour. Pupae are creamy white and 40–60 mm long. Adult beetles are 45–60 mm long, black with fine ashy or yellowish gray pubescence. Pronotum has two elongated yellowish white spots. The elytra have numerous shining black tubercles at base and several rounded and broken white marks extending upto apex. A female lays 55–60 eggs which hatch in 8–15 days. The grub stage lasts for 20–25 months, prepupal stage for 50–182 days and the pupal stage for 40–90 days (Rahman and Khan, 1942). Adult longevity is 4–5 months (June to November) and a single life cycle occupies 23–32 months. The full grown grubs overwinter from October to March but smaller, less than a year old continue to feed even during winter.

Quetta borer *Aeolesthes sarta* Solsky is a major pest of walnut in Kashmir and Baluchistan in Pakistan. Other well known hosts of this pest are apple, peach, fig, apricot, almond, quince, *chinar*, elm, poplar and willow. It makes holes surrounded by small rotting areas on stem and branches. The larval galleries are filled with fibrous excreta and chewed wood is visible after the bark has fallen off. The dark brown (3–5 cm long) beetles having fine satin grey pubescence mainly emerge during May- June and lay about 50 milky white elliptical eggs (4 mm long) within 7–10 days of mating, in groups of 5–10 in wounds on the bark, or on broken ends of branches or in pits on stem. Egg hatch in 10–14 days and the neonate larva (6 mm long) bores inside, initially feeding on soft sappy bark and subsequently makes galleries in the bark and sapwood. In winters, it bores inside the heart of the tree, makes pupal chamber in spring and pupates. In most cases, pupation takes place during May-June and pupal stage lasts 4–5 months. Beetles rest for several months in pupal chambers and start coming out in May. The life cycle is completed in two years (Janjua and Samuel, 1941).

Cherry stem borer *Aeolesthes holosericea* Fab. occasionally bores the walnut trees in Himachal Pradesh and Jammu and Kashmir. It is a minor pest of walnut.

Walnut tree borers can be managed by pruning and destruction of newly attacked branches and killing the older larvae by inserting flexible wire into the live tunnels. In the case of severe infestation, grubs are killed by inserting solid fumigant (0.05% paradichlorobenzene or 0.2% aluminium phosphide) in the larval gallery already cleaned by iron spoke and plugging the hole with mud plaster, as mentioned under cherry for the control of stem borer.

Root Borer *Dorysthenes hugelii* Redt. (Cerambycidae: Coleoptera), a major pest of apple, occasionally damages the roots of walnut trees (Butani, 1979; Sharma and Khajuria, 1997). The affected trees become weak and may even get uprooted in due course if feeding grubs are not controlled.

The caterpillars of the leopard moth, *Zeuzera* spp. (Cossidae: Lepidoptera) attack shoots and stems of walnut trees in India. It is a polyphagous pest and attacks several fruit trees. Eggs are laid singly in cuts and wounds. On hatching, the young caterpillars feed just below the bark and later bore into the tender shoots and feed within for 8 to 9 months, come out and bore the thick stems and main branches to feed for another 10 to 11 months. Consequently, young trees may succumb while in older ones the leaves may wither and branches may die thereby affecting the growth of the tree. The life cycle is completed in two years or longer depending upon the climatic conditions. For detailed biology and control refer to apple and cherry.

### Walnut Weevil, *Alcidodes porrectirostris* Marshall

The walnut weevil is a serious and specific pest of walnut. It has been reported from all the walnut growing areas in the Himalayan and sub mountane region (Gaffar and Bhatt, 1990). Damage is caused by grubs as well as adults. Female weevils excavate pits on fruits for laying eggs. One to 4 eggs are laid per fruit. Newly emerged grubs bore into the fruits, feed on the kernels and reduce them into a black mass. Feeding grubs also cause premature dropping of fruits from April to August, maximum being during June-July. The affected fruits also show dark brown spots which are infact dried resinous excretions. Full-fed grubs pupate inside the fruits and the adults emerge from the fruits by bitting circular holes. The weevils feed on petioles, female floral buds, tender shoots and even on young fruits (Husain and Khan, 1949).

The eggs are 1.0–1.5 mm long and transluscent. Full grown grubs and pupae are creamy white in colour and measure 12–16 mm and 9–12 mm in length, respectively. Adult weevils are jet black when young turning dark brown with age, 9–12 mm long, with a prominent snout bending downwards. The incubation period is 3–5 days, larval period 13–22 days and pupal period 9–17 days. Pupation takes place inside the fruit. The first generation grubs feed on developing kernels during April to July and do more damage than second generation (June-September) grubs which feed inside the rind and fail to penetrate the hard shell. Adult longevity is 4–6 weeks during active season whereas hibernating adults live for 6–7 months. After completing two generations between April and September, the weevils continue to feed as long as twigs are green and hibernate from November to March by hiding under the bark of trees, debris, clods or in cracks and crevices in the soil.

Cost effective and satisfactory control of this pest is difficult to achieve unless fallen fruits are collected and destroyed every 10 days to kill the grubs. This operation is useful and efficient only if all the walnut tree growers in a particular area co-operate and carry out the destruction of fallen fruits simultaneously. If feasible, spray the affected trees with quinalphos (0.04%)or dichlorvos (0.05%). Grubs of this weevil are parasitized by a tachinid fly, *Hemidegera villeneuvei* Bar.

**Codling Moth, *Cydia pomonella* (L.) (Olethreutidae)**

Codling moth is a widely distributed pest throughout the world wherever apple is grown. However, in India this pest is confined to Ladakh region of Jammu and Kashmir (Pawar *et al.*,1981). Besides apple, it attacks walnut, pear, apricot, hawthorn, quince and crab apple. Three generations of the pest damage walnut fruits in India. The first generation larvae enter the growing fruit through the blossom end. The infested immature fruits drop down from the tree and serve as breeding site. The second and third generation larvae enter the nut through the side of the husk. The damaged nuts are unmarketable. This pest is categorised as the most serious pest of walnut in Turkey, the third largest producer of walnuts, after USA and China (Barut, 1996; Guclu *et al.*, 1995).

## REFERENCES

Barnett, W.W. and Hendricks, L.C. 1992. Comparison of pheromone trap catches to adult emergence and oviposition of peach twig borer, *Anarsia lineatella* Zeller, in California almond orchards. *Acta Phytopathologica et Entomologica Hungrarica,* **27** (1–4): 85–88.

Barut, E. 1996. Overview of walnut culture in Turkey. *Chronica Horticulturae,* **36**(1): 14–14.

Beeson, C.F.C. 1941. *The Ecology and Control of Forest Insects.* F.R.I., Dehradun. pp. 1007.

Behdad, E. and Barooti, S. 1978. Biological studies on *Eulecanium coryli* L., the pest of fruit trees in Erfahan. *Entomologie et Phytopathologie Appliquees,* **46** (1–2): 55–65.

Bhagat, K.C., Kaul, V.K., Masoodi, M.A. and Kaul, V. 1994. Development and predation of *Metasyrphus corollae* (Fabr.) (Diptera: Syrphidae) on dusky veined walnut aphid, *Callaphis juglandis* Kalt. *Pest Management and Economic Zoology,* **2** (2): 151–152.

Bhalla, O.P. and Gupta, P.R. 1993. Insects Pests of Temperate Fruits. *In: Advances in Horticulture.* K.L. Chadha and O.P.Pareek (eds). Vol. 3. *Fruit Crops.* Part 3. Malhotra Publishing House, New Delhi. pp. 1567–1589.

Bhalla, O.P. and Pawar, A.D. 1977. *A survey study of insect and non-insect pests of economic importance in Himachal Pradesh.* Department of Entomology and Zoology, College of Agriculture, Solan.

Bhardwaj, S.P. 1982. Note on a new record of leopard moth, *Zeuzera multistrigata* Moore (Lepidoptera: Cossidae), on cherry. *Indian Journal of Agricultural Sciences,* **52**(12): 881–882.

Butani, D.K. 1979. *Insects and Fruits.* Periodical Expert Book Agency, New Delhi. pp. 316–330.

Chander, R. and Dogra, G.S. 1983. Lepidopterous insects infesting apple in Himachal Pradesh. *In: Proceedings of Workshop on High Altitude Entomology and Wildlife Ecology.* Zoological Survey of India, Calcutta.

El-Tariki, A.W. and El-Sharif, R. 1987. A survey of the important insects, diseases and other pests affecting almond trees in Tunisia. *Arab Journal of Plant Protection,* **5**(1): 1–7.

Gaffar, S.A. and Bhatt, A.A. 1990. Record of walnut weevil, *Alcidodes porrectirostris* (Marshall) (*Neomecyslobus porrectirostris*): A new insect pest on walnut in Karnah, Kashmir. *Bulletin of Entomology,* **31**(2): 231.

Guclu, S.; Hayat, R. and Ozbek, H. 1995. An investigation on phytophagous insect species on walnut (*Juglans regia* Linnaeue) in Erzurum and its neighbouring provinces. *Turkiye Entomoloji Dergisi,* **19**(2): 137–145.

Gupta, B.P.; Joshi, N.K. and Joshi, L.D. 1977. Some important defoliating and fruit eating beetles of temperate fruit trees: A survey and control. *Progressive Horticulture,* **8**(4): 53–60.

Hogmire, Jr. H.W. 1995. *Mid Atlantic Orchard Monitoring Guide.* Northeast Regional Agricultural Engineering Service, Ithaca, New York. pp. 361.

Hoy, M.A.; Cave, F.E. and Caprio, M.A. 1991. Guthion resistant parasite ready for implementation in walnuts. *California Agriculture,* **45** (5): 29–31.

Husain, M.A. and Khan, M.A.W. 1949. Bionomics and control of the walnut weevil (*Alcidodes porrectirostirs* Mshll.) (Alcidinae: Coleoptera). *Indian Journal of Entomology,* **11**: 77–82.

Janjua, N.A. and Samuel, C.K. 1941. *Fruit Pests of Baluchistan.* ICAR, New Delhi, *Misc. Bull.No. 42.*

Kehat, M.; Anshelevich, L.; Dunkelblum, E. and Greenberg, S. 1994. Sex pheromone traps for monitoring the peach twig borer, *Anarsia lineatella* Zeller: effect of pheromone components, pheromone dose, field aging of dispenser, and type of trap on male captures. *Phytoparasitica,* **22**(4): 291–298.

Kozer, F. and Konstantinova, G.M. 1981. San Jose scale in deciduous fruit orchards of some European countries—survey of scale insect infestations in European orchards (No. IV). *Bulletin Organisation Europeenne et Mediterraneenne pour la Protection des Plantees,* **11**(2): 127–133.

Kuprii, A.V. and Damrose, I.P. 1986. Aphids in cherry orchards in the conditions of the southern Ukaraine. *Trudy Latviiskoi Sel skokhozyaistvennoi Akedemii,* **234**: 12–15.

Masoodi, M.A.; Kaul, V.K.; Bhagat, K.C. and Bhat, O.K. 1989. Bioecology of dusky veined walnut aphid, *Callaphis juglandis* on walnut in Kashmir. *Indian Journal of Plant Protection,* **17**(1): 139–141.

Masoodi, M.A.; Kaul, V.K.; Bhagat, K.C. and Singh, T. 1987. Control of *Lymantria obfuscata* Walker on walnut in Kashmir. *Indian Journal of Plant Protection,* **15**: 15–17.

Mohamad, M.A. and Al-Mallah, N.M. 1990. Preliminary field observations on the ecology and biology of the apricot leaf aphid, *Hyalopterus pruni* G. (Homoptera: Aphididae) in Mosul region -Iraq. *Arab Journal of Plant Protection,* **8**(1): 1–5.

Mustafa, T.M. and Hamdan, A.S. 1989. Flight activity of the short tailed almond aphid, Brachycaudus amygdalinus (Hompotera: Aphididae) in the central highlands of Jordan. *Bollettino della Societa Entomologica Italiana,* **121**(3): 176–180.

Pawar, A.D. and Parry, M. 1989. Record of natural enemies of important fruit pests in Ladakh (J&K). *Indian Journal of Plant Protection* **17**(2): 291–292.

Pawar, A.D.; Tuhan, N.C.; Balasubramanium, S. and Parry, M. 1981. Distribution, damage and biology of codling moth, *Cydia pomonella* (L.). *Indian Journal of Plant Protection,* **9**: 111–114.

Rahman, K.A. and Khan, A.W. 1942. Bionomics and control of *Aeolesthes holosericea* F. (Cerambycidae: Coleoptera). *Proceedings, Indian Academy of Sciences (B),* **15**: 181–185.

Reil, W.O.; Johnson, T.W.; Frofita, J.C.; Davis, C.S.; Hendricks, L.C. and Rough, D. 1981. Monitoring peach twig borer in almonds with sex pheromone traps. *California Agriculture,* **35**(9–10): 19–21.

Rice, R.E. and Jones, R.A. 1988. Timing post bloom sprays for peach twig borer (Lepidoptera: Gelechiidae) and San Jose scale (Homoptera: Diaspididae). *Journal of Economic Entomology,* **81**(1); 293–299.

Rolli, K. 1974. Aphid attack on almond trees in Tunisia. *Gesunde Pflanzen,* **26**(1): 15–16.

Roltsch, W.J.; Zalom, F.G.; Barry, J.W.; Kirfman, G.W. and Edstrom, J.P. 1995. Ultra-low volume aerial applications of *Bacillus thuringiensis* variety *Kurstaki* for control of peach twig borer in almond trees. *Applied Engineering in Agriculture,* **11**(1): 25–30.

Schonbeck, H. 1988. Biological control of aphids on wild cherry. *Allgemeine Forstzeitschrif,* **34**: 944.

Sharma, H.K.; Chander, R. and Thakur, J.R. 1990. Incidence of flat headed peach tree borer *Sphenoptera lafertei* Thomson (Coleoptera: Buprestidae) on stone fruits in Himachal Pradesh. *Indian Journal of Agricultural Sciences,* **60**(10): 705–707.

Sharma, H.K. and Gautam, D.R. 1994. Evaluation of some almond rootstocks for tolerance to the flat headed peach tree borer *Sphenoptera lafertei* Thomson (Coleoptera: Buprestidae). *Pest Management and Economic Zoology,* **2**(2): 153–155.

Sharma, J.P. and Khajuria, D.R. 1997. Bioecology and management of apple root borer *Dorysthenes hegelii* Redtenbacher (Cerambycidae: Coleoptera) in HP. *In National Seminar on Plant Protection Towards Sustainability*. Rajendranagar, Hyderabad. Plant Protection Association of India. Abstract R-13.

Sharma, P.L.; Attri, B.S. and Aggarwal, S.C. 1969. Beetles causing damage to pome and stone fruits in Himachal Pradesh and their control. *Indian Journal of Entomology,* **31**: 377–379.

Sharma, P.L.; David, A.C. and Bhalla, O.P. 1963. Flat-headed borer, *Sphenoptera lafertei* Obb. and its control in Himachal Pradesh. *Himachal Horticulture,* **4**: 1–5.

Singh, C. 1964. Temperate Fruits. *In: Entomology in India*. Entomological Sociaty of India, New Delhi. pp. 213–276.

Stimmel, J.F. 1982. Seasonal history of the white peach scale, *Pseudaulacaspis pentagona* (Targ. Tozz.) (Homoptera: Diaspididae) in northwestern Pennsylvania. *Proceedings of the Entomological Society of Washington,* **84**(1): 128–133.

Triggioni, O. 1973. Contribution to knowledge of the activity of the natural enemies of aphids on almond (*Amygdalus communis*) in the countryside round Basi. *Entomologica,* **9**: 119–135.

Tripathi, G.M. and Gupta, B.P. 1985. Studies on collection of beetles on light trap at Chaubattia in UP Hills. *Progressive Horticulture,* **17**(4): 347–353.

Veeresh, G.K. 1988. White grubs as pests of forest and fruit trees in India and their management. *In: V.K. Gupta and N.K. Sharma (eds). Tree Protection.* Indian Society of Tree Scientists, Solan.

Yadav, C.P.S.; Saxena, R.C.; Mishra, R.K. and Dadheech, L.N. 1977. Population management of white grub, *Holotrichia consanguinea* Blanch. in an agro-eco-system. *Indian Journal of Entomology,* **39**: 205–210.

Youngman, R.R. and Barnes, M.M. 1986. Monitoring peach twig borer by standardized trapping methods. *California Agriculture,* **40**(3–6): 29–30.

*Pest Management in Horticulture Crops*
L.R. Verma, A.K. Verma and D.C. Gautam (eds)


*Chapter 14*

# Pests of Mango

*R.P. Srivastava*

## INTRODUCTION

DESPITE THE LARGEST production base of mango in the world, with respect to number of varieties grown and area under cultivation, the productivity of mango fruit in India is rather low. Besides some inherent limitations on high yielding susceptible cultivars and incidence of diseases, this might be attributed to high incidence of insect pests. Also, many insects hitherto of minor importance, have attained the status of major pests.

Mango is highly susceptible to a number of insect pests which include hoppers, mealy bug, stem borer, shoot borer, fruit flies, stone weevil, leaf and flower feeding caterpillars, fruit borers, fruit sucking moths, leafminers, termites, scale insects, thrips, aphids and ants. A brief account on the biology, nature of damage and management strategies of important pests, are given in this chapter.

## MANGO HOPPERS

***Idioscopus Clypealis* Lethierry, *I. nitidulus* (Walker) and *Amritodus atkinsoni* (Lethierry)**

The infestation by hoppers causes colossal loss to the mango and they are considered as the most serious pests of this fruit. Three species of hoppers namely, *Idioscopus clypealis* Lethierry, *I. nitidulus* (Walker) and *Amritodus atkinsoni* (Lethierry) are generally noticed in mango orchards with varied degrees of infestation in different agro-climatic regions of the country. Apart from India, *I. clypeallis* has also been reported from Pakistan, Bangladesh, Taiwan, Myanmar, Srilanka and the Philippines (Alam, 1962; Kato, 1928; Baker, 1915; Palo and Garcia, 1935).

There is no authentic data available regarding the seriousness of this pest and the losses it causes. However, according to Ramachandra Rao (1930), these hoppers cause a loss of 20 to 100 per cent to inflorescence.

### Biology

All the three species of hoppers have a wedge shaped body with a broad head and narrow abdomen towards the back. The adults of *A. atkinsoni, I. nitidulus* and *I. clypealis* measures 4.2–5 mm, 4–4.8 mm and 3.5 mm in length respectively. *A. clypealis* has been reported to be the most prolific breeder but it breeds only during January-February, *I. nitidulus* breeds throughout the flowering season while *A. atkinsoni* is a shy breeder and generally confines to vegetative shoots but may also be found on inflorescence.

The hopper lays minute eggs singly, embedding them into the plant tissues. Nymphs hatch out, move actively and suck sap from inflorescence and shoots. The nymphs pass through three instars to become adult. The egg period ranges from 4 to 7 days, nymphal period from 8 to 13 days and the life cycle is completed in 15 to 19 days. Srivastava and Butani (1972a) reported that hoppers have two or more broods in a year with two peak periods, i.e. spring generation in February to April and summer generation from June to August.

### Management

#### *Biological control*

***Predators:*** Srivastava *et al.* (1979) recorded twelve species of predatory spiders namely, *Phidippus* sp., *Rhene indicus* Tikader, *Marpisa* sp., *Oxyopes shweta* Tikader, *Cyrtophora* sp., *Cyrtophora cicatrosa* Stoliczka, *Araneus sinhagadensis* Tikader, *Cheiracanthium donieli* Tikader, *Linphia* sp., *Stegodyphus sarasinorum* Karsch, *Uloborus* sp., *Hersilia savigryi* Lucas and *Theridion indica* Tikader. They also observed *Coccinella septempunctata* Linnaeus, *C. transversalis* Fabricius and *Menochilus sexmaculatus* Fabricius, preying upon *I. clypealis*. The other important predators reported are *Mallada boninensis* (Okanoto) and *Chrysoperla lacciperda* preying upon *I. nitidulus* (Fasih and Srivastava, 1990). A mite, *Bochartia* sp. was also observed predating on the nymphs.

***Parasitoids:*** The parasitoid, *Pipunculus annulifemer* Brun was found parasitizing the adults of all the three species of mango hoppers during November and December. Other species, *Pyrillosenos paracompactus* Pierce and *Epipyrops fulginosa* (Tams.) were also observed parasitizing the hoppers in many orchards in Uttar Pradesh. The egg parasites, *Aprostocetus* sp., *Gonatocerus* sp., *Polynema* sp. and *Tetrastichus* sp. were recorded on *I. nitidulus* and *A. atkinsoni* (Fasih and Srivastava, 1990; Chou and Chou, 1990; Herting and Simmonds, 1972).

***Pathogens:*** The entomogenous fungus, *Verticillium lecanii* (Zimmerman) and *Beauveria bassiana* (Balsamo) were found to be pathogenic to *I. clypealis* (Srivastava and Tandon, 1986; Tripathi *et al.*, 1990; Shukla and Prashad, 1984). Natural occurrence of the fungus, *Isaria tax* on *I. clypealis* and *A. atkinsoni* was reported from Tripura (Kumar *et al.*, 1985). The fungus, *Hirsutella aversicolor* was recorded on *I. clypealis* and *Isaria stellata* on *I. nitidulus* (Nayar *et al.*, 1979). Srivastava (1997 c) and found that Naturalis-L (Boverine) (0.5%) and *B. bassiana* (0.2%) were effective against *I. nitidulus*.

***Plant products:*** Essential oils like *Mentha spicata* Linn., *Cymbopogan winterianus* Jowitt, *C. flexuosus* (Steud.) wats. and *Mentha arvensis* (Linnaeus) at 0.125 per cent and *Mentha citriata* Ehrhart at 0.25 per cent were found effective in field against *I. nitidulus* (Anonymous, 1995).

***Insect growth regulator:*** Buprofezin was found to be effective against *I. nitidulus* in field trial at 0.0125% (Srivastava, 1997 b).

***Chemical control***

Tandon and Lal (1979) found carbaryl 0.15%, monocrotophos 0.04%, phosphamidon 0.05% and methyl parathoion 0.05% to be effective against these pests. Srivastava and Verghese (1985) reported that fogging with mixture of malathion and diesel (1.5: 8.5) caused drastic reduction in hopper population. Kumar *et al.* (1985) found methyl demeton and monocrotophos both at 0.05% effective against these pests. Carbaryl 0.2%, quinalphos 0.06%, monocrotophos 0.05%, chlorpyriphos 0.04%, fenitrothion 0.075% and dimethoate 0.06% were reported to be effective against *I. nitidulus* (Anonymous, 1995). Srivastava *et al.* (1987) reported quinalphos (0.063%), carbaryl (0.2%) and chlorpyriphos (0.04%) effective against *A. atkinsoni*. Pingle and Patel (1988) observed carbaryl (0.1%), permethrin (0.01%) and dimethoate (0.03%) effective against *A. atkinsoni* and *I. nitidulus*. Srivastava and Verghese (1989) recommended monocrotophos (0.04%), cypermethrin (0.002%) and permethrin (0.05%) for the control of *I. Clypealis*.

***Integrated pest management programme of mango hoppers***

Srivastava (1997a) suggested the following programme for the management of mango hoppers:

- July-August: Avoid dense plantation in mango orchards. Plant grafted plants at proper distance.
- September-December: The orchards should be kept clean by ploughing and removal of weeds. Pruning of overcrowding and overlapping branches should be done.
- January: Fogging with mixture of malathion and diesel in the ratio of 1.5: 8.5 should be done.
- February: Spray 3% alcohol extract of *Alpinias galanga* rhizomes in the first fortnight and buprofezin 0.0125% (Applaud 25 WP) in second fortnight, if required.
- March: Spray 0.5% alcohol extract of neem seed kernels in the first fortnight and 0.125% spearmint oil, citronella oil or lemon grass oil in second fortnight, if required.
- April: Spray 0.0125% buprofezin in the first fortnight, if required.

## MANGO MEALYBUG

***Drosicha mangiferae* (Green)**

Mealybugs are other serious pests of mango. The giant mealy bug, *Drosicha mangiferae* (Green) is the most commonly occurring species. Besides, *Perissopneumon ferox* Newst. also infests mango. *D. mangiferae* is found in all the mango growing regions of India. It is quite serious in Punjab,

Uttar Pradesh, Bihar and Delhi. It is also reported as a serious pest of mango in Bangladesh, Pakistan and China (Alam, 1962). The nymph and adult females suck the sap from inflorescence, tender leaves and shoots. Excessive desapping leads to wilting and drying of foliage, ultimately resulting in poor fruit set.

### Biology

The fertilized females start migrating from 3rd week of April till beginning of May below the soil and lay eggs which remain in diapause in the soil from May to December. The eggs are oval in shape and when freshly laid, are shining pink in colour but gradually change to yellow. A female lays 300 to 400 eggs in second and third week of May and the hatching occurs in following January. There are three nymphal instars. The adult males are morphologically distinct from the females as they are crimson coloured with brownish-black fore wings while the females are wingless, mealy white, elliptical in shape and body covered with numerous minute hair.

### Management

#### *Biological control*

***Parasitoid:*** A species of *Phygadenon* was found to oviposit in the second and third instar nymphs. Thompson and Simmonds (1964) reported a parasitoid, *Gitonoides persipicax* Knab. (Diptera, Drosophillidae) from India while Rafiq and Ghani (1972) reported *Tetrastichus* sp. from Rawalpindi, Pakistan. Srivastava and Verghese (1987) reported *Gitona* (*Gitonoides*) *magnipalpis* Duda parasitizing the nymphs. Fasih and Srivastava (1990) reported *Cryptochetum* sp. parasitizing the mealybug.

***Predators:*** A large number of predators control mealybug. These are *Adonia doubleayi* Mulsant, *Aulis vestitus* Mulsant, *Coccinella transversalis* Fabricius, *C. septempunctata* Linnaeus, *C. undecimpunctata* L., *Hippodamia variegata* (Goeze), *Menochilus sexmaculatus* Fabricius, *Rodalia fumida* Mulsant, *R. guerini*, *Sumnius cardoni* Welse, *S. renardi* Wiesl., *Chrysoperla scelestis* Banks, *Bochartia* sp., *Leptus* sp., *Cyrtophora cicatrosa* Stolicza, *Stegodyphus sarasinorum* Karsch, *Hersilia savigryi* Lucus, *Oxyopes shweta* Tikadir, *Marpissa* sp., *Phiddipus* sp., *Rhene indicus* Tikadir, *Linphia* sp. and *Argya caudata caudata* (Rehman and Latif, 1944; Srivastava *et al.*, 1979; Herting and Simmonds, 1972).

***Pathogens:*** The entomogenous fungi reported on this pest are *Beauveria bassiana* (Balsamo) and *Aspergillus parasiticus* Speare (Fasih and Srivastava, 1988).

#### *Plant products*

Neem products, 1 per cent oil-based and 2 per cent kernel-based concentrates were reported to cause 83.6 and 77 per cent mortality of the nymphs, respectively. Buprofezin at 0.0625% and 3% alcohol extract of *Alpinia galanga* Willd. were also found effective against mealybug nymphs (Srivastava, 1997d).

*Mechanical control*

The first instar nymphs climb the tree by crawling through the main trunk. To stop the nymphs from climbing, various banding methods have been suggested from time to time. Srivastava and Butani (1972b) recommended alkathene sheet (400 gauge) 30 cm wide, wrapped over the tree trunk about 30 cm above the ground level. However, Chandra *et al.* (1991) found alkathene band, 20 cm wide and 200 gauge thick, to be most effective. Due to the slippery nature of the alkathene sheet the nymphs are unable to crawl over the band. Srivastava (1980) reported that the double girdle bands were better in effectiveness as they stopped the nymph movement to crawl on the tree completely.

*Chemical control*

Srivastava and Butani (1972b) suggested sprays of 0.05% diazinon, malathion or dichlorvos for the control of mealybugs.

*Integrated pest management programme for Drosicha mangiferae (Green)*

Srivastava (1997a) suggested the following programme for the management of *Drosicha mangiferae*:

- October: Flood the orchard with water to kill mealybug eggs.
- November: Plough the orchard to expose the eggs to solar radiation.
- December: In the middle of month, 25 cm wide 400 gauge alkathene sheet may be fastened to the tree trunk. Also rake the soil around the tree trunk and mix 2% methyl parathion dust @ 200 g per tree in the last week.
- January: Spray spore suspension of *Beauveria bassiana* containing 1 x 107 viable spores/g in the last week.
- February: Spray 2% alcohol extract of *Alpinia galanga* rhizomes in the first week or 2% alcohol extract of neem before opening of flowers, if required. Release predators like *Sumnius renardi* Wiesl. or *Rodalia fumida* Mulsant, etc.
- March: Spray 0.125% spearmint oil, citronella oil or lemongrass oil. Avoid spraying on open flowers.
- April: Spray diazinon 0.1% to control adults, if necessary.

## MANGO GALL MIDGES

Mango suffers from about 20 species of cecidomyids infesting various parts of the plant. Of these, four species attack flowers and the remaining feed on leaves and twigs. Mani (1952) gave a comprehensive account of the gall midges affecting mango crop and described three new species, *viz.*, *Asynapta mangiferae* on mango bark, *Dasyneura mangiferae* (Procystiphora) Felt on flower buds, *Erosomyia mangiferae* (Felt.) in leaf-blister galls from West Indies and *Mangodiplosis mangiferae* causing galls on inflorescence axis from Brazil. Earlier in 1935, he reported *Rhabdophaga mangiferae*

Mani causing galls on shoots. Grover (1965) described three new species of mango midges namely, *Dasyneura amaramanjarae* Grover, *Procystiphora indica* Felt. and *Erosomyia indica* Grover.

### *Erosomyia indica* Grover

The midge attacks the preflowering shoot buds, buds and axis of inflorescence, panicles, newly formed fruits and the post-flowering shoot buds (Prashad, 1971). The adult is a minute yellowish bifiline midge with greyish back and broad wings. The males are bigger in size than females. The genitalia in male is darker than the body. The females have a long ovipositor. The eggs are laid in between the leaves and buds either singly or in groups of 4–6 in one sitting and the female lays about 150 eggs in its life time. The newly hatched larvae penetrate the tissues and form small blister like galls. The larvae feed within the galls and at the time of pupation come out with a swift jump and pupate in the soil. Black spots of varying sizes are the characteristic symptoms of their attack. Also, the inflorescence gets bent at an angle at the site of infestation. The larvae get entry into the developing ovary and develop inside the fruit. The attacked fruit turns pale, becomes deformed, stops growing and finally drops down. The exit holes are usually observed in infested fruits. The pupal stage lasts for 4 to 6 days.

### *Dasyneura amaramanjarae* Grover

This species is a a blossom feeder. The larvae feed on filaments of the stamen, and the style of the gynoecium and other internal organs. The female midge lays 40 to 50 eggs in many sittings. The number of eggs in a bud varies from 3 to 6 although as many as 12 eggs have also been recovered from a single bud. The eggs are minute, cylindrical, structured, rounded at one end while the other end is drawn out to form a delicate stalk. The first instar larva takes 40 to 45 hr to develop, while the second and third instar take 48 to 50 hr each in February and 72 hr in March and the fourth instar takes 50 to 60 hr to develop. The pupation takes place in the soil and pupation period varies from 4 to 6 days depending upon relative humidity and temperature. The life cycle is completed in 15 to 16 days.

### *Procystiphora mangiferae* Felt. and *P. indica* Grover

*P. mangiferae* was first reared on mango flowers at Coimbatore, Tamil Nadu. This species was found to attack 60 to 70 per cent of mango buds in Cochin while *P. indica* was found attacking mango blossom in Uttar Pradesh. The various phases of life cycle of the two species are similar. The eggs are translucent, elongate and cylindrical. The first instar maggots feed inside the buds. There are four larval instars. The fully-fed maggot of *P. mangiferae* spins a coccon and pupates inside the bud while in *P. indica*, pupation takes place in the soil. The pupal period varies from 4 to 6 days.

### Management

#### *Cultural control*

Tillage and cultivation of the soil proves effective in controlling the midge. Deep ploughing helps in killing pupae. Gangwar (1974) observed in Bihar that cultivars Jardalu and Bombay were highly susceptible to *E. indica* and *P. mangiferae* whereas, Safeda was least susceptible while Kalipad and Rumani were resistant to these pests. However, the two prized varieties of North India, Dusehri and Langra, were highly susceptible.

***Biological control***

Prashad and Grover (1974) recorded three parasitoids on *E. indica,* viz., *Platygaster* sp., *Systasis dasyneurae* Mani and *Eupelmus* sp. Other parasitoids recorded are *Aphanogmus* sp. on *Procontarinia* spp. (Austin, 1984); *Aprostocetus sankarani* on *Procontarinia* sp. (Boucek, 1985); *Aprostocetus* sp. on *E. indica* (Grover, 1986). The predators reported by Grover (1986) are *Camponotus* sp., *Formica rufa* L. and *Oecophylla* sp. on *E. indica*, *D. amaramanjarae* and *Procystiphora* spp.

***Chemical control***

Prashad (1972) recommended sprays of phosphamidon (0.02%) + diazinon (0.03%). Srivastava and Tondon (1983) found fenitrothion (0.05%), dimethoate (0.04%) or diazinon (0.04%) effective against the gall midges when sprayed twice at 15 day interval at the time of bud burst.

***Integrated pest management programme for gall midges***

Cultural practices, such as summer ploughing, raking etc. have been found useful in reducing the midge population in the following year. Cutting of heavily infested inflorescence at the early stage of infestation and destroying them is also useful. Monitoring of the population with the help of rectangular wooden traps placed under the mango tree helps in synchronising the control measures with initiation of pest attack. Foliar sprays of fenitrothiion (0.05%), dimethoate (0.04%) or diazinon (0.04%) may be done at the bud burst stage in the case of *E. indica* and at the time of hatching of larvae in other species. Release of biocontrol agents, such as the parasitoid, *Platygaster* sp. and others may help in the management of gall midges on mango (Srivastava, 1997a).

## MANGO SHOOT GALL PSYLLA

***Apsylla cistellata*** **(Buckton)**

The mango shoot gall psylla, *Apsylla cistellata* (Buckton), is a serious pest of mango causing galls in the leaf axils which inhibit the formation of inflorescence. Besides, another species namely, *Pauropsylla brevicornis* Crawford was reported by Chandrika and Nair (1971) from Kerala, India and *P. nigra* by Lee and Wen (1980) from China. Gupta (1953) reported that the pest was localized to tarai area, upto 30 to 60 km from the base of Himalayas. However, presently it is a common pest in Uttar Pradesh, Bihar and West Bengal and also in Bangladesh (Srivastava, 1997a). The pest seems to have spread to the major mango growing areas of North India.

The scaly leaves formed around the central axis give rise to galls due to feeding by nymphs inside the buds. The nymphs suck the sap and secrete whitish sticky material in drops. These secretions bring about the swelling of the buds and ultimately gall formation.

The infestation usually starts in October. The female lays upto 166 eggs, inserting them in the mid rib of the leaf in March-April in two parallel rows. The eggs are transparent and white, and egg hatch in the second week of August. Freshly hatched nymphs are minute creatures having yellowish coloration with red markings all over the body. There are five nymphal instars and the nymphal period extends from August to February. The adult psyllid measures 3 to 4 mm in length. The thorax and head are black and abdomen is light brown in colour. The females are slightly longer than males. The adults are short lived and die soon after oviposition (Mathur, 1946).

### Management

#### *Mechanical control*

Pruning of infested shoots bearing galls is very useful.

#### *Biological control*

Two species of predators namely, *Baccha pulchrifrons* and *Micromus timidus* Hagen, were reported to suppress the population of mango gall psylla (Beeson, 1941; Srivastava and Verghese, 1987).

#### *Chemical control*

Monocrotophos (0.04%) or dimethoate 0.06% give effective control of the psyllid when sprayed during mid of August (Singh and Mishra, 1978). Srivastava *et al.* (1982) recommended quinalphos 0.05%, monocrotophos 0.05%, phosphamidon 0.075% + urea 1% or dimethoate (0.06%) for the control of gall psylla.

## FRUIT FLIES

Fruit flies are the most serious pests of mango throughout the world. In view of the embargo enforced on India by many countries, the export potential of mango has reduced tremendously. Eight species of fruit flies infesting mango in India include *Bactirocera dorsalis* (Hendel), *B. zonatus* Saunders, *B. [Hemigymnodacus] diversa* Coquillett, *B. hageni* deMeijere, *B. cucurbitae* Coquillett, *B. (Zeugodacus) tau* Walker, *Dacus correctus* Bezzi and *D. incisus* Walker (Kapoor, 1970). Apart from these, a large number of other mango fruit flies occur elsewhere in the world but fortunately these species are not found in India and hence, strict quarantine measures are required to avoid the introduction of these species in the country.

### *Bactrocera (Dacus) dorsalis* (Hendel)

This species, called the Oriental fruit fly, is considered to be the main pest of mango affecting the export of fresh mango fruits from India. Besides, it is reported from Pakistan, Ceylon, Myanmar, Siam, Java, Malaya, Mauritius, the Philippines, East Indies, Amboina, China, Formosa, Taiwan, Bonin Islands, Hawaii and Australia.

The adult fly punctures the fruit with the help of its highly chitinized ovipositor. The oviposition punctures are not easily visible on the fruits. The eggs are laid in clusters of 2–15 eggs inside the fruit tissues.

### Management

#### *Cultural and mechanical*

Ploughing of fields, 2–3 times in a year, reduces the pest population by exposing the pupae to natural enemies and to sun's heat. Collection and destruction of infested fruits also help in reducing the population. Early harvesting of fruits before ripening is helpful in reducing the incidence of fruit flies.

#### *Chemical control*

Good control of *B. dorsalis* and *B. cucurbitae* using bait spray of 25 g sugar and 10 ml malathion in one litre water was achieved by recommended sprays of carbaryl (2%) or dimethoate (0.06%) for effective control of fruit flies on mango reported successful control of fruit flies using bait sprays of 20 ml malathion or 50 ml diazinon and 200 g *gur* or molasses in two litre water.

Shukla and Prashad (1984) reported that deltamethrin at 0.0025% gave the best results followed by fenthion (0.05%), carbaryl (0.1%) and dimethoate (0.3%). Fenthion was also reported to be effective who also found that malathion and protein hydrolysate mixture was better than other bait combinations. Recently, a vapour heat equipment has been developed in Japan in which mango fruits are treated at 110 to –130°F (43.3 to 46.1°C) and the larvae are killed within 30 minute treatment. This treatment is acceptable for quarantine purposes.

#### *Integrated pest management programme for fruit flies*

Ploughing the orchards and exposing the diapausing pupae to sun's heat, parasitoids and predators during December-February is helpful in reducing the pest population. Monitoring of adult emergence with methyl eugenol traps can also be done.

Early harvesting of matures fruits, collection and proper disposal of infested and fallen fruits are also helpful.. Bait sprays of malathion (0.1%) + molasses (0.1%), given thrice at fortnightly intervals before ripening of fruits are effective. Spraying of fenthion 0.5% or carbaryl 0.1% is also effective. Control the fruitfly infestation *in situ* by hot water treatment or vapour heat treatment is also possible (Srivastava, 1997a).

## MANGO STONE WEEVIL

***Sternochetus mangiferae*** **Fabricius**

The mango stone weevil is a very important pest in view of its being the major constraint in the export of fresh fruits to foreign countries. It is a specific pest of mango and prefers sweet varieties like Alphonso, Bangalora, Neelum etc. The pest is more sensitive to low temperature and low humidity and hence is more common in South India where late varieties suffer the most. Two species of stone weevil are found namely, *S. mangiferae* and *S. frigidus* [*gravis*] Fabricius.

Shukla *et al.* (1985) observed that the adult weevils emerge from the fruits during June-July and hibernate till March next year.

Field sanitation by removing and destroying all fallen fruits and seeds helps in reducing the infestation. A baculovirus has been reported on *S. mangiferae* and a fungus, *Aspergillus* sp. has been reported on *S. gravis* found carbaryl (0.2%), monocrotophos (0.05%) and dimethoate (0.06%) effective against the stone weevil recommended vapour heat treatment at 46°C for 280 min, or 50°C for 240 min. This treatment besides killing the stone weevil, also reduced the rate of fruit softening and mesocarp colour development.

## MANGO STEM BORER

***Batocera rufomaculata*** **deGeer**

The mango stem borer was not a serious pest earlier but in the last 10 years, it has become an important pest of mango. Five species, viz, *B. rufomaculata* deGeer, *B. rubus* Linnaeus, *B. roylei* Hope, *B. numitor* Newman and *B. titana* Thompson occur in mango, these first two are more common and destructive.

### Biology

This pest is quite serious when it attacks the main stem and the branches of the trees. When the stem is attacked, the sap exudes from the hole and mass of frass is expelled by the larva. The adult beetle is a large prominent, greyish brown beetle having white triangular patches at the junction of the elytra. The antennae are long with 11 segments. The pre-oviposition period is 24 hr and the oviposition period ranges from 20 to24 days. The eggs were laid at a depth of 2.5 to 4 mm inside the bark. The freshly laid eggs are white and shiny with oval, narrowly rounded ends. The freshly formed first instar larva is slender, pale white turning creamy white yellowish tinge on the thorax. The larval development is completed through six instars occupying 71 days. The pupa is ivory coloured. The average length of the pupa is 5.27 cm. The life cycle of the beetle from egg to adult emergence takes 182 days on mango and 124 days in semi-synthetic diet (Palaniswami *et al.*, 1979).

### Management

The infested branches should be cut and burnt. Beetles should be collected and destroyed. The grubs should be extracted through hooks and wires or destroyed by injecting dilute kerosene oil, petrol, naphthalene, chlorosol, carbon disulphide or tar water and the holes covered with wet clay soaked in tar recommended sprays of monocrotophos (0.1%) or phosphamidon (0.1%) at fortnightly intervals reported·that this insect can be controlled by injecting the hole with dichlorvos (0.02%) or endosulfan (0.05%) at the rate of 5 ml/hole and sealing the treated hole with mud reported a species of actoparasitic mite, *Proctolaclaps bickleyi* (Bram.) and a rhabditid nematode *Brevibucca* sp., parasitic on grubs of the beetle.

## MANGO LEAF WEBBERS

The leaf webber species recorded on mango trees in India are: *Orthaga euadrusalis* Walker, *O. exvinacea* Hampson and *O. mangiferae* Mishra. Other species of *Orthaga* recorded on mango are: *O. chinonalis* and *O. icarusalis* from Malaysia and Thailand and *O. leucatina* reported from Egypt and Sri Lanka.

***Orthaga euadrusalis*** **Walker**

This species was recorded for the first time on mango trees. It has now become a major limiting factor to mango production in Uttar Pradesh, Bihar and West Bengal. Earlier, this species was reported on *Shorea tolura* Roxb.

The webber infestation begins from June and continues upto December. There are five generations in a year. The adult moth is sombre coloured and lays eggs on leaves. The eggs are greenish dull in colour and hatch in 4–7 days. The first instar larvae feed on leaf chlorophyll while second instar onwards, start webbing the leaves and feed on the entire leaf, leaving behind the mid rib and veins. The webbing gives a tent-like appearance and so it is popularly called as tent caterpillar. A full grown caterpillar measures between 2.5 and 3 cm and its colour is brownish blue with white striations dorsally. There are five or seven instars, each of 2 to3 days duration in July. Larval period varies between 15 to 33 days. Larvae pupate within the web but the last generation larvae pupate in the soil. The pupae are dark black in colour and pupal period lasts for 4 to 5 days at 35°C and 16 to18 days at 25°C.

**Management**

Pruning of infested shoots and burning them in June-July helps to reduce the pest infestation. Raking of soil in tree basin tree in January helps in killing the hibernating pupae. The webbed parts of the plant should be removed. For this purpose a web removing equipment has been developed at the Central Institute for Sub Tropical Horticulture, Lucknow.

Isolated a red pigmented bacterium, *Serratia marcescens* Bizio and a fungus, *Aspergillus flavus* Link and found that both these were highly pathogenic to this pest. Another fungus, *Beauveria brassiana* (Balsama) was also found highly pathogenic to the larvae. Srivastava and Verghese (1987) reported the parasitoids, *Brachymeria lasus* (Walker), *Hormius* sp., *Pediobius bruchicida* (Rondani), *Eucepsis* sp. and *Exorista* sp. Out of these species, *Hormius* sp. showed promise for the control of the pest. Srivastava *et al*. (1982) recommended three sprays of quinalphos (0.05%), monocrotophos (0.05%) or carbaryl (0.2%) at fortnight intervals, starting from the last week of June.

## LEAF EATING CATERPILLARS

The important leaf eating caterpillars are *Orgyia postica* Walker, *Euthalia garuda garuda* Moore, *Lymantria marginata* Walker, and *Latoia lepida* Cramer.

***Orgyia postica*** **Walker**

Large scale defoliation of mango trees by *O. postica* caterpillars occurs in Uttar Pradesh during June-July. The young larvae feed on new flush, nibbling the leaf buds and affecting growing points. In the event of severe attack, fruit stalks and fruits are also scrapped. The female moth lays eggs in groups. Larval development takes place in 3 to 4 weeks. The full grown larva is a light brown hairy caterpillar, 3 to 5.5 cm in length. Pupation takes place on the plant in a thin papery cocoon covered with larval hair. Pupal period lasts for 7 to 10 days.

The larval parasitoid, *Exorista* sp. and pupal parasitoid, *Brachymeria lasus* (Walker) were recorded on this pest.

### *Euthalia garuda garuda* Moore

The species of leaf caterpillars is also known to cause severe damage to mango foliage in Uttar Pradesh. Severe defoliation affects vigour and vitality of the trees eventually affecting flowering and fruiting. The adults are medium sized butterflies of dark brown colour. The caterpillars are bright green about 3 cm in length. Pupa is also bright green in colour.

The larvae were found to be parasitized by *Apenteles euthaliae* Bhatnagar and *Ceraphron manilae* Ashmead and pupae by *Brachymeria lasus* (Walker) and *Charops optusus* Morlay.

### *Lymantria marginata* Walker

This species is a serious pest in western Uttar Pradesh and appears in the major mango growing areas of the state.

About 400 eggs are laid by a single female. Eggs are oval in shape and grey in colour. The newly hatched caterpillars congregate near the egg mass and feed on empty egg shells. The larval development is completed in 36 days. The female has seven instars while male has six instars. Fully matured larvae are of ash colour. The larval period lasts for 25 to 56 days and the pupal period lasts for 8 to 21 days. The adults have brown wings, speckled with white spots.

The larvae of *L. marginata* were also parasitized by *Eupelmus* sp., *Brachymeria lasus* (Walker) and *Cantheconidea furcillata* (Wolff).

### *Gastropacha pardale* Walker

This pest was reported for the first time feeding on mango leaves. The adults have hairy bodies with dark brown fore wings, thickly covered with scales. The mature larva is about 7 cm in length, dull brown with creamish tufts of hairs at the sides. The life cycle is completed in 38 to 51 days during September-October. The larvae are voracious feeders, feeding on the whole leaf except the mid rib.

The leaf eating caterpillars can be controlled by sprays of quinalphos (0.05%), monocrotophos (0.04%) or carbaryl (0.2%).

## SHOOT BORER

### *Chlumetia transversa* Walker

The shoot borer, *Chlumetia transversa* Walker, is found in India while *C. alternans* Moore and *C. brevisegna* Hollaway are reported from Japan and *C. transversa* Walker from Bangladesh, Taiwan, Java, Philippines and China. Shoot borers cause damage to mango shoots during the period when there is a new flush on the trees.

The female lays eggs singly on the tender leaves which hatch in 2 to 3 days. The newly hatched larva bores into the mid rib of the leaf and feeds for 2 to 3 days before boring into the tender shoots. The full-grown larvae are pink with black head and prominent legs. The larval stage lasts for 10 to 12 days and pupal stage for 15 to 18 days. The adult longevity is 15 days. The forewings are dark grey and beautifully patterned with wary designs. The life cycle is completed in about 45 days.

The affected shoots should be collected and burnt. The bioagents reported are *Bracon greeni* Ashmead, *Gorypus* sp., *Meteorus* sp., *Araneus singhagadensis* Tikader and *Stgodyphus sarasinorum* Karsch.

## SOFT SCALES

Scale insects infest mango trees and feed on the cell sap from flowers, leaves, stem and twigs. Over 70 species of scale insects are reported on mango trees but only a few cause severe damage.

***Chloropulvinaria polygonata* Cockarell**

*C. polygonata* is one of the serious pests in western Uttar Pradesh. The insect has a bright green colour but after oviposition, a distinct cottony sac is observed behind the female. The eggs are laid beneath the body of the mature female in conspicuous egg sac. A single female lays 1500 to 3000 eggs in her life time. The eggs hatch during June. The young crawlers move to the leaves and suck the sap along the mid rib or veins. The females crawl back to the twigs and small branches where they pass the winter.

***Aspidiotus destructor* Sign.**

This scale is a serious pest in mango nurseries. The pest prefers grafted varieties of mango. The damage starts on the underside of leaves with the advent of summer season. Due to sap sucking, the leaves first become yellow and then drop down. The infestation is more severe in areas with high rainfall. The body of female is bright yellow and nearly circular in outline. The eggs are yellow, tiny and the incubation period is 7 to 8 days. The crawlers spread out on the leaves and start sucking sap. The life cycle is completed in 31 to 35 days and there are 3 to 4 generations in a year.

***Rastrococcus iceryoides* Green**

This scale insect is a common pest on mango trees. The crawlers infest leaves, twigs and fruits. The female lays 300 to 600 eggs in compact, waxy sac beneath the rear end of its body. The egg laying continues for 1 to 2 weeks. The young nymphs crawl over the plant. They are flattened, oval and light yellow in colour. The pre-oviposition period lasts for 7 to 9 days. The male nymphs moult four times while female only thrice. There are 6 to 9 generations in a year and the insect overwinters as adult females. The young nymphs suck sap from the plant tissues. The nymphal period of male and female lasts for 24 to 77 and 25 to 42 days, respectively.

Pruning of affected plant parts such as leaves, twigs and branches helps in reducing the infestation. A large number of parasitoids species have been reported by various workers parasitizing coccids on mango. Srivastava and Verghese (1987) reported *Anicetus* sp. and *Comperiella aspidiotiphaga* Subba Rao on *C. polygonata*.

## REFERENCES

Anonymous, 1995. *Annual Report.* Central Institute for Tropical Horticulture, Lucknow, India.

Alam, M. Z. 1962. *Insect and mite pests of fruits and fruit trees in East Pakistan and their control.* Department of Agriculture, East Pakistan, Dacca, pp. 115.

Austin, A.D. 1984. New species of Platygastridae (Hymenoptera) from India which parasitise pests of mango particularly *Procontarinia* spp. (Diptera: Cecidomyiidae). *Bulletin of Entomological Research,* **74**(4): 549–557.

Baker, C.F. 1915. Studies of Philippine Jassoidea. IV. The Idiocerini of the Philippines. *Philippine Journal of Science,* **10**(6): 317–342.

Beeson, C.F.C. 1941. *The Ecology and Control of Forest Insects of India and Neighbouring countries*. Vasant Press, Dehra Dun, India. pp. 1007.

Boucek, Z. 1985. Taxonomic studies of chalcidoid wasps (Hymenoptera) associated with gall midges (Diptera: Cecidomyiidae) on mango trees. *Bulletin of Entomological Research,* 76(3): 393–407.

Chandra, A.; Bhati, D.P.S. and Singh, K.M. 1991. Mechanical control of mango mealybug (*D. mangiferae* Green) by tree trunk banding. *Indian Journal of Entomology,* **53**(1): 76–79.

Chandrika, S. and Nair M.R.G.K. 1971. Biology and control of *Pauropsylla brevicornis* Crawford, a pest of mango in Kerala. *Indian Journal of Entomology,* **33**(4): 360–362.

Chou, K.C. and Chou, L.Y. 1990. Survey of the natural enemies of mango leaf hoppers in Taiwan. *Journal of Agricultural Research China,* **39**(1): 70–75.

Fasih, M. and Srivastava, R.P. 1988. Natural occurrence of *Beauveria bassiana*, an entomogenous fungus on mango mealybug, *D. mangiferae* Green. *Indian Journal of Plant Pathology,* **6**(1): 8–10.

Fasih, M. and Srivastava, R.P. 1990. Parasites and predators of insect pests of mango. *International Pest Control,* **32**(2): 39–41.

Gangwar, V.S. 1974. The mango inflorescence midges and their economic importance. *Cecidologia Indica,* 7: 117–118.

Grover, P. 1965. Studies on Indian gall midges (Diptera: Cecidomyiidae). XII: The mango blossom midge, *Dasyneura amaramanjarae* n. sp. *Annals of the Entomological Society of America,* **58**(2): 202–206.

Grover, P. 1986. Population fluctuation of *Erosomyia indica* Gover and *Dasyneura amaramanjarae* Grover and extent of damage. *Cecidologia Internationale,* **7**(1–2): 43–57.

Gupta, R.L. 1953. Distribution of *Apsylla cistelata* Buckton (Psyllidae: Homoptera) causing bud galls in mango in Uttar Pradesh. *Indian Journal of Entomology,* **15**(4): 378–379.

Herting, B. and Simmonds, F.J. 1972. *A catalogue of parasites and predators of terrestrial arthropods section- A host or prey/enemy.* Vol. II *Homoptera*, Commonwealth Agricultural Bureaux, England, pp. 210.

Kapoor, V.C. 1970. Indian Tephritidae with their recorded hosts. *Oriental Insects,* **4**(2): 220–251.

Kato, M.C. 1928. On the homopterous insects infesting mango. *Formosan Agricultural Review*, **22**: 47–52.

Kumar, D.; Roy, C.V.; Yazdani, S.S.; Hameed, S.F. and Khan, Z.R. 1985. Efficacy of some insecticides against hopper complex on mango (*Mangifera indica* Linn.). *Pesticides,* 19(11): 42–43.

Lee, H.S. and Wen, H.C. 1980. Seasonal occurrence of mango psyllid and the experiments of control. *Journal of Agricultural Research of China,* **29**(3): 219–224.

Mani, M.S. 1952. Some gall midges (Itonichadae: Diptera) associated withh *Mangifera indica* Linn. in India and their galls. *Research Journal of Agra Univercity,* **1**(Science): 47–54.

Mathur, R.N. 1946. On immature stages of some Psyllidae. *Indian Journal of Entomology,* **8**: 224–236.

Nayar, K.K., Ananthakrishnan, T.N. and David, B.V. 1979. *General and Applied Entomology*. McGraw Hill Publishing Company Limited, pp. 588.

Palaniswami, M.S., Sundra Babu, P.C. and Subramaniam, T.R. 1979. Studies on the biology of mango stem borer, *Batocera rufomaculata* de Geer (Cerambycidae: Coleoptera). *South Indian Journal of Horticulture,* **27**(3–4): 100–106.

Palo, M.A. and Garcia, C.E. 1935. Further studies on the control of leafhoppers and tip borers on mango inflorescence. *Philippines Journal of Agriculture,* **6**(4): 425–464.

Prashad, S.N. 1971. The mango midge pests. *Official Bulletin of the Cecidological Society of India* **6**: 174.

Prashad, S.N. 1972. On the control of mango midge pest. *Cecidologia indica,* 7(2): 51–63.

Prashad, S.N. and Grover, P. 1974. Population fluctuation of *Erosomyia indica* Grover. *Cecidologia Indica,* **9**(1–2): 1–38.

Rafiq, A. and Ghani, M.A. 1972. Coccoidae and their natural enemy complexes in Pakistan. *Technical Bulletin of the Commonwealth Institute of Biological Control*, Rawalpindi, **50**: 104.

Ramchandra Rao, Y. 1930. Mango hopper problem in South India. *Agricultural Journal of India,* **25**: 17–25.

Rehman, K.A. and Latif, M. 1944. Description, bionomics and control of the giant mealybug, *Drosicha stebbingi* Green (Homoptera: Coccidae). *Bulletin of Entomological Research,* **35**(2): 197–209.

Shukla, R.P. and Prashad, V.G. 1984. Evaluation of some insecticides against mango hopper, *Idioscopus clypealis* (Leth.) (Homoptera: Cicadellidae). *Indian Journal of Agricultural Sciences,* **54**(8): 677–668.

Shukla, R.P. and Prashad, V.G. 1985. Population fluctuations of the fruit fly, *Dacus dordalis* (Hendel) in relation to hosts and abiotic factors. *Tropical Pest Management,* **31**(4): 273–275.

Singh, G. and Mishra, P.N. 1978. The mango shoot gall psyllid, *Apsylla cistellata* Buckton and its control. *Pesticides,* **12**(9): 15–16.

Srivastava, R.P.1980. Efficacy of alkathene bands to prevent the ascent of mango mealybug nymphs on mango trees. *Indian Journal of Entomology,* **42**(1): 122–129.

Srivastava, R.P.1997a. *Mango Insect Pest Management.* International Book Distributing Co., pp. 272.

Srivastava, R.P. 1997b. Field efficacy of Buprofezin, an insect growth regulator in the control of mango hopper, *Idioscopus nitidulus* (Walker). *Proceedings of International Symposium on Toxicity, Safety and Proper Use of Biopesticides*, October 27–31, 1996, Phitsanulok, Thailand. pp. 137–139.

Srivastava, R.P. 1997c. Field efficacy of products of *Beauveria bassiana,* an entomogenous fungus against mango hopper, *Idioscopus nitidulus*. *Proceedings of International Symposium on Toxicity, Safety and Proper Use of Biopesticides*, October 27–31, 1996, Phitsanulok, Thailand. pp. 161–162.

Srivastava, R.P.1997d. Laboratory screening of Buprofezin and alcoholic extract of *Alpinia galanga* against mealybug nymphs, *Drosicha mangiferae*. *Indian Journal of Entomology,* **59**(4): 366–368.

Srivastava, R.P. and Butani, D.K. 1972a. Mango hopper menace. *Entomologists Newsletter,* **2**(2): 10–11.

Srivastava, R.P. and Butani, D.K. 1972b. A method to prevent the mango mealy bug (*Drosicha mangiferae* Green) damage. *Entomologists Newsletter,* **2**(5): 35–36.

Srivastava, R.P. and Tandon, P.L. 1983. Field evaluation of some insecticides for the control of mango blossom midge, *Erosomyia indica. Pesticides,* **17**(3): 21.

Srivastava, R.P. and Tandon, P.L. 1986. Natural occurrence of two entomogenous fungi pathogenic to mango hopper, *Idioscopus clypealis. Indian Journal of Plant Pathology,* **4**(2): 121–123.

Srivastava, R.P., Fasih, M. and Abraham Verghese 1982. Evaluation of insecticides for the control of mango shoot gall psylla, *Apsylla cistellata* (Buckton) (Psyllidae: Homoptera). *Entomon,* **7**(3): 281–284.

Srivastava, R.P., Tandon, P.L. and Lal, B. 1979. Natural control of important insect pests of mango. *Research Report on mango*, *Mango Workers Meeting* held at Panji, Goa, 2–5 May, 1979: 275–277.

Srivastava, R.P. and Verghese, Abraham 1985. Fogging as a new method of hopper control in mango. *Indian Journal of Entomology,* **47**(1): 113–115.

Srivastava, R.P. and Verghese, Abraham 1987. Records of new parasites and predators of some insect pests of mango. *Bulletin of Entomology,* **28**(2): 162–164.

Srivastava, R.P. and Verghese, Abraham 1989. Chemical control of mango leaf hopper, *Idioscopus clypealis* Lethierry. *Indian Journal of Entomology,* **51** (3): 358–359.

Tandon, P.L. and Lal, B. 1979. Studies on the chemical control of the mango hopper, *Idioscopus clypealis* Leth., population in relation to physical environmental factors. *Entomon,* **8**(3): 257–261.

Thompson, W.R. and Simmonds, F.G. 1964. A Catalogue of Parasite and Predators of Insect Pests, *Section-3, Predator host catalogue*, Commonwealth Agricultural Bureau, England. pp. 204.

Tripathi, G.M.; Tripathi, N.L.M. and Mishra, R.S. 1990. New records of *Beauveria bassiana* (Bals.), an entomogenous fungus on mango hopper, *Idioscopus* spp. *Indian Journal of Plant Protection,* **18**(1): 140.

*Pest Management in Horticulture Crops*
L.R. Verma, A.K. Verma and D.C. Gautam (eds)


# *Chapter 15*

# Pests of Citrus

*Des Raj and Nirmala Devi*

## INTRODUCTION

OVER 250 SPECIES OF insects and mites have been reported to infest citrus in India and elsewhere with over one dozen species being the major pests (Pruthi and Mani, 1945; Bodenheimer, 1951; Jeppson and Carman, 1960; Wadhi and Batra, 1964; Rajput and Hari Babu, 1985; Tandon, 1993; Wait, 1992; Broeksma *et al* 1993). Various pests attacking citrus in Himachal Pradesh (Sharma and Bhalla, 1964), Punjab (Bindra, 1966; Sethi and Jawanda, 1964; Batra and Sidhu 1981); Jammu and Kashmir (Koul, 1994), Uttar Pradesh (Gupta, 1969), South India (Ayyar, 1963) and North Eastern Region (Sachan, 1984) have been documented. Citrus psylla, white flies, scales, mealy bugs and lemon butterfly are major pests in many areas whereas leaf roller, bark eating caterpillars, stem borers, fruit sucking moths and fruit flies are sporadic causing substantial losses. In addition, aphids, thrips, leaf eating beetles/weevils, blossom midges, flower moths, fruit sucking bugs, fruit borers, mites and termites are minor pests (Tandon, 1993). Need for their management with bias for utilization of biotic agents has been emphasized by Rozen (1993) and Singh (1993a, 1994, 1997).

## CITRUS PSYLLA

### *Diaphorina citri* Kuwayama

Citrus psylla, *Diaphorina citri* Kuwayama (Pyslidae: Homoptera) is a serious pest of citrus, widely distributed throughout the Oriental regions. It has been reported from China, Japan, Burma, Sri Lanka, East Indies, New Gunea and India. In Himachal Pradesh, it has been listed as an important pest (Awasthi 1990; Azad and Chauhan,1990). The pest also attacks other plants of Rutaceae family. Nymphs and adults suck sap from leaves, buds and tender shoots as a result of this leaves get curled and buds and shoots dry up. Severe infestation may result in complete crop failure. It has been reported to be a vector of greening mycoplasma causing the dreaded, citrus decline (Bindra, 1966). In addition to it continuos feeding by myraids of insects devitalise the plants and because of phototoximia the twigs die off tip backwards. Death of branches and ultimately complete trees my result into a loss of 90 to 95% the yield (Garg, 1978).

The pest has been reported breeding throughout the year (Bindra, 1957; Mangat, 1966; Pruthi and Mani, 1945; Atwal, *et al.*, 1970; Lakra *et al.*, 1983, Khan *et al* 1984). Atwal *et al.*, (1970) reported it to be inactive in severe cold of January in North India but in Karnatka, a peak has been reported in January by Bhumannavar (1987) under favourable conditions. Population peaks in February to April have been recorded at different places viz., U.P. (Lal, 1950), Allahabad (David, 1964), Rajasthan (Maheshwari and Sharma, 1978) and that in July-August by Dharmaraju and Reddy (1975) in Andhra Pradesh. The maximum activity of the pest has been reported during spring and autumn and least in monsoon and winter (Lal, 1950; Atwal and Verma, 1968; Koli *et al.*, 1981; Sachan *et al.*, 1982; Doharey and Butani, 1985). Sandhu (1978) recorded serious damage of *D. citri* during flowering and fruit set as a key (regular) pest on kinnow in Punjab. Aggregate distribution patterns in eggs and nymphs and appearance of adults in random to aggregate form has been observed by Dharajothi *et al.* (1989).

Almond shaped eggs are layed in the fold of half open leaves, on leaf petioles or between flower buds. Single female lays about 800 to 900 eggs. On hatching, the nymphs crawl on terminal shoots and buds for feeding. The development from eggs to adults takes 15 days in summer and 47 days in winter and there are 9 to 10 overlapping generations in a year (Mangat, 1966). Hibernating adults may live up to 189 days.

**Management**

- In biocontrol, parasitoid, Tetrastichus radiatus Waterston has been found effective in regulating the pest population. Tetrastichus phyllocnistoides and Diaphorencyrtus aligharensis are other natural enemies of D. citri. Tamarixia radiata a nymphal parasitoid has been found promising for the control of this pest. (Shivankar and Potineni, 1995). Conservation and augmentation of natural enemies (Predators) such as Chilocorus nigritus Fbr. Coccinella septempunctata Linn., C. rapanda Thumb, Menochilus sexmaculatus Fabr. Brumus suturalis Fabr., Chrysopa spp. and syrphids will play important role in its control.
- Soil application with phorate 15 or 20 g ai or dimethoate @ 16 ml ai/tree (Bindra *et al.*, 1970) or spray of (0.05%) monocrotophos or 0.045% dimethoate, or 0.05% phosalone is very effective. Endosulfan or malathion (0.05%) has also been recommended (Butani and Jotwani, 1975; Maheshwari and Sharma, 1978). Batra *et al.* (1990) obtained satisfactory control with monocrotophos, dimethoate, oxydemeton methyl or phosphamidon.
- Before opening of flowers on spring flush, spray of dimethoate (0.03%) (Rogor 30 EC) or methyl demeton (0.025%) (metasystox 25 EC) or monocrotophos (0.04%) (Nuvacron/ Monocil 36 WSC) or fenitrothion 0.05%(Folithion/Sumithion 50 EC) has been recommended (Anonymous,1995). If necessary, the spray should be repeated after the fruit set. Second flush of growth should be covered in August-September with one of these insecticides. All plants including hedges should be sprayed and fruit should not be picked before three weeks of spray. Fenvalerate (0.01%), monocrotophos (0.04%) and cypermethrin (0.01%) in linear order have proved highly effective against citrus psylla (Koul, 1994). Patel *et al.* (1998) found methyl- o-demeton @ 350 a.i./ha as the most effective against psylla.

## CITRUS BLACK/WHITE FLIES

***Aleurocanthus woglumi* (Quaintance)**

It is native of India and is found all over the world. Whiteflies are widely distributed in India, China and America. They cause heavy losses in quality and quantity of citrus fruits (Butani and Srivastava, 1983). The species commonly causing damage to citrus include *Dialeurodes* citri Riley and Howard, *D. citrifolia* (Morgan), *D. elongata* Dozier, *Aleurocanthus husaini* Corbett, *A. woglumi* (Quaintance), *Aleurolobus citrifoli* Corbett, *A. marlatti* (Quaintance), etc. *D. citri* is the most serious pest while *A. woglumi* and *A. husaine* sometimes become serious pest in certain areas. A very low population (of 1 to 4%) unidentified white flies was recorded in Jammu region of J&K state (Koul, 1994).

Both nymphs and adults suck sap from tender leaves and adversely affect plant vigour. Severely infested foliage turns yellowish green, becomes curled and finally drops down. Sooty mould develops on honeydew excreted by the white flies and gives black appearance to the foliage. Infested trees suffer in chlorophyll content, nitrogen, crude protein, flower production which may fall down and the set fruits are insipid. The white flies infest different species of citrus, cotton, coffee, castor and some ornamental plants. In South and East, in addition to citrus, it has also been recorded on coffee, mango and several other hosts.. *D. citri* is a small pale yellow insect with red eyes. The adults are found in abundance during spring and early autumn while the nymphs are observed during March-August. The life cycle is completed in 149 to 250 days under north Indian conditions. *A woglumi* is a serious pest of citrus in Maharastra, lays eggs in spiral rings on the lower side on fresh leaves. The nymphs from eggs laid, hatch in 7 to 14 days in different months and the nymphal stage is completed in 38 to 60 days. The pupal stage lasts 99 to 161 days. The pest hibernates in nymphal stage(Singh, 1994). The nymphs are black in colour. Freshly emerged adults are reddish and their bodies get covered with a heavy pulverulence within 24 hours of emergence giving them slaty bluish look. There are 4 to 5 generations in ayear and maximum adult emergence occurs in March, July and November. These periods coincide with initiation of new flush of *ambia, mrig* and *hastha bahar* respectively (Kuchanwar *et al.,* 1984). Recurrent drought conditions, dense planting in heavy soils and indiscriminate use of broad spectrum insecticides have been found responsible for its flare up.

**Management**

- Parasitioids, *Prospaltella lahorensis* How. and *Eretmocerus* sp. have been reported parasitizing nymphs of whiteflies. Besides, predators like *Brumus suturalis, Chrysopa* sp., *Catana paracesetosa, Scymnus* sp, *Lasius*, sp. and *Amystus* sp. have been recorded predating on eggs and nymphs. Entomopathogenic fungs, *Cladosporium* sp. also destroys large proportion of its population (Saini, 1964; Sharnagat *et al*. 1984). On *A woglumi,* some native natural enemies viz., *Amitus* sp. *Encarsia, E. citrifolia*, *Chrysoperla* sp., *Pullus xerampelinus*, *Aschersonia* sp., and *Verticillium lecanii* have been recorded from Kodagu (Karnataka), where it has not attained pest status. 90% parasitism of *A. woglumi* has been recorded due to *Encarsia divergens* and *E. merceti* in Gauhati whereas parasitism was 80–90 % due to *E. clypealis* in Pune and 80% due to *E. clypoealis* and *Amitus hesperidium* in Bombay and Bangalore respectvely (Singh, 1991a).The predator *Mallada boninensis* and aphelinid parasitoid *Encarsia opulenta* have also been found promising against *A. woglumi* (Potineni and Shivankar 1995).

- Close planting and waterlogging encourge black/whitefly infestation. Hence these should be avoided. For better management, initial infestation can be checked by clipping off and destroying infested shoots.
- Spray of 0.03% dimethoate, or 0.05% phosphamidan, or 0.05% acephate is very effective (Potineni, 1993). Spraying should be commenced at the initiation of new flush and repeated at 10 to 12 days interval. Other chemicals mentioned for control of citrus psylla can also be effectively used for their control.

## LEMON BUTTERFLIES

*Papilio demoleus* Linn., *P. helenus* Linn; *P. helenus* Moore, *P. polytes* Linn, *P. polyctor* Boisd, *P. polymnestor* Cramer, *P. machaon* Menetrius, *P. memmon* Linn and *P. protenor* Cramer (Papilionidae: Lepidoptera) have worldwide distribution and have been reported feeding on citrus leaves in India (Butani, 1979). *P. demoleus* and *P. polytes* are widely distributed throughout India while *P. polymenstor* and *P. helenus* are confined to submountain areas with milder climate (Ayyar, 1963; Singh, 1994) *P. demoleus* attacks almost all citrus species but prefers *Citrus sinensis* and *C. grandis*. Among various insect pests infesting citrus, the damage caused by *P. demoleus* ranks next to citrus leaf miner *Phyllocnistis citrella* Stainton in Jammu division. Its epidemics occurred in 1940, 1969 and 1982–83 (Radke et al; 1984).

The newly emerged caterpillars feed on young leaves biting them from edges towards midrib and feeding throughout the larval stages. The nature and extent of damage has also been studied by Rahman (1946), Ganguli and Ghosh (1967) and Brar and Rataul (1973). In event of severe attack entire plant was found defoliated. Foliar damage by caterpillars was observed to synchronize with the fresh growth in April, August and September. Severely attacked plants did not bear any fruit. Various aspects of its biology have been studied by Fletcher(1914), Pruthi and Mani(1945), *Talgeri et al.* (1972), Atwal (1964), Doharey and Butani (1985), Bhan(1994).The eggs are laid singly or in groups of 2 to 3 on tender leaves. Each female is capable of laying upto 183 eggs. Incubation period is 3 to 4 days, larval period 23 to 26 days and pupation takes place on twigs. The adults from non-diapausing pupae emerge in 7 to 24 days (Singh, 1994). Life cycle of *P. demoleus* is completed in 3 to 6 weeks during summer and 13 to 15 weeks during winter. The pest hibernates in pupal stage. There are 4 to 6 overlapping generations depending upon different locations. Atwal (1964) observed 10 while Bhan (1994) observed 9 generations of the pest under laboratory conditions. Bindra (1969) stated that the pest passes through 8 to 10 overlapping generations annually in Punjab.

### Management

- Larval stages of *Apanteles paplionis* and *Melalophacharops* sp. and pupal stages of *Pteromalus puparium* and *Holcojoppa coelopyga* are the potential parasitiods. Entomopathgenic-bacterium, *Serratia marcescens* and fungus *Fusarium* sp. killed substantial population of *Papilio* spp. (Singh, 1994). *Trichogramma chilonis* and *Telenomus* sp. *Incommodus* parasitised 75 and 78%

eggs (Krishnamoorthy and Singh, 1986; 1988). Trichogrammatids and *Bacillus thuringiensis* var. Kurstaki have been found promising (Singh 1991a). Entomophilic nematode, *Steinernema* sp.is effective against this pest (Singh, 1993b).

- Spray of *Bacillus thuringiensis* (Dipel 0.05%) or 5% neem seed extract is effective. In severe infestation, spray of 0.1% carbaryl, or 0.05% endosulfan or 0.05% quinalphos should be done. Bhan (1994) reported cypermethrin 0.01%, decamethrin (0.0025%) and quinalphos 0.04% more effective.

## CITRUS LEAF MINER

***Phyllocnistis citrella* Stainton**

It is distributed from Africa to Australia. And is reported to infest citrus in Japan, China, Malaya, Formosa, Philippines and East Indies (Rajput and Hari Babu, 1985). It is a serious pest of citrus all over India. In Himachal Pradesh, it is listed as common and important pest (Awasthi, 1990; Azad and Chauhan,1990) The larvae mine tender leaves and feed on epidermis making zigzag galleries characteristic of miner damage. The infested leaves turn pale, curl and get dried off. The infestation is also suspected of aiding infection by bacteria which cause citrus canker. Sometimes larvae also mine epidermis of tender shoots. The incidence has been reported as high as 87.41% during *ambia* flush in Maharashtra. The larvae have also been reported to mine the leaves of jasmine and cinnamon. The work on seasonal abundance and population dynamics has been done by Atwal and Josan(1962), Koli *et al.* (1981), Singh, (1984), Maheswari and Sharma (1986), Batra (1990) and Bakshi (1993). The pest is active throughout the year except during extreme cold. The peak activity coincides with the emergence of new flushes in February-March, June –July and October-November. Koul (1994) found that in Jammu region the peak incidence periods of *P. citrella* synchronized with the emergence of new flush on the host plants and commenced earlier under rainfed conditions.,

The greenish yellow eggs are usually laid singly on the underside of tender leaves near the mid rib. Incubation period is 3 to 6 days. On hatching young larvae enter the epidermis at once and start mining. Larval period is 1 to 2 weeks. Depending upon the season, peak period may last for 3 to 4 weeks. The total life cycle may be completed in 20 to 60 days and there are 9 to 13 generations in a year.

**Management**

- The varieties Carrizo, Sacaton, Savage, Troyer, Yama citrange, Citrumelo(*Poncirus trifoloiata* x grape fruit), Cambell, Valencia, Pomary and Rubidoux, and *Muraya koenijii* have been found resistant to *P. citrella* on the basis of leaf infestation (Batra *et al.* 1992). *Poncirus trifoliata* (L) Raf; Carrizo, Citrumelo, Morton, Sacaton, Savage and Troyer were found least susceptible among citrus germplasm screened by Batra and Sidhu (1981). *Citrus reticulata, C. reshni* and *P. trifoliata* have been reported to register less incidence of leaf miner as compared to others (Singh, 1994).
- Natural enemies such as *Tetrastichus phyllocnistoides*, *Cirrospilus quadristriatus, Bracon phyllocnistoids, Tetrastichus* sp., *Simpiesis purpurea, Kratosysma* sp., *Pediobius sp. Sympiesis* sp.

have been found commonly parasitizing *P. citrella* (Singh, 1994). Atwal (1964) reported that *Ageniaspis* sp. parasitized 80% larvae of this pest in Punjab.

- Fenvalerate (0.01%), cypermethrin (0.01%), permethrin (0.01%) and, decamethrin (0.0025%) have been recorded highly effctive (Radke and Kandalker, 1990; Singh and Sharma, 1990).
- Spray of (0.05%) monocrotophos, (0.05%), dimethoate, 0.003% decamethrin or 0.005% cypermethrin has proved effective in managing this pest (Tandon, 1993). Spray of neem seed extract (2%) has also been found quite effective and safe. Lefnuron (14 g a.i./ha) has been found effective against *P. citrella* upto 7 days and was superior to methyl-o-demeton (Patel *et al.* 1998).

## SCALE INSECTS AND MEALY BUGS

Coccids infesting citrus include *Aonidiella aurantii* (Maskell), *A. citrus citrina* Coquillett, *A. orientalis* (Newstead), *Chrysomphalus aonidum* (L), *Parlatoria oleae* Colvee, *P. pergandii* Comstock, *P. zizyphis* (Lucas), *Coccus hesperidum* Linnaeus, *C. viridis* (Grem), *Saissetia coffeae* (Walker), *Rastrococcus iceryoidea* (Green) and *Planococuss citri* (Risso*), P. pacificus* Cox, *Icerya purchasi* Mask (Pruthi and Mani, 1945; Tandon, 1993, Singh, 1994). *A auranti* is the most common. Citrus scale has been found to infest some orchard in Nalagarh area of Solan (Sharma *et al.,*1992). It mainly attacks leaves and tender shoots but in severe infestation even fruits are also attacked. They have 5 to 6 generations in a year. The green scale *C. viridis* (Green) is quite serious in southern and eastern regions. It has been recorded on several other fruits, ornamental plants, coffee, tea and weeds (Tandon, 1985). The scale insects also exude honey dew on which sooty mould grow and it attracts ants also. Infested plants lose vigour and become unproductive. *Planococus citri* (Risso) commonly attacks citrus in India. In addition to citrus, it has been recorded on guava, grape, mango and coffee (Tandon, 1993; Singh, 1994). They infest leaves, tender shoots and fruits. Under severe infestation, plant growth is arrested and fruit drop prematurely. Continuous drought causes outbreak of this pest. The nymphs and adult females are ovate in shape and light brown in colour. The body remains covered with dense white mealy secretion. Each female lays 600–800 yellow eggs in loose cotteny masses. There are 2 to 3 generations in a year.

### Management

- The natural enemies regulate the population of *A. aurantii* and *C. aonidium* such as *Aphytis mytilaspidis* (LeBron), *Aphytis* sp., *Colophora biplagiata* Swartz, *Chilocorus circumaduts* Sch., *C. nigritus* (F.)*, Chrysopa* sp., *Pharoscymnus horni Wse, Microterys* sp. *and Tetrastichus sp.* should be encouraged. Numerous natural enemies viz. *Aneristus ceroplastae* How, *Anicetus annulatus* Timb., *Coccophagus* sp., *C. cowperi, C. bogoriensis, Cheloneuromyia javensis, Encyrtus lecaniorum, Microterys flavus, Tetrastichus* sp. *Cryptoblabus* sp., *Chilocorus circumdatus, C. nigritus, Chrysopa* sp. *Cryptolaemus montrouzieri, P. hori, Amblysieus* sp., *Neoseiulus* group, *Tydecus caudatus, Entomopthora* sp. and *Verticillium lecanii* have been found regulating green scale populations. However, indiscriminate use of pesticides has reduced their population (Tandon, 1985, 1993). *Mallada boninensis, Plesiochrysa (Chrysopa) lacciperda, Anisochrysa basalis, Chrysoperla* *carnea, Triommata* sp., *Scymnus* sp. and *Spalgis epius* have been recorded on *P. citri* at Bangalore. *Coccidoxenoides peregrina* has been recorded for the first time on *P. citri* in India (Singh, 1994).

- To control mealy bugs, release of coccinellid *C. motrouzieri* @ 10 beetles per plant has been recommended. Inoculative releases of an exotic parasite, *Leptomastix dactylopii* have also been recommended (Krishnamoorthy and Singh 1987, Nagarkatti *et al.* 1992, Tandon, 1993).
- Removal of weeds which act as additional hosts of coccids and proper sanitation by pruning and destroying infested twigs is helpful in management of coccids. The ant colonies should be destroyed by ploughing the soil around trees and by mixing 5% carbaryl dust around plant bases.
- In event of severe infestation, spray of 2% neem oil, or 2% Pongamia oil, or 2% neem seed extract should be repeated twice at 7 days interval.

## APHIDS

They are distributed through out the world but are not known to cause serious direct damage but act as vectors of tristeza virus. The infested foliage get curled up and fruits drop prematurely. They excrete honey dew on which sooty mould develops. Various species include *Aphis pomi*(DeGeer), *A. gossypii* Glov., *Myzus persicae* (Sulz), *Toxoptera aurantii* (Fonscolomibe), *T. citricidus* (Klrk), *A. craccivora* (Koch.), *Dactynotus jaceae* (L.) and *A. citriela*(Pruthi and Mani, 1945; Verma et al, 1965, Singh, 1994).Except *Toxoptera* spp. others are polyphagous and may attack custoudapple, Jackfruit, Tamarind etc. Aphids attack citrus during flowering spell but peak population level is attained when wet season is followed by dry weather they normally reproduce parthenogenetically and have a life cycle of 6–8 days. Temperature beyond 30°C is not favourable for their development.

**Management**

- Need based initial spray of 2% *mahua* or neem oil or neem seed extract can be done prior to flowering.
- The role of natural enemies like *Coccinella septempunctata* L., *C. repanda* Thumb. *Menochilus sexmaculata* (F.), *Chrysopa* spp., *Ischiodon scutellaris* (Fabr.), *Chrysoperla* sp. and *Cardiastethus pygmaeus* may be exploited subsequently in the season.
- Spray of malathion (0.05%), methyl demeton (0.025%) or phosphamidon (0.025%) or dimethoate (0.03%) are recommended.

## FRUIT SUCKING MOTHS

A number of species of fruit sucking moths have been reported on citrus by Butani (1979) but *Othreis fullonia* Linnaeus, *O. materna* L, *Calpe emerginata* Fab, *Achaea janata* L. cause considerable loss. The caterpillars feed on forest plants like *Cocculus hirsutus*, *Erythrina* sp.*Tinospora cordiforlia*. Due to destruction of forests these insects have now adapted to the cultivated plants.

The larval stage of the moths is not destructive to citrus but moths migrate and puncture the rind of fruits. Injury on fruits make way for bacteria which hasten fruit rotting.

**Management**

- Egg parasitoid *Trichogramma chilonis* and larval parasitoid *Microplitis maculipennis* which are important natural enemies of *Achaea janata* (Singh 1990).
- Alternate hosts such as weeds should be destroyed.
- At the advent of fruit setting, generate smoke for 2 hrs after sunset. Dispose off fallen and decaying fruits and harvest the fruits early.
- Trapping and killing moths using 1kg sugar + 50 ml dichlorovos in 10 litres water is also effective.

## LEAF ROLLER

***Psorostica zizyphi*** **(Stainton)**

It is also called as orange caterpillar and has been reported as serious pest in Punjab, Maharashtra and Tamilnadu. In Punjab it is gaining importance of a serious pest on young citrus plantations and nursery. Peak activity of this pest is during August-October. The caterpillars initially feed on tender leaves or sprouting buds. Later on they secrete silken strands and role up the leaves longitudinally. They initiate feeding from tip downwards. The life cycle is completed in 18 to 21 days.

**Management**

- Important natural enemies include *Apanteles machaeralis*, *Brachymeria euploea, Mesochorus* sp. and *Pristomerus marginicollies* (Tandon, 1993).
- Fish oil rosin soap has been found quite effective to control this pest. Besides 0.01% cypermethrin or fenvalerate has also been recommended.
- Methyl parathion alongwith quinalphos and monocrotophos each @ 0.05% gave significantly better results for the control of citrus leaf roller in Punjab (Batra, 1990).

## THRIPS

Several species of thrips viz. *Frankliniella dampfi* Priesner, *Thrips florum* Schmutz, *T. niligiriensis* Ramakrishna, *T. subnudula* Karny, *Heliothrips haemorrhoidalis* (Bouche), *Ramaswamiahiella subnudula* Karny, *Rhopalandrothrips* (Moulton) have been recorded damaging leaves, flowers and fruits of citrus. *H. haemorrhoides* also attacks tea, banana, coffee and pines.

**Management**

i)Spray of 0.05% phosphamidon or 0.03% dimethoate or 0.1% carbaryl is recommended.

## SHOOT AND TRUNK BORERS

### Bark eating caterpillar

*Indarbela quadrinotata* Wlk. (Inderbellidae: Lepidoptera) is important pest of citrus and its host plants include guava, pomegranate, mango and casuarina. Newly hatched larvae bore into shoots and bark. They make galleries and as a result, plants present sickly look and ultimately die. Infested plants exhibit frass-covered areas on bark. Shoot borers, *Oberea lateapicalis* Pic. and many bupresid and longicon beetles have been reported as pests of citrus (Singh *et al.*1983). By and large they are categorized as minor pests but some of them cause damage regularly in certain areas. *O. lateapicalis* is mainly important in southern and north eastern hilly regions. Attack is more when citrus is inter planted with coffee. It appears in March and oviposition continues till May-June. On hatching caterpillars bore into young twigs and feed on soft tissues. Later on they bore downwards and feed inside the shoots. Infested shoots break and dry. Orange stem borer, *Cheliodonium argentatum* is a serious pest of orange in Kodagu and Wynad (Karnataka). The grubs make spiral tunnels in the shoots and trunk.Trunk borer, *Anoplophora varsteegi* has been reported causing 20–30% mortality of young trees in Meghalaya and Arunachal Pradesh (Sachan, 1984).

### Management

- Prune and destroy infested shoots.
- Apply 0.1% chlorpyriphos on main trunk.
- Spray 0.05% endosulfan or 0.2% carbaryl from second fortnight of April as per need.
- Inject 0.1% emulsion of dichlorovos or methyl parathion and plug the holes with wet clay.
- To kill the grubs apply phorate (Thimet 10 G) as per age of the plant and to combat adults, dust carbaryl or phosalone on trees during evening time.

## MITES

They are cosmopolitan in distribution and have been recorded as serious pest in USA, Java, Philippines, Iraq, Iran, Egypt, Sri Lanka and India (Rajput and HariBabu, 1985). As many as 54 species of mites have been reported on citrus in India (Garg, 1978). *Eutetranychus orientalis* feeds on the upper surface of the leaves and does not form webs while *Tetranychus fijiensis* which feeds on the lower surface of the leaves, makes profuse webbing (Singh, 1994) Both are serious pests of citrus during dry period. They suck sap and devitalise the plants. Infestation by *T. figiensis* results in curling of leaves in downward direction. In the event of severe infestation, orchard presents dull yellow look and leaves start falling thereby plant vigour and fruit production is adversely affected. Other species include *E. banki*, *Panonychus citri*, *Paratetranychus citris*, *Phyllocoptruta oleivora* Ashm, *Schizotetranychus hindustanicus*, *Tetranychus cinnabarinus*, *Tetrarychus sexmaculatus* Riley sometimes attain the status of major pests damaging leaves, flowers and fruits. Their peak incidence has been recorded during January-May and August-September in different parts of the country (Tandon, 1993).

## MINOR INSECT PESTS

Low intensity attack of weevil, *Lepidospyris demisea* Marshall, jassids/plant hoppers *Oxyrhachis* sp., *Paramesodes lineaticollis*; plant bug, *Bagrada hilaris* Burmeister and severe attack of termites have been reported on Kinnow madarin in Jammu region by Koul (1994). Termite *Odo ntotermes obesus* Rambur infests citrus in drier areas as also in sandy or sand-loam soils and may kill trees if not checked (Butani and Srivastava, 1983). Application of pesticide like chlorpyriphos in soil or trunks can check the infestation. Red ant *Oecophylla samaragdina* F. although does not feed on plants yet it spreads all over trees and builds nests of leaves and hence causes nuisance. Other inhabited trees include mango, citrus, chikoo and guava. They protect noxious insects like coccids, aphids etc. For obtaining honeydew from these harmful insects. Dusting with 5% sulphur dust has been found effective against red ant. Other minor insect-pests attacking citrus include weevil, *Myllocerus discolor* and hairy caterpillar *Euproctis fraterna* attacking foliage, citrus flower moth *Prays citri* whose larvae web the flowers and feed inside on the floral parts and blossom midge, *Sayneura citri*. In India, several species of fruit flies viz., *Dacus correctus* Bezzi, *D. curcurbitae* Coq., *D. diversus* Coq, *D. dorsalis* Hendel, *D. tace* Walker and *D. zonatus* Saunders have been found damaging citrus. *D. diversus* sometimes becomes major pest of sore oranges. The incidence can be reduced by placing 20–100 traps baited with 1% methyl eugenol and 0.5% malathion in a groove. *Poekilocerus pictus* F and Chaffer beetle, *Adoretus* have also been recorded as minor pests in citrus nurseries and can be controlled by spraying the area with carbaryl 0.15% or endosulfan 0.05%.

### Management

- Among different varieties of mandarin screened for mite infestation, highest infestation (64%) was found in Nagpur variety while lowest (4%) was found in kinnow (Anonymous,1978).
- Coccinellid (*Scymnus gracilus* Motsch), thrip (*Scolothrips indicus* Prieshner) and four species of mites (*Pronematus* sp. *Amblyseius cucumeris*, *A. hibisci* and *Agistemus* sp.) have been reported predating on *E. orientalis* (Sadana and Kanta, 1971). Predatory mites *Amblyseius aerialis* and *A. finlandicus* were most common (Rao, 19ω9). In addition, *Arthrocnodax* sp., *Chrysopa* sp. and *Hemerdieus* sp. have also been reported as predators (Rajput and Hari Babu, 1985).
- Spray 0.04% dicofol or 0.04% dimethoate or 0.05% phosalone or thiometon 0.1%.

## REFERENCES

Anonymous, 1978. Incidence, biology and control of citrus fruit mite. *Annual Report,* Department of Entomology-Zoology, HPKV, Solan. pp. 26–29.

Anonymous, 1995. *Package of Practices for Horticultural Crops*. Directorate of Extension Education, Dr.Y.S.Parmar University of Horticulture and Forestry, Solan (H.P.). pp. 186.

Atwal, A.S. 1964. Insect pests of citrus in Punjab. VI. Biology and control of citrus leaf miner, *Phyllocnistis citrella* Station (Lepidoptera: Phyllocnistidae). *Punjab Hort. J.* 4(2): 100–103.

Atwal, A.S; Choudhary, J.P. and Ramzan, M. 1970. Studies on the development and field population of citrus psylla, *Diaphorina citri* Kuwayama (Psyllidae: Homoptera). *J. Res. PAU,* 7(3): 333–338.

Atwal, A.S. and Josan, G.S. 1962.Citrus pests round the year. *Punjab Hort J.* 2(1): 8–16.

Atwal, A.S. and Verma, G.C. 1968.Studies on the control of citrus psylla, *Diaphorina citri* Kuwayama (Hemiptera: Psyllidae) by foliar spray and soil application. *J.Res. PAU,* **5**(2): 240–243.

Awasthi, R.P.1990. Problems and prospectus of kinnow in Himachal Pradesh. *In: Citriculture in North Western India, Procedings* of a *Show-cum-Seminar* on *Prospects and Problems of Kinnow Cultivation*. J.J. Gill; J.S.Kanwar and S. Raghbir (Eds). Mohindra Art Press Ludhiana. pp. 93–96.

Ayyar, R.T.V. 1963. Insects affecting fruits and palm trees. *In*: *Hand Book of Economic Entomology for South India.* Govt Press Madras. pp. 280–281.

Azad, K.C. and Chauhan K.C.1990. Development of citrus production in Himachal-Prospects and Problems. *In: Citriculture* in *North-Western India, Proceedings* of a *Show-cum-Seminar* on *Prospects and Problems of Kinnow Cultivation.* K.S.Gill, J.S.Kanwar and S.Raghbir Mohindra Art Press, Ludhiana. pp. 97–108.

Bakshi, C.L.1993. *Population studies, relative incidence and chemical control of citrus leaf miner Phyllocnistis citrella Stainton(Gracilariidae: Lepidoptera) on citrus.* Unpublished. M.Sc.(Ag) thesis submitted to Sher-e-Kashmir University of Agricultural Sciences and Technology R.S.Pura, Jammu.

Batra, R.C.1990. Insect pest problems in citrus nursery with special reference to kinnow. *In: Proceedings of a Citrus Show-cum- Seminar on Prospects and Problems of Kinnow Cultivation*. 6–7 January, 1989 held at PAU Ludhiana. pp. 189–195.

Batra, R.C. and Sidhu GS 1981. Screening of germplasm for citrus leaf miner in Punjab. *Punjab Hort. J.,* **6**: 108–116.

Batra, R.C.; Sharma D.R. and Chanana Y.R. 1992. Screening of citrus germplasm for their resistance against citrus leaf miner, *Phyllocnistis citrella* Stainton. *J. Insect Sci,* **5**(2): 150–152.

Batra, R.C., Sharma D.R., Singh, R. and Singh, S.N.1990. Status of citrus insect and mite pests and their natural enemies in different agroclimatic zones of Punjab. *Indian J. Hort.,* **43**(3): 331–336.

Bhan, R.1994. Studies on the bionomics and management of lemon butterfly (*Papilio demoleus* Linnaeus) on citrus M.Sc. thesis submitted to Sher-e-Kashmir University of Agricultural Sciences and Technology (J&K) pp. 85.

Bhumannavar, B.S.1987. Evaluatioin of pyrethroids compounds against citrus leaf miner, *Phyllocnistis citrella* Stainton(Lepidoptera) on Coorg mandarin. *Entomon,***12**: 183–185.

Bindra, O.S. 1957. Insect pests of citrus and their control. *Indian J. Hor.* **14: 89–98.**

Bindra, O.S. 1966. Role of insects and other animals in relation to citrus decline. *Punjab Hort. J.,* **6**: 108–116.

Bindra, O.S. 1969. Controlling citrus pests. *Indian Horticulture,* **13**: 28–31.

Bindra, O.S.; Verma, G.C.; Sandhu, G.S. and Chhabra, S.K. 1970. Studies on control of citrus pests by soil application of systemic insecticides. *J. Res. PAU,* **7**(2): 197–202.

Bodenheimer, F.S.1951. *Citrus Entomology*. Hoitsema Bros. Groningen, Holland. pp. 663.

Brar, T.S. and Rataul, H.S. 1973. Bionomics of citrus butterfly in the Punjab. *Punjab Hort. J.* **13**: 208–210.

Broeksma, A.; Robtertse, E. and Saba, F. 1993. Field trials with Confidor(Imidacloprid) for control of various insect species on citrus in the Republic of South Africa. *Pflanzenschutz-Nachrichten* Bayer (English ed), **46**(1): 5–32.

Butani, D.K.1979. Insect pests of citrus and their control. *Pesticides,* **13**: 15–21.

Butani, D.K. and Jotwani, M.G. 1975. Trends in the control of insect pests of fruit crops in India. *Pesticides Annual,* 139–149.

Butani, D.K. and Srivastava, R.P. 1983. Insect pests of tropical and subtropical fruits. *In: Agricultural Entomology.* P.D.Srivastava, M.G. Jotwani, R.A. Agarwal, S.R. Wadhi, R.K. Bhanotar and R.K. Bhatnagar (eds.). All India Scientific Writers, Society, New Delhi. pp. 189–211.

David, B.A. 1964. Insect pests of *Citrus* spp. and varieties grown in the Allahabad Agricultural Institute Orchards. *Allahabad Farmer,* **38**(4): 170–177.

Dharajothi, B., Verghese, A. and Tandon, P.L. 1989. Ecological studies of citrus psylla with special reference to spatial distribution and sampling plan. *Entomon,* **14**(4–4): 319–324.

Dharmaraju, E. and Reddy, C.S.1975. Insect pests associated with citrus decline. *In:. Citrus decline in Andhra Pradesh.* pp. 46–55.

Doharey, K.L. and Butani, D.K. 1985. Newer approaches in the management of citrus pests. *Pesticides,* 19(4): 38–40.

Fletcher, B.T. 1914. Some South Indian Insects. Superintendent, Govt Press, Madras. pp. 565.

Ganguli, R.N. and Ghosh MR. 1967. Biology of *Papilio demoleus* pest of citrus in Tripura state. *Indian Agriculturist,* **11**: 13–18.

Garg, D.O.1978. Major citrus pests. In: *Insects Pests of Citrus Fruits.* Agri-Horticultural Publishing House, Nagpur, India. pp. 15–25.

Gupta, B.P.1969. Pests of citrus in hilly regions of Uttar Pradesh. *Prog. Hort.,* **1**(37): 55–63.

Jeppson, L.R. and Carman, G.E.1960. Citrus insects and mites. *Annu. Rev. Entomol.* **5**: 353–378.

Khan, K.M.; Radke, S.G. and Borle, M.N. 1984. Seasonal activity of citrus psylla, *Diaphorina citri* Kuwayama in the Vidarbha region of Maharashtra. *Bull.Ento.,* **25**(2): 143–146.

Koli, S.G.; Maker, P.V. and Choudhari, K.G. 1981. Seasonal abundance of citrus pests and their control. *Indian J. Ent.* **43**(2): 183–187.

Koul, P.K.1994. *Studies on inportant insect pests of citrus and control of citrus psylla in Jammu(J&K).* M.Sc. Thesis submitted to Sher-e- Kashmir University of Agricultural Sciences and Technology (J&K). pp. 95.

Krishnamoorthy, A. and Singh, S.P.1986. Record of the egg parasite, *Trichogramma chilonis* on *Papilio* spp. in citrus. *Curr. Sci.* **51**: 461.

Krishnamoorthy, A. and Singh, S.P.1987. Biological control of citrus mealybug with an introduced parasite *Leptomastix dactylopii* in India. *Entomophaga,* **32**: 143–148.

Krishnamoorthy, A. and Singh, S.P.1988. Observatioinal studies on the occurrence of parasites on *Papilio* spp. in citrus./*Indian J. Plant Prot.* **16**: 79–81.

Kuchanwar, D.B.; Hardas, M.G. and Sharnagat, B.K. 1984. Agrotechnique in the management of citrus pests. *In: Proc. National Seminar Progress and Problems.*PKVK(MS). Nagpur, 5–7 February 1984. pp. 51–52.

Lakra, R.K.., Singh, Z. and Kharub, W.S.1983. Population dynamics of citrus psylla, *Diaphorina citri* Kuwayama in Haryana. *Indian J.Ent.* **45** (3): 301–310.

Lal, K.B.1950. Insect pests of fruit trees grown in the plains of Uttar Pradesh and their control. *Journal of Agriculture and Animal Husbandry,* **1** (4): 30–45.

Maheshwari, A.C. and Sharma, L.S.1978. Preliminary studies on seasonal incidence, chemical control and screening of citrus species against citrus psylla in Rajasthan. *Haryana J. Hort.Sci.* 7(3–4): 125–129.

Maheshwari, A.C. and Sharma, L.S. 1986. Seasonal incidence and control of citrus leaf miner *Phyllocnistis citrella* Statinton at Udaipur. *Pesticides,* **20**(1): 37–38.

Mangat, B.S. 1966. Biology and control of citrus psylla,, *Diaphorina citri* Kuwayama *Plant Protection Bulletin,* **18**(2): 18–20.

Nagarkatti, S.; Singh, S.P.; Jayanth, K.P. and Bhummanavar, B.S. 1992. Introduction and establishment of *Leptomastix dactylopi* How. against *Planococcus* spp. in India. *Indian J.Plant Prot.,* **20**: 102–104.

Nayar, K.K.; Ananthakkrisnan, T.N. and David, B.V. 1985. *General* and *Applied Entomology*. Tata MC Graw Hill Publishing Company Limited, New Delhi. pp. 589.

Patel, J.J.; Patel, J.R.; Valand, V.M.; Patel, B.H. and Patel, M.J. 1998, Bioefficacy of some of the new insecticides against leaf miner *Phyllocnistis citrella* and Psylla, *Diaphorina citri* infesting citrus. *Indian J. Ent.* **60**(1): 50–56.

Potineni, K. 1993. Comparative efficacy of conventional and new insecticides for control of citrus blackfly (*Aleurocanthus woglumi* and its economics. *Indian J. agri Sci.* **63**(8): 526–528.

Potineni, K. and Shivankar, V.J.1995. Integrated mangement of citrus blackfly, *Aleurocanthus woglumi* Ashby using natural enemies and selective insecticides. *In: Annual Report, National Research Centre for citrus ICAR, Seminany hills Nagpu.* pp. 47–50.

Pruthi, H.S. and Mani, M.S. 1945. Our knowledge of insect and mite pests of citrus in India and their control. *Scientific Monograph*. Imperial Council of Agricultural Research, New Delhi:**16**: 42.

Radke, S.G. and Kandalkar, H.G. 1990. Chemical control of citrus leaf miner. *Indian J. Ent.* **52**(3): 397–400.

Radke, S.G.; Borle, M.N. and Vawalkar, K.S. 1984. *National Seminar Pest Management Citrus, Cotton, Sugarcane and Sorghum. Progress and Problems,* PKVK(MS). Nagpur, 5–7 February 1984. pp. 112.

Rahman, K.A. 1946. Insect pests of fruit trees. *Punjab Farmers Journal,* **10**(40): 141–144.

Rajput, C.B.S. and Hari Babu, R. Sri. 1985. *Citriculture*. Kalyani Publishers, New Delhi-Ludhiana. pp. 364.

Rao, V.P.1969. India as a source of natural enemies of citrus pests. *Proc. Ist Int. Citrus Symposium,* **2**: 785–792.

Rozen, D.1993. Role of biological control in IPM programmes in Israel. *Entomologicheskoe obozrenie,* **72**(3): 537–543.

Sachan, A.S., Gangawar, S.K. and Sachan, J.N. 1982. Seasonal incidence of major insect pests of Khasi mandarin (*Citrus reticulata* Blanco) at high altitudes in Meghalaya. *Indian J. Plant Prot.* **10**: 46–48.

Sachan, J.N. 1984. Management of Citrus pests in NE. hilly region. *Proc. National Seminar Pest Management Citrus, Sugarcane and Sorghum Progress and Problems,* PKVK(MS) Nagpur, 5–7 February 1984. pp. 57.

Sadana, G.K and Kanta. 1971. Predators of citrus mite *Eutetranychus orientalis* in India. *Sci. Cult. 37*: 530.

Saini, B.S. 1964. Bionomics and control of *Dialeurodes citri* (Raley and Howard) (Homoptera: Aleurodidae). M.Sc. thesis PAU, Ludhiana.

Sandhu, G.S. 1978. Insects pests of Kinnow and their control in Punjab. *Punjab Hort.J. 18(1–4):* 166–170.

Sethi, S.L. and Jawanda, J.S. 1964. Control of citrus pests of the arid irrigated regions of the Punjab. *Punjab Hort. J.* **4**(2): 67–69.

Sharma, N.K.; Sharma, R.; Badiyala, S.D. and Pathania, N.K. 1992. *Status Report on the Sub-montane and low hills sub-tropical zone (Zone-1). Horticulture. National Agril. Research Project.* Dr.Y.S.Parmar Unoiversity of Horticulture and Forestry, Solan. pp. 64.

Sharma, P.L. and Bhalla, O.P. 1964. Survey study of insect pests of economic importance in Himachal Pradesh. *Indian J. En,.* **26**: 318–331.

Sharnagat, B.K., Kuchanwar, D.B. and Hardas, M.S. 1984. Management of citrus blackfly on Nagpur Mandarin in Vidarbha region. In: *Proc. Seminar: Pest Management Citrus, Sugarcane, Sorghum Progress and Problems*. PKVK(MS), Nagpur, 5–17 February, 1984. pp. 12.

Shivankar, V.J. and Potineni, K. 1995. Integrated Management of citrus psylla(*Diphorina citri* Kuwayama). *In: Annual Report, National Research Centre for Citrus,* ICAR, Seminary Hills, Nagpur. pp. 50–51.

Singh, K. and Sharma, V.K. 1990. Chemical control of citrus leaf miner(*Phyllocnistis citrella* Stainton). *Advances in Horticulture and Forestry,* **1**: 92–94.

Singh, S.P. 1984. Biological control of insect pests of natioinal importance. *In: Integrated Pest and Disease Management*. S.Jayaraj (ed). TNAU, Coimbatore. pp. 309–318.

Singh, S.P. 1990. Biological suppression of pests in fruit crops. *In: Proceedings of the Indo-USSR Joint Workshop on Problems and Potentials of Biocontrol of Pests and Diseases.* Biological control Centre, Bangalore. pp. 91–165.

Singh, S.P. 1991a. Natural enemies of *Papilio* spp. *Indian J. Hort.* **48**: 237–242.

Singh, S.P.1991b. Potential of two commercial formulations of *Bacillus thuringiensis* against citrus butterfly caterpillar *Papilio demoleus* Linnaeus (Lepidoptera: Papilionidae). *Indian J. Hort.* **48**: 135–138.

Singh, S.P.1993a. Integrated pest management in horticultural crops. *Indian Horticulture,* **38**(1): 25–28, 37, 39–40.

Singh, S.P. 1993b. Effectiveness of an indigenous entomophilic nematode against citrus butterfly.. *Insect Sci.* 6(1): 107–108.

Singh, S.P. 1994. Biological suppression of fruit pests. *In: Fifteen Years of AICRP on Biological Control.* Project Directorate of Biological Control, Bangalore. pp. 139–171.

Singh, S.P.1997. Biological control in India. *In: IPM System in Agriculture: Biocontrol in Emerging Biotechnology.* R.K. Upadhyay., K.G. Mukerji, and R.L.Rajak (eds.). Aditya Books (P) Ltd. New Delhi. pp. 216–328.

Singh, S.P., Rao, N.S. and Kumar, K.K.1983. Studies on the borer pests of citrus. *Indian J. Ent.* **45**:286–294.

Talgeri, G.M., Borle, M.N., Mohalkar, D.R. and Kalbhor, S.E.1972. Pests of fruit crops. *In: Crop Pests and How to Fight them*, Directorate of Publicity, Govt. of Maharashtra, Bombay. pp. 131–157.

Tandon, P.L. 1985. Spatial distribution, sampling technique and population dynamics of citrus green scale, *Coccus viridis* (Green) on mandarin Ph.D thesis, UAS, Bangalore. pp. 278.

Tandon, P.L. 1993.. Insect and mite pests of tropical fruits. *In Advances in Horticulture-Fruit crops.* Malhotra Publishing House, New Delhi. **3**(3): 1530–1550.

Verma, P.M.; Rao, D.G. and Capoor, S.P. 1965. Transmissioin of tristeza virus by*Aphis craccivora* Koch. and *Dactynotus jaceae* L. *Indian J. Ent.* **27**: 67–71.

Wadhi, S.R. and Batra, H.N. 1964. Pests of tropical and sub-tropical fruit trees. *In: Entomology in India.* N.C. Pant (ed.).Entomology Society of India, New Delhi. pp. 251–252.

Wait, A.J. 1992. *Citrus in Home Gardens: a guide for tropical Australia.* Brisbane, Australia, Queensland of Primary Industries. pp. 40.

*Pest Management in Horticulture Crops*
L.R. Verma, A.K. Verma and D.C. Gautam (eds)


*Chapter 16*

# Pests of Grapes

*Abraham Verghese*

## INTRODUCTION

GRAPE CULTIVATION OR viticulture is gaining momentum in India. The main grape growing belts are found in Andhra Pradesh, Karnataka, Maharashtra and Tamil Nadu. North India also has some grapes but not to a large extent. Insect pests of grapes were not of much consequence till about sixties. But after that by eighties, insecticides formed major input of viticulture, proving that insect pests have become a force to reckon with (Azam, 1983; Manjunath, 1985). In fact, insecticide pressure has been so high that residues in grapes have become a bottle neck in our grape export.

Insects in grape or vineyard ecosystem are subjected to a major 'ecological shock' at the time of pruning. This results in a sudden loss of food and shelter, and a drastic change in the microenvironment, all of which affect the insects. This chapter mainly deals with the insect pest problems associated with grape cultivation in India and the integrated pest management programme (IPM) developed at the Indian Institute of Horticultural Research, Bangalore (IIHR) between 1985 and 1997. Roving surveys, on farm studies (mainly in Karnataka, Maharashtra, Tamil Nadu and Andhra Pradesh) and field experimentation served as basis for categorising the seriousness and distribution of insect pests, and the efficacy of control measures applied by the growers.

## PEST COMPLEX

Over 85 species of insects have been reported from grapevines in India (Tandon and Verghese, 1994a). Of these, the major pests are mealybugs, thrips, flea beetle, shot hole borer, scale insects and berry webber. The lesser important ones include, cockchafer beetles, girdler beetle, stem boring beetle, leaf roller and termites; while grasshoppers, whitefly, leaf sucking bugs, leaf eating caterpillars, fruit sucking moths, wasps, etc. are of minor importance (Butani, 1979).

### Mealybug *Maconellicoccus hirsutus* (Green)

The pink mealybug *M. hirsutus* has become a serious pest of grapevine in recent years in peninsular India. It poses serious threat in summer months and also in the fruiting phase. According to Azam (1983), the grape mealybug alone caused yield loss ranging from 50 to 100 per cent in the field. They are called as 'hard to kill' insects due to concealed habitat and waxy coating over their bodies. Each mealybug produces 300–750 orange coloured eggs in an ovisac formed by female. The crawlers hatch in 6–7 days and colonise ventrally on new leaves, on rachis of fruits or on any convenient site among the foliage. Crawlers spread in an orchard by their own movement or wind drift.

Normally it takes 8 to 11 days for the mealy covering to appear conspicuously on a bug. Once mealy covering envelops a nymph, it becomes difficult for insecticides to penetrate its body. The males are winged and have four instars, while the females have three instars. Nymphs are pinkish brown in colour. On an average the mealy bug takes 25 to 40 days to complete its life cycle, depending on prevailing ecological conditions (Verghese and Tandon, 1989). There are several overlapping generations every year (Mani, 1986).

### Grapevine thrips *Rhipiphorothrips cruentatus* Hood

It is a major pest of grapevine and is highly polyphagous. The nymphs and adults rasp the ventral surface of tender leaves and flower stalks and suck the oozing cell sap. Nymphs are reddish in colour and adults bright yellow (males) to dark brown (females). Both sexual and parthenogenetic reproduction occur side by side; the progeny of the latter being always males. Eggs are laid @ 2 to 6 per day. Incubation, pre-imaginal, pupal and total lifecycle durations occupy 3 to 8, 11 to 22, 2 to 5 and 14 to 33 days, respectively (Butani, 1979). There are 5 to 8 generations in a year and overwintering is in the pupal stage, up to 15 cm deep in the soil.

### Grape Flea Beetle *Scelodonta strigicollis* (Motschulsky)

The flea beetle is becoming a more regular and serious pest on grapevines. The beetle is especially serious when it attacks buds following pruning. The affected buds or sprouts soon dry out. Many growers fail to detect it as the adults are nocturnal in habit and hide during day under loose bark or clods. But when the populations are fairly high, adults can be detected in the orchards easily. As the beetles feed on leaves also, due care must be taken to prevent flare up. Characteristic long slits in the lamina of the leaves is another indication of the presence of the beetle in the field. In such cases, the growers have to be alert about flea beetles in the next season (Verghese and Tandon, 1996). The grubs which feed on roots of weeds live in the soil and have not been observed to cause any major damage to grape roots (Verghese, 1997). A female lays 200 to 600 cigar shaped, creamy-white eggs in her life time of about 8 months (mainly from March to October), in clusters of 20 to 40 eggs. Incubation period is 4 to 8 days, the grub stage lasts for 34 to 45 days and pupal period is 7 to 11 days (Butani, 1979). There are four overlapping generations in a year (Butani, 1979). Hibernation is in adult stage during November-December to March. Adults are small, 4 to 5 mm long, copper coloured beetles with 3 black spots on each elytra. Mating takes place a month after emergence and the eggs are laid 4 to 7 days thereafter beneath the bark or crevices in the stems.

### The shot-hole borer *Xyleborus crassiusculus*

The shot-hole borer is a tiny beetle, measuring 3 to 4 mm in length and is brownish to brownish-black in colour. It belongs to the group of insects popularly called 'Ambrosia beetles'. Ambrosia is a fungus 'cultivated' by these beetles in their shot-hole nests; and the grubs of the beetle feed on this fungus. It is the nest-building activity of the beetle on the main trunk that actually damages the vine.

The adult beetles many attack the main trunk, starting from the base of the plant. The external hole is the initial nest excavation by the adult beetle. Next, it tunnels into the main trunk, cutting through xylem and phloem vessels. These tunnels can be seen as galleries, if a cross-section of the trunk is cut and examined. The holes are almost uniformly distributed along the main trunk. The foremost symptom of the pest injury is the pinholes on main trunks, more conspicuous, if the loose bark is removed. Powdery wastes due to the tunneling can be seen falling from the holes. As infestation progresses, gummy exudates can be seen along the trunk. Besides, presence of beetle themselves is sure indication of the pest. Each hole opens as a gallery extending between 1–2 cm. In these galleries eggs are laid. The grubs which hatch out feed on the fungus cultured by the beetles in the galleries. Pupation takes place within the galleries. Adults which emerge, further mate and reinfest fresh vines and orchards

### Defoliating beetles

Defoiating beetles, cockchafer beetles or white grubs are widely distributed al over India damaging a variety of economic crops including fruit trees. Grape is defoliated by several beetles which attack vineyard in May/June, following the early rains. The more important ones are: *Holotrichia serrata* (Fabricius) and other *Holotrichia* spp.; *Adoretus bengalensis* Blanchard and other *Adoretus* spp. These beetles do not attack grape regularly, but sporadically these can be very serious (Verghese and Tandon, 1994b). Eggs are laid in soil. On hatching, the grubs feed on rootlets of various grasses and even economic crops including grapevines. Pupation takes place in the soil. Adult beetles are nocturnal, come out of the soil after dusk and feed on leaves making innumerable holes in leaf lamina. In case of severe infestation, the entire leaf lamina may be eaten away. The adult beetles appear soon after the first heavy shower (May-June) and are active for about two months.

### Girdler beetle *Sthenias grisator* Fabricious

The girdler beetle, *S. grisator* has been reported as a serious pest from South India and Sri Lanka. Besides grapevine, the beetles also damage almond, apple, citrus, jack-fruit, mango, mulberry, ornamentals, etc. The adult beetles emerge usually around July-August and maximum damage is caused during August to October. The beetles with their strong mandibles, ring or girdle the young green branches of the host plant. This is a unique device that invariably precedes egg laying and allows the girdled branch to dry up so as to enable the grubs to bore and tunnel the dry wood. For egg laying, the females cut transverse slits under the bark of girdled branch and thrust in between the bast and sap wood, one to four eggs. On hatching, the grubs straight away tunnel into the wood. Pupation takes place in an elongate chamber inside the tunnel and the beetles emerge out by circular exit holes.

Eggs are elongated oval in shape, about 4 mm long and whitish in colour. Freshly formed grubs are 2 to 3 mm long, having dark brown head, globular thorax and cylindrical abdomen sparsely covered with hair. When full grown, grubs are 11 to 14 mm long. Adult beetles are about 24 mm long, greyish-brown with white and brown irregular markings resembling the bark colour; elytra have an elliptical greyish median spot and an eye shaped patch. Incubation period is 8 to 10 days but the duration of, grub stage varies considerably according to dryness and nutritive value of food material and competition. When the infestation is severe the grub stage is usually 7 to 8 months with only one generation in a year.

**Stem boring beetle *Sinoxylon anale* Lesne**

*S. anale* is an important pest reported from Punjab (Butani, 1979). It has also been recorded from France, Italy, former USSR, China and Japan. In addition to grapevines, the beetle attacks a number of forest trees in India. The adults appear with the onset of winter, bore inside the stems and feed within, making longitudinal tunnels and a number of exit holes. Chewed up wood particles are thrown out of these holes. The eggs are laid in these tunnels and on hatching, the grubs feed therein causing further damage. As a result of the attack, the plant portion above the point of attack gradually dries and dies away. The pest is active only in dormant vines and prefers late-sprouting varieties like Anab-e-shahi. The breeding of the pest continues in the dead vines throughout the year. Grubs are thick, slightly curved and yellowish-white in colour. Adults are dark brown, stout beetles, 3 to 5 mm long.

**Leaf roller *Sylepta lunalis* Guenee**

*S. lunalis* is found all over India and though a specific pest of grapevine, it has been found attacking eleven species of Vitaceae in South India (Butani, 1979). A female lays 100 to 150 eggs, usually on ventral surface of leaves. On hatching, the young caterpillars feed on epidermis of leaves and skeletonize the same. Later, these caterpillars roll the leaves and feed within. Pupation takes place within rolled leaves. Generally each leaf-roll contains only one caterpillar or pupa. Sometimes the pupae may fall down from the rolled leaves and are seen lying on the ground on fallen leaves and debris. The activity of the pest is accelerated with the onset of monsoon (July-August) and continues till October.

Eggs are oval, about 1.5 mm long and creamy-white in colour. Caterpillars are hairy, cylindrical, 18 to 24 mm long when full grown and pale green in colour. Pupae are 8 to 12 mm long and brown in colour. Adults are dirty brown moths with white spots and wavy patterns or fore and hind wings. Egg, larval and pupal stages last for 2 to 3, 14 to 20 and 5 to 7 days, respectively. The total life cycle during the active period lasts for 3 to 4 weeks.

**Fruit borer *Oxyptilus regulus* Meyrick**

*O. regulus* is a grape boring plume moth, reported from India, Sri Lanka and Australia. In India, it is more common in Karnataka and other southern states than in the North. Eggs are laid singly at night on fruits, around fruit stalks or peduncles. On hatching, the caterpillars bore into the ripening berries and tunnel through the pulp up to the seeds. A single caterpillar may damage number of berries. Pupation generally takes place on peduncles of healthy berries. Some of the larvae that fall down pupate on weeds or dry leaves lying on the ground.

## VERTEBRATE PESTS

Among the vertebrate pests, bats and birds are reported pestilent on the ripe grape fruits. The short-nosed fruit bat, *Cynopterus sphinx* Vahl, seriously damages grapes in South India (Verghese and Tandon, 1994a).

### Fruit-eating bat *Cynopterus sphinx* Vahl

The short-nosed fruit bat, *C. sphinx* is widely distributed in Peninsular India and South-East Asia. The length of the bat from head to tail ranges from 10 to 12 cm and the wing span ranges from 33 to 34 cm. According to Prator (1971), this species feeds on nectar, fruit juice and tiny nectar-rich flowers. In forest ecosystems, it plays a significant role in pollination and seed dispersal of wild fruits, like other fruit bats, but in horticultural ecosystems it causes marginal yield losses (Isacc *et al.*, 1993).

In grapes, the incidence and economic loss caused by this bat is increasing probably due to decline in the number of wild fruits and flowers and an increase in the area under grape cultivation. The bat sucks the juice from berries after plucking individual fruit off the bunch and then expels the pulp, seed and skin which litter the base of the grape vine. Grape bunches with bare stalks are also an indication of bat damage. Bats are difficult to control, as they are nocturnal, and the day roosting sites are invariably located away from vineyards. They are not easily amenable to chasing or scaring because of their nocturnal habits. As *C. sphinx* is also a useful pollinator of forest trees, it is not desirable to kill them. Hence, non-destructive control strategies are advocated in areas where fruit damage by bats is serious.

### Frugivorous birds

The major avain species which feed on grapes are: jungle myna (*Acridotheres fuscus*(Wogler)), Indian myna (*A. tristis* (L.), jungle crow (*Corvus macrorhynchos*) and house crow (*C. splendens* Vieillot). Besides, two species of minor consequence are the rose-ringed parakeet (*Psittacula kramer* Scopoli) amd red-vented bulbul (*Pycnonotus cafer* Blyth). Beak gnawing mark, presence of faecal matter on adjacent leaves, pillars, etc., part of pulp/skin remains on stalklets and presence of birds on perches near vineyards are characteristic symptoms useful in diagnosis of bird pestilence (Verghese and Tandon, 1993).

## MANAGEMENT STRATEGIES

Insecticides which seemed potential as control agents, were either laboratory or field tested, and then recommended. Use of eco-friendly botanicals as insecticides were also investigated. The subsequent orchard assessment and/or grower feed back and analysed data formed basis for drawing up control strategies. The integrated pest management (IPM) for grapes, has been drawn up, following crop phonology, modelled after typical October fore pruning and April back pruning system. These have been divided into three phases, A, B and C, and definite directions for IPM have been serially outlined into a decision making model pattern, for the convenience of researchers and growers. This model is available as a software package on request from the Director, IIHR, Bangalore.

**Phase-A: Fore pruning to berry formation**

This phase is approximately between October and February. After pruning, the vineyard is bare, but insect communities are gearing to get active. The flea beetle, *Scledonta strigicollis* Mostchulsky would be the first to emerge, to feed on the new buds, turning it into a 'dried sprout'. Later on, it takes to full grown leaves, causing characteristic longitudinal slit feeding on the lamina.

The two main species of thrips, *Rhipiphorothrips cruentatus* Hood, and *Scirtothrips dorsalis* Hood would emerge when tender leaves are ready as ovipositional sites. There are overlapping generations, and over a period of time, would migrate and rasp mustard sized berries. Thrips injury at this stage manifest as 'thrips scab' on full grown berries, considerably reducing market value. There are several other species of thrips on grapes, but of only minor status. During this phase, especially immediately after pruning, the defoliators *Helicoverpa armigera* Hubner and *Spodoptera litura* (Fab.), are serious in some regions.

As the crop progresses (approximately by December) the eggs masses of the mealybug, *Maconellicoccus hirsutus* (Green) from under the bark or a clod begin to hatch and crawlers ascend the vines. Or, it may immigrate as wind drift from alternate hosts or adjacent vineyards. These settle on leaves, causing diagnostic, 'marginal cup curling'. If left unchecked, they multiply and migrate to the fruit bunches, where they are very difficult to control.

***Strategies of IPM during Phase-A***

- Immediately after pruning remove all pruned twigs and burn, as it may harbour stages of thrips and mealybug.
- After pruning, remove all loose bark from main trunk and primary arms. Then swab the trunk and arms with double dose of spray concentration of contact insecticide using like carbary (l0. 4%) or malathion 0.2% a hard brush. For swabbing, the swab mixture should have neem oil 1% plus neutral soap solution which should be mixed with the above insecticides. This strategy eliminates egg masses of mealybugs, stages of scales (e.g., *Hemiberlesia lataniae* (Signoret), etc. If shot-hole borer (*Xyleborus* sp.) is present, the swabbing mixture should contain dichlorvos (0.15%). Swabbing has to be repeated once or twice till powdery 'exudates' stop. Shot hole borer attack is mainly on the variety Bangalore Blue, and is characterised by pin holes and powdery exudates on the main trunk.
- Rake the soil immediately after pruning to expose and kill immature stages of flea beetle, thrips and mealybugs.
- After A-3, rake in to the soil neem cake (mechanically pressed) @ 250 to 500 g/vine on the ground immediately around the trunk to kill any residual egg mass of mealybug. This needs to be done only in mealybug prone orchard. In addition, it has some manurial value.
- At bud sprout, spray carbaryl (0.2%) or quinalphos (0.06%). After one week a spray of neem cake extract (8%) may be given whenever flea beetle is serious. This prevents flea beetle feeding, and limits thrips establishment. Repeat carabryl or quinalphos after two weeks from first spray.
- At post-bloom (approximately a month after pruning) spray dimethoate 0.06% followed by a single spray of azadirachtin (10 ppm) to prevent thrips, berry webber and mealybugs

eventually establishing or damaging the bunches. Defoliators, if any, at this stage will also be controlled.

- Prune infested leaves with mealybugs and burn.

Note: *All the above measures, except last one, are prophylatic and have to be followed, especially in orchards with history of infestation and considering the commercial nature of viticulture.*

Caution: Over use or excess concentration of insecticides recommended, and unreasonable use of synthetic pyrethroids can set off several chains of secondary resurgence.

**Phase B: Berry formation to harvest**

This phase is approximately from February to April/May. During this, mealybugs not checked in phase A will firmly establish on berries beyond control. Birds and/or bats could cause damage during this time. Sometimes, berry moth, *Oxyptilus regulus* Meyr, may cause some pestilence. The mite (*Oligonychus* sp.) infestation may occur in this phase, but is more pronounced in the next phase.

***Strategy of IPM during phase B***

- During early berry stage, spray as in sixth strategy of phase A, if there has been an infestation of mealybug in the orchard the previous year.
- If mealybug is serious, spray dichlorvos (0.1%) at least twice before harvest. There should be at least 15 days waiting period prior to harvest. Spray only if infestation is sighted as berry formation progresses. All spraying should stop a fortnight before harvest for safety reasons. If on an average less than one mealy bug is found per bunch, there is no need to spray, as natural enemies in the ecosystem will suppress this population. Bioagents which naturally suppress mealybugs inchude: *Spalgis epius* (Westwood), *Mallada boninensis* (Okamato), *Cryptolaemus montrouzieri* (Mulsant), *Anagyrus dactylopii* (How.), spiders and insectivorous birds. Release of *C. montrouzieri* @ 10 grubs per vine controls moderate infestation of mealybug. However, no insecticidal spray should be given while the predators are active.
- Spread nylon nets around the orchard and cover canopy gaps. With brier or twigs to prevent bird/bat damage.

Note: *The above strategies are supervisory than prophylatic.*

**Phase C: Back pruning to fore pruning**

This phase is approximately between May and October. During this phase, the plant is in a 'rest' phase. Following back pruning, like in post fore pruning, flea beetles and thrips can infest. The general lush vegetative growth invariably coincides with onset of monsoon. This attracts defoliating battles like *Holotrichia* sp., *Adoretus* sp. on to the leaves. The stem girdler, *Sthenias grisator* Fab. may infest the stem, after hatching from eggs laid below the bark. Sometimes, the leaf roller, *Sylepta lunalis* G., sphingid, *Hippotion celerio* L., and aleyrodid, *Aleurocanthus spiniferus* Q. may mildly infest the grapevines. The scale, *H. lataniae* and the mealybug *Ferrisia virgata* may also infest, the former mainly on stems and trunk and latter on shoots and leaves.

***Strategies on IPM during phase C***

- Repeat first two strategies of phase A. This, in addition will prevent stem girdler infestation. If girdler persists, or is serious, a spray of quinalphos (0.06%) should be given.
- In case defoliating beetles, leaf—roller, sphingid or aleyrodid is noticed, a spray as in A-5 may be given.
- In case of thrips or mealybugs, one or two sprays as in A-6 may be given.
- If mites are a problem, spray diamethoate (0.06%) or micronised sulphur (0.08%) (Normally sulphur sprays for disease also control mites).

Note: *During the phase, the strategies except for first strategy, are supervisory and entirely need-based.*

## REFERENCES

Azam, K.M. 1983. Losses due to pests in grapes. *Indian Journal of Entomology* (special issue), 2: 387–389.

Butani, D.K. 1979. *Insects and Fruits.* Periodical Expert Book Agency, New Delhi. pp. 415.

Isacc, S.S.; Kumaraswamy, P. and Marimuthu, G. 1993. Observations on pollinating behaviour of a tropical fruit bat, *Cynopterus sphinx* (Chiroptera: Pteropodidae). *In: Pollination in Tropics*. G.K. Veeresh, R. Umashanker and K.N. Ganeshaiah (eds.). International Union for the Study of Social Insects. Indian Chapter, Bangalore. pp. 151–152.

Mani, M. 1986. *Distribution, bioecology and management of Maconellicoccus hirsutus with special reference to natural enimies.* Ph.D. Thesis, UAS, Bangalore.

Manjunath, T.M. 1985. *Maconellicoccus hirsutus* on grapevine. *FAO Plant Protection Bulletin,* **33**: 74.

Prator, S.H. 1971. *The Book of Indian Animals*. Bombay Natural History Society, Bombay. pp. 160–176.

Tandon, P.L. and Verghese, A. 1994. Present status of insect and mite pests of grapes in India. *Drakshavrita Souvenir.* Maharashtra Growers Association, Pune, India. pp. 149–157.

Verghese, A. 1997. Crop loss due to flea beetle in grapes. *Drakshavrita Souvenir*. Maharashtra Growers Association, Pune, India. pp. 147–148.

Verghese, A. and Tandon, P.L. 1989. Management of mealybug in grapevines with special reference to Thompson seedless. *Drakshavritta Souvenir*. Maharashtra Growers Association, Pune, India, pp. 79–82.

Verghese, A. and Tandon, P.L. 1993. Avian depredators in vineyard ecosystems. *Drakshvritta Souvenir.* Maharashtra Growers Association, Pune, India. pp. 117–119.

Verghese, A. and Tandon, P.L. 1994a. Foraging model and other observations on grape eating bat, *Cynopterus sphinx* (Vahl). Drakshavritta Souvenir, *Maharashtra Growers Association*, Pune, India. pp. 141–142.

Verghese, A. and Tandon, P.L. 1994b. Management of defoliating beetles in grapevine. Drakshavritta Souvenir, *Maharashtra Growers Association*, Pune, India. pp. 145–147.

Verghese, A. and Tandon, P.L. 1996. Management of flea beetle damage in grapes. Drakshavritta Souvenir, *Maharashtra Growers Association*, Pune, India. pp. 107–108.

*Pest Management in Horticulture Crops*
L.R. Verma, A.K. Verma and D.C. Gautam (eds)


*Chapter 17*

# Insect Pests of Fruits of Arid and Semi-Arid Regions of India

*R.K. Lakra*

## INTRODUCTION

DAMAGE BY INSECT PESTS, *inter alia,* is one of the major constraints in enhancing the production of fruits, especially the minor fruits, which mainly occupy the neglected, poor and marginal lands in India. The pests are known to cause 8 to 35 per cent production losses in fruit plants of India. The estimate is well within the realm of practicability and could be on an even higher scale since these are grown under constraint and harsh conditions of arid and semi-arid areas. Intensification (over diversified agroclimatic areas having variations in soil type, water, plant material, etc.) of some hardy fruit plants like *aonla/amla*, *bael*, *ber*/jujube, date palm, fig, *jamun/jamon*, *karonda/corinda*, mulberry, *phalsa/falsa* and pomegranate in fragile environment of arid and semi-arid regions of India has led to increase in pest problems and also brought some newer ones to the wake. However, despite the fact that these minor fruits contribute significantly towards fruit production at the national level, they have not yet received due attention of scientists and policy planners.

In the recent decades, area under fruit trees has increased considerably, and so have their pest problems. Cultivation of high yielding cultivars responsive to fertilizers, irrigation, dense planting and new technological packages including use of broad spectrum insecticides, has further aggravated the pest problems in these plants. As a consequence, if systematic studies on their pests are not initiated timely, these might thwart the likely benefits of their cultivation and more so of their sustainability in sparsely inhabited arid and semi-arid regions. In this chapter, an attempt has been made to briefly deal with the major, minor, sporadic and new pests of some of the under-exploited fruit trees of these regions. The information focussing on the taxonomic position, distribution, host plants, life and seasonal history, nature and extent of damage, natural enemies and control strategies of these pests is included. It is high time to accelerate systematic studies on the pest complexes of these horticultural plants and devise economically feasible and ecologically viable pest management strategies so as to reduce production losses. Research should focus on low input IPM approaches that encompass farmers' current practices, host plant resistance, natural biological control, optimum tree density, minimum use of insecticides and good crop husbandry.

## JUJUBE/BER (*ZIZIPHUS MAURITIANA* LAMK.)

*Ber/Bor* or jujube, *Ziziphus mauritiana* Lamk., being an indigenous fruit is an integral part of Indian culture and has no rivals in arid fruits. *Ziziphus*, with many forms (cultivated and wild; tall trees, shrubs and bushes; fruit trees and weeds, etc.) and genotypes grow extensively throughout the Indian subcontinent. Being a hardy tree, it requires little post planting care and tolerates high salinity, prolonged water-logging as well as drought conditions, and grows well even in marginal and poor lands. Its fruits are rich source of vitamin C and are eaten by the languqrous natives fresh and ripe, or are dried and used as dessert, candied or used in making a very refreshing drink. Its fruit is considered 'poor man's apple' as it is highly nutritious with medicinal value. The twigs are used in religious rites and ceremonies. Its foliage is used as fodder, wood as fuel and timber, and live branches for rearing lac insect and tassar silkworms. A wide spectrum of as many as 225 pest species has been found attacking different *Ziziphus* spp. in the world, of which, nearly 150 species are reported from India alone. In addition, several species of birds, bats, squirrels, rodents and monkeys also feed on its luscious fruits, while many species of four footed animals damage its foliage. However, most of these are mere records of occurrence of a species in a given locality and only about 20 species are of economic importance. The major ones are as follow:

### *Ber* fruit fly *Carpomyia vesuviana* Costa

Of the various insect pests of *Ziziphus* spp., *C. vesuviana* is undisputedly the most destructive and well known. In fact, the depredations caused by this fruit fly are a limiting factor in successful cultivation of *ber* in several regions. The pest is found in all the *ber* growing areas in India and is widely distributed in the Indian subcontinent, several countries of Oriental Asia, Indian Ocean Islands, Middle-East, Temperate Asia and South Europe. The fruit fly is monophagous and breeds only in fruits of *Ziziphus* spp.

The adult fly is brownish-yellow in appearance, slightly smaller than the common house fly, and possesses a distinctive pattern of black spots with brownish longitudinal stripes on the thorax and four yellowish or greyish-brown characteristic cross bands on the hyaline wings. The female lays an average of 22 eggs in its life time, in newly set ("pea" size) to ripe fruits. Eggs are deposited in a small cavity, made with the movement of its protrusive ovipositor, just below the epidermis of the fruit, usually singly, occasionally 2 to 5 and exceptionally more than 8. The eggs hatch in 1 to 4 days.

The newly hatched legless, tiny, whitish maggot burrows its way through the pulp towards the stone of the fruit, surreptitiously and comes out only when fully grown. The maggots undergo two moults inside the reddish brown irregular galleries, which become conspicuous around stone with the increase in larval size and age, and take 7 to 24 days to become full-fed. Initially the tunnels are confined to a portion of pulp but gradually they furrow the entire pulp and latter instar maggot restricts its feeding around stone of the fruit. The larval burrows are filled with brownish or dirty white maggoty frass. On becoming full grown, the maggot carves out a typical circular hole in the epicarp and wriggles out slowly to drop on ground in a peculiar spring-like fashion and pupates in soil at a shallow depth of 3 to 6 cm. The pupal stage lasts for 5 to 42 (average 14) days. However, some pupae hibernate during December-January, and aestivate during May-June and may last upto 43 to 122 (average 78) days. All the maggots which complete development towards the end of fruiting season in April, and those which complete development by end of November, undergo long term pupation so as to cope with the inclement weather. The adults are good fliers and live for 2 to 55 days. They feed on sugary fluids secreted by homopterans.

The fly remains active from July to April and completes 6 to 9 overlapping generations a year. It passes 3 to 5 generations in off-season and seasonal fruits of *jhar-ber, Z. nummularia* (Burm-f.) Wight and Arn. and *bordi, Z. mauritiana var. rotundifolia* Lamk., from July to October, and another 3 to 4 broods from November to April in cultivated *ber* fruits. However, in areas where only the cultivated fruit trees are available, the tephritid completes 2 to 4 generations during the main fruiting season. Under favourable conditions a single life cycle is completed in 15 to 39 (average 24) days, while under unfavourable ones it takes as long as 35 to 80 (average 56) days.

Only the larval stage is harmful. However, the oviposition punctures made by the female also result in arresting growth of the surrounding tissues giving the fruits a rugged appearance in the form of protuberances and/or depressions. Younger the fruits punctured, more conspicuous are the deformities produced. Similarly, more the oviposition punctures per fruit more are the deformities and vice versa. Crinkling is least observed when fruits are infested after attaining full size and ovipuncture is more distinguishable as a dark dot in such fruits. Off-season infested fruits exhibit more uneveness in shape than seasonal ones. Infested fruits become unfit for consumption and when the number of maggots per fruit is more, the entire pulp is riddled and hollowed out. In such fruits, invariably bacterial and fungal infection set in, leading to decay of pulp and emission of a peculiar offensive and alcoholic smell. Externally rotting first becomes visible around ovipuncture or exit hole (Lakra and Singh, 1983). The fly infestation hastens ripening and dropping of fruits, and such fruits are prone to bird attack. On an average, the tephritid destroys 24 (5 to 61) per cent fruits of *Z. mauritiana* and 31 (6 to 51) per cent of wild bushes (*Z. nummularia*) (Lakra and Singh, 1984). However, in years of severe infestation and/or in susceptible genotypes/cultivars, as high as 80 per cent fruits are riddled by the fly maggots. Normally in a given region, 20 to 33 per cent fruits are spoiled by infestation of *C. vesuviana* every year. Fruits of early and late ripening cultivars, bigger size, and those possessing thin skin, sweeter and higher pulp stone ratio are more susceptible to attack (Batra, 1953).

***Management***

In no other family of insect pests, management is as difficult as in the case of fruit flies. Complete protection of fruits against the attack of tephritids is neither possible nor required. The adoption of the following strategies will help in reducing or minimising the infestation of fruit flies in *Ziziphus* fruits:

- Collect all the fallen and infested fruits at regular intervals (twice a week from fruit setting till havest) and destroy them promptly, feed such fruits to sheep, goats, camels or other farm animals, or bury them at least one metre deep in compact soil, lest the adult flies escape.
- Deep and thorough raking up of soil in the basin area of trees twice during peak summer and winter months, exposes the pupae to inclement weather and natural enemies. A large

number of aestivating/overwintering pupae is also destroyed by mechanical injury during this operation.

- Do not allow wild bushes, trees, isolated growing or abandoned jujube plants of *Z. nimmularia* or *Z. mauritiana var. rotundifolia* or other *Ziziphus* species to grow in the vicinity of *ber* orchard, as their fruits serve as good source for multiplication and help in carry over of fruit flies to cultivated *ber* fruits.
- Left over, off-season and early set *ber* fruits on trees and bushes provide a continuum for *Carpomyia* to breed, multiply and switch over to the main crop. Ensure that no jujube fruits are available to the fruit flies during off-season, i.e. after harvest of the main crop till next fruiting.
- Early harvesting of mature fruits (at colour change stage) avoids infestation. Do not allow the fruits to over ripe on the trees as they are more prone to attack of fruit flies.
- Some insecticides, especially those with systemic or penetrant activity and longer residual effect, when applied as cover sprays, synchronising with the oviposition activity of fruit flies, protect the developing fruits from egg deposition and also provide effective control of the maggots. In areas of pest damage history, when majority of newly set *ber* fruits attain the size of "pea", spray the trees with 120 ml oxydemeton methyl (Metasystox 25 EC) or 100 ml dimethoate (Rogor 30 EC) in 100 litre water. Repeat the spray, if necessary, after 30 to 45 days.
- In order to protect the mature/ripe fruits, spray (if needed) 100 ml malathion (Cythion 50 EC) + 1 kg *gur* or sugar in 100 litre water. Bait sprays consisting of brewery waste and malathion have been found effective against *Bactrocera* species. Hydrolysed protein is the usual choice for bait (20 g per litre solution), however, it has not been evaluated against *C. vesuviana* (Joshi and Shinde, 1971).
- Methyl parathion (Metacid 2% dust) or fenvalerate (Fenval 0.4% dust) when mixed thoroughly in pulverised soil in basin area of the tree, at the rate of 25 kg per hectare, provides satisfactory control of the falling mature larvae, pupating maggots and prepupae of the fruit fly.
- Early ripening, sweet, soft and thin skinned, more juicy and flavoured pulpy varieties with moderate to bigger sized fruits are more susceptible to attack. There is no identified resistant *ber* variety to the fruit fly though pronounced variations in infestation level among various genotypes have been reported (Saxena and Rawat, 1968; Mann and Bindra, 1976; Sachan, 1984; Singh, 1984). Where possible it should be used as a tool for the control of fruit flies.
- A good number of parasitoids and predators are associated with fruit flies. Singh (1989) reported a braconid *Biosteres vandenboschi* Fullaway as a parasitoid of *C. visuviana* from India. However, their populations remain in negligible proportion and thus as a tenet of IPM, the biotic agents alone are not effective in checking the pest. Nevertheless, efforts should be made to conserve and augment their populations to our advantage.
- Post-harvest disinfestation of fruits with fumigants, though does not seem convincing and logically practical in small sized fruits like *ber*, yet is very effective in causing mortality of eggs and larvae of fruit flies. Treat the plucked *ber* fruits wih 0.45 kg ethylene dibromide per 28 $m^3$ for 6 hr or with 0.25 kg for 12 hr in a sealed earthenware or air tight space.

- Many chemical lures are known to attract fruit flies belonging to different genra like *Bactrocera*, *Ceratitis*, etc. But none of them are effective again *C. vesuviana*. This aspect needs exploration for developing chemicals that can be used against jujube fruit fly.

### Cockchafer beetles or May/June beetles

***Holotrichia spp., Adoretus spp. Schizonycha spp. and Anomala spp.***

Several species of scarabacid beetles have been recorded damaging foliage of jujube, *phalsa, jamun*, fig, grapevine, guava, pomegranate, *karonda*, mulberry and many other agroforestry trees, ornamentals and shrubs, found growing in arid and semi-arid regions throughout the country. In fact, white grubs are the most destructive of soil insects and are important national pests. The adults, commonly known as chafers or cockchafers, are harmful to fruit trees, whereas, the soil dwelling larvae, commonly called white grubs, chiefly destroy the roots of several field crops.

The beetles emerge from the soil after first good pre-monsoon or monsoon showers in May or June and hence are commonly called as May/June beetles. The adults gather in good numbers on trees, mate and feed voraciously on the leaves and tender shoot tips during night till dawn, and hide during the day. They make characteristic round holes in the leaves. In the case of severe attack only the skeletons of the leaves are left and the whole tree is stripped. The growth and fruiting capacity of the trees are adversely affected. The lamellate beetles are usually brownish-black or brown, stoutly built, and possess hardened elytra. Male beetles are smaller than the females and have longevity of about a month.

Eggs are laid in soil and on hatching the cream coloured grubs feed on decaying organic matter, dung and roots of almost all *kharif* field crops, grasses and even on rotting underground parts of trees from July to October. Crops having tap root system and poorly developed roots are more susceptible than those with adventitious root system and fully established roots. The full-fed grub is curved, C-shaped and dingy white in colour and pupates deep in soil in an earthen cell. The grubs move deeper in soil in search of food, moisture and for pupation. During winter the pupae enter into hibernation within the pupal cells and resume activity as soon as good showers of rain are received after April. In case there is no rain during this period, the beetles perish inside the soil. There is a single brood of the pest in a year and the life cycle is completed in 4 to 6 months.

#### *Management*

- Collect the congregated beetles at night by vigorously shaking the host trees on a sheet of cloth and kill them in kerosenised water. Beetle collection should be started from the day first good premonsoon or monsoon rain is received and continued daily after sunset for 5 to 6 days. The beetles are easily dislodged from the trees by jerking and can be easily and safely picked by hand. Conscientious collection of adults on a large scale in the form of a campaign is the cheapest and safest method of control.
- Frequently plough up the orchard soil during July to September and again during December-January, so as to expose the grubs, diapausing pupae and inactive adults to predatory birds and other natural enemies. Ploughing before the grubs go below plough depth is more effective.

- A good number of the newly emerged adults is attracted towards light at night. The use of light traps having low intensity light source should be deployed at suitable places in the orchard to attract the beetles collected and killed in kerosenised water.
- The beetles exhibit differential response towards various host trees. Wherever feasible, some highly attractive host trees like drumstick or *senjana* (*Moringa oleifera*), *karonda* (*Carissa carandas* Auct.), neem (*Azadirachta indica* A. Juss.), jujube (*Ziziphus mauritiana* Lamk.), khejri (*Prosopis cineraria*), *Ficus* spp., etc. should be planted at the border of orchard so as to attract the cockchafer beetles and avoid damage to fruit trees. The congregating beetles on such hosts should be collected and killed.
- The adults are attracted to manure, dung heaps, plant debris and rotten organic matter heaps for egg laying. Such heaps made at a few places near fruit trees help greatly to concentrate the eggs and grubs to a limited area where they can be killed effectively either by mechanical means or by application of methyl parathion dust (Metacid 2% dust) @ 25 kg per hectare. Since the freshly turned soil with high organic content is attractive to adult beetles, only well rotten organic matter should be applied in tree basins.
- Several natural enemies including parasitoids, predators, entomopathogenic fungi, bacteria and virus, etc. contribute to mortality of the scarabaeids in nature but none of them are effecient biocontrol agents. Likewise, several species of birds, lizards, toads and mammals (pig, hedgehog) also feed upon the scarabaeids but with no regularity. However, the use of fungi, *Beauveria brongniartii* (Sacc.) Petch and *Metarrhizium anisopliae* (Metchnikoff) and bacteria, *Bacillus* spp. has shown some promise and has been recommended against the pest.
- Since the adult beetles congregate on the host trees for feeding from dusk to dawn, this provides an opportunity to kill them by application of insecticides. Spraying the trees in the evening with 100 ml monocrotophos (Monocil 36 WSC) or 200 ml quinalphos (Ekalaux 25 EC) or 300 g carbaryl (Sevin 50 WP) in 100 litre water, one day after emergence of the adults is useful. Repeat the spray after 5 to 7 days, if the attack continues or the insecticidal deposits are washed away by rain. All host trees in the vicinity of orchard should also be treated. Avoid spray of insecticides during flowering or ripening of fruits (Bindra and Singh, 1971; Khan and Ghai, 1974; Tandon, 1994).

### Hairy caterpillars

***Euproctis fraterna* (Moore), *E. lunata* Walker, *E. subnotata* Walker and *E. flava* (Bremer)**

The caterpillars of *Euproctis* spp. are polyphagous and sporadic pests of several forest trees, fruits, vegetables and field crops. They are widely distributed in the Palaearctic, African, Oriental, Australian regions, and the Indian subcontinent. More than 50 species of *Euproctis* occur in India and Pakistan, of which, *E. fraterna*, *E. lunata*, *E. flava* and *E. subnotata* are common on *ber*, but the first one is predominant. The caterpillars cause considerable damage to several arid and semi-arid fruit trees, viz., plum, jujube, *jamun*, phalsa, pomegranate, *aonla*, fig, mulberry, custard-apple, loquat, grapevine, guava, peach, pear, citrus, mango, etc. but plum and castor are the preferred hosts (Sah *et al.*, 1972).

The females are robust, yellowish moths with brownish forewings which lay 92 to 241 eggs in clusters on the under surface of tender leaves. The egg masses are covered with yellow hair from anal tuft of the female. The egg, larval and pupal stages last for 9 to 14, 25 to 56 and 7 to 15 days, respectively. The larval development is faster during the spring and summer, slows down during winter and again becomes active in February. The larva pupates in thin silken cocoon in leaf fold or on a branch. The adults live for 5 to 7 days and the life cycle is completed in 44 to 84 days. On certain arid fruit trees, *E. fraterna* completes six generations in a year, and the duration of development varies considerably on different hosts (Kuswaha and Bhardwaj, 1968).

The newly hatched yellowish hairy caterpillars feed gregariously on ventral surface of tender leaves by scrapping the chlorophyll, consuming the leaf margins and eventually skeletonizing them completely. After 3rd instar the larvae spread out and feed voraciously, resulting in defoliation of twigs and branches and sometimes the whole trees. Quite often, the catepillars also nibble the epidermis of fruits resulting in corky, deformed surface and sometimes fungus develops on the injured portion and renders the fruits unfit for human consumption. The losses are more pronounced when the larvae feed on the developing fruits. The pest is more destructive to tender and succulent leaves than the older ones. The full grown larva is stout, reddish brown, thickly covered with poisonous hair. At harvest, body contact with urticating hair leads to skin irritation and itching to workers. The attack of the pest usually remains confined to trees on which eggs are laid and do not spread in the entire orchard. Another lymantriid, *Dasychira mendosa* (Hubner) often appears in considerable numbers on jujube.

*Management*

- Collect and destroy mechanically or in kerosenised water the congregating larvae and egg masses.
- In the case of severe infestation, spray the affected trees with 100 ml monocrotophos (Monocil 36 WSC) or 200 ml endosulfan (Thiodan 35 EC) or quinalphos (Ekalaux 25 EC) or 300 g carbaryl (Sevin 50 WP) or 25 ml fenvalerate (Fenval 20 EC) or 30 ml fluvalinate (Mavrik 25 EC) in 100 litre water (Lakra, 1997a). Repeat the spray after 2 to 3 weeks, if necessary.

**Lac insect *Kerria lacca* (Kerr.)**

The pestilent aspect of this insect has largely and truely been overshadowed by its usefulness. The tachardiids are mostly confined to the tropics and sub-tropics, and the predominant genus *Kerria* (syn. *Laccifer*) best develops in the Oriental region. It is widely distributed in the Indian subcontinent and occurs all over the country where *ber* is grown. The insect infests about 230 host plants of widely differing families. Besides *ber* (*Z. mauritiana*), *ghont* (*Z. xylopyra* Willd.), *Ficus* spp., *Acacia* spp., palas and kusum are also the preferred hosts. It also attacks grapevine, *aonla*, pomegranate, *lasoda*, litchi, sugar apple, tamarind and several forest and avenue trees. Old branches harbouring lac encrustations of the previous generation serve as source of infestation (brood lac) for the next brood (Glover, 1930; Roonwal, 1953).

The brood lac famale lays about 100 eggs underneath its body in lac encrustation. The eggs hatch in 2 to 4 days in July-August (*katki* brood) and 4 to 7 days in November—December (*baisakhi* brood) (Kher and Lakra, 1989). The newly hatched nymphs are ovate, soft bodied, crimson red and minute (about 0.6 mm long and 0.25 mm wide). The nymphs, soon after coming out of the mother lac cell, wander/crawl on host twigs in search of a suitable spot to settle. The swarming activity is generally restricted between 2nd week of July and 1st week of August in *katki*, and 3rd week of November and 1st week of December in *baisakhi* brood (brood names are after Indian months in which the lac crop is harvested). However, an individual nymph crawls for 1 to 4 days before settling. Innumerable tiny nymphs settle very close to each other (16 to 32 nymphs/cm$^2$) and insert their proboscis into the tender succulent shoots and feed on the sap from phloem tissues.

Once settled, the nymphs do not move and moult thrice to become sexually mature. The nymphal period is completed in 42 to 51 days in *katki* and 124 to 130 days in *baisakhi* brood depending on the climatic conditions and nutritional status of the host (Kher and Lakra, 1989). The females lose their legs, eyes and antennae at first moult once for all. The males are both pterous (winged) and apterous (wingless), while the females are only apterous. The dimorphic males live for 5 to 14 days and mate discriminately with unmated females. Nymphs after settling, start secreting lac from almost all over the body except mouth parts, breathing pores and anus, and slowly gets encased in a cell of lac, which keeps on increasing with increase in the size of the insect. As the nymphs settle very close together, the lac secretion from adjacent cells coalesces to form a continuous encrustation. Males produce smaller quantity of lac in comparison to females which produce the commercial lac. As already mentioned, there are two broods of *K. lacca* in a year.

De-sapping by large population of lac insect over a long period of time reduces the vitality and vigour of the tree resulting in substantial loss in fruit yield and quality. The cottony appearance of certain healthy encrustations is due to long wax filaments, while the sooty appearance on the leaves is due to black mould growth on the honeydew which imparts ugly appearance to the trees. The severely infested branches dry up. All trees are not evenly infested. 'Umran' and 'Kaithli' cultivars of *ber* are preferred over Gola. Neglected *ber* orchards having close plantations, intermixed with other host trees, and devoid of regular summer pruning are more prone to lac infestation. Some other species of genus *Kerria* also occur on *ber* but are less common than *K. lacca* (Kapur, 1962).

### *Management*

- After fruit harvest or when trees are in dormancy, prune all severely infested branches and burn without delay.
- Avoid overcrowding and growing other lac insect host plants in the vicinity of fruit trees.
- Spray the infested trees with 100 ml monocrotophos (Monocil 36 WSC) or 100 ml dimethoate (Rogor 30 EC) or 150 ml oxydemeton methyl (Metasystox 25 EC) in 100 litre water, in August and again in December coinciding with hatching/settling of motile crawlers.

## Bark eating caterpillars

### *Indarbela tetraonis (Moore) and I. quadrinotala (Walker)*

These are described under aonla.

### *Ber* leaf webbers/rollers

#### *Synclera univocalis (Walker) and Pagyda traducalis Zeller*

Of about a dozen species of leaf webbers/rollers, *Synclera univocalis* (Walker), *Pagyda traducalis* Zeller, *Ancylis* spp., *Acroclita corinthia* Meyrick, *Phycita* sp., *Limaecia* sp., *Pleurota* sp., *Psorosticha zizyphi* (Stainton), have been reported damaging *ber* leaves from different countries. However, at a given time in an area, normally one or two species predominate. It appears that different species avoid intra- and inter-species competition. In India, *S. univocalis, P. traducalis* (*Synclera* earlier synonymised with *Pagyda*), *A. corinthia, Ancylis* spp. and *Phycita* sp. are widespread, but the first one is predominant (Mann and Singh, 1981). Besides India, *S. univocalis* has been reported from Sri Lanka, Myanmar, Pakistan, Palestine, Syria, South Africa and South America under different synonyms. In India, it occurs widely in Punjab, Haryana, Delhi, Uttar Pradesh, Rajasthan and Gujarat. Cultivated *Z. mauritiana* trees are the most preferred while wild, *Z. mauritiana* var. *rotundifolia* trees are the least preferred and the bushy *jhar ber* (*Z. nummularia*) is not attacked at all. The leaf webbers are more destructive to tender leaves of new flush, especially on young plants and trees receiving frequent irrigation and nitrogenous fertilizers.

The symptoms of damage by various species broadly look similar. The newly hatched caterpillar attacks the unopened or partially opened leaf and the later instars keep the leaf folded along the long axis with the help of silken threads. The larva consumes the green matter by scrapping, leaving behind the papery epidermis. Sometimes, the advanced stage larva may join the two adjacent terminal leaves together longitudinally. Often, the larvae of some species web together the terminal leaf buds and unopened leaves and feed within. Usually a single larva is present in each web, but at times 2 to 3 larvae may be present. Severely attacked leaves curl up (margins turn upwards or downwards), turn brown and later dry up, imparting an unhealthy appearance to the tree. Due to caterpillar attack, small plants remain stunted and feeding on growing points adversely affects flowering and fruit set.

These insects generally remain active throughout the year but inflict greater damage during summer and rainy seasons, i.e. between May and November when trees put forth new flush. During peak vegetative flush, sometimes more than 50 per cent tender leaves are attacked by one or the other species and there may be more than 50 larvae and/or pupae per 100 attacked leaves. On an average, *S. univocalis* lays 75 eggs, usually singly or in groups of 2 to 5, on tender leaves, which hatch in 2 to 4 days. Larval and pupal stages are completed in 13 to 17 and 8 to 14 days, respectively. Adult moths live for 4 to 11 days. The pest completes as many as 14 generations per year. Likewise, the female of *A. corinthia* lays about 60 eggs, usually singly on upper surface of young tender leaves. On an average, the egg, larval and pupal stages take 3.5, 15.2 and 6.7 days, respectively. Females live slightly longer (5.3 days) than the males (4.5 days). The life cycle is completed in about 26 days and there are more than 12 generations in a year. Its peak incidence occurs during August-September.

#### *Management*

- On appearance of the caterpillars on new vegetative flush, spray 25 ml fenvalerate (Fenval 20 EC), 200 ml quinalphos (Ekalaux 25 EC), 25 ml cypermethrin (Cymbush 25 EC) or 200 g carbaryl (Sevin 50 WP) in 100 litre water.

- Repeat the spray after 3 to 4 weeks if there is recurrence of pest infestation (Pajni and Rose, 1978; Mann and Brar, 1980; Mann, 1981; Mann and Singh, 1981; Singh and Mann, 1982; Batra *et al.*, 1988).

**Miscellaneous pests**

Dacine fruit flies namely, the Oriental fruit fly *Bactrocera dorsalis* (Hendel), peach fruit fly *B. zonatus* (Saunders), and guava fruit fly *B. correcta* (Bezzi) (Tephritidae: Diptera) being *polyphagous,* occasionally attack *ber* fruits (Agarwal and Kapoor, 1985). The fruit borers, *Meridarchis scyrodes* Meyrick, and *M. reprobata* Meyrick (Carposinidae: Lepidoptera) bore into developing *ber* fruits and often inflict serious losses in South India. Among the lepidopteran leaf feeders, the castor semilooper, *Achaea janata* (Linn.) and *Thiacidas postica* Walker (Noctuidae), the tassar silkworm, *Antheraea mylitta* Drury (Saturniidae) and *ber* butterfly, *Tarucus theophrastus* (Fabr.) (Lycaenidae) are minor pests of *ber*. The grey weevils, *Myllocerus* spp. and *Xanthochelus superciliosus* Gyllenhal (Curculionidae: Coleoptera) feed on the tender leaves from July to October and are sporadically serious in nurseries and young *ber* plantations. The termites, *Microtermes obesi* (Holmgren) and *Odontotermes obesus* (Rambur) (Termitidae: Isoptera) are highly polyphagous, more common and widespread in arid and semi-arid areas. Their damage is more pronounced in ber nurseries and new plantings, in light and virgin soils.

Several species of homopterans/hemipterans, viz., spittle bugs, *Machaerota* spp. (Machaerotidae); mealybugs, *Perissopneumon* (=*Drosichiella*) spp. and *Drosicha mangiferae* (Green) (Margarodidae); armoured scales, *Aonidia zizyphi* Rahman and Ansari, *Aonidiella aurantii* (Maskell), *Aspidiotus destructor* Signoret (Diaspididae); soft scales belonging to *Planococcus, Coccus* and *Ceroplastes* genera (Coccidae); and the treehopper, *Otinotus oneratus* Walker (Membracidae) suck the sap and are of minor importance. Among these, the spittle bugs are gaining importance especially in irrigated areas where nitrogenous fertilizers are increasingly used. The *ber* leaf gall midge, *Phyllodiplosis jujubae* Grover (Cecidomyiidae: Diptera) is a sporadic minor pest of nurseries and young plants (gangwar (1983). Among the 20 species of mites recorded from jujube, the gall mites, *Eriophyes cernuus* Massee (Eriophyidae), *Larvacarus transitans* (Ewing) (Tenuipalpidae) and *Eutetranychus* spp. (Tetranychidae) are quite common but of minor importance (Sadana *et al.*, 1982).

## AONLA/AMLA (*EMBLICA OFFICINALIS* GAERTN.)

The emblic or Indian gooseberry, as *aonla/amla* is often referred to, is a sacred Hindu tree believed to have originated in India. It is highly prized fruit in Asia and is cultivated throughout the Indian subcontinent. The graceful tree is of great importance in Indian 'ayurvedic' medicines and its fruits are considered as the "richest known natural source of vitamin C". Its plant extracts work as synergists and are used to increase the residual toxicity of insecticides besides being known to possess juvenile hormone activity. Being a comparatively neglected fruit, not much research work has been done on its insect pests. Till date, as many as 30 insect and mite species have been recorded feeding on this tree from India, but only a few cause serious damage.

**Bark eating caterpillars**

The bark eating caterpillars namely, *Indarbela tetraonis* (Moore) and *I. quadrinotata* (Walker) are widely distributed in the tropical and sub-tropical regions of the Indian subcontinent comprising India, Sri Lanka, Bangladesh, Pakistan and Myanmar. These are highly polyphagous species, attacking a wide range of fruit, forest and avenue trees, especially the soft-wooded ones. While the fruit trees like *aonla*, *jamun*, pomegranate, loquat, falsa, pear and guava are the preferred hosts; *ber*, fig, mulberry, *bael*, *karonda*, custard apple, litchi, citrus, mango, etc. are also severely attacked by these insects. In any given locality, usually one or the other species is predominant, though their combined incidence is not uncommon and they show affinity to certain host trees. Generally, the incidence of *Indarbela* spp. is more severe in neglected orchards where older, overcrowding trees in mixed plantations are subject to greater infestation. The biology, life and seasonal history and nature of damage of both the bark species are more or less similar.

The nocturnal females lay eggs in groups of 15 to 25, in cracks and crevices or under loose bark of trunk, stem or main branches, usually near the forks. Upto 2000 eggs are laid by a single female mainly during summer (May-June). The incubation period is 8 to 11 days and the newly hatched dirty brown caterpillar nibbles the live bark superficially for 2 to 3 days before boring into the branch/stem. The larvae take as many as 9 to 11 months to complete development. Larval feeding and development slows down during winter months. The full grown larva is light brown or dirty brown with strong biting mouth parts. It pupates in the bored hole in April and pupal stage lasts 3 to 4 weeks. Adult moths emerge in May-June leaving a portion of empty pupal case protruding out of the bore hole. The moth is stout, pale brown with rufous head and thorax. The forewings are pale brown with numerous brown spots and bands/streaks, while the hind wings are fucous. Males are slightly smaller than females. Adults have a short life span of less than a week. There is only one generation in a year (Sontakay, 1945; Patel *et al.*, 1966).

The larvae, while feeding on bark, destroy the food transporting phloem vessels and thus interrupt the flow of nutrients and sap. Further, boring into the branches and stems by the larvae also weakens the trees. All these, in turn, affect the vigour, vitality and fruiting capacity of the trees. Severely infested branches/stems or even main trunks often break off under the pressure of strong winds. Trees of all ages are attacked by the bark caterpillars, but generally older and soft wooded trees are more prone. The catterpillars are usualy not seen and their presence is indicated by the winding ribbon-like webbings, made of their excreta, pieces of bark and frassy material of varying colour on the infested trees. The caterpillars construct these silken galleries while feeding on the bark. Such tunnels extend around the feeding sites and bore holes in a zig-zag manner usually near the forks of branches and may be a few to more than 65 cm in length, and thus can be easily located from a distance. The larva bores a hole in the stem or branch and remains in it during the daytime, tunnelling downwards, and feed on the bark sheltered under the cover of webbing during night. There is only one larva in each hole but several holes or webbings may be present on a tree (David *et al.*, 1963).

***Management***

The pest is indeed not very difficult to control provided timely and appropriate management practices are adopted, like

- Keep the orchard clean and avoid overcrowding of trees.
- Kill the caterpillars mechanically by inserting a thin wire/iron spike into the larval holes. The mechanical destruction of the larvae, rather tedious on tall trees and in large area, is quite effective.
- In early stage of infestation, i.e. during September-October, clean the webbed areas and paint the bark around borer holes with emulsion prepared by adding 10 ml monocrotophos (Monocil 36 WSC) or 10 ml methyl parathion (Metacid 50 EC) in 10 litre water.
- During February-March, remove the webbings and give the following treatment: Inject into each borer hole 5 ml emulsion prepared by diluting 2 ml dichlorvos (Nuwan 76 EC) or 5 ml methyl parathion (Metacid 50 EC) or 30 ml endosulfan (Thiodan 35 EC) or 40 g carbaryl (Sevin 50 WP) in 10 litre water or 10 per cent kerosene emulsion (1 litre kerosene + 100 g soap + 9 litre water) or push into each borer hole a cotton swab soaked in any of the above mentioned insecticide emulsions and plug the treated hole with mud. Again retreat the re-opened holes.
- Treat all the alternate host plants in the vicinity of fruit trees. The entomogenous fungi, *Aspergillus candidus* cause considerable larval mortality during the rainy season at some places. Older, overcrowded trees are more susceptible to attack and should be given special care in control programmes (Sandhu *et al.*, 1987; Lakra, 1997a).

**Apical twig gall maker, *Betousa stylophora* Swinhoe (= *Hypolamprus stylophora*)**

The apical twig gall maker *H. stylophora* is a sporadic and specific pest of *aonla*. The new growth of branches suffer damage temporarily as a result of feeding by the sluggish caterpillars which form galls at the apices of growing twigs. The blackish larva while feeding, secretes some metabolites which induce typical galls resembling "snake charmer's flute". The growth of heavily affected branches is temporarily arrested. However, from the apices of these galls, normal growth of twig is restored in subsequent years with not much reduction in bearing. The larva forms a single gall to complete development which is very slow. The pest is univoltine.

***Management***

Prune the twigs with galls and destroy the larvae inside the gall by crushing them mechanically or by burning the twigs. Being a minor pest, usually no other control measure is required.

**Other lepidopterous pests (Noctuidae, Lycaenidae, Lymantriidae, Oecophoridae, Geometridae, Zeuzeridae, Gracillariidae)**

In addition to bark eating caterpillars and apical twig gall maker, 12 species belonging to seven families are known to feed on various parts of *aonla* trees. Of these, the pomegranate butterfly, *Deudorix* (= *Virachola*) *isocrates* (Fabr.) (Lycaenidae) and fruit sucking moths, *Achaea* sp. (Noctuidae) are the ones that cause damage to fruits (Srivastava and Jain, 1973). The larva of pomegranate butterfly enters into developing fruit by making a conspicuous hole and feeds on the hard yet green seeds, without eating the fleshy portion, but otherwise destroying the fruit completely. The larval excreta is pushed out through the entrance hole. The stout noctuid moth (*Achaea* sp.) with strong, well built proboscis pierces the skin of soft ripened fruits and sucks the juice during night and thus damages them in the process. The larvae defoliate wild plants especially the creepers but also occasionally devour *aonla* leaves. The attack of fruit sucking moths is more pronounced in orchards raised within or near the forest area or wild vegetation.

Among the leaf defoliators, the hairy caterpillars, *Euproctis fraterna* (Moore), *Euproctis* sp. (Lymantriidae) and *Selepa celtis* Moore; leafroller, *Gracillaria theivora* Welsm, leaflet twister, *Caloptilia acidula* (Meyr.) (Gracillariidae), and citrus leafroller, *Psorosticha zizyphi* (Stainton) (Oecophoridae) cause minor damage and seldom attain serious form (Joshi and Meshram, 1989). The larvae of looper, *Buzura* (=*Biston*) *suppressaria* (Guenee) (Geometridae) voraciously devour the tender leaves leading to defoliation of tree but they are not regular in occurrence (Gupta and Singh, 1988). The red larvae of the borer *Zeuzera coffeae* (Nietner) and *Zeuzera* sp. bore into the stems and branches feeding and tunnelling within. Their attack is mostly directed towards the base of the trees. The red borers appear only as minor pests (Velavadi and Shah, 1987).

**Sap sucking insects—Mealybugs, soft scales, aphids, whiteflies, treehoppers and bugs**

Three species of pseudococcids, *Ferrisia virgata* (Ckll.), *Nipaecoccus vastator* (Mask.) and *Pseudococcus* sp. (Pseudococcidae); three species of mealybugs, *Drosicha mangiferae* (Green), *Drosicha* sp. and *Drosichiella* (=*Perissopneumon*) *phyllanthi* (Green) (Margarodidae); and four species of aphids, *Cerciaphis emblica* Patel and Kulkarni, *Setaphis viridis* Van der Goot, *S. bougainvilliae* Thunberg and *Schoutedonia emblica* Patel and Kulkarni (Aphididae) have been reported feeding on *aonla* trees (Patel and Kulkarni, 1954). The treehopper, *Oxyrhachys tarandus* (Fabr.) (Membracidae) and the whitefly *Aleyrodes sizoukinesis* Kuw. (Aleyrodidae) are occasional minor pests of *Emblica* sp. (Chatterjee, 1914; Babu, 1971). The nymphs and adults of the green plant bug, *Scutellera perplexa* (Westw.), *S. nobilis* (Fabr.) (Scutellaridae) inflict damage to ripe fruits by sucking the sap. The raw fruits are spared. They are occasional pests of minor importance and hardly cause economic damage. Majority of these homopterans are polyphagous pests that suck cell sap from leaves, tender shoots and even fruits and devitalize them. They also secrete large quantity of honey dew which attracts ants and favours development of sooty mould which interfere with photosynthetic activity. These insects prefer dry weather. Heavy populations of these sap sucking insects, though uncommon, reduce the vitality and vigour of the trees.

***Management***

Management of these sap sucking insects s discussed under the heading 'pomegranate'.

**Miscellaneous pests**

The termites, *Odontotermes obesus* (Rambar) and *Microtermes* sp. (Isoptera: Termitidae) are cosmopolitan in arid and semi-arid regions and attack underground parts and base of trunks of several fruit trees including *aonla*. Young saplings when raised in virgin and light soils are more severely damaged and often die. However, well established *aonla* trees are rarely attacked and moreover the full grown trees withstand the termite attack extremely well. Feeding by gall midge, *Asphondylia phyllanthi* Felt (Cecidomyiidae) causes small galls on leaves and tender shoots in some regions. Its infestation damages the auxillary buds inhibiting growth and normal branching of *aonla* trees (Agarwal, 1980 a, b). Likewise, an unidentified small wasp (Chalcididae) has also been observed causing galls on the leaves and shoots of *aonla* (Agarwal and Jain, 1981). The gall mite, *Calacarus channabasavannae* Lakkundi (Eriophyidae) also causes small galls on the leaves and tender twigs (Channabasavanna, 1966). All these, however, are of minor importance.

## POMEGRANATE (*PUNICA GRANATUM* L.)

The original home of pomegranate lies somewhere between Iran and Himalaya in North India. The ancient fruit is now widely cultivated in most parts of the tropics and sub-tropics where the climate is not too humid. The attractive shrub is grown throughout the Indian subcontinent and in drier areas of South-East Asia. However, the best quality fruits come from areas with hot dry summers and cool winters having irrigation facilities. The small tree is liberally branched, more or less spiny and possesses extremely long life. Its roots, rinds and seeds are used in medicines. Despite being a hardy plant, its fruits are very delicate and are prone to attack by insect pests, though more species have been recorded feeding on the foliage which incidentally is not serious. Of about 5 dozen insect and mite species recorded on pomegranate, less than a dozen cause serious damage.

### Pomegranate butterfly (Anar butterfly)

***Deudorix isocrates (Fabr.), Deudorix epijarbas (Moore)***

The anar butterfly, *D. isocrates* is an economically important and regular pest among all the butterflies. It is found all over India and in the adjoining countries, wherever pomegranate is grown. It is decidedly the most destructive pest of this fruit in the plains, whereas, the Cornelian butterfly, *D. epijarbas*—a closely related fruit borer species, occurs mainly in hills. Both these fruit borers limit the cultivation of pomegranate in certain areas, the former being widely distributed and predominant species (Zaka-ur-rab, 1980; Kakar *et al.*, 1987). Although pomegranate is the most preferred host, yet fruits of aonla, ber, tamarind, mulberry, guava, loquat, sapota, litchi, citrus, peach, pear, plum, apple, etc. are also attacked. The mode of feeding of both the lycaenid borers is almost identical and there is resemblance in their biology, behaviour, life and seasonal history.

The medium sized colourful butterflies lay shiny white, oval eggs, singly, mostly on calyces of flowers and on newly set fruits from button size onwards, and sparingly on leaves. The incubation period varies from 7 to 10 days depending upon the season. The newly hatched caterpillar bores into the fruit where the rind is thin and soft, and feeds mainly on the developing seeds and reluctantly on the pulp just below the rind. The larva feeds inside the fruit for 18 to 47 days to become full grown. The larvae are stoutly built, 17 to 20 mm long, dirty brown in colour and have short hair and whitish patches all over the body. Before pupation, the larva ties the fruit stalk with the branch and repeatedly reinforces the junction of the fruit with the twig with the help of silken strands. This is ostensibly done to prevent the fruit from falling from the tree. The larva then re-enters the fruit and pupates therein. Occasionally pupation takes place on the branch/stem and in fallen fruits also. The pupal duration ranges from 7 to 34 days depending upon the prevailing climatic conditions. The lycaenid breeds throughout the year on various host plants. It completes 2 to 3 generations from March to August on the main pomegranate crop and often 1 to 2 broods on winter fruiting season from November to February. Males of these glossy violet blue butterflies are comparatively smaller and females have a conspicuous orange spot on the fore wings. The life cycle is completed in one to two months. The characteristic symptoms of damage are offensive smell and excreta of larva coming out of the entry hole and stuck around the hole. Such holes predispose the fruits to bacteria, fungi and scavenging insects and birds, which lead to rotting. Often the anal segments of larva can be seen plugging the borer hole and at times more than 3–4 larvae could be seen feeding inside a single fruit. Affected fruits generally show conspicuous holes on the surface and drop prematurely and are not fit for consumption. The chalcid, *Brachymeria euploeae* Westwood (Chalcididae) frequently parasitizes the larval population in nature.

***Management***

- Remove and destroy all infested fruits, fallen as well as those still on the plant, at regular intervals.
- Wrap individual fruit with paper envelopes, muslin or polythene bags soon after fruit formation/beginning of butterfly incidence. Bagging of fruits is very effective but laborious and expensive, and often results in colonization of mealybugs.
- In areas where bagging of fruits cannot be practised, give 4 cover sprays at monthly intervals with 100 ml monocrotophos (Monocil 36 WSC), 130 ml dimethoate (Rogor 30 EC), 160 ml oxydemeton methyl (Metasystox 25 EC) or 200 ml endosulfan (Thiodan 35 EC) alternated with 20 ml cypermethrin (Cymbush 25 EC), 25 ml fenvalerate (Fenval 10 EC) or 100 ml deltamethrin (Decis 2.8 EC), in 100 litre water. The cover sprays should be started from setting of fruits/beginning of butterfly attack. Avoid application of persistent insecticides towards ripening of fruits (Awate *et al.,* 1977; Shukla and Prasad, 1983; Dubey *et al.*, 1993).
- Avoid growing highly susceptible cultivars like Kagzi Anar, Achik Dana, Jalore Seedless in butterfly endemic areas.

**Bark eating caterpillars**

***Indarbela tetraonis (Moore) and I. quadrinotata (Walker)***

These are discussed under the heading 'aonla'.

**Sap sucking insects—Mealybugs, scales, aphids, whiteflies and bugs**

More than 25 species of sap sucking homopterans/hemipterans have been recorded on pomegranate in India. Majority of these are polyphagous and widely distributed. Besides sucking the plant sap from leaves, twigs, branches and even fruits, they produce large quantity of honey dew which attracts ants and enables sooty mould to grow on foliage. In case of severe infestation, plants get devitalized and flower and fruit production is hampered.

Of these *Parasaissetia nigra* (Nietner) (Coccidae), *Planococcus lilacinus* (Cockerell), *P. citri* (Risso) and *Ferrisia virgata* (Cockerell) (Pseudococcidae), *Icerya purchasi* Mask. and *Drosicha mangiferae* (Green) (Margarodidae), *Aonidiella orientalis* (Newstead) and *Aspidiotus* sp. (Diaspididae) are more common in India (Mani and Krishnamoorthy, 1990). *Aphis punicae* Passerini (Aphididae) and whiteflies (*Siphoninus* spp.) (Aleyrodidae) are specific pests of pomegranate and their occurrence coincides with the new flush, the aphids being more regular (Bhagat, 1986). The leaf-footed plant bug, *Leptoglossus membranaceus* Fabr. (Coreidae) is becoming increasingly common in some areas (Jadhav *et al.*, 1980). At any one time of the year usually one or the other species of these homopterans predominate while others occur occasionally though they remain active throughout the year.

*Management*

- Prune and destroy the affected parts in initial stages to prevent further spread and build up.
- Ants, which protect these homopterans (in order to obtain honey) from their natural enemies, should be kept away from the infested trees. Destroy the ant colonies by digging/ploughing the soil around trees frequently.
- In case of severe infestation spray the infested trees with 100 ml monocrotophos (Monocil 36 WSC) or 200 ml diazinon (Dimecron 20 EC) or quinalphos 200 ml (Ekalux 25 EC) (against mealybugs and scales), and 35 ml phosphamidon (Dimecron 85 WSC) or 100 ml dimethoate (Rogor 30 EC) or 120 ml oxy-demeton methyl (Metasystox 25 EC) (against aphids and whiteflies), in 100 litre water. Repeat spray after three weeks, if necessary (Noor *et al.*, 1978; Ghule and Dhumal, 1992).
- Weeds, which act as alternate hosts, should be removed. Orchard sanitation is advantageous in keeping the pest population low.

**Miscellaneous pests**

Pomegranate fruits suffer occasionally by the attack of fruit sucking noctuid moths, *Othreis* spp. and *Achaea janata* Linn., pyralid fruit borer, *Dichocrocis punctiferalis* (Guen.) and fruit flies, *Bactrocera zonatus* (Saunders) and *B. dorsalis* (Hendel) (Mote *et al.*, 1991). Similarly, species of hairy caterpillars, *Euproctis* spp., bagworm, *Clania crameri* Westw.; thrips, *Scirtothrips dorsalis* Hood, *Rhipiphorothrips cruentatus* Hood and *Retithrips syriacus* (Mayet); weevils, *Myllocerus* spp., chafer beetles *Anomala* spp. and *Holotrichia* spp. and mites, *Brevipalpus* spp. and *Tenuipalpus* spp. also inflict damage to pomegranate leaves (Menon *et al.*, 1971; Dhooria and Sandhu, 1973; Bagle, 1993). Termite, *Odontotermes obesus* (Rambur) infests newly planted saplings and established plants in light soils. However, most of these are of minor importance and seldom require control measures.

## MULBERRY (*MORUS ALBA* L.)

There are hardly any organised orchards of mulberry, but it is frequently grown as a shade-cum-fruit tree in almost every orchard, and is used extensively for rearing silkworms and for certain other purposes. Its wood is used in making sports goods such as hockey sticks and tennis racquets. Its long flexible tender twigs are used in making baskets. It is a small tree, upto 5 m in height, and the fruit is a syncarp. Mulberry trees are subjected to the attack of nearly 85 species of insects and mites, of which, only a handful are of economic importance while the rest are minor pests.

**Stem borers**

About a dozen stem boring beetles have been recorded feeding on mulberry trees. These are mostly polyphagous and are widely distributed all over the country. Their attack is more conspicous on trees which are not looked after well. Less vigorous, old, unhealthy, injured and dying trees are heaven for these beetles. The major species are:

- Fig stem borers, *Batocera rufomaculata* (deGeer) and *Batocera* spp. (Cerambycidae: Coleoptera).
- Other stem borers are *Apriona rugicollis* Chev., *A. cinerea* Chev., *A. germarii* (Hope) and *Aeolesthes holosericea* Fabr. (Cerambycidae: Coleoptera).

The adult beetles have a habit of cutting leaves. They feed on bark of tender shoots and later girdle the shoot which gradually withers away. The damage is much more than what the adults actually consume. The grubs, however, are more destructive than the adults. After hatching the grub bores inside the shoot and makes circuitous galleries. From the main tunnel several offshoot galleries arise which open outside as an exit hole, through which oily sap, excreta and chewed wood is thrown out. A single grub makes 8 to 10 such exit holes spaced at 10 to 12 cm apart. Several grubs tunnel down from branches to main stem and hollow it out. The vitality and productivity of the infested trees are greatly impaired but such trees do not die quickly. The pest is active from March to October, remains quiescent through winter, and resumes feeding in March. A single life cycle is completed in 2 years. Fig trees are the preferred hosts while local mulberry plants are less preferred than the grafted ones.

*Management*

The management of stem borers is discussed under the heading 'fig'.

**Bark eating caterpillars**

***Indarbela quadrinotata (Walker) and I. tetraonis (Moore)***

These are discussed under the heading '*aonla*'.

**Sap sucking insects—Mealybug, soft scale, and armoured scale**

Of about two dozen species of mealybugs and scales that have been recorded feeding on mulberry trees, Pseudococcidae is more common pest of sporadic occurrence. It is a polyphagous species that infests custard apple, grapevine and citrus in addition to mulberry. The nymphs and adult females cluster on ventral surface of leaves and tender shoots and suck sap from there. In case of severe infestation, leaves exhibit typical curling symptoms and tree loses vigour and becomes unproductive. It also excretes large amount of honey dew, which encourages growth of black fungus (*Capnodium* sp.) and also attracts black ants which in turn protect the mealybugs from their natural enemies (Rao *et al.,* 1993).

In addition, *Pseudococcus* sp. and *Nipaecoccus viridis* (Newstead) (Pseudococcidae), *Drosicha mangiferae* (Green), *D. contrahens* (Walker) and *Drosichiella* (=*Perissopneumon*) *tamarindus* (Green) (Margarodidae) and about half a dozen species of coccid genera *Coccus, Pulvinaria, Chloropulvinaria, Ceroplastes,* etc. (Coccidae) have been reported feeding on mulbery plants. Some species of armoured scales, viz., *Aonidiella aurantii* (Maskell), *A. orientalis* (Newstead), *A. citrina* (Coquillett), *Quadraspidiotus perniciosus* (Comstock), *Chrysomphalus* spp., *Parlatoria oleae*

(Colvee), most of which are polyphagous, also attack mulberry plants in various regions (Chatterjee, 1960).

*Management*

The management of sap sucking insects is discussed under the heading 'pomegranate'.

**Mulberry bug**

***Halys dentatus (Fabr.)***

The mulberry or shield bug, *H. dentatus* is an important and common pest of mulberry. Besides mulberry, its other known hosts are *jamun*, litchi, mango and several forest and wild trees. The pentatomid is widely prevalent in the plains and hills all over the country. The adult bug is shield shaped, stoutly built and measures about 20 mm in length. It is more active during summer months when swarms of nymphs and adults are seen sitting on tree trunks and main stem. The nymphs and adult bugs suck sap from the main stems and tree trunks through their strong piercing and sucking mouth parts resulting in general weakening of trees. Their deceptive colouration, which closely resembles the bark of host tree provides protection from natural enemies especially predators. Not much is known about its biology, life and seasonal history.

*Management*

- Do not grow mulberry, *babool, Acacia arabica,* and wild casuarina trees, near *jamun* trees.
- If need be, spray the tree trunks and main stems with 150 ml endosulfan (Thiodan 35 EC), 100 ml fenitrothion (Folithion 50 EC) or 100 ml methyl parathion (Metacid 50 EC) in 100 litre water during summer when population is high.

**Miscellaneous pests**

Various species of birds feed on ripe mulberry fruits, especially on unattended trees, Red and yellow wasps feed on the ripened fruits. The girdler beetle, almond chrysomelid beetle, leaf weevils and thrips occur quite frequently on mulberry trees and are relatively important. One or the other species of homopterans is invariably present in mulberry orchards and their collective damage is generally substantial. However, serious outbreaks are not common.

## FIG (*FICUS CARICA* L.)

Fig (*Ficus carica*), is a native of western Asia from where it spreaded to semi-arid tropical and sub-tropical areas of the world. It does not grow well in low wet tropics but does well at higher altitudes and drier parts of tropics, and mild temperate climates. There are very few orchards exclusively of fig in India, but it is grown in almost every orchard either as a shade or avenue tree or as a fruit tree. Fig is attacked by about 100 species of insects and other arthropods, of which the stem borers, leaf-rollers, fruit flies, midges, mealybugs and scales are important.

**Fig stem borer *Batocera rufomaculata* (deGeer)**

This cerambycid has wide distribution in India and north-western part of the Indian subcontinent. Outside India, it is known to occur in South China, S.E. Asia, Far East, Sri Lanka, eastern Mediterranean region, South Africa, Mauritius, Madagascar, etc. It is a highly destructive pest of mango, fig, jack fruit, mulberry, apple, papaya and rubber, *jamun*, guava, pomegranate, etc. are among the less preferred hosts. Although the borer is not very common and regular in occurrence on fig, yet whenever it appears in the main trunk or a branch, it invariably kills the host within a few years. The grubs are far more destructive than the adults which emerge between May and August. The female deposits eggs singly, usually under the loose bark or in slits or cavities and injured, diseased or decling portion of the stem or branch is preferred over the vigorous parts. The incubation period is 1 to 2 weeks.

The newly hatched grub makes zig-zag burrows beneath the bark and slowly tunnels inward by feeding on the inner wood of the stem or trunk. The burrows are directed upward and are filled with frassy material consisting of fibrous matter and excreta. The grubs, which are large, stoutly built and fleshy with well developed mandibles, cut more fibres than they actually consume. Because of their large size they make large extensive tunnels in the sap wood which interfere with the sap flow. Consequently the attacked branches or stems become weak, shed leaves and finally die and dry up. The infested tree trunks break off from the tree during windy days. The vitality and fruiting capacity of the severely attacked trees is greatly impaired and the tree dies within a few years. Usually the grub stage lasts for about 4 to 7 months but may continue for more than a year. Before pupation the grubs enter diapause during winter and resume feeding as soon as the weather warms up in the spring season. The full grown larva excavates an elliptical chamber within the stem and pupates 5 cm below the surface. Pupal stage lasts for about a month but may extend upto 6 months. The stoutly built, large longicorn beetle with long antennae, cuts its way out of pupal chamber and appears during the summer. The adults feed on the bark of tender twigs and petioles and may live for 8 months. The life cycle is completed in one year but may take upto two years if the larval and pupal stages get prolonged.

***Management***

The cerambycid stem borers of fruit trees are some of the most difficult pests to control. Usually no single method of control is effective, but a combination of several methods can provide adequate protection to the trees against these beetles.

- Since vigorously growing trees are less attacked than declining ones, promote growth and vigour of trees by providing balanced nutrients and timely irrigation.
- Orchard sanitation by way of cutting the infested branches/trunks, scrapping loose bark of stems, pruning the diseased plant parts and burning the pruned twigs promptly, helps in destroying the pest.
- Remove the weak, injured, dying, dry branches/stems, uprooted trees and broken trunks and burn, as these are the likely targets of borer attack.
- Do not allow the wild *Ficus* spp. and *Morus* spp. to grow in the vicinity of the orchard as these are preferred hosts of longicorn beetles.

- Do not stock freshly fallen tree trunks, logs and pruned twigs etc. near the orchard as these may be harbouring the borer larvae.
- Kill the grubs inside the burrow by inserting stiff iron spikes. Adult beetles, whenever sighted, should be killed mechanically.
- Locate the borer hole, remove the frass and inject or pour insecticide emulsion, prepared by mixing 2 ml methyl parathion (Metacid 50 EC) or 15 ml dichlorvos (Nuvan 76 EC) per litre water and plug it with mud. Alternatively, injecting dilute kerosene oil, petrol, crude oil or formalin in borer holes and subsequently plugging these with mud is also effective. Live re-opened holes if any, should be treated again.
- Use of aluminium phosphide @ 0.5 to 0.75 g tablet per hole has also been found effective. Insert the tablet in the hole and plaster with mud immediately.
- Provide suitable breeding and nesting sites for wood peckers and other insectivorous birds which dig out the grubs and pupae from infested trees.

### Fig leaf rollers *Phycodes minor* Moore and *P. radiata* (Ochs.)

Both the leaf-rollers are sporadic pests and cause severe damage in some areas. *P. minor* occurs commonly in North India, whereas, *P. radiata* is more common in Southern India. The larvae feed on the epidermis of young terminal leaves by webbing them together or by rolling individual leaf with the help of silken threads. At times they fold one leaf over another and feed within. The folded leaves are eaten partially or completely under the cover of white cottony web which eventually drop off. When the defoliation is serious, the plant withers and takes long time to put on fresh growth which leads to stunted growth. The pest is particularly severe in young plantations, remains active from March to October and hibernates in leathery silken cocoons in cracks and crevices of the bark from November to February. There are 4 overlapping generations in a year (Sinwat and Sidhu, 1974; Lakra and Kharub, 1977).

#### *Management*

- Clip off and destroy all the leaf-roller webs/folded leaves along with caterpillars regularly.
- Spray 200 ml endosulfan (Thoidan 35 EC), chlorpyriphos (Durmet 20 EC) or 200 ml carbaryl (Sevin 50 WP) in 100 litre water.

### Cockchafer beetles

For details see under the heading 'jujube'.

### Fig midges

### *Anjeerodiplosis peshawarensis* Mani and *Udumbarie nainiensis* Grover

The midge species have been reported from Punjab, Haryana, Delhi and Uttar Pradesh. *A. peshawarensis* is remarkably host specific and most abundant. The female midge deposits eggs in raw and tender fruits. On hatching, the maggots feed on the pulp inside. Infested fruits become elongated, shrivelled, wrinkled, soft, fail to ripe and drop prematurely. More than 200 maggots can be seen in a single fruit.

*U. nainiensis* on the other hand, damages the flower buds by ovipositing in them. On hatching, the maggots feed on internal contents of inflorescence and cause discoloration, wrinkling and formation of small protuberances. Pupation in both the species occurs in soil and adults are short lived. There are several overlapping generations in a year.

*Management*

- All infested flowers and fruits should be removed and destroyed.
- Plough the soil around the trees frequently so as to expose and kill the pupae.
- Spray 200 ml malathion (Cythion 50 EC) 200 g carbaryl (Sevin 50 WP) or 200 ml endosulfan (Thiodan 35 EC) in 100 litre water at flower bud formation and again when the fruits are 'pea' size. Repeat the spray, if necessary, after 2 weeks.

**Scales, mealybugs and lac insects**

More than 35 species of homopterans, mainly scales and mealybugs, have been recorded feeding on fig trees in India. Of these, *Drosicha mangiferae* (Green) and *Icerya purchasi* Mask., (Margarodidae); *Parlatoria oleae* Colvee, *Aonidiella aurantii* (Maskell) and *Aspidiotus destructor* Sign. (Diaspididae); *Kerria lacca* (Kerr.) and *K. fici* (Green) (Tachardiidae), *Planococcus citri* (Risso) and *Pseudococcus* sp. (Pseudococcidae), *Ceroplastes* spp., *Chloropulvinaria psidii* (Maskell), *Coccus discrepans* (Gr.) and *Saissetia oleae* (Bernard) (Coccidae) are more prevalent. One or the other species is invariably present in almost every orchard. Of all these sap sucking insects, *P. oleae* and *K. lacca* are relatively economically important. The fig spittle bug, *Cosmoscarta niteara* Dist. (Cercopidae) and the bug, *Riptortus linearis* Fabr. (Coreidae) devitalise the trees by sucking the sap from succulent twigs, but are of minor importance. (For further detail refer to pomegranate and jujube).

**Miscellaneous pests**

The fruit flies, *Bactrocera dorsalis* (Hendel), *B. correcta* (Bezzi) and *B. zonatus* (Saunders) being highly polyphagous, also attack ripe fig fruits in some regions. Unripe fruits are rarely attacked as their latex is not ideally suited for the development of tephritid maggots (Bezzi, 1916). Besides these pests, a diverse group of about 50 species of insects and mites belonging to more than two dozen families, spreading over the orders Orthoptera (1 species), Hemiptera (2), Homoptera (10), Thysanoptera (2), Lepidoptera (15), Coleoptera (8), Hymenoptera (1) and Acarina have been reported feeding on fig plants in different regions. A vast majority of these are not regular in occurrence and seldom appear in serious form. Being of minor importance, usually no control measures are required against them (Simwat and Sidhu, 1974).

## BAEL (*AEGLE MARMELOS* CORREA)

*Bael/Bel*, also called Bengal quince or Indian quince is more valued for its medicinal virtues than for edible fruit quality. It is a subtropical species having the reputation of thriving successfully where most other fruit trees cannot survive. It is cultivated throughout India and does best in well drained soil. However, the trees fruit well only in areas where there is a long dry season. It also grows wild in dry forests in the hills as well as plains of India, Pakistan, Bangladesh and Myanmar, and also in drier areas of Sri Lanka, Thailand and Java. *Bael* is a hardy tree and does not suffer much from ravages of insect pests, although more than 30 species of insects and mites have been reported feeding on bael in India. Majority of these arthropods are polyphagous and have been recorded primarily as pests of other fruit trees growing nearby and bael is generally an alternate/collateral host of several of these. Moreover, leaf extract of *bael* is used as protectant against stored grain pests, as antifeedant against some crop pests, and also shows fairly good insect growth regulatory (juvenoid) activity against certain insects.

**Leaf eating beetles and weevils**

The chrysomelids, *Clitea picta* Baly and *C. indica* J. are specific pests of *bael* and sometimes cause noticeable damage to leaves and shoots. The leaves are riddled with patches eaten by small black shining adult beetles, while the yellowish white larvae tunnel inside the petiole and midrib of tender leaves. The affected young shoots become stunted as a result of larval injury. The damage is more significant on young plants and succulent leaves which are severely defoliated.

The epilachna beetle, *Henosepilachna* 12- *punctata Epilachna duodecimpunctata* (Coccinellidae) scrap the leaves in characteristic manner. The affected leaves are typically skeletonised. The grubs also feed on leaf tissues and present lace like appearance. Among the curculionids, *Myllocerus discolor* Boh. and *M. undecimpustulatus* Faust are commonly found feeding on *bael* leaves. The grey weevils nibble the leaves from edges and their grubs feed on roots of weeds and other crops. The beetles and weevils appear when the trees put forth new leaves. The losses caused by coleopterans are usually not much and as such no control measures are required.

**Lepidopterous pests**

Six species of moths and butterflies are known to be associated with *bael*. Three of them namely, *Phyllocnistis citrella* (Stn.) (Phyllocnistidae), *Papilio demoleus* Linn. (Papilionidae) and *Psorosticha zizyphi* (Stn.) (Oecophoridae) are well known and established pests of citrus in India, and *bael* serves as an alternate host. However, when *bael* is grown near citrus orchards, these pests show preference for the latter. The larvae of *Prays endocarpa* Meyr. (Yponomeutidae) and *Euzophera* sp. sometimes bore into the pulp of bael fruit but do not cause much damage. The larvae of *Cryptophlebia illepida* (Butler) also bore into the fleshy portion of *bael* fruit but only scarcely. Most of these species generally do not require any control measures (Latif and Yunus, 1951).

**Homopterous pests**

About a dozen species *(Myzus persicae* (Sulzer) and *Aphis craccivora* Koch (Aphididae), *Aleurocanthus woglumi* Ashby (Aleyrodidae), *Coccus viridis* (Green) (Coccidae) and Diaspididae *Aonidiella aurantii, A. orientalis, Fiorinia* sp., *Parlatoria* spp., *Aspidiotus* sp. (Diaspididae) have been recorded on *bael* trees in various parts of India. No control measures are usually needed against these as all of them are minor pests of *bael* (Ramakrishna, 1919).

**Miscellaneous pests**

The newly planted saplings of *bael* in dry areas are attacked by termites, *Odontotermes* sp. and *Microtermes* sp. leaves by gall mites, *Tetra aegleis* (Eriophyidae: Acarina), fruits by fruit fly, *Bactrocera* spp. and tender bark by the snail, *Rachis bengalensis* Lamarck (Enidae: Gastropoda) in different areas. Except termites, which devour the root and bark of new plantations, all other pests are of minor importance (Ramakrishna Ayyer, 1940).

## JAMUN (*SYZYGIUM CUMINI* SKEELS)

Jambolan, *Syzygium cumini*, known by various colloquial names, viz., *jamun, jaman, jambul, jambu*, Indian black-berry, Indian black plum, etc. is native to India, Myanmar and Sri Lanka. It grows abundantly along roadsides, borders of fields and orchards, on river banks, near temple ponds, etc. throughout the country from sea level to 1800 m. This highly nutritious fruit has immense medicinal properties and religious value. The multipurpose tree is very hardy and can withstand both flood and drought equally well. Thouogh this tree flourishes on sandy soils, hardly any serious/regular cultural practices are adopted for its cultivation. Jamun is known to be attacked by 150 species of insects and nearly a dozen species of mites, besides some species of birds and higher animals. However, only a handful of these are important pests. Those of common occurrence in arid and semi-arid areas are bark eating caterpillars, thrips, leaf-miners, whiteflies/blackflies, mealybugs, leaf-rollers, fruit borers and mulberry bug.

**Bark eating caterpillars**

Bark eating caterpillars are discussed under the heading 'aonla'.

**Thrips**

Six species of thrips, one belonging to family Thripidae and five to family Phlaeothripidae have been reported to infest *jamun*. While *Thrips florum* Schmutz. infests flowers, the others feed on leaves and developing fruits. The long tailed *jamun* thrips, *Leeuwenia ramakrishnae* Ananthakrishnan (Phlaeothripidae) is specific pest of jamun, confined to India especially in the South. The grapevine thrips, *Rhipiphorothrips cruentatus* Hood (Thripidae), though distributed all over the country, predominantly occurs in arid and semi-arid regions of North India. It is highly polyphagous but has a distinct preference for anacardiaceous and myrtaceous plants like cashew, mango, guava, rose, pomegranate and some wild plants and avenue trees.

The female, on an average, lays 50 eggs singly in small slits made on the undersurface of leaf tissues usually 2 to 6 per day. The egg stage lasts for 3 to 8 days and the small pale coloured (with reddish tinge) nymphs are just visible to the unaided eyes. The nymphs pass through two stages before entering pre-pupal and pupal stage in 9 to 22 days. The prepupal and pupal stages are capable of moving/crawling a bit, if disturbed. During active breeding season, pupation occurs on leaves but otherwise the nymphs fall on to the ground to pupate in soil and change to adults in 2 to 5 days. The black-brown adults which live among nymphs, have yellowish feather-like wings. The females live for a longer duration (upto 18 to 20 days) than males (2 to 6 days) and a single generation is completed in 14 to 33 days. There are 5 to 8 generations in a year. In areas, where there is a distinct winter as in the North, the pest hibernates in pupal stage, 8 to 18 cm deep in soil. Where winter is not severe, the thrips remains active throughout the year. Both sexual and parthenogenetic reproduction occurs side by side, the fertilized eggs hatching into females and unfertilized ones into males.

Both nymphs and adults feed on the ventral surface of leaves where they lacerate the tissues and imbibe the oozing cell sap. The population usually remains confined in between the veins where they feed motionlessly. The nymphs far outnumber the adults in a colony and are far more destructive too. The nymphs are often seen carrying a drop of excretory liquid at the raised anal end of their body. Such black shiny globules of excreta are seen scattered in the feeding zone. Due to intensive feeding by thrips, the leaves develop silvery white scorchy patches which coalesce and turn brown. The attacked leaves curl up, turn yellowish and give withered appearance, become brittle and drop prematurely, which in turn reduces the bearing capacity of plant. Attack on inflorescence induces shedding of flowers and newly set fruits. The surface of infested fruits become rough and corky and thus their market value is adversely affected. Severe incidence of thrips occurs during hot, dry and dusty spells from April to June and August to November, after the rains stop (Rahman and Bhardwaj, 1937; Patel and Patel, 1955).

*Management*

- Remove grasses and weeds underneath the trees.
- Spray 100 ml malathion (Cythion 50 EC), 150 ml endosulfan (Thiodan 35 EC) or 30 ml fenvalerate (Fenval 20 EC) in 100 litre water.

**Leafminers**

In India, four species namely, *Acrocercops phaeospora* Meyrick, *A. loxias* Meyrick, *A. syngramma* Meyrick and *A. telestis* Meyrick have been reported on *jamun*. Of these, *A. phaeospora* is specific to *jamun* and is comparatively more common and regular in occurrence. *A. loxias* and *A. telestis* on the other hand, are sporadic, while *A. syngramma* is widely distributed in Indo-Pakistan plains. However, these are polyphagous in nature and attack *jamun*, mango, litchi, sweet potato, etc. (Lakra, 1997b).

The newly hatched, delicate, pale yellow caterpillar mines a narrow thread like silvery gallery on the leaf along the mid rib. The mine is slowly transformed into a tubular blister-like swelling which extends from 3 to 6 cm in length on the dorsal surface of the leaf. The caterpillar feeds within the lacerated gallery. When full-fed, it forms a circular flat cocoon in depressed portion or cavity in the mine, usually near the raised conspicuous leaf vein and pupates there. The moth, on emergence, cuts a circular hole in the swollen fragile mine to get an easy exit. There may be a single blister covering both halves of the leaf blade or one blister on each half of the leaf blade. After emergence of moths, the swollen leaf tissues start drying, and eventually rupture and drop off, leaving conspicuously big longitudinal holes in the infested leaves. Heavily mined leaves later appear as if scortched due to heat. As high as 66 per cent leaves are sometimes found riddled by *Acrocercops* larvae. The gracillariid remains active from April to October and passes through several overlapping generations. Most of the damage takes place during summer and rainy season, coinciding with the onset of new vegetative flush and also fruiting stage of the crop. The pest overwinters from November to February as pupa inside the tubular mine. Tiny moths emerge during March and lay eggs singly, along the midrib on the underside of the tender leaves. The moths are short lived (Dahiya *et al.,* 1995).

### *Management*

- Clipping off and burning the mined leaves during winter reduces the incidence of leafminers on new growth in March-April.
- Spraying the trees with 150 ml oxydemeton methyl (Metasystox 25 EC), 120 ml dimethoate (Rogor 30 EC) or 50 ml phosphamidon (Dimecron 85 WSC) per 100 litre water effectively checks the leafminer in the active season (Kanta *et al.*, 1966; Singh and Singh, 1979; Lakra, 1997). Avoid insecticide sprays in areas where the *Acrocercops* larvae/pupae are found heavily parasitized by a small black wasp, and protect insect pollinators by avoiding insecticidal sprays during flowering.

### Mulberry bug, *Halys dentatus* (Fabricius)

The mulberry bug is discussed under the heading 'mulberry'.

### Whiteflies/blackflies

Whiteflies and blackflies are important pests of *jamun* in the tropics and sub-tropics. As many as 7 species of aleyrodid are known to infest *S. cumini* in different regions of India. Some of the species are rather highly polyphagous, while others are specific to *jamun*. The *jamun* whitefly, *Dialeurodes eugeniae* (Maskell) is known to occur practically all over the country (Lamba, 1949). It is a specific pest of *S. cumini* and rose apple (*Syzygium jambos* (L.) Alston). Although widely distributed yet it hardly causes any serious damage, except in certain pockets. The adults measure 1.6 to 2.0 mm, ash grey coloured males are smaller in size than females which have orange abdomen. There is a profuse deposition of white waxy material on the wings and body. The hind wings are cinereous and similar coloured bands are found on the abdomen. Eggs are deposited in clusters, arranged in a semi-circle on tender leaves. The newly hatched sienna coloured nymph moves about for a short period before settling at a suitable spot for feeding. The young nymphs are located in a peculiar form in little gall-like pits on the upper surface of fresh leaves. Once fixed, into the nymphs do not move for rest of their life and undergo three months to enter into 4th instar or transform into pupae at the same spot. The pupa is sub-circular, bluish-black with an opaque white rim, the male pupa is slightly smaller than female. The pre-imaginal stages occupy about two months in summer and slightly longer in winter. The whitefly completes two generations in a year. The nymphs are more harmful than the adults. When in good number, feeding causes grey areas on the leaf as a result of sucking of vital sap. The nymphal feeding causes characteristic punctuations on both the surfaces of leaves, but more often on the dorsal surface. Excessive desapping may also affect the fruiting capacity of the tree.

Besides this, *Dialeurodes vulgaris* Singh, *Singhiella bicolor* (Singh), and a blackfly, *Aleurocanthus rugosa* Singh are also fairly common and occur all over the country. All these are specific pests of *jamun* and their feeding behaviour, symptoms of damage, biology and life history are more or less similar to that of *D. eugeniae*.

*Management*

The adoption of following management strategies will help in controlling these pests on *jamun*. Avoid overcrowding in the vicinity of *jamun* trees to enable penetration of sunlight and air which will prevent creating suitable habitat for whitefly.

- Phytosanitation: Keeping the orchard clean by removal of weeds and other plants reduces pest population and prevents the initiation of infestation.
- Avoid water-logging, persistent drought situation and other stress conditions which predispose the trees to whitefly infestation.
- Avoid excessive irrigation and application of nitrogenous fertilizers.
- In case of localized infestation clip off and destroy the affected shoots.
- Parasitoids, predators and other natural enemies which keep the whitefly population under check in nature, should be preserved/protected and, if need be, augmented.
- If the infestation is severe, spray the trees with 200 ml endosulfan (Thiodan 35 EC), 100 ml monocrotophos (Monosil 36 WSC), or 130 ml dimethoate (Rogor 30 EC) in 100 litre water. Spraying should be done at the initiation of new flush and pest infestation and may be repeated after 2 to 3 weeks, if necessary (Rao, 1958).

**Leaf gall psylla, *Trioza jambolanae* Crawford**

The leaf gall psylla, *T. jambolanae* is a specific pest of *jamun* in all *jamun* growing areas of Asia. The psyllid occurs throughout the country but hardly came any serious harm, though it produces malformation of tender shoots and foliage. The female deposits about 500 eggs, usually singly but sometimes in clusters, scattered on underside of young leaves. The eggs hatch in 3 to 6 days. The young scale-like nymphs settle on the ventral surface and suck the plant sap. While feeding, the nymphs also inject some secretion in the leaf tissues which stimulates cell multiplication leading to gall formation. The galls are hemispherical, unilocular, isolated or conglomerate at the site of stylet insertion mainly on dorsal surface of leaves. Initially the galls are smooth and greenish-yellow, but later harden and turn brownish. Finally, irregular radiating cracks appear on the surface splitting it into several lobes which in turn facilitate the adult emergence. The nymphs remain sheltered within the galls and feed for 13 to 40 days in summer, and 160 to 170 days in winter. Overwintering occurs in nymphal stage inside the gall. The galls are often globose or somewhat oblong and measure about 7 mm in length and 5 mm in diameter. In severe infestation the leaves are conspicuously laden with galls and get badly distorted/disfigured, dry up and drop off. Young plants are more prone to attack as compared to full grown trees.

*Trioza vitiensis* (Kirkaldy), *T. fusca* Mathur, *T. spinulata* Mathur, *T. eugenioides* (Crawford), *Euphyllura caudata* Mathur and *E. concolor* Mathur are the other psyllids reported feeding on *jamun* leaves in India, but all of them are of little economic importance (Chandrasekaran and Balasubramanian, 1972).

**Miscellaneous pests**

As many as 40 species of butterflies and moths have been reported feeding on *jamun*. The larvae of *Polychrosis cellifera* Meyrick, *Argyroploce* spp., *Lepidogma* sp., *Oenospila flavifusata* Wlk. and

*Homona coffearia* (Nietner) roll or fold the tender leaves with the help of silken threads and feed within. Some notable leaf eating caterpillars namely, *Carea subtilis* Wlk., *Trabala vishnou* (Lefevere), *Metanastria* spp. and *Euproctis* spp. are sporadic/minor pests which become occasionally serious in some regions. The fruit borers, *Meridarchis* spp. and fruit sucking moths, *Calyptra bicolor* (Moore) and *Anua triphaenoides* Wlk. attack the fruits but are seldom serious. Of more than 3 dozen coleopteran species recorded attacking *jamun* only a few are important pests. The chafer beetles, *Holotrichia* spp., *Anomala bengalensis* Blanchard, and the grey weevils, *Myllocerus* spp., *Apoderus* spp. are polyphagous and feed on *jamun* foliage also. Among the fruit and seed weevils, *Curculio c-album* (Fabr.), *Balaninus c-album* Fabr. and *Apotomorhinus cribratus* Sch. are minor pests recorded feeding on fruits and seeds of *jamun* at few locations. Likewise, some long-horn cerambycid stem borers, viz., *Batocera rufomaculata* (deGeer) and *Aeolesthes holosericea* Fabr., being pests of woody vegetation also bore into trunks of old *jamun* trees. Similarly, more than a dozen species of mealybugs, scales, aphids, termites and some species of tetranychid mites are also known to feed on *jamun* foliage. Usually these pests do not cause serious losses and, therefore, no control measures are used against them (Lakra, 1977b).

## DATE PALM *(PHOENIX DACTYLIFERA* L.)

Date palm, often called edible date, is native to Persian Gulf and is believed to be one of the first fruit trees ever cultivated. The date is more suited for cultivation in arid areas than any other fruit. It is remarkably tolerant to alkali soils and long periods of drought. It likes full sun, high temperature and low humidity. The fruit is known for its long keeping quality and high food value (about 54% sugar content and 7% protein) and is a staple food in hot deserts. Economic damage to date palm is caused by about 20 pest species in India, although it serves as an alternate host to all insects attacking coconut and other palm trees. Of these, the date palm scale, black palm beetle, red palm weevil, black headed caterpillar, termites and some birds are important pests.

**Date palm scale *Parlatoria blanchardi* (Targ.)**

Of the 6 species of armoured scales reported infesting date palm, *P. blanchardi* is the most serious pest. It is specific to palms, especially *P. dactylifera,* and has a wide range of distribution from Iraq to India spreading over to Iran, Afghanistan and Pakistan. It occurs all over the plains of India. On hatching, the light violet colour crawlers leave the maternal scale and take up position to settle on the under surface of leaves, pinnules and main veins of the fronds, and suck the sap. A severe infestation forms a continuous crust of yellowish brown scales on the underside of leaves. Due to sap loss the vitality of tree, fruit setting and development is affected. Attacked fruits become dry and deformed and do not mature. Closely planted, frequently irrigated young plants grown under shade suffer the most. Fruits covered with cloth bags are more severely infested. Hot and dry climate is favourable for this diaspidid (Batra and Sohi, 1974).

The female lays from 6 to 15 rose coloured eggs underneath its body which hatch in 8 to 15 days. The rose coloured newly hatched tiny crawlers leave the maternal scale to settle on the leaves. Once settled they do not move. The female nymphs become adults in 30 to 40 days while male nymphs take 20 to 25 days and then pass through prepupal and pupal stages in another 6 to 8 days. The winged males are short lived, mate soon after emergence and do not feed. The life cycle is completed in 50 to 60 days in summer and 75 to 85 days in autumn and spring. The pest hybernates during winter.

Another diaspidid, the coconut scale, *Aspidiotus destructor* Sign., is also widely prevalent on date palm. The semi-transparent scale is polyphagous and infests citrus, mango, fig, *ber*, pomegranate, *jamun,* etc. and remains active even during winter. *Parlatoria ziziphus* (Lucas), *P. pergandii* Comstock, and *Chrysomphalus ficus* Ashmead, being polyphagous, sometimes infest date palm but the infestation is not serious.

*Management*

- Plant healthy off shoots free from scales.
- Remove and burn all the infested leaflets and fronds in initial stage of infestation.
- Sanitation and good crop management practices check pest development and favour the coccinellid predators (Murlidharan *et al.*, 1992).
- Spray 125 ml monocrotophos (Monsil 36 WSC), 150 ml dimethoate (Rogor 30 EC) or 200 ml diazinon (Basudin 25 EC) in 100 litre water. Repeat the spray after 3 weeks, if necessary (Bhargawa *et al.*, 1990).

**Rhinoceros beetle *Oryctes rhinoceros* (Linn.)**

The Indian rhinoceros beetle is an extremely important pest of coconut and occurs in all coconut growing areas of the Indian subcontinent. It causes considerable damage to other palms including date palm. Adult is the only destructive stage that feeds during night on the crown. The beetle bores into unopened tender frond and feeds on the growing point which is damaged and palm tree dries up. Infested fronds when open give rise to leaflets having V shaped cuts. The beetle throws out fibrous mass while feeding in the burrow made in young plants which die due to destruction of growing point. The growth and yield of older palms suffer badly. Maximum damage is caused during the monsoon. Isolated trees and palms of irregular heights are more prone to attack. Grubs feed on decaying organic matter and are harmless. Damaged palms are often infected secondarily by fungal pathogens.

The life and seasonal history of rhinoceros beetle has been studied to a considerable extent on coconut and only a little information is available on date palm. The female on an average, lays 90 to 150 eggs in decaying organic matter, manure/compost heaps and dead rotting palm trunks. Eggs are laid usually singly at 5 to 15 cm below the soil surface. Hatching takes 8 to 18 days. The grubs feed on the decaying matter, fairly deep in soil, and become full grown (7 to 10 cm in length) in 75 to 190 days passing through three instars. Pupation occurs deep in soil and pupal period lasts for 2 to 4 weeks. The adult remains within the pupal chamber for 5 to 30 days to attain sexual maturity. Adults live for 75 to over 200 days and are stout and robust, 3 to 5 cm long. The beetles are shiny black dorsally and reddish brown ventrally. They are nocturnal in habit. The life cycle is completed in 105 to 260 days depending upon climatic conditions. There are one or two generations per year.

*Management*

- Orchard sanitation checks the pest development. Burn all dead palm trunks, leaves and logs.

- Kill the beetles in the bore hole with a pointed wire or iron spoke during rainy season.
- Destroy all potential breeding sites (manure/compost pits, refuse dumps, decaying material, etc.) in and around the plantation. All rubbish and rotten vegetation should be dried and burned. Spraying of insecticides is not practicable on large scale and is uneconomical.

**Red palm weevil *Rhynchophorus ferrugineus* (Olivier)**

The palm weevil is an important pest of date palm, although its preferred hosts are coconut and oil palm. It is widely distributed in Indian subcontinent and S.E. Asia to China. The grub is the chief damaging stage and is a voracious feeder. The larvae bore into the trunk near the growing point and tunnel downward. Once inside, they do not come out and hence it is difficult to detect the infestation early in the season. The outer leaves turn chlorotic and die and the symptoms soon spread to innermost leaves. Young palms are preferred. The larvae burrow in hard woody portions in all directions making tunnels of up to 1 m length and feed on soft and succulent tissues. As the galleries become more extensive, the trunk weakens and the tree may be easily decapitated in strong wind. The trunk is completely hollowed out and is filled with decaying rubbish. Viscous plant sap, unpleasant smell and chewed frassy material are thrown out from the holes in the stem. Growing points of infested young plants wither, dry, rot and give offensive smell. In older plants top portion of trunk bends and breaks just below the crown. The pest is multivoltine and breeds throughout the year. Several grubs (40 to 50) could be found in a single trunk.

Eggs are laid singly in small holes scooped out by the adult weevil on soft regions of young palms, usually at the base of leaf sheaths, in cuts or wounds or injured portion of the stem made by other insects (*Oryctes* spp.) and man. The female lays on an average about 200 eggs in its life span of 2 to 3 months. Hatching takes place after 2 to 5 days and the yellowish-white apodous grubs penetrate the crown and feed on the internal tissues for 24 to 78 days. Pupation occurs in oval cocoons made of fibrous threads within the infested palm trunk. The pupal stage lasts for 15 to 35 days. The entire life cycle is completed in 50 to 90 days.

***Management***

- Orchard sanitation, especially prompt destruction of infested, dying and dead palms at initial stage of attack suppresses the pest development.
- Avoid the occurrence of cuts, wounds, scars, etc. on the plants as female weevils search such sites actively for oviposition.
- This pest is very difficult to control in advance stage of infestation. Hence, the affected plants should be destroyed promptly to prevent further spread.
- Inject insecticidal emulsion in borer holes as suggested for the control of bark eating caterpillars under *aonla*.

**Black-headed caterpillar *Opisina arenosella* Walker**

The black-headed caterpillar is a major pest of coconut in India, Sri Lanka and Myanmar. The xyloryctid, though not a serious pest of date palm, is unquestionably the most important of lepidopterous pests attacking date in India. Being polyphagous it also attacks other palms and occurs commonly along the East and West coast of India. The caterpillars feed gregariously on the green matter within the galleries of silk and frass, on the underside of leaflets. The gallery is extended, as the larva grows, covering under it major portion of ventral surface of the leaflets. Attacked leaflets turn brown and dry up affecting the photosynthetic activity. The leaflets are reduced to papery tissues, severely infested trees give a burnt appearance, and could be recognized from a distance by dried up patches in the fronds. The fruiting capacity of the tree is adversely affected.

The female lays an average of 137 scale-like whitish eggs, in batches, on old leaflets. The incubation period varies from 3 to 7 days, averaging 5 days. The larva is light green with red brown stripes, black head and prothorax, and reddish mesothorax. The full grown larva is elongate, slender and measures about 15 mm in length. The caterpillar becomes full grown in 32 to 55 (average 42) days and pupates within thin tough cocoon inside the silken web on leaves. Pupal period ranges from 8 to 14 days and ash-grey moths live for 3 to 15 days. A single generation is completed in 50 to 60 days and there are 4 to 5 generations in a year. The incidence of the pest becomes serious during dry months but the population declines during rainy season.

*Management*

- Remove and burn the infested plant parts alongwith silken galleries and larvae as and when located.
- Spray the infested trees with 100 ml monocrotophos (Monosil 36 WSC), 200 ml endosulfan (Thiodan 35 EC) or 200 g carbaryl (Sevin 50 WP) in 100 litre water.
- Release of some parasitoids, viz., *Bracon brevicornis* Nesmael, *Elasmus nephantidis* Rohwer, *Trichospilus pupivora* Ferrieri and *Perisierola nephantidis* Mues. in areas where infestation uccurs regularly.

**Miscellaneous pests**

Some insects and mites recorded as minor pests of date include coleopterans like the date stone weevil, *Coccotrypes dactyliperda* (Fabr.) (Scolytidae), *Haptoncus luteolus* (Erichs.) and *Carpophilus mutilatus* Er. (Nitidulidae) which feed on the ripening fruits and typical circular holes are seen drilled on the infested fruit causing decay. The fruit clusters at 'dang' and 'dhoka' stages are often damaged by hymenopterans like the yellow wasp, *Vespa orientalis* Linn., red wasp, *Polistes* sp. (Vespidae) and honey bees (Apidae). These species feed on the soft, ripe and injured fruits. Likewise, the red tree ants, *Oecophylla smaragdina* (F.) (Formicidae) make aerial nests on the trees and feed on ripened fruits. Being ferocious, they may gave a bite when disturbed, and are a source of nuisance. Red ants also provide protection to sap sucking coccids. The thrips, *Haplothrips ceylonicus* and *Adihetrothrips jambudvipae* Ramakrishna (Thripidae) feed on flowers and developing fruits. The nymphs and adults of banana lacewing bug, *Stephanitis typicus* Dist. (Tingidae) occasionally devitalize the date trees by sucking sap from tender leaves but their population remains fairly low. The coconut white grub, *Leucopholis coneophora* Burm. (Scarabaeidae) has also been observed feeding on roots of date palm in some white grub endemic areas. The termite, *Odontotermes obesus* (Rambur) (Termitidae) is fairly widespread in arid and semi-arid regions and attack several fruit trees including newly planted date palms. The trunk of date is often seen covered with mud galleries made by the workers. The scarlet mite, *Raoiella indica* Hirst

(Tenuipalpidae) kills the new foliage in some places, but the attack is seldom severe. Beside arthropod species, several species of birds (common myna, bulbul, parakeet, sparrow, purple sun bird, etc.) and some species of squirrels and monkeys also feed on ripening date fruits at 'dang' or 'doka' stage. Likewise, date fruit kept in storage or during transit is subject to invasion by storage pests, viz., fig moth, *Ephestia cautella* (Hb.), (Pyralidae), cigarette beetle, *Lasioderma* sp. (Anobiidae) and tineid moth, etc. Being pests of minor importance control measures are not generally needed against them (Batra, 1972).

## PHALSA/FALSA (*GREWIA SUBINAEQUALIS* DC.)

*Phalsa* or *Falsa*, as the scraggy shrub is indigenous to India. It is grown mostly as a catch crop amidst various commercial fruit trees. The shrubby tree is drought tolerant and grows well in poor and marginal lands and requires little post planting care. The ripe fruits keep poorly and are eaten fresh as dessert, the fresh leaves are valued as fodder and its strong flexible shoots/canes are used for basket marketing. The plant possesses certain medicinal qualities too. *Falsa* is one of the most hardy Indian fruit plants, least attacked by insect pests. However, over 3 dozen species of insects and mites have been observed feeding on this plant. In general, no single species causes severe damage but the overall loss inflicted by various pests is sometimes substantial. Majority of the species noted feeding on *falsa* are polyphagous.

### Plum hairy caterpillars

These are discussed under the heading 'jujube'.

### Bark eating caterpillars

These are discussed under the heading 'aonla'.

### Sap sucking insects

Twenty species of insects recorded on *Grewia* spp. are sap suckers belonging to 11 families of Homoptera/Hemiptera. Majority of these are mere records of occurrence of the pest on *falsa* as an alternate host. However, the mango mealybug *Drosicha mangiferae* (Green) and *ber* mealybug, *Perissopneumon tamarindus* (Green) (Margarodidae), leafhoppers, *Zygina rubronotata* (Dist.) and *Tautoneura manica* Thapa (Cicadellidae), *phalsa* bug, *Gargara mixta* Buckton and the membracid bug, *Leptocentrus taurus* (Fabr.) (Membracidae), the shield bug, *Scutellera nobilis* (Fabr.) (Pentatomidae), tingid bugs and psyllids, cause considerable damage to the shrub, while the rest of the pests are of minor importance. Of these, *G. mixta* is widely distributed and specific pest of *falsa* while the lygaeid bug, *Oxycarenus hyalinipennis* Costa, infests the fruits. Nymphs and adults of all these species suck sap from leaves, shoots, inflorescence and fruits. Normally no control measures are required against these pests.

### Leaf eating beetles and weevils

As many as 12 species of beetles, belonging to family Scarabaeidae, Curculionidae and Chrysomelidae have been reported feeding on *falsa* leaves and flowers. The scarabaeids,

*Holotrichia consanguinea* (Blanchard), *H. insularis* Brenske, *Adoretus* sp., *Anomala bengalensis* (Blanchard), *Apogonia uniformis* Blanchard, and *Schizonycha* sp. are polyphagous pests and defoliate the *falsa* leaves during night (Khan and Ghai, 1974). Another scarabaeid, *Oxycetonia versicolor* (Fabr.) feeds on leaves and occasionally on flowers, whereas, *O. albopunctata* feeds on flowers only. The attacked flowers wither and die.

Weevils like *Myllocerus* sp., *Colobodes tetramaculata* Sch., and *Attelabus octomaculatus* Jek., etc. though of minor importance, feed on leaves of wild and cultivated *Grewia* sp. (Arya *et al.*, 1987). Likewise the almond beetle, *Mimastra cyanura* Hope, also causes minor damage to *falsa* leaves and as such no control measures are required against them. However, among the coleopterans, the cockchafer beetles are important and cause serious damage.

(For more details see under *ber*.)

**Leaf eating caterpillars**

The plum hairy caterpillars, *Euproctis fraterna* (Moore) and *E. lunata* (Wlk.) (Lymantriidae) and *falsa* caterpillar, *Giaura sceptica* Swinhoe (Noctuidae) gnaw the leaves and often cause serious damage. Occasionally the pyralid, *Lygropia quaternalis* (Zeller) and the sphingid, *Agnosia microta* (Hampson) caterpillars are also noticed feeding on the foliage. The leafminer, *Bucculatrix univoca* Meyr. is an important sporadic pest. The gracillariid makes transparent mines on tender and middle leaves (1 to 320 mines/leaf) between July and October when it passes through 3 to 4 generations (Lakra and Dahiya, 1988). The severely mined leaves curl up and drop. Another leafminer, *Phyllocnistis citrella* (Stn.) also occasionally infests *phalsa* leaves.

**Miscellaneous pests**

The fruit fly, *Bactrocera dorsalis* (Hendel) being highly polyphagous also damages *phalsa* fruits. The termites, *Microtermes* sp. and *Microcerotermes beesoni* Snyder often damage the roots and stems of *Grewia* spp. in arid areas (Sen-Sarma and Gupta, 1979). Among the mites, *Eutetranychus orientalis* (Klein) (Tetranychidae) and the gall mite *Tetra coimbatorensis* Mohansundaram (Eriophyidae) (Mohansundaram, 1994) have also been recorded feeding on *Grewia* spp. but are of little economic significance.

## KARONDA/CORINDA (*CARISSA CARANDAS* AUCT.)

*Karonda/karanda/corinda*, as this straggly shrub is known by various colloquial names is native of India. The thorny shrub is primarily cultivated as a hedge or ornamental, but the sweet-subacid flavoured fruits are edible. The plant often grows wild in dry sandy or rocky terrain, and thrives even in poorest and rockiest soils. Its chief requirement is said to be exposure to sun. *Karonda* is perhaps the most hardy fruit plant and is not severely attacked by many insect pests. In all, 15 species of hexapods have been recorded infesting *Carissa* spp., none of which except the scale insects, is serious.

Scale insects, both armoured and soft, are among the principal pests of which *Aspidiotus* sp. and *Aonidiella* sp. are widely distributed, more common in occurrence and severe. Heavy infestation of this pest has been reported from many areas leading to heavy reduction in fruit formation.

Among the soft scales, *Coccus* sp. and *C. viridis* (Green) have been noted to occur sporadically. All these homopterans are polyphagous and infest several fruit plants. They suck cell sap motionlessly and devitalize the plants. In case of severe infestation, the growth of the tree is arrested and fruits may fall off prematurely.

*Karonda* also serves as alternate/collateral host of the citrus blackfly, *Aleurocanthus woglumi* (Ashby) but the incidence of the aleyrodid on *C. carandas* is negligible. Fortunately the psylla, *Diaphorina carissae* Kuwayama (Psyllidae: Homoptera) has not so far been recorded on this host from Indian. The cockchafer beetles, *Holotrichia consanguinea* (Blanchard), *H. insularis* Brenske and *Anomala bengalensis* Blanchard (Scarabaeidae) have been observed damaging foliage of *karonda*. The scarab adults feed by making characteristic round holes on the leaves during night especially after the first premonsoon or monsoon rains. Being polyphagous, most of the fruit trees grown in arid and semi-arid areas are prone to their attack. Sometimes complete defoliation may occur and the losses could be substantial.

The larvae of castor semi-looper, *Achaea janata* (Linn.), a polyphagous pest, feed on leaves of several fruit trees. It punctures and feeds on the juice of ripening fruits causing rotting and premature fruit fall. It is a sporadic pest of minor importance (Pandey, 1967). The conspicuous caterpillar of a sphingid, *Nepheles hespera* Fabr. voraciously feeds on the leaves of *karonda* in some regions but its population remains very low and no significant damage is caused. The bark eating caterpillars, *Indarbela* spp. (Metarbelidae) and termites *Odontotermes obesus* (Rambur) and *Microtermes obesi* (Holmgren) (Termitidae) are other important polyphagous pests of arid zone fruits including *karonda*. Their attack is more serious in the neglected orchards. The fruit fly, *Bactrocera correcta* (Bezzi) (Tephritidae), being highly polyphagous, also infests fruits of *karonda* in some areas but rarely.

**Termites or white ants**

About a dozen species of termites belonging to three families have been reported damaging arid and semi-arid fruit trees. These are *Microtermes* sp., *M. obesi* (Holmgren), *Microcerotermes beesoni* Snyder, *Odontotermes sp. O. obesus* (Rambur), *O. wallonensis* Wasmann and *Trinervitermes biformis* (Wasm.), *Bifiditermes beesoni* (Gardner), *Neotermes* sp., *Coptotermes heimi* Wasmann, and *Heterotermes indicola* (Wasmann).

Termites are polymorphic, mostly photophobic, eusocial insects which live in small to large colonies, in nests (termitaria) of their own construction, the larger ones may contain over a million individuals. Each colony consists of members of different castes which are morphologically and functionally distinct. They are exceedingly abundant and are found throughout the tropical and subtropical areas. Fruit tree/wood tree eating termites tend to be polyphagous and certain species in semi-arid and arid areas attack all types of plants including those which otherwise are poisonous to mammals. In fact termites are one of the most destructive pests in fruit tree savannah of arid regions. Of the 260 species of termites reported from India, among more than 2100 species known the world over, a dozen have hitherto been recorded to attack the aforesaid fruit trees. However, the most common species widely attacking arid fruit plants, are *Odontotermes obesus* (Rambur) and *Microtermes obesi* (Holmgren).

The destructive capacity of termites lies in their number. Workers, which form the majority in a colony, cause the most damage. Cellulose, the most abundant organic compound in terrestrial habitats is the decisive nutritional requirement for termites. Workers gather it by feeding rapaciously on all kinds of plant material, both living and dead. The workers become more active at night and collect food over a large area surrounding the nest through a network of subterranean tunnels. The warmth loving, heliophobic workers withdraw into their termitarium during daytime. While foraging they live mostly underground but some even forage on the surface over well-trodden trails. Subterranean plant portions are greatly at risk. The isopterans first invade the dead or dying plant parts and then attack the living tissues. The attack usually begins at roots and then moves upwards and a slow hollowing out of the stem occurs, or they build mud galleries on trunks during night and feed on the bark beneath. The workers move along these roofed runways to feeding places, protected from light and dryness. The presence of mud galleries on tree trunk is a clear sign of termite attack. In severe cases the entire tree/shrub is covered with mud galleries and such plants die after sometime.

As a result of gnawing of bark, roots or heartwood by the workers of termites, tree trunk is often fully hollowed out leading to drying and death of whole tree. The infested trees become prone to lodging in rain and storm because of loosening of soil under the damaged roots. The termites cause havoc in orchards raised on virgin lands in warmer regions. Also, their damage is more pronounced in nurseries having sandy or sandy-loam soils and newly planted saplings. Old, weak, injured or unhealthy trees are liable to be attacked more by termites. Fruit trees are usually more susceptible to termite attack than majority of indigenous forest and wild trees. Some species of termites form mounds above the ground which rise to a height of a couple of metres, and are among the largest structures constructed by animals (except man). Besides attacking the living fruit, forest and avenue trees, termites also damage dry wood, timber and such material as textile, paper, etc. The termites remain active all the year round though their incidence and activity slow down during monsoons and in severe winter. The formation of earthern galleries on tree trunks is stepped up after the monsoons. Drought seems to play a major part in enhancing their incidence.

### *Management*

The control of termites attacking fruit trees in the arid regions is by no means easy, because of their subterranean habit, prolific rate of multiplication, absence of any effective predator or parasitoid and uninterrupted availability of host trees all the year round. Obviously, it will be more economical to prevent termite attack than to control its infestations. The following methods appear to be feasible to formulate a viable management programme against these pests.

#### *Cultural control*

- Major growth providing practices, viz., balanced nutrients, timely irrigation, hoeing, etc. are suitable means for avoiding termite damage, as healthy and vigorous trees are rarely attacked by termites.
- Termite control largely centres around orchard sanitation and tree hygiene. All dead wood, timber, crop residues, dry stubbles or decaying matter which attract termites should be removed from the orchard.
- Unnecessary injuries to plants which facilitate entry of termites should be avoided. Whenever noticed, prune the injured parts and scars. Termite attack is also favoured in weak,

unhealthy, fungus infected, weed infested trees/orchards, uprooted or dieing plants and drought. Such trees and situation should be avoided.

- Deep ploughing around/beneath trees helps in destroying the subterranean gallery system of termites. Well cultivated/pulverised soil followed by frequent copious irrigations check the termite damage and should be practised wherever possible.
- Avoid use of green manure and partially decomposed farm-yard manure, as these attract termites. However, termite damage to fruit trees can be reduced remarkably if humus content of soil is maintained at high level. So only fully decomposed manure should be applied.
- Most consistent and satisfactory control of termites can be achieved by the destruction of termite colony. This can be done easily in the case of mound forming species by breaking open the termitarium and killing the occupants by using insecticides, together with large amount of water. Interculture, hoeing and re-sowing of bald patches and planting new saplings in empty pits prevent the formation of termite nests in the orchard.

*Biological control*

Despite the large number of predators (dragonflies, ants and several other hexapods; frogs, lizards, birds, bats, large mammals, viz., spiny ant eater, great ant eater, antbear, pangolins, monkeys, man, etc.), parasitoids and pathogens, successful biological control of termites under field conditions still continues to be the cherished goal. All these biotic agents contribute in regulating the termite population in nature, though most species are able to maintain relatively high colony density. Nevertheless, the role and importance of the battery of natural enemies in keeping the population of termites under check should not be under-rated. The chances of success of biological control of termites in future stand more bright in fruit trees than the field crops.

*Chemical control*

Like other pest insects, complete control of termites is still a far cry. However, satisfactory control of termite infestation in orchards was earlier accomplished through soil application of persistent organochlorine insecticides like HCH, aldrin, dieldrin, chlordane, heptachlor, etc. as emulsion or dust. Of late, most of these soil insecticides have either been banned for use or their use has been restricted and few others are on the verge of being phased out because of certain undesirable effects and ecological considerations. Alternatively, chlorpyriphos and endosulfan have been found effective. However, the insecticides should only be used when possibilities of economic losses in an orchard are suspected. At times treatment of whole orchard is not required. The following practices, if adopted timely, will provide an effective control of termite infestation in the orchard.

*Treatment of termitarium*

In the case of mound forming species, locate the termitarium, scrap upper crust of the nest and destroy the occupants by applying 0.2 per cent chlorpyriphos, lindane (g -HCH), or kerosene oil @10 litre per cubic metre. The insecticide solution should be sprinkled so as to drench thoroughly each of the vertical shafts of the nest. This exercise should be undertaken preferably towards the end of rainy season each year.

*Pre-plantation treatment:* The pit should be filled 7 to 10 days before planting the sapling incorporated with soil 30 g methyl parathion (Fol: dol 2% dust), or 20 g fenvalerate (Fenval 0.4% dust) in 3 or 4 basket fulls of soil + fully decomposed manure or drench thoroughly (5–10 litre/pit) the soil in the pit, with 0.2 per cent chlorpyriphos (Durmet 20 EC), endosulfan (Thiodan 35 EC), or lindane (g-HCH; Kanodane-20EC) emulsion, so as to safeguard the newly planted sapling against the ravages of termites.

*For newly planted/established trees:* To protect the new plantations till they become vigorous and established enough to withstand the termite attack (usually after 2 to 3 years of planting), the following practices should be adopted:

- Mix thoroughly methyl parathion (Folidol 2% dust) @ 25 kg/ha (usually 200 to 300 g dust/tree), or fenvalerate (Fenval 0.4% dust) @ 15 kg/ha (about 120 to 150 g dust/tree) in pulverised soil covering whole basin area of tree, once a year, or
- Soil application of insecticides around tree trunk gives long lasting protection and is not phytotoxic at the recommended concentrations. The usual procedure is to apply chlorpyriphos, lindane or endosulfan emulsion @ 5 litre formulation/ha alongwith irrigation water. Wherever possible, drench the plants uniformly with insecticidal emulsion once a year, preferably in September.

*Treatment of mud galleries*

If the mud galleries appear on tree trunks, remove the same by breaking or scrapping the mud, and sponge/spray the affected parts with 0.2 per cent chlorpyriphos, lindane or endosulfan emulsion. Appearance of mud galleries on tree trunks is a clear indication of termite attack and also of neglected situation of the orchard.

## REFERENCES

Agarwal, M.L. and Kapoor, V.C. 1985. On a collection of Trypetinae (Diptera: Tephritidae) from northern India. *Annals of Ent*omology, **3**(2): 59–64.

Agrawal, B.D. 1980a. Morphology and anatomy of the galls of *Asphondylia phyllanthi* Felt on the leaf and shoots of *Emblica officinalis* Gaertn. *Cecidologica Internationale,* **1**(1–3): 93–99.

Agrawal, B.D. 1980b. Biology and bionomics of *Asphondylia phyllanthi* Felt forming galls on the leaves and stem of amla (*Emblica officinalis* Gaertn.). *Cecidologica Internationale,* **1**(1–3): 109–122.

Agrawal, B.D. and Jain, M.K. 1981. Studies on the leaf and shoot galls of *Emblica officinalis* Gaertn. formed by Chalcidoidea (Hymenoptera). *Cecidologia Internationale,* **2**(1): 39–44.

Arya, L.S., Singh, J. and Pajni, H.R. 1987. Two new species of genus *Colobodes* Schonherr (Coleoptera: Curculionidae: Cryptorynchinae) from India. *Entomon,* **12**(3): 201–206.

Awate, B.G., Naik, L.M., Dhumal, V.S. and Khandge, S.V. 1977. Chemical control of pomegranate fruit borer, *Virachola isocrates* Fabricius. *Indian Journal of Plant Protection,* **5**(2): 211–212.

Babu, P.C.S. 1971. A note on the occurrence of the Aleyrodid, *Aleyrodes sizoukinensis* on star goose-berry, *Phyllanthus acidus* and its control. *Madras Agricultural Journal,* **58**(5): 413–414.

Bagle, B.G. 1993. Seasonal incidence and control of thrips *Scirtothrips dorsalis* Hood in pomegranate. *Indian Journal Entomology,***55**(2): 148–153.

Batra, H.N. 1953. Biology and control of *Dacus diversus* Coquillett *and Carpomyia vesuviana* Costa and important notes on other fruit flies in India. *Indian Journal of Agricultural Sciences,* **23**(2): 87–112.

Batra, R.C. 1972. Insect pests of date-palm at Abohar and their control. *Punjab Horticulture Journal,* **12**(1): 44–45.

Batra, R.C., Sandhu, G.S. and Singh, S.N. 1988. Note on chemical control of ber leaf webber. *Entomon,* **50**(1): 53–54.

Batra, R.C. and Sohi, B.S. 1974. A new record of *Parlatoria blanchardi* (Targ.) from India. *Indian Journal of Entomology,* **36**(3): 236.

Bezzi, M. 1916. On the fruit-flies of the genus *Dacus* (s.l.) occurring in India, Burma and Ceylon. *Bulletin Entomological Research,* 7(2): 99–121.

Bhagat, R.C. 1986. Aphid pests of fruit-trees and their natural enemies in Kashmir valley. *Indian Journal of Agricultural Science,* **56**(7): 532–534.

Bhargawa, M.C., Khatri, N.K. and Dhaka, R.S. 1990. Efficacy of different insecticides against scale insects, *Parlatoria blancharidii* Targ. of date palm, *Phoenix dactylifera* L. *National Symposium on Pests, Parasites and Environment,* Aligarh Muslim University, Aligarh, (Abstract).

Bindra, O.S. and Singh, J. 1971. Biological observations on the white-grub *Lachnosterna (Holotrichia) consanguinea* Blanchard (Coleoptera: Scarabaeidae) and the relative efficacy of different insecticides against the adults. *Indian Journal of Entomology,* **33**(2): 225–227.

Chandrasekaran, V. and Balasubramanian, M. 1972. Morphology, height-diameter relationship and reduction in leaf area of leaf galls on *Eugenia jambolana* Lam. and on *Pongamia glabra* Vent. *Madras Agrictural Journal,* **59**(8): 411–414.

Channabasavanna, G.P. 1966. *A Contribution to the Knowledge of Indian Eriophyid Mites*. University of Agricultural Sciences, Bangalore. pp. 153.

Chatterjee, H.C. 1914. A note on *Oxyrhachys tarandus* Fabr. *Indian Forester,* **40**(2): 75–79.

Chatterjee, S.N. 1960. Some insect pests of mulberry in Kalimpong, Darjeeling district, West Bengal. *Indian Journal of Entomology,* **22**(4): 258–260.

Dahiya, K.K., Lakra, R.K. and Ombir, 1995. Studies of population dynamics, and damage caused by jamun leafminer, *Acrocercops syngramma* Meyrick on jamun, *Syzygium cumini* Skeels. *Haryana Journal of Horticulture Sciences,* **24**(3&4): 193–196.

David, B.V., Vijayaraghavan, S. and Narayanaswamy, P.S. 1963. *Indarbela tetraonis* Moore—A new bark borer pest on curry leaf tree and its control. *Madras Agricultural Journal,* **50**: 195–199.

Dhooria, M.S. and Sandhu, G.S. 1973. Occurrence of *Tenuipalpus punicae* Pritchard & Baker (Tenuipalpidae: Acarina) on pomegranate in India. *Current Science,* **42**(5): 179–180.

Dubey, J.K., Nath, A. and Thakur, J.R. 1993. Chemical control of pomegranate fruit borer(s) *Virachola isocrates* (Fabr.) and *Deudorix epijarbas* (Moore). *Indian Forester,* **119**(11): 928–931.

Gangwar, V.S. 1983. Biology of *Phyllodiplosis jujubae* Grover infesting leaves of *Ziziphus jujuba* Lamb. (Cecidomyiidae: Diptera). *Cecidologia Internationale,* **4**(1/2/3): 57–73.

Ghule, B.D. and Dhumal, V.S. 1992. Chemical control of scale insects on pomegranate. *Journal of Maharashtra Agricultural University,* **17**(2): 322–323.

Glover, P.M. 1930. Entomological aspects of lac research in India. *Bulletin Entomological Research,* **21**(3): 261–266.

Gupta, B.P. and Singh, Y.P. 1988. New record of *Buzura* (=*Biston*) *suppressaria* (Guenee) (Geometridae: Lepidoptera) as a pest of mango and some other fruit trees. *Progressive Horticulture,* **20**(3–4): 362–364.

Jadhav, L.D., Kadam, M.V., Ajri, D.S. and Pokharkar, R.N. 1980. Observations on the biology of leaf footed plant bug *Leptoglossus membranaceus* Fabricius a pest of pomegranate. *Bulletin Entomology,* **21**(1/2): 79–82.

Joshi, H.C. and Shinde, V.K.R. 1971. Control of ber fruit fly, *Carpomyia vesuviana* Costa (Tephritidae: Diptera). *Indian Journal of Entomology,* **33**(2): 142–147.

Joshi, K.C. and Meshram, P.B. 1989. A new report of *Selepa celtis* Moore (Lepidoptera: Noctuidae) as a pest of babul (*Acacia nilotica*) and *aonla (Emblica officinalis*). *Indian Forester,* **115**(11): 847.

Kakar, K.L., Dogra, G.S. and Nath, A. 1987. Incidence and control of pomegranate fruit-borers, *Virachola isocrates* (Fabr.) and *Deudorix epijarbas* (Moore). *Indian Journal of Agricultural Science,* **57**(10): 749–752.

Kanta, S., Rai, B.K. and Lal, R. 1966. Comparative toxicity of insecticides to jamun leafminer, *Acrocercops phaeospora* (Lepidoptera: Gracillaridae). *Indian Journal of Entomology,* **27**(4): 393–397.

Kapur, A.P. 1962. The lac insect. *In: A Monograph on Lac*. Mukopadhyay and Muthana (eds). Indian Lac Research Institute, Namkum, Ranchi, Bihar, India. pp. 59–89.

Khan, K.M. and Ghai, S. 1974. White grubs and their control in India. *Pesticides,* **8**(12): 19–25.

Kher, Samir and Lakra, R.K. 1989. Bioecological studies on lac insect, *Kerria lacca* (Kerr) in Haryana. *Journal of Insect Science,* **2**(2): 124–128.

Kushwaha, K.S. and Bhardwaj, S.C. 1968. Biology and external morphology of forage pests. 1. Tussock caterpillars, *Euproctis* spp. (Lymantriidae: Lepidoptera). *Indian Journal of Agricultural Science,* **37**(2): 93–107.

Lakra, R.K. 1997a. Some important pests of fruit crops of arid regions and their management. *Proceedings of the Silver Jubilee National Symposium on Arid Horticulture*, 5–6 December, held at CCS Haryana Agricultural University, Hisar. pp. 140–171.

Lakra, R.K. 1997b. Insect pests of some under-exploited fruits and their management 1. Jamun (*Syzygium cumini* Skeels) A. Lepidopterous Pests. *Haryana Journal of Horticultural Science,* **26**(1&2): 35–48.

Lakra, R.K. and Dahiya, A.S. 1988. Biology of *Bucculatrix univoca* Meyr. A new leafminer pest of phalsa. *Indian Journal of Plant Protection,* **16**(2): 237–239.

Lakra, R.K. and Kharub, W.S. 1977. Fig leaf roller, *Phycodes minor* Moore (Glyphipterygidae: Lepidoptera)-A new pest of fig in Haryana and its control. *Haryana Journal of Horticultural Science,* **6**(1&2): 35–36.

Lakra, R.K. and Singh, Z. 1983. Oviposition behaviour of ber fruit fly, *Carpomyia vesuviana* Costa and relationship between its incidence and ruggedness in fruits in Haryana. *Indian Journal of Entomology,* **45**(1): 48–59.

Lakra, R.K., and Singh, Z. 1984. Calender of losses due to 'ber' fruit-fly, *Carpomyia vesuviana* Costa (Diptera: Tephritidae) in different *Zizyphus* spp. in Haryana. *Indian Journal of Entomology,* **46**(3): 261–269.

Lamba, K.S. 1949. On the anatomy of the fourth instar larva (prepupaphare) of the jamun whitefly, *Dialeurodes eugeniae* (Maskell) (Homoptera, Aleyrodidae). *Indian Journal of Entomology,* **11**(1): 25–46.

Latif, A. and Yunus, C.M. 1951. Food plants of citrus leaf-miner (*Phyllocnistis citrella* Stn.) in the Punjab. *Bulletin Entomological Research,* **42**(2): 311–316.

Mani, M. and Krishnamoorthy, A. 1990. Outbreak of mealybugs and record for their natural enemies on pomegranate. *Journal of Biological Control,* **4**(1): 61–62.

Mann, G.S. 1981. Chemical control of *Synclera univocalis* Walker on jujube at Ludhiana. *Indian Journal of Agricultural Science,* **51**(7): 517–520.

Mann, G.S. and Bindra, O.S. 1976. Evaluation of different jujube cultivars for resistance to the ber fruit-fly. *Punjab Horticulture Journal,* **16**(1&2): 64–67.

Mann, G.S. and Brar, J.S. 1980. Record of *Synclera* sp. nr. *traducalis* Led. and *Pleurota* sp. as a pest of jujube (*Zizyphus mauritiana* Lmk.) at Ludhiana (Punjab). *Indian Jpournal of Horticulture,* **37**(1&2): 106–107.

Mann, G.S. and Singh, Dharampal. 1981. Population studies of jujube leaf-webbers on *Zizyphus mauritiana* Lamk at Ludhiana (Punjab). *Entomon,* **6**(4): 351–355.

Menon, M.G.R., Ghai, S. and Katiyar, K.N. 1971. New records of Tenuipalpid mites from India. *Entomologists' Newsletter,* **1**(2): 13.

Mohanasundaram, M. 1994. Three new species of eriophyids from Tamil Nadu (Eriophyidae: Acari). *Entomon,* **19**(3/4): 95–99.

Mote, U.N., Tambe, A.B. and Patel, C.S. 1991. Observations on incidence and extent of damage of fruit sucking moths on pomegranate fruits. *Journal of Maharashtra Agricultural Universities,* **16**(3): 438–439.

Muralidharan, C.M., Sodagar, N.N., Ramdevputra, M.V. and Patel, P.K.1992. Date palm scale (*Parlatoria blanchardi*), a new host of black beetle (*Chilocorus nigritus*). *Indian Journal of Agricultural Sciences,* **62**(10): 720–721.

Noor, A., Kavadia, V.S. and Bhatnagar, V.S. 1978. Evaluation of various insecticides against citrus mealy bug *Nipaecoccus filamentosus* Cockerell. *Pesticides,* **12**(10): 23–23.

Pajni, H.R. and Rose, H.S. 1978. Revision of the genus *Pagyda* Walker for the revalidation of its synonym *Synclera* Lederer alongwith the description of a new species (Lepidoptera: Pyralidoidea: Pyraustidae). *Entomon,* **3**(2): 215–220.

Pandey, Y.D. 1967. Countering the castor semi-looper. *Indian Farming,* **17**(6): 23, 25.

Patel, G.A. and Kulkarni, H.L. 1954. *Cerciaphis emblica sp. nov*. (Fam: Aphididae)—A new aphid pest on *Emblica officinalis*. *Journal of the Bombay Natural History Society,* **1**(2): 435–438.

Patel, N.G. and Patel G.A. 1955. Host plants, distribution and abundance of thrips (Thysanoptera) of Bombay State. *Journal of the Bombay Natural History Society,* **51**: 597–607.

Patel, R.M., Sahay, G. and Patel, V.M. 1966. Biology of bark—eating caterpillar (*Indarbela quadrinotata* Wlk.) on guava in Gujarat. *Indian J. Hort.* **23**: 120–124.

Rahman, K.A. and Bhardwaj, N.K. 1937. The grape-vine thrips (*Rhipiphorothrips cruentatus* Hood) [Thripidae: Terebrantia: Thysanoptera]. *Indian Journal of Agricultural Science,* **7**(4): 633–651.

Ramakrishna Ayyar, T.V. 1919. Some South Indian coccids of economic importance. *Journal of the Bombay Natural History Society,* **26**(2): 621–628.

Ramakrishna Ayyar, T.V. 1940. *Handbook of Economic Entomology for South India*. Superintendent Govt. Press, Madras. pp. 516.

Rao, A.A., Teotia, R.S., Chauhan, S.S., Chakraborty, S. and Rao, G.S. 1993. Studies on the seasonal incidence of the mealybug *Maconellicoccus hirsutus* (Green) causing 'tukra' on mulberry in West Bengal. *Indian Journal of Sericulture,* **32**(1): 111–113.

Rao, A.S. 1958. Notes on Indian Aleurodidae (Whiteflies), with special reference to Hyderabad. *Proceedings of the 10th International Congress of Entomology* Vol. **1**: 331–336.

Roonwal, M.L. 1953. A phylogenetic analysis of the host-plants of the lac insect *Laccifer lacca* (Kerr) (Order Hemiptera, Family Coccidae), with a view to determine host preference. *Indian Journal of Entomology,* **15**(4): 283–292.

Sachan, J.N. 1984. Varietal susceptibility of ber fruits to the damage of fruitfly (*Carpomyia vesuviana* Costa). *Bullein Entomology*, **25**(2): 198–199.

Sadana, G.L., Chhabra, S.C. and Gupta, B.K. 1982. Tetranychoid mites infesting fruit trees in the Punjab together with records of some new hosts and description of a new species. *Entomon,* 7(3): 397–401.

Sah, B.N., Mehra, B.P. and Naqvi, A.H. 1972. Bionomics of *Euproctis fraterna* Moore (Lepidoptera: Lymantriidae) on *Ziziphus xylopyra* Willd. *Indian Journal of Agricultural Science,* **42**(7): 630–636.

Sandhu, G.S., Sohi, A.S. and Batra, R.C. 1987. Comparison of different control methods for the management of bark eating caterpillar, *Indarbela quadrinotata* Walker (Metarbelidae: Lepidoptera) infesting fruit and avenue trees in Punjab. *Punjab Horticulture Journal,* **27**(3&4): 255–259.

Saxena, D.K. and Rawat, R.R. 1968. A study on the incidence, varietal preference, and biology of the ber fruit fly in Madhya Pradesh. *Madras Agricultural Journal,* **55**(8): 328–333.

Sen-Sarma, P.K. and Gupta, B.K. 1979. Food preference in laboratory colonies of *Microcerotermes beesoni* Snyder. *Indian Journal of Entomology,* **41**(3): 283–285.

Shukla, R.P. and Prasad, V.G. 1983. Comparative efficacy of various treatments for controlling pomegranate fruit borer, *Virachola isocrates* (Fabricius). *Entomon,* **8**(4): 381–383.

Simwat, G.S. and Sidhu, A.S. 1974. Biology of fig leaf roller, *Phycodes minor* Moore (Glyphipterygidae: Lepidoptera). *Indian Journal of Entomology,* **36**(2): 149–150.

Singh, D. and Mann, G.S. 1982. Behavioural studies of the mobile forms of *Synclera univocalis* on jujube, *Zizyphus mauritiana. Phytoparasitica,* **10**(3): 201–204.

Singh, H. and Singh, B. 1979. Control of jamun leafminer, *Acrocercops phaeospora* Meyr. (Gracillaridae: Lepidoptera). *Indian Journal of Entomology,* **41**(3): 276–277.

Singh, M.P. 1984. Studies on the field resistance of different jujube cultivars to the fruit fly, *Carpomyia vesuviana* Costa. *Madras Agricultural Journal,* **71**(6): 413–415.

Singh, M.P. 1989. *Biosteres vandenboschi* Fullaway—a new braconid parasite of *Carpomyia vesuviana* Costa from the Indian desert. *Entomon,* **14**(1–2): 169.

Sontakay, K.R. 1945. The bark-eating borer of orange. *Indian Farming,* **6**(2): 74–75.

Srivastava, O.S. and Jain, A.C. 1973. Occurrence of *Virachola isocrates* Fabr. (anar caterpillar) (Lepidoptera: Lycaenidae) on *Emblica officinalis* (amla or aonla) in Madhya Pradesh. *Indian Journal of Entomology,* **35**(4): 352–353.

Tandon, P.L. 1994. Problems and prospects of insect pest management in fruit trees. *In:* G.S.Dhaliwal and R. Arora (eds.). *Trends in Agricultural Insect Pest Management* (eds.). Commonwealth Publishers, New Delhi. pp. 373–416.

Velavadi, V.V. and Shah, N.K. 1987. *Zeuzera* sp. (Lepidoptera: Zeuzeridae)—a new record on amla from South Andaman. *Journal of Andaman Science Association,* **3**(1): 56–58.

Zaka-ur-Rab, M. 1980. The cornelian, *Deudorix epijarbas* Moore (Lepidoptera: Lycaenidae) as a serious pest of pomegranate fruits in Kashmir. *Journal of Entomological research,* **4**(2): 233–235.

*Pest Management in Horticulture Crops*
L.R. Verma, A.K. Verma and D.C. Gautam (eds)


*Chapter 18*

# Pests of Potato

*V.K. Chandla and K.D. Verma*

## INTRODUCTION

POTATO IS A HIGHLY nutritious food. It produces more protein per hectare per day than wheat, maize or rice. The production of potato in India in 1995–1996 was 19.24 million tonn from 1.14 million ha with an average yield of 16.9 t/ha. in 1995–96. It is proposed that by year 2020, the total annual potato production in the country would be raised to 49 million tonnes. For this the area has to be doubled and average yield has to be increased to 24.5 t/ha. This could be achieved by the use of inputs- efficient high yielding cultivars, appropriate plant protection and agro-techniques and ensuring availability of quality seed. According to recent estimates worth 6,000 crores a year is lost due to pests damage every year which account for 10 to 20 percent of the annual produce. In fact, indications reveal that the annual demand of pesticides will rise from 80,000 tonnes to 1,44,000 tonnes by the turn of century. But trends indicate that heavy application of chemical pesticides will not increase productivity. In the year 2000, the amount of consumable agricultural produce will fall by a drastic 26.6 per cent. Most of the agriculturist including potato scientists feel that the solution to this big problem lies in a "back to basics" approach, i.e. turning to economically sound pest control techniques, such as biological agents and natural pesticides.

A great variety of insect pests attack potato crop in India. These pests damage and potato plants by feeding on leaves reducing photosynthetic area and efficiency by attacking the stems thus weakening the plant and inhibiting potato tubers destined for consumption or use as seed. Potato pests can be grouped into soil pests, foliage feeders or defoliators, sap feeders, storage and nematode pests. This chapter deals with their status and distribution, nature and extent of damage, population dynamics, and biology and management strategies.

## SOIL PESTS

## CUTWORMS

### Distribution, nature of damage, losses caused and biology

Cutworms are polyphagous insects of cosmopolitan distribution. Five species viz., *Agrotis ipsilon* (Hfn), *A segetum* Schiff., *A. flammatra* Schiff., *A. interacta* Wek. and *A. spinnifera* (Hb.) infest potato crop in India (Srivastava, 1958; Saxena, 1974). The first two species are widely distributed in plains and hilly regions, respectively. Their larvae damage the crop in its initial stages by cutting the young plants at their base and later by feeding on shoots and leaves. After tuberisation they feed on underground tubers and thus reduce their market value. In infested fields, tuber damage by cutworms has been reported to vary from 12–40% (Chaudhuri, 1953; Chandla, 1983). They also feed on barley, maize, mustard, linseed, cabbage, pea, gram, tobacco etc.

The females of cutworms deposit their eggs in cluster of 15 to 35 on the outside of potato leaves or on wet soil of potato fields. On an average a female lays about 300 eggs in its whole life. The egg, larval and pupal period range, 2 to 13 days, 12 to 35 days and 10 to 68 days, respectively. The total life cycle is completed in 30 to 68 days.

Moths are generally not found in North Indian plains during May to September but appear during October and remain active till March-April. Venkataraman (1954) indicated the possibility of their migration from plains to hills during summer months. Mass appearance of moths is of common sight during April in hills.

### Management

#### *Cultural and mechanical control*

Cutworms either go in the soil for aestivation during summer months or for hibernation in winters while completing their life cycles. As such, deep ploughing of fields during summer months in plains exposes the immature stages of these pests to high temperatures and predatory birds resulting in decrease of population to a great extent (Chandla,1985, Misra and Agarwal, 1988, Misra *et al.*,1995). Similarly, deep ploughing of potato fields during autumn in hill helps in minimizing the population of cutworms. Light traps placed in around potato fields attract the adults of cutworms. Hence these could be used as a tool in IPM programme for mass collection and destruction of individuals.

#### *Chemical control*

In the past, several persistent chlorinated hydrocarbons as pre-sowing treatments in soil, were being used for controlling the cutworms (Chaudhry, 1953, Nirula and Kumar, 1963). However most of these chemicals are banned now because of their toxic residue problems. The present trend is to use less persistent insecticides like endosulfan, chlorpyrifos, carbaryl, dicrotophos, phosphamidan etc., at concentrations ranging from 0.03 to 0.05 % (Chandla, *et al.,* 1977, Rizvi and Chandla, 1977).

***Biological control***

Fletcher (1916) had reported *Broscus punctatus* Dist and *Liogryllus bimaculatus* Linn parasitising the larvae of cutworms. Later, Chandla *et al.*, (1989) and Singh (1993) reported *Macrocentrus collaris* (Spin), *Netelia ocellaris* (Thomson) sp. complex, *Coelichneumon* sp. (nr) *Truncatulus* Thomson, *Periscepsia carbonaria* Panzer and *Turanogonia chinensis* Wiedemann parasitising *A. ipsilon* and A. *segetum* under natural field conditions. Misra *et al.,* (1995) also reported entomopathogen fungus, *Metarrhizium anisopliae* (Meld), pathogenic bacteria *Bacillus thuringiensis* and entomophilic nematode *Stinernema (Neoplectana)* sp. as a dominant regulatory factor for cutworm population in various parts of the country.

## WHITE GRUBS

### Distribution, nature of damage, losses and biology

White grubs are distributed throughout the country. The species infesting potato are: *Lachnosterna (Brahmina) coriacea* (Hope), L. *(B.) longipennis* (Blanch) and *Melolontha indica* Blanch. Damage to potato is caused by the grubs which feed on root lets, roots and tubers. The first stage grub can survive on the organic matter present in the soil but roots are preferred. The damage to potato tubers is mainly caused by the second and third instar grubs which eat by making large circular holes in the potato tubers. The severe tuber damage may occur before the grower comes to know about pest problem. Damage is most severe in years following peak adult flights. In Shimla hills, loss to potato crop was reported upto 85% (Misra and Chandla, 1989).

The larvae of white grubs feed on the roots of the cultivated crops.The adult feed upon the foliage and fruits of orchard trees. In the hills no crop is completely free from the attack of these grubs.

The biology of important specics, L. *(B.) coriacea* has been studied. The adult start emerging from soil immediately at dusk soon after good monsoon showers during June -July. The emerging beetles coud be seen settling on nearby bushes/trees such as *Acacia spp.* peach, plum, apricot, pear, and apple, primarily for mating and later on for feeding. The mated females re-enterthe soil for egg laying at dawn.The eggs are laid singly in 2–7 installments. A single beetle may lay 4 to 41 eggs in its whole life. The incubation period ranges from 7 to 12 days.There are three larval instars and full grown grubs measure about 35–38 mm in length. The larval period ranges from 98 to152 days. These pass their winter as grubs in earthen cells.The full grown grubs afterwards spend 5–7 days in earthen cells before entering into pupal stage during April-May. The pupal period varies from 12 to 20 days. The longevity of beetles ranges between 15 and 145 days.

All larval stages live in the soil from July to November and initially feed upon living plant roots ingesting at the same time quantities of soil and dead organic matter and later on tubers. On completion of development, larvae construct a cell in the soil in which they may undergo a prolonged resting (hibernation or a estivation) period before pupating (December to April-May). Adult emergence from the soil is in response to environmental conditions i.e. with the onset of first monsoon showers in June onwards (Chandla, 1986).

### Management

It is too difficult to control white grubs by adopting a single control measure. The management programme should be practised in the endemic localities on an area comparing basis by involving all the recommended control techniques viz., autumn ploughing, application of insecticide on the adult host plant, installation of light traps to monitor adult beetles, conservation of natural enemies and their field application.

### *Cultural and mechanical control*

Cultural techniques are helpful in reducing the grubs as well as adult population when adopted in a scientific way. In the endemic areas, autumn ploughing exposes the different stages of grubs which are picked up by the predatory birds. The use of nitrogenous fertilizers, especially ammonia and urea, at higher doses kill the first instar grubs. (*Misra et al.*, 1995)

The light trap may be used for collecting the adult beetles during nights. The beetles can also be collected by shaking or jerking their host plants. The fallen beetles should be collected and destroyed by putting them in kerosinized water or by burning them. The host trees should be chopped or prunned the branches of these host plants can be transplanted in the endemic area to attract the adult beetles these could then be picked up from the branches and destroyed. At the time of beetle emergence, spraying of host trees with carbaryl or fenitrothian or monocrotophos or endosulfan or Quinalphos results in reduction in the beetle population considerably(Yadav *et al.*, 1978)

### *Chemical control*

Insecticides act as important tool to get immediate relief from the white grub infestation. The chemical method should be followed in the endemic localities to bring the population below the economic threshold level. Effective control is possible if chemical controlis applied when the grub are tiny or young. The older grubs are very hardy to controlas they move to a great depth in the soil. Soil treatment with granules containing fensulfothion or quinalphos or phorate or carbofuran or isofenphos, is possible when half dose is applied at planting and half after 70 days later, coinciding with larval activity to control the pest significantly (Misra and Chandla, 1989).

### *Biological control*

A number of bioagents i.e. parasites, predators and pathogens, which attack white grubs in different parts of the country, have been reported by several workers. The population of these natural agents should be encouraged in the endemic localities either by conserving the existing population or by introducing, establishing and colonizing the new kinds of parasites, predators and pathogens. Some of the potential bio-agents are scoliids, *Campsomeris collaris*, F *Scolia aureipennis* Lepeletier and *S. pustulata* Magr. (Gupta and Awasthy, 1960, Singh, 1993) the pathogensare *Bacillus popillae* Dulky, *Metarrhizium anisopliae* (Metch.), *Beauveria brongniartii* (Sacc.) Petch, Isyr. *B. tenella* (Del.) and entomophilic nematode, *Steinernema* sp. (Singh, 1993).

## MINOR SOIL PESTS

Other minor soil pests recorded on potato crop are termites, viz., *Odontotermas obesus* Ramb and *Eremotermes* sp.; lunate fly, *Eumerus* sp.; red ant, *Dorylus oriental is* Wastwood; Wireworm, *Agroites* sp; grubs of Myllocerus weevil, *Myllocerus subfasciatus* (Guerin).

All these pests chew and make shallow to deep pits in tubers, there by reducing the grade of produce. Young injured tubers often become misshaped. Plant gets reduced if seed pieces and young shoots are attacked. Feeding sites are the entry points for other insects and diseases organism. For the management of these pests practices suggested for cutworms and white grubs will take care of these pests too.

## FOLIAGE FEEDERS OR DEFOILATING PESTS-

### DEFOLIATING CATERPILLARS

#### Distribution, nature of damage, losses and biology

Several leaf eating caterpillars such as the pea semilooper, *Plusia orichacea* (f.) and *P. ni* (Hb); lucerne caterpillars, *Spodoptera exigua* (Hb); gram pod borer, *Heliothis armigera* (Hb); Bihar hairy caterpillars, *Dicrisia obliqua* Walk; and brinjal shoot and fruit borer, *Leucinodes orbonal is* Guen have been reported feeding on potato foliage from different regions of the country. Of these, *H. armigera* and *P. orichalcea* are important species (Lall 1964; Saxena and Rizvi, 1974; Misra *et al.,* 1995).

*H. armigera* caterpillars feed on potato foliage. In case of heavy population load of caterpillars, the crop may suffer with great losses. It is widely distributed in the country and foind on red gram, bengal gram, groundnut, tomato, potato, tobacco, maize, hemp, linseed, safflower, cotton, peas, opium, lucerne, indigo, arhar, caster seeds, and numerous other hosts (Misra *et al.,*1995).

In plains, it migrates from gram to potato crop in spring season and feeds on its foliage in patches. In northern hills, its moths appear in very large numbers ornamental plants by the end of March. On hatching, caterpillars feeds on potato foliage and other alternate host plants (Chandla, 1986).

Female moths of *H.armigera* after mating lay thier eggs singly on the lower side of potato leaves in the fields. Initially the colour of eggs remain yellowish white which later on becomes darker at hatching. The incubation period ranges from 2 to 5 days. The newly hatched caterpillars are sluggish but become active within 3–4 hours. There are five larval instars of this pest and entire larval period ranges between 17 and 22 days. Prior to pupation, the full grown larvae suspend all thier activity and remain in pre-pupal stage for 1 to 2 days. Pupation takes place between the folded leaves or in the soil. The pupal period ranges from 6 to 10 days. The moths mate soon after emergence and lay 40 to 112 eggs in its life span of 4 to 6 days.

*P. orichalcia* cause severe foliage damage to potato crop in the summer in Meghalaya and the spring in Punjab. It is also a polyphagous insect having wide range of host plants including peas, cabbage, cauliflower, radish, turnip, carrot, potato, celery, indigo and linseed. The biology of *P. orichalcia* has been studied on other crops also. The mated female normally lays 350–450 eggs which hatch in 3 to 5 days. The larval and pupal periods occupy respectively 20 to 28 and 8 to 15 days.

#### Management

*Cultural and mechanical control*

Hand picking of various stages of defoliators and their destruction, if undertaken properly and in time, is the best way of controlling the defoliators on potato crop particularly when their population is manageable. Besides this, suitable crop rotation with non-host crops and clean cultivation are some other effective contıol components which could be adopted without much expenses (Misra *et al.*, 1995).

### *Chemical control*

Many chemicals viz., acephate, chlorfenvinphos, endosulfan, formothion, monocrotophos and methyl-parathion at 0.05% have been reported as effective against lepidopterous pests (Chandla and Rizvi, 1977; Chandla, 1985; Misra et al., 1995).

### *Biological control*

The polyphagous lepidoterous pests of potato have a long list of natural enemies on their key host plants, but not on potato. They appear on potato occassionally Some of the records made and attempts to utilize them are prestented here. Eleven species of parasitoids have been recorded in Banglore district from H.armigera infesting different crops including potato (Divakar and Pawar, 1979) The parasitoids includes Ichnemonids Campoletis chlorideae Uchida, Eriborus sp and Xanthopimia punctata Fabr., the Braconids Bracon hebetor Say, B. greeni Ashm, and Apanteles sp., the Bathylid Goniozus (= Parasierola) sp., the Trichogrammatid, Trichogramma chilonis, The Tachinids Carcelia illota Curq, Palexorista laxa Curr. And Goniophthalmus halli Mesnil and the Mermithid nematode Hexmerimis sp.Three Hymenopterons, Chelones narayani Gupta, Horogene fenestralis Opius and Ecphoropsis sp parasitosed the larvae of H.armigera in Bihar [Lall,1964]. Ecphoropsis sp has also been found parasitising the larvae of H. armigera in Punjab and Himachal Pradesh (Chandla,1986]. At Patna, a nuclear polyhyedrosis virus and Bracon sp from semilooper, P. orichalcea are commonly recorded in February- March (Singh, 1974). Bracon sp. nr. chinensis Szepl; Shirakia sachoenobii Vier. and Iphiaulax sp. have been reared on the larvae of L. orbonalis, Bracon sp.nr. chinensis has been found parasisting 35 percent of the larvae at Chickballapur, Bangalore (Ramachandran Nair, 1967). A large number of predators have been reported on *H. armigera* and *P. orichalcea*, such as Chrysopa sp. feeds on eggs; Micromius nilghiriousm (Novas), Hemerabiuin onontanus Kimmins, Henricohania spinosa Dist. and Sphedanolestes variabillis feed on the larvae in Shimla hills (Chandla, 1986).

## EPILACHNA BEETLES

### Distribution, nature of damage, losses and biology

*Henosepilachna ocellata* Red ten-bacher cause serious damage to the potato crop in the Northern Higher Hills and in the years of severe attack, the crop gets completely defoliated (Kapoor, 1950). The activity of other species, i.e. *H. vigintioctopunctata* (Fabricius) is restricted to plains and mid-hills. It has been reported to feed on a wide range of host plants including potato, tomato, brinjal, datura, and many *Solanum* spp. (Lall, 1964). Both the adults and grubs voraciously feed on leaves by scrapping the chlorophyll thus resulting in pre-mature destruction of foliage.

The biology of *E. ocellata* has been studied in depth. The beetles and grubs of this pest feed on potato foliage from May to September in highier hills and pass through two overlapping generations during the crop season. The female lays eggs in clusters of 5–57 on the underside of potato leaves and total number of eggs laid per female varies from 5 to 450. Freshly laid eggs are cigar shaped and yellow before hatching. The incubation period ranges from 4 to 5 days. There are four larval instars. The larval period ranges between 21–28 days. The full grown grubs stop feeding and spend 2–3 days in pre-pupal stage. Pupation normally takes place on exposed leaves. The pupal period range between 8 and12 days. The longevity of female beetles varies from 29 to 295 days while the male survive for 17 to 225 days (Chandla, 1986).

### Management

#### *Cultural and mechanical control*

The hand picking of epilachna beetles especially egg clusters and their destruction is the best way of controlling the epilachna beetles on potato crop. This practice should be adopted when its population is manageable (Misra *et al.,* 1995)

#### *Chemical control*

Several insecticides, like methyl-parathion, carbaryl, chlorpyrifos, diazinon, dimethoate, endosulfan, formothion, monocrotophos etc. of 0.05% concentration have been reported as effective against *Epilachna* beetles (Verma and Misra, 1975 and Misra *et al.,* 1995)

#### *Biological control*

An eulophid egg parasitoid, *Tetrastichus ovulorum* Ferriere, has been recorded from *H. vigintioctopunctata* around Bangalore and Ranchi (Krishnamurti, 1932; Lall, 1964). A Chalcid parasitoid, *Ugna menoni* Kerrich, has been recorded from *H. vigintioctopuncta* in Andhra Pradesh (Azam *et al.,* 1974) and from *H. ocellata* from Shimla hills. The parasitism of *H. vigintioctopunctata* in the field by *Pediobius foveolatus* Crawford reaches as high as 77 percent (Appanna, 1948) and subsequent report shows that, in untreated fields the parasitism reaches 70–74 percent in Feburary-March and 68 percent in July (Pattarudriah and Krishnamurti, 1954), but after 40 years of the first report, only 9 percent parasitism has been recorded (Trivedi, 1988). This may be due to indiscriminate use of pesticides. A number of predators have also been recorded. Two Reduviids, *Coranus* sp. and *Henricoclania* sp. nr. *Spinosa* Dist. and a Nabid *Aptus mussooriensis* (Dist.) have been found feeding on the grubs and adults of *E. ocellata* at Shimla (Saxena and Singh, 1977). Reduviid *Rhinocoris fusipes* in Karnataka and *Cantheconia (Eocanthecona) furcellata* (Wolf.) in Kerala on *E.vigintioctopunctata* (Shiela and Abraham, 1981) have also been reported.

## MINOR DEFOLIATING PESTS

There are many more minor or sporidic defoliating pests which have been recorded on potato crop from different parts of the country. Some of them are flea beetles, *Psylloides plana* Maulik, weevils, *Myllocerus subfasciatus* Guerin and *Alcidodes westermanni* (Bohemann); bluster beetles, *Epicauta hirticorris* Hagg; army worm, *Mythimna seperata* (Walker) and grass hopper locust, *Chrotogonus* sp., *Heiroglyphus banian* (Fabr.) and *Schistocerca gregaria* (Forsk.) (Chandla,1986; Misra *et al.,* 1995) All these pests damage the potato crop by feeding on the green foliage and tender shoots thus affecting the photosynthesis process in the potato plants. These pests can also be effectively managed on potato crop by adopting same IPM practices recommended for Epilachna beetles and defoliating caterpillers (Misra, 1993).

## SUCKING PESTS OR SAP FEEDERS

### *APHIDS*

**Distribution, nature of damage, losses and biology**

Out of the various insect pests, aphids constitute an important group infesting potato crop. Though the direct damage caused by them is marginal compared to most other pests but they are extremely important as vector of many potato viruses. Both the adults and nympus of aphids suck the sap from potato leaves. Heavily infested leaves curl downward, turn yellow and some times die. In addition to causing direct damage, two species viz., *Myzus presicae* (Sulzer) and *Aphis gossypii* Glover also transmit potato virus Y (PVY) and potato leaf roll virus (PLRV) (Nagaich et al., 1970 and Singh *et al.,* 1982). Other important species colonising on potato crop are, *Rhopalosiphum rufiabdominalis* Sasaki, *Aphid fabae* Scopoli and O*tysiphon* Davidson (Verma, 1993).

*M. persicae* feeds on several ornamental plants, weeds, fruit trees, the important one are peach, cherry, spinach, lettuce, beets, tomato, potato, egg plant, various cruciferous and cucurbitaceous plants and tobacco (Rizvi *et al.,*1976). Under inoculated condition the losses in crop yields due to PLRV can be 20–50% and by PVY 40–85% (Nagaich and Agrawal,1969).If tubers from infested crop are used as seed, the losses in subsequent crops can even increase considerably.

Realising the importance of aphids as vectors of viral diseases, the population dynamics of *M. persicae* was studied under different potato growing regions of the country.It starts appearing on potato crop in the North- Western plains of India from second week of November and in the Eastern plains in early December, reaching the critical level (20 aphids per 100 compound leaves) in the end of December in former area and between 1–2nd week of January in the latter area. In the North-westren higher hills, i.e. Kashmir to Kumaon and North-Eastern hills the critical level is generally reached by the end of July. In the plateau region of Maharashtra aphid infestation remains almost negligible in Kharif crop while in Rabi crop it appears in the second week of November and critical level is crossed by the third week of December.

This pest has a complicated life-cycle due to different morphs and methods of reproduction. These morphs include, winged, wingless parthenogenetic forms, oviparous males, males etc. Reproduction is both by asexual and sexual methods.In case of asexual reproduction winged viviparous females starts appearing on potato crop and other secondary hosts from the middle of November onwards in the Indo-Gangetic plains. These forms reproduce parthenogenetically. and gives birth to living young ones which pass through four nymphal instars. Each instar takes about 1 to 5 days and one generation is completed in about a fortnight. This asexual reproduction continues for many generations on the secondary hosts including potato. The parthenogenetically produced aphids reach to their peak in Februaryand populations starts declining from March onwards. In April/May these migrate to the hills where they start their reproduction on secondary hosts including potato. In addition to asexual reproduction in the Northern plains during December-February some sexually developed adults fly to primary host (peach) and produce oviparous females. The male also fly to these trees and start mating. Mated females lays eggs in the small crevices of axillary buds of peach trees. Which hatch in spring season.(Verma, 1985, 1993; Ghosh and Verma,1990; Verma and Chauhan 1993a).

### Management

The seed accounts almost 50% of total cost of cultivation in potatoes and this could be curtailed by practising integrated approach for managing the aphid vectors.

#### *Cultural and Mechanical control*

Planting of clean seed free from tuber-borne infection is the most important step in the control of disease. Potatoes being grown from vegetative parts, there is a direct relationship between the buildup of disease and the years of continuous cultivation. The percentage of 'seed' infected must be kept at a minimum to satisfy commercial requirement of epidemics. Seed plot technique ensure the low infection rate of seed crop.

It is best to grow the crop in periods in which vectors are absent or when their population density is low. The useful information on the (I) time of first arrival of *M. persicae* on potato crop, (ii) time of approaching or crossing the critical level of 20 aphids/100 compound leaves, (iii) peak activity period and (iv) disappearance from the crop has been worked out (Verma, 1993)

The intensity of virus spread is affected not only by the number of vectors, but also by their distance from the target crop. It is, therefore, advisible to keep the potato fields and surroundings free from alternate or collateral weed/crops. Such plants, if available in the vicinity, the aphid will feed, multiply quickly under favourable environmental conditions and would migrate to seed crops being grown in neighbourhood.

#### *Chemical control*

Several systemic insecticides in their granular/emulsifiable concentrate formulations are being used effectively for managing aphid vectors on potato crops. In higher hills (rainfed crop) only foliar systemic insecticides such as demeton-methyl, dimethoate, thiometon or monocrotophos etc at 0.03% concentration @ 1000–1250 of spray/ha/spray, depending upon the age/canopy of the crop, are recommended for controlling aphids on seed crop (Misra and Chandla, 1979; Misra, 1993 and Misra *et al.*,1995). In the plains, under irrigated conditions, application of granular systemic insecticides phorate 10G or carbofuran 3G @ 1.0–1.5kg a.i./ha has been found effective for 45–65 days. Granular insecticides applied at planting will not be sufficient to protect the crop throughout the season. It is, therefore, suggested that 1–2 need based sprays with any of the earlier mentioned foliar systemic insecticides at 10–15 day intervals should be given (Verma *et al.*, 1976 and Misra *et al.*,1995)

#### *Biological control*

Natural enemies of potato aphids have been reviewed by Singh (1988). *M persicae* is attacked by over 24 predators, 22 parasitoids and 5 entomopathogenic fungi, but only few of these, viz., predators *Allograpta favana* (Wiedemnn) (Pruthi, 1946); *Episyiephus balteatus* (De Gear) (Ray-Chaudhuri et al., 1979; Ghosh *et al.,* 1985); *Ischiodan scutellaris* (Fabricius), *S. scripta* (Linneus) (Ghosh *et al.,* 1985), *Sphaerophoria indiana* Bigot (Pruthi, 1946); *Leucopis fumidilarva* Tanas (Saxena, 1981); *Menochilus sexmaculatus* (Fabricus) and parasitoids, *Aphelinus* sp. (Verma *et al.,* 1976) and *Aphidius colemani* Viereck (Trivedi, 1988) have been recorded attacking *M. persicae* on potato crop. On *A. gossypii*, over 36 predators and 21 parasitoids have been recorded

(Singh, 1988). In Shimla, *Aphelinus sp* has shown its maximum capability by parasitising 100 per cent of *M.persicae* alone in glass houses (Verma *et al.*,1976). Similarly, in Karnataka, A. *colemani* parasitized up to 70 per cent individuals of this species in certain potato fields (Trivedi, 1988).Seven species of coccinellid predators, two species of syrphid and chrysopiid were recorded on potato aphids around Farrukhabad, Agra and Meerut districts of Uttar Pradesh, of which *Coccinella septempuncata* Linn and *M. sexmaculatus* Fabr. were the ten species of coccinellids, species of chrysopiids and species of predacious bugs were recorded feeding on aphids in Shimla during May-October (Chandla, 1986).It is observed that the polyphagous predators control aphids on a number of crops including potato. Some systematic studies are necessary on all India basis on all the major predators, parasitoids and pathogens for using these in potato pest management programme.

## LEAF HOPPERS

### Distribution;nature of damage, losses and biology

*Alebroides nigroscutulatus* Dist and *Seriana equata* Singh are reported as the vectors of yellows viz., Purple Top Roll (PTR) and Marginal Flavescence (MF) in potato crop. Besides several species viz., *Empoasca devastans* Dist, *Balclutha hebe* kirkadly and *Amrasca biguttula* Ishida cause direct damage to crop. Hopper burn is a physiological condition resulting from potato leafhopper feeding. Both nymphs and adult feed by sucking the sap from the leaves and stems of potato plants, and while feeding, they introduce into the potato a toxin that causes hopper burn. The damage is first evident as a small brown triangular spot at the leaf top. The margin of the leaf turns brown and rolls inward. Severe infestation affects all leaves, through older leaves are affected first (Prasad, 1961). Potato leafhopper feeding increases plant respiration, depleting reserves available for growth and tuber development. More importantly, there is considerable loss of photosynthetic tissue and of vascular elements. Early planted crop in the plains suffers severe losses due to hopper burn (Misra *et al.*, 1995, Verma *et al.,* 1993b).

They feed on large number weeds growing in or around potato fields, including potato. In hilly regions they feed on potato plants throughout the crop season i.e. May- September, while in Northern plains they cause serious damage to early planted potato crop.

The females of *A.nigroscutulatus* deposit white and elongated eggs singly inside the lamina of leaves just below the epidermis. With the progress of embryonic development, colour of eggs changes to yellow and finally to dark yellow before hatching. The incubation period ranges from 6 to 7 days. There are five nymphal instars and the total nymphal period is of 18 to 28 days. The longivety of females ranges from 10 and 17 days, whereas of males ranges from 3 to 7 days. On an average single female lays 20 eggs in its life time. Under field conditions, 4 to 5 overlapping generations could be seen during May-August at/around Shimla. After harvest of potato crop, the adult migrate on alternate host like dahalia, wild potato, mulberry and calendula plants and survive throughout the winter on them (Chandla, 1986).

### Management

Leafhoppers do not have any effective natural enemies. So chemical control should beapplied. In areas where insecticides are used to control aphids, the leaf hoppers are usually not a problem.

Healthy, vigorously growing plants with optimum soil moisture withstand damage better than stressed plants. Potato leaf hoppers populations build on certain crops and after harvesting or maturity of those crops leave to colonize on potato. Therefore, potatoes should not be planted next to cotton or other host crop.

## BROAD MITE

### Distribution, nature of damage and losses

The broad mite, *Polyphagotarsonemus* Banks, is a polyphagous mite found throughout the country on potato crop. Its damage is very high on kharif potato crop in Maharashtra Karnataka and Himachal Pradesh and on early potato crop in Uttar Pradesh, Haryana and Punjab. Due to mite infestation, the leaves of infested plants quickly wither and dry up. This malady is locally called " Tambra" in Maharashtra. In heavily infested fields, the mite infestation may result in about 60 per cent loss in yield (Saxena and Rizvi, 1974; Channa Basavanna and Puttarudrish, 1959, and Raj *et al.,* 1979).

On potato, its population remains high in the early planted crop in northern plains during September-October months. In Maharashtra and Karnataka higher infestation in kharif crop while in hills high population the main potato crop occurs during May-June months.

### Management

Mites can be effectively controlled by spraying the crop with Kelthane or tranid or micronized wettable sulphur @ 2.0litre, 0.5 litre and 2.5 kg per hectare, respectively. Endosulfan, quinalphos, carbaryl and Mancozeb are also reported to be effective against mites @ 1.0 litre, 0.5 litre, 2,0 litre, 2.0 kg and 1.5 kg per hactare respectively. Besides chemical control, suitable crop rotations, clean cultivation and not growing of alternate or collateral host plants of this mite in the vicinity of potato crop are some of the other effective ways of its management (Misra, 1993).

### Other minor sucking pests or sap feeders

Some of the minor sucking pest of sporadic in nature infesting potato crop are green bug, *Nezara viridula* Linn, whiteflies, *Bemasia, tabaci* Geendiius and *Trialeurodes vaporariorum* (westwood); thrips, *Thrips* Spp. and bulb mite, *Rhizoglyphus echinopus* F and R. The damage is caused to potato crop by their nymphs/immature stages and adults by sucking the sap from the tender shoots and thus affects plant vigour (Misra *et al.*, 1995).

# STORAGE PEST

## POTATO TUBER MOTH (PTM)

### Distribution, nature of damage, losses and biology

*Phthorimaea operculella* (Zeller) (Lepidoptera Gelechiidae) a native of South America, is the most obnoxious insect pest of potatoes both in fields and country stores. In areas where the pest remains unchecked, losses in country stores may go up to 70%. Inadequate cold storage facilities, practice of taking more than one crop of potatoes in a year, non-adoption of recommended control-measures and prolonged dry and hot weather prevaiing during last 3–4 years in different regions of the country, have favoured PTM to became key pest of potato crop. It was first introduced into India in 1906 alongwith the seed potatoes imported from Italy (Lefroy, 1907). It was afterwards reported from Ajamgarh, Uttar Prasad, Coimbatore (Tamil Nadu), Dharwad (Karanataka), Pune (Maharashtra) and Purnea (Bihar) Fletcher (1919). By the nineteen fifties, PTM spread in most of the potato growing states of country (Lal, 1993).The larvae (caterpillars) of PTM damage the potato plants by feeding on leaves and hence the photosynthetic area and efficiency is reduced drastically. The larvae also attack the stems which results in the weakening of plants and inhibition of nutrients flow/movements. In addition, the larvae also attack on the tubers (potatoes) exposed in the fields and those stored in ordinary/country stores. When such infested tubers are examined by cutting them, the tunnels packed with black excretory pellets alongwith larvae could be seen easily inside such infested tubers. Sampling techniques developed by Trivedi (1990) have indicated that the losses to potato foliage in the fields and to tubers stored in country stores are to the tune of 26 and 56 per cent respectively.

PTM is primarily a pest of potato but its larvae also feeds on the leaves of egg plant, tobacco, tomato, sugarbeet and *Solanaceous* plants. Under favourable weather conditions 8–13 generations in a year are completed. Egg laying is inhibited and moths die at 36°C and no emergence takes place at temperature below 13°C and above 40°C.

The female moths lay their eggs on the underside of leaves or on exposed tubers, near the eyes. On an average, a single female lays about 150 eggs. The young larvae are grey- yellowish white in colour with brown head. The full grown larvae come out of tubers/foliage for pupation. They pupate in greyish silken cocoon and cover themselves with soil for comouflage. Pupation occurs either in dried leaves, soils over stored potatoes or in the cracks and crevices of country stores. The egg, larval and pupal periods range from 3 to 7, 10 to 25 and 5 to 9 days, respectively depending upon the environmental temperatures. The duration of one complete life cycle varies from 20 to 30 days at optimum temperature i.e. 20°C to 25°C. As many as 8 to 9 generations in northern plains, 10–13 generations in peninsular India and 11 generation in North eastern hill region of India are reported to occur in a year (Chandla,1986).

## Management

Studies carried out in the country and abroad have clearly indicated that no single control measure can provide adequate protection to potatoes from PTM attack, especially when the population of the pest is high, because the infestation initially starts from the field and later on reaches to the stores. As such, integrated management is the only solution for suppressing the PMT population and minimising its incidence Raman and Booth (1983) have also advocated the integrated management programme (IPM) for protecting the potatoes from PTM attack.

### *Cultural and mechanical control*

The cultural practices described below could be utilised successfully for reducing the PTM population and its damage on potatoes in the fields.

*Depth of planting:* Tuber damage due to PTM attack could be reduced to great extent by deep planting. Planting of potatoes at a depth of 10cm reduces the PTM infestation when compared with the traditional planting depth of 6cm (Akhade etal.,1970). It is further recommended that the field should be ridged after 6 to 7 weeks of planting so that the tubers are burried at least 25cm below the soil surface. In areas where PTM population remains quite high ridging should be done twice so that the tubers are not exposed for PTM infestation.

*Irrigation:* Timely and adequate irrigations minimise soil cracking and thereby reduce the exposure of PTM tubers for infestation. Hence, if timely and adequate irrigations are not given to the crop, it will cause compaction and cracking of soil. These problems are common in the areas where potatoes are grown in heavy soils. Such cracks expose the tubers for PTM attack. It is, therefore, essential to apply timely irrigations of required level for maintaining the good soil tilth and mositure contents in the fields necessary for avoiding the cracking of soil and exposure of tubers for egg laying by PTM moths.

*Sanitation:* Harvested potatoes must be removed from the fields as early as possible and in any case, the harvested potatoes should not be kept for overnight in the fields. Left over tubers, after harvest, should positively be collected and used either for table purposes or destroyed as these left over tubers rough plants and acts as source of infestation. Besides, the plants debris should also be collected and destroyed. The weeds belonging to family '*Solanaceae*' should immediately be destroyed because they act as reservoirs for PTM populations. Further, the crops like tomato, tobacco, chillies and brinjal etc. should also not be grown in the vicinity of potato fields. These precautions are more important for PTM prone areas (Lal, 1993).

### *Host plant resistance*

As control tactics, varietal resistance remains the first line of defence in the control of insect pests. Although immunity to PTM has not been identified so far in potato cultivars but various levels of genetic resistance have been reported in potato cultivars.

### *Sex pheromones*

Presently PTM sex pheromones consisting of trans 4, Cis-7-tridicadien-1-01 acetate (PTM-1) and trans-4, cis-1-cis-10-tridectrien-1-01 acetate (PTM-2) (0.4mg of PTM 1 + 0.6mg of PTM-2) are commercially available with CIPA-LIMA and Laboratory for Research on Insecticides, Marijkeweg 22,6700 PG, Wageningen, the Netherlands. These sex pheromones are mainly used for (I) monitoring and detecting the first occurrence of PTM, (ii) facilitating the proper use of insecticides where ever necessary and (iii) mass trapping in direct control of PTM (Saxena *et al.,* 1984, and Chandla *et al.,*1987). For using the sex pheromones effectively either of the traps viz., funnel, water and sticky trap could be used.For monitoring purposes, either of the trap designs could be used by placing two traps/field at a distance of 50 meters apart within the crop. Normally, the traps should be checked at 7 day intervals but when the PTM population is high, checking should be done at alternate days instead of weekly intervals. The number of traps to be used for direct control of PTM depends on the population density of the moths. For example in Lima (Peru) as and when PTM population reaches 2000 moths/trap/day, it becomes essential to use 42 traps/ha (Raman and Booth, 1983).

*Biological control*

Attempts made on the biological control of PTM in the country were sporadic and had given variable results in term of controlling this pest in fields and country stores. However, some of these are discussed below:

(i) *Parasites*: In Karanataka, indigenous parasites viz., *Chelonus curvimaculatus, Bracon gelechiae, Apanteles* sp., *Pristomerus vulnerator* and other Braconid were parasites in abundance and had registered 4–17% parasitism under field conditions during 1965–67 (Nair and Rao, 1972). *Melanips* sp. and *Diadgma molliplum* could give only 2.5 to 5.0% parasitism in hilly tracts at Shimla (Saxena and Raj,1980). On the other hand, field releases of *Bracon hebator* gave 12% parasitism to PTM larvae under field conditions in Bangalore (Divakar and Pawar, 1979).

(ii) *Parasidoids:* A number of exotic parasitoids were also introduced into India between 1970 and 1980. Of these, *Copidosoma koehleri*, an exotic egg/larval parasite gave 28.4–60.8% parasitism in Maharashtra (Dalaya and Patil,1973). Besides, *Orgilus jennieae, Apanteles subandinus, Chelonus blackburni* and *C.kellieae,* which are the ovo-larval parasities, were also tested for their potency in suppressing the PTM population at Rajgurunagar district of Pune (Maharashtra) for crop seasons. Of these, *C. jennieae* and *A. subandinus* registered 59.6 and 16.6% parasitism, respectively. (Chaudhary *et al.,* 1983)

(iii) *Pathogens: Bacillus thuringiensis* (Berliner) a gram positive, flagellated, rod shaped bacterium, which produces a crystal during sporulation, has been reported to be effective in controlling PTM in Peru, Tunisia, and India (Amonkar *et al.,* 1979; Chandla *et al.,* 1993: Lal, 1987). Recently, a baculovirus of the PTM has also been in collected in Peru by the CIP. The diseased larvae, in this case, were opaque, milky white. Such diseased larvae do not move violently when disturbed and also fail to pupate. This virus has been reported quite effective in suppressing the PTM population and its damage both in fields and stores. Long-term studies have indicated that this virus remains effective upto a period of 120 days under storage condition (Chandla *et al.,* 1993, Lal, 1993).

(iv) *Plant Products having repellent properties*: For last so many centuries, the farmers of the country have been protecting their crops in fields and produce in the stores with the help of natural insect-repellents found in the leaves, stems, flowers and fruits of various plants. Several experiments have been carried out, both in the country and abroad, for determining the repellent effect in locally available plants/plant products against the PTM for protecting the potatoes from its attack. Raman and Booth (1983) have reported that three plant species viz., *Lantana camara*, *Eucalyptus globulus and Minthostachys* sp. both in dried shredded leaves and powder forms, were effective in preventing PTM damage to potatoes stored for four months in ordinary/country stores. In India, Lal (1987) has reported that out of the several plant species tested for their repellency against PTM, *Lantana aculeata* provided the best protection to the control to below the 5%. *Lantana aculeate* provided the best protection to the tubers by reducing PTM damage which was more than 70% in the control to below the 5%. Likewise the sprout damage was also reduced from 45% in control to below the 3%.

Chopped and dried leaves in 2cm thick layers are pread below and above the stored potatoes. *Lantana* leaves can be collected fresh and air dried at room temperatures. These dried leaves should be crushed either by hand or by placing it in a cloth/hesian bag and beating with stick. Afterwards spread the crushed leaves in a thin layer of about 2cm thick over and below the stored tubers for physical barrier and for its repellent action against PTM. These layers protect the potatoes stored in ordinary stores from PTM firstly by acting as physical barriers and secondly through repellent action.

***Chemical control***

Chemical control of PTM under field condition with the help of insecticides is still considered to be of paramount importance and as such it is very common in practice. Spraying the crop with chlorfenvinphos (0.4 kg a.i./ha) acephate (0.5 kg a.i:/ha), quinalphos (.375kg a.i./ha), methamidophos (0.9 kg a.i./ha), phosalone (0.525 kg a.i./ha)and monocrotophos (0.6kg a.i./ha). have been found to be effective against PTM in fields (Raj *et al.,* 1986). In country stores, the potatoes meant for seed purposes can also be saved from tuber moth's attack by dusting them either with fenvalerate 2% or malathion 5% or quinalphos 1.5% dusts @ 125 g dust/100kg of potatoes. Besides, the dipping of seed potatoes in insecticidal solution containing 0.1% Carbaryl 50 WP or 0.05% fenitrothion 50EC or 0.025% decamethrin for 2 minutes. The treated (dipped) potatoes must be dried in shades before storing them in stores (Saxena *et al.,* 1984 and Lal, 1993).

*Important:* The potatoes meant for table purposes should not be treated, in any case, with any insecticides because these insecticides may cause serious health hazards to the cosumers.

Under situations where PTM population is low, the adoption of indivisual control measures described above may be effective even if used alone. However, in areas where the pest population is very high, the different control components should be integrated together for evolving economically effective IPM programmes against this pest because single control components under these conditions do not serve the purpose of management of PTM, both in fields and country stores.

## NEMATODE PESTS

Phytonematodes are known to damage potatoes for more than a century. Now, about 140 species belonging to 45 genera are known to be associated with the crop and causing 12% loss in global potato yield. In our country, nearly 90 different species of plant parasitic nematodes are reported to be associated with potato crop. Among these, potato cyst nematodes (PCN) and root-knot nematodes (RKN) are considered to be the problems in potato, not only in our country but elsewhere also (Gill, 1974).

## POTATO CYST NEMATODES

### Distribution, nature of damage, losses and biology

***Globodera pallida (Stone) Mulvey and Stone and Globodera rostochiensis (Wollenweber) Mulvey and Stone***

Infestation of this nematode on potato was observed in 1881 in erstwhile West Germany by Julius Kahn who recorded the occurrence of the cyst nematodes, now popularly known as Golden cyst nematode of potato. There are two species i.e. *Globodera pallida* (Stone) Mulvey and Stone, and *G. rostochiensis* (Wollenweber) Mulvey and Stone which infest the potatoes. The former is dominant and able to adapt itself under subtropical climatic conditions although basically it is a pest of temperate region.

In western countries, nematological research started only after the discovery of the golden nematodes from potato field in Long Island of New York, USA in 1941 necessitating a quarantine enforcement regulations.

Dr. F.G.W. Jones, then the Head of Nematology, Harpenden, UK, on visit to India, observed for the first time cyst of the golden nematode of potato from Nilgiri hills (Jones, 1961). In view of the potentially serious nature of the cyst nematodes, Govt. of India applied a provision of the Destructive Insect pest Act in 1971 and banned the movement of the seed potatoes from Tamil Nadu to other parts of the country.

Preliminary survey in Tamil Nadu (T N) conducted in 1962 revealed that PCN was distributed in many fields around Ootacamund, though its infestation level was low (Seshadri and Sivakumar, 1962) An organized survey in T.N. hills i.e. Nilgiris, Kodaikanal and Karnataka state was undertaken. Data collected from different locations revealed that both the species i.e. *G. pallida* and *G. rostochiensis* were widely distributed on potato in T.N. hills. In locations like Mynali, and Thumanhatti, the population of *G. pallida* is dominant while in places like Adasholi and Yellenhalli, *G. rostochiensis* is predominant. Similarly in Kodaikanal hills, both the species of PCN are found in mixed population in all the potato growing belts (Singh and Kishore 1982, Singh 1983). Studies conducted at Masanagudi in the foot-hills of Nilgiris indicated that the cyst nematode could also develop at lower altitudes of 900—1000 mt. amsl where the climatic conditions of the substropics prevail. Potato growing districts of Karanataka like Bangalore, Belgaum, Chickmagalur, Dharwar, Hasan, and kolar were surveyed to see if PCN was present. The survey report indicated that presence of PCN ranged between 1 and 26 cysts/100 ml soil (Prasad and Singh, 1986, Trivedi *et al.*, 1990). However, the viability test conducted at Ootacamund revealed that the cysts collected from these areas (Karnataka State) were non viable. The occurrence is of serious nature since potato cultivations there in Karnataka is done in rainfed as well as the irrigated areas at an altitudes ranging between 600 and 1000 mt. ams1. Report of occurrence of PCN is Kerala points to the need of stricts quarantine measures to avoid further spread of these nematodes. (Raman and Mohandas, 1988).

PCN infestation appears in patches in the initial stages and then gradually spreads to the entire field. The nematodes normally spread through irrigation channel, wind, agricultural implements, humus, and seed movement from infested fields. Infested plants lose vigour and become sick in appearance with stunted growth, dull appearance and unhealthy foliage. Plants show tufting of leaves on the top as the other leaves turn yellow and die. Root system is smothered, secondary roots are induced at the collar region and the plants can be easily pulled out. Tubers formed are less in number and are reduced in size. In extreme cases, tuber formation remains arrested. Nematode infested plants wilt during the day when sun is bright and take more water per unit weight than healthy ones. Field experiments conducted from 1982–86 during summer and autumn season to estimate the avoidable tuber yield loss in potato cultivar "Kufri Jyoti". Which was estimated to be 99.93 and 99.68% in autumn and summer season, respectively (Prasad,1993).

The life cycle and development of cyst nematodes are same like root-knot nematode except that the root and tubers do not form galls. The adult females of nematodes are round in shape with protruding neck which they keep remain attached to the potato roots. The females are 0.7 to 0.8 mm. in diameter. The young females are white in colour and then change to cream in *G.pallida* and golden yellow in colour *G.rostochiensis* and finally both the species change to form coloured cysts. These cysts are easily dislodged at the times of harvest and lie free in the soil. The dead females called cyst, contains 200–600 eggs. These nematodes hibernate inside the cysts in the form of egg and larval stages and can survive for 10–15 years in unfavourable weather conditions.

The root exudates stimulate the eggs inside the cyst to hatch out. Larvae move actively in soil in search of food, invade the roots and lie parallel to the vascular system. They undergo four successive moulths increasing in size each time to reach round shaped adult female. Males have no other role except mating. PCN takes 31 days to complete its life-cycle on susceptible cultivar "Kufri Jyoti". Multiplication of this nematodes on potato partly depends on initial population density as there is competition for root-space. Dormant period after browning of cysts is found to be 75 to 90 days during April-September.in *G. pallida* while in *G. rostochiensis* it is 45–60 days. A conservative estimate shows that the PCNs are distributed in about 6000 ha in T.N. (Prasad, 1993).

**Integrated Management**

As *G. pallida* and *G. rostochiensis* spread through the soil adhered on to tubers, farm machinery, tools, wind, water, and seedlings raised in infested soil, it is essential to follow the preventive as well as prophylactic measures. Some of the measures practised and recommended are mentioned below:

- Deep summer ploughing at 10–15 days intervals to expose the cysts to weather factor. Preplant application of carbofuran 3G @ 3kg a.i./ha.
- Use of healthy, truthful, certified and cysts free seed potatoes for raising potato crop.
- It has been observed that PCN spreads through infested soil adhered to tubers, irrigation water, nursery seedlings, farm machinery and tools, caution should be taken to avoid these things.
- Crop rotation of 7 to 8 years with non-hosts like plantation crops, wheat, maize, beans, peas, cabbage, cauliflower etc. helps in suppressing the PCN population.
- Raise potato crop by adopting resistant cultivar "Kufri Swarna".
- Potato grown in infested soil should not be used as seed.
- Potato, tomato and brinjal should not be grown for 3–4 years nor their volunteer seedlings be allowed to grow in infested fields as the host range of *G. pallida* and *G. rostochiensis* is restricted to solanaceae.
- Clean cultivation and frequent removal of volunteer plants and unharvested tubers should be strictly followed.

## ROOT-KNOT NEMATODE (RKN)

**Distribution, nature of damage, losses and biology**

The root-knot nematode causing typical root galls is the most widely distributed plant parasitic nematode of crop plants. The most commonly occurring species that infect potato are *Meloidogyne incognita, M. javanica, M hapla, and M. arenaria*. In India, Thirumalachar (1951) recorded the root-knot nematode infestion on potato tubers from shimla hills.Root-knot nematodes are almost thoroughly spread in the world. The important species on a world wide basis are *M. incoginita M. javanica* and *M. arenaria* distributed in the warm localities while *M. hapla* occurs in cooler zones. Similarly in India, the dominant species is *M. incognita*. Root-knot nematodes have wide host range including cereals, pulses, oilseeds, vegetables, ornamental plants, horticultural and cash crops. Most of the weeds found in potato fields also serve as alternate host. The sedentary female deposits about 300–400 eggs into a gelatinous matrix which is usually found adhering to the root galls. The eggs hatch readily and at the time of tuber formation, the juveniles usually enter tubers the root system starts decaying (Prasad, 1993). At Shimla and other North-western hilly regions, *M. incognita* completes its life cycle in 25–30 days during summer (April -September), whereas in winter (October-March) it takes about 65–100 days. In Bihar plateau region, it takes 28–35 days in main season (June-July) and 50–56 days in winter (November-March) to complete one generation. In the hilly regions, two generations of this nematode are completed by the time tuber development takes place hence tuber infestation is invariably observed even in low infested fields. The eggs hatch in to first stage juveniles that remain in eggs. The second stage juveniles emerge from eggs and they are the infective stage in the whole life cycle. The second stage juveniles penetrate roots at root elongation and root tip region, establish themselves there and continue to feed on the metabolically active cells there. They complete 3rd instar and then become swollen females. Under favourable conditions, there is high rate of parthenogenesis but amphimictic reproduction also takes place under fovourable conditions. There is a sex reversal mechanism by which the females turn into males.

Nematodes of *Meloidogyne* genus usually induce gall on host roots. Root galls contain certain pearly white swollen female adults. Eggs hatch without requiring root exudates or diffusates. Second stage juveniles invade roots at close to the tip and migrate to the differentiating vascular tissues. Symptoms are not recognizable on the aerial parts of the plants but heavily infected plants are stunted with yellow leaves. The typical galls on roots often go unnoticed. Wart like structures are found on tubers (Thirumalachar, 1951). The tissues under wart like pimples are glossy and show slight necrosis. *Meloidogyne* spp are obligate sedentary parasites of higher plants, generally in the roots but occasionally on the aerial parts of the stem and leaves. Root-knot nematodes have also got their role in spread and development of bacterial wilt and early blight (Bombawala *et al.*, 1982).

**Integrated management**

***Breeding for resistance***

The work on resistance breeding started at CPRI in 1961. Screening of accessions of tuber bearing wild *Solanum* indicated that a high degree of resistance was available in *S. spegazinii* which were used as potential material for producing several crosses. Low infestation was observed in several lines of *S. tuberosum andigena* during testing of commercial varieties and advance cultures. Many resistant resources are available in several tuber bearing *Solanum* spp and these could be utilized in the breeding programme.

### *Cultural control*

Effect of crop rotation with four locally existing crops at Shimla, namely garlic, onion, peas, and turnip potato and discouraged the growth of Root-knot nematodes (Nagesh *et al.*, 1994). Mustard crop has been found to decrease the nematode population in Shimla hills.

### *Biological control*

Indigenous pathogenic bioagent (*Bacillus.*) was isolated at Shimla from infested RKN females and egg masses. Laboratory studies showed 25–30% mortality and 20–25% nematode (juveniles) static effect in 42–48 hrs. of exposure.

### *IPM Strategies*

Since potato cultivation varies in hills as well as in plains, it is required that separate approaches should be applied for managing RKN.

### *RKN management in plains*

- Summer ploughing, 2–3 times at 10–15 days intervals should be done.
- Use of uninfested certified/truthful seed should be used. Only seeds multiplied from uninfested fields should be planted.
- Clean cultivation and frequent removal of volunteer plants and unharvested leftover tuber from the fields is required.
- Localized pre-plant application of Carbofuran 3G or Phorate 10G @ 2–3kg a.i/ha is highly appreciable.
- Crop rotation with cereals like wheat, maize, and rice and vegetable Crops like onion, turnip, garlic, and peas should be done.
- Several rotation for different agroecological regions of the country should be denitrified. For instance, In Jalandhar region crop rotations combining maize-potato-wheat-vegetables or paddy-potato-maize/wheat vegetables are highly feasible.
- In Patna region, Paddy-Potato-Maize-Vegetable or Paddy-Potato-Maize/Paddy-Potato-Paddy or Maize-Potato-Paddy aare highly feasible for checking RKN.
- Solanaceous crops should be avoided for raising in infested fields.
- Clean harvest of leaf-out tubers and crop rotation must be strictly followed/practised.

### *RKN management in hills*

- Uninfested, certified and seed multiplied from uninfested fields should be used.
- Localized preplant application of Carbofuran 3% or phorate 10G @ 2–3kg a.i./ha should be done.

## REFERENCES

Akhade, M.N.; Tidke P.M. and Patkar, M.S. 1970. Control of potato tubermoth *(Gnorimoschema operculalla* Zell.) in Deccan Plateau through insecticides and depth of planting. *Indian J. Agric. Sci.*, **40**: 1071–1076.

Amonkar, S. V.; Pal, A.K.; Vijayalakshmi, L. and Rao, A.S.1979.Microbial contol of potato tuber moth (Phthorimaea *operculella* Zell.).*Indian J.Exptl.Biol.* **17**: 1127–1133.

Apanna, M. 1948. The larval parasite *Pleurotropis faveolatus* C. of the potato beetles Epilachna 24- punctata. *Curr. Sci.* **17**: 154–155.

Azam, K.M.; Aziz, S.A. and Ali, M.M. 1974. A new record of *Ugna menoni* Kerrich as parasite of *Epilachna vigintioctopunctata* (F) in India. *Curr. Res.* **3**: 88.

Bombawala O.M.; Bedi, P.S.; Sharma, N.K. and Sharma, S.K. 1982, B.B. *In: Developing Countries.* Indian Potoato Association CPRI, Shimla. pp. 374–378.

Chandla, V.K. 1983. Extent of damage by pests. *In: Potato production storage and utilization.* B.B. Nagaich (ed.). Central Potato Research Institute, Shimla. pp. 425–427.

Chandla, V.K.1985. Potato pests and their management. *Indian Farming,* 36 **(12)**: 31–32 and 44.

Chandla, V.K.1986. *Insect pest complex of potato crop in Shimla hills and their managmenet.* Ph. D. thesis. Department of Entomology-Apiculture, Dr. Y.S. Parmar University of Horticulture. and Forestry, Solan (H.P). pp. 194.

Chandla, V.K.; Bhalla, O.P and Dogra, G.S. 1987. Monitoring of adult potato tubermoth, *Phthorimaea operculella* Zeller with sex pheromone. *Nat. Acad. Sci. Letter* **10** (11): 397–399.

Chandla, V.K.; O.P.Bhalla and K.S., Negi,1989. Natural Enemies of cutworm, *Agrotis* spp in Shimla hills. Abstract. Presented in "X *Annual., General., Body Meeting of IPA*" held at HAU., Hisar on 20th August,1989.

Chandla, V.K.; Misra, S.S.; Verma, K.D. and Bist, B.S., 1977. Insecticidal control of cutworms, *Agrotis* spp infesting potato crop. *Pesticides*, **11**(2): 29–30.

Chandla, V.K. and Rizvi, S.M.A 1977, Chemical control of *Heliothis armigera* Hb. (Noctuiidae: Lepidoptera) on potato. Pesticides, **11**: 65

Chandla, V.K.; Trivedi, T.P.; Misra, S.S.; Sharma, D.C. and Kashyap, N.P. 1993. Potato tubermoth menace in Himachal Pradesh and its management (Abstract). *J. Indian Potato Assoc.* **20** (I): 60–61.

Channa Basavanna, G.P. and Puttarudriah, 1959. *Hemitarsonemus latus (*Banks), a potential mite pest of cotton in Mysore. *Sci. Cult.* **25**: 322–323.

Chaudhary, R.; Trivedi, T.P. and. Raj B.T. 1983. Field evaluation of some exotic parasitoid of potato tuber moth, *Phthorimaea opercelella* (Zell.). *Indian J. Ent.* **45**: 504–506.

Dalaya, V.P. and Patil, S.P. 1973. Laboratory rearing and field releases of *Copidosoma koehleri* Blachard, an exotic parasite for the control of *Gnorimoschema operculella* Zeller. *Research J. Mahtma Phule Agric. Univ.*, **4**: 97–107.

Divaker, B.J. and Pawar, A.D. 1979. Field recovery of *Chelonus blackburni* and *Bracon hebator* from potato tubermoth. *Indian J. Plant Pro.* 7: 214.

Fletcher, T.B., 1916 Annotated list of Indian Crop pests. *Rep. Agric. Res.Inst. and Coll.Pusa.*1915–16. pp. 58–77.

Fletcher, T.B., 1919. Life historice of Gelechides. *Mems Dept. Agric.India* (Entomological Series), **6**: 75–76.

Ghosh, D.; Debnath; Chakrabarti, S. 1985. Predators and parasites of Aphids (Hom., Aphididae) from North West Himalaya: Ten species of syrphids (Diptera -Syrphidae) from Garhwal range. *Entomon,* 10(**4**): 301–303.

Ghosh, L. and Verma, K.D.1990. Discovery of sexual females of *M.persicae* (Sulzer) with redescription of alate male from India. *J. Aphidol.* **4**: 30–35.

Gill, J. S.,1974. Advances made in the potato nematology in India. *JIPA,* **1**(1–2): 51–55.

Gupta, B.D. and Awasthy, P.N.1960. Insect pests of sugarcane and their control. Indian Central Sugarcane Committee, New Delhi. pp. 9–34.

Jones, F.G.W. 1961. Potato root eelworm *Heterodera rostochiensis* (Woll). in India. *Current Science,* **30**: 187.

Kapoor, A.P. 1950. A note on *Epilachna ocellata* Redt. (Coleoptera: Coccinellidae) with description of three species hitherto confused with it.Rec. *Indian Mus.* **48**: 17–19.

Krishnamurti, B.1932. The potato epilachera beetle, *Epilachna vigintioctopuncta* (Fabr.). *Bull. Dept. Agric Mysore. Ent. Ser.* (**9**): pp. 16.

Lall, B.S., 1964. Vegetable pests in Entomology in India. Dr. NC Pant, *Entomological., Society of India. pp.* 184.

Lal, L.1987. Studies on natural repellents against potato tuber moth (*Phthorimaea operculella* Zeller) in country stores. *Potato Res.* **30**: 329–334.

Lal, L. 1993. Potato tuber moth-Bionomics and Management. *In: Advances in Horticulture–potato.* K.L.Chandha and J.S. Grewal, (eds). **7**: 591–602.

Lefory, H.M. 1907. The potato tubermoth. *Indian Agric. J.* **2**: 294–295.

Misra, S.S., 1993. Integrated pest management. *In*: *Advances in Horticulture potato*. K.L. Chadha and J. S.Grewal (eds). Malthora Publishing house, New Delhi, India. **7**: 603–613.

Misra, S.S. and Agarwal, H.O.1988. Potato Pests in India and their control. *Tropical Pest Management,* **35**: 199–209.

Misra, S.S.and V.K. Chandla, 1979. Take care of potential vector for raising disease-free seed potatoes. *IPA News* letter, **1**: 5–6.

Misra, S.S. and Chandla, V.K. 1989. White grubs infesting potatoes and their management. *J. Indian Potato Assoc.* **16**: 29–33.

Misra, S.S.; Chandla, V.K. and Singh, A.K. 1995. Potato pests and their management. Technical Bulletin No. 45. Central Potato Research Institute, Shimla. pp. 1–13.

Nagiach, B. B., and Agrawal, H.O.1969. Research on potato viruses in India. *Indian J. Agric. Sci.* **30**: *286–296.*

Nagaich, B.B.; Verma, K.D. and Upreti, G.C. 1970. Hereditary variation in the ability of *Myzus persicae* to transmit potato leafroll and virus. *Y. Final., Tech. Rept. PL 480 Scheme*, CPRI.

Nagesh, M.; Negi, K.S. and Kumar, V. 1994 Effect of crop rotation on nematode multiplication and potato yield in Shimla. *In: Potato Persent and Future.* G.S.Shekhawat *et al.,* (eds). IPA Shimla. pp. 255–260.

Nair, Ramchandran and Rao, V.P.1972. Results of the suvey for natural enemies of potato tuber moth, *Phthorimaea operculella* Zeller(Leoidoptera: Gelechiidae) in Mysore state and the parasites reared from Maharashtra state.*Tech.Bull. Commonwealth Institute of Biological control.* **15**: *115–130.*

Nirula, K.K.and Kumar, R. 1963. Studies on the rate and mode of application of heptachlor and aldrin for the control of *Agrotis ypsilon* Rott. *Indian Potato J.* **5**: 58–63.

Prasad, K.S.K.1993. Nematodes-Distribution, Biology and Management. *In: Advances in Horticulture.* K. L. Chadha and J. S. Grewal. Potato. Malhotra Publishing House, New Delhi. **7**: 635–647.

Prasad, K.and Singh, D.B. 1986, Note on the nematode associated with potato in Karnataka state of India. *Intrn. Nematol. Net work News letter,* **34**: 11–13.

Prasad, S.K. 1961. Quantitative estimation of damage to potatoes caused by *Emposca fabae* (Harris). *Indian Potato* J. **3**: 105–107.

Pruthi, H.S. 1946. *Report of the Imperial Entomologist*. Abrigd. Sci. Rep. Agric Res. Inst. New Delhi. 1941—44 Delhi. pp. 64–71.

Puttarudriah, M. and Krishnamurti, B. 1954. Problems of *Epilachna* control in Mysore.Insecticide control found undesirable when natural incidence of parasite is high. *Indian. J. Ent.* **16**(2): 137–141.

Raman, K.V. and Booth, R.H. 1983. Evaluation of Technology for Integrated control of potato tube rmoth in field and storage. *Technology Evaluation Series No. 10.CIP, Lima*, Peru. pp. 18.

Raman K.V. and Mohandas.,1988. Occurrence of potato cyst nematode *(Globodera pallida* C.Stone 1973) in Kerla. *Indian. J. Nematol.* **18**: 141.

Ramchandran Nair, K. 1967. *Leucinodes orbonalis* Guen (Lepidoptera: Pyralidae) as a serious pest of potoato plants in Mysure state. *Indian J. Ent*. **29(1)**: 96–97.

Raj, B.T.; Saxena, A.P. and Ansari, M.A.1979. Chemical control of tarsonemid mite *(Polyphagotarsonemus latus* Banks) on potato crop. *J. India Potato Assoc*. **6**: 194–198.

Raj. B.T.; Trivedi, T.P. and Saxena, A.P.1986. Evaluation of some insecticides for control of potato tuber moth *J. Indian Potato Assoc. 13: 73–76.*

Raychaudhuri, D.N.; Dutta, S.; Agarwal, K.; Raha, S.K. and Raychaudhuri.D. 1979. Some parasites and predators of aphids in North-east India and Bhutan. *Entomon*, **4(2)**: 163–166.

Rizvi, S.M.A. and Chandla, V.K. (1977). Chemical control of cutworm on potatoes. *Indian J.Pl. Prot*. 5(1): 105–107.

Rizvi, S. M. A.; Lal, L. and Khurana, S.M.P. 1976. Host plants of *Myzus persicae* Sulzer in Jullunder and Simla.*Indian J. Pl. Prot. 4: 32–34.*

Saxena, A.P. 1974. Save potato crop from the ravages of insect pests. *Indian Fmg*. 24 : 25–28.

Saxena, A.P.1981. *Leucopis fumidilarva* Tans (Diptera: Chamaemycides) a new predator of aphids and leaf hoppers on potato crop. *J. Indian Potato Assoc*. **8(2)**: 96–98.

Saxena, A.P., and Raj, B.T.,1980. Studies on the biological control of potato pests. *Proc. Workshop on Biol. Control of Crop Pests and Weeds*. PAU, Ludhiana. October 27–30: 121–125.

Saxena, A.P., Raj, B.T. and Rawat, P.S. 1984. Chemical control of potato tuber moth in country stores. *J.Indian Potato Assoc*. **II**: 113–115.

Saxena, A.P. and Rizvi, S.M.A.1974. Insect pest problems of potato in *India. J. IndianPotato Assoc*. **1(1–2)**: 43–50.

Saxena, A.P. and Singh, V. 1977. Predacious bugs on *Epilachna ocellata* Redt.(Coccinellidae: Coleoptera). *J. Indian Potato Assoc*. **4**: 70.

Seshadri A.R. and Sivkumar, C.V.1962. Golden nematode of potatoes (*Heterodera rostochiensis* Woll.): A threat to potato cultivation in the Nilgris. *Madrass Agric. J*. **49**: 281–288.

Sheila, M.K. and Abraham, G.C.1981. *Cantheconidea furcellata* (Wolf.) (Pentatomidae, Hemiptera) as a predator of *Henosepilachna vigintioctopunctata* (Fabr.) *Agric. Res. J. Kerala,* **19**: 142.

Singh D.B; 1983, Survey of phytonematodes associated with potato in Kodaikanal and Nilgris hills. *Third Nematol. Symp*. Solan. pp. 25.

Singh, D.B., Kishore, V.1982.Nematode pests of potato in India and fungal diseases in Nilgiris. *Seminar on Tuber borne Diseases and pests in potato seed production.* At Thimpu, Bhutan, Oct. 12–14 pp. 43–56.

Singh, S.P.1974. Observations on the biology of *Trichoplusia orichalcea* Fabr. on potato. *Indian. J. Plant Prot.* **2**: 127–128..

Singh, S.P. 1988. Natural enemies of aphids. *Fourth National Symposium of Aphidology*, CPRI Shimla. pp. 66.

Singh, S.P. 1993. Biological control of potato pests. *In: Advances in Horticulture potato.* K.L. Chadha, and J.S.Grewal (eds). Malhotra Publishing house, New Delhi. **7**: 615–634.

Singh, M.N.; Khurana, S.M.P.; Nagaich, B.B. and Agarwal, H.O.1982. Efficiency of *Aphis gossypii* and *Acyrthosiphon pisum* in transmitting potato viruses X and leafroll. *In: Potato in Developing countries.* B.B. Nagaich *et.al.*, (eds). CPRI, Shimla.

Srivastava, B.K. 1958. A new cutworm pest of potato in Rajasthan. *Curr. Sci*. **27**: 504.

Thirumalachar, M. J.1951.Root-knot nematode in potato tubers in Simla. *Curr.Sci. 20: 104.*

Trivedi, T.P. 1988. Natural enemies of potato pests in Karanataka. *J. Indian Potato Assoc*. **15**: 159–160.

Trivedi, T.P. 1990. *Spatial distribution, phenology, life table and assessment of losses due to potato tuber moth. Phthorimaea operculella* (Zeller). Ph.D. Thesis of University of Agriculture Sciences, Bangalore (India). pp. 182.

Trivedi, T.P., Prasad, K.S.K. and Singh, D.B.1990. Pest problem of potato in Karnataka. In *National., Sym.on stategies forpotato production, marketing storage and utilization.* IARI.New Delhi (Abstract.).

Venketraman, T.V.1954. A note on the possibility of *Agrotis ypsilon* Redt. migrating from the plains to the hills during summer months in parts of India. *Indian J. Entomol.* **16**: 197–98.

Verma, K.D.1985. Potato aphids with special referance to *Myzus persicae* (Sulzer). *Proc. 2nd Nat.Symp.on recent trends in Aphidol studies.* S.P.Kurl (ed). pp. 39–46.

Verma, K.D. 1993. Aphid–Biology and virus transmission. *In*: *Advances in Horticulture*. K.L. Chadha, and J.S. Grewal (eds). Malhotra Publishing House, New Delhi. **7**: 575 : 589.

Verma, K.D. and Chauhan, R.S. 1993a. The life-cycle of potato vector, *Myzus persicae* (Sulzer). *Curr. Sci.* **65**: 488–89.

Verma, K.D. and Misra, S.S.1975. Chemical control of *Epilachna ocellata* Redt. grubs infesting potato crop. India *J. Pl. Prot*. **3**: 214–215.

Verma, K.D.; Misra, S.S. and Saxena, A.P.1976. Role of *Aphelinus* sp. in the natural control of *Myzus persicae* on potato. *J. Indian Potato Assoc*. **3**(**1**): 40.

Verma. K.D., Parihar, S.B.S. and Khan, I.A.1993b. Field reaction of some potato varieties/germplasm lines to leafhopper, *Empoascae fabiae*(Harris). *J.Indian Potato Assoc 20: Abstr.49.*

Verma, K.D.; Chaudhry, R.P.; Chandla., V.K.; Misra., S.S. and Tandon, M.L.1976a. Further studies with systemic insecticides against the green peach aphid on potato. *Indian J. Pl. Proct.* **4**: 65–169.

Yadava, C.P.S.; Saxena, R.C.; Mishra, R.K.; Dadheech, L.N. and Poonia, F.S. 1978. Integrated control of white grubs—ICAR Scheme. *White grub News letter* 2(**23**): 23–26.

*Pest Management in Horticulture Crops*
L.R. Verma, A.K. Verma and D.C. Gautam (eds)


*Chapter 19*

# Insect Pests of Beans, Pea and Okra

*K.C. Sharma*

THE INSECT PESTS of beans, pea and okra and their management are described as follows:

## BEANS

French bean (*Phaseolus vulgaris* L.) commonly known as bean, string bean, snap bean or field bean is an important vegetable crop whose immature, delicate and tender pods are used as fresh vegetable while dry seeds are used as pulses. The growth period of beans coincides with the onset of monsoon season. Prevalence of congenial temperature coupled with high humidity during this season favours not only luxuriant growth of the crop but also that of many insect pests which reduce the yield to a considerable extent. Following table lists the insect and mite pests which are commonly found on bean crop.

**Insect and mite pests of beans, *Phaseolus vulgaris* L.**

| *Common Name* | *Scientific name* | *Order* | *Family* |
|---|---|---|---|
| Black bean bugs | *Chauliops nigrescens* Distant | Heteroptera | Lygaeidae |
| | *C. fallax* Scott | Heteroptera | Lygaeidae |
| Blister beetles | *Mylabris pustulata* Thunberg | Coleoptera | Meloidae |
| | *M. phalerata* Pallas | Coleoptera | Meloidae |
| | *M. mecilenta* Marshal | Coleoptera | Meloidae |
| | *M. tiflensis* Bills | Coleoptera | Meloidae |
| Whitefly | *Bemisia tabaci* (Gennadius) | Homoptera | Aleyrodidae |
| | *Trialeurodes vaporariorum* (Westwood) | Homoptera | Aleuirodidae |
| Bean stemfly | *Ophiomyia phaseoli* (Tryon) | Diptera | Agromyzidae |
| Bean aphid | *Aphis craccivora* Koch | Homoptera | Aphididae |
| Surface grasshopper | *Chrotogonus trachypterus* Blanch. | Orthoptera | Acridiidae |
| Bean bug | *Riptortis pedestris* (F.) | Heteroptera | Coreidae |
| | *R. linearis* (F.) | Heteroptera | Coreidae |
| Red spider mite | *Tetranychus telarius* (L.) | Acarina | Tetranychidae |

**Black bean bug *Chauliops nigrescens* Distant**

The black bean bug is an important pest of various types of beans. It is widely distributed in Ceylon, Japan and South-eastern Australia. In India, the pest is distributed in the northern states on various beans. This species has since been reported to feed on soybean, cowpea, moth bean, french bean, horse gram, and green gram (Kashyap and Adlakha, 1971; Rawat and Sahu, 1973; Lal, 1974; Bhalla and Pawar, 1977; Rustogi, 1981; Singh *et al.*, 1987; Kakar *et al.*, 1990; Sharma *et al.*, 1994a).

Both adults and nymphs suck sap from the plant usually from the lower surface of the leaves. As a result of sap sucking, the chlorophyll content gets reduced which ultimately affects the quality and yield of the crop. Severely attacked leaves show several minute, yellowish specks and small, black pustules of excreta. The damaged leaves gradually wither and fall off. Adults appear in June (with the onset of monsoon) and remain on the crop until final harvest.

***Life cycle***

The eggs are cylindrical, dark brown and glued to the leaf surface. The chorion exhibits fine reticulations. The eggs are generally laid singly but sometimes in pairs near the mid-rib or veinlets. Before hatching the eggs become dark. The incubation period averages 8.7 days.

The bug has five nymphal instars. Just after hatching the nymph is light orange in colour but after feeding, the thoracic parts of the body become dark red while the abdominal parts turn red having light greenish tinge. Antennae are four segmented. Stylets are transparent and the abdominal portion is slightly compressed. The average length and breadth of the first instar nymph is 0.073 and 0.022 mm, respectively, and the average duration is 3.2 days. The second instar nymph is light red but changes to red before moulting. The body shape is similar to first instar except the size. The length and breadth of this instar averages 1.025 and 0.560 mm, respectively, and the average duration is 3.5 days. The general body colour of the third instar nymph is dark red. During this stage rudimentary wing pads are seen through a microscope. The shape of the body is similar to the second instar. The length and breadth of the third instar averages 1.309 and 0.761 mm, respectively, and a mean duration of 3.1 days. The fourth instar nymph is dull red and the wing pads are in developing stage. The abdominal segments are distinct. The average length and breadth of the fourth instar nymphs is 1.878 and 0.932 mm, respectively, and the mean duration is 4.3 days. The body colour of the fifth instar is dark brown. Wing pads are visible. Abdominal segments are more distinct than the fourth instar. This instar resembles the adults. The length and breadth of the fifth instar averages 2.386 and 1.185 mm, respectively. This stage takes 5.5 days to become adult. The total nymphal development takes 18–23 days. Females are larger than the males. The average length and breadth of male is 2.77 and 1.00 mm while that of females 2.86 and 1.25 mm, respectively. Pre-oviposition period averages 8.7 days while the oviposition period averages 15.5 days. Repeated matings are observed. Adult females live for an average duration of 25.4 days and males for 20.9 days. The average fecundity is 45 eggs. The bug has three overlapping generations in a year. It overwinters in the adult stage (Bhardwaj and Bhalla, 1975; Lal, 1981; Sharma *et al.*, 1994a).

***Natural enemies***

Mermethids (Nematoda) were reported to attack lygaeid bugs in China (Xu, 1987).

*Management*

Sprays of phosphamidon, dichlorvos, monocrotophos, endosulfan, lindane, dimethoate or oxydemeton methyl at 0.05% and malathion at 0.1% were recommended for the control of this pest (Lal, 1981). Two sprays at fortnight interval were found to give full protection from the bug until harvest (Lal, 1992).

**Blister beetles**

***Mylabris pustulata* Thunb., *M. macilenta* Marshell and *M. phalerata* Pall.**

Blister beetles are serious pests of beans, cowpea, soybean, mung bean, urd bean, okra, cucurbits, maize and pigeon pea, etc. Kàkar *et* al. (1990) reported that blister beetles were the regular pests of french bean in the mid-hills of Himachal Pradesh. Beetles appear in mid July at the time of flower bud formation and flowering of french beans and remain on the crop till harvest.

The blister beetles are so called because they are used for making blister plaster. The elytra of the beetles contain a pharmaceutical product called Cantharidin. Cantharidin preparations are prescribed by physicians for the removal of warts from the body. The damage is caused by adults which feed on foliage, flowers and developing pods. The larval stages are predatory on the eggs of grasshoppers while some other eat eggs of solitary bees and their food stores.

*Life cycle*

The life cycle of the blister beetles is very complicated. Eggs are laid in the hard soil singly or in batches upto 20. The female digs holes in the ground and lays 24 to 60 eggs at a time. The egg laying occurs at the same period as that of the grasshoppers/locusts. The incubation period varies from 20 to 22 days.

The newly hatched larva, known as triungulin, is very active. Immediately after hatching, it moves in search of food. On finding the egg pods of grasshoppers, the larva enters into the pod by making a hole in the upper half of the pod. Feeding larva destroys the egg membranes and emits excreta, by which the infested egg pod can be easily recognised. The larva leaves the egg pod after becoming full-fed and burrows in the ground at a depth of about 10 cm. It then transforms into sixth instar which is immobile. This stage lasts for 10 to 30 days. Each larva requires one egg pod for its development thus destroying 40 to 70 eggs (Verbeek, 1930). Pupation takes place in soil. Pupal stage lasts for 2 months. The adults are large beetles (25 to 35 mm long) and are sluggish in behaviour. A female lays 10,000 to 20,000 eggs during its life span. It overwinters in the egg stage (Patnaik *et al.*, 1993).

*Management*

Spray of cypermethrin (30 g ai/ha) or fenvalerate (75 g a.i./ha) at flowering stage in July-August gives good control of the beetles (Kakar *et al.,* 1990). The beetles can also be controlled by spraying endosulfan (0.1 %), dimethoate (0.03 %) or a mixture of dimethoate and endosulfan (0.1%) (Chhabra *et al.,* 1993).

**Bean fly *Ophiomyia phaseoli* (Tryon)**

*Ophiomyia phaseoli* is a polyphagous pest that infests almost all leguminous crops. The pest has been reported as one of the most destructive pests of legumes in South India. It is a minute insect that causes extensive damage to beans. In Maharastra, the infestation occurs immediately after germination, becomes severe during the 10th week and gradually declines thereafter (Mote, 1983). The damage is caused by the larva which mines the leaf lamina, veins, petiol and stem. It pupates at the base of the stem thus blocking the food uptake. As a result, plants show sudden wilting at the time of flowering. In Andhra Pradesh, upto 90 per cent damage has been reported to french bean crop by this pest (Singh *et al.*, 1991).

***Life cycle***

The flies are seen in the field as soon as the seeds germinate. The eggs are laid by the female singly on the stem just above the ground level by inserting its ovipositor in superficial tissues or on the upper surface of young leaves. The eggs hatch in 48 to 72 hours. The newly hatched maggot makes tunnel underneath the epidermis of the leaf and then in leaf stalk and stem. There are three larval instars. The first instar takes 24 to 48 hours to develop. The second instar maggot is white and is distinguished by the presence of prominent head with mandibular hooks at the anterior portion and spiracles at the posterior side. This is the most injurious stage. This stage lasts for 8 to 48 hours. On entering the third instar, the colour turns to creamy white. During this stage the segmentation is clearly visible with prominent head, anterior spiracles at the cephalic region, mandibular hooks are fully developed with three projections at the lower side. This stage lasts for 24 to 56 hours. Pupation takes place inside the stem. Pupa is dark brown with visible segments and is barrel shaped. The pupal stage lasts for 6 to 13 days. The adults are small active flies. The female is bigger than male. The total life cycle is completed in 11 to 22 days. During the active season upto seven generations have been reported (Singh *et al.*, 1991).

***Management***

Two sprays of endosulfan (700 g a.i./ha) at unifoliate stage, first at the appearance of puncture marks (12 days after sowing) and second at petiole mining stage (16 days after sowing), have been found to give good control of the pest (Krishnamoorthy and Srinivasan, 1991). Sprays of 4 per cent neem seed kernel extract also gives adequate control under moderate insect pressure when timed at appearance of puncture marks and later on petiole mining stage (Srinivasan and Krishnamoorthy, 1992).

**Whitefly *Bemisia tabaci* (Gennadius)**

Whitefly is a polyphagous insect and is very destructive pest in the northern and western regions of the Indian subcontinent (Atwal, 1986). It is distributed throughout India and is considered as the major pest in Punjab, Rajasthan, Andhra Pradesh, U.P., Bihar, Maharastra, Tamil Nadu and Kerala. Whitefly problem has accentuated in the recent past in countries like Israel, Thailand and India. *B.* tabaci is known as serious pest of cucurbitaceous and solanaceous crops. Mound and Halsey (1978) reported more than 320 host plants of this pest. The Damage is caused by nymphs as well as adults which cluster on the underside of leaves and suck the cell sap. Besides inflicting direct damage due to depletion of cell sap, it also acts as a vector of many viruses and transmits a number of viral diseases. Females are more efficient vectors than males. Heavy infestation can cause death of seedlings and wilting of older plants, shedding of leaves and flowers. Continuous feeding affects the physiology of the plant leading to deterimental effect on yield. Additionally, the honeydew excreted by the insects attracts sooty mould growth that interferes with photosynthetic activity in leaves.

*Life cycle*

Whitefly adults are small insects with two pairs of wings and white waxy appearance. The louse-like nymphs are sluggish creatures, clustered on the underside of the leaves and their pale-yellow bodies make them standout against the green background. Females lay spindle shaped eggs on the underside of young leaves and the egg is anchored on to the leaf surface by a short pedicel on the broader end. The minute eggs measure less than 0.2 mm, are pale or yellowish green initially but turn greyish black before hatching. The incubation period varies from 3 to 33 days. Sometimes the females resort to parthenogenetic reproduction.

The nymphs on emergence look elliptical and soon fix their mouth parts into the plant tissues. After passing through three stages, the nymphs convert into pupae. The entire development takes place in 14 to 22 days under favourable conditions, however, during winter it may extend upto 107 days. A female lays about 120 eggs within a span of 18 days. Breeding takes place almost all the year round and there are 10 to 12 overlapping generations in a year. High temperature (>33°C), associated with scanty rainfall are conducive for rapid development and breeding of whiteflies.

*Management*

*Cultural/Agronomic practices:* Cultivation of whitefly prepared 6 alternate hosts like brinjal, tomato, sunflower and tobacco may be avoided. In case their cultivation is unavoidable, plant protection measures should be extended to these crops by spraying appropriate insecticides. Judicious application of recommended doses of fertilizers particularly nitrogenous fertilizers may be followed. Field sanitation may be given attention.

*Chemical control:* Monocrotophos (0.4%) and phosphamidon (0.4%) have been reported to give efficient control of whiteflies in french bean (Azmi and Rathi, 1992). Rao *et al* (1996) reported that profenofos (0.05%), was effective in reducing the population of whitefly. Endosulfan (0.5%) and neem seed kernel extract (5%) also give good control of this pest.

*Biological control:* Coccinellids, chrysopids and spiders are the important predators and *Encarsia* sp. are the major parasitoids of whiteflies. However, the parasitoids are effective only at low populations. Besides a fungus, *Beauveria bassiana* has been found to be effective for the control of *B.* tabaci under field conditions.

**Bean aphid *Aphis craccivora* Koch**

The bean aphid, also known as groundnut aphid, blue aphid, cowpea aphid, etc. is cosmopolitan in is distribution and occurs as a serious pest of several leguminous and non-leguminous plants in

India, Africa, Argentina, Australia, Chille and the U.S.A. (Srikanth and Lakkundi, 1988). Sharma and Bhalla (1964) reported this aphid as a pest of soybean, cowpea and beans and minor pest of groundnut in Himachal Pradesh. The damage is caused by nymphs as well as adults which suck sap from the underside of leaves and other aerial parts resulting in distortion and withering of leaves thereby reducing the vigour of the plant which ultimately affects the yield. In severe infestation, the plant may die. This species is an important vector of groundnut rosette virus and 13 other viruses. The virus is brought into the crop by winged adults and is transmitted within the crop by both wingless and winged forms.

### *Life cycle*

The adults reproduce parthenogenetically and viviparously. There are four nymphal instars. The total nymphal period is about 6 days. Adults live for 10 days of which 8 days are utilized by the females to lay youngones. An average of 36.5 nymphs are laid by the female during its life time. Under favourable environmental conditions, the life cycle is completed in about 15 days (Srikanth and Lakkundi, 1988).

### *Management*

Cultural control can be effective through early planting and close spacing.

*Chemical control:* Cypermethrin (0.01%), fenvalerate (0.01%) and dichlorvos (0.05%) were reported to be effective in controlling this pest on moong bean (Gaffar *et al.*, 1990). Khurana and Kaushik (1991) recommended sprays of oxydemeton methyl (0.025%), monocrotophos (0.025%) and endosulfan (0.05%) on chickpea. Deltamethrin (0.0028%) and dimethoate (0.03%) are also effective against this aphid.

*Biological control:* The coccinellid and syrphid predators are the important agents which could suppress the population of this aphid. The carabid beetle, *Chlaenius panagaeoides* (Laferte) was found to be the most important predator in Bangalore which could check the initial population of this aphid under field conditions (Rajgopal and Kumar, 1988).

## Red spider mite *Tetranychus telarius* (L.)

The red spider mite is a polyphagous pest feeding on a wide range of plant species. It was described for the first time by Linnaeus (1758) under the name *Acarus telarius* L. The nymphs and adults feed on the underside of leaves and breed near the petiole in between the main veins. Under heavy infestation mites are found all over the leaves. The mite lives gregariously and is protected by silken web. Feeding is accomplished by the chenceral stylets which penetrate the leaf tissues and the liberated fluids is sucked near the tip of rostrum. Due to extraction of chlorophyll, leaves show characteristic blotching. Sucking on the underside of the leaves results in uneven growth of the two surfaces, exhibited by crinkling of leaves from below. The infested leaves show yellowish speckled appearance which later turns brown (Pal, 1966; Srivastava and Mathur, 1962; Bindra and Singh, 1970).

*Life cycle*

The eggs are smooth, round, pearl white when freshly laid but later change to light yellow. Before hatching the colour becomes opaque and pinkish. Egg hatches in 5 to 8 days. The newly hatched larva is slender light brown in colour with two red eye spots. This stage lasts for 2 to 3 days. Before moulting the larva becoms quiscent for few hours and is elliptical and dark brown in colour. It then converts to protonymph with four pairs of legs. Protonymph is very active stage and changes to deutonymph without any quiscent period. During this stage, sexes can be distinguished. Lall *et al.* (1965) reported the duration of protonymph and deutonymph as 2.45 and 2.78 days, respectively. The total life cycle is completed in 10 to 11 days. Females are bigger than males and have round abdomen. A female lays about 75 eggs during its life span of 14 to 17 days.

*Management*

Malathion (0.05%) or dimethoate (0.09%) spray is effective for the control of this mite (Bindra and Singh, 1970; Patel and Patel, 1968). A single pre-flower application of dicofol (250 g a.i./ha) or phophamidon (140 g a.i./ha) was recommended for the control of this mite on french bean (Kakar *et al.*, 1990).

## PEA

Pea is an important vegetable crop that plays a major role in improving the economy of the farmers. It is attacked by a number insect pests at one or the other stage of crop growth and reduces the yield to a considerable extent. The following table lists important insect pests that are reported to attack the pea crop.

**Insect pests of pea, *Pisum sativum* L.**

| *Common Name* | *Scientific name* | *Order* | *Family* |
|---|---|---|---|
| Pea leafminer | *Chromatomyia horticola* (Goureau) | Diptera | Agromyzidae |
| Pea pod borer | *Etiella zinckenella* (Treitshke) | Lepidoptera | Phycitidae |
| Pea blue butterfly | *Lampides boeticus* (Linnaeus) | Lepidoptera | Lycaenidae |
| Black cutworm | *Agrotis segetum* (Denis & Schiffer) | Lepidoptera | Noctuidae |
| Greasy cutworm | *Agrotis ipsilon* (Hufnagel) | Lepidoptera | Noctuidae |

**Pea leafminer *Chromatomyia horticola* (Goureau)**

The pea leafminer is a polyphagous species and is reported on 20 plant species belonging to 20 families (Ahmad and Gupta, 1941). It was reported as a serious pest in Punjab (Singh and Sohi, 1957) and as a regular pest causing heavy damage to pea and *Brassica* crops in the Northern India (Atwal *et al.*, 1969).

The maggots mine into leaves, eat through the mesophyll and leave the two epidermal layers intact. The mines start from the periphery and end towards the mid rib. The mines become wider as the larva grows in size and the faecal matter left by the larva can be seen in these galleries. The minute slender larvae can be seen feeding within the mines if the attacked leaves are held against bright light. The photosynthesis is considerably reduced due to larval feeding. Under severe infestation leaves wither away and are shed as a result of which flowering and fruiting in reduced considerably (Kaurava *et al.,* 1970).

### *Life cycle*

The eggs are oval, spherical, tapering towards bluntly rounded end, translucent white when freshly laid but turn dull white before hatching. The incubation period ranges from 1 to 2 days. The larva is translucent white and the inverted 'Y'-shaped oral hook is clearly visible having dark brown colour. The second instar larva is cylindrical, tapering anteriorly. Both anterior and posterior spiracles are visible. The third instar is difficult to differentiate from the second. The duration of instars is about 6 days.

The pupa is fusiform with distinctly defined segments. Initially it is light brown but turns dark toward maturity. The pupal stage occupies 9.5 days. Adults are two winged flies having grayish black mesonotum. Adult females are black with light brown head. Mating occurs after 1 to 8 days of emergence and lasts for 15 to 90 minutes (Ahmad and Gupta, 1941; Tandon, 1971; Sharma *et al.,* 1994b). For egg laying, the female moves over the leaf surface and on finding a suitable place bends its abdomen and inserts the ovipositor into leaf tissue. One egg is laid in one insertion. Most of the eggs are laid near the leaf margins. A female on an average lays 294 eggs during its life span of 27 days. Males live for about 10 days.

### *Management*

*Chemical control:* Application of phorate granules @ applied at the time of sowing is effective against this pest (Chopade, 1975) whereas Verma *et al.* (1980) recommended fenthion (0.05 %) or chlorpyriphos (0.05%) sprays for effective control of the leafminer. From Karnataka, Srinivasan and Shukla (1980) reported that granular formulation of carbofuran phorate or aldicarb applied at 500 g ai/ha at sowing and 40 days after germination reduced the leafminer damage. Mavi and Singh (1988) reported two sprays of dimethoate at 300 g ai/ha to keep the level of infestation below the economic threshold. For effective control of pupae, thiodicarb and carbaryl were suggested by Singh *et al.,* (1986). Dash (1990) reported that methyl demeton at 500 g ai/ha was effective in reducing the incidence of this pest. Khajuria and Sharma (1995) reported that monocrotophos (0.05%) treated plots gave maximum seed yield followed by chlorphyrifos whereas, Mehta *et al.,* (1995) reported that methyl demeton treated plots gave the highest net profit.

*Biological control:* The population of the pea leafminer is naturally kept under check by a large number of larval and pupal parasitoids in India and elsewhere. These include braconids, eulophids, chalcids and other hymenopterans (Kumar, 1985a, b; Stenentite, 1986; Darvas *et al.,* 1988; Sheng *et al.,* 1989). From Ludhiana, Atwal *et al.,* (1969) reported *Solenatus guptai* Subba Rao, *Rhopalosus thakerei* Subba Rao, and *Opius* sp. as important larval parasitoids of this pest. They further reported that parasitiation was low (2.1to 6.3%) upto mid February but reached 72.8 to 81.8 per cent by the end of March, whereas, Kaurava *et al.* (1970) reported larval parasitization from 2.85 per cent by *Neochrysochavi* sp. and pupal parasitization upto 40 per cent by *Opius* sp. Tandon (1971) reported 97.2 per cent parasitization of the pupae by chalcids and braconids in the second week of April. Lal and Nazi (1979) reported parasitization by *Diglyphus* isaea (Walker) to the extent of 25 per cent, whereas Kumar (1985 a, b) reported *Chrysonotomyia formosa*, *D. isaea* as larval parasitoids and *Opius turcicus,* (Fischer), *O. exiguus* (Wesrn.) and *Sphegigaster* sp. as pupal parasitoids showing important controlling influence on agromyzid population. Darvas *et al.* (1988) identified *Dacnusa sibrica*, *Dapsilartha rufiventis* and *D. isaea* as the larval parasitoids and the parasitization was reported to be 35, 12 and 37 per cent, respectively.

## Pea pod borer *Etiella zinckenella* (Treitschke)

The pea pod borer is a major pest of pea which causes severe damage to flowers and pods. Besides India, the pest is reported from Mynamar, Sri Lanka, Hungary and Poland. It also attacks lentil gram, chickpea, etc. The larvae of this pest feed on flowers, pods and seeds. In the advanced stages of larval development, 2 or more pods may be webbed together. In Punjab, 50 to 90 per cent pod infestation and 23.9 per cent reduction in the seed weight due to this pest was reported by Singh and Dhooria (1971).

### *Life cycle*

Adults are greyish brown with distint pale white band along the front margin of forewings. Moths become active by the last week of February. The eggs are laid singly or in groups of 2–27 on pea pods. Immediately after hatching, the larvae feed on floral parts and green pods but later bore inside the pods and feed on seeds within. Generally one caterpillar is found in one pod. The caterpillars are green in colour with 5 black spots. The larval development is completed in 12.6 days. Pupation takes place in the soil. The pest overwinters in pupal stage and the adults emerge in February-March. Adult lives for 35.8 days. There are five generations in a year.

### *Management*

Srivastava and Singh (1974) and Verma *et al.,* (1980) reported that fenthion (0.05%), chlorpyriphos (0.02%) and phosphamidon (0.03%) were effective in controlling the pest, whereas Chopade (1975) recommended monocrotophos (0.04%) to reduce the damage caused by the pea pod borer.

## Pea blue butterfly *Lampides boeticus* L.

The pea blue butterfly is widely distributed in India, Mynamar, China, Japan and Europe. Besides pea, it feeds on lima bean, hyacinth bean, snap bean, urd bean, mung bean and pigeon pea, etc. The damage is caused by the larva which bores into the buds, flowers and green pods causing reduction in yield to a considerable extent.

### *Life cycle*

The female lays eggs on buds, flowers and green pods which are round and light green in colour. The eggs hatch in 4–6 days. The newly hatched larva is green with yellowish tinge, black head, a dark-brown patch on the prothorax, cylindrical body and scattered hair. The brownish mid-dorsal and yellowish lateral lines are well marked. The full grown larva is yellowish green to yellowish red. The larva passes through five instars. The larval period varies from 22 to 35 days. Pupa is dark/brown in colour. Pupation takes place on leaves or in the infested pod itself. The pupal stage lasts for 6 to 7 days. Adults live for 2 to 5 days (Cheng and Chan, 1989).

### *Management*

The pest can be controlled by spraying endosulfan (0.05%) or carbaryl (0.1%). Spray can be repeated after 15 days if the attack persists.

## Cutworms *Agrotis segetum (Denis & Schiffermuller)* and *A. ipsilon* (Hufnagel)

Cutworms are prolific, polyphagous insects that infest a large number of cultivated crops throughout the world and are widely distributed pests of agricultural crops. They feed on potato, chickpea, wheat, cabbage, cauliflower, knol khol, okra, broccoli, chinese cabbage, sugar beet, pepper, maize, soybean, onion, lettuce, chicory, tomato, chilli, cotton, etc. In recent years, the pest has been found severely infesting pea crop in some localities in the mid-hills of Himachal Pradesh. The larvae cut seedlings at the base and feed on leaves. They generally feed between dusk to dawn and keep hiding among the crop refuge or in the soil during day time. The caterpillars cut many more plants than they actually consume.

### *Life cycle*

The cutworms are active from October to April in the plains and during summer in mountainous regions of the country. Adults of *A. segetum* are stout and has a wing span of 40 mm. The forewings are pale, greyish-brown with dark markings which include rings (stigmata) and lines (fascia). The hindwings are white in case of males and suffused with brown, particularly towards the edges, in females. The robust body and head of the adult moth are brown in colour and the collar region has a black band. The moths of *A. ipsilon* have a wing span of 45 mm. The forewings are dark purple brown to pale brown in colour, suffused with a distinctive pattern of brown lines and black streaks and rings. The hindwings are white to translucent with brown lines and veins.

Females lay creamy white dome-shaped eggs in clusters on the under surface of leaves or in soil. The round, creamy white eggs hatch in 3 to 6 days. Just after hatching, the light grey larva remains hidden in the soil and feeds on the egg shell. There are six larval instars. The larvae of *A segetum* are hairless, the body is plump, greyish brown in colour, sometimes tinged with green and has a dark dorsal line and lighter spiracle line. The thoracic legs of the larvae are yellowish brown, the abdominal prolegs are pale, greyish brown whereas, the larvae of *A. ipsilon* are grey or purple brown with dark lines along the spiracles and back. The head is dark brown with yellow brown frons. The larvae become rich purple brown dorsally and yellow or brown ventrally in the last two instars and have a greasy appearance. In general, the larvae of *A. segetum* are smaller in size than that of A. ipsilon. Cannibalism is common in advanced larval instars. in both the species. The larval stage lasts for about a month. Pupation takes place in earthern chambers in the soil. The newly formed pupa is pale white but becomes dark brown later. The pupal stage lasts for 10 to 15 days. A female lays upto 700 eggs during its life span The oviposition takes place during night and continues upto 10 to 12 days. The life cycle is completed in eight weeks.

*Management*

Otto *et al.* (1985) achieved 29 to 40 per cent control of the larvae with trichlorfon baited with a mixture of bran, water and sugar, broadcasted at the rate of 30 kg/ha (1.5 kg a.i./ha). A bait formulation containing 0.25% quinalphos applied at 15 and 20 kg/ha (37.5 and 50 g a.i./ha) by broadcasting gives good control of this pest (Tarraborrelli *et al.,* 1989). When the infested area is small, the larvae can be collected and killed. Cutworm larvae can be controlled to some extent by flooding the field. In the evening, heaps of dry grass are made at different places of the affected field, the cutworms cluster themselves beneath the heaps during night, and can be killed by burning the grass.

## OKRA

Okra, *Abelmoschus esculentus* L., is an important vegetable grown in the tropical and sub-tropical areas in different parts of the world. The crop is attacked by a large number of insect pests which not only reduce the yield but also affect the quality of crop. The abundance of the pests varies from one region to another. The following table lists the major species that attack okra crop in India.

**Insect pests of okra, *Abelmoschus esculentus* L. in India**

| *Common name* | *Scientific name* | *Order* | *Family* |
|---|---|---|---|
| Jassid | *Amrasca biguttula biguttula* (Ishida) | Homoptera | Cicadellidae |
| Shoot and fruit borer | *Earias vitella* (F.) | Lepidoptera | Noctuidae |
| Blister beetle | *Mylabris pustulata* Thunberg | Coleoptera | Meloidae |
| | *M. phalerata* Pallas | Coleoptera | Meloidae |
| | *M. mecilenta* Marshall | Coleoptera | Meloidae |
| | *M. tifiensis* Bills | Coleoptera | Meloidae |
| Okra aphid | *Aphis gossypii* Glover | Homoptera | Aphididae |
| Red cotton bug | *Dysdercus koengii* (F.) | Homoptera | Pyrrhocoridae |
| Green semilooper | *Anomis flava* (F). | Lepidoptera | Noctuidae |
| Fruit borer | *Helicoverpa armigera* (Hubner) | Lepidoptera | Noctuidae |
| Whitefly | *Bemisia tabaci* (Gennadius) | Homoptera | Aleyrodidae |

### Jassid *Amrasca biguttula biguttula* (Ishida)

Jassids are serious pests of okra and came damage not only but also by acting as potential vectors of many virus diseases. They also feed on cotton, brinjal, potato, cucurbits, sunflower, mung bean, urd bean, soybean and many other malvaceous crops. The species is widely distributed in India. The nymphs and adults suck sap from the plant. The infested leaves turn yellow near the margins and show upward cupping. Under severe infestation, yellowish patches develop which later turn brownish thus showing symptoms of hopper burn (Bindra and Mahal, 1979).

*Life cycle*

Eggs are inserted into the leaf veins on the underside of the leaf which hatch in 4 to 14 days. The newly hatched nymphs are wedge-shaped, extremly active and suck sap on the under surface of the leaves. After passing through six instars in 5 to 19 days, they are transformed into adults. The adults are minute, greenish yellow insects, flying away at the slightest disturbance and live for about two weeks. A female lays an average of 50 eggs. There are 3 to 4 generations in a year.

*Management*

Sprays of thiodicarb (0.15%), oxydemeton methyl (0.03%), monocrotophos (0.05%) or phosphamidon (0.4%) give good control of this pest. Dahiya *et al.* (1990) reported that cypermethrin, fenvalerate and flucithrinate (0.006%), deltamethrin (0.0028%) and endosulfan (0.07%) were effective in checking the population for 15 days whereas fenthion, diazinon and phenthoate (0.05%), malathion (0.075%) and carbaryl (0.01%) checked the population for only a week. Okra seeds treated with carbofuran or phorate granules at 150 and 300 g a.i./ha or foliar application of malathion 1000 ml/ha in combination with seed treatment and exclusive foliar sprays of malathion or carbaryl at the same rate increased the fruit yield by reducing the population of this pest (Kumar *et al*., 1996).

**Shoot and fruit borer *Earias vitella* (F.)**

The larva, the shoot and fruit borer is a widely distributed pest of malvaceous plants like okra, cotton, hollyhock, etc. The larva bores into tender shoots, moves downwards and blocks the passage for movement of nutrients. The infested shoots ultimately wither and the growing tip or the branch is killed. The larva also bores into the flower buds, flowers and fruits remaining in shedding of fruits. Fruits become distorted and are rendered unfit for human consumption due to contamination by excreta.

*Life cycle*

The pest is active throughout the year and prefers high temperature. The eggs are laid on bracts, leaf axils and veins on the under surface of leaves. The hairy parts of the plant are preferred for oviposition. The eggs are crown shaped, sculptured and deep sky blue in colour which hatch in three days. The larva passes through six instars and the larval stage is completed in 10 to 12 days. The full grown larva is brownish with longitudinal white stripe on the dorsal side. Pupation takes place on plants or on fallen buds or leaves. Pupal cocoon is brown. The pupal period lasts for 7 to 9 days. Adult moth is yellowish green in colour. A female lays an average of 385 eggs. The life cycle is completed in 20 to 22 days. The species completes several generations in a year.

*Management*

*Chemical control*: Krishnakumar and Srinivasan (1984) recommended sprays of fenvalerate 50g a.i./ha, deltamethrin 10 g a.i./ha or cypermethrin 30g a.i./ha at flowering (40 days after sowing) and later at 20–25 days intervals for the control of this pest. Sarkar and Nath (1989) reported fenvalerate (0.5 ml/l) spray in reducing the fruit borer infestation and increasing the seed yield.

Foliar spray of carbaryl 500–600 a.i./ha was also found to give good control of this pest on okra (Kumar *et al.*, 1996).

*Biological control:* The eggs of *E. vitella* are parasitized by *Trichogramma evenescens* Westwood, larvae by *Actia aegyptica* Vill., *A. hyalinata* Mall., *Bracon brevicornis* (Wesmael), *B. greeni* (Ashmead), *B. hebetor* (Say), *Rhogas aligharensi* Qadri, *R. testaceus* Grav., and pupae by *Chelonus rufus* Lyle, *Brachymeria respomsonator* (Walker), *B. tachaidiae* (Cameron) and *Goryphus nursci* (Camerson). Clean cultivation, destruction of stubbles, infested shoots and fruits help in reducing the incidence.

**Okra aphid *Aphis gossypii* Glover**

*Aphis gossypii* is a cosmopolitan, polyphagous species found on a number of host plants all over India. Among the important cultivated crops, it is reported on capsicum, papaya, okra, plum, *Pyrus* and cucurbits (Bhalla and Pawar, 1977), whereas, Roy and Behura (1983) reported 220 host plants of this aphid throughout the world. The nymphs and adults suck the sap from tender leaves and twigs. Under severe infestation, plants wither away and affect the yield. Besides causing direct damage, the aphid acts as a vector and transmits several virus diseases.

*Life cycle*

The species remains active from the fourth week of July on okra and reproduces parthenogenetically. The biology of this pest was studied by Kandoria and Jamwal (1988) on okra. There are four nymphal instars. The duration of the respective instar being 2.4, 1.9, 1.9 and 2.0 days, respectively, and the total duration is 8.3 days. After a brief pre reproductive period, the females produce young ones for 7 to 8 days during which an average of 15.76 nymphs are produced by a female.

*Management*

Applications of monocrotophos (0.05%), dimethoate (0.03%) and endosulfan (0.07%) give satisfactory control of this aphid.

**Red cotton bug *Dysdercus koengii* (F.)**

The red cotton bug a minor pest of okra. The pest is also reported on cotton, hollyhock, bajra, maize, wheat, gooseberry, mulberry, simbal and silk cotton, etc. The nymphs and adults cause damage by sucking the sap from leaves and fruits. The plant becomes weak and stunted which ultimately affects the yield. The infested fruits may become distorted. The bugs also deprive the plants of carbohydrates, free amino acid and proteins.

*Life cycle*

The eggs are laid in clusters of 80 to 100 in cracks of moist soil or dry leaves near the plant. The eggs are ovoid, yellow but turn to orange colour before hatching. The incubation period varies from 5 to 8 days. There are five nymphal instars. The first instar does not feed but requires moisture and congregates near the empty chorions while the second and third instar feed gregariously.

Nymphs sometimes become cannibalistic. The latter instars feed freely and suck the sap from fruits and seeds. The nymphal period varies from 21 to 35 days. The adult bug is elongated, slender having crimson colour. A female on an average lays 170 eggs during its life span. The life cycle is completed in 40 to 90 days.

### *Management*

The pest can be managed by spraying malathion (0.05%). Hand picking and killing of insects in kerosinized water is also effective. Deep ploughing exposes the eggs and checks the population of this pest. A tachinid fly parasitizes the adults. Birds namely, *Crioius melanocephalus, Oriolus kundoo* Sykes, *Hierococcoyx varius* Vahl. and *Sitta ferontalis* Swainson feed on these bugs.

## Green semilooper *Anomis flava* F.

Green semilooper is a sporadic pest of okra. It also feeds on cotton, hollyhock, *Hibiscus*, cowpea, etc. The caterpillars feed on leaves leaving behind only the veins. In severe infestation, the caterpillars also damage the tender shoots and buds causing complete defoliation.

### *Life cycle*

The eggs are laid singly on the underside of leaves. The eggs are round and green in colour but change to pale green before hatching. Incubation period varies from 3 to4 days. The larvae feed voraciously from the initial instar. The full grown larva is pale yellowish green with five longitudinal white lines on the dorsal surface. The larval stage lasts for 15 to 20 days. Before pupation the larva spins a cocoon and pupates in the soil or debris. The pupae pass the winter in hibernation and the moths emerged in 9 days. The life cycle is completed in 4 to 6 weeks.

The pest can be managed by spraying endosulfan (0.05%). The incidence can also be minimized by deep ploughing.

## REFERENCES

Ahmad, T. and Gupta, R.L. 1941. The pea leafminer, *Phytomyza atricornis* (Meigen) in India. *Indian Journal of Entomology,* **3**: 37–49.

Atwal, A.S. 1986. *Agricultural pests of India and South-east Asia.* Kalyani Publishers, New Delhi, Ludhiana. pp. 508.

Atwal, A.S., Choudhary, J.P. and Ramzan, M. 1969. Studies on the biology and control of pea leafminer (*Phytomyza atricornis* Meigen). *Journal of Research, Punjab Agricultural University,* 6(Suppl.): 163–169.

Azmi, O.R. and Rathi, Y.P.S. 1992. Efficacy of insecticides against whitefly-a vector of french bean crinkle stunt virus. *Plant Disease Research,* 7(2): 257–258.

Bhalla, O.P. and Pawar, A.D. 1977. *A survey study of insect and non-insect pests of economic importance in Himachal Pradesh.* Tiku and Tiku, Kitab Mahal, Bombay. pp. 80.

Bhardwaj, S.P. and Bhalla, O.P. 1975. Laboratory observations on the biology of bean bug, *Chauliops nigrescens* Distant (Hemiptera: Lygaeidae). *Indian Journal of Entomology,* 37(4): 408–410.

Bindra, O.S. and Singh, G. 1970. Mite pests of ornamental plants in India. *Pesticides,* 3(6): 17–24.

Bindra, O.S, and Mahal, M.S. 1979. Investigation on varietal resistance in okra, *Abelmoschus esculentus* (L.) to the cotton jassid, *Amrasca biguttula biguttula* (Ishida). *Indian Journal of Horticulture,* **36**(2): 212–219.

Cheng, T.C. and Chan, C.C. 1989. Observations of three lepidopterous pests attacking leguminous vegetables in Taiwan. *Bulletin of Taichung District Agricultural Improvement Station,* **24**: 21–29.

Chhabra, K.S.; Lal, S.; Kooner, B.S. and Verma, M.M. 1993. *Insect pests of pulses: Identification and control manual.* Punjab Agricultural University, Ludhiana and Directorate of Pulses Research, Kanpur. pp. 88.

Chopade, H.M. 1975. Chemical control of some major pests of field pea, *Pisum sativum. Entomologists Newsletter,* **5**(2): 17.

Dahiya, A.S.; Sharma, S.S.; Verma, A.N. and Ombir. 1990. Comparative efficacy of different insecticides against jassid, *Amrasca biguttula biguttula* (Ishida) on okra. *Journal of Insect Science,* **3**(1): 83–87.

Darvas, B.; Szarukan, I.; Kaminszky, M.E. and Dulinafka, G. 1988. Occurrence of *Chromatomyia horticola* (Goureau) in sunflower. *Novenyvedlem,* **24**(9): 399–404.

Dash, A.N. 1990. Evaluation of some insecticides for their efficacy against the pea leafminer, *Phytomyza atricornis* (Meigen). *Indian Journal of Plant Protection,* **18**(2): 295–297.

David, B.V. and Jesudaasan, R.W.A. 1986. Status of the cotton whitefly, *Bemisia tabaci* (Genn.) excluding its vector biology. *Pesticides,* **20**(7): 42–45.

Distant, W.L. 1904. The Fauna of British India including Ceylon and Burma. Taylor and Frances, Fleet Street, London. *Rhynchaeta*- **II** (1): 35–36.

Gaffar, S.A.; Rather, A. and Mattoo, M.S 1990. Evaluation of various insecticides for the management of cowpea aphid, *Aphis craccivora* (Koch) infesting moong. *Journal of Aphidology,* **4**(1&2): 36–38.

Kakar, K.L.; Bhalla, O.P. and Singh, A.K. 1990. Studies on pest complex of french bean and their control under mid hill regions of Himachal Pradesh. *Indian Journal of Plant Protection,* **18**(1): 71–75.

Kandoria, J.L. and Jamwal, R. 1988. Comparative biology of *Aphis gossypii* Glover on okra, brinjal and chilli in the Punjab, India. *Journal of Aphidology,* **2**(1&2): 35–39.

Kashyap, N.P. and Adlakha, R.L. 1971. New records of insect pests of soybean crop. *Indian Journal of Entomology,* **33**(4): 467–468.

Kaurava, A.S.; Odak, S.C. and Dhamdhere, S.V. 1970. Studies on the biology of *Phytomyza atricornis* Meigen (Agromyzidae: Diptera). *Journal of Bombay Natural History Society,* **67**(3): 597–601.

Khajuria, D.R. and Sharma, J.P. 1995. Efficacy of insecticides in controlling pea leafminer (*Chromatomyia horticola*) on seed crop of pea (*Pisum sativum*). *Indian Journal of Agricultural Sciences,* **65**(5): 381–384.

Khurana, A.D. and Kaushik, H.D. 1991. Bioefficacy of insecticides against *Aphis craccivora* Koch and *Agrotis ipsilon* (Hufn.) on chickpea. *Journal of Insect Science,* **4**(2): 193–194.

Krishnamoorthy, P.N. and Srinivasan, K. 1991. Diagnosis and management of beanflies in beans. *In: Horticulture New Technologies and Applications*. J. Prakash and R.L.M. Pierik (eds). Kulwer Academic Publishers, Netherland. pp. 363–366.

Krishnakumar, N.K. and Srinivasan, K. 1984. Extending spray interval with synthetic pyrethorids for the control of fruit borer, *Earias vitella* Fab. on okra. *Madras Agriculture Journal,* **71**: 478–480.

Kumar, A. 1985a. Estimation of damage caused to *Chromatomyia horticola* population on *Brassica campestris* by hymenopteran parasites. *Entomon,* **10**(1): 49–53.

Kumar, A. 1985b. Incidence of parasitism of *Diglyphus isaea* (Walk.) of *Chromotomyia horticola* (Goureau) on *Pisum sativum. Entomon,* **10**(1): 55–58.

Kumar, R.; Singh, S.K.; Khokhar, K.S. and Aggarwal, N. 1996. Comparative efficacy of different insecticides applied against fruit borer infestation as seed treatment and foliar sprays on fruit yield of okra. *Annals of Agricultural Research,* **17**(3): 241–243.

Lal, O.P. 1974. Occurrence of *Chauliops fallax* Scott (Hemiptera: Lygaeidae) on french bean and horse gram in H.P. *Indian Journal of Entomology,* **36**(1): 67–68.

Lal, O.P. 1981. A contribution to the knowledge of ecology, biology, host range and control of the lygaeid bug, *Chauliops fallax* Scott (Hemiptera, Lygaeidae), a pest of pulse crop in Kullu Valley. *Rivista di Agriculture Subtropicale e Tropicale,* **75**(4): 381–403.

Lal, O.P. 1992. Evaluation of certain insecticides against the pest complex of french bean seed crop under field conditions. *Journal of Entomological Research,* **16**(1): 57–61.

Lal, O.P. and Nazi, A.H. 1979. Observations on some new pests and parasites from Socialist Peoples, Libyan Arab Jamahiriya. *Rivista di Agricoltura Subtropicale e Tropicale,* **73**(3/4): 219–232.

Lall, B.S.; Dutta, C.P. and Jayaswal, A.P. 1965. Bionomics and control of red spider mite. *Indian Journal of Entomology,* **27**: 389–392.

Linnaeus, C. 1758. *Acarus telarius. Systema Naturae,* **10**: 616.

Mavi, G.S. and Singh, Sukhminder. 1988. Assessment of losses by pea leafminer, *Chromatomyia horticola* (Goureau) on pea. *Journal of Insect Science,* **1**(1): 61–64.

Mehta, P.K., Vaidya, D.N. and Kashyap, N.P. 1995. Efficacy and economics of some insecticides against pea leafminer, *Chromatomyia horticola* (Goureau) on pea. *Journal of Insect Science,* **8**(1): 116–117.

Mote, U.N. 1983. Seasonal incidence and chemical control of stemfly on french bean and pea. *Journal of Maharashtra Agricultural Universities,* 8(2): 159–161.

Mound, L.A. and Halsey, S.H. 1978. *Whitefly of the world. A systematic catalogue of the Aleyrodidae (Homoptera) with host plant and natural enemy data.* British Museum (Natural History) 1978 and John Wiley and Sons. Chickester-New York-Brisbane-Toranto. pp. 340.

Otto'D., Karabensch, K., Muller, A. and Suss, A. 1985. Bait formulations for the control of the cutworm (*Scotia segetum* Schiff.). *Nachrichtenblatt fur den pflanzenschutz in der DDR,* **39**(9): 194–195.

Pal B.P. 1966. *Rose in India.* Indian Council of Agricultural Research, New Delhi, pp. 265.

Patel, H.K. and Patel, U.C. 1968. Control of red spider mite on soybean. *Pesticides,* **1**: 21–23.

Patnaik, H.P., Mohapatra, H.K. and Senapati, B. 1993. Blister beetle incidence and some observations on its ovipositional behaviour on pigeonpea in Orissa, India. *International Pigeonpea Newsletter,* **18**: 27–30.

Rajagopal, D. and Kumar, P. 1988. Predation potentiality of *Chlaenius panagaeoides* (Laferte) (Coleoptera: Carabidae) on cowpea aphid *Aphis craccivora* Koch (Homoptera: Aphididae). *Journal of Aphidology,* **2**(1&2): 93–99.

Rao, C.V.; Punnaiah, K.C.; Prasad, V.D. and Krishnaiah, P.V. 1996. Studies on the efficacy of certain newer insecticides against brinjal whitefly, *Bemisia tabaci* Gennadius. *Pestology,* **20**(7): 17–19.

Rawat, R.R. and Sahu, H.R. 1973. Estimation of losses in growth and yield of okra due to *Empoasca devastans* and *Earias* spp. *Indian Journal of Entomology,* **35**(3): 252–254.

Roy, D.K. and Behura, B.K. 1983. Notes on host plants, feeding behaviour, infestation and ant attendances of cotton aphid, *Aphis gossypii* Glover. *Journal of Bombay National History Society,* **80**: 654–656.

Rustogi, K.B. 1981. Record of *Chauliops fallax* Scott on urd bean in Haryana. *Haryana Agricultural University Journal of Research,* **11**(3): 418.

Sarkar, B.B. and Nath, S.C. 1989. Field evalulation of various insecticides against fruit borer (*Earias vitella* F.) and effect on yield of lady's finger seed. *Pestology,* **12**(4): 19–21.

Sharma, K.C.; Chauhan, U.; Verma, A.K. and Sood, A.K. 1994a. Biology of the black bean bug, *Chauliops nigrescens* Distant (Hemiptera: Lygaeidae) infesting French bean, *Phaseolus vulgaris* L. *Journal of Entomological Research,* **17**(4): 305–307.

Sharma, K.C.; Chauhan, U. and Verma, A.K. 1994b. Biology of pea leafminer, *Chromatomyia horticola* (Goureau) (Diptera: Agromyzidae) on pea, *Pisum sativum* L. *Indian Journal of Agricultural Sciences,* **64**(1): 72–73.

Sharma, P.L. and Bhalla, O.P. 1964. A survey study of insect pests of economic importance in Himachal Pradesh. *Indian Journal of Entomology,* **26**(3): 318–331.

Sheng, J.K.; Zhong, L. and Wu, Q. 1989. Studies on the parasite complex of garden pea leafminer, *Phytomyza horticola* (Diptera: Agromyzidae) in Nanchang area. *Chinese Journal of Biological Control,* **5**(4): 164–167.

Singh, H. and Dhooria, M.S. 1971. Bionomics of the pea pod borer, *Etiella zinckenella* (Tr.). *Indian Journal of Entomology,* **33**(2): 123–130.

Singh, O.P.; Neha, K.K. and Verma, S.N. 1987. Outbreaks and new records. India: new insect pests of soybean (*Glycine max)* in Madhya Pradesh, India. *FAO Plant Protection Bulletin,* **35**(3): 100–102.

Singh, O.P.; Dhamdhere, S.V. and Misra, U.S. 1986. Toxicity of insecticides against pupae of pea leafminer, *Phytomyza horticola* (Meigen.) infesting vegetable pea. *Indian Journal of Entomology,* **48**(4): 372–374.

Singh S and Sohi GS. 1957. List of insect pests of economic importance in the Punjab. *Government Agriculture College Magazine, Ludhiana,* **5**: 19–35.

Singh, T.; Kumar, V.; Satyanarayana, P. and Goud, T.R. 1991. Studies on the biology of french bean stemfly, *Ophiomyia phaseoli* (Tryon.). *Journal of Insect Science,* **4**(2): 172–173.

Srikanth, J. and Lakkundi, N.H. 1988. Host preference studies of cowpea aphid, *Aphis craccivora* Koch. *Indian Journal of Plant Protection,* **16**(1): 103–107.

Srinivasan, K. and Krishnamoorthy, P.N. 1992. Changing scenario of pest problems in solanaceous, malvaceous and leguminous vegetables and their management. Paper presented at National Seminar on Changing scenario in pests and pest management in India held at CPPTI, Hyderabad, Jan-Feb., 1992. Abst. p.6.

Srinivasan, K. and Shukla, R. 1980. Control of pea leafminer (*Phytomyza atricornis* Meigen) and aphids (*Acyrthosiphon pisum* Harris) with certain granular insecticides. *Pesticides,* **14**(12): 19–20.

Srivastava, A.S. and Singh, Y.P. 1974. Survey, damage, life history and control of pea pod borer, *Etiella zinekenella* Tr. (Pyralidae: Lepidoptera). *Science and Culuture,* **40**(8): 370–372.

Srivastava, B.K. and Mathur, L.M.L. 1962. Bionomics and control of castor mite. *Indian Journal of Entomology,* **24**: 229–235.

Stanentite, A.P. 1986. 14 species of Hymenoptera Chalcidoidea new to the hithuanian SSR found in 1969–1985. *In: Novye i redkii dlya utovskoi SSR Vidy nasekmykh, Soobschcheniyon i.,* pp. 43–49.

Tandon, S.K. 1971. Notes on the biology of *Phytomyza atricornis* Meigen, a common leafminer of cruciferous plants (Agromyzidae: Diptera). *Agra University Journal of Research* (Sc.), 20: 47–67.

Tarraborrelli, L.; Talame, M.; Principe, P. and Corvi, F. 1989. Control of noctuiids in horticulture with bait formulation based on quinalphos. *Difesa delle Piante,* **12**(1–2): 75–80.

Verbeek, F. 1930. Meloid beetles in Java. *Tectona,* **23**: 304–308.

Verma, R.S.; Prakash, N. and Nath, V. 1980. Note on chemical control of pea leafminer, *Phytomyza atricornis* Meigen and pea pod borer, *Etiella zinckenella* Tr. *Indian Journal of Applied Entomology,* **42**(2): 824–826.

Xu, G.J. 1987. Hemipteran, homopteran, lepidopteran and dipterans insects parasitized by mermithid nematodes. *Natural enemies of Insects,* **9**(2): 118–121.

*Pest Management in Horticulture Crops*
L.R. Verma, A.K. Verma and D.C. Gautam (eds)


*Chapter 20*

# Insect Pests of Radish and Turnip

*G.C. Sachan*

## INTRODUCTION

Radish (*Raphanus sativus* Linnaeus) and turnip (*Brassica rapa* Linnaeus) are root crops, rich in minerals, vitamin A, fibre, ascorbic acid and oher essential nutrients required by the human body (Gopalan and Balasubramanium, 1966). Roots are used as salad, cooked as vegetable or used in pickles whereas leaves are generally used as cooked products. These crops are attacked by several insect pests (Table 1), out of which mustard sawfly, aphids including mustard aphid, cabbage aphid and green peach aphid are of major concern.

**Table 1.** Important insect pests of radish and turnip

| *Order/Family* | *Scientific name* | *Common name* |
|---|---|---|
| Hemiptera | | |
| Aleyrodidae | *Bemisia tabaci* (gennadius) | Cotton Whitefly |
| Aphididae | *Lipaphis erysimi* (kaltenbach) | Mustard aphis |
| | *Brevicoryne brassicae* (Linnaeus) | Cabbage aphid |
| | *Myzus persicae* (Sulzer) | Peach aphid |
| Pentatomidae | *Bagrada cruciferarum* Kirkaldy | Painted bug |
| Thysanopera | | |
| Thripidae | *Thrips tabaci* Lindemann | Onion thrip |
| Diptera | | |
| Agromyzidae | *Chromatomyia horticola* Goureau | Leafminer |
| Hymenoptera | | |
| Tenthredinidae | *Athalia proximz* Klug | Mustard sawfly |
| Lepidoptera | | |
| Plutellidae | *Plutella xylostella* (Linnaeus) | Diamond back moth |
| Pyraustidae | *Hellula undalla* (Fabricius) | Cabbage borer |

| | | |
|---|---|---|
| Noctuidae | *Agrotis ipsilon* (Hufnagel) | Cut worm |
| | *Spodoptera litura* Fabricius | Tobacco caterpillar |
| Arctiidae | *Spilosoma obliqua* (Walker) | Bihar hairy caterpillar |
| Pyralidae | *Crocidolomia binotallis* zeller | Greater moth |
| Coleoptera | | |
| Chrysomelidae | *Phyllotreta cruciferae* Goeze | Flea beetle |
| | *P. chotanica* Duvier | |
| | *P. downset* Baly | |
| | *P. vittata* Fabricius | |
| | *Menolepta signata* oliver | |
| | *Chaetocnema basalis* Baly | |

## SAWFLY

The mustard sawfly, *athalia proxima* Klug, is one of the few hymenopterous insects that cause damage to field crops. The mustard sawfly is an oligophagous species feeding on the plants of family Cruciferae because of the presence of sinigrin in them. Detailed account of identification of different stages, biology and seasonal history of this insect has been given by Bakhetia and Sachan (1997). The greenish black larvae look like caterpillars but have eight pairs of prolegs and measure 16–20 mm. They defoliate the crop completely particularly at the seedling stage necessitating resowing. Owing to their extensive potential for defoliation of host plants, a very low economic threshold level of 2 larvae per square metre area has been proposed by them. Adults inflict damage to cotyledonary stage of the plant by the act of egg laying with the help of their sawlike ovipositor.

The pest is active throughout the year except during April to June when it remains under aestivation. On radish it becomes active from mid July and completes five generations upto end of October. Later, it completes five more generations on other cruciferous hosts till the end of February. Tenth generation larvae descend into the soil for pupation in earthen cells till June end (Shirke *et al.*, 1968; Nayar *et al.*, 1979). In North India, it appears in September and remains active till the second week of March with peak period of activity during November to January on various crucifers and completes 3 to 4 generations (Rai, 1976). It aestivates in soil during summer and hibernates in severe winter (below 7°C) (Dhillon, 1966). Hibernation takes place in prepupal stage in earthen cocoon.

The larvae prefer sandy soil for pupation with soil moisture varying from 10 to 15 per cent (Sachan and Tripathi, 1986; Verma and Sachan, 1993, 1996).

Sachan and Chhibber (1986) indicated that larvae have the habit of basking in the sun. They come out at 6.30 AM from their hiding places and get collected on the top of the plant reaching to peak congregation at 10.30 AM in winter. They feed there till 3.30 PM and start descending. On cloudy days, only a few larvae are seen on plants. Larval population is negatively correlated with bright sun shine. Similarly morning vapour pressure, evening relative humidity and bright sun shine are responsible for variation in insect population to the extent of 67.4 per cent under

Gujarat conditions (Patel, 1995). A positive association with temperature and sawfly population was observed by Patel *et al*. (1996).

Under laboratory conditions, the food preference of the larvae was found to be: turnip > radish > mustard > rai, whereas in the field, turnip was found to have the maximum infestation (Srivastava, 1957). Grubs consumed more leaf area of mustard and turnip than radish, cauliflower and cabbage (Bindra and Mehra, 1964). Sharma (1979) also food preference for radish than toria. Similarly, Patel (1995) reported that radish was the most preferred host followed by mustard, cauliflower and cabbage. The insect successfully completed development on radish (Kapadia *et al*., 1980; Singh, 1993) and turnip (Sehgal and Ujagir, 1977; Sharma, 1979). Some parameters in the life cycle of this insect on turnip and radish as studied by Bogawat (1967) are presented in Table 2.

**Table 2**: Duration of *A. proxima* on turnip and raddish

| *Parameters* | *Turnip* | *Radish* |
|---|---|---|
| Larval period (days) | 12.00 | 13.75 |
| Pupal period (days) | 10.00 | 11.25 |
| Pupation (%) | 80.00 | 66.66 |
| Adult emergence (%) | 91.66 | 100.00 |
| Adult longivity (days) | | |
| Male | 08.50 | 08.00 |
| Female | 10.50 | 11.00 |
| Preoviposition period (days) | 01.00 | 02.00 |
| Oviposition period (days) | 06.00 | 06.00 |
| Average Fecundity | 31.25 | 29.25 |

The larvae which pupated after 4 moults developed into males and those after five into females. Unmated female progeny undergoing five moults developed into males only. The duration of total life cycle varied with weather conditions. It took 29.9 days during July-October and 32.19 days during November-March in Kolhapur, Maharashtra where females outnumbered males (Jagtap and Kadam, 1978; Kapadia *et al*., 1980). The optimum temperature for larval development was worked out to be 28oC but at 30oC all the larvae died (Bogawat, 1968). Temperature below 19oC could bring down the pupation rate drastically (Wahla *et al*., 1979). Mating behaviour of *A. proxima*has also been studied in detail (Verma and Sachan, 1991; Shirke *et al*., 1968; Jagtap and Kadam, 1978). Mustard, turnip, radish, *rai*, cauliflower and cabbage are more preferred for egg laying (Bindra and Mehra, 1964; Dhillon, 1966; Bogawat, 1967).

### Natural enemies

Two larval parasitoids namely, *Exacrodus populans* Morley and Perilissus cingulator Morley and a predator, *Canthoconia furcellata* Distant have been recorded. The adults prey on sawfly larvae.

**Management**

- Hand picking of larvae and their destruction.
- Irrigation results in drowning of larvae.
- Application of quinalphos (0.025%,) endosulfan(0.05%,) malathion (0.05%) (Patel *et al*., 1994); dichlorvos (0.05%), (0.1%) malathion (Butani *et al*., 1977): carbaryl 5% dust, (Butani and Jneja, 1984); Spark (0.036%) (Triazophos + deltamethrin (35 : 1), quinalphos (0.05%) (Patel, 1995) have been recommended. In addition, phosphamidon, monocrotophos, phosalone and methyl parathion have also been recommended for its control (Patil and Pokharkar, 1973; Jagtap and Kadam, 1981; Srinivasan and Krishnakumar, 1986).
- Use of neem has been found very effecitve in the management of sawfly. This is mainly due to antifeedant, repellent and insecticidal properties of neem (Pandey *et al*., 1977). Petroleum ether extract of seed kernels of neem is very effective against sawfly larvae (Singh *et al*., 1991). On the other hand, extract of neem leaves was reported to possess antifeedant activity against this pest (Verma, 1992). Among the various products based on neem, Repelin (1%), Margocide CK (0.1%), Neemark (1%) and NSKS (5%) were reported to possess high oviposition deterrent action, antifeedant and/or larvicidal action on sawfly. *Acous calamus* extract also has antifeedant as well as insecticidal properties (Pandey *et al*., 1979), bitter gourd seed oil emulsion (2%) (Kumar *et al*., 1979), alcohol extract of Derris indica seeds (0.0%), petroleum ether extract of *A. calamus* (0.5–2.5%) and alcohol extract of *Allium sativum* (2.5%) provided complete protection to radish crop (Banerji *et al*., 1982). Ethyl acetate extract of *Lantana camara* leaves and benzene extract of *Parthenium hysteophorus* and neem leaves were more potent antifeedants than their methanol and water extracts (Verma and Sachan, 1994).
- *Bacillus thuriengiensis* Berliner suspension when applied @ 0.05—0.2% indicated 50 per cent larval mortality in 3 days and complete mortality after 4 days of treatment but no toxicity after 7 days of application under field conditions (Saxena, 1979). Three sprays of DD136 (based on *Steinernema feltiae* (Filipjev)) at 1.1 x 103 nematodes/ml at weekly intervals checked the population of sawfly on radish (Narayanan and Gopalakrishnan, 1988).

## APHIDS

Aphids species reported on radish and turnip include the mustard aphid, *Lipaphis erysimi* Kalt., cabage aphid, *Brevicoryne brassicae* (L.) and green peach aphid, *Myzus persicae* (Sulzer). The important identification characters, nature and extent of damage of these species have been described in detail by Bakhetia and Sachan (1997). The workers have also given a detailed account of biology and seasonal history of mustard aphid on rape seed mustard.

Since the arrival of mustard aphid on early and mid season radish is late, the leaves are least affected and there is no remakable loss in quality and quantity. This is also true for roots which are used for culnery purposes. In the case of heavy infestation the plants are completely devitalized, leaves and shoots curl up, become yellowish and finally die. This is due to diminished sap flow and increased growth inhibiting substances with concomitant decrease in the content of cytokinins, gibberellins and auxins in aphid infested seedlings (Forrest *et al.*, 1973; Hussain *et al.*, 1973). The seed crop is badly affected resulting in tremendous loss in qualilty and quantity (60–70%) of seed (Sinha *et al.*, 1977). This is one insect which directly affects the radish seed industry.

In Orissa the mustard aphid appears in late November and disappears in early March on radish (Rout and Senapati, 1968), whereas in West Bengal this aphid appears on seed crop in early December reaching to its peak in January and starts declining in February (Ghosh and Mitra, 1983).

Singh *et al.* (1993) summarised the information on various abiotic factors responsible for population regulation of mustard aphid on rapeseed mustard and indicated that temperature range from 17 to 32oC with 61 to 85 per cent relative humidity was conducive for the insect. Rainfall more than 10 mm is detrimental whereas, mild shower (about 2 mm) and cloudy weather favour its multiplication. A positive correlation exists between sunshine and aphid population during January/February which becomes negative during March/April. They further pointed out that rainfall and relative humidity, and temperature account for 68 and 53 per cent influence on the population, respectively.

**Management**

- Plucking and destruction of infested twigs at the initial stage of attack will help in checking the spread and build up of aphid population.
- Seed crop of variety S-6 and Pusa Chetki whose stocklings are planted earlier than mid December, suffers less attack but the best time was the third week of November (Singh *et al.*,1987).
- Among the radish varieties Punjab Safed, White-5, Pusa Chetki and S-6, the last one is less attacked by the mustard aphid in early and mid planted groups (Singh *et al.*, 1987). Bhalla and Sandhu (1973) reported that turnip (White-4) is preferred by the aphids than radish. Srinivasachar and Malik (1972) indicated that light green leaves of turnip having low wax content were less susceptible than cauliflower, cabbage and knol khol having higher wax content and blue green and waxy green leaves.

*Insecticides:* Application of carbaryl (0.1%), dimethoate (0.05%), DDVP 0.04% or monocrotophos (0.04%) at 10 to 12 days interval has been recommended (Butani and Juneja, 1984). Spraying of oxydemeton methyl, 0.34 kg a.i./ha (Sekhon *et al.*, 1980); monocrotophos 300 g a.i./ha (3 applications at 21—day interval) (Singh and Singh, 1987; Srivastava *et al.*, 1991); fluvalinate 50 g a.i./ha, 50 g a.i./ha cypermethrin (Umapathy and Bhaskaran, 1991) were also found effective in controlling the mustard aphid.

The economic threshold level of mustard aphid on seed crop of radish was prescribed by Bath and Singh (1989). They pointed out that 50 aphids per plant was the critical number for the application of oxydemeton methyl @ 300 g a.i./ha by middle of February. Foliar application of insecticides for the control of aphid on seed crop of radish is generally not found suitable because it causes damage to grown up crop during spraying, kills natural enemies and other beneficial insects especially honey bees.

Soil application of aldicarb granule @ 1 kg a.i./ha (half at planting and half at aphid appearance) (Bhanot and Verma, 1978; Kashyap *et al.*, 1985), full dose at 40 to 50 days after planting the stecklings (Sinha *et al.*, 1983), side dressing of aldicarb @ 1 kg a.i./ha, or phorate @2 kg a.i./ha (Sekhon *et al.*, 1983) are effective treatments against the mustard aphid. Stecklings dipped in 0.5% thiometon solution before transplanting protects the seed crop upto 58 days (Yadav and Jotwani, 1969). Treatment of stecklings with 0.2% aldicarb slurry is also an efficient and cost effective treatment (Sinha *et al.*, 1983; Bhanot and Verma, 1978; Kashyap *et al.*, 1985).

*Botanical pesticides:* Initial laboratory experiments to assess the effectiveness of application of neem/neem products against the mustard aphid resulted in delay in reproduction, reduction in fecundity and aphicidal action (>80% mortality) when fed on leaves treated with aqueous suspension of seed powder (Sharma *et al.*, 1986) or 1–1.5% leaf extract (Pandey *et al.*, 1987). Aqueous extract (salanin), ethanolic extract (salanin derivatives) and hexane extract (non terpenoids) of neem seed kernel indicated LC50 values of 0.055, 0.0255 and 0.14 per cent, respectively (Singh *et al.*, 1988). Among the several plant extracts the extract from neem seed was found very effective to repel mustard aphid (Kabir and Mia, 1987).

*Natural enemies:* Among the coccinellids, *Coccinella septempunctata L., C. repanda Thub., Brumoides suturali, Hippodamia variegata (Goeze)* and *Menochilus sexmaculatus* (F.) were effective predators of mustard aphid. Two species of chrysopids, *Chrysoperla carnea* (Stephens), and *Anisochrysa boniennsis* Ckomata, the syrphid, *Episyrphus balteatus* (De Gear) and Chamaemyid, *Leucopis*sp. are also found associated with colonies of the mustard aphid. The parasitoid, *Diaeretiella rapae* (McIntosh) parasitises the aphid and regulates the musard aphid-population on seed crop of radish (Dhiman and Kumar, 1985).

## FLEA BEETLES

The important flea beetles attacking cruciferous crops especially radish and turnip include *Monolepta signata* Oliver, *Phyllotreta chotanica* Duvier, *P. cruciferae* Goeze, *P. downsei* Baly, *P. vittata* Feb. and *Chaetocnema basalis* Baly. They cause extensive perforations (shot holes) and pitting of the foliage. Severe damage may be done to the developing seedlings before they appear above ground as well as to cotyledons as they break through the soil. Injury is also done during rough leaf stage but worst injury is caused to the seedlings in cotyledonary state. The larvae of *P. neumorum* L. (large striped flea beetle) mine the leaves. These mines are linear at first but later become extended to appear like blotches.

*Phyllotreta downsei* and *M. signata* are common on radish in South India whereas *P. cruciferae* is more common in North India.

The more severe white spotted beetle, *M. signata* is is 2–3 mm in length with reddish brown body and pale elytra having two big white spots. The adults of *P. chotanica* measure 2 mm and are metallic bronze in colour while that of *P. cruciferae* are 2 mm in size, shining dark blue and very active. The biology and seasonal history of flea beetles on rapeseed mustard have been discussed by Bakhetia and Sachan (1997).

Flea beetles can be controlled by carbaryl (5%) or endosulfan (4%) dust or spraying with carbaryl (0.2%) DDVP (0.05%) or endosulfan (0.05%) (Butani and Juneja, 1984), or dichlorvos (0.1%) (Srinivasan, 1993).

Srinivasan and Krishnakumar (1986) suggested that there is no need to control flea beetles and other lepidopterous insects as market value of roots is not affected. However, control is needed when flea beetles as well as sawfly are infesting the crop and suggested the use of endosulfan (0.07%) monocrotophos (0.05%), fenvalerate (0.01%) or deltamethrin (0.004%).

## LEAFMINER

The pea leafminer, *Chromatomyia horticola* Goureau, is a polyphagous and serious pest of pea which also attacks radish (Ahmad and Gupta, 1941. Zaka-Ur-Rab (1941) reported this insect for the first time from Kashmir. Detailed marks of identification, nature of damage, biology and seasonal history have been described by Bakhetia and Sachan (1997). In North India infestation on radish is initiated from mid February and remains till maturity of the crop with peak period of activity during mid March (Omvir and Bhanot, 1997). They further reported that the crop planted on December 25, had more population than those planted 10 and 20 days later. However, non-significant correlation was worked out between leafmoner population and abiotic factors of the environment. Collection and destruction of infested leaves is useful. The pest can be controlled by the use of insecticides suggested for aphid control.

## LEAF EATING CATERPILLARS

Five species of lepidopterous larvae are found feeding on the radish foliage (Butani and Juneja, 1984; Srinivasan and Krishnakumar, 1986). Among these, *Plutella xylostella* L., *Crocidolomia binotalis* Zeller and *Hellula undalis* (F.)are oligophagous, restricting their activity only on crucifers while *Spilosoma obliqua* (Walker)and *Spodoptera litura* F. are polyphagous.

The caterpillars of *P. xylostella* first make pit-like feeding marks when young but later the entire leaf is devoured. *C. binotalis* larvae feed gregariously on chlorophyll just after hatching but segregated later. After segregation they spin webbs and feed within the leaf tissues and when grown up these larvae web the terminal growing points having inflorescence and feed gregariously. Due to this the growth of the plant is checked, the plant becomes stunted and flowers are shed. Larvae also act as pod borers and feed on seeds. The young larvae of *H. undalis* mine the leaves along the side veins and the affected leaves turn papery.

The Bihar hairy caterpillar and tobacco caterpillar feed gregariously on the leaves and when grown up disperse throughout the field and defoliate the crop.

Since these are the minor pests of radish no separate efforts are required to control these pests in North India but sometimes *P.xylostella* and *C. binotalis* pose threat to seed crop in South India. Collection and destruction of infested leave/plant parts particularly when larvae are in gregarious form is recommended. Several chemicals like 5% carbaryl or malathion (5%) dust, DDVP (0.05%) or endosulfan (0.05%) (Butani and Juneja, 1984), deltamethrin (0.0014%), fenvalerate (0.01%), phosalone (0.07%), monocrotophos (0.05%) and endosulfan (0.07%) (Srinivasan and Krishnakumar, 1986) have been recommended.

## REFERENCES

Ahmad, T. and Gupta, R.L. 1941. The pea leafminer, *Phytomyza atricornis* (Meigen), in India. *Indian Journal of Entomology,* 3: 37–49.

Bakhetia, D.R.C. and Sachan, G.C. 1997. *Identification and Management Strategies for Insect-pests of Rapeseed and Mustard*. Directorate of Extension, Ministry of Agriculture, Govt. of India, pp. 43.

Banerji, R.; Mishra, G.; Nigam, S.K.; Prasad, N.; Pandey, R.S. and Mathur, Y.K. 1982. Indigenous plants as antifeedants. *Indian Journal of Entomology,* **44**: 71–76.

Bhalla, J.S. and Sandhu, G.C. 1973. Varietal susceptibility of different turnip and radish varieties to mustard aphid, *Lipaphis erysimi* Kalt. *Punjab Vegetable Grower,* **8**: 28–29.

Bhanot, J.P. and Verma, A.N. 1978. Effectiveness of three granular systemic insecticides applied by different methods and malathion sprays on aphid control and seed yield of radish. *Seed Research,* **6**: 48–57.

Bath, D.S. and Singh, D. 1989. Studies on the economic threshold level of mustard aphid, *Lipaphis erysimi* on the radish seed crop in India. *Tropical Pest Management,* **35**: 154–156.

Bindra, O.S. and Mehra, C.K. 1964. A note on host preference by Athalia proxima Klug the mustard sawfly. *Indian Journal of Entomology,* **26**: 284–288.

Bogawat, J.K. 1967. Biology of mustard sawfly on different host plants. *Indian Journal Entomology,* **29**: 270–274.

Bogawat, J.K. 1968. Effect of temprature on the development of sawfly, *Athalia proxima* Klug (Hymenoptera: Tenthredinidae). *Indian Journal of Entomology,* **30**: 80–82.

Butani, D.K.; Jotwani, M.G. and Verma, S. 1977. Insect pests of vegetables and their control: Cole crops. *Pesticides,* **11**: 19–24.

Butani, D.K. and Juneja, S.S. 1984. Pests of radish in India and their control. *Pesticides,* **18**: 10–12.

Dhillon, S.S. 1966. On Indian insect types. VII Morphology and biology of *Athalia proxima* Klug (Tenthredinidae: Hymenoptera). Aligarh Muslim University Publication (Zoology. Series). pp. 166.

Dhiman, S.C. and Kumar, V. 1985. Biological control of mustard aphid, *Lipaphis erysimi* (Kalt.) by using the parasitoid, *Aphidius (Diaeretiella)* rapae (Curtis). I. Influence of parasitism on aphid. *In: Proceedings of National Symposium on Pesticides Residue and Environmental Pollution.* Muzaffarnagar, India 2–4 October, 1985.

Forrest, J.M.S.; Hussain, A. and Dixon, A.F.G. 1973. Growth and wilting of radish seedlings, *Raphanus sativus*, infested with the aphid, *Myzus persicae. Annals of Applied Biology,* **75**: 267–274.

Ghosh, M.R. and Mitra, R. 1983. Incidence pattern and population composition of *Lipaphis erysimi* (Kaltenbach) on mustard and radish. *Pranikee,* **4**: 43–51.

Gopalan, C. and Balasubramanium, S.C. 1966. *Nutritive value of Indian Foods*. (6th revised Edn.), Indian council of medical Research, New Delhi.

Hussain, A.; Forrest, J.M.S. and Dixon, A.F.G. 1973. Changes in growth regulator content of radish seedlings, Raphanus *sativus*, infested with the aphid, *Myzus persicae*. *Annals of Applied Biology,* **75**: 275–84.

Jagtap, A.B. and Kadam, M.V. 1978. Biology of mustard sawfly, *Athalia proxima* Klug on radish (*Raphanus sativus*) *Journal of Maharashtra Agricultural Universities,* **3**: 107–108.

Jagtap, A.B. and Kadam, M.V. 1981. Studies on chemical control of mustard sawfly on radish. *Pesticides,* **15**: 9–10.

Kabir, K.H. and Mia, M.D. 1987. Effectiveness of some indigenous materials as repellentsagainst mustard aphid. *Bangladesh Journal of Zoology,* **15**: 87–88.

Kapadia, M.N.; Bhadoria, R.K. and Vora, V.J. 1980. Biology and larval and post larval development of *Athalia proxima* Klug on different host plants. *Gujarat Agricultural University Research Journal,* **6**: 13–16.

Kashyap, R.K.; Bhanot, J.P.; Pandey, U.C. and Lal, S. 1985. Evaluation of aldicarb (Temik 10G) and malathion sprays against aphids in seed crop of radish. *Haryana Agricultural University Journal of Research,* **15**: 414–417.

Kumar, A.; Tewari, G.C. and Pandey, N.D. 1979. Antifeeding and insecticidal properties of bitter gourd, *Momordica charantia* Linn. against Athalia proxima Klug. *Indian Journal of Entomology,* **41**: 103–106.

Narayanan, K. and Gopalakrishnan, C. 1988. Effect of DD-136 (*Steinernema feltiae*) for the control of mustard sawfly, *Athalia proxima* on radish and its persistence in soil. Abstr. 10th International Soil Zoology Colloquium, 7–13 August, Bangalore, India. pp. 15.

Nayar, K.K.; Ananthakrishnan, T.N. and David, B.V. 1979. *General and Applied Entomology*. Tata McGraw Hill, New Delhi.

Ombir and Bhanot, J.P. 1997. Population dynamics of pea leaf miner, *Chromatomyia horticola* on seed crop of radish planted on different dates. Abstr. International Conference on Ecological Agriculture, 15–17 November, 1997, Chandigarh, India. pp. 82.

Pandey, N.D.; Singh, I.; Singh, Y.P. and Tripathi, R.A. 1987. Effect of certain plant extracts against *Lipaphis erysimi* Kalt. under laboratory conditions. *Indian Journal of Entomology,* **49**: 238–242.

Pandey, N.D.; Singh, M. and Tewari, G.C. 1977. Antifeeding, repellent and insecticidal properties of some indigenous plant material against mustard sawfly. *Athalia proxima* Klug. *Indian Journal of Entomology,* **39**: 60–64.

Pandey, N.D.; Sudhakar, T.R.; Tewari, G.C. and Pandey, V.K. 1979. Evaluation of some botanical antifeedants under field conditions for the control of *Athalia proxima* Klug. *Indian Journal of Entomology,* **41**: 107–109.

Patel, C.C.; Patel, G.R. and Patel, M.G. 1994. Bioefficacy of insecticides against mustard sawfly, Athalia *lugens proxima* Klug. *Gujarat Agricultural University Research Journal,* **19**: 50–52.

Patel, G.P. 1995. *Bionomics and chemical control of sawfly, Athalia lugens proxima Klug (Tenthredinidae: Hymenoptera) on radish, Raphanus sativus L.* M.Sc. Thesis Gujarat Agriculture University, Anand, pp. 180.

Patel, K.I.; Patel, M.G.; Sheikh, A.M.; Patel, J.R.; Savani, M.B. and Patel, C.T. 1996. Effect of temperature on population build up of mustard sawfly, *Athalia lugens proxima* and mustard aphid, *Lipaphis erysimi* Kalt. *Gujarat Agricultural University Research Journal,* **22**: 85–89.

Patil, S.P. and Pokharkar, R.N. 1973. Bionomics and control of *Athalia proxima* Klug. *Mahatama Phule Agricultural University Research Journal,* **4**: 44–50.

Rai, B.K. 1976. *Pests of Oilseed crops in India and their Control. Indian Council of Agricultural Research*, New Delhi, pp. 121.

Rout, G. and Senapati, B. 1968. Biology of mustard aphid, *Lipaphis erysimi* in India. *Annals of Entomological Society of America,* **61**: 259–261.

Sachan, G.C., Chhibber, R.C. 1986. Basking behaviour of *Athalia proxima* Klug. *Cruciferae Newsletter,* **11**: 103.

Sachan, G.C. and Tripathi, N.L.M. 1986. Influence of soil type on pupation of *Athalia proxima* Klug. *Cruciferae Newsletter,* **11**: 104–105.

Saxena, R.C. 1979. Efficiency of Bacillus thuringiensis Berliner for the control of *Athalia proxima* Klug. *Indian Journal of Ecology,* **6**: 33–35.

Sehgal, V.K. and Ujagir, R. 1977. Plant natural defense in Cruciferae and Tropaeolaceae against mustard sawfly, *Athalia proxima* Klug. *Indian Journal of Ecology,* **4**: 199–205.

Sekhon, S.S.; Bindra, O.S. and Chahal, B.S. 1983. Studies on the control of aphids infesting seed crop of radish with soil granular systemic insecticides. *Indian Journal of Plant Protection,* **11**: 9–12.

Sekhon, S.S.; Sajjan, S.S. and Kanta, U. 1980. Chemical control of mustard aphid. *Lipaphis erysimi* on seed crop of radish. *Indian Journal of Plant Protection,* **8**: 151–153.

Sharma, M.M. 1979. *Consumption and utilization of six plants by mustard sawfly grubs, Athalia proxima Klug and their effect on growth and development.* M.Sc. Thesis G.B. Pant University of Agriculture and Technology, Pantnagar, pp. 106..

Sharma, R.K.; Saxena, R. and Nathani, S. 1986. Aphicidal action of neem (*Azadirachta indica A. Juss.*) on mustard aphid, *Lipaphis erysimi* (Kalt.). *Neem Newsletter,* **3**: 1–2.

Shirke, D.B.; Rawat, R.R. and Dhamdhere, S.V. 1968. Studies on the biology of *Athalia proxima* Klug (Hymenoptera: Tenthredinidae). *Madras Agricultural Journal,* **55**: 228–234.

Singh, D. and Singh, H. 1987. Effectiveness of granular and foliar application of systemic insecticides controlling mustard aphid, *Lipaphis erysimi* Kalt. on radish crop. *Indian Journal of Entomology,* **49**: 453–456.

Singh, D.; Singh, H. and Brar, H.S. 1987. Influence of varieties and dates of planting of radish stecklings on infestation by mustard aphid, *Lipaphis erysimi* (Kaltenbach). *Punjab Agricultural University, Journal of Research,* **24**: 396–398.

Singh, H.; Rohilla, H.R. and Ahlawat, D.S. 1993. Ecology of major pests of rapeseed mustard with reference to forecasting of incidence of mustard aphid. Abstract Group Discussion on IPM Strategies in Oilseeds in India. Punjab Agricultural University, Ludhiana. December 23–24, pp. 23–24.

Singh, R.P.; Devakumar, C. and Dhingra, S. 1988. Activity of neem seed kernel extract against mustard aphid, *Lipaphis erysimi*. *Phytoparasitica,* **16**: 225–229.

Singh, S.P. 1993. *Developmental behaviour of Athalia proxima Klug on crucifers.* Ph.D. Thesis, G.B. Pant University of Agriculture and Technology, Pantnagar. pp. 147.

Singh, Y.P.; Pandey, S.; Guddewar, M.B.; Shukla, A. and Saini, M.L. 1991. Efficacy of some plant extracts against mustard sawfly (*Athalia proxima*Klug). *Plant Protection Bulletin India*, **43**: 26–30.

Sinha, S.N.; Chakrabarti, A.K. and Randhawa, K.S. 1983. A note on the control of aphid in radish seed crop. *Indian Journal of Entomology,* **54**: 481–484.

Sinha, S.N.; Peshwani, K.M.; Chakrabarti, A.K. and Saxena, A.N. 1977. Control of mustard aphid, *Lipaphis erysimi* Kalt. in radish seed crop. *Seed Research,* **5**: 49–51.

Srinivasachar, D. and Malik, R.S. 1972. An induced aphid resistant non-waxy mutant in turnip, *Brassica rapa*. *Current Science,* **41**: 820–821.

Srinivasan, K. 1993. Pests of vegetable crops and their control. *In: Advances in Horticulture.* K.L. Chadha and G. Kalloo (eds). *Vol. Vegetable Crops Part II* Malhotra Publishing House, New Delhi. pp. 868.

Srinivasan, K. and Krishnakumar, N.K. 1986. Evaluation of insecticides for the control of major pest complex of radish. *Madras Agriculture Journal,* **73**: 134–139.

Srivastava, K.M.; Srivastava, S.K. and Singh, D.R. 1991. Efficacy of selected insecticides against mustard aphid, *Lipaphis erysimi* Kalt (Homoptera: Aphididae). *Entomon,* **16**: 241–244.

Srivastava, P.D. 1957. Studies on the choice of food plant and certain aspects of the digestive physiology of the larvae and adults of *Athalia proxima* (Klug) and Epilachna vigintioctopunctata (F.). *Bulletin of Entomological Research,* **48**: 289–297.

Umapathi, G. and Bhaskaran, P. 1991. Insecticidal control of major pests of radish. *Madras Agriculture Journal,* **78**: 94–96.

Verma, R.K. 1992. *Population dynamics of Athalia proxima Klug with special reference to efficacy of some plant products.* Ph. D. Thesis, G.B. Pant University of Agriculture and Technology, Pantnagar pp. 116.

Verma, R.K. and Sachan, G.C. 1991. Studies on eclosion and mating behaviour of *Athalia proxima* Klug. *Indian Journal of Entomology,* **53**: 552–558.

Verma, R.K. and Sachan, G.C. 1993. Effect of soil type on pupation behaviour of *Athalia proxima* Klug. *New Agriculturist,* **4**: 77–78.

Verma, R.K. and Sachan, G.C. 1994. Studies of some plant products as antifeedants against the larvae of mustard sawfly, *Athalia proxima* Klug. Abstract National Symposium on pests of Agricultural importance and their management. Allahabad 19–21 February, 1994. p 5.

Verma, R.K. and Sachan, G.C. 1996. Effect of moisture regimes and soil types on pupation behaviour of *Athalia proxima Indian Journal of Entomology,* **58**: 235–238.

Wahla, M.A.; Asghar, M.A. and Ahmad, T. 1979. Effect of different temperatures on the development of *Athalia proxima* Klug grubs. *Pakistan Journal of Zoology,* **11**: 337–340.

Yadav, T.D. and Jotwani, M.G. 1969. Protection of seed crop of radish against mustard aphid, *Lipaphis erysimi* Kalt. by insecticidal root dipping during transplanting. *Indian Journal of Entomology,* **31**: 32–35.

Zaka-Ur-Rab M. 1981. Studies on the agromyzids (Diptera) of Kashmir. I. Some infesting palearectic species. *Filippo Silvastri,* **38**: 133–137.

*Pest Management in Horticulture Crops*
L.R. Verma, A.K. Verma and D.C. Gautam (eds)


*Chapter 21*

# Insect Pests of Cucurbit Vegetables

*Divender Gupta*

## INTRODUCTION

CUCURBITS CONSTITUTE THE largest group of summer vegetables grown all over the world. This group consists of wide range of vegetables which are used as salad, cooked vegetable, pickle and dessert and serves as a good source of vitamin A and C and various minerals. The plants of these crops possess the trailing habit except for summer squash, which is bushy in growth. In Indian market, cucurbit vegetables are available almost throughout the year due to diversity in climate prevailing in different regions. Due to the advent of polyhouse technology, the crops can be taken almost at will to make them more remunerative.

Cucurbits are infested by a number of insect pests but fortunately few of these namely, red pumpkin beetle, fruit flies and hadda beetle are of serious concern. The losses caused by these insects can be minimised by proper management practices. Various insect pests infesting cucurbit vegetables are discussed hereunder:

## MAJOR PESTS

### Red pumpkin beetle *Aulacophora foveicollis* (Lucas)

Red pumpkin beetle is an important pest of cucurbits in the tropic, subtropical and temperate regions of the world. It causes severe damage to cotyledonary and tender leaves of many cucurbit vegetables, becoming thereby a limiting factor in commercial cultivation of these crops. Two allied chrysomelids, namely, blue pumpkin beetle, *A. lewisii* (Baly) and grey pumpkin beetle, *A. cincta* (F.) have also been reported infesting cucurbits, the former being more common in northern India while the latter is prevalent in the southern parts of the country. However, none of these is as destructive as *A. foveicollis* (Butani and Verma, 1977).

The beetles feed mainly on cucurbits though they have been reported feeding on sunflower, okra, sweet potato (Chaudhary and Alikhan, 1990) and even on Ak (*Calotropis gigantea* L.) leaves under scarce food conditions (Vishvanathan *et al*., 1982).

Among the cucurbits cucumber is the most and bitter gourd the least preffered whereas bottle gourd and sponge gourd are equally preferred by the beetles (Khan and Hajela, 1987; Mehta and Sandhu, 1989). Cucurbitacins widely distributed in Cucurbitaceae, play a major role in host acceptance, as the extracts of preferred hosts when applied on relatively non preferred hosts elicited greater feeding response indicated by injury (Mehta and Sandhu, 1992). Bitter gourd also contains cucurbitacin but in addition it contains a triterpenoid glucoside which acts as a feeding deterrent (Chandravadana and Pal, 1983).

The adult of *A. foveicollis* is an orange coloured, 6 to 8 mm long beetle with black ventral surface clothed with short hairs. *A. cincta* is almost similar except for the colour of the elytra which is greenish yellow to brownish grey with distinct pale margins. The adults of *A. lewisii* are slightly smaller with blackish blue elytra. The grubs are dirty white, 10–12 mm long and pupae are pale white.

With the warming up of the weather, females start laying yellowish pink and oval eggs. A female lays 150 to 300 eggs, singly or in batches of 8 to 9, in moist soil around the base of the plants. The egg, larval and pupal period vary from 6 to 15, 13 to 25 and 7 to 17 days, respectively, depending upon the weather conditions. The pupation takes place in oval, and water proof earthern cells in the soil and the life cycle is completed in 1–2 months. There are 3–4 generations in a year. Most of the adults hibernate in winter in the soil while some may overwinter among weeds, grasses and dried creepers (Dutt and Dalpati, 1977) meaning thereby that hibernation is not obligatory.

Both the grubs and adults cause damage, grubs feed on underground parts whereas adults feed on cotyledons, flowers and leaves. Grubs enter into the root and underground stem or sometimes the fruit touching the ground. The underground portion starts rotting due to attack of saprophytic fungi at the injury site and the fruits become unfit for human consumption. Adults are very destructive particularly in the initial stages of the crop growth and sometimes the attack is so severe that the entire crop requires resowing.

The beetles resume activity in March and remain in the field till October. The peak period of activity is April-June and from September onward, the population starts declining (Shinde and Purohit, 1978; Butani and Jotwani, 1984). However, in eastern India, high population is found in December and less in August and the population density is positively correlated with temperature and rainfall (Roy and Pande, 1991).

### *Management*

The pest can be managed by using combination of various methods like collection of beetles in the early morning hours when the insects are sluggish and their killing by dipping in water containing insecticide. Early sowing of the crop can help in infestation escape particularly in the initial stages, as by the time beetles resume activity, the plants will already cross the cotyledonary stage. For this, the seeds can be sown in plastic bags kept in the polyhouse for germination and growth which can later be transferred to the field. The crop can also be grown in polythene cages in the field itself. Chaudhary (1995) reported that a cage made up of transparent polythene sheet (200 gauge) of 30 cm height and 120 cm perimeter (open from top) effectively protected cucumber seedlings upto one month after germination and the beetles attacked plants only after the crop crossed the height of the cage and at this stage it can tolerate the infestation due to the sufficient foliage. The creepers should be removed and burnt at the end of the season so that the adults are unable to overwinter among them.

The plant products like leaf extract of *Ageratum conyzoides* L. @ 0.06% proved quite effective (Pande *et al.*, 1987) and the oil of basil, *Ocimum basilicum* L. @ 1.5 ml per litre of water (Dube *et al.*, 1989) acted as a repellent against the beetle but since they are not available readily, their use is restricted.

The carbaryl poisoned bait with cucurbitacin as an attractant can be used but it remains effective upto 24 hours (Mehta and Sandhu, 1990). In case of heavy infestation, spraying of 0.1% carbaryl is recommended. The grubs can be controlled by drenching the soil around the base of the plants with methyl parathion 50 EC @2ml/litre of water or phorate granules (10G) @ 3kg/acre can be incorporated in soil around the base of vines (Sandhu and Bhalla, 1974).

**Fruit flies *Bactrocera cucurbitae* Coquillett and *B. tau* (Walker)**

The fruit flies constitute an important group of pests infesting cucurbit vegetables. Two species, *Bactrocera cucurbitae* and *B. tau* commonly called as melon fruit flies, are the major species found on cucurbits. Beside, they have been reported feeding on solanaceous crops like tomato and brinjal (Kapoor and Agarwal, 1983). Among cucurbits, these flies seem to prefer cucumber over other hosts while sponge gourd is least preferred (Gupta and Verma, 1993; Gupta and Verma, 1995). The damage by *B. tau* resembles the one by *B. cucurbitae* so closely that there is every possibility of mix up in their studies. Two allied species common in India are *B. ciliatus* Loew and *B. dorsalis* Hendel, but none of them is as serious as the melon fruit flies.

The adults of *B. cucurbitae* are medium sized (6 to 8 mm) flies, reddish brown with lemon yellow curved vertical markings on the thorax. The fore wing bears a dark brown band at its apex and in between radio-medial vein. The adults of *B. tau* are light brown in colour and the fore wing is devoid of dark band across radio-medial vein. However, the maggots of both the species are creamy white, taper anteriorly and measure 9–10 mm when full fed. The puparium is barrel shaped (5–7 × 2–2.5 mm), light brown, turning to dark brown as it develops. Eggs, maggots and pupae of different species look alike and are very difficult to distinguish.

Before becoming sexually active, adults go through maturation period during which they require carbohydrates and water to survive and protein sources for gonad maturation. Such nutrients are obtained in nature from honeydew, plant exudates, extrafloral nectaries, pollen, fruit juice and even bird droppings (Fletcher, 1987). After a preovipostion period of 9–14 days, the female lays upto 300 white, cylindrical eggs (1.15–1.30 × 0.25–0.27 mm), slightly curved on one side, in batches of 1–10 by thursting its ovipositor 4 to 5 mm deep inside the fruit. Young and tender fruits are preferred owing to their soft rind which facilitates penetration of the ovipositor. From the point of puncture, where the fly has laid eggs, oozes out a thin but profuse watery fluid which solidifies into a brown resinous deposit, giving protection to eggs. Presence of larva inside a fruit has deterrent effect on ovipositing females of *B. cucurbitae* (Prokopy and Koyama, 1982). At the time of oviposition, the fly also transfers some symbiotic bacteria present in the alimentary canal, to eggs, some entering the micropyle and others remaining on the egg chorion. The bacteria seek entry in larva at the time of hatching and multiply inside it. These bacteria hydrolyse fruit tissue proteins for feeding by the maggots while in adults they help in hydrolysing leaf surface proteins and also synthesize amino acids which are essential in the diet of both maggots and adults.

Besides, they may also have other functions including detoxification of plant defence chemicals (Fletcher, 1987). The egg, larval and pupal periods vary from 1–3, 3–6 and 7–12 days, respectively, during summer and rainy season, and are prolonged during winter.

The full-fed maggots leave the host prior to pupation in soil, by peculiar jumping habit. Before jumping, it doubles its body so that its mouth hooks hold the posterior end of the body and then by sudden contraction of longitudinal muscles and release of mouth hooks it springs into the air covering a distance of 10–20 cm of ground. Due to this limitation, the puparia are generally found in the vicinity of the plants. The life cycle is completed in 2 weeks but is prolonged in winter upto three months. There are a number of generations in a year and the pest overwinters as an adult.

Environmental moisture is of special importance as a determinant of abundance of the pest. The population expands when the rainfall is adequate and contracts during dry periods. The flies start appearing in the field from March onwards and the peak population is observed during rainy season i.e. July-August/September, after which the population starts declining (Agarwal *et al.*, 1987; Gupta and Verma, 1992).

Maggots feed on the pulp of the fruit by making gallaries and simultaneously the secondary infection also comes in, resulting in rotting of fruits. Fruit fly attack may reduce the yield drastically. As per some recent estimates nearly 50 per cent of cucurbits are partially or completely damaged by fruit flies in India (Atwal, 1986; Agarwal *et al.*, 1987; Gupta and Verma, 1992). The late maturing varieties are more prone to attack.

Life stages of fruit flies which are most exposed to the action of natural enemies are those associated with the soil i.e mature larvae (after they have left the fruit for pupation) and pupa. The eggs and early larval stages are relatively well protected inside the host. In nature, the pupae are parasitized by *Opius fletcheri* Silvestri, *O.compensans* Silvestri, *O. incisus*, *O. longicaudatus* (Ashmead), *Spalangia phillipinensis* Fullaway, *Dirhinus giffardi* Silvestri, *D. luzonensis* Rohwer etc. (Narayanan and Batra, 1960; Kapoor and Agarwal, 1983) but the extent of parasitization generally remains low.The ants and earwigs predate upon the pupae to a lesser extent.

### *Management*

The management of this pest is quite difficult as its economic threshold (ET) is very low (Bateman, 1976), rate of development is high and also the eggs are laid inside the fruits. Our efforts for the management of the pest should be Concentrated on the period before the eggs are laid inside the fruit. Since once the eggs are deposited inside, there is no use of applying control measures and any insecticide applied at this stage will merely poison the fruit.

Various methods for the management of fruit flies are being tried throughout the world as these are the pests of international importance. One such method is the "sterile insect technique" (SIT) which has been shown to be quite effective at least in Island situations (Bateman, 1976). The melon fly, *B. cucurbitae* was eradicated from islands of Kume, Miyako, Okinawa and Yayeyama of Okinawa prefecture Japan in 1978, 1987, 1990 and 1993, respectively, by using sterile insect technique through the release of gamma irradiated pupae. Prior to the release, suppressive control was carried out with male annihilation technique by distributing cotton string soaked with cue-lure (a male attractant) and an insecticide, from a helicopter to reduce the wild populaton density. Thereafter, a release of 30 million pupae was made every week with the help of a helicopter (Kakinohana *et al.*, 1997). But this technique is cumbersome and involves cost as it requires mass production facilities for the release. Further it is feasible particularly in island situations as the population is isolated. The application of SIT in southeast Asian countries weather for eradication or for control would face some unique problems, as often two or even three species may infest a single fruit and elimination of one species may result in resurgence of another.

The synthetic male lures have been used successfully to eradicate isolated populations and also to detect outbreaks and new introductions. A recent pilot test in Island of Rota in USA demonstrated melon fly suppression of about 98 per cent by male annihilation technique utilizing aerial distribution of cue-lure and malathion. This technique provide an effective adjunct of SIT by reducing the males in a population to a level that can be more easily flooded with sterile releases (Vargas, 1993). In India this lure has been tried for monitroring purposes. Pawar *et al.* (1991) monitored populations of *B. cucurbitae* in bitter gourd using cue-lure as an attractant and observed peak catch during early October. By monitoring the population, we can pin point the period when control measures should be applied.

Attempts have been made by various workers to formulate a strong attractant or bait for both male and female fruit flies using assortment of substances ranging from fruit juices, sugar, mollases and ammonia. However, none have met with as much success as protein baits (both hydrolysed vegetable protein and autolysed yeast). The protein bait sprays have become a major method of suppressing fruit fly population in many parts of the world but its use is not as widespread as as it should be, in Asian countries. A major factor is that protein baits in Asian countries have to be imported from foreign sources, thus making them expansive and inacessible to a large number of growers. Further, most of the growers probably apply bait spray in high volumes much similar to the insecticide cover spray thus accruing little additional benefit (Vijaysegaran, 1993). The research in Malaysia has shown that a protein bait formulated from a locally available industrial byproduct when applied in very low volumes as spot sprays provided an excellent control of fruit flies infesting carambola fruit (Vijaysegaran, 1989). The research should be conducted on these lines and use of protein bait should be promoted so that many undesirable side effects of insecticide cover spraying may thus be avoided.

Under Indian conditions, the management of this pest can be done by combining various practices like regular removal and detruction of infested fruits by burying them deep in pits in order to reduce adult emergence. A light ploughing should be given after lifting the crop which will help in killing the pupae by mechanical injury and also by exposing them to predators. Cultivation of early maturing varieties is useful as these normally escape infestation. Spraying of the crop with malathion 0.05% + 1% gur/sugar in water, will attract the flies and kill them, or a protein hydrolysate baits can also be tried. A male attractant cue-lure mixed with some contact insecticide can be used for monitoring and trapping of the adults (Pawar *et al.*, 1991). In case of heavy infestation, spraying of the crop with fenitrothion (0.05%) or fenthion (0.05%), repeated at 15 days interval is quite effective. However, care should be taken that fruits are harvested after observing a waiting period of 15 days.

**Hadda Beetle *Henosepilachna vigintioctopunctata* (F.)**

This is a serious pest of the bitter gourd and other cucurbit vegetables. An other species *Epilachna duodecastigma* (Weid.) is also associated with these crops. These beetles are polyphagous pests and their major hosts are solanaceous crops like brinjal, potato and tomato.

The adult of *H. vigintioctopunctata* is medium sized (8–10 mm), copper colored beetle which bears 7 to 14 black spots on each elytron, while *E. duodecastigma* has 6 spots on each elytron. The full-grown grubs (6–7 mm) are yellow, hump shaped and spiny. The yellow coloured pupa is 6–7 mm in length and hemispherical in shape.

The female lays upto 450 cigar shaped eggs in batches of 5–40 on the underside of leaves, which hatch in 3–5 days to produce yellow spiny grubs. The grubs start feeding on the lower surface of leaves, become full- fed in 10–15 days and pupate on the plants. The life cycle is completed in 2–3 weeks in summer and upto two months in winter. There are a number of generations in a year and the adults hibernate in winter under straw, creepers or cracks and crevices in the soil.

The beetles resume activity in March-April and remain active upto October but highest damage is done during April-May and August-September. During hot, dry months the population declines but again builds up in August (Singh, 1984).

Both grubs and adults cause damage by scrapping green matter from leaves which die and dry up. The severe infestation in early stages can kill the plant, whereas the older vines show stunted growth and poor yield.

In nature, immature stages of these beetles are parasitized by *Tetrastichus ovulorum* Ferriere, *Achrysocharis appannai* L., *Pleurotropis fovelolatus* Crawford, *P. epilachnae* Rohwer, *Pediobius foveolatus* (Crawford), *Microbracon brevicornis* Wesmael, *Campoletis perdistinctus* (Viereck), *Chelonus narayani* Subba Rao, and *Horogenes fenestralis* Holmgren but the extent of parasitization is low (Butani and Jotwani, 1984).

### *Management*

The infestation can be checked to a certain extent by hand picking and mechanical destruction of egg masses and adults in the early stage of attack. In case of severe infestation, dusting of crop with 5% carbaryl or spraying 0.1% dichlorvos or 0.05% malathion is effective.

### Cotton or melon aphid *Aphis gossypii* Glover

It is a major pest of melons and other cucurbit crops widely distributed throughout the country. Being polyphagous, it also feeds on malvaceous and solanaceous plants and is even reported feeding upon medicinal plants like basil, *Ocimum* sp.(Ghosh, 1975), mentha, *Mentha arvensis* L. (Bhalla and Pawar, 1977) and forest trees like sal, *Shorea robusta* L.(Ghosh, 1975).

Nymphs and adults are small, soft bodied, pear shaped and greenish brown insects. There are alate and apterous forms, the latter being more common and abundant. The apterous forms are robust while the alates are tender and fragile. The reproduction generally is by parthenogenetic viviparity and sometimes by sexual oviparity.

In Europe, *A. gossypii* is treated as anholocyclic i.e it breeds only parthenogenetically whereas in Japan and China the aphids go through sexual and parthenogenetic phases (holocyclic) and the primary hosts on which the sexuales are found in east Asia include *Hibiscus syriacus* Willd., the buckthorns *Rhamnus dahurica*, *R. japonica* and pomegranate, *Punica granatum* L. (Blackman and Eastop, 1994). A single apterous female produces 8 to 22 nymphs per day which mature in 3 to 4 days and the life cycle is completed in 7–9 days. (Butani and Jotwani, 1984).

The pest is active throughout the year except during winter and the maximum activity is observed during April-June High humidity and cloudy weather favours multiplication whereas the heavy rains reduce the population. Singh *et al.*(1994) found a peak population of *A. gossypii* on cucumber in first fortnight of May (145 aphids/3 leaves/plant) in Manipur.

As a result of sucking of cell sap by nymphs and adults from underside of leaves and tender parts, the leaves become cupped or otherwise distorted which ultimately dry up. The clusters of aphid can be seen on young shoots and underside of leaves. They also exude honey dew on which sooty mould develops which hinders the photosynthetic activity of the plants, resulting in stunted growth and loss in yield. The fruits covered by sooty mould become unattractive and loose their market value. This aphid also acts as a vector of cucumber mosaic virus (Andotra *et al.*, 1995) due to which large number of plants require uprooting to prevent spread of the disease.

### *Management*

A large number of natural enemies are reported to suppress aphid population in nature (Ghosh, 1975). These include parasites, predators and fungi. Singh and Rao (1995) reported a braconid parasitoid, *Trioxys indicus* Subba Rao and Sharma controlling *A. gossypii* on bottle gourd, sponge gourd and pumpkin. Two initial releases of parasitoid @ 10 mummies per square meter reduced population upto 20 to 25 per cent after first release and upto 65 to 70 per cent after the second release. In USA, *Aphidius colemani* Viereck is being used against *A. gossypii*. The predators like *Menochilus sexmaculatus* F. and *Chrysopa sclestis* Banks were found effective in regulating the population (Krishnamoorthy, 1988).

In case of heavy infestation, the aphid can be effectively controlled by spraying malathion (0.05%), methyl demeton (0.025%), phosphamidon (0.05%) and monocrotophos (0.04%). Repeat the application at 15 days interval, if required.There should be a waiting period of atleast 15 days between the harvest and last spray application.

## Pumpkin caterpillar *Diaphania indica* (Saunders)

The pumpkin caterpillar is reported as an irregular and sporadic pest of cucurbits (Patel *et al.*, 1993) and is widely distributed throughout the country. The adults are medium sized moths and their forewings are white and transparent with big brown marginal patches. Females have a tuft of orange coloured hair at the anal end. The caterpillar is elongated, bright green and bears a pair of narrow longitudinal white streaks along the mid dorsal line. The female lays upto 360 eggs singly or in groups on the lower surface of leaves. The egg, larval and pupal period lasts for 3–6, 9–14 and 5–13 days, respectively. The pupation takes place in cocoons spun among the leaves. The temperature has a positive influence on development of all stages of the pest and minimum threshold of development is 12.05°C while the development from egg to adult require 455.44 degree days (Peter and David, 1992a).

The caterpillar, the damaging stage of the pest, feeds on the leaves by binding them together. Sometimes ovaries of flowers and young developing fruits are also eaten up (Nair, 1986; Patel *et al.*, 1993). The affected flowers bear no fruit and infested fruits become unfit for consumption.

A gregarious endoparasitoid, *Apanteles taragamae* Viereck is reported to feed upon *D. indica* and the extent is as high as 80 per cent in Tamil Nadu, but still the pest is not under control due to attack of hyperparasite *Eurytoma braconidis* Ferriere on the host pupa (Peter and David, 1992b). Another endoparasitoid, a braconid, *Phanerotoma hendecasisella* is recorded for the first time from India as a solitary egg larval parasitoid on *D. indica* larvae. Kuruvilla and Jacob (1980) tested the pathogenicity of an entomogenous fungus *Paecilomyces farinosus* and reported it to cause 90 per cent mortality of *D. indica* larvae in the laboratory.

### *Management*

The pest can be effectively managed by hand picking and mechanically destroying tIhe caterpillars in early stage of infestation which will check the population build up. The entomogenous fungus can also be tried in the field conditions. In case of heavy infestation, dusting with 5% carbaryl or spraying with 0.1% carbaryl or 0.05 % endosulfan is effective.

### Stink bug *Aspongopus janus* F.

The stinkbug *A. janus*, a sporadic pest of cucurbits, is found all over India and its associated species namely, *A. observus* F. and *A. brunneus* Thunberg are found mainly in the North-East and South India, respectively (Butani and Jotwani, 1984). The desending order of host preference by *A. janus* as studied by Mishra and Sharma (1989) is cucumber, pumpkin, bitter gourd, brinjal, etc.

The adults are flat, medium sized insects. *A. janus* measures 30 mm in length. Its pronotum and base of elytra are bright red while head and wing membrane are black. *A. brunneus* adults are pale brown and smaller in size. The female lays eggs in rows on lower surface of leaves, tendrils, stems or other parts of the host plant. The egg and nymphal duration varies from 9–10 and 24–28 days, respectively.

The pest is active from March to September and is found only occasionally in winter. It is more active in the morning and evening and shelters under the leaves in intense sunlight. Both nymphs and adults suck the cell sap and can be seen in groups clinging to the leaves and tender shoots. Attacked plants loose vigour which ultimately affects the fruit set and yield. The bugs emit a stinking smell due to the presence of an odoriferous gland (Mishra Sharma, 1989).

### *Management*

This pest can be effectively managed by collecting and destroying the leaves and twigs bearing congregating bugs and spraying with 0.1 % carbaryl or 0.05% endosulfan or dusting with 5% carbaryl or malathion in case of heavy infestation.

### Red vegetable mite *Tetranychus neocaledonicus* Andre (Tetranychidae: Acarina)

The vegetable mite is one of the most destructive pests of cucurbits and is present throughout the country. Apart from cucurbits, it infests a variety of crops including brinjal, okra, potato, tomato and many ornamental plants.

The red colour adult measures 0.3–0.5 mm in length. The abdominal segmentation is quite distinct and the abdomen bears 12 pairs of hair on the dorsum (Butani and Jotwani, 1984).

The pest reproduces both sexually as well as asexually and in the latter case the progeny is only of males. In sexual reproduction, both sexes are produced though the females dominate. Eggs are laid at random on the webbings on the ventral surface of leaves. A fertilized female lays 61–93 eggs whereas the unfertilized female lays 39 to 59 eggs (Nair, 1986), which are spherical, minute and translucent. The egg hatches into six legged pinkish larva in 2–3 days in summer and upto 30 days in winter. The larval period for male and female varies from 2–6 and 2–9 days, respectively, during May to October, and may prolong to 20–22 days in January- February (Butani and Jotwani, 1984). The life cycle is completed in 5–10 days during summer and upto 60 days in winter. The mite hibernates as gravid females in winter.

The mites become active with the warming up of the season in April and remain active upto July. Heavy rains kill all the stages except eggs. After rains they again become serious in September-October. Thereafter, the population declines with the drop in temperature.

Nymphs as well as adults lacerate the underside of leaves and suck the cell sap causing small yellowish specks visible on the upper surface of leaves. The colonies of mite can be seen on the ventral surface of leaves which are protected under fine silken webs. These webbings interfere with the plant growth. The infested leaves dry up, the plant vigour is reduced, which ultimately affect the fruit set and yield.

Besides *T. neocaledonicus*, the other species such as *T. angolensis* Meyer, *T. cinnabarinus* (Bois du val), *T. utricae* Koch; *Bravipalpus phoenicus* (Geijskes), *Eutetranychus anneckei* Meyer, and *E. maximae* (Klein) have been reported damaging different cucurbit vines (Butani and Jotwani, 1984).

### *Management*

Mites can be effectively managed by spraying dimethoate (0.025%), methyl demeton (0.025%) or wettable sulphur (0.05%), repeated at 20 day intervals. The spray should be directed more towards lower side of leaves as the mites feed on the lower side of the foliage.

## MINOR PESTS

The pests of minor importance associated with cucurbits are the blister beetle *Mylabris pustulatus* Thunberg, jassids *Empoasca binotata* Pruthi and *Hishimonus phycitis* Distant, whitefly *Bemisia tabaci* (Gennadius), American serpentine leafminer *Liriomyza trifolii* (Burgess), plume moth *Sphenarches caffer* (Zeller), flea beetle *Phyllotreta cruciferae* (Goeze), gall fly *Lasioptera falcata* Felt, leaf footed plant bug *Leptoglossus australis* (F.), white grub *Holotrichia insularis* Bern etc., but they generally do not assume serious proportions.

Since cucurbit vegetables are not high valued crops and only a few pests are of serious concern, the insecticide use should be limited to the bare minimum on these crops, keeping in view the economic considerations. Offensive management strategies especially those designed to control early season populations are preferable to defensive strategies that are used after the pest population has expanded. From the consumer safety angle, farmers should observe the required waiting period after the last application of insecticides for picking and marketing the produce. The use of systemic insecticides particularly in the granular form should cease about 30 to 40 days before the harvest.

## REFERENCES

Agarwal, M.L., Sharma, D.D. and Rahman. O. 1987. Melon fruit fly and its control. *Indian Horticulture*, **32**(2): 10–11.

Andotra, P.S., Garg, I.D. and Chowfla, S.C. 1995. Characterisation of viruses causing cucumber mosaic in Himachal Pradesh. *Journal of Tropical Plant Diseases,* **13**(1): 43–50.

Atwal, A.S. 1986. *Agricultural pests of India and southeast Asia*. Kalyani Publishers, Ludhiana. pp. 529.

Bateman, M.A. 1976. Fruit flies. *In: Studies in Biological Control*. V.L. Delucchi, (ed). Cambridge University Press. pp. 11–49.

Bhalla, O.P. and Pawar, A.D. 1977. *A survey study of insect and non insect pests of economic importance in Himachal Pradesh*. Tiku and Tiku Kitab Mehal, Bombay. pp. 80.

Blackman, R.L. and Eastop, V.F. 1994. *Aphids on the world trees. An Identification and information guide*. CAB International. pp. 987.

Butani, D.K. and Verma, Shashi. 1977. Pests of vegetables and their control: Cucurbits. *Pesticides,* **11**(3): 37–41.

Butani, D.K. and Jotwani, M.G. 1984. *Insects in vegetables*. Periodical Expert Book Agency, New Delhi. pp. 356.

Chandravadana, M.V. and Pal, A.B. 1983. Triterpenoid feeding deterrent of *Raphidopalpa foveicollis* Lucas (red pumpkin beetle) from *Momordica charantia* L. *Current Science,* **52**(2): 87–88.

Chaudhary, Z.A. and Alikhan, M.A. 1990. Effects of temperature and relative humidity on the development and fecundity of the red pumpkin beetle, *Aulacophora foveicollis* Lucas (Chrysomelidae: Coleoptera). *Indian Journal of Entomology,* **52**(2): 274–278.

Chaudhary, R.N. 1995. Management of red pumpkin beetle, *Aulacophora foveicollis* (Lucas) using polythene cages on cucumber. *Pest Management in Horticultural Ecosystems,* **1**(1): 55–57.

Dhaliwal, G.S. and Arora, R. 1996. *Principals of insect pest management*. National Agricultural Technology Information Centre, Ludhiana. pp. 374.

Dube, S.; Upadhyay, P.D. and Tripathi, S.C. 1989. Antifungal, physiochemical and insect repelling activity of the essential oil of *Ocimum basilicum*. *Canadian Journal of Botany,* **67**(7): 2085–2087.

Dutt, N. and Dalpati, A. 1977. Dormancy in red pumpkin beetle, *Raphidopalpa foveicollis* (Lucas). *Indian Journal of Entomology,* **39**(1): 70–80.

Fletcher, B.S. 1987. The biology of dacine fruit flies. *Annual Review of Entomology,* **32**: 115–44.

Ghosh. A.K. 1975. *Aphids (Homoptera: Insecta) of economic importance in India*. The Agricultural Society of India, Calcutta. pp. 134.

Gupta, Divender and Verma, A.K. 1992. Population fluctuations of the maggots of fruit flies (*Dacus cucurbitae*) Coquillett and *D. tau* (Walker) infesting curbitaceous crops. *Advances in Plant Sciences,* **5**(2): 518–23.

Gupta, Divender and Verma, A.K. 1993. Biology of *Dacus tau* (Walker) on different vegetable hosts. *Journal of Insect Science,* **6**(2): 299–300.

Gupta, Divender and Verma, A.K. 1995. Host specific demographic studies of the melon fruit fly, *Dacus cucurbitae* Coquillett (Diptera: Tephritidae). *Journal of Insect Science,* **8**(1): 87–89.

Kakinohana, H.; Kuba, H.; Kohama, T.; Kinjo, K.; Taniguchi, M.; Nakamori, H.; Tanahara, A. and Sokei, Y. 1997. Eradication of the melon fly, *Bactrocera cucurbitae* Coquillett, by mass release of sterile flies in Okinawa Prefecture, Japan. *Japanese Agricultural Research Quarterly,* **31**: 91–100.

Kapoor, V.C. and Agarwal, M.L. 1983. Fruit flies and their natural enemies in India. *In: Fruit flies of economic importance*. Cavalloro (ed). Balkema, Rotterdam. pp. 104–105.

Khan, M.Z. and Hajela, K.P. 1987. Studies on *Aulacophora foveicollis* Lucas (Coleoptera: Chrysomelidae) -food preference and extent of damage. *Indian Journal of Entomology,* **49**(4): 457–59.

Krishnamoorthy, A. 1988. Biology of *Chrysopa sclestis* Banks (Neuroptera: Chrysopidae). *Bulletin of Entomology,* **29**(1): 69–72.

Kuruvilla, Suma and Jacob, Abraham. 1980. Pathogenicity of the entomogenous fungus *Paecilomyces farinosus* (Dickson Ex Fries) to several insect pests. *Entomon,* **5**(3): 175–176.

Mehta, P.K. and Sandhu, G.S. 1989. Studies on host preference and rate of feeding by red pumpkin beetle on different cucurbits. *Vegetable Science,* 16(1): 66–74.

Mehta, P.K. and Sandhu, G.S. 1990. Monitoring of red pumpkin beetle (*Aulacophora foveicollis*) by cucurbitacins as kairomones through poison baits. *Indian Journal of Agricultural Sciences,* **60**(3): 225–226.

Mehta, P.K. and Sandhu, G.S. 1992. Feeding behaviour of red pumpkin beetle, *Aulacophora foveicollis* (Lucas) on leaf extracts of different cucurbits. *Uttar Pradesh Journal of Zoology,* **12**(2): 87–94.

Mishra, R.K. and Sharma, R.N. 1989. Host preference by adult *Aspongopus janus* F. *Indian Journal of Entomology,* **51**(4): 480–481.

Nair, M.G.K. 1986. *Insect and mites of crops in India*. Indian Council of Agricultural Research, New Delhi. pp. 408.

Narayanan, E.S. and Batra, H.N. 1960. *Fruit flies and their control*. Indian Council of Agricultural Research, New Delhi. pp. 68.

Pande, Y.D., Gosh, D. and Guha, S. 1987. Toxicity of leaf extract of *Ageratum conyzoides* L. (Compositae) to red pumpkin beetle, *Raphidopalpa foveicollis* (Lucas) (Coleoptera: Chrysomelidae) in Tripura. *Indian Agriculturist,* **31**(4): 251–255.

Patel, J.R., Mehta, D.M. and Shah, B.R. 1993. Field evaluation of insecticides for the control of pumpkin caterpillar *Margaronia (Diaphania) indica* Saunders. *Indian Journal of Entomology,* **55**: 446–448.

Pawar, D.B., Mote, U.M. and Lawande, K.E. 1991. Monitoring of fruit fly population in bitter gourd crops with the help of lure traps. *Journal of Maharashtra Agricultural University,* **16**(2): 281.

Peter, Clement and David, B.V. 1992a. Studies on the thermal requirement for development of *Diaphania indica* (Saunders) (Lepidoptera: Pyralidae). *Journal of Insect Science,* **5**(2): 172–74.

Peter, Clement and David, B.V. 1992b. Biology of *Eurytoma braconidis* Ferriere (Eurytomidae), A hyperparasite of *Apanteles taragamae* Viereck (Braconidae). *Journal of Insect Science,* **5**(2): 203–205.

Prokopy, R.J. and Koyama, J. 1982. Oviposition site partitioning in *Dacus cucurbitae*. *Entomologia Experentia et Applicata,* **31**: 28–32.

Roy, D.C. and Pande, Y.D. 1991. Seasonal incidence, host preference and feeding rate of red pumpkin beetle (*Raphidopalpa foveicollis*) in Tripura. *Indian Journal of Agricultural Sciences*, **61**(8): 603–607.

Sandhu, G.S. and Bhalla, J.S. 1974. Red pumpkin beetle- an important pest of cucurbits. *Punjab Horticultural Journal,* **14**(1–2): 70–74.

Shinde, C.B and Purohit, M.L. 1978. Population study of red pumpkin beetle (*Aulacophora foveicollis* Lucas)—a serious insect pest of cucurbits. *Science and Culture,* **44**(7): 335.

Singh, Harcharan. 1984. *Household and kitchen garden pests*. Kalyani Publishers, Ludhiana. pp. 420.

Singh. K.G.; Singh, L.S. and Singh, T.K. 1994. Population dynamics of *Aphis gossypii* Glover (Homoptera; Aphididae) on cucumber (*Cucumis sativus* L.) in Manipur. *In: National Symposium on Emerging Trends in Pest management.* June, Nauni (Solan) India. pp. 28–30.

Singh, Rajendra and Rao, S.N. 1995. Biological control of *Aphis gossypii* Glover on cucurbits by a parasitic wasp *Trioxys indicus* Subba Rao & Sharma. *Biological Agriculture and Horticulture,* **12**(3): 227–236.

Surulivelu, T. and Kumaraswami, T. 1990. Effect of different formulations and methods of application of insecticides on some sucking pests and their predators in cotton. *Journal of Biological Control,* 4(1): 25–30.

Vargas, R.I. 1993. Fruit fly eradication research in Hawaii. *In: Fruit Flies: Biology and Management*. M.A. Aluja and P Liedo (eds). Springer Verlag, New York, Inc. pp. 416–418.

Vijaysegaran, S. 1989. An improved technique for fruit fly control in carambola cultivation using spot sprays of protein baits. *National Seminar on Carambola: Developments and Prospects*, 18–19 July, K. Lumpur, Malaysia.

Vijaysegaran, S. 1993. Control of fruit flies in tropical regions of Asia. *In: Fruit Flies: Biology and Management*. M.A. Aluja and P. Liedo, (eds). Springer Verlag, New York, Inc. pp. 453–463.

Vishvanathan, T.; Abdul Kareem, A. and Sasitharan, P. 1982. *Calotropis gigantea* Linn. as a new host for the red pumpkin beetle, *Raphidopalpa foveicollis* (Lucas). *Pesticides,* 16(7): 22.

*Pest Management in Horticulture Crops*
L.R. Verma, A.K. Verma and D.C. Gautam (eds)


*Chapter 22*

# Insect and Mite Pests of Ornamental Plants

*Usha Chauhan*

## INTRODUCTION

CULTIVATION OF FLOWERS has gained much recognition in India in the past few years. In fact, Indian floriculture industry is still young but is making rapid strides in production of flowers for export purpose. This is basically due to benefits of high density planting, controlled conditions of greenhouse, regular irrigation and fertilization, etc. Although India's share in the global flower trade is small yet recently its export earnings from floriculture has registered a growth of nearly 300 per cent. Thus, India is expected to play a major role in world flower market in the near future. One of the major limitations of flower cultivation in India is that most of the farmers grow them in the open fields where they are subjected to vagaries of nature and attack of various pests. This is totally in contrast to the western world where floriculture is carried out essentially under glasshouse conditions and are highly mechanised and intensively managed enterprises.

Ornamental crops are highly entomophilous in nature and attract a large number of insect species for food, nectar, pollen and shelter. The injury inflicted by these pests reduces the aesthetic and market value of flowers. Sometimes, these pests act as vectors of virus diseases in these plants. All these factors necessitate use of a number of management practices against these pests, the effectiveness of which depends upon the exact identification of the species, its biology, damage potential, symptoms of attack and weak points in life cycle. In this chapter, various aspects of the biology and management practices of major insect and mite pests infesting important flowering plants such as rose (*Rosa* spp.), gladiolus (*Gladiolus* sp.), chrysanthemum (*Chrysanthemum morifolium* Ramat), carnation (*Dianthus caryophyllus*), lilies (Tiger lily and Easter Lily), snapdragon (*Amtirrhinum majus* L.) and fuschia (*Fuschia* sp.) are highlighted.

## APHIDS

The aphids are one of the serious pests of ornamental plants because of their high reproductive capacity, polymorphism, parthenogenetic mode of reproduction, wide host range and high damage potential. There are more than 4000 species of aphids known world over, out of which about 250 species are of economic significance. According to Behura (1986) in India, about 653 species are known out of which 150 are reported to be serious pests on 1185 plant species of 691 genera falling under plant families. Many aphid species are found on one or at the most a few host plants but some have very large host range, e.g. *Myzus persicae* (Sulzer) is reported to occur on more than 400 plant species throughout the world. The more common species that occur on flower crops include the green peach aphid, *M. persicae,* melon/cotton aphid, *Aphis gossypii* Glover; chrysanthemum aphid, *Macrosiphoniella sanborni* Gillette, rose aphid *Macrosiphum rosae* L., potato aphid, *Macrosiphum euphorbiae* Thomas and peach leaf curling aphid, *Brachycaudus helichrysi* Kalt. The important flowering plants infested by aphids include China aster, carnation, cacti, chrysanthemum, dahlia, gladiolus, antirrhinum, cineraria, larkspur, lupins, petunia, sweet pea, lilies, rose, many shrubs and ornamental trees.

### *Aphis gossypii* Glover

It is commonly found on glasshouse plants and measures 1.8 to 2.9 mm in length. Its colour varies from yellow, dark green to almost black. The shape of the body in apterae is pyriform, broadest behind the middle. The antenae are at least half the length of the body. The frontal lobes are almost absent in such a way that outline of the front of the head between the antennae is sinuate. The cornicles taper slightly from the base and have a apical flange. The dark green or black cauda is narrow with three hairs on each side.

### *Brachycaudus helichrysi* (Kaltenbach)

The primary host of this species is *Prunus* species. It migrates to secondary hosts including cineraria, chrysanthemum, etc. This aphid is small, 1.3 to 2.00 mm, pale-green to orange in colour. The colour depends upon the type of food it eats. Dorsum of the aphid is pale brown. The cauda is short, semi-circular with a slight basal constriction. Cornicles are pale, short and distinctly flanged. It attacks both shoots and flowers of chrysanthemum leaving dense deposits of crystalline honeydew.

### *Macrosiphoniella sanborni* (Gillette)

This species is a small, 2 to 2.5 mm long, dark, shining brown aphid. The cauda is considerably longer than the cornicles. The latter are stout, black and enlarged at the base with a coarsely reticulated pattern on the apical two-third of their length. During summer and early autumn, stems harbour dense colonies which distort the developing buds and flowers. This aphid is also a vector of two important viruses, vein mottle and chrysanthemum virus B.

### *Macrosiphum euphorbiae* Thomas

This species migrates from Rosaceae to a wide range of herbaceous plants including potato, carnation, cineraria, tulip, etc. It is a large aphid, 4.0 mm long, green in colour with brownish lobes and covered with a fine white pubescence on the thoracic disc. Antennae are longer than the body. Cornicles are long, yellowish, slender and apically reticulate. The pink cauda is half the length of the cornicles. This species is an active, restless and vector of many virus diseases including anemone mosaic, chrysanthemum virus B, iris mosaic and tulip break, etc.

***Macrosiphum rosae* L.**

This species is comparatively larger than other aphids, about 3 mm long. Its colour varies from pink to green. It has prominent frontal tubercles and long, slender and entirely black cornicles. Under glasshouse conditions, the reproduction may be viviparous throughout the year but outdoors, small reddish females, lay eggs on young shoots during autumn.

***Myzus persicae* (Sulzer)**

This polyphagous species is the most important glasshouse aphid. Its colour varies from shades of pink to yellow or grey-green. It is about 2 mm long. The cornicles are long and thin, slightly swoollen in the middle, the extreme tip is characteristically darker than the rest and bears a distinct flange. The cauda is triangular, less than half the length of the cornicles, pale in the wingless form (apterae) and darker in winged form (alatae). Both the forms possess between the antennae, well developed, convergent and lateral roughened frontal prominences. The development is rapid at higher temperatures. It takes 7 days for development from nymph to adult at 24°C and 43 days at 6°C. The fecundity varies from 30 to 50 nymphs per female. *M. persicae*, unlike many other aphid species, usually does not form dense colonies on plants. It is a restless species whose populations are in a continuous state of movement. On carnation, pale areas often develop around the individual feeding punctures, giving leaves a mottled appearance. Among the more important viruses transmitted by *M. persicae* are carnation latent, carnation mosaic, chrysanthemum virus B, dahlia mosaic, orchid mosaic and tulip breaking, etc.

**Detection**

Aphid infestation can be detected on crops by several ways, viz., winged aphids are attracted to yellow colour and hence yellow sticky traps or yellow tray water traps can be used for population monitoring. Presence of ants or flies on the infested plants, which visit the aphid colonies for honeydew and inturn may guard the colonies from natural enemies, presence of ladybird beetles and syrphid flies which are aphid predators around the infested plants also indicate the presence of aphid colonies.

**Management**

Clean culture and hygienic practices are important for preventing aphid attack arising from infestation on weeds which often serve as hosts to aphid species and act as reservoirs for aphid infestation. Such plants should be removed in and around the cropped area. Screens and such other physical barriers should be used against spread of aphids in areas where stock plants are being maintained and to avoid spread of virus infection. Fertilizers should be used in appropriate quantity, avoid excessive use of nitrogenous fertilizers, as high nitrogen level promotes aphid multiplication. Monitoring of aphid population with yellow sticky traps or yellow water traps should be done regularly to pinpoint the time of initiation of aphid infesttion. Following indications of aphid infestation, the plants may be sprayed with soap solution. Water rinse may be given after each spray to wash off excess soap from the plants after a few hours. It is important to control aphids on flowering crops before the buds open because once they get inside the inflorescence it becomes difficult to control them with any insecticide. On severe infestation spraying with methyl demeton 0.05 per cent, phosphamidon 0.03 per cent, dimethoate 0.03 per cent or monocrotophos 0.05 per cent should be done at 10 to 15-day intervals. Do not use dimethoate on chrysanthemum as it may prove phytotoxic.

In general, most of the organo-phosphorus insecticides are phytotoxic to chrysanthemum, but carbamates like methomyl and aldicarb are effective against aphids (Worthing, 1967). Phosphamidon and methiocarb may cause damage and spotting on leaves and infloresence in gerbera. Thus, for aphid control on gerbera spray monocrotophos or propoxur at 0.05 per cent (Romankow *et al.*, 1972). Aphids are controlled effectively on cut flowers of chrysanthemum by dipping them in azodrin at 250–1600 ppm for 48 hours (Poe and Marousky, 1972).

Biological control agents for the suppression of insect pests of ornamental crops have not been used extensively in India. However, in other countries several biocontrol agents like *Chrysoperla* spp., coccinellids *Harmonia axyridis* and hymenoptrous parasites like *Aphelinus abdominalis* Dalm., etc. are effectively used against aphids in glasshouses (Hassan, 1974; Ferran *et al.*, 1996). By using *A. abdominalis* along with insecticides, there was at least 75 per cent reduction in insecticide usage on roses (Blumel and Hausdorf, 1996).

## THRIPS

Thrips are polyphagous insects that infest almost every kind of flowers. There are more than 6000 species of thrips of which several hundred are reported to attack cultivated plants but only a few cause problems on ornamental plants. The commonly occurring species worldover on flower and foliage crops are the tobacco or onion thrips, *Thrips tabaci* Lind., greenhouse thrips, *Heliothrips haemorrhoidalis* (Bouche), eastern flower thrips, *Frankliniella tritici* (Fitch), western flower thrips, *F. occidentalis* (Pergande) and gladiolus thrips, *Taeniothrips simplex* (Morison). In India, important thrips species reported to infest gladiolus, carnation, rose, chrysanthemum and bulbous plants include *Rhipiphorothrips cruentatus* Hood, *Thrips tabaci* Lind., *T. flavus* (Schrank), *T. coloratus* (Priesner), *T.* simplex, *T. traegardhi* (Trybom), *Frankliniella dampfi* (Priesner), *Microcephalothrips abdominalis* (Crawford), *Aelothrips collaris* (Priesner), *A. meridionalis* Priesner (Ananthakrishnan, 1971, 1973; Thakur *et al.*, 1973; Bhalla and Thakur, 1974; Chandra and Lal, 1973; Sood, 1988; Hole and Salunkhe, 1997a). Thrips become problematic in hot and dry weather conditions when they develop and reproduce very fast. However, high humidity does not seem to reduce their number. Thrips belong to two suborders namely, Terebrantia and is Tubulifera. Most of the flower thrips except lily flower thrips belong to the former suborder. In Terebrantia, the female possesses a saw-like ovipositor with which it slits the plant tissues and the eggs are laid within. On the other hand, in Tubulifera, females lack modification in ovipositor and the eggs are laid on the surface of the plant. Thrips are gregarious in nature and large number of adults of the same or even different species may be found together on a leaf or within a single flower. Thrips rarely fly, though some are able to jump when disturbed.

### Biology

Thrips are small, slender bilaterally asymmetrical insects, 1–2 mm in length and their body colour varies from yellow to brown or blackish brown. They have two pairs of narrow wings fringed along the margins with long, fine hairs. When at rest, the wings are laid parallel along the back. In a few species, the wings may be either fully developed (macropterous), much reduced in size (brachypterous) or absent altogether (apterous). The antennae are 6 to 8 segmented. Females are usually more numerous than males and are distinguished by their large sized pointed apex of the abdomen. In few species, however, males are rare or unknown, so the reproduction is generally parthenogenetic. The metamorphosis is complete and the duration of different stages of life cycle is directly influenced by temperature and relative humidity. The life cycle of most of the species is almost similar. Each female lays about 60 whitish reniform eggs singly, through a slit cut by the ovipositor, within leaf or flower tissues. One end of the egg protrudes slightly from this slit to facilitate the escape of newly hatched larva. Eggs hatch after about 8 days into translucent white or pale yellow larva about 0.4 mm long. Its head is abnormally large and the eyes are bright red. At first gregarious, the larvae later become scattered over the leaf surface. After 10 to14 days the fully fed larvae conceal themselves in the soil to undergo a short (1–2 days) prepupal stage before changing to pupa. The pupa changes to adult in 4 to7 days. Under glasshouse conditions the life cycle is completed in 23 to31 days and at least 6 to7 generations are completed in summer. Larvae or adult females overwinter in sheltered sites, in the soil or under the bark of trees and shrubs. The duration of hibernation depends on temperature and may last 4 to5 months.

The larvae and adults cause damage by rasping the epidermal tissues and sucking the oozing cell sap from tender leaves, leaf axils and flowers. Maximum injury is inflicted by the larval feeding within developing leaves and flowers but the damage is seen only when the affected parts expand and by that time thrips may not be present on the plant. On leaves, the damage generally appears as silvery streaks due to removal of cell contents and the affected portion, later turns brown. Sometimes the margins of infested leaves roll upwards, leaf and petal size gets reduced and mottled, deformed flowers having slivery patches or burnt margins are produced (Ananthakrishnan, 1971; Butani, 1974; Hole and Salunkhe, 1997a). *Heliothrips* and some other species deposit large amounts of faecal matter on the foliage which attract sooty mould growth and accentuate the damage to ornamental plants.

**Detection**

A monitoring system is required for the successful management of thrips. This is done by installing yellow sticky traps to which thrips are generally attracted. Traps may also be placed outdoors near glasshouses which provide an early warning of thrips invasion. Gladiolus corms infested with thrips can be detected as they become sticky due to sap that oozes out as a result of feeding injury. Sometimes indicator plants like petunia and beans are grown around the main crop. These plants can be removed as soon as virus symptoms appear on them in order to avoid the disease spread to the main crop. Flowers can be shaken over a piece of moistened filter paper to detect the presence of thrips.

**Management**

No other flowering plant should be allowed to grow near the main crop so that they do not serve as a source of attraction for thrips. Grow indicator plants around the greenhouse to trap thrips and reduce their number before they move on to the main crop. Reflective aluminized tape, cloth and screen can also be used in greenhouses and their vents to reduce entry of thrips into the glasshouse. Mild infestation may be controlled by spraying soap solution but for heavy infestation spray malathion (0.05%) or fenitrothion (0.05%) on rose (Pirone *et al.*, 1960; Hessan, 1981). According to Hole and Salunkhe (1997a) the rose thrips, *Rhipiphorothrips cruentatus* Hood, can be controlled by monocrotophos (0.05%), methidathion (0.05%) or quinalphos (0.05%). In Poland, Labanowski and Mynett (1990) recommended acephate (0.1%), pirimiphos methyl or cypermethrin (0.05%) against the gladiolus thrips, *T. simplex*. Gladiolus corms infested with *T. simplex* may be stored in a room provided with dichlorvos strips (1 strip/100 ft3) space at 26.7°C. Storing of infested gladiolus corms at 2o C for 6 weeks and later treating them in hot water at 46°C completely killed the thrips. In the USA, during transportation, cut flowers of carnation are wrapped in polyethylene bags containing 1%chlorpyriphos as fumigant to control the western flower thrips, *Frankliniella occidentalis* (Fitch) (Tjosvold and Ali, 1989).

## LEAFMINERS

There are about 150 species of leafminers known to feed on cultivated plants, but only about a dozen cause problems to greenhouse and outdoor plants. In India, two genra are commonly found, viz., *Chromatomyia* and *Liriomyza*. The later is reported to feed on flowering plants including China aster, zinnia and nasturitium in South India (Udayagiri, 1987). Besides, leafminers infest several ornamental plants like chrysanthemum, gerbera, sweet pea, larkspur, marigold, etc.

### Biology

The leafminers are small, black or yellow flies that inhabit the crop canopy and are active fliers in adult stage. The female makes small feeding or egg laying punctures in leaves with the ovipositor. Both the sexes feed on the fluid that oozes out from the leaves. Each female produces several hundred white translucent eggs depending on the host and its nutritional condition. Eggs are embedded in the leaf tissues singly beneath the epidermis. They hatch in 4 to 5 days into small legless maggots. The maggots feed in between leaf epidermis forming characteristic narrow trails or mines. The larvae pass through three instars in 4 to 7 days to become full-fed. Pupation generally occurs within the mine in *Chromatomyia* whereas in *Liriomyza*, it occurs mostly in the soil or on lower leaves. The pupal stage lasts for about nine days and the life cycle is completed in 21 to 30 days. Besides causing aesthetic injury to plant foliage, leafminer attack also reduces the photosynthetic activity of the plant to a considerable extent.

### Detection

The planting material should be checked for leafminer infestation and culled out before planting. The adult activity can be monitored with the help of yellow sticky traps placed a few centimeter above the crop canopy (Parrella and Lindquist, 1983; Chandler, 1987; Robin and Mitchell, 1987; Tatara and Furuki, 1994).

### Management

Screens or other barriers are effective in preventing leafminer invasion in greenhouses. To some extent, adults can be removed from the greenhouse by using hand vacuum suction devices which will reduce the future larval injury. Leafminer attack is more on plants given higher dose of nitrogen fertilizers. Therefore, its application should be regulated to optimum. Remove and burn all infested leaves from the crop on weeds growing nearby and destroy the refuge of cutflowers immediately after harvest. Steam or fumigate beds before planting another crop. Inter planting with field bean (*Vicia faba* L.) as a trap crop is an effective method to check leafminer infestation. Use of yellow sticky traps coated with chlorpyriphos or permethrin reduces adult population (Sanderson *et al.*, 1989).

Cold treatment of infested rooted cuttings of chrysanthemum at 1 to 2°C followed by fumigation is effective against leafminers. Storage of chrysanthemum cuttings at 1.18°C for 15 days was advocated by Mora *et al.* (1985) for obtaining complete larval mortality and reduction in egg hatching. They also suggested dipping of cuttings in pyrazophos (0.02%) for 1 to 5 minutes for the prevention of larval damage. Under field conditions severe infestation can be checked by spraying the crop with pyrazophos (0.02%) or triazophos (0.04%) (Lindquist *et al.*, 1984). Aldicarb was also found effective on potted plants of chrysanthemum at 25 mg/pot while in marigold and zinnia, treatment with cyromazine at 20 mg/pot gave good control (Lindquist and Casey, 1988). Several insecticides have been found effective against *L. trifolii* (Burgess) such as avermectin B1, cyromazine, methomyl, diflubenzuron (Parrella, 1983; Smith, 1986). A method of stem injection with systemic insecticide was developed in Israel and used effectively against leafminer on ornamental trees (Halperin, 1981). Treatment of chrysanthemum cuttings with neem extracts (Margosan-D) resulted in the disruption of life cycle of *L. trifolii* as a result of which reinfestation can be avoided (Sanderson *et al.*, 1989a). Resistance to pyrethroids in *L. trifolii* and *L. huidobrensis* was reported from United Kingdom (Macdonald, 1991).

Among the promising biocontrol agents of leafminers, the parasitoid, *Diglyphs* sp. (Eulophidae) is used effectively for their control cither alone or in combination with insect growth regulators like methoprene and cyromazine, or entomopathogenic nematode, *Steinernema* sp. (Parrella *et al.*, 1982; Powell and Lindquist, 1996; Sher *et al.*, 1996). However, certain chemicals like permethrin, methomyl, etofenprox and prothiofos, etc. are found to destroy most of the parasitoids of *Liriomyza* sp. (Saito *et al.*, 1996). Release of *Encrsia formosa* Gahan, a parasitoid of *Bemisia argentifolii* on poinsettia reduced the number of insecticide applications by 75% but the cost of using parasitoids was 9.5 times more than that of insecticides (Hoddle *et al.*, 1996).

## MEALYBUGS AND SCALE INSECTS

Mealybugs and scale insects have a wide host range and the ornamental plants attacked by them include rose, gladiolus, orchids, clerodendrum, geranium, amaryllis, fuchsia, begonia, bignonia, jasmine, cacti, gardenia, crossandra, hibiscus, verbena, bird of paradise, and many others. These insects have piercing and sucking type of mouth parts. In case of severe infestation, owing to continuous draining of plant sap flowering and yield is affected adversely and the plant succumbs to injury. These insects, while feeding inject toxic saliva into the plant due to which the affected parts turn yellow and wither away. All scale insects and mealybugs (except armoured scale) excrete honeydew on plants on which sooty mould fungus develops which gives unsightly look to the plant which in turn interferes with the photosynthetic activity and the vitality of the plant is greatly impaired.

Scales are of two types, armoured (Diaspididae) and unarmoured or soft scales (Coccidae). Adults of armoured scales produce hard, waxy covering over their soft and yellow coloured bodies which helps in protecting them from harsh weather, predators and parasites. This hard covering can be separated from the adult body. The shape of the scale varies between male and female even within the same species. The female produces several hundred yellow eggs which are kept beneath its body. After hatching the yellow coloured crawlers come out and move towards new growth in search of food. After finding suitable site they loose their legs and settle there in sessile state and start secreting waxy covering over their bodies. There are three nymphal instars in males and five in females before becoming adult. Adult males are winged insects. There may be 4 to 6 generations in a year.

Unarmoured scales also secrete a coverning over their bodies but in adult stage. Unlike armoured scales, the covering cannot be reported from their bodies. Soft scales are comparatively bigger in size than the armoured scales and there are 4 to 6 generations in a year. The important species which attack various ornamentals are: the citrus red scale, *Aonidiella aurantii* (Maskell), San Jose scale, *Quadraspidiotus perniciosus* (Comstock), black scale, *Saissetia nigra* Nietn, cottony cushion scale, *Icerya purchasi* Maskell, and rose red scale, *Lindingospis rossi* (Maskell), etc.

Mealybugs (Pseudococcidae) are soft bodied insects which produce a waxy powder over their bodies. Some mealybugs also produce waxy projections around their bodies and many other have long, filamentous growth consisting of waxy strands. They are found mostly in groups around nodes and stem branchings. Female produces a mass of waxy threads called ovisac in or near the soil or in which eggs are laid. After hatching the young crawlers move about in search of suitable feeding site and on finding one, immediately start feeding. But unlike scale insects, crawlers of mealybugs retain their legs throughout their life. Females of mealybugs look like scale insects and adult males are winged. The important species are: *Planococcus citri* (Rosso), *Ferrisia virgata* (Cockerell), *Pseudococcus latipes* Green, *P. longispinus* Targ. and *Nipaecoccus nipae* (Maskell).

### Detection

Early detection is of great significance for efficient management of scales and mealybugs. Plant inspection is the only effective method of detection. Therefore, new plant material should be thoroughly checked and infested material should be discarded. Presence of black fungal growth and ants, wasps, bees, etc. is a sure sign of the presence of these insects in the greenhouse.

### Management

Use of insect-free plant material and cutting and burning of infested parts reduces further spread of the pest. Spray of monocrotophos, phosphamidon, endosulfan and malathion at (0.05%) provides effective control of the scales (Butani, 1974; Nandakumar *et al.*, 1988; Hole and Salunkhe, 1997c). Soil application of carbofuran or phorate @ 1.0 kg a.i./ha after pruning effectively reduces population of *A. auranti* on rose. Sprays of methyl demeton 0.025 per cent or dimethoate (0.03%) cent can effectively check infestation of soft scale and mealybug. Natural enemies, viz., *Aphytis* sp., *Chilocorus* sp. and *Pharascymnus* sp. are common in occurrence on red scale. In China, inundative releases of coccinellid predator, *Cryptolaemus montrouzieri* Mulsant reduced the mealybug population on glasshouse ornamentals to the extent of 75 to 90% (Dong, 1993). In the United States, scales and mealybugs were controlled on harvested commercial flowers and foliage by giving 5 minute dip in soap solution consisting of potassium salts of fatty acids (0.98 %) and an insecticide fluvalinate (0.025 %) (Hansen *et al.*, 1992).

## WHITEFLIES

There are about 1100 species of whiteflies out, of which the important species infesting flowering plants include the greenhouse whitefly, *Trialeurodes vaporariorum* (Westwood) and cotton whitefly, *Bemisia tabaci* Guen. The flowering plants that are attacked by whiteflies include geranium, fuchsia, ageratum, hibiscus, hollyhock, coleus, begonia, primula, salvia, etc. Some information on the biology of greenhouse whitefly is given hereunder:

### *Trialeurodes vaporariorum* (Westwood)

The adults of this species are about 1 mm long, gregarious, collecting in vast numbers on the upper foliage of plants. Wings and yellow body is covered with minute spirals of wax giving the wings a pure white colouration and yellow body a mealy appearance. Reproduction is largely parthenogenetic but sometimes males are also present which are smaller than females. Mated females produce both the sexes. The reproductive rate is affected by the nutrient status of the host and the environmental conditions. Under favourable condition, females lay between 150 to 500 eggs at an average rate of 25 per day. Freshly laid eggs are 0.2 to 0.25 mm long and yellowish in colour but become black after 2 days. They are attached to the leaf surface by a short stalk normally on the underside of the leaves, though a few may be laid on the upper surface or even on the main stem. While ovipositing the female usually keeps her rostrum fixed in the tissues and pivots her abdomen around in a circle. Hence, on smooth leaved plants, where such movements are uninterrupted, a complete circle of 20 to 40 eggs may be laid without ceasing feeding. The incubation period is of 7 to 10 days. The newly hatched larva is flat, 0.3 mm long and light green in colour with bright red eyes. It has functional legs and moves about for 2 to3 days in search of suitable feeding site before settling down after which it becomes sedentary. The larva moults after 3 to 4 days into second instar nymph which flattens out on the leaf and becomes even more transparent than the first instar, legs and antennae become vestigial. Within 3 days the nymphs moult into third instar which is slightly larger (0.5 mm) than the previous instar. Within next 3 to 4 days the nymph again moults to whitish-green pupa. As the pupa matures, it is lifted from the leaf surface by submarginal wax rods which form a striated palisade. Unlike other stages the pupa has 5 to 8 pairs of prominent dorsal wax hairs and shorter marginal processes. Within 6 days, adult emerges through a T-shaped opening in the dorsal surface of the pupa. The life cycle is completed in 3–4 weeks at 15 to 20°C. Adults and immature stages have piercing and sucking type of mouth parts, which are inserted into the phloem tissues for removal of cell sap. The infested plants loose their vigor due to sap sucking, wilt, turn yellow or may even die. The leaves are covered with a coating of glazed sticky honeydew on which black sooty mould (*Cladosporium* sp.) develops and reduces the photosynthetic activity of the plant besides reducing the aesthetic value drastically. Whiteflies also transmit virus diseases in ornamental plants like geranium, hibiscus, zinnia, anthurium, etc.

### Detection

The whitefly infestation can be easily detected by shaking the plants and dislodging the adults. Yellow sticky traps can also be used to trap the adults. Greenhouse whitefly, *T. vaporarium* can be differentiated from *B. tabaci* in the pupal stage as the pupae of the former have vertical sides with spines and fringes of setae around the top perimeter while in the later pupae are convex or flat and without spines and fringes.

**Management**

Whiteflies are polyphagous pests which can thrive on many hosts including weed plants. Therefore, all weeds and left over plants in and around the cropped area should be removed after each crop. Screening can prevent the movement of whiteflies into the greenhouse from outdoors. On outdoor crops, mulching with aluminium or yellow polyethylene sheets reduces virus transmission by whiteflies. Yellow polythene mulch helps in attracting and killing the whiteflies due to high temperature whereas aluminium mulch acts as light reflector, repelling the insects. Removal of infested leaves from the plants also reduces further build up of population. At low infestation level, spray mild soap solution at 3 to 4 days intervals with fine mist sprayer so that the material also reaches the undersides of leaves (Ramesh, 1994).

On heavy infestation, spray deltamethrin (0.0028%), dichlorvos (0.04%) or trizophos (0.08%) at 5 to 7 day intervals. Among the promising biocontrol agents the parasitic wasp, *Encarsia formosa* Gahan, has been found effective in controlling *Bemisia tabaci* on poinsettia along with soap sprays in the USA (Parrella *et al.*, 1991). On poinsettia and chrysanthemum, when used along with growth regulators abamectin or Enstarse (kinoprene) against *T. vaporariorum*, this parasite lowered the population to great extent throughout the season (Fisher and Shanks, 1979; Zchori *et al.*, 1994). Entomopathogenic fungi, *Verticillium lacanii* is also used in the form of a wettable powder in the Netherlands against the glasshouse whitefly which gives upto 90 per cent reduction in infestation (Ravensberg *et al.*, 1990). This fungus is compatible with conventional insecticides and miticides but not with fungicides.

## CATERPILLARS

Caterpillars of many species of moths and butterflies are known as pests of ornamental plants. Some of the important caterpillars pests are: the pod borer, *Helicoverpa armigera* (Hubner), hairy caterpillars, *Euproctis fraterna* Moore, *E. virgincula* Walker, *E. scintillans* Walker, castor semilooper, *Achaea janata* Linn., leafroller, *Sylepta derogata* Fabr. on hollyhock, tobacco caterpillar, *Spodoptera litura* (Fabr.), cabbage caterpillar, *Pieris brassicae* (L.), and cabbage semilooper, *Thysanoplusia orichalcea* Fab. And *T. nigrisigna* Walker (Ayyar, 1963; Santhanarman, 1965; Suba Rao *et al.*, 1973; Singh, 1983; Sood, 1988; Kaushal, 1997). The flowering plants attacked by caterpillars include carnation, chrysanthemum, gladiolus, poppy, gerbera, hibiscus, hollyhock, antirrhinum, fuschia, kolenchoe, petunia, sweet william, tuberose zinnia, sunflower, sweet pea, spanish ivy and many shrubs and trees.

**Biology**

Caterpillars have chewing type of mouth parts and cause conspicuous damage by consuming leaves and flowers by tying or rolling leaves or boring into stems and buds or cutting plants at the base. The biology of some of the more common species is discussed as under:

***Achaea janata* L.**

The moth is stout, greyish brown with black blotches on hind wings. Young larvae are greenish with light brown posterior margins. The full grown caterpillars are 5.0 to 5.5 cm long, smooth, dull greyish-brown with white or brown stripes along the body. Caterpillars are voracious feeders. Eggs are green in colour and are laid singly on the underside of leaves. The chorion is sculptured with ridges.

### *Agrotis segetum* (D. & S.), *A. ipsilon* (Huff.)

Cutworms are polyphagous insects that infest a variety of cultivated plants including ornamentals. The cutworm larvae, as the name implies, cut the seedlings at the base of the stem and feed on leaves. They generally feed on seedlings between dusk to dawn and keep hiding among the crop refuge or in the soil during daytime. Among the important species of cutworms found in India, five species of the genus *Agrotis* including *A. ipsilon* (Hufnagel), *A. segetum* (D.&S.), *A. interacta* Walker, *A. flammatra* Schiff.and *A. spinifera* (Hubner) have been reported to cause extensive damage to various field crops, the first two being major pests of potato (Saxena and Rizvi, 1974) and cole crops (Sachan and Gangwar, 1990).

The females lay upto 1000 dome-shaped eggs in clusters either on the underside of leaves or on the moist soil in the field. The eggs hatch in 2 to 13 days depending upon temperature. The young larvae feed on leaves touching the ground and remain hidden in the soil during daytime. Older larvae cut the plants at ground level during night. There are six instars and the larval development is completed in 21 to 25 days. Pupation occurs in earthen cells (cocoons) in the soil and pupation period lasts for 10 days in summer and 30 days in winter. Cutworms damage more than they actually consume, often necessitating replanting of seedlings.

### *Euproctis fraterna* Moor

The adults are medium sized moths with orange yellow wings having light wavy markings on them. The female lays eggs in clusters on the underside of leaves covering them with a dense tuft of hair. The grown up caterpillar is 3.5–4.0 cm long, reddish brown in colour and has tufts and tussocks of hair on the head and posterior end of the body. Pupation takes place in the soil.

### *Helicoverpa armigera* (Hubner)

It is a polyphagous pest of cosmopolitan occurrence feeding on almost every type of plants available in the field. Among the ornamental plants, carnation, chrysanthemum, sweet pea, rose, marigold and gerbera, etc. are its preferred hosts. The moth is stoutly built, yellowish brown with fore wings marked with greyish wavy lines and black spots on the underside. The hind wings are lighter in colour with a broad blackish band along the anal margin. A female can lay upto 750 eggs singly on tender parts of the plant and on flower buds. The eggs are round and shining greenish yellow. They hatch in 2 to4 days. Generally, the caterpillars are greenish in colour with dark brown/grey lines along the side of the body. Just after hatching, the larvae feed on the tender foliage but later move and bore into lower buds and flowers. The caterpillar is about 3.5 cm when full-fed and its colour changes according to the food. If not checked in time, carnation crop is completely destroyed by this pest. A single larva may destroy many flowers before reaching maturity. Larvae are highly cannabalistic and feed on fellow caterpillars if food is not available. The larvae become full-fed in 13 to19 days. Pupation occurs in soil. Depending upon the temperture, 3 to 8 generations may be completed in a year.

***Spodoptera litura* (Fabr.)**

The moth is stout, dark with wavy white markings on the fore wings and hind wings are white with brown margins. The full grown caterpillars are 3.7 cm in length and are greenish brown with dark markings. The caterpillars feed gregariously on the foliage in the early stages of development while older ones are voracious feeders and remain hidden in the soil during daytime.

***Thysanoplusia orichalcea* Fab.**

The adults are medium sized mouths with golden markings on the fore wings. The young caterpillar is green in colour. A female lays about 400 to 500 eggs singly on leaves during night. The life cycle is completed in 19 to-34 days. The pupation occurs within the folded leaves. The pupal period is of 7 to 15 days.

**Detection**

The presence of moths can be monitored with the help of light traps or black light traps by attracting then to these traps. However, these traps are useful only in open fields and not in greenhouses as the attracted females may start laying eggs on the otherwise protected crop. Pheromone traps are species-specific and can be used for monitoring important polyphagous pests in the open fields. These traps provide an early warning of a possible infestation of a particular species and help in timely application of insecticides. Thorough scrutiny for injury symptoms like small round holes, presence of caterpillars and focal pellets on the foliage, etc is useful. It is important to detect and treat infestation early as grown up caterpillars that disperse to uninfested plants, are voracious feeders and are difficult to control even with most potent insecticides. Early infestations can be checked with endosulfan 0.05%, fenvalerate 0.1%, cypermethrin 0.0075% or deltamethrin 0.0028% at 10 to 15 days intervals (Sood, 1988; Kaushal, 1997). Among the biocontrol agents the egg parositoid, *Trichogramma* sp., bacterial pathogen, *Bacillus thuringiensis kurstaki* Berliner and certain specific nuclear polyhedrosis viruses are registered for insect control in some countries (Foschi and Deseo, 1987; Bene and Del, 1983) the later two have resulted in complete mortality of the caterpillars on rose and other ornamentals under glasshouse conditions (Burges and Jarrett, 1978; Smits *et al.*, 1987; Fritzsche *et al.*, 1991).

Foschi and Deseo (1987) reviewed the work on the prospects of using microbial control of phytophagous pests in ornamental plants and listed biocontrol agents like *B. thuringiensis kurstaki*, *Verticillium lecanii*, *Steinernema* (=*Neaplectana*) and *Heterorhabditis* spp. out of which some are registered and are being used for the control of caterpillars. On cut flowers of carnation infested with tortricid caterpillars, ionizing radiation applied at the rate of 6 to 9 Krad proved effective them (Bestagno *et al.*, 1973).

## BEETLES

Beetles are common pests of ornamental crops and account for about 40% of the known insect species. Most of the pest species belong to families, Elateridae, Scarabaeidae, Chrysomelidae, Curculionidae, Meloidae, etc. These insects have chewing and biting type of mouth parts and cause severe injury to foliage and flowers. The plants attacked by beetles include China aster, marigold, narcisus, cyclamen, celosia, morning glory, petunia, lupin, primula, rose and many other ornamental shrubs and trees.

Adults of most of the beetles like cockchafer beetle come out at dusk time on the plant to feed on them by making small round holes or irregular notches from the margins of the leaves. Some species of beetles and weevils feed on flowers during daytime by hiding either in them or among the leaves. During day time they may also hide in crevices or in the soil among grasses or leaves. The grubs are white, fleshy with wrinkled bodies, brown head and three pairs of strongly developed thoracic legs. These grubs feed on the roots of the plants including grasses and in severe infestation, the entire root system may be destroyed. In cyclamen, cavities are made in the corms by the grubs. Beetles mostly lay eggs in the soil and depending on the species, grubs may take 1 to 2 years to mature. Pupation also occurs in soil and adults generally emerge and come out with the first monsoon showers. Some of the commonly found species are *Adoretus versutus* Harold, *Anomala transversa* Burmeister, *Adoretus* spp., *Myllocerus* spp., *Mylabris* spp., *Altica* spp., *Oxycetonea* spp. and *Hyperstylus* sp., etc.

**Management**

Deep ploughing exposes grubs, pupae and eggs to natural enemies and sun. Soil application of phorate or carboforan @ 1.5 kg a.i./ha controls grubs. Setting up of light traps helps in collecting adults and monitoring their incidence. Spraying of quinalphos (0.07%), chlorphyriphos (0.05%)or carbaryl 0.2%, once in a fortnight in the evening hours helps in controlling them on rose. Ethanol extract of neem seed kernel at 2.5% inhibits feeding by chafer beetles on rose. In such pests with long life cycle and typical behaviour, timing of insecticide application is more important than the number of applications (Fregonese and Zandigiacomo, 1992). Many biocontrol agents are being used in the European countries to control beetle and weevil pests like, *Bacillus thuringiensis*, *tenebrionis*, entomogenic nematodes like *Neoaplectana* and Heterorhabditis, and Entomogenous fungus, *Metarhizium anisopliae* (Metsch.) to some extent but complete success has not been achieved so far (Deseo and Costanzi, 1987; Garnham and McNab, 1990; Moorhouse *et al.*, 1993; Eckberg and Cranshaw, 1994). However, about 90% control of the weevil was achieved on potted plants of begonia and cyclamen in the UK by applying conidia of entomogenous fungus, *M. anisopliae*.

## MITES

Mites are important pests of agricultural crops and being polyphagous cause extensive damage to many crops of economic importance including ornamentals. Feeding by mites results in loss of chlorophyll, stipping or bronzing of foliage, stunting of growth, production of various types of deformities and reduction in yield. Apart from direct damage, many species belonging to Eriophyidae and Tetranychidae are known as vectors of virus diseases causing extensive damage to flower plants.

Among the various species, the two spotted spider mite, *Tetranychus urticae* Koch, and reddish carmine mite, *T. cinnabarinus* (Bois du val), are the most serious pests attacking ornamental plants in glasshouses and out doors. The mite species recorded from various ornamental plants in India include *Tetranychus telarius* (Linnaeus), *T. ludeni* Zacher, *T. cinnabarinus*, *T. neocaledonicus*

Andre, *T. urticae* (Koch), *T. cucurbitae* Rahman and Sapra, *Paraphytoptus chrysanthemi* Keifer, *Brevipalpus obovatus* Donn., *B. phoenicis* (Geijskes), *B. rugulosua* Chaudhri, Akbar and Rasool, *B. californicus* Banks, *B. karachiensis* Chaudhri, Akbar and Rasool, *Typhlodromus confuscus* Narayanan, Kaur and Ghai (Channabasavanna, 1966; Menon and Ghai, 1968; Ghai, 1964; Bindra and Singh, 1970; Sadana *et al.*, 1981; Sharma and Pande, 1984; Sood, 1988).

***Brevipalpus*** **spp.**

These are minute, red or yellowish mites with body length 250 u. The palpi are 4-segmented, the body flat and egg-shaped, and the cuticle has a reticulate pattern. The legs are usually wrinkled. Unlike most true spider mites, these mites do not produce a silken web. These mites crowd under the leaves producing a brown, scurfy discolouration extending along either side of the midrib to the base of the leaf. The damage often induces premature senescence. Reproduction is mainly parthenogenetic. Each female lays about 60, bright red, elliptical eggs with the long axis perpendicular to the leaf in groups, usually in crevices or along the groove of the mid rib. These mites are recorded on *Ageratum*, fuchsia, campanula, coleus, orchids, peperomia and many other ornamentals.

***Bryobia*** **sp.**

These species occasionally cause damage to cultivated ivy and chrysanthemum and are distinguished from other tetranychids by their hairy, 6-segmented legs, the first pair being extremely long. The body is flattened and covered with clavate, hair and has a distinct peripheral step on its dorsal side. The eggs are smooth, dark red, 0.2 mm in diameter, slightly sticky and covered with fine dust particles. Newly hatched larvae are orange-red but gradually become greener. Both protonymph and deuteronymph stages are brownish at first but later become greenish. In the dormant stages the first pair of legs is folded together underneath the body and appear greatly shortened. In general, *Bryobia* has a striking preference for the upper side of the leaf and unlike other tetranychids, does not produce silken web. Males are generally absent and the reproduction is parthenogenetic. Infested leaves become mottled with greyish yellow spots and in severe cases, plant growth may be stunted.

***Eriophyid mites***

Eriophids are minute, almost wormlike mites and are either free living on plants, or gall makers. The long body is annulate and bears in all developmental stages only two pairs of anteriorly placed legs. Outside India these mites are recorded on chrysanthemum and carnation, etc. (Vernon, 1957; Miles, 1964). In India, 12 species of eriophyid mites belonging to two sub-families were recorded by Channabasavanna (1966) on different ornamental plants.

***Rhizolyphus echinopus*** **Fumouze and Robin**

This species is found on hyacinth lilies, tulips, narcissus and other kinds of ornamental bulbs. It is not a primary pest of bulbs, but attacks and breeds very fast in slightly injured bulbs. Once the mite is established on the bulbs, the latter are reduced to a rotten pulp. During storage the mites wander from bulb to bulb and may help in spreading various pathogenic fungi. The mite has a smooth, colourless and glistening body, about 0.7 mm in length with reddish brown appendages. Each female lays about 50 eggs. The species is generally found in large groups on the plant and is frequently associated with *Fusarium* spp.

***Stenotarsonemus laticeps*** **Halbt.**

On amaryllis, this mite causes red streaks and spots on the leaves as they emerge from the neck region of the bulb. The reddish lesion elongates as the leaf or flower stem grows and ultimately a scar results. In severe cases the flowers shrivel or the flower stem remains short. Storage of bulbs at low temperature and humidity in a well lit storage room reduces the serious mite infestation.

Infested bulbs of narcissus may be recognized by their abnormal dryness. The outer scales adhere tightly to the infested bulb. Severely affected bulbs loose vigour and foliage become senescent prematurely and may be killed in extreme cases. Yellow streaks and blotches appear on the basal portions of the leaves and flower stems while many afffected leaves develop a saw-tooth disfiguration. The rapidly increasing mite population congregates in the neck of the bulb where they become visible to the naked eye at the bases of young leaves. As the foliage dies, the mites wander along the dying leaves to adjacent bulbs.

***Tarsonemus pallidus*** **Banks**

These are minute, less than 0.3 mm long with shining, oval and segmented bodies. The females of this mite are pale, amber coloured which congregate within unopened buds and folded leaflets. Hundreds of white, opaque smooth and oval eggs are laid in dense groups. To the naked eye, such groups appear like a fine coating of dust. The foliage, expanding from infested buds, is usually curled, twisted, brittle, and covered with small scabs. Infested flower buds are discoloured, flecked with dark spots even when small number of mites are present and ultimately wither and die. These mites produce toxins during feeding which induce distortion of plant growth.

***Tetranychus urticae*** **(Koch)**

This species is greenish or yellowish in colour with large spots on either side of the body. The body is soft, pear-shaped and arched behind the second pair of legs by a transverse groove. The full grown females are 0.45 mm long and males are 0.3 mm long. Males have slender and rather pointed abdomen. Females produce about 100 spherical creamish white eggs, 0.1 mm in diameter, on the underside of leaves or silken strands produced by the adults. As the eggs develop, the colour changes to red. At 25°C, the incubation period is 3 days after which it changes to six-legged larva. After one day, it enters into a quiscent stage for a day or two. This resting stage is firmly attached to the leaf surface, usually near a vein or silken fibre. During this period the legs are rigid, the front two pairs held straight forward parallel to the palps. The next stages, protonymph and deuteronymph, also have active and quiscent stages each lasting 1 to 1.5 days. Sexes can be differentiated at deuteronymph stage as the males are more slender and elongate than females. Adult males are more active than females and copulate immediately after female emergence. Egg laying begins 24 to 48 hours after fertilization and the females produce both the sexes but slightly more number of females. Unfertilized eggs develop into males. The adults as they move over the leaves, form a web of fine silken strands which help in moving from one place to another in search of food. When the population is abundant, the entire plant may be enveloped in fine silken threads. The silk is produced from the glands discharging the secretion on the upper surface of the stylets. The silken webbing does not increase the damage but on flowering plants, such as carnation and chrysanthemum, the flowers are rendered unsightly and unsalable. The mite enters into hibernation during winter months.

**Detection**

The mite infestation is monitored through frequent inspections. Spider mite infestation is indicated by extensive webbing on leaves, and mottling or yellowing of leaves. Bulbs infested with mite become sticky and dry. Indicator plants like beans, grown around the main crop may help in detecting spider mite infestation before their further spread to the main crop.

**Management**

The best way to avoid loss due to bulb scale mite (*Stenotarsonemus* laticeps) is to ensure that the bulbs are free from infestation. Hot water treatment of the narcissus bulbs for 4 hours at 43.3°C or 3 hours at 44.4°C is an established treatment of mite control. Fumigation of the bulbs with methyl bromide is also recommended in some countries but this may damage the bulbs in some instances.

Weeds and infested plants growing around the main production area should be removed and burnt to reduce further multiplication of mites. Spider mites develop very fast on moisture-stressed plants and in areas where soil is rich in nitrogen. Therefore, water and nitrogen supply should be regulated in order to avoid excessive build up of population. Bulbs of ornamental plants should be freed from mite infestation before moving them to the field. Weekly spray with plain water or soap solution may take care of the mild infestation of spider mite. For heavy infestation, spray malathion (0.05%)or dicofol (0.037%) (Ramesh, 1994). Among the more recent chemicals, fluvalinate and fenpropathrin, both at 0.02% were found to give effective control of the mite on rose (Gough, 1990; Hole and Salunkhe, 1997b).

The phytoseiid mite, *Phytoseinulus persimilis* Athias-Henriot is the most effective biological control agent of mite pests on rose, chrysanthemum and gerbera (Vacante and Garzia, 1987). However, use of pesticides for the control of other pests might kill the predatory mite as well.

## REFERENCES


Ananthakrishnan, T.N. 1971. Thrips (Thysanoptera) in agriculture, horticulture and forestry: diagnosis, bionomics and control. *Journal of Scientific and Industrial Research,* **30:** 113–146.

Ananthakrishnan, T.N. 1973. *Thrips: Biology and Control*. McMillan Company of India, Delhi. pp. 120.

Ayyar Ramakrishna, T.V. 1963. *Handbook of Economic Entomology for South India*. Government Press, Madras. pp. 516.

Behura, B.K. 1986. Biology of aphids. *In: Proceedings of the Second National Symposium in Recent Trends in Aphidological Studies.* S.P. Kurl (ed.). M.M. P.G. College, Modinagar, India. pp. 21–28.

Bene, G-del and Del, B.G. 1983. Effects of *Bacillus thuringiensis* Berliner on *Epichoristodes acerbella* (Walk.) Diok. *Redia,* **64:** 229–236.

Bestagno, G.; Piana, S.; Roberti, L. and Rota, P. 1973. Ionizing radiations for controlling tortrix moths on carnations. Notiziario sulle Malattie delle Piante, **88–89:** 195–220.

Bhalla, O.P. and Thakur, J.R., 1974. A note on thrips fauna of Himachal Pradesh. *Himachal Journal of Agricultural Research,* **2**: 2.

Bindra, O.S. and Singh, G. 1970. Mite pests of ornamental plants in India. *Pesticides,* **4**(6): 17–24.

Blumel, S. and Hausdorf, H. 1996. Greenhouse trials for the control of aphids on cut-roses with the chalcid *Aphelinus abdominalis* Dalm. (Aphilinidae: Hymenoptera). *Anzeiger fur Schadlingskunde Pflanzenschatz Umweltschutz,* **69**: 64–69.

Burges, H.D. and Jarrett, P. 1978. Caterpillar control with *Bacillus thuringiensis*. *Grower,* **90**: 13, 589–590, 593–595.

Butani, D.K. 1974. Pests damaging roses in India and their control. *Pesticides,* **8**(12): 40–42.

Chandler, L.D. 1987. Evaluation of yellow trap sizes and within-field densities for monitoring *Liriomyza trifolii* in bell peppers. *Southwestern Entomologist,* **12**: 245–251.

Chandra, J. and Lal, O.P. 1973. Record of thrips on some vegetables and ornamental plants from Kullu Valley, Himachal Pradesh. *Indian Journal of Entomology,* **35**(2): 164–166.

Channabasavanna, G.P. 1966. *A contribution to the knowledge of Indian triophyid mites (Eriophyoidaea: Trombidiformes: Acarina*). University of Agricultural Sciences, Hebbal, Bangalore. pp. 153.

Deseo, K.V. and Costanzi, M. 1987. Use of nematodes pathogenic to insects against the larvae of weevils injurious to flowering and ornamental crops. *Difesa delle Piante,* **10**(1): 127–132.

Dong, H.F. 1993. A preliminary study on the occurrence of *Pseudococcus comstocki* (Homoptera: Pseudococcidae) on *Clivia minuta* and its control with inundative release of *Cryptolaemus montoazieri* (Coleoptera: Coccinellidae). *Chinese Journal of Biological Control,* **9**: 12–14.

Eckberg, T.B. and Cranshaw, W.S. 1994. Larval biology and control of the rabbitbush beetle, *Trirhabda nitidicollis* Leconte (Coleoptera: Chrysomelidae). *Southwestern Entomologist,* **19** 249–256.

Ferran, A.; Niknam, H.; Kabiri, F.; Picart, Je.; Herce-C-de; Brun, J.; Iperti, G.; Lapchin, L. and De-Herce, C. 1996. The use of *Harmonia axyridis* larvae (Coleoptera: Coccinellidae) against *Macrosiphum rosae* (Hemiptera: Sternorrhyncha: Aphididae) on rose bushes. *European Journal of Entomology,* **93**(1): 59–67.

Fischer, S.J. and Shanks, J.B. 1979. Whitefly infestation on chrysanthemum and poinsettia treated with plant and insect growth regulators. *Journal of the American Society for Horticultural Science,* **104**(6): 829–830.

Foschi, S. and Deseo, K.V. 1987. Prospects of microbiological control of phytophagous pests in flowering and ornamental crops in a protected environment. *Difesa delle Piante,* **10**(1): 121–125.

Fregonese, A. and Zandigiacomo, P. 1992. *Otiorhynchus sulcatus* (Fabr.), a cureulionid beetle damaging ornamental plants in nurseries. *Informatore Agrario,* 48: 73–77.

Fritzsche, R.; Geissler, K. and Schliephake, E. 1991. Results of experiments with granulosis and nuclear polyhedrosis virus preparations in fruits, vegetables and ornamental plants. *Nachrichten blatt des Deutschen Pflanzensch utzdienstes,* **43**(5): 92–95.

Garnham, M. and McNab, H. 1990. The use of entomogenous nematodes for the contro of black vine weevil in cyclamen. *Proceedings of the 43rd New Zealand Weed and Pest Control Conference.* Lincolu, Canterbury, New Zealand. pp. 70–72.

Ghai, S. 1964. Mites. *In:* N.C. Pant (ed.). *Entomology in India.* N.C. Pant (ed.). Silver Jubilee of Entomological Society of India, New Delhi. pp. 385–396.

Gough, N. 1990. Evaluation of miticides for the control of two-spotted mite, *Tetranychus urticae* Koch on field roses. *Crop Protection,* **9**(2): 119–127.

Halperin, J. 1981. Use of trunk injections for pest control on ornamental trees in Israel. *Mitteilungen der Deutschen Gesellschaft fur Allgemeine und Angewandte. Entomologie,* **3**(1–3): 294–297.

Hansen, J.D.; Hara, A.H. and Tenbrink, V.L. 1992. Insecticidal dips for disinfesting commercial tropical cut flowers and foliage. *Tropical Pest Management,* **38** (3): 3, 245–249.

Hassan, S.A. 1974. Mass-culturing and utilisation of *Chrysopa* spp. (Neuroptera, Chrysopidae) in the control of insect pests. *Zeitschrift fur pflanzenkran kheiten und pflanzenschutz,* **81**(10): 620–637.

Hoddle, M.S.; Driesche, R.V. and Van, D.R. 1996. Evaluation of *Encarsia formosa* (Hymenoptera: Aphelinidae) to control *Bemisia argentifolii* (Homoptera: Aleyrodidae) on poinsettia (*Euphorbia pulcherrima*): a life table analysis. *Florida Entomologist,* **79**: 1–12.

Hole, U.B. and Salunkhe, G.N. 1997a. Evaluation of certain insecticides against rose thrips. *Rhipiphorothrips cruentatus* Hood. *Pestology,* **21**(4): 34–36.

Hole, U.B. and Salunkhe, G.N. 1997b. Efficacy of different pesticides against the rose mites, *Tetranychus urticae* (Koch). *Pestology,* **21**(5): 9–12.

Hole, U.B. and Salunkhe, G.N. 1997c. Field efficacy of some insecticides against red scale, *Aonidiella aurantii* Maskell on rose. *Pestology,* **21**(6): 24–27.

Kaushal, Preetee. 1997. *Studies on Helicoverpa armigera (Hubner) on carnation and its chemical control*. M.Sc. Thesis, Dr YS Parmar University of Horticulture and Forestry, Nauni, Solan, India. pp. 49.

Labanowski, G. and Mynett, M. 1990. Control of the gladiolus thrips, *Taeniothrips simplex* (Morison), on gladiolus crop. *Acta Horticulturae,* **266**: 547–551.

Lindquist, R.K. and Casey, M.L. 1988. Evaluation of cyromazine formulations and aldicarb 10 G for leafminer control on potted chrysanthemums, marigolds and zinnias. *Ohio Florists Association Bulletin,* **70**: 7–8.

Lindquist, R.K.; Casey, M.L.; Helyer, N. and Scopes, N.E.A. 1984. Leafminers on greenhouse chrysanthemum: control of *Chromatomyia syngenesiae* and *Liriomyza trifolii*. *Journal of Agricultural Entomology,* **1**: 256–263.

Macdonald, O.C. 1991. Responses of the alien leafminers *Liriomyza trifolii* and *Liriomyza huidobrensis* (Diptera: Agromyzidae) to some pesticides scheduled for their control in the U.K. *Crop Protection,* **10**: 509–513.

Menon, M.G.R. and Ghai, S. 1968. Three new records of Bryobiinae from India (Acarina: Tetranychidae). *Indian Journal of Entomology,* **30**: 88–89.

Miles, P.M. 1964. Chrysanthemum semi-bad mite in Wales. *Entomologists' Monthly Magazine,* **100**: **97**.

Moorhouse, E.R.; Gillespie, A.T. and Charnley, A.K. 1993. Application of *Metarhizium anisopliae* (Metsch.) Sor. conidia to control *Otiorhynchus sulcatus* (F.) (Coleoptera: Curculionidae) larvae on glasshouse pot plants. *Annals of Applied Biology,* **122**: 623–636.

Mora, M.H.R.; Mosquera, P.F. and Medina, H.R.M. 1985. Evaluation of physical and chemical methods of control for the chrysanthemum leafminer *Liriomyza trifolii* (Burgess) (Diptera: Agromyzidae). *Revista Colombiana Entomologia,* **11**: 36–38.

Nandakumar, C.; Reghunath, P. and Sheela, K.R. 1988. Control of the red scale, *Aonidiella aurantii* (Maskell) on rose. *Entomon,* **13** (3–4): 275–277.

Parrella, M.P. 1983. Evaluation of selected insecticides for control of permethrin-resistant *Liriomyza trifolii* (Diptera: Agromyzidae) on chrysanthemum. *Journal of Economic Entomology,* **76**: 1460–1464.

Parrella, M.P.; Christie, G.D., Robb, K.L. and Bethke, J.A. 1982. Control of *Liriomyza trifoli* with biological agents and insect growth regulators. *California Agriculture,* **36**: 11–12, 17–19.

Parrella, M.P. and Lindquist, R.K. 1983. Research on biology and control of leafminers (Diptera: Agromyzidae) in the genus *Liriomyza* Mik. *Proceedings 10th International Congress of Plant Protection, Brighton, England. pp.* 1117.

Parrella, M.P.; Paine, T.D.; Bethka, J.A.; Robb, K.L. and Hall, J. 1991. Evaluation of *Encarsia formosa* (Hymenoptera: Aphelinidae) for biological control of sweet potato whitefly (Homoptera: Aleyrodidae) on poinsettia. *Environmental Entomology,* **20**(2): 713–719.

Pirone, P.P.; Bernard, O.D. and Rickett, H.W. 1960. *Diseases and Pests of Ornamental Plants*. Constable & Co. Ltd., London.

Poe, S.L. and Marousky, F.J. 1972. Control of aphids and mites on cut chrysanthemums by post-harvest absorption of azodrin and demeton. *Proceedings of the Florida State Horticultural Society,* **84:** 432–435.

Powell, C.C. and Lindquist, R.K. (eds.) 1996. *Ball Pest and Disease Manual*, 22nd Edition, Ball Publishing Co., Illinois, USA. pp. 426.

Ramesh, P. 1994. *Pests of Floriculture Crops and their Control*. Kalyani Publishers, Ludhiana, pp. 205.

Ravensberg, W.J.; Malais, M.; Van-der-Schaaf Da and Van-der-Schaaf, D.A. 1990. *Verticillium lacanii* as a microbial insecticide against glasshouse whitefly. *Proceedings of the Brighton Crop Protection Conference, Pests and Diseases*., Brighton, England. pp. 265–268.

Robin, M.R. and Mitchell, W.C. 1987. Sticky trap for monitoring leafminers *Liriomyza sativae* and *L. trifolii* (Diptera: Agromyzidae) and their associated hymenopterous parasites in watermelon. *Journal of Economic Entomology,* **80:** 1345–1347.

Romankow, W.; Baranowski, T. and Kowalewski, T. 1972. Results of studies on the control of the green peach aphid (*Myzodes persicae* Sulz.) and red spider mite (*Tetranychus urticae* Koch) on *Gerbera jamesonii* Bolus by selected chemical prepartions with reference to their phytotoxicity. *Biuletyn Instytutu Ochrony* Roslin, **54**: 223–238.

Sachan, J.N. and Gangwar, S.K. 1990. Seasonal incidence of insect pests of cabbage, cauliflower and knol-khol. *Indian Journal of Entomology,* **52**(1): 111–124.

Sadana, G.L.; Chhabra, S.C. and Kumari, N. 1981. New records of tetranychoid mites of ornamental and medicinal plants and their hosts from the Punjab State. *Entomon,* **6**(4): 325–327.

Saito, T.; Iked, F. and Ozawa, A. 1996. Effect of pesticides on parasitoid complex of serpentine leafminer, *Liriomyza trifolii* (Burgess) in Shizuoka Prefecture. *Japanese Journal of Applied Entomology and Zoology,* **40**(2): 127–133.

Sanderson, J.P.; Parrella, M.P and Trumble, J.T. 1989. Monitoring insecticide resistance in *Liriomyza trifolii* (Diptera: Agromyzidae) with yellow sticky cards. *Journal of Economic Entomology,* **82:** 1011–1018.

Santhanaraman, T. 1965. Record of two new Lymantriid pests on rose in Coimbatore. *Science and Culture,* **31**(12): 647–648.

Saxena, A,P. and Rizvi, S.M.A. 1974. Insect pest problems of potato in India. *Journal of Indian Potato Association,* **1**(1–2): 43–50.

Sharma, B.L. and Pande, Y.D. 1984. Host range of cucurbit mite, *Tetranychus neocaledonicus* Andre (Tetranychidae: Acarnina). *Science and Culture,* **50:** 29–30.

Sher, R.B.; Parrella, M.P. and Lenteren, Jc. Van. 1996. Integrated biological control of leafminers, *Liriomyza trifolii* on greenhouse chrysanthemums. *Proceedings of the meeting on Integrated control in glasshouses, Vienna, Austria*, 20–25 May, 1996. Bulletin OILB SROP, **19**(1): 147–150.

Singh, S.P. 1983. *Heliothus armigera* (Hubner) as a pest of ornamental plants in Simla Hills. *Indian Journal of Plant Protection,* **11**(1/2): 95.

Sood, A.K. 1988. *Insect pest complex of ornamental plants in Himachal Pradesh*. M.Sc. Thesis, Dr YS Parmar University of Horticulture and Forestry, Nauni, Solan, India. pp. 112.

Smith, R.F. 1986. Efficacy of selected insecticides against *Liriomyza trifolii* (Burgess) (Diptera: Agromyzidae), a leafminer of chrysanthemum. Canadian Entomologist, **118:** 761–766.

Smits, P.H.; Vrie, M.V.D.; Vlak, J.M.; Van D.V.M. and De, V.M.V. 1987. Nuclear polyhedrosis virus for control of *Spodoptera exigua* on glasshouse crops. *Entomologia Experimentalis et Applicata,* **43**: 1, 73–80.

Subba Rao P.V.; Mohansaundram, M. and Subramaniam, T.R. 1973. Occurrence of *Spodoptera litura* (Fabr.) (Noctuidae: Lepidoptera) on rose flowers in Coimbatore. *Indian Journal of Entomology,* **35**(1): 70.

Tatara, A. and Furuki, T. 1994. Efficiency of yellow flat sticky trap for monitoring the leagume leafminer, *Liriomyza trifolii* (Burgess), in greenhouses. *Proceeding of the Kanto Tosan Plant Protection Society,* **41**: 235–237.

Thakur, J.R.; Bhalla, O.P. and Bhardwaj, A.K. 1973. Thrips fauna of Solan Area of Himachal Pradesh. *Current Science,* **42**(7): 256.

Tjosvold, S.A. and Ali, A.D. 1989. Fumigation of western flower thrips using banana bags (polyethylene-D) during simulated shipment of cut flowers. *Flower and Nursery Report, Summer*. pp.3–4.

Udayagiri, S. 1987. Frist record of *Liriomyza compositella* Spencer and *L. brassicae* (Riley) (diptera: Agromyzidae) as pests of ornamental crops in South India. *Entomon*,. **12**: 85–88.

Vacante, V. and Garzia, G.T. 1987. Present possibilities of control against *Tetranychus urticae* Koch by means of *Phytoseiulus persimilis* Athias-Henriot on rose and gerbera in a protected environment. *Difesa delle Piante,* **10**(1): 139–147.

Vernon, J.D.R. 1957. An eriophid mite causing damage to chrysanthemums. *Plant Pathologist,* **6**: 111.

Worthing, C.R. 1967. Observations on the chemical control of pests on year round chrysanthemums. *Proceedings of 4th Britain Insecticides and Fumigation Conference*. pp. 63–75.

Zchori, Fein E.; Roush, R.T. and Sanderson, J.P. 1994. Potential for integration of biological and chemical control of greenhouse whitefly (Homoptera: aleyrodidae) using *Encorsia formosa* (Hymenoptera: Aphelinidae) and abamectin. *Environmental Entomology,* **23**(5): 1277–1282.

*Pest Management in Horticulture Crops*
L.R. Verma, A.K. Verma and D.C. Gautam (eds)


# *Chapter 23*

# Nematode Pests of Fruit Crops

*N.K. Sharma*

## INTRODUCTION

NEMATODES, THOUGH, have been co-existing with plants since times immemorial and doubtlessly causing unrecognised damage to plant health, yet the work on fruit nematology was started towards the end of 19th century when Neal (1889) reported root-knot nematode infestations in peach and oranges from Florida (USA). Larsen (1910) in Hawaii in 1910 gave yet another account of the injury caused by nematodes in pineapple and suggested that these should be controlled. The discovery of citrus nematode, *Tylenchulus semipenetrans* in 1912 (Thomas, 1913; Cobb, 1913) accelerated the research on fruit nematology. The continuous investigations on *T. semipenetrans* over a decade demonstrated its pathogenic effects on citrus, both under controlled and field conditions (Thomas, 1923). However, it was only in the middle of this century that the destructive role of nematodes, in general, could be amply demonstrated following the discovery of highly potent soil fumigant nematicides like DD, EDB, DBCP (Nemagon) etc. These soil fumigants also made it possible to demonstrate that elimination of plant parasitic nematodes often removed certain disease symptoms which were formerly attributed to fungi or bacteria.

The problems like insidious "slow decline", or "dieback" of walnut, cherry, prune, apple, citrus and other orchard trees which baffled the investigators for many years were also found to be associated with large populations of plant parasitic nematodes. The investigations carried out throughout the world during the last 50 years have further revealed that these tiny organisms are not only the major limiting factors in our horticulture economy by way of their feeding and root destruction but they are also the basic causes in most diseases of complex nature (root and rhizome rots, replant diseases, shortening of tree productive life, "dieback" etc.), disorders (leaf fall, winter injury, leaf yellowing) and other problems like, graft incompatibility and bacterial cankers. The nematodes have also been demonstrated to transmit a number of nepo-viruses in fruit trees.

Inspite of the well established and recognised role of nematodes in leading fruit producing countries of the world, the awareness about the occurrence of nematodes in fruit nurseries and orchards in India is still lacking. In this chapter, nematode problems in fruit industry have been highlighted and effective management strategies have been suggested.

## ABOUT FRUIT NEMATODES

Nematodes are multicellular, non-segmented, bilaterally symmetrical, vermiform, colourless organisms which are microscopic in size. Typically, the nematode body is elongate, spindle shaped or fusiform, tapering towards both ends and circular in cross section. The length may vary from 0.2 mm (*Paratylenchus* =pin nematode) to about 11 mm (*Paralongidorus maximus* =needle nematode) but majority are of 1.0 mm in length. The body width may vary from 0.01 to 0.5 mm. In a few plant parasitic genera, the body may be pear, kidney or saccate shaped as in females of *Meloidogyne, Heterodera, Rotylenchulus* and *Tylenchulus.*

The life cycle consists of an egg stage, four larval stages and adult (both sexes). Generally, the life cycle extends from 15 to 30 days. Plant nematodes are obligate parasites and possess a protrusible stylet for feeding. They occur chiefly in soil where they attack plant roots. They are broadly classified into above ground and below ground forms.

### Above-ground forms

In general these forms are less than one per cent of the total plant parasitic species described so far. Even these forms have to pass some part of their life cycle in soil and plant debris. They attack the above ground plant parts by moving through a thin film of water present on their hosts. Classical example of above ground nematode species infesting fruit plants is *Aphelenchoides fragariae* on strawberry, mainly during spring and rainy season.

### Below-ground forms

The below ground nematode forms feed and complete their life cycle on plant roots only. They are broadly classified into;

#### *Ectoparasites*

These do not penetrate into the roots, but insert their stylets into the root tissue and suck cell sap. Examples of ecoparasites are *Criconemella, Paratylenchus, Xiphinema, Longidorus,* etc.

#### *Semi-endoparasites*

Semi-endoparasites such as *Tylenchulus semipenetrans, Rotylenchulus reniformis,* and some species of *Hoploaimus, Tylenchorhynchus, Helicotylenchus,* etc. normally feed with the front part of the body burried in the root.

#### *Endoparasites*

Endoparasites enter completely into the root tissues and feed directly either on cell cytoplasm or modify the cell forms like giant cells or syncytia, e.g., species of *Meloidogyne, Heterodera, Nacobbus, Radopholus, Pratylenchus*, etc.

In both the groups, migratory and sedentary forms are found. The migratory forms frequently shift their feeding sites from one place to another but in the case of sedentary forms, the nematode development from infective larval to adult stage is completed at one feeding site.

## NEMATODE DISEASES: CONCEPT AND DIAGNOSIS

Nematode diseases are not the diseases we generally refer to in the case of other pathogens, where we have separate nomenclature for each disease with the help of a known casual agent and definite symptomatology it produce. Whenever we talk of a nematode disease, it generally means nematode association with the plant roots in soil. The adverse effects of nematodes on plant become apparent only when there are sufficient number of individuals of the same species on the same plant or a population of parasitic nematodes may become pathogenic when the host ability to resist or tolerate the disturbance is inadequate (Wallace, 1973). These effects get enhanced when secondary soil micro-organisms (mainly bacteria and fungi) are involved with nematode. Therefore, nematodes are known to cause decline in orchards. "Decline" according to McElroy (1972), is a situation whereby fruit trees that have previously produced profitable yields, no longer grow or produce satisfactorily. Under such conditions a "replant disease" may arise when a second or a following planting of the same or closely related species, planted at the same site fails to become established. Factors known to be involved in a replant disease include parasitic nematodes, weak parasitic fungi and bacteria, toxic chemicals from old roots, unbalanced soil nutrition etc. On these bases, it can be suggested that the nematode disease should always be taken in totality and not in isolation, since, they are the main contributors in creating such situations in orchards over a period of time.

Investigating the exact role of nematodes in tree diseases etiology is a cumbersome process. However, their involvement in causing a particular disease, disorder or situation can be fully established under field conditions by using the technique as elaborated herein.

### Suspecting a nematode problem

This is an initial but the most important step towards detecting a nematode problem in tree crops. Fruit nematodes, though root parasites, do express their presence by producing some non-specific symptoms on above ground plant parts. Symptoms like retarded and/or unthrifty shoot growth, nutritional imbalances, leaf yellowing, increased plant susceptibility to foliar fungal diseases and pest attack, occurrence of bacterial canker, graft incompatibility, other plant disorders and winter injuries, die-back and decline in yields are known to be produced by nematodes. These symptoms if persist or reappear even after the adoption of curative control measures, the problem should be suspected to be due to nematode attack.

### Analysis of nematode population

Once the problem has been suspected on the basis of non-specific symptoms, the presence and population build up of nematodes should be recorded to establish them as a possible cause. For making proper assessment of nematode population, the root and soil samples should be drawn from 30 to 90 cm soil depth, 30 to 60 cm away from tree trunk. In nursery, the sampling is to be planned in such a way so as to cover the entire area. The composite samples should be collected in polythene bags and tied properly to avoid any loss of moisture. To ascertain the populations of both endo- and ecto-parasitic nematode forms on realistic basis, standard extraction techniques prescribed for individual groups should be employed.

### Confirming nematode involvement

After establishing presence of nematode species, the sites should be treated with nematicide(s)/organic amendment(s)/saw dust mulch, etc., to control nematode populations. In case the fruit trees start restoring normal growth/yield behaviour and the suspected symptoms of nematode attack are reduced/disappear and, there is marked reduction in nematode populations, the involvement of nematodes in causing the disease or disorder is confirmed. For determining the exact role of nematodes in the disease etiology suitable experiments should be laid out under controlled conditions.

## NEMATODE PESTS OF FRUIT CROPS

### Apple

Over 70 nematode species belonging to 30 genera have been reported to occur in apple tree rhizosphere from various parts of the world. Of these, the associations of five nematode groups, viz., *Pratylenchus, Meloidogyne, Xiphinema, Paratylenchus* and *Longidorus* with apple are of major economic importance as these cause pronounced deleterious effects on plant growth and productivity (Hoestra, 1967; McElory, 1972; Inagaki, 1978; Mai and Abawi, 1978; Tiev, 1983; Melton *et al.*, 1985). Besides, the nematodes have also been reported to act as vectors of apple viruses (Utkhede *et al.*, 1992). These nematode groups (except *Longidorus*) have been reported to occur in high number in apple nurseries and orchards in all the apple growing regions of India (Sindham, 1976; Kaul, 1985; Sharma and Kaur, 1985; Sethi *et al.*, 1988; Sharma and Sharma, 1988).

### *Pratylenchus* species (= Lesion nematode)

These are highly destructive migratory endoparasitic nematode forms which cause: decline and replant diseases. Out of the 12 species of this genus reported from apple, the association of *P. penetrans* is world-wide and of major concern. Ten nematodes of this species per 100 g of soil is the threshold level in apple (Zepp and Szczygiel, 1980). In India, this species is most frequently encountered in apple orchards in high number (upto 500 nematodes/250 g soil) (Sharma and Kaur, 1985)

The pathogenicity of *P. penetrans* on apple has been proved under controlled conditions in the USA (Mai and Abawi, 1978) and Poland (Zepp and Szczygiel, 1986) and is known to play a major role in apple replant problem (Mai and Abawi, 1978). The other species of *Pratylenchus* which occur in apple orchards and can be important locally are *P. coffeae, P. vulnus, P. thornei, P. crenatus, P. scribneri* and *P. pratensis* (McElory, 1972; Kir'yanova & Krall, 1980; Anonymous, 1985–93).

*Meloidogyne* species (= Root-knot nematode)

These are sedentary endoparasitic nematode forms which and Cause: unthrifty plant growth, yield reduction and enhance multiplication of fungi and bacteria in root tissues are known to exert deleterious effects on apple in some parts of the world. Root-knot nematodes, Two species of namely, *Meloidogyne mali* and *M. incognita* have been reported to occur on apple. Inagaki (1978) correlated poor fruit quality of apple to the abundant population of *M. mali* in apple orchards in Japan. Koch and Covey (1980) reported that an apple orchard raised in *M. mali* infested soil after pre-plant application of methyl bromide, the fruit yield increased atleast by 20 lb/tree in eight years.

In India, on the occurrence of root-knot nematode in apple for the first was reported in 1987 by Sharma and Kaur (1987). The infestation was particularly heavy in the mid hill and valley areas of Himachal Pradesh and the nematode infested trees exhibited symptoms of unthrifty growth. These nematode assist and enhance the entry and multiplication of various plant pathogens and thus aggravating and causing diseases of complex nature. Sethi *et al*., (1988) recorded heavy infestation of root-knot nematode and crown gall from severely damaged apple trees.

*Paratylenchus* species (= Pin nematode)

These are ectoparasitic species which feed on growing points of the roots and prevent them from functioning in a normal manner, thus Cause: quick drying, "leaf drop" and decline symptoms in fruit and forest trees. This is probably the least understood nematode group inspite of the fact that *Paratylenchus hamatus* was identified as the cause of "leaf drop" in figs in the USA (Thorne, 1961).

The high populations of this nematode have been found to be invariably associated with quick drying, "leaf drop" and decline symptoms in fruit and forest trees (Jenkins, 1960; Thorne, 1961; Corbett, 1966). In India, four species namely, *P. hamatus, P. manaliensis, P. nainianus* and *P. prunii* are known to occur in apple orchards (Sharma and Kaur, 1985; Khan, *et al*., 1988; Sharma and Sharma, 1988; Khan and Sharma, 1991). Sharma and Sharma (1990) recorded maximum population of *P. prunii* when soil samples were drawn from 20 cm depth, 30 cm away from tree trunk. In order to detect and have realistic assessment of its population the soil samples should not only be drawn from the site of its maximum concentration but also be examined carefully as the immature stages of these nematodes occur in great numbers and are not easily recognizable. Heavy infestation of *Paratylenchus* is also known to cause dwarfism in apple seedlings (Braun *et al*., 1966; Fisher, 1967).

*Xiphinema* species (= Dagger nematode)

An ectoparasitic form armed with long dagger shaped spear in its mouth. The nematode feeds on newly emerging rootlets thus causing extensive destruction of feeder roots, resulting in decline symptoms in orchords and tolso transmit viruses.

Out of the seven species of this genes reported from apple orchards, *X. americanum*, *X. diversicaudatum* and *X. rivesi* are of economic importance in apple throughout the world. In India, *X. americanum* is reported to occur in apple orchards and nurseries (Kaul, 1985; Sharma and

Sharma, 1988). The frequency of occurrence of this nematode varies in soil samples drawn from different apple orchards.

*Xiphinema* spp. were found to be extensively prevalent in declining apple orchards in the USA (Thorne, 1961) and are known to play an important role in apple replant disease (Mai and Abawi, 1978). Besides causing direct damage, these nematodes are known to transmit and enhance viral infections. Forer *et al*. (1984) confirmed the transmission of tomato ring spot virus (TmRSV) disease in apple by *X. rivesi* and correlated the prevalence and severity of TmRSV disease in various apple orchards in the USA with the incidence and population intensity of *X. americanum* and *X. revesi*.

### *Longidorus species* (=Needle nematode)

It is a large nematode form which possesses a long needle-like spear in its mouth and feeds ectoparasitically on roots. Swelling and curling of root tips, necrosis and retorted plant growth are common symptoms produced by Longidorus (*L. elongatus*) on apple. These species also transmit viruses.

*Longidorus* species are known to transmit plant viruses. In USSR, 20 year old apple trees were debilitated by the parasitism of *L. elongatus* resulting in yield reduction to one third (Tiev, 1983). The decline in an established apple orchard in the UK was also attributed to the prevalence of *L. profundorum* 1982). *Longidorus* species though of common occurrence in India, yet have not so far reported on apple.

## Pear

Many plant parasitic nematodes are prevalent in pear orchards, but they do not appear to cause economic losses to this fruit crop. At some places, however, the nematode populations have been correlated with pear diseases and decline. An important example is the association of *P. penetrans* with pear diseases in the Netherland (Oostenbrink, 1954), Germany (Decker, 1960), and *X. americanum* and *Paratylenchus hamatus* with pear decline in California, USA (French *et al*., 1964).

## Banana

Nematodes constitute one of the major limiting factors to banana production. According to Blake (1969), nematode parasitism is the most serious and widely distributed cause of root damage through out the banana growing areas of the world. Among a number of nematodes reported in association with banana roots, the burrowing nematode, *Radopholus similis*, is the most important pest. *Helicotylenchus multicinctus, Pratylenchus coffeae* and *Heterodera oryzicola* are the other species of economic importance. The association of root-knot though reported from banana from various places yet has not been considered of much significance.

### *Radopholus similis* (=Burrowing nematode)

This species has a wide geographical distribution and is known to occur in most areas of the world where banana is grown commercially. The disease caused by *R. similis* is known by different names like Radopholus root rot, black head, black-head toppling disease and decline. The nematode was first described by Cobb (1893) from Fiji where it was known to cause extensive necrosis on the roots of *Musa sapientum* since 1890 (Cobb, 1915). From India, the first report of its occurrence on banana was made from Kerala. Since then it has been reported from Tamil Nadu, Karnataka, Maharashtra, Andhra Pradesh and Madhya Pradesh (Parvatha Reddy and Singh, 1980; Tiwari and Dave, 1985; Parvatha Reddy and Khan, 1987).

The nematode after entering the root, mostly through root tip, takes up feeding position between parenchyma cells in the cortex, where it feeds on the cytoplasm of nearby cells, destroying them and causing cavities to develop. These cavities coalesce and are continually enlarged by the nematode feeding to form tunnels laterally and towards endodermis. The characteristic reddish brown lesions are also formed throughout the cortex. Three to four weeks after infection, when extensive cavities have been formed, one or more deep cracks with raised margins appear on the root surface. The stele is not invaded. Eventually the root system may be reduced to a few short stubs (Blake, 1966). Reddish brown cortical lesions are diagnostic of nematode infection. The falling or toppling of mature banana plants especially during wet and windy weather is because of the nematode damage to primary and those roots which anchor the plant in the soil.

This species hastens the onset of Panama wilt caused by *Fusarium oxysporum* and *F. cubense*. The incidence of Panama wilt was shown to be doubled in the presence of *R. similis* in three months (Newhall, 1958). From the available evidences it is construed that the nematodes create a food base in the root cortex for rapid colonization of the fungus, which in turn helps causing larger lesions and even grow through the endoderious to colonize stelar region and cause necrosis of vascular cells.

### *Helicotylenchus multicinctus* (=Spiral nematode)

It is well known pathogenic species. However, because of its concomitant infestation with *R. similis* in most banana growing areas of the world, its role in banana cultivation is not fully understood. The only area where *H. multicinctus* infests banana in the absence of *R. similis* is Jordan valley in Israel, and there it is considered as the principal factor limiting the profitability of banana production (Minz *et. al.*, 1963). Its pathogenicity has been proved under aseptic conditions (Black, 1966). Like *R. similis*, this species is also a cortical feeder and destroys parenchymatous cells by partially and/or fully penetrating into the roots. The lesions or necrosis caused by *H. multicinctus* are usually discrete but smaller and superficial than those caused by *R. similis*.

### *Pratylenchus coffeae* (=Lesion nematode)

This nematode is a very destructive species known to cause yellowing of leaves in various field, tree and plantation crops and is reported as an endoparasite of banana but little is known about its status. Both adults and larval forms destroy the cortex through feeding and migration within the root tissues. The symptoms caused by *P. coffeae* are similar to that of *R. similis* and its presence along with *R. similis* has also been associated with black head topping disease of banana.

### *Other nematodes*

Banana is also known to be infected by sedentary nematode forms like root-knot (*M. incognita, M. javanica, M. arenaria, M. hapla*), cyst (*H. oryzicola*) and reniform (*R. reniformis*). However, their occurrence is not wide and dominant.

## Grapes

Grapes are grown in various agro-climatic conditions. Consequently, the nematode fauna should also differ greatly particularly between tropics and areas with heavy snowfall but, no such information is available. On the basis of nematode species found associated with grape cultivation in various parts of teh world it appears that the species of *Pratylenchus, Paratylenchus, Xiphinema* and *Meloidogyne* are mainly prevalent in temperate and sub tropical zones, while *Meloidogyne incognita* and *Rotylenchulus reniformis* are the predomient in the tropics. Some of the important nematode pests of grapes which pose serious threat to grape vineyards are discussed hereunder:

### *Meloidogyne* species (= Root-knot nematode)

Grapes are *Meloidogyne incognita, M. javanica, M. arenaria* and *M. hapla*. The major species of root-knot nematodes associated with grapes. They cause stunted plant growth, yield reduction and increased plant sensitivity to stress and nutritional deficiencies mainly in sandy soils. Root symptoms are distinctive, with small swellings or galls on feeder and young secondary rootlets, where single nematode invasion occurs. Larger galls may be produced by multiple infections. Root-knot nematodes can also kill feeder roots and generally debilitate plant growth (Raski, 1986).

### *Pratylenchus* species (=Pin nematode)

There are seven *Pratylenchus* species reported to occur on grapes, viz., *P. vulnus, P. brachyrus, P. pratensis, P. scribneri, P. neglectus, P. coffeae* and *P. thornei*. Among these, the most damaging is *P. vulnus*. Though there are no marked above-ground symptoms, there may be slight indication of potassium deficiency. Affected plants exhibit poor growth and low yields. The root system is similarly affected, many feeder roots are killed, more are produced and also killed eventually, giving rise to a watted witch's broom effect in some parts and total lack of roots in other parts of the root system (Raski 1986). *Pratylenchus coffeae* on the other hand has been found to cause wilt in grapes in Australia (Colbran, 1966).

### *Xiphinema* species (=Dagger nematode)

About nine species are known to occur on grapevine roots under this group. The important ones are *X. index, X. italiac, X. mediterraneum* and *X. pachtaicum*. Their feeding on roots mostly confines to root tips, where even a single nematode feeding for a short time can stop further development of that rootlet. Growth does not resume even if the feeding stops. Continuous feeding on roots and newly emerged rootlets produces witch's broom symptoms. Roots swell slightly due to proliferation of cells near the feeding site and often form a characteristic bend or crook near the tip, reminiscent of *Phylloxera* damage. Prolonged feeding results in blackened necrotic areas that may extend over the entire swollen tip. The root system is ultimately destroyed and the vine starves for water and nutrients. *X. index* and *X. italiac* are also known as vectors of fan leaf virus, highly pathogenic disease agent. The combined effects of nematode and viral damage shorten the life of a vineyard more than the sum of the virus or nematode acting separately (Raski, 1986).

### *Tylenchulus semipenetrans*

This species is a primary pathogen of citrus but also attacks grapevines mainly happens at places where either vineyards are planted adjoining to citrus orchards or at sites previously used for citrus plantation. Heavy root infestation with *T. semipenetrans* was demonstrated by inoculating grapevines artificially in the then USSR in 1957 (Kir'yanova and Krall, 1980).

### *Rotylenchulus reniformis*

This species is mainly prevalent on grapes in South India. Being polyphagous, it has adapted itself on this crop in the absence of other hosts in these areas. *R. reniformis* has also been reported as most prevalent species on grape in Cebu in the Philippines (David *et al.* 1979).

## Strawberry

The nematode problems of strawberry are slightly different from those on other fruit plants. Both above and below ground species are found associated with this crop.

### *Aphelenchoides* fagariae (= Leaf and bud nematode)

*Aphelenchoides fragariae* attacks the above-ground parts of strawberry in spring and rainy season. Therefore, it is usually termed as foliar, leaf or bud nematode. It is known to cause "Spring dwarf" or "Crimp" disease in this crop. The nematodes remain in quiescent form usually in soil, plant debris or propagating material (mother plants) from where these are transmitted to daughter plants. The infected plants are deformed and reduced in size. Petioles are short and leaflets rather elongate in comparison to their width. Leaf margins, showing cupping usually upwards in young leaves and downwards in older ones. In extreme cases, leaves are merely rudimentary and there is no further growth. Petioles, veins and underside of leaflets are often reddish purple. Infested leaves are distinctly more brittle than normal ones. If the main bud is killed, the plant may fail to produce more leaves and often dies. In some instances, adventitious buds develop to perpetuate a deformed, weak plant (Thorne, 1961).

In India, the occurrence of *A. fragariae* on strawberry was reported from Punjab (Sharma, 1977) and Himachal Pradesh (Sharma and Kaur, 1985). Kir'yanova and Krall (1980) reported that a strawberry race of *Ditylenchus dipsaci* (*D. fragariae*) was widespread on strawberry in the then USSR and caused severe crop losses. The symptoms reported however, resembled that caused by *A. fragariae*.

### *Other nematodes*

On the roots, strawberry is known to be attacked by nematodes belonging to 12 genera namely, *Pratylenchus*, *Meloidogyne*, *Xiphinema*, *Tylenchorhynchus*, *Tetylenchus*, *Helicotylenchus*, *Criconemoides*, *Hoplolaimus*, *Trichodorus*, *Belonolaimus*, *Longidorus* and *Ditylenchus* from USA and Europe (Braun, 1958; McElory, 1972). Out of these *Pratylenchus, Meloidogyne* and *Xiphinema* are of wide occurrence and importance. Some of the important Species of these two genera associated with strawberry are *P. penetrans, P. pratensis, P. scribneri, M. hapla* and *M. javanica* (McElory, 1972).

Among *Xiphinema* spp., the pathogenicities of *X. diversicaudatum, X. americanum* and *X. chambersi* have been proved on strawberry (Schindler and Braun, 1957; Perry, 1958). Nematode species of *Xiphinema* and *Longidorus* are also known to transmit tomato ring spot, tomato black ring spot, raspberry ring spot, strawberry latent ring spot, yellow bud mosaic and Arabic mosaic viruses in strawberry plants (Jha and Posnette, 1959; Harrison *et al.*, 1961; Frazier and Maggenti, 1962; Taylor, 1962; Harrison, 1964; Lister, 1964; Teliz *et al.*, 1967).

**Citrus**

Citrus is an important fruit widely grown in tropical and sub-tropical climates, occupying a belt extending around the world at both sides of the equator to a latitude of 35oN and 35oS. Over 200 nematode species belonging to 44 genera have so far been reported from citrus rhizosphere throughout the world. However, only about 20 species are known to cause pathogenic injury to citrus (Raski 1986). Some of the important nematodes which occur on citrus are discussed hereunder.

***Tylenchulus semipenetrans* (=Citrus nematode)**

This species is undoubtedly the most important nematode associated with citrus. The distribution of this species almost coincides with citriculture in the world. In India, it occurs in all citrus growing areas ranging from sub-tropical hilly tracts to tropical areas in the plains. The adult female is a sedentary semi-endoparasite, much smaller than most sedentary types such as cyst and root knot nematodes. An established infection site usually contains many females closely aggregated and covered with gelatinous egg masses containing several hundred eggs. All larval stages, young female and males are slender. The posterior portion of the female becomes enlarged only in the later part of the life cycle, probably to accommodate the gonad development. The stylet and digestive system in the males are not well developed and can be said to be degenerated compared to females.

The above ground symptoms, which are not very specific, include reduction in plant growth and vigour, yellowing of leaves and decline in yield. The nematode makes the citrus orchards uneconomical shortly after it has reached a level of enormous build up. This generally happens in about 10 years duration. The decline symptoms produced by the nematode are more pronounced in the upper most part of the tree. Since, it causes slow reduction in tree growth and yield (both in quality and quantity) parameters, the disease caused by this species is known as "Slow decline" of citrus.

On roots the nematode infection hinders spread of new roots initiated by the plant. The root biomass is reduced and is much less compared to the roots of uninfested trees. The soil particles usually cling tightly to the gelatinous egg masses and do not get detached even after washing. The capability of root system to draw nutrients from the soil is adversely affected. This in turn leads to nutritional deficiency symptoms in the above ground parts. Histopathological changes induced by this species are confined to the cortical cells. Six to ten cortical cells known as nurse cell, with dense cytoplasm, enlarged nucleus and nucleolus and without vacuole are produced by the nematode for its feeding and wriggles its way deeper in root cortex for further feeding.

Studies on spatial distribution of *T. semipenetrans* undertaken in Punjab (Sharma and Sharma, 1981) revealed that the maximum concentration of larvae was recorded 30 cm away from the tree trunk and in top 30 cm column of soil, irrespective of age and growth of the tree. The larval density is reduced drastically from 30 to 90 cm depth. As the tree grows, there is a slow shift in nematode population from inner circle of rhizosphere indicating deterioration of feeder roots.

The association of this nematode is not only confined to citrus species but several other plants of related genera are also attacked. There are few unexpected hosts of this nematode like olive, grapevine and loquat. Among these grapevine is the most important as this nematode is known to cause economic losses to this crop.

### *Radopholus citrophilus* (= Burrowing nematode)

Earlier known as a biotype of *R. similis*, has now been established as a separate species. It causes "spreading decline" disease of citrus in Florida.

*R. citrophilus* is a migratory endoparasite which penetrates the roots completely and kills cells by feeding and migration. It causes cavities and tunnels in the cortex and stele of roots. Once inside, the nematode may spend its entire life within the root. The life cycle takes about 18–20 days at 24–26°C. It may survive in soil without feeding for about 6 months (Raski, 1986). On roots, it causes brown-black lesions. The feeder roots fail to develop and cause wilting symptoms on above-ground parts during the period of water stress.

### *Other nematodes*

*Pratylenchus coffeae, P. vulnus, Meloidogyne* spp. and *Hemicycliophora arenaria* are other important species of nematodes which infest citrus as root parasites. A recent study by Sharma *et al.* (1986) revealed that decline was more common in a citrus orchard infested with other tylenchids (mainly *Helicotylenchus indicus)* than the orchard that harboured *T. semipenetrans* as the major population. *Fusarium oxysporum* was the main fungal pathogen isolated from declined trees in both the orchards.

## Papaya

Root-knot nematode, *Meloidogyne incognita* and reniform nematode, *Rotylenchulus reniformis* are the major species associated with papaya cultivation. The root-knot nematode is a vascular feeder and causes galling on root while the reniform nematode feeds mainly on phloem (Sivakumar and Seshadri, 1972). Plant dwarfism, leaf yellowing, dropping, reduction in yield and premature dropping of fruits are the symptoms general of attack. The yield losses in papaya of 38.4 and 50 per cent due to reniformis and root knot nematode, respectively (Rajendra and Naganathan, 1981; Parvatha Reddy, 1993).

## Pineapple

On pineapple, *Meloidogyne incognita, Rotylenchulus reniformis, Trichodorus christiei* and *Criconemella xenoplax* are the major species of nematodes. The first two being the most serious on pineapple in India (Parvatha Reddy, 1993). Ayala *et al.* (1969) reported yield losses to the tune of 217 tonnes per hectare due to nematodes. Since nematode infestation do not allow proper development and functioning of the roots, the symptoms like, stunted plant growth and conditions of general degeneration of crop are exhibited on above-ground parts.

## Peach

### *Pratylenchus* species (= Lesion nematode)

These are migratory endoparasitic nematode forms. Both young and adult stages enter roots by forcing their way through epidermis and then feed on cortical cells as they migrate through the cortical layer. The openings made on roots provide entry sites for pathogenic fungi and bacteria. The combined action of these associated organisms produces conspicuous lesions on roots and hence the name. However, it is not unusual to find considerable number of *Pratylenchus* in root tissues without any evidence of necrosis.

Among the stone fruits, peach is the most seriously decline and replant disease in which nematodes play an important role. The associated nematode species with these diseases in many cases belong to genus *Pratylenchus*. Out of six species viz., *P. penetrans, P. vulnus, P. zeae, P. coffeae, P. brachyurus and P. scribneri* reported from this fruit tree, the importance of *P. penetrans* and *P. vulnus* has been substantiated (Barker and Clayton, 1969). Mountain and Patrick (1959) demonstrated the pathogenicity of *P. penetrans* in the absence of bacteria and fungi. Under aseptic condition, it invaded and killed peach roots by hydrolysing the cyanophoric B-glucoside and amygdalin which released hydrogen cyanide and benzaldehyde, both of which are phytotoxic. In non-sterile growth room tests, peach seedlings were more severely damaged than under sterile conditions. Hence, it was suggested that the main role of *P. penetrans* in peach replant failure was its ability to incite root degeneration by providing extensive infection courts for soil microorganisms. In a survey conducted in Himachal Pradesh about 30 per cent peach trees were found infested with this species (Sharma and Kaur, 1985). All the soil samples collected from a declined peach orchard near Solan contained *P. criconemella species (= Ring nematode)*.

According to Weaver *et al.* (1974) *C. xenoplax* was found to have relationship with peach tree short life. Nyczepir *et al.*, (1983) and Nyczepir (1990) also reported that trees parasitized by *C. xenoplax* in the South-easern USA were predisposed to cold injury and/or bacterial canker (*Pseudomonas syringae* cv. *syringae*), a replant or production disease complex termed as peach tree short life.

### *Meloidoyne* species (= Root-knot nematode)

Four *Meloidogyne* species namely; *M. incognita, M. javanica, M. arenaria* and *M. hapla* are known to parasitize peach of which the first two are predominant. Root galls are the typical symptoms of root-knot nematode attack. Affected trees show signs of reduced vigour, growth and yield. Such trees may also show signs of defoliation, which becomes more pronounced when trees are exposed to drought conditions. Heavy infestation of *Meloidogyne* species often leads to stunted growth of young trees and occasionally death of the tree. Lownsbery (1959) reported severe stunting of Lovell peach seedlings within three months with an initial inoculum of five or more *M. incognita* juveniles. Sharma and Sharma (1988b) reported 66 per cent reduction in shoot weight and 44 per cent reduction in shoot length after six months of inoculation @ 20 juveniles per gram soil.

### *Paratylenchus prunii* (= Pin nematode)

This minute nematode species is predominantly associted with peach cultivation in Himachal Pradesh (Sharma and Sharma, 1987b). It does not produce any specific symptom but retards plant growth when its population is more than 20 per gram soil (Sharma and Sharma, 1987a). It is present in higher density during early spring season in peach rhizosphere than during summer months indicating that it attacks new root growth initiated during spring (Sharma and Sharma, 1988a). It is a cortical feeder and can cause complete destruction of the cortical region.

*Other nematodes*

Feeding on peach roots and causing reduction in root and top growth has also been reported in the case of *Paratrichodorus lobatus* (Sirling, 1976), *X. americanum* and *X. vuittenezi* (Gobor and Astavaan, 1980; Bonsi *et al*., 1984). *Xiphinema* species are economically more important as virus vectors. In a region where nepo-virus occur, the damage threshold level has been fixed at one nematode per 100 cc soil (Bird, 1989), compared to 100 nematodes per 100 cc soil of other species.

### Plum

Ring nematode, *Criconemella xenoplax* is the most dominant species associated with plum throughout the world. In Himachal Pradesh, this country, this nematode frequently occurs in high number on plum trees (population upto 3 nematodes/g of soil). The role of *C. xenoplax* in plum diseases is still not fully understood. Majtahedi *et al.* (1975) however, reported that bacterial canker caused by *Pseudomonas syringae* in plum trees was more severe on trees whose roots were inoculated with *C. xenoplax*.

The other nematode species which are of some concern in this fruit are species of *Paratylenchus, Meloidogyne, Helicotylenchus, Xiphinema* and *Pratylenchus*.

### Apricot

Apricot has long been considered practically resistant to nematode attack. In recent years, however, some of the economically important nematode species namely, *Meloidogyne incognita, Paratylenchus neoamblycephalus, Pratylenchus vulnus, P. penetrans, Criconemella xenoplax, Rotylenchulus macrodorus*, etc. were recorded in association with apricot (Siniscalco *et al*., 1976; Koliopanos and Vovlas, 1977; Sharma and Kaur, 1985, which may be because of apricot orchards being planted on the sites previously occupied by peach or plum.

### Almond

All nematode species which are reported from peach, plum and apricot also attack this crop. The severity of problem generally depends on the use of rootstock in Himachal Pradesh, almond trees were observed to have severe infection of crown gall (*Agrobacterium tumefaciens*) in the presence of heavy nematode populations (Sharma and Sharma, 1988).

### Cherry

Cherry plantations suffer from severe nematode attack. Several genera are found in established cherry orchards but the root injuries in most instances are attributed to *Pratylenchus* spp. mainly

*P. penetrans* (McElroy, 1972). Hoestra and Oostenbrink (1962) reported *P. penetrans* as an important cause in the cherry replant problem from the Netherland. Parker and Mai (1956) from USA proved its pathogenicity on Mazzard cherry in greenhouse tests. According to these authors, *P. penetrans* is highly pathogenic to cherry seedlings. Severely damaged roots may either lack feeder rootlets or may develop Witch's broom symptoms appearing as tufts of short, partially dead roots.

The other species that damage the cherry plantations are *Meloidogyne hapla, Tylenchorhynchus claytoni, Xiphinema americanum, X. diversicaudatum, X. vuittenezi* and *Longidorus macrosoma*. Among these, heavy populations of *M. hapla* were found to completely destroy the feeder roots (Szczygiel, 1980) and enhance the cherry wilt due to *Verticillium dahliae*. *Tylenchorhynchus claytoni* is also known to assist this fungus in causing more severe growth reduction (Ndubizu 1977). *Xiphinema* and *Longidorus* species are mainly associated with transmission of cherry leaf roll virus (Harrison, 1964; Fritzsche and Keglar, 1964; Teliz *et al.*, 1967; McElory 1972). In India, cherry plantations have been found to harbour heavy populations of *P. penetrans* (upto 3200/250 g soil) and *Xiphinema* species (upto 1700/250 g soil) in Kullu area of Himachal Pradesh (Sharma and Sharma, 1988).

### Walnut

#### *Pratylenchus vulnus* (=Root lesion nematode)

This species reduces growth and yield of English walnut. Being highly pathogenic and worldwide in distribution, this species is a potential threat to walnut industry. It is primarily a cause of replant problem in walnut. Young trees are devitalized when the root damage exceeds the plant ability to regenerate roots, but mature trees tolerate high populations without any adverse effects (Lownsbery, 1956; McGranahan and Catlin, 1987).

#### *Other nematodes*

*Criconemella xenoplax* and *Cacopaurus pestis* have been reported as pests of walnut but their importance has not been documented in recent years. Thorne (1943) reported that parasitism of *Juglans regia* (Persian walnut) by *C. pestis* resulted in extensive lesions on roots (2.4 cm in diameter) and death of smaller roots. Nematode infested trees initially exhibited smaller leaves followed by reduction in number of twigs and leaves, reduced yield and died when sudden temperature drop occurred in January.

### Pecannut

*Criconemella rucium*, commonly known as the ring nematode is known to cause necrosis on pecan roots in the presence of *Pythium irregulare* and *Fusarium solani* (Hsu and Hendrix, 1973).

## NEMATODE MANAGEMENT

In fruit crops, nematodes mainly spread through the use of infested plant material. Even the ectoparasitic forms like *Paratylenchus* species which are attached to young roots were found to survive after the nematode infested plant material was transported from California to Salt Lake

City Utah in the USA (Thorne, 1961). Further, nematodes are known to inflict economic losses in fruit crops only when their populations reach a threshold level. Thus, preventive and curative measures are the basic components of management strategy for nematode control.

**Preventive measures**

***General***

*Pre-planting treatment:* The land on which the nursery stock is to be grown should be fumigated before planting, preferably with formaldehyde. The other potent fumigants for nematode control are DD, EDB, Telone, chloropicrin, DBCP (Nemagon) etc. However, these nematicides are not easily available. Being highly phytotoxic (except Nemagon) their application is recommended 2–3 weeks prior to planting.

*Post-planting treatment*: In case the nursery area has not been fumigated prior to sowing, and the nematode problem is detected on the seedlings, use of granular formulation of non-phytotoxic chemicals like, aldicarb, carbofuran or phorate @ 3 kg a.i./ha is recommended. The application of these granular nematicides though helps in reducing infestation and boosts seedling growth yet it does not eliminate nematodes completely. Therfore, such seedlings should be subjected to hot water and/or chemical root dip treatment to ensure complete denematization before planting in the orchard. However, hot water treatment and chemical root dips should not be applied unless standardized.

*Shifting of nursery site*: The nursery should not be continuously grown at the same site. The old site should not only be kept weed free but planted with crops like pangola grass, tegetes, crotolaria, mustard etc., which are antagonistic to nematodes.

*Use of certified planting material:* Use only certified nematode-free planting material.

***Specific***

*Paring infected sets:* Banana roots arise from the medulla of the rhizome and pass through the thick, fleshy tissues to the cortex before becoming external. *R. similis* invades the endogenous parts of these roots, having migrated from the root to the adjacent tissues of the rhizome. The resulting reddish-brown to black lesions make the infected areas conspicuous. Loos and Loos (1960) described a successful method for disinfecting banana sets, this involving peeling the rhizome and cutting away all discoloured internal roots and rhizome tissues. Rhizomes with diameter greater than 10 cm at their base are preferred for paring because smaller rhizomes have higher mortality rate.

*Pralinage:* It is the process of soaking the rhizome sets of banana for a few seconds in nematicide mixed clay slurry so as to disinfect banana sets against nematode infection.

*Hot-water treatment of planting material*: Plant parasitic nematodes generally do not survive at tempratures above 40°C but when embedded deep into plant tissues higher temprature is required to kill them. Very high temperature however, is lethal to the plant tissues. Therefore, time-temprature combinations for different planting materials should be carefully worked out.

Time-temperature combinations of hot water treatment for denematizing planting stock (Parvatha Reddy, 1983)

| *Planting stock* | *Nematode* | *Temp. (°C)* | *Time (min)* |
|---|---|---|---|
| Banana suckers | *Rodopholus similis* | 45.0 | 10 |
| Banana rhizome sets* | *Rodopholus similis* | 55.0 | 20 |
| Citrus rooted | *Tylenchulus semipenetrans* | 46.7 | 10 |
| | *R. citrophilus* | 50.0 | 10 |
| Cherry rootstocks | *Meloidogyne* spp. | 50–51 | 5–10 |
| Peach rootstocks | *Meloidogyne* spp. | 50–51 | 5–10 |
| Strawberry roots | *Meloidogyne* spp. | 52.8 | 5 |
| Strawberry (dormant plants) | *Pratylenchus penetrans* | 51.0 | 7.5 |

* Blake (1961).

*Nematicidal dip of planting materials:* The planting material can be made nematode free by dipping in chemicals having nematicidal properties. Some examples of chemical dips to plants are presented in Table below.

**Nematicidal dip treatment for different planting material**

| *Planting stock* | *Nematode* | *Chemical (conc.) & Time* | *Reference* |
|---|---|---|---|
| Cherry rootstock Spp. | *Pratylenchus* | Phorate (1%) for 30 min. | Sher, 1960 |
| Cherry seedlings | *P. penetrans* | Diazinon (10%) for 30 min. | Sher, 1960 |
| Cirtus seedlings (1000 ppm) for 30 min. | *Rhodopholus similis* | Zinophos, B-25141 & mocap | O'Bannon & Taylor, 1967 |
| Tylenchulus semipenetrans | -do- | for 60 min. | |
| Peach (Lovell) | *Meloidogyne* spp. | Phorate (2,4,6 & 8%) for 5, 30, 60 & 120 min., respectively) | Reed, 1963 |

## Curative measures

Nematode populations in orchards can be effectively kept below threshold levels employing cultural, biological and chemical methods. The threshold levels worked out for some tree fruit under pot culture and field conditions, are given in Table below.

**Threshold levels of nematode population for some tree fruits**

| *Nematode species* | *Fruit plant* | *Threshold level (no./g soil)* | *Reference* |
|---|---|---|---|
| *Cricnemella xenoplax* | Peach | | |
| Walnut (black) | | > 42 | Lownsbery *et al.*, 1978 |
| *Meloidogyne incognita* | Peach (Lovell) | > 5 | Lownsbery, 1959 |
| *Paratylenchus neoamblycephalus* | Peach | 13 | Braun *et al.*, 1975 |
| *P. prunii* | Peach | 20 | Sharma and Sharma, 1987a |
| *Pratylenchus penetrans* | Apple | 0.15 | Crossa-Raynaud and Audergon, 1987 |
| | Cherry | 0.80 | -do- |
| | Pear | 0.30 | -do- |
| | Plum | 3.20 | -do- |
| | Pome fruits | 0.25–1.50 | Abawi and Mai,1990 |
| *Tylenchulus semipenetrans* | Citrus | 2 | Chandel, 1986 |
| *Xiphinema** spp. | Apple | 1 | Bons *et al.*, 1984 |
| | Peach | 1 | -do- |

* one nematode/100 cc is the threshold level for nepo-virus transmission in fruit crops.

## Cultural control

Application of oilseed cakes like castor, neem, mustard, karanj etc., use of saw dust, wood chips in tree basins, and use of black polythene sheet as mulch are some of the commonly followed cultural methods for nematode control (Blizzard, 1975; Colbran, 1979; Vadivelu *et al.*, 1987; Bhattacharya and Rao, 1985). Recently, Parvatha Reddy *et al.* (1997) recorded three times more yield of banana by controlling *Radopholus similis* with combined use of endomycorrhiza (*Glomus mosseae*) and Karanj cake.

Crop rotation seemes impracticable in fruit plants. However, Ayala *et al.* (1969) reported complete elimination of nematodes when pineapple was rotated with pangola grass.

## Chemical controls

Chemicals are one of the potent agents to bring down nematode populations but are cost prohibitive. Hence, these should be used only when the nematode infestations are very heavy. Nemagon, aldicarb, carbofuran, phenamiphos, ethoprop, phorate etc., are some of the nematicides which can be applied in the orchards.

## Biological control

The biological control aimes at increasing the population of parasites, predators or pathogens of plant-parasitic nematodes in the plant rhizosphere. Some of the important biological agents of common occurrence for nematode control are: *Arthrobotrys oligospora, Dactylaria brochophaga, D. candida, Dactylella ellipsopora, D. bembicodes*, predaceous fungi); *Paecilomyces lilacinus* (endoparasitic fungus); *Bacillus penetrans, Pseudomonas dinitrificans* (endoparasitic bacteria) and *Mononchus papillatus, Tripyla, Mononchoides potohikus, Diplogaster* sp., *Seinura tenuicaudatus* (predatory nematodes), tardigrades, etc. The population of parasites and predators present naturally or inoculated artificially in soil can be enhanced for effective control of nematodes through application of organic amendments and green manures. Kauri Paesake (1973) reported that soil sickness in apple nurseries caused by *P. penetrans* was eliminated by green manuring or adding milled peat in the soil. In the former case, the nematophagous fungi were activated, while in the later nematode preying tardigrades were found to be stimulated.

Use of resistant plant material is often suggested for nematode control in fruit crops. Nemagourd and Nemared in peach, Trifoliate orange and Grape fruit in citrus, Dogridge and Salt Creek in grapes are some of the known resistant rootstocks against important nematode species attacking these crops. However, there is always a danger of breakdown of resistance due to development of new races in nematodes.

## REFERENCES

Abawi GS and Mai WF. 1990. Dagger nematodes. *In: Compendium of Apple and Pear Diseases*. A.L. Jones and H.S. Aldwinckle (eds). APS Press, St. Paul, Minnesota. pp. 73–74.

Anonymous. 1982. *East Malling Research Station Britain. Report for 1981*. pp. 227.

Anonymous. 1985–90. Annual Reports, AICRP-on plant parasitic nematodes with integrated approach for their management. UHF-Solan Centre, ICAR, New Delhi.

Ayala A, Gonzalez-Tejere E and Irizarry H. 1969. Pineapple nematodes and their control. *In: Nematodes of Tropical Crops*. J.E. Peachey (ed.). Tech. Comm. No. 40, C.A.B. St. Albans, Herts, England.

Barker, K.R. and Clayton, C.N. 1969. Relative host suitability of peach cultivars to six species of lesion nematodes. *Phytopathology*, **59**: 1017.

Bhattacharya RK and Rao VNM. 1985. Soil cover increases banana production *Indian Hort*. **30**(1): 5–8.

Bird G.W. 1989. Avoidance and management of nematode problems. Tree fruit production in Michigan. Michigan State University Ext. Bulletin. pp. 8–9.

Blake C.D. 1961. Root rot of banana caused by *Radopholus similis* (Cobb) and its control in New South Wales. *Nematologica*, **6**(4): 259–310.

Blake C.D. 1966. The histopathological changes in banana roots caused by *Rodopholus similis* and *Helicotylenchus multicinclus*. *Nematologica*, **12**(1): 129–137.

Blake C.D. 1969. Nematodes of banana and their control. *In:* J.E. Peachey (ed.). *Nematodes of Tropical Crops*. Technical Communication No. 40, C.A.B., St. Albans, Herts, England.

Bons, C., Stonffer, R. and Shaffer, R. 1984. Effect of different initial population densities of *Xiphinema americanum* and *Xiphinema rivesi* on growth of apple and peach seedlings. *Phytopathology*, **74**: 626.

Braun, A.J. 1958. Plant parasitic nematodes food in association with strawberry roots in the United States *Pl. Dis. Reptr*. **42**: 76–83.

Braun, A.J., Palmiter, D.J. and Keplonger, J.A.. 1966. Nematodes find in apple orchards in the Hudson Valley 1956- 1965. *Pl. Dis. Reptr*. **42**: 76–83.

Braun AL, Mojtahedi H and Lawnsbery BF. 1975. Separate and combined effects of *Paratylenchus neoamblycepalus* and *Criconemoides xenoplax* on "Myradalan" plum. *Phytopathology*, **65**: 328–330.

Chandel Y. S. 1986. *Studies on nematodes associated with citrus in Himachal Pradesh* Ph.D. Thesis U.H.F., Solan (H.P.) India.

Cobb, N. A. 1913. Notes on *Mononechus* and *Tylenchulus*. *Journal of Washington Academry of Science,* **3**: 288–289.

Cobb, N.A. 1915. *Tylenchus similis*; the cause of root disease of sugarcane and banana, *J. agric. Res,* **4**(6): 561–568.

Colbran, R.C. 1966. Studies on plant and soil nematodes: 13 *Tylenchulus claricaudatus* n.sp. (Nematoda: Tylenchulidae) a parasitic of the liana *Deeringia arborescens* (RBr). *Druce. J. agric. Anim. Sci,* **23**: 421–427.

Colbran, R. C. 1979. Problems in tree replacement. The effect of cultural practices and soil fumigation on root lesion nematode and on the growth of apple replant. *Aust. J. agric. Res.,* **301**): 113–123.

Corbett, D.C.M. 1966. Central African Nematodes. II *Paratylenchus crenatus* n.sp. (Nematoda: Criconematidae) from Malawi. *Nematologica,* **12**(1): 101–104.

Crossa-Raynand, P. and Audergon, J. M. 1987. Apricot rootstocks. *In: Root stocks for fruit crops.* R.C. Rom and R F Carlson (eds.). John Wiley & Sons, New York. pp. 295–320.

David Dand, R.G.; Sana, R.M. and Maranrn, L.R. 1979. Plant parasitic nematodes associated with grapes in Cabu and Batanga. *Phillipine Phytopathology,* **15**: 1–8.

Decker, H. 1960. *Pratylenchus penetrans* as the cause of soil sickness in East Germany. *Nemeatolgica Suppl.,* **11**: 68–75.

Fisher, J.M. 1967. Effect of temperature and host on *Paratylenchus neoambly cephalus* and effect of the nematode on the host. *Aust. J. agric. Res.*, **18** (6): 921–929.

Forer, L.B.; Powell, C.A. and Stouffer, R.E. 1984. Transmission of tomato ring spot virus to apple root stock cuttings and to cherry and peach seedlings by *Xiphinema rivesi*. *Pl. Dis.* **61**(12): 1052–1052.

Frazier, N.W. and Maggenti, A.R. 1962. Nematode transmission of yellow bud mosaic virus to strawberry. *Pl.Dis. Reptr.* **46**: 303–304.

French, A.M.; Lownsberry, B.F.; Ayoub, S.M.; Weiner, A.C. and Elgholu, N. 1964. c.r.f. Nematodes of tree fruits and small fruits. *In: Economic Nematology.* J.M. Webster (ed). Academic Press, New York. pp. 335–376.

Fritzsche, R. and Keglar, H. 1964. The transmission of cherry leaf-roll virus by nematodes. *Naturwiss,* **51**: 299.

Gobar, J. and Astvaan, S. 1980. Study on the pathogenicity of *Xiphinema vuittenezi* to the roots of fruit trees. *Novenyvedelem,* **16**: 14–17.

Harrison, B.D. 1964. Specific nematode vectors for serologically distinctive forms of raspberry ring spot and tomato black ring virus. *Virology,* **22**: 544–550.

Harrison, B.D.; Mowat, W.P. and Taylor, C.E. 1981. Transmission of a strain of tomato black ring virus by *Longitorus elongatus* (Nematoda). *Virology,* **14**: 480–485.

Hoestra H. 1967. Specific apple replant disease problem and prospects. Meded. *Rigksfec Lamb. wet. Gent.* **32**: 427–432.

Hoestra, H. and Oostenbrink, M. 1962. Nematode in relation to plant growth IV *Pratylenchus penetrans* (Cobb) on orchard trees. *Neth. J. agric. Sci* **10**: 286–296.

Hsu, D. and Hendrix, F.F. Jr. 1973. Influence of *Criconemoides quadricornis* on pecan feeder root necrosis caused by *Pythium irregulare* and *Fusarium solani* at different temperatures. *Can. J. Bot.* **51**: 1421–1424.

Intagaki, H. 1978. Apple root-knot nematode, *Meloidogyne mali* its taxonomy, ecology damage and control. *Kasetsari J.* **12**: 25–30.

Jha, A. and Posnette, A.F. 1959. Transmission of a virus to strawberry plants by a nematode (*Xiphinema* spp.). *Nature,* **184**: 962–963.

Jenkins, W. R. 1960. *Paratylenchus marylandicus* n.sp. (Nematoda: Criconemetidae) associated with roots of pine. *Nematologica*, **5**(3): 175–177.

Kaul, V. 1985. Seasonal fluctuations of phytonematodes in two fruit nurseries in Kashmir. *Indian J. Helminth.,* **37**(1): 4–11.

Kauri Paesake, M. 1973. Biological control of replant diseases in nurseries. *Lantbrukshogskolen Meddelanden Serie*, **190**; pp. 11.

Khan, M. L; Sharma, G.C. and Sharma, N.K. 1988. New record of *Paratylenchus curvitatus* and *Gracilacus peperpotti* (Nematoda: *Paratylenchidae)*. *Nematol. medit*. **17**: 11–12.

Khan, M.L. and Sharma, N.K. 1991. A description of *Pratylenchus manaliensis* sp.n. (Nematoda: Pratylenchinae) associated with apple in India. *Nematol. medit.* **19**(2): 275–277.

Kir'yanova, E.S. and Krall, E.L. 1980. *Plant parasitic nematodes and their control* Volume II. Amerind Publishing Co., pp. 731.

Koch, B.L. and Covey, R.P. 1980. Preplant soil fumigation increased apple tree growth and fruit production. Good fruit Grower, **31**: 15–16.

Koliopanos, C.N. and Vovlas, N. 1977. Records of some plant parasitic nematodes in Greece with morphometric descriptions *Nematol. medit.* **5**: 207–215.

Larsen, L.D. 1910. Diseases of the pineapple. *Hawaii Sug. Plvs. Ass., Exp. Stat. Path. Ser. Bull.* **10**.

Lister, R.M. 1964. Strawberry latent ring spot; a new nematode borne virus. *Ann. appl. Biol.* **54**: 167–176.

Loos, C.A. and Loos, S.B. 1960. Preparing nematode-free banana "Seed". *Phytopathology,* **50**(5): 383–386.

Lownsbery, B.F. 1956. *Pratylenchus vulnus*, primary cause of the root-lesion disease of walnuts. *Phytopathology*, **46**: 376–379.

Lownsbery, B.F. 1959. Studies of the nematode *Criconemoides xenoplax* on peach. *Pl. Dis. Reptr.* **43**: 913–917.

Mai, W.F. and Abawi, G.S. 1978. Determining the cause and extent of apple, cherry and pear replant disease under controlled conditions. *Phytopathology*, **68**(11): 1540–1544.

Majtahedi, H. Lownsberg, B.F and Moody, E.H. 1975. Ring nematodes increase development of bacterial canker in plums. *Phytopathology,* **65**: 556–559.

McElroy, F.D. 1972. Nematodes of tree fruits and small fruits. *In:* J.M.Webster (ed.), *Economic Nematology.* Academic Press, New York, pp. 335–373.

McGranahan, G.H. and Catlin, P.B. 1987. *Juglans* rootstocks. *In:* R.C.Rom and R.F. Carlson (eds.). *Rootstocks for Fruit Crops*. John Wiley and Sons, New York. pp. 411–428.

Melton, T.A., Ries, S.M., Noel, G.R. and Doll, C.C. 1985. The peach and apple nematodes of Illinois. Trans. *Illinois State Hort. Soc.* **119**: 78–91.

Minz, G., Strich-Harari, D. and Cohn, E. 1963. [*Plant parasitic nematodes in Israel and their control*] Tel-Aviv: Hassaden Publishing House. p. 84. [In Hebrew: English Summary pp. 3–7].

Mountain, W.B. and Patrick, Z.A. 1959. The peach replant problem in Ontario: VII the pathogenicity of *Pratylenchus penetrans Can. J. Bot*. **37**: 153–161.

Nair, M.R.G.K., Das, N.M and Menon, M.R. 1966. On the occurrence of the burrowing nematode, *Redopholus similis* (Cobb 1915) Thorne 1949 on banana in Kerala, Indian. *J. Ent.* **28**: 553–554.

Ndubizu, T.O.C. 1977. Effect of earthworm, nematode, cultivations and host plants on verticillium wilt of peach and cherry. *Ann. appl. Biol.* **86**: 153–161.

Neal, J.C. 1889. The root knot disease of the peach, orange and other plants in Florida due to the work of *Anguillula*. *Bull. Bur. Entomology. U.S. Dep. Agric. No.* **20**: 31.

Newhall, A.G. 1958. The incidence of Panama disease of banana in the presence of the root knot and burrowing nematodes (*Meloidogyne* and *Radopholus*). *Pl. Dis. Reptr.* **42**(7): 853–856.

Nyczepir, A.P. 1990. Influence of *Criconemella xenoplax* and pruning time on short life of peach trees. *J. Nematol*. **22**: 97–100.

Nyczepier, A.P, Zehr, E.I., Lewis, S.A. and Harshman, D.C. 1983. Short life of peach trees induced by *Criconemella xenoplax*. *Pl. Dis*. **67**: 507–508.

O'Bannon, J.H. and Taylor, A.C. 1967. Control of nematodes on citrus seedlings by chemical bare-root dip. *Pl. Dis. Reptr.* **49**(1): 51–53.

Parker, K.G. and Mai, W.F. 1956. Damage to tree fruits in New York by root lesion nematodes. *Pl. Dis. Reptr.* **40**: 694–699.

Parvatha Reddy, P. 1983. *Plant Nematology*. Agricol publishing Academy. pp. 287.

Parvatha Reddy, P. 1993. Nematodes of fruits. *In:* K.L Chadha and O.P Pareek (eds.) *Advance in Horticulture. Vol. 3-Fruit Crops* Malhotra Publishing House, New Delhi, India.

Parvatha Reddy, P. and Khan, R.M. 1987. First report on the occurrence of *Radophilus similis* (Cobb 1893) Thorne, 1949 in Andhra Pradesh, India. *Int. Nematol. Network Newsl.*, **4**(1): 11.

Parvatha Reddy, P., Nagesh, M., Rao, M.S. and Deveppa, V. 1997. Integrated management of the burrowing nematode, *Redopholus similis* using endomycorrhiza, *Glomus mosseal* and oil cakes. *Pest Management in Hort. Ecosystems*, **3**(1): 25–29.

Parvatha Reddy, P. and Singh, D.B. 1980. Plant parasitic nematodes associated with banana in South India. *Proc. National Seminar on Banana Prod. Technol.* Coimbatore. pp. 161–163.

Perry, V.G. 1958. Parasitism of two species of dagger nematodes (*Xiphinema americanum* and *X. chamberoi)* to strawberry. *Phytopathology,* **48**: 420–423.

Rajendra, G. and Naganathan, T.G. 1981. Evaluation of systemic chemicals for the control of *Rotylenchulus reniformis* on papaya. *Indian J. Nematol.*, **11**: 130.

Raski, D.J. 1986. Plant parasitic nematodes of banana, citrus, coffee, grapes and tobacco. *Union Carbide Agricultural Products Company Inc. Publication, USA.* pp. 43–57.

Reed, H.E. 1963. Chemicals dips for eradication of root-knot nematodes for plant root. *Phytopathology,* **53**(6): 625.

Schindler, A.F. and Braun, A.J. 1957. Pathogenicity of an ectoparasitic nematode *Xiphinema diversicandatum* on strawberry. *Nematologica,* **11**: 91–93.

Sharma, G.C. and Sharma, N.K. 1987a. Effect of initial inoculum levels and multiplication potential of *Paratylenchus prunii* on peach. *Indian J. Nematol*. **17**(2): 211–213.

Sharma, G.C. and Sharma, N.K. 1987b. Plant parasitic nematodes associated with peach in Himachal Pradesh. *Him. J. Agric. Res.* **13**(1): 27–29.

Sharma, G.C. and Sharma, N.K. 1988a. Effect of environmental temperature on the population build up of *Paratylenchus prunii* in the peach rhizosphere. *Indian Phytopath*. **41**: 469–470.

Sharma, G.C. and Sharma, N.K. 1988b. Effect of *Paratylenchus prunii* and *Meloidogyne incognita* on peach seedlings. *Nematol. medit.* **16**: 117.

Sharma, G.C. and Sharma, N.K. 1990. Distribution of three important nematode species in the rhizosphere of plum trees. *Indian J. Hort.* **47**(2): 207–209.

Sharma, N.K. 1977. Incidence of spring dwarf disease of strawberry from India. *Curr. Sci.* **46**: 566.

Sharma, N.K. and Kaur, D.J. 1985. Nematodes associated with temperate crops in Himachal Pradesh. *In:* T.R. Chadha; V.P. Bhutani and J.L. Kaul (eds). *Temperate Fruits*. Y.S. Parmar University of Horticulture and Forestry, Solan. pp. 358–362.

Sharma, N.K and Kaur, D.J. 1987. Parasitism of root-knot nematode, *Meloidogyne* sp. on apple in Himachal Pradesh. *Indian J. Hill Fmg*. **1**(1): 69.

Sharma, N.K.; Mangat, B.P.S. and Thind, T.S. 1986. Possible role of nematodes other than *Tylenchulus semipenetrans* in citrus decline and their relationship with fungal flora of citrus rhizosphere. *J. of Tree Sci.* **5**: 30–35.

Sharma, N.K. and Sharma, G.C. 1988. Nematode pests of temperate and sub-tropical fruits and their management. *In:* V.K. Gupta and N.K. Sharma (eds.) *Tree Protection* Indian Society of Tree Scientist, Publication, UHF, Solan. pp. 192–210.

Sharma, N.K. and Sharma, S.K. 1981. Spatial distribution of citrus nematode, *Tylenchulus semipenetrans* in relation to tree age. *Indian. J. Nematol.* **11**(2): 226–228.

Sethi, C.L.; Gaur, H.S,; Kaushal, K.K.; Srivastava, A.N. and Khan, E. 1988. Occurrence of root-knot nematodes in fruit plants in association with *Agrobacterium tumefasciens. Int. Nematol. Network Newsl.* **5**(2): 12–13.

Sher, S.A.. 1960. Chemical control of plant parasitic nematodes in plant roots. *Phytopathology*, **50**: 654.

Shivakumar, C.V. and Seshadri, A.R. 1972. Histopathology of infection by the reneformis nematode, *Rotylenchulus reniformis* Linford & Oliveira 1940 on castor, papaya, tomato. *India J. Nematol.* **2**: 173–181.

Sindham, G.C. 1976. Plant parasitic nematodes associated with fruit trees in Kumaon. *India. J. agric. Res.* **10**(1): 69–72.

Siniscalco, A; Lamberti, F. and Inserra, R. 1976. Reaction of peach root stocks to Italian populations of root-knot nematodes, *Meloidogyne* spp. *Nematol Medit*. **4**: 79–84.

Stirling, G.R. 1976. *Paratrichodorus lobatus* associated with citrus, peach and apricot trees in South Australia. *Nematologica,* **22**: 138–144.

Szczygiel, A. 1980. Pathogenicity of *Meloidogyne hapla* to some seedling root stocks of stone fruits. *Zeszyty Problemowe Postepow Nauk Rolniczych*, **232**: 39–44.

Taylor, C.E. 1962. Transmission of rasberry ring spot virus by *Longidorus elongatus* (de Man)(Nematoda: Dorylaimidae). *Virology,* **17**: 493–494.

Teliz, D. Lownsberg, B.F., Grogam, R.G. and Kimble, K.A. 1967. Transmission of peach yellow bud mosaic virus to peach apricot and plum by *X. americanum. Pl. Dis. Reptr.* **51**: 841–843.

Thomas, E.E. 1913. A preliminary report of a nematode observed on citrus roots and its possible relation with the mottled appearance of citrus trees. *Circ. Calif. Agric. Exp. Stn. No.* **85**: 14.

Thomas, E.E. 1923. The citrus nematodes, *Tylenchulus semipenetrans. Circular, Univ. Calif. Agr. Expt. Sta.*

Thorne, G. 1943. *Cacopaurus pestis* nov. gen. nov. spec. (Nematoda: Criconematinae), a destructive parasite of the walnut, *Juglans regia* Linn. *Proceedings of the Helminthological Society, Washington*, **10**: 78–83.

Thorne, G. 1961. *Principles of Nematology*. McGraw Hill Book Company Inc., New York. pp. 553.

Tiev, R.A. 1983. Diseases of fruit trees caused by nematode vectors of virus from the family Longidoridae. *Byull. Vsesoy. Inst. Gelmintol, Steryabine,* **33**: 80.

Tiwari, S.P. and Dave, G.S. 1985. Burrowing nematode. *Radophulus similis* (Cobb) Thorne 1949 on banana in Madhya Pradesh. *Curr. Sci*. **54**: 1286.

Utkhede, R.S; Vrain, T.C and Yorston, J.M. 1992. Effects of nematodes, fungi and bacteria on the growth of young apple trees grown in apple replant disease soil. *Plant and Soil,* **139**(1): 1–6.

Vadivelu, R.S; Rajendran, G.; Naganathan, T.G. and Jayaraj, S. 1987. Studies on the nematodes pests of banana and their management-A Review. *Proc. Third Group Discussion on the Nematological problems of plantation crops.* Sugarcane Breeding Institute, Coimbatore. pp. 32–33.

Wallace, H.R. 1973. *Nematode ecology and plant disease*. Edward Arnold(Publishers) Ltd., London. pp. 228.

Weaver, D.J.; Wehnut, E.J. and Dowler, W.M. 1974. Association of tree site *P. seudomonas syringae, Creconemoides xenoplax* and pruning data with short life of peach trees in Georgia. *Pl. Dis. Reptr.* **58**: 76–79.

Zepp, L. and Szczygiel, A.C. 1986. Pathogenicity of five populations of *Pratylenchus penetrans* to three fruit tree seedling root stocks. *Zeszyty Problemowe Postepow Nank Roliniczych*, **323**: 73–90.

*Pest Management in Horticulture Crops*
L.R. Verma, A.K. Verma and D.C. Gautam (eds)


## *Chapter 24*

# Nematode Pests of Vegetable Crops and Ornamental Plants

*R.K. Walia*

## INTRODUCTION

NEMATODE PESTS OF CROPS, the minute unseen and subterranean creatures, have been stealthily taking the toll of agricultural produce since the time immemorial. The undiagnostic symptoms are largely responsible for the damage unnoticed by agriculturists and farmers. However, it was only after World War II that systematic global research projected their role in agricultural production, the horticultural crops notwithstanding. According to an estimate, plant parasitic nematodes cause 12.3% average loss on global basis in a few major crops (Sasser and Freckman, 1987).

## NEMATODE PESTS OF VEGETABLE CROPS

Several plant parasitic nematodes are associated with vegetable crops but the economically important groups belong to the root-knot, reniform and cyst nematodes.

### Root-knot nematodes (*Meloidogyne* spp.)

*Meloidogyne* spp. are considered most important worldwide attacking almost every vegetable crop. Four species namely, *Meloidogyne incognita* (Kofoid and White) Chitwood, *M. javanica, M. arenaria* and *M. hapla* are economically important. The first two constitute 95% of the total *Meloidogyne* species recorded throughout the world, whereas *M. hapla* is restricted to temperate climate only.

#### *Biology and nature of damage*

The vermiform second-stage juveniles present in the soil infect the plant roots, establish feeding sites (multinucleate giant cells) in the vascular tissues, and block the transportation of water and nutrients to the shoots, resulting in stunted plant growth, yellowing of leaves and reduced yields. The plants are seldom killed outrightly. The juveniles develop into saccate, sedentary endoparasitic females, each laying 200–400 eggs in an egg sac, mostly on the root surface. The vermiform adult males wriggle out into the soil and are generally not required for reproduction which is mostly by parthenogenesis. The eggs hatch into second-stage juveniles, which reinfect the roots and several generations are completed in a crop season, each lasting about one month under favourable conditions.

Root galls at the sites of infection are the characteristic symptoms of nematode attack. The size of the galls vary with the intensity of infection and susceptibility of the crop. Tomato, brinjal and okra are the most favourable hosts showing very prominent galls, the cucurbitaceous crops bear very big and compound galls and generally contain egg masses within. Tuberous crops like potato bear pimple-like outgrowths on the tuber surface thereby suffering qualitative losses as well. Crops like radish, turnip, etc., often get forked, and galls are visible on the finer feeder roots.

Root-knot nematode infected plants are often more vulnerable to infection by secondary pathogens, that accentuate the overall damage. *M. incognita* present along with *Pseudomonas solanacearum* greatly increases the incidence of bacterial wilt of brinjal. Resistance in brinjal to bacterial wilt is also broken by root-knot nematode (Parvatha Reddy *et al.*, 1979b). Root-knot nematodes reduce nodulation due to *Rhizobium* in leguminous vegetables (Singh and Parvatha Reddy, 1981) and decrease the leghaemoglobin content of the root nodules (Sharma and Sethi, 1975). *M. javanica* was reported to increase the extent of damage in pre-and post-emergence phases of damping off caused initially by *Fusarium oxysporum* f. sp. *lycopersici, Rhizoctonia solani* and *Pythium debryanum* in tomato (Nath *et al.*, 1984). The growth of tomato plants is also reduced significantly in simultaneous inoculations with *M. incognita* and tobacco mosaic virus or when nematode infection precedes the virus inoculation (Goswami and Chenulu, 1974). In the case of brinjal and bottlegourd, root-knot development is enhanced by the presence of mosaic virus (Mahmood *et al.*, 1974).

### *Crop losses*

The extent of crop loss varies with the initial nematode population, host susceptibility, edaphic factors and ambient climatic conditions. In the USA alone, the vegetable crops suffer 11% annual loss due to nematodes (Anonymous, 1971). In India, several empirical estimates on crop losses are available. Sen (1958) reported a loss of 70% in chillies, brinjal, okra and tomato from Bihar. In Mahendergarh (Haryana), losses in the yields of okra, tomato and brinjal were 90.0, 46.2 and 27.3%, respectively, in a field having initial population of 3 to4 juveniles of *M. incognita* per g soil (Bhatti and Jain, 1977). The per cent avoidable losses in yield of okra, brinjal, french bean and cowpea due to *M. incognita* in Bangalore (Karnataka) were 28.1, 33.7, 43.5 and 28.6, respectively (Parvatha Reddy, 1986). Darekar and Mhase (1988) reported that yield losses in tomato, brinjal and bittergourd due to *M. incognita* in Maharashtra were 46.9, 32.7 and 36.7%, respectively.

### *Management*

*Cultural practices:* Fallowing, deep ploughing, polythene mulching, crop rotations, etc., provide effective and low cost/no cost control of root-knot nematode. In tropical areas, 2 to 3 deep ploughings with clean fallow during the months of May and June kill most of the root-knot nematode juveniles in soil due to desiccation, heat and starvation. Planting non-hosts or poor host crops like onion, garlic, mustard or cereal crops in rotation keeps the root-knot nematode population below damaging levels.

Intercropping with certain plants in vegetable fields reduces the incidence of root-knot nematode. Root-knot development in tomato and okra was low when interplanted with *Targetes erecta* (Alam *et al.*, 1980) due to presence of α-terthienyl, the active principle in *Tagetes* spp., which is toxic to nematodes (Gommers and Voor in't Holt, 1976). Crotalaria root system exerts a toxic effect on *Meloidogyne* populations and when planted in advance, reduces subsequent damage (Bijloo, 1968). Atwal and Mangar (1969) showed that root exudates of sesame have nematicidal properties against *M. incognita*.

*Chemical control:* Synthetic nematicides, though most effective in controlling root-knot nematodes, are generally discouraged to cut down costs, and to avoid residue problems and environmental pollution. These are generally recommended when nematode populations are very high and economic returns are ensured. In transplanted vegetable crops like tomato, brinjal and chillies, application of carbofuran or phorate @ 2 kg a.i./ha in nursery beds only, gives effective control of root-knot infection at the seedling stage which is most vulnerable to nematode damage (Jain and Bhatti, 1983). In bold-seeded crops like leguminous vegetables (pea, cowpea, french bean) and cucurbits, seed coating with Posse 25 ST @ 3% w/w is highly effective in giving initial protection against root-knot infection (Gupta *et al.*, 1986). Carbofuran and oxamyl at 1000 ppm for 10–15 minutes (Parvatha Reddy and Singh, 1979a); fenamiphos at 250–750 ppm for 30 minutes (Thakar and Patel, 1985); dimethoate at 500 ppm for 6 hours, phosphamidon and dichlofenthion at 1000 ppm for 8 hours (Jain and Bhatti, 1978) and thionazin at 500 ppm for 15 minutes (Reddy and Seshadri, 1975) gave effective control of root-knot nematode on tomato when used as bare-root dip of seedlings.

*Host resistance:* Significant progress has been made in the recent past in the development of nematode resistant varieties. Over 235 cultivars of 15 major crops are reported to carry resistance to one or more species of root-knot nematodes. Most of these crops have resistance to *M. incognita* with a large number of cultivars occurring in tomato, cowpea and sweet potato among vegetable crops.

In tomato, the source of resistance to *Meloidogyne* spp. was *Lycopersicon peruvianum*. Because of high levels of resistance obtained from Mi gene, F1 hybrids derived from crosses of homozygous resistant and homozygous susceptible lines can be grown in root-knot nematode infested soils. Although many of these express incomplete dominance, these can be economically valuable for planting in infested soils. In Haryana, an advanced tomato breeding line, NRT-8 (Hisar Lalit) has been recommended for cultivation in root-knot nematode problem areas (Jain *et al.*, 1991). Higher concentrations of phenolics and ortho-dihydroxy phenolic compounds and more active IAA-oxidase and peroxidase systems have been observed in resistant tomato varieties (Masood and Husain, 1976; Bajaj and Mahajan, 1977; Ganguly and Dasgupta, 1981).

*Biological control: Arthrobotrys irregularis* was developed as a commercial biocontrol agent ('Royal 350') against root-knot nematodes in tomato (Cayrol and Frankowski, 1979). Field trials using this fungus on granules of oat seeds @ 140g/m2 incorporated one month in advance of cropping resulted in good protection of tomato against nematodes. In case of heavy populations of *Meloidogyne,* however, Cayrol and Frankowski (1979) indicated that 'Royal 350' fungus would be more appropriate as a secondary control measure after an initial nematicide treatment.

The discovery of *Paecilomyces lilacinus* as an effective parasite of *M. incognita* and *Globodera pallida* eggs was reported by Jatala *et al.* (1979). *P. lilacinus* has several attributes of a successful biocontrol agent (Jatala, 1986). Its adaptability to a wide range of soil pH and temperature (Bansal *et al.*, 1991) makes it a competitive organism. The successful results obtained in the Philippines have been the major reason for its commercial production under the name of 'Biecon' in that country.

Mankau (1972) provided the initial information on the effectiveness of a bacterial parasite, *Pasteuria penetrans* against root-knot nematode on tomato. The bacterium can survive in air dry soil for several years (Mankau and Prasad, 1977). Stirling (1984) found that galling of tomato roots and the number of nematodes in soil at harvest were reduced significantly when the parasite preparation was incorporated @ 212–600 mg/kg soil into root-knot infested soil. Walia (1994) reported that treatment of nematode infested nursery beds with small amount of parasite preparation may facilitate introduction of bacterium to the main field through infected nursery transplants. Using this technique, Walia and Dalal (1994) reported an increase of 17–20% in the yield of tomato crop.

*Organic amendments:* Aqueous extracts of leaves of several plants *in vitro*, and oil cakes of several plants such as neem, groundnut, castor, mahua, etc., have been reported to be toxic to root-knot nematodes (Bhatti and Nandal, 1994; Mishra and Vijayalakshmi, 1994). Dried leaf powder of *Chenopodium album, C. murale* and *Azadirachta indica*, when incorporated @1–2 g/kg soil before seeding to the root-knot nematode infested nursery soil, suppressed root galling and boosted the growth of tomato seedlings (Walia and Gupta, 1995; Tabil and Walia, 1997).

*Integrated control:* Under Indian conditions, deep summer ploughing can be effectively employed in vegetable growing areas to reduce soil population of root-knot nematodes. Further, protection to crops like tomato, brinjal and chillies can be provided by nematicide application or nursery soil amendments with leaf extracts of several plant species. In okra, seed treatment with cabofuran can be used along with summer ploughing. Cultivation of root-knot nematode resistant varieties/non-hosts can be introduced on alternate years.

### Reniform nemotode (*Rotylenchulus reniformis* Linford and Oliveira)

*R. reniformis* is widely distributed throughout the world and is considered to be a serious problem in the tropics, semi-tropics and warmer areas of temperate zone. Several important crops have been recorded as hosts in India (Sitaramaiah, 1984) including vegetable crops (tomato, brinjal, okra, sweet potato, cucumber, potato, cowpea, etc.).

#### *Biology and nature of damage*

*R. reniformis* is a sedentary semi-endoparasite. Pre-adult females infect the roots and establish a permanent feeding site. Young females swell to kidney-shaped adults, with most of the posterior portion protruding out of the roots. Each female deposits 39–120 eggs on vegetable crops in an egg mass on the root surface. According to Sivakumar and Seshadri (1971), on okra it may take

24–29 days to complete one life cycle. The second-stage juveniles hatch from eggs and develop to pre-adult females/males through a series of superimposed moults in soil without feeding. Males are not infective.

The pre-adult females establish a permanent feeding site in the cortex, endodermis, pericycle or phloem tissues depending upon the host. The tissues involved were reported to be phloem in tomato (Sivakumar and Seshadri, 1972), and pericycle in cauliflower and cowpea (Razak and Evans, 1976; Khan and Khan, 1985). According to Khan *et al.* (1985), in tomato feeding sites are established in the pericycle cells in the first four weeks but later phloem cells are also invaded. Unlike the root-knot nematode, the feeding zone of *R. reniformis* is restricted to pericycle or periphery of stele. Hypertrophy, hyperplasia, thickening of cell walls, dense granular cytoplasm, enlargement of nucleus, giant cells/syncytial formation, and feeding peg and feeding tube formation within the syncytium are the cellular modifications that result following *R. reniformis* infection in the host roots. A semi-lunar arrangement of modified cells is observed in cauliflower and tomato roots.

The pathogenically of *R. reniformis* has been established in several crops, including vegetables (Verma and Prasad, 1969). The inoculum level at and above 100 nematodes per 1500 cc soil resulted in significant reduction in the growth of brinjal plants (Singh and Khera, 1979). However, the extent of damage is influenced by host plant, environmental factors and initial nematode density. The above-ground symptoms are not characteristic but general unthrifty growth, yellowing and stunting is observed in the infected plants.

*R. reniformis* interacts with a variety of micro-organisms, including other nematodes, fungal pathogens, mycorrhyzae, bacteria and viruses. The incidence of wilt caused by *Rhizoctonia solani* is advanced in okra (Kumar and Sivakumar, 1981). Increase in nematode population was recorded in virus infected okra (Sivakumar and Meerzainuddin, 1973) and brinjal (Naqvi and Alam, 1975).

### *Crop losses*

Heald (1979) reported that yields from fumigated plots of okra, tomato, cowpea and lettuce were 19, 20, 15 and 52% higher, respectively, than non-fumigated *R. reniformis* infested plots in Texas, USA. *R. reniformis* caused maximum reduction in okra yields in poor soils not supplemented with fertilizers (Sivakumar and Meerzainuddin, 1974). Nath *et al.* (1976) observed that *R. reniformis* infested *parwal* (*Trichosanthes dioica*) plants had slower growth, reduced leaf size, thinner veins, reduced fruit size and significantly lower yields as compared to non-infested plants.

### *Management*

Cropping sequences including mungbean, sesbania, marigold, wheat, barley or fallow period reduced the population of *R. reniformis* to very low level (Khan *et al.*, 1984). Organic amendments in the form of green manure or oil cakes are also effective in managing the reniform nematode population. Balasubramanian and Ramakrishnan (1983) observed that Kalyanpur Selection, Kalyanpur III, CA-121 of tomato, and *Lycopersicon pimpinellifolium* exhibited resistant reaction against *R. reniformis*. Additional resistant cultivars of vegetable crops are listed in Table 1.

Table 1. Nematode resistant cultivars of vegetable crops

| *Crop* | *Nematode* | *Resistant cultivar* | *References* |
|---|---|---|---|
| Tomato | *Meloidogyne* sp. | Nematex, SL-120, NTR-1, SL-12, Patriot, VFN-8, VFN-Bush, Piersol, Ronita, Radiant, Nemared, Anahu | Rao and Tikoo,1975; Patel *et al.*,1979 |
| | | Bresch, Helani, Punuui, Campbell-25 | Anonymous, 1985a |
| | | Pelican, Hawaii-7746, Hawaii-7747 | Jain *et al.*,1983 |
| | | NRT-8 (Hisar Lalit) | Jain *et al.*,1991 |
| | | PNR-7 | Nandpuri *et al.*, 1988 |
| | *Rotylenchulus reniformis* | Radiant, SL-120, Patriot, Bresch, Ronita, Piersol, F-38-E-2 | Anonymous, 1983 |
| Brinjal | *M. incognita* | Giant of Banaras, Gola, Black Beauty | Alam *et al.*, 1974 |
| | *M. javanica* | Ghatikia White, Vijaya Hybrid, N(17)-4, Gacha Baigan, PBr-91–2, Dingra's Multiple Purple | Anonymous,1983 |
| Chilli | *M. javanica* | 579, CAP-63, Pusa Jwala | Jain *et al.*, 1983; Anonymous., 1983 |
| | *R. reniformis* | Pusa Jwala | Anonymous, 1983 |
| Potato | *M. incognita* | Kufri Dewa | Raj and Gill, 1983 |
| | *Globodera rostochiensis* | Kufri Swarna | Anonymous, 1985b |
| Carrot | *M. incognita* | Black | Arya and Tiagi, 1982 |
| Cowpea | *M. incognita* | Barsati Mutant, Iron, New Selection, C-152, 82–1-B, IC 9642-B, TVU 24-O-P, Dixie Cream | Sharma and Sethi, 1976; Darekar and Patil, 1981; Singh and Parvatha Reddy, 1982 |
| French bean | *M. incognita* | Banat, Blue Lake, Stringless, Bountiful Flat, Brown Beauty, Cambridge Countless, Gallaroy, Kenya-3, Pinto W5–114, Seafarer, Sutton's Masterpiece | Singh *et al.*, 1981 |
| Okra | *Meloidogyne* spp. | Harichikni, Kanki, Local Green, Vaishali, Badher | Anonymous, 1985a |
| Ridge gourd | *M. incognita* | Panipati, Meerut Special | Khan *et al.*, 1971 |
| Ash gourd | *M. incognita* | Jaipuri | Khan *et al.*, 1971 |
| Pumpkin | *M. incognita* | Jaipuri, Dasna | Khan *et al.*, 1971 |

Bare-root dip treatment with aldicarb/carbofuran/terbufos @ 500 ppm or 1000 ppm for 15–30 minutes (Krishna Prasad and Krishnappa, 1981) or seed dressing with carbofuran 3% or 6% (w/w) in okra (Sivakumar *et al.*, 1973) were found effective in reducing nematode population. Sivakumar *et al.* (1979) obtained 89% reduction in *R. reniformis* population by growing *Capsicum annum* (non-host) and application of carbofuran/aldicarb/phorate @ 1 kg a.i./ha, 15 days after transplanting. Deep ploughing followed by fallowing for one month integrated with aldicarb treatment @ 0.8 kg a.i./ha at sowing, or deep ploughing together with seed treatment with carbofuran in okra was found very effective (Parvatha Reddy, 1983). Combined application of organic amendments and nematicide has been recorded to reduce *R. reniformis* population and increase tomato yield (Badra and Mohamed, 1979).

### Potato cyst nematode (*Globodera rostochiensis* Woll. and *G. pallida*)

Potato cyst nematode is vastly distributed throughout Europe and presently is reported from 60 countries. In India, it was first recorded in the Nilgiri hills by Jones (1961) and is believed to have been introduced along with soil adhering to tubers imported from Scotland. It has remained restricted to 3050 ha in Nilgiri hills and about 200 ha in Kodaikanal hills of Tamil Nadu. However, its incidence was also recorded from potato fields in the adjoining areas of Karnataka and Kerala (Krishna Prasad, 1992).

The devastation caused by potato cyst nematode particularly in Europe has been responsible in giving boost to the science of nematology in general. Its record in Long Islands (USA) led to enforcement of strict quarantine regulations in that country. Originally thought to be constituting one species—*Globodera rostochiensis*, it was subsequently split into two, the second one *G. pallida* being distinguished by white or cream coloured females, in contrast to golden yellow females of *G. rostochiensis*. Besides potato, tomato and brinjal are also preferred hosts of potato cyst nematode, in addition to a few solanaceous weeds.

#### *Biology and nature of damage*

Eggs within the cysts can remain viable in soil upto 15 years. Larval hatching occurs in the presence of potato root diffusates. The second-stage larvae infect the potato roots and induce formation of multinucleate syncytia in the vascular tissues on which they feed. The larvae grow in size, moult thrice and become adult females (0.7–0.8 mm in diameter), while vermiform males emerge from larval cuticle, wriggle out into the soil and inseminate the females which are protruding out of the roots. After laying eggs within their bodies, the white females turn creamish/golden yellow and ultimately brown, when they get dislodged from the roots and remain in soil till next potato crop. Under Indian conditions, the second-stage larvae take 37–39 days during April-July and 40–42 days during September-December to become cysts, and generally one generation is completed in a season (Krishna Prasad, 1992).

According to European scheme, using standard *Solanum* genotypes, eight pathotypes of potato cyst nematode (5 of *G. rostochiensis* and 3 of *G. pallida*) have been distinguished (Kort *et al.*, 1977). In India, mixed populations of pathotypes Ro 1, Ro 2, Pa 1, Pa 2 and Pa 3 are indicated (Amalraj and Rao, 1992). The nematode spreads over long distances through infested soil adhering to the tubers.

Low nematode populations may not exhibit any detectable symptom. However, when populations build up to appreciable levels, nematode damage may manifest itself as patchy growth of stunted plants, yellowing of foliage, temporary wilting during hotter parts of the day, poor root development, reduction in the size and number of tubers.

*Management*

*Crop rotation:* Considering the narrow host range of potato cyst nematode, growing non-solanaceous vegetables such as beet root, cabbage, carrot, cauliflower, french bean, garlic, radish, turnip, etc., during autumn brings down the nematode population considerably. The rotational sequence of potato-french bean-pea-potato in a 4-year rotation decreases cyst population upto 98 per cent. The adoption of potato-cabbage rotation by Nilgiri farmers is best for the management of potato cyst nematode (Krishna Prasad, 1992).

*Host resistance:* Breeding potato varieties resistant to potato cyst nematode is difficult since the nematode species and pathotypes occur as mixed populations. However, sources of resistance are available in several non-tuber forming wild *Solanum* species. High degree of resistance to several populations of cyst nematode was exhibited in *Solanum vernei* clone VTn2 62–33–3. Since this hybrid was highly susceptible to late blight disease, Kufri Jyoti, a late blight resistant variety, was used as female parent with this clone to obtain several genotypes. One of these selections was subsequently released as Kufri Swarna in 1985, which presently occupies about 40% of potato area in the Nilgiris (Krishna Prasad, 1995).

*Chemicals:* A massive chemical control attempt was made under the Indo-German Nilgiri Development Project during 1971–75. Nematicidal treatment was made mandatory under the Tamil Nadu Pest Act, 1971 and all the infested fields at that time in the Nilgiris were treated with fensulfothion @ 30 kg a.i./ha in the first year and @ 15 kg a.i./ha in the next year. Later, however, after standardisation of use of systemic pesticides, aldicarb or carbofuran @ 2 kg a.i./ha was found effective for the economical management of potato cyst nematode under Nilgiri conditions.

*Quarantine:* Several European countries have enforced strict quarantine regulations against potato cyst nematode. The spread of this nematode from Long Islands to the mainland USA was checked successfully due to legal enforcement. In India, domestic quarantine was introduced under the Destructive Insect Pests Act, 1971 whereby the movement of potato tubers from infested areas of Nilgiris to other areas was banned, and mandatory chemical treatment was enforced. The records of this nematode in Karnataka and Kerala, however, warrant strict vigilance in enforcing domestic quarantine.

*Integrated control:* In view of the protective nature of cyst nematodes and long persistence in soil in the absence of host, it is practically impossible to eradicate this nematode. However, the damaging effect can be reduced considerably by using a combination of management practices. Following economically viable crop rotations including non-hosts, cultivation of resistant varieties, and strict enforcement of domestic quarantine can help in keeping the potato cyst nematode populations restricted to limited areas below damaging levels.

## NEMATODE PESTS OF ORNAMENTAL PLANTS

The nematode pests of ornamental plants have not been studied extensively in India. However, considering the export potential of cut flowers and plant material of ornamentals, the area under ornamental plants has shown tremendous increase in the recent years. Consequently, the importance of nematode pests in successful cultivation of ornamental plants has also been realised. The major nematode pests of ornamental plants in India are dealt with hereunder.

### Root-knot nematodes (*Meloidogyne* spp.)

*Meloidogyne* spp. have been recorded on crossandra in Tamil Nadu, Karnataka and Andhra Pradesh (Rajendran *et al.*, 1976; Khan and Parvatha Reddy, 1992b); tuberose in Tamil Nadu (Jayaraman *et al.*, 1975; Sundarababu and Vadivelu, 1988b); chrysanthemum (Chandwani and Reddy, 1967; Sen and Dasgupta, 1977); rose from Delhi and Uttar Pradesh (Prasad and Dasgupta, 1964; Alam *et al.*, 1973); gladiolus (Parvatha Reddy *et al.*, 1979c) and jasmine (Rajaram and Rajendran, 1979).

Crossandra (*Crossandra undulaefolia*) plants infected with *M. incognita* show stunting, dried peripheral branches bearing chlorotic leaves turning white at later stages (Rajendran *et al.*, 1976). Inflorescences are undersized and sometimes fail to produce flowers. Roots exhibit severe galling. Premature death of plants due to synergistic action of *M. incognita* on *Fusarium oxysporum* and *F. solani* has also been reported (Khan and Parvatha Reddy, 1992a). The nematode is responsible for 25.6 and 21.6% loss in number and weight of flowers, respectively (Khan and Parvatha Reddy, 1992b).

Root-knot nematode infected tuberose plants show rotting of bulbs, besides shoot malformation (Jayaraman *et al.*, 1975). In severe infestation upto 65% reduction in top growth may occur and emergence of spikes may be suppressed (Sundarababu and Vadivelu, 1988b). Khan and Parvatha Reddy (1992b) reported reduction in plant weight, number of flowers, spike length, spike weight and number of bulblets.

In gladiolus, severe galling on the roots results in yellowing of leaves and stunted growth, nematode infection spreads to daughter corms and cormels where they survive and serve as source of inoculum for the next season (Parvatha Reddy *et al.*, 1979b).

### Lesion nematodes (*Pratylenchus* spp.)

Among the lesion nematodes, *Pratylenchus delattrei* is considered important on crossandra in Tamil Nadu (Srinivasan and Muthukrishnan, 1975); *P. coffeae* on chrysanthemum in Karnataka (Rashid and Khan, 1975); *P. zeae* on rose in Tamil Nadu (Sundarababu and Vadivelu, 1988a) and *P. pratensis* also on rose in Delhi (Prasad and Dasgupta, 1964).

Lesion nematodes invariably infect cortical tissues, destroy parenchymatous cells and form lesions which ultimately lead to root rot due to secondary infection by other pathogens. Infected crossandra plants exhibit chlorosis and wilting, leaves turn brown and pinkish, dry at tips and are shed off. No side shoots are formed. Heavily infected plants do not produce tertiary spikes and flower yield is reduced (Srinivasan and Muthukrishnan, 1975). Association of *Fusarium* spp. with root lesions and premature wilting on crossandra has also been reported (Srinivasan and Muthukrishnan, 1975; Khan and Parvatha Reddy, 1992a). Rashid and Khan (1975) reported heavy root damage to chrysanthemum due to *P. coffeae* causing reduction in the size of flowers. The nematode is also known to aggravate disease severity caused by *Pythium aphanidermatum* and *Rhizoctonia solani* (Hasan, 1988). *P. zeae* reduces shoot length, shoot weight, root length and root weight of rose by 69.6. 36.4, 59.6 and 33.3%, respectively (Sundarababu and Vadivelu, 1988a).

### Other nematodes

High incidence and density of *Rotylenchulus reniformis* was reported on tuberose causing reduction in flower yield (Khan and Parvatha Reddy, 1992a). It was also recorded in chrysanthemum (Swarup *et al.*, 1967). *Longidorus africanus* was attributed to stunting, swelling and darkening of root tips of crossandra (Muthukrishnan *et al.*, 1977). In Himachal Pradesh, the incidence of bud and leaf nematode, *Aphelenchoides ritzemabosi* was reported on several ornamental plants including chrysanthemum, zinnia, salvia, aster and dahlia (Gill and Sharma, 1976). Characteristic symptoms of interveinal leaf necrosis leading to browning and withering of leaves were noticed. *Xiphinema diversicaudatum* and *X. basiri* have been recorded on rose causing stunting and unproductive flowering (Prasad and Dasgupta, 1964; Muthukrishnan *et al.*, 1975). *Ditylenchus dipsaci* is rated as the most important nematode pest of bulbous ornamentals in the whole of Europe.

### Nematode management

Soil and root populations of root-knot and lesion nematodes on crossandra were significantly reduced by interplanting marigold (Khan and Parvatha Reddy, 1992b). Ohkawa and Saigusa (1981) found that *Rosa indica* (Major) and *R. multiflora* (60-S) were resistant to *Pratylenchus penetrans* and *P. vulnus*. They also reported that rose cv. Mavetti was resistant to *Meloidogyne hapla*. Application of *Paecilomyces marguandi* (a fungus parasitic on nematode eggs) on paddy seeds @ 2g/kg soil effectively controlled root-knot nematode on crossandra (Khan and Parvatha Reddy, 1992b).

Various methods of applying chemical nematicides have been tested. Nursery bed treatment with phorate, aldicarb, fensulfothion and carbofuran each @ 5 g/m$^2$ significantly reduced infection of *P. delattrei* (Vadivelu and Muthukrishnan, 1979). Dale (1973) achieved significant reduction in root-knot nematode infection on rose by dipping bare roots in 0.1% fenamiphos for 30 minutes. Dipping of gladiolus corms in thionazin or fensulfothion (0.5 g a.i./l solution) reduced root-knot nematode significantly (Overman, 1969). Vadivelu and Muthukrishnan (1979) used 0.05–0.1% concentration of aldicarb, carbofuran, fensulfothion fenamiphos and phorate as root dip treatment for 8 hours and achieved protection of crossandra seedlings from *P. delattrei* for 30 days after transplanting. Foliar spray of methyl parathion, chlorpyriphos and quinalphos each at 0.05% was effective to control *A. ritzemabosi* in zinnia and chrysanthemum (Gill and Walia, 1980; Gill, 1981).

Integration of aldicarb or neem cake and biocontrol agent (*Trichoderma herzianum* or *Verticillium chlamydosporium*) was found promising for controlling root-knot nematode on crossandra. Combination of VAM fungi- *Glomus mosseae* or *G. fasciculatum* with neem cake and/or aldicarb also gave effective management of root-knot nematode on tuberose (Khan and Parvatha Reddy, 1992b).

## REFERENCES

Alam, M. M.; Khan, A. M. and Saxena, S. K.1973. Some new host records of root-knot nematode, *Meloidogyne incognita* (Kofoid & White, 1919), Chitwood, 1949. *Indian Journal of Nematology,* **3**: 159–160.

Alam, M. M.; Khan, A. M. and Saxena, S. K. 1974. Reaction of some cultivated varieties of egg plant, pepper and okra to the root-knot nematode, *Meloidogyne incognita. Indian Journal of Nematology,* **4**: 64–68.

Alam, M. M.; Siddiqui, Z. A.; Saxena, S. K. and Khan, A. M. 1980. Effect of different cropping sequences on the population of plant parasitic nematodes. *Indian Journal of Nematology,* **10**: 35–39.

Amalraj, S.F.A. and Rao, G.R. 1992. Potato cyst nematode pathotypes in the southern hills of India. *Indian Journal of Nematology,* **22**: 82–85.

Anonymous 1971. Estimated crop losses due to plant parasitic nematodes in the United States. *Journal of Nematology,* (Supplement, Special Publication No. 1), pp. 7.

Anonymous 1983. *Proceedings of the Fifth Workshop of All India Coordinated Research Project on Nematode Pests of Crops and their Control*. Himachal Pradesh Krishi Vishvavidyalaya, Solan.

Anonymous 1985a. *Proceedings of the Fourth Biennial Workshop of All India Coordinated Research Project on Nematode Pests of Crops and their Control*. Mohan Lal Sukhadia University, Udaipur, pp. 129.

Anonymous 1985b. Kufri Swarna, a high yielding late blight and cyst nematode resistant cultivar suitable for cultivation in Tamil Nadu hills. *Extension Bulletin* No. **21**(E). Central Potato Research Institute, Shimla, pp. 4.

Arya, M. and Tiagi, B. 1982. Changes in total proteins in three carrot cultivars infested with *Meloidogyne incognita. Indian Journal of Nematology,* **13**: 398–400.

Atwal, A.S. and Mangar, A. 1969. Repellent action of root exudated of *Sesame orientale* against the root-knot nematode, *Meloidogyne incognita* (Heteroderidae: Nematoda). *Indian Journal of Nematology,* **31**: 286.

Badra, T. and Mohamed, M.I. 1979. Influence of combinations of organic amendments and nematicides on tomato infected with *Rotylenchus reniformis* and associated populations of predacious and saprophytic arthropods and microphagous nematodes. *Indian Journal of Nematology,* **9**: 101–110.

Bajaj, K.L. and Mahajan, R. 1977. Phenolic constituents of tomato resistant and susceptible to *Meloidogyne incognita* (Kofoid & White) Chitwood. *Nematologia Mediterranea,* **5**: 329–333.

Balasubramanian, P. and Ramakrishnan, C. 1983. Resistance to the reniform nematode *Rotylenchulus reniformis* in tomato. *Nematologia Mediterranea,* **11**: 203–204.

Bansal, R.K.; Walia, R.K. and Bhatti, D.S. 1991. Studies on the effect of pH and temperature on the nematode destroying fungus, *Paecilomyces lilacinus. Journal of Biological Control,* **5**: 36–38.

Bhatti, D.S. and Jain, R.K. 1977. Estimation of loss in okra, tomato and brinjal yield due to *Meloidogyne incognita. Journal of Nematology,* **7**: 37–41.

Bhatti, D.S. and Nandal, S.N. 1994. Nematode management through phytotherapeutic substances. In: D.S. Bhatti and R.K. Walia (eds) *Nematode Pests of Crops.* CBS Publishers and Distributors, Delhi. pp. 148–164.

Bijloo, J.D. 1968. Substances in plants toxic to nematodes. *First International Congress of Plant Pathology*, London.

Cayrol, J.C. and Frankowski, J.P. 1979. Une methode de lutte biologique contre les nematodes a galles des racines appartenant au genre *Meloidogyne. Revista Horticulturae,* **193**: 15–23.

Chandwani, G.H. and Reddy, T.S.N. 1967. The host range of root-knot nematode *Meloidogyne javanica* in tobacco nurseries at Rajamundry, Andhra Pradesh. *Indian Phytopathology,* **20**: 383–384.

Dale, P.S. 1973. Elimination of root-knot nematodes from roses by chemical bare-root dips. *New Zealand Journal of Experimental Agriculture,* **1**: 121–122.

Darekar, K.S. and Mhase, N.L. 1988. Assessment of yield losses due to root-knot nematode, *Meloidogyne incognita* Race 3 in tomato, brinjal and bittergourd. *Indian Journal of Nematology Network Newsletter,* **5**(4): 7–9.

Darekar, K.S. and Patil B.D. 1981. Reaction of some cowpea varieties to root-knot nematode. *Indian Journal of Nematology,* **11**: 83–84.

Ganguly, S. and Dasgupta, D.R. 1981. Protein pattern in resistant and susceptible tomato varieties inoculated with the root-knot nematode, *Meloidogyne incognita. Indian Journal of Nematology,* **11**: 180–188.

Gill, J.S. 1981. Fighting the foliar nematode of chrysanthemum. *Indian Horticulture,* **25**: 21–22.

Gill, J.S. and Sharma, N.K. 1976. Additional hosts of the foliar nematode, *Aphelenchoides ritzemabosi* from India. *Indian Journal of Nematology,* **6**: 169–171.

Gill, J.S. and Walia, R.K. 1980. Control of chrysanthemum foliar nematode, *Aphelenchoides ritzemabosi* on *Zinnia elegans. Indian Journal of Nematology,* **10**: 86–88.

Gommers, F.J. and Voor in't Holt, D.J.M. 1976. Chemotaxonomy of Compositae related to their host suitable for *Pratylenchus penetrans. Netherlands Journal of Plant Pathology,* **82**: 1–8.

Goswami, B.K. and Chenulu, V.V. 1974. Interaction of root-knot nematode, *Meloidogyne incognita* and tobacco mosaic virus in tomato. *Indian Journal of Nematology,* **4**: 69–80.

Gupta, D. C.; Jain, R. K. and Paruthi, I. J. 1986. Chemical control of root-knot nematode (*Meloidogyne javanica*) infecting okra. *Proceedings of the National Academy Symposium on Short term and Long term Hazards of Pesticides and their Safe Use, New Delhi.* pp. 68–69.

Hasan, A. 1988. Interaction between *Pratylenchus coffeae* and *Pythium aphanidermatum* and/or *Rhizoctonia solani* on chrysanthemum. *Journal of Phytopathology,* **123**: 227–232.

Heald, C.M. 1979. Effect of the reniform nematode on vegetable yields. *Plant Disease Reporter,* **62**: 902–904.

Jain, B.K. and Bhatti, D.S. 1978. Bare root treatment with systemics for controlling root-knot nematode in tomato transplants. *Indian Journal of Nematology,* **8**: 19–24.

Jain, R.K. and Bhatti, D.S. 1983. Chemical control of *Meloidogyne javanica* in tomato through nursery bed treatments. *Third All India Nematology Symposium.* H.P. Krishi Vishvavidyalaya, Solan. pp. 37–38.

Jain, R.K.; Bhatti, D.S.; Bhutani, R.D. and Kalloo, G. 1983. Screening of germplasm of some vegetable crops for resistance to root-knot nematode, *Meloidogyne incognita. Indian Journal of Nematology,* **13**: 212–215.

Jain, R.K.; Bhutani, R.D.; Kalloo, G.; Bhatti, D.S. and Singh, K. 1991. Do not fear root-knot nematode, sow tomato Hisar Lalit. *Indian Horticulture,* **36**: 24–28.

Jatala, P. 1986. Biological control of plant parasitic nematodes. *Annual Review of Phytopathology,* **24**: 453–489.

Jatala, P.; Kaltenbach, R. and Bocangel, M. 1979. Biological control of *Meloidogyne incognita acrita* and *Globodera rostochiensis* on potatoes. *Journal of Nematology,* **11**: 303.

Jayaraman, V.; Rajendran, G. and Muthukrishnan, T.S. 1975. Occurrence of root-knot nematodes in *Polianthes tuberosa* L. in Tamil Nadu. *Indian Journal of Nematology,* **5**: 101–102.

Jones, F.G.W. 1961. The potato root eelworm *Heterodera rostochiensis* Woll. in India. *Current Science,* **30**: 187.

Khan, A.H.; Saxena, S.K. and Mahmood, I. 1984. Effect of different cropping sequences on the population of plant parasitic nematodes. *Pakistan Journal of Nematology,* **2**: 29–36.

Khan, A.M.; Saxena, S.K. and Siddiqui, Z.A. 1971. Reaction of some cultivated and wild species of *Lycopersicon* and cucurbits to *Meloidogyne incognita* (Kofoid and White) Chitwood. *Indian Phytopathology,* **24**: 197–199.

Khan, R.M.; Khan, A.M. and Khan, M.W. 1985. Cellular changes in tomato roots following infection of reniform nematode. *Pakistan Journal of Nematology,* **3**: 73–82.

Khan, R.M. and Khan, M.W. 1985. Histopathology of cauliflower roots infected with root-knot nematode. *Pakistan Journal of Nematology,* **3**: 83–85.

Khan, R.M. and Parvatha Reddy, P. 1992a. Nematode pests of ornamental plants. *In:* D.S. Bhatti and R.K. Walia (eds). *Nematode Pests of Crops.* CBS Publishers and Distributors, Delhi.

Khan, R.M. and Parvatha Reddy, P. 1992b. *Proceedings of the National Seminar on Commercial Floriculture in India.* Bangalore.

Kort, J.; Ross, H.; Rumpenhorst, H.J. and Stone, A.R. 1977. An international scheme for identifying and classifying pathotypes of potato cyst nematodes *Globodera rostochiensis* and *G. pallida. Nematologica,* **23**: 333–339.

Krishna Prasad, K.S. 1992. Nematode pests of potato. *In:* D.S. Bhatti and R.K. Walia (eds). *Nematode Pests of Crops.* CBS Publishers and Distributors, Delhi. pp. 99–108.

Krishna Prasad, K.S. 1995. Nematode problems of potato. *In: Nematode Pest Management- An Appraisal of Ecofriendly Approaches.* G. Swarup, D.R. Dasgupta and J.S. Gill (eds). Nematological Society of India, New Delhi. pp. 179–185.

Krishna Prasad, K.S. and Krishnappa, K. 1981. Effect of bare root-dip treatments with pesticides on *Rotylenchulus reniformis* affecting brinjal. *Indian Journal of Nematology,* **11**: 154–158.

Kumar, S. and Sivakumar, C.V. 1981. Disease complexes involving *Rotylenchulus reniformis* and *Rhizoctonia solani* on okra. *Nematologia Mediterranea,* **9**: 145–147.

Mahmood, K.; Kumar, S.; Naqvi, A. and Alam, M.M. 1974. The influence of bottlegourd mosaic virus and brinjal mosaic virus on the development of root-knot caused by *Meloidogyne incognita* (Kofoid and White, 1919) Chitwood, 1949. *Indian Phytopathology,* **27**: 610–611.

Mankau, R. 1972. Utilization of parasites and predators in nematode pest management. *Proceedings Annual Tall Timbers Conference on Ecology. Animal Control and Habitat Management* **4**: 129–143.

Mankau, R. and Prasad, N. 1977. Infectivity of *Bacillus penetrans* in plant parasitic nematodes. *Journal of Nematology,* **9**: 40–45.

Masood, A. and Husain, S.I. 1976. Phenolic and ortho-dihydroxy phenolic changes and their role in the resistance and susceptibility of three tomato varieties to *Meloidogyne incognita. Indian Journal of Nematology* **6**: 86–93.

Mishra, S.D. and Vijayalakshmi, K. 1994. Management of plant parasitic nematodes through organic amendments of soil. In: D.S. Bhatti and R.K. Walia (eds). *Nematode Pests of Crops.* CBS Publishers and Distributors, Delhi. pp. 133–147.

Muthukrishnan, T.S.; Chandrasekaran, J.; Lakshmanan, P.L. and Chinnarajan, A.M. 1975. Occurrence of the nematode *Hemicycliophora labiata* Colbran, 1960 on roses in Tamil Nadu. *South Indian Horticulture,* **23**: 75–76.

Muthukrishnan, T.S.; Vadivelu, S. and Rajendran, G. 1977. *Longidorus africanus* and its pathogenic effect on *Crossandra undulaefolia. Indian Journal of Nematology,* **7**: 69–72.

Nath, R.P.; Haider, M.G.; Akhtar, S.W. and Prasad, H. 1976. Studies on the nematodes of vegetables in Bihar-I. Effect of the reniform nematode, *Rotylenchulus reniformis* on *Trichosanthes dioica. Indian Journal of Nematology,* **6**: 175–177.

Nath, R.P.; Khan, M.N.; Wanshi, R.S.K. and Dwivedi, R.P. 1984. Influence of root-knot nematode, *Meloidogyne javanica* on pre-and post-emergence damping-off of tomato. *Indian Journal of Nematology,* 14: 135–140.

Naqvi, S.Q.A. and Alam, M.M. 1975. Influence of brinjal mosaic virus on the population of *Tylenchorhynchus brassicae* and *Rotylenchulus reniformis* around egg plant roots. *Geobios,* **2**: 120–121.

Ohkawa, K. and Saigusa, T. 1981. Resistance of rose rootstocks to *Meloidogyne hapla, Pratylenchus penetrans* and *Pratylenchus vulnus. HortScience,* **16**: 559–560.

Overman, A.J. 1969. Gladiolus corm dips for root-knot nematode control. *Proceedings of the Florida Station Horticultural Society,* **82**: 362–366.

Parvatha Reddy, P. 1983. *Plant Nematology*. Agricole Publishing Academy, New Delhi. pp. 287.

Parvatha Reddy, P. 1986. Analysis of crop losses in certain vegetables due to *Meloidogyne incognita. International Nematology Network Newsletter,* 3(4): 3–5.

Parvatha Reddy, P. and Singh, D.B. 1979a. Control of root-knot nematode *Meloidogyne incognita* on tomato by chemical bare-root dips. *Indian Journal of Nematology,* **9**: 40–43.

Parvatha Reddy, P.; Singh, D.B. and Ram Kishun. 1979b. Effect of root-knot nematodes on the susceptibility of Pusa Purple Cluster brinjal to bacterial wilt. *Current Science,* **48**: 915–916.

Parvatha Reddy, P.; Singh, D.B. and Rao, V.R. 1979c. The root-knot nematode on *Gladiolus* from India. *Current Science,* **48**: 82.

Patel, G.J.; Shah, H.M. and Patel, D.J. 1979. Reaction of some tomato cultivars to root-knot nematode. *Indian Journal of Nematology,* **9**: 60–61.

Prasad, S.K. and Dasgupta, D.R. 1964. Plant parasitic nematodes associated with roses in Delhi. *Indian Journal of Entomology,* **26**: 120–121.

Raj, D. and Gill, J.S. 1983. Reaction of some potato varieties to root-knot nematode, *Meloidogyne incognita. Proceedings of Third All India Nematology Symposium*, H.P. Krishi Vishvavidyalaya, Solan. pp. 27–28.

Rajaram, B. and Rajendran, G. 1979. Report of *Meloidogyne incognita* (Kofoid and White, 1919) Chitwood 1949 in *Jasminum sambac* and *J. flexile. South Indian Horticulture,* **26**: 70.

Rajendran, G.; Vadivelu, S. and Muthukrishnan, T.S. 1976. Pathogenicity of *Meloidogyne incognita* on crossandra. *Indian Journal of Nematology,* **6**: 115–116.

Rao, V.R. and Tikoo, S.K. 1975. Resistance of tomato cultivars to the root-knot nematode, *Meloidogyne incognita* (Kofoid and White, 1919) Chitwood, 1949. *Current Science,* **44**: 282–283.

Rashid, A. and Khan, A.M. 1975. Studies on the lesion nematode, *Pratylenchus coffeae,* associated with chrysanthemum. *Indian Phytopathology,* **28**: 22–28.

Razak, A.R. and Evans, A.A.F. 1976. An intracellular tube associated with feeding by *Rotylenchulus reniformis* on cowpea root. *Nematologica,* **22**: 182–189.

Reddy, D.D.R. and Seshadri, A.R. 1975. Elimination of root-knot nematode infestation from tomato seedlings by chemical bare-root dips or soil application. *Indian Journal of Nematology,* **5**: 170–175.

Rohde, R.A. and Jenkins, W.R. 1958. The chemical basis of resistance of *Asparagus* to the nematode, *Trichodorus christiei. Phytopathology,* **48**: 463.

Sasser, J.N. and Freckman, D.W. 1987. A world perspective on nematology: The role of the society. In: J.A. Veech and D.W. Dickson (eds). *Vistas on Nematology*. SON, Maryland. pp. 7–14.

Sen, A.C. 1958. Nematode attacking vegetable crops. *Indian Journal of Entomology,* **20**: 311–312.

Sen, K. and Dasgupta, M.K. 1977. Additional hosts of the root-knot nematode, *Meloidogyne* spp. from India. *Indian Journal of Nematology,* **7**: 74.

Sharma, N.K. and Sethi, C.L. 1975. Leghaemoglobin content of cowpea nodules as influenced by *Meloidogyne incognita* and *Heterodera cajani. Indian Journal of Nematology,* **5**: 113–114.

Sharma, N.K. and Sethi, C.L. 1976. Reaction of certain cowpea varieties to *Meloidogyne incognita* and *Heterodera cajani. Indian Journal of Nematology,* **6**: 99–102.

Singh, D.B. and Parvatha Reddy, P. 1981. Influence of *Meloidogyne incognita* infestation on *Rhizobium* nodule formation in french bean. *Nematologia Mediterranea,* **9**: 1–5.

Singh, D.B. and Parvatha Reddy, P. 1982. Cultivars of cowpea resistant to root-knot nematode, *Meloidogyne incognita. Tropical Pest Management,* **28**: 185–186.

Singh, D.B., Parvatha Reddy, P., Rao, V.R. and Rajendran, R. 1981. Reaction of some varieties and selections of french bean to *Meloidogyne incognita. Indian Journal of Nematology,* **11**: 81–83.

Singh, R.V. and Khera, S. 1979. Pathogenicity of *Rotylenchulus reniformis* on brinjal (*Solanum melongena* L.). *Indian Journal of Nematology,* **9**: 117–124.

Sitaramaiah, K. 1984. *Plant Parasitic Nematodes of India*. Today & Tomorrow's Printers and Publishers, New Delhi. pp. 292.

Sivakumar, C.V.; Kuppuswamy, S. and Meerzainuddin, M. 1973. Evaluation of carbofuran seed treatment for the control of root-knot on okra. *Indian Journal of Nematology,* **3**: 71–72.

Sivakumar, C.V.; Lakshmanan, P.L. and Balasubramaninan, M. 1979. Interaction of non-host plant (*Capsicum annum* L.) and certain systemic nematicides in population decline of *Rotylenchulus reniformis. Revue de Nematologie,* **2**: 165–168.

Sivakumar, C.V. and Meerzainuddin, M. 1973. The effect of yellow vein mosaic disease of okra on the reniform nematode, *Rotylenchulus reniformis. South Indian Horticulture,* **21**: 68–69.

Sivakumar, C.V. and Meerzainuddin, M. 1974. Influence of N, P and K on the reniform nematode *Rotylenchulus reniformis* and its effect on the yield of okra. *Indian Journal of Nematology,* **4**: 243–244.

Sivakumar, C.V. and Seshadri, A.R. 1971. Life-history of the reniform nematode, *Rotylenchulus reniformis* Linford and Oliveira, 1940. *Indian Journal of Nematology,* **1**: 7–20.

Sivakumar, C.V. and Seshadri, A.R. 1972. Histopathology of infection by the reniform nematode *Rotylenchulus reniformis* Linford and Oliveira, 1940 on castor, papaya and tomato. *Indian Journal of Nematology,* **2**: 173–181.

Srinivasan, A. and Muthukrishnan, T.S. 1975. A note on nematode fungal complex in crossandra (*Crossandra undulaefolia* Salisb.) in Coimbatore. *Current Science,* **44**: 743–744.

Stirling, G.R. 1984. Biological control of *Meloidogyne javanica* with *Bacillus penetrans. Phytopathology,* **74**: 55–60.

Sundarababu, R. and Vadivelu, S. 1988a. Studies on the nematodes associated with rose. *Indian Journal of Nematology,* **18**: 139.

Sundarababu, R. and Vadivelu, S. 1988b. Pathogenicity of *Meloidogyne* species to tuberose (*Polyanthes tuberosa* L.). *Indian Journal of Nematology,* **18**: 146–148.

Swarup, G.; Nath, R.P. and Sethi, C.L. 1967. The plant parasitic nematodes of Genus *Rotylenchulus* in India. *Indian Phytopathology,* **20**: 118–123.

Tabil, M.A. and Walia, K.K. 1997. Control of root-knot nematode, *Meloidogyne incognita* by amendment of nursery soil with *Chenopodium* species. *Annals of Plant Protection Sciences,* **5**: 61–65.

Thakar, N.A. and Patel, C.C. 1985. Control of root-knot nematode, *Meloidogyne javanica* on tomato by chemical bare-root dips. *Indian Journal of Nematology,* **15**: 110–111.

Vadivelu, S. and Muthukrishnan, T.S. 1979. Control of root lesion nematode, *Pratylenchus delattrei* on crossandra. *Indian Journal of Nematology,* **9**: 86.

Verma, S.K. and Prasad, S.K. 1969. The reniform nematode *Rotylenchulus reniformis* Linford and Oliveira 1940. I. Bio-ecological studies. *Indian Journal of Entomology,* **31**: 36–47.

Walia, K.K. and Gupta, D.C. 1995. Neem- an effective biocide against *Meloidogyne javanica* attacking vegetable crops. *Plant Disease Research,* **10**: 59–61.

Walia, R.K. 1994. Assessment of nursery treatment with *Pasteuria penentrans* for the control of *Meloidogyne javanica* on tomato in green-house. *Journal of Biological Control,* **8**: 68–70.

Walia, R.K. and Dalal, M.R. 1994. Efficacy of bacterial parasite, *Pasteuria penetrans* application as nursery soil treatment in controlling root-knot nematode *Meloidogyne javanica* on tomato. *Pest Management and Economic Zoology,* **2**: 19–21.

*Pest Management in Horticulture Crops*
L.R. Verma, A.K. Verma and D.C. Gautam (eds)


*Chapter 25*

# Rodent Pests of Horticultutal Crops

*Chander Sheikher*

## INTRODUCTION

THE ASSOCIATION OF THE pests with man dates back to time immemorial, and with the introduction of agriculture, the menace of pests has increased manifold. Among the various vertebrate pests, the rodents have attained the status of one of the most serious pests of agro-forest ecosystem the world-over. These mammals inflict damage at practically each stage of the development of agricultural, horticultural and forest crops as well as their products in storage, distribution and actual consumption by man. These animals, though considered primitive, have been so successful and abundant worldwide that they alone account for more than half of the known mammalian fauna (Goodwin, 1964).

The success of rodents lies in their number, their excellent capabilities of reproduction and adaptations. Generally, these animals have (with a few exceptions) a short estrous cycle, large litter size and number. They can subsist on a variety of food and even without water for a long time. Due to their prowess of adaptability, they have successfully adapted to varied ecological and physiological niches, and therefore are inhabitants of practically each and every latitude and altitude. They are globally present right from the areas at the sea level to snow line. They can withstand a temperature of sub-zero to over 50°C due to their magnificent burrowing habit. They have not only adapted to fossorial and terrestrial mode of life but have exploited arboreal mode of life too, though all rodents are not good climbers. They are able to swim vast stretches of water bodies and thus have invaded islands also. Their neophilic and neophobic behaviours have also contributed towards their success as these lead to exploration of newer and alien environments followed by either avoidance or rejection of unsuitable new items or adaptation of useful items of food and shelter.

All the known rodents can be divided into three major groups (sub-orders) as:

i. Sciumorpha, i.e., squirrel-like which include arboreal squirrels such as tree squirrels and flying squirrels as well as ground squirrels and their kins like woodchucks, chipmunks, prairie dogs, pocket gophers, marmots, etc.,

ii. Myomorpha, i.e., mouse-like such as rats, mice, voles, jerboas, lemmings, gerbils, etc., and

iii. Hystricomorpha, i.e., porcupine-like including tree porcupines, capybara, agoutis, etc.

All the rodents have enlarged pre-maxillae, maxillae and lower jaws to accommodate the long roots of incisors, and to provide for the attachment of powerful muscles used in gnawing and mastication. Owing to their gnawing habit, they damage not only the food items but also other utility items such as furniture articles, boxes, cartons, jars, pipes, wires, etc. Besides, they render still larger quantities of food unfit for human consumption by contaminating it with their droppings, urine and hair. In addition, they are the known carrier of some of the most dreadful diseases of man and live-stock.

## DAMAGE

The rodents have diverse food habits and can subsist practically on anything and everything, though they are basically herbivores. Due to this factor, they have attained the status of major vertebrate pests in all ecosystems. Though their devastation has been amply felt and documented in cereal crops and stored grains, yet they are equally important as pests in orchards, vegetable crops, grasslands and pastures, and forests, as well as residential premises and poultry farms.

Rodents damage plants during their fossorial, feeding and gnawing activities, and the damage is spread to all parts of the plants. The rodent damage to a plant can be categorized as below ground damage, which is restricted primarily to root system and to underground rhizome and stem and above ground damage which may either be restricted to some specific part or widely spread, i.e., to stem, leaves, flowers, fruits, etc. Depending upon the severity and nature of damage, the plant may be uprooted and killed (if the roots are completely destroyed), or the vitality/vigour of the plant may be adversely affected as reflected in slowed growth or reduced yield of the plant. The extent of rodent damage is directly related to the density of the rodent population (or their number).

### Orchards and plantation crops

Rodents inflict heavy damage to a variety of fruit and plantation trees under diverse ecological conditions. The species responsible for damage also vary in diverse niches and continents. Tree squirrels and some arboreal rats and mice (all rats and mice are not good climbers) are responsible for heavy losses in orchards as they gnaw at the fruits throughout the crop cycle, though there may be some special preference for some particular stage of the fruit maturity, and thus incur direct loss to the crop, whereas other rodents (ground squirrels, rats, mice, porcupines, etc.) cause indirect loss by damaging trees at or below the ground level by either girdling or debarking the stem above ground level or gnawing and girdling of the collar or other finer roots below ground level.

A direct correlation exists between the extent of damage and the frequency and specific location of the active burrows of the rodents. Sheikher and Jain (1991) emphasized that the damage was more pronounced in trees which were 1.5 m or less away from the surface opening of active burrow of rodents (*Bandicoata bengalensis* and *Rattus meltada*) in the pecan nut orchards of Himachal Pradesh (India) and that the age of the plant was also an important factor as the severity of the damage was more in plants which were less than 10 years of age; the incidence of death being as high as 17.4%. Apparently the older trees could withstand the damage due to their more elaborately and strongly developed root system since the rodent damage is restricted to roots only. A similar relationship was recorded by Swihart and Picone (1994) in apple orchards where 96% of the trees growing >1.5m from active woodchuck (*Marmot monax*) burrows were gnawed at, and the incident of death in such trees was 17% as against nil in the trees growing 12-15 m away from the burrows. The trees growing nearer to burrows recorded a reduced growth in terms of height, maximum canopy, width, stem diameter as well as apple size and weight. The fact that the rodents in orchards adversely affected the yield was highlighted by Askham (1988) who recorded 31% production loss of fruit in Red Delicious and 53% in Golden Delicious cultivars in terms of weight due to the severe attack and high population of mountain vole (*Microtus montanus*) in mixed apple orchards. Elimination of vole from the orchards resulted in 3.2-fold increase in production. Earlier, Pearson and Forshey (1978) and Richmond *et al.* (1987) also determined and showed reduction in growth, yield and fruit size of apple in the trees damaged by pine vole population of known size maintained in enclosures around the trees. Ferguson (1980) estimated that the voles alone caused $50 million loss to farming community in eastern United States.

Rodents inflict damage to citrus orchards also. Ciampolini *et al.* (1985) recorded that the vole, *Pitymis savii*, gnawed at and damaged the roots below the collar of young trees whereas in the older trees only the root cortex was damaged, and that the trees showed a loss of growth vigour in the subsequent year inasmuch as the shoot number and fruit set was concerned, depending upon the severity of damage. In Himachal Pradesh, *Bandicota bengalensis* and *Rattus meltada* dig their elaborate burrows in citrus orchards and are responsible for the death of young plants or a loss of vigour due to extensive damage to the root system. Elias and Fall (1988) have reported girdling of citrus fruit trees by *Oryzomys*, *Sigmodon* and other rodent species.

Rodent damage is not restricted to only apple and citrus but is wide spread in other orchards too. Marsh rats (*Holochilus brasiliensis*) is the most important pest of banana cultivation in Latin America (Argentina, Brazil and other countries), though other rodents such as *Oryzomys, Calomys, Akodon* and *Ctenomys* species are also known to be destructive to fruit trees, primarily damaging the underground parts of the plants (Manjon *et al.*, 1983; Bruggers and Zaccagnini, 1994). Rodents, especially *Oryzomys* and *Punomys*, have been held responsible for causing even a total loss of pineapple crop in Peru during seventies and *Holochilus* and *Sigmodon* are very severe on pineapple in Venezuela and Nicaragua (Elias and Fall, 1988), whereas the marsh cane rat, *Thryonomys swinderianus*, has been reported to damage these plants in southern Africa (Fiedler, 1988a). Singh *et al.* (1994) reported 8.53% damage to pine-apple in India, primarily to ripening or ripened fruit.

Reports on the rodent damage to other fruit trees also abound. In Africa, *Cricetomys gambianus* damages avocado, pear and mango from Sudan to Senegal, and *Atherurus africanus* (the African brush-tailed porcupine) severely damages the banana and guava plantations particularly in the areas near forests or the newly opened agricultural or horticultural areas from forests (Delany, 1975). Squirrel, *Sciurus carolinensis,* damages peaches, grapes and almonds in South Africa, however, the damage is minimum to young trees (Fiedler, 1988a). Almond and apple are also damaged by arboreal rats and mice in Himachal Pradesh. Rodent damage in guava orchards was also recorded in India by Malhi and Sheikher (1989). Five striped squirrel, *Funambulus pennanti*, shows special liking for ripe fruits of pomegranate, though unripe fruits are also damaged, and Patel *et al.* (1993) recorded 4.6% damage in this crop in the arid region of Rajasthan (India). Plants of *Zizyphus mauritiana* (*ber*), a cash crop in tropics in India, face rodent damage (Jhala *et al.*, 1989). *Meriones shawi* attack lowers the fruit yield in olive, almonds and pistachio nut orchards due to debarking and consumption of the nuts (Fiedler, 1988a).

Rodent damage to fruit trees is not restricted to only established orchards but also in the nurseries. Parshad *et al.* (1984) reported that 26.8% plants in almond nursery were damaged 3 days after sprouting and additional 14.9% suffered damage 15 days after sprouting due to the activity of ground rodents (*Bandicota bengalensis, Tatera indica, Golunda ellioti and Mus spp.*) which seemed to have been attracted by the exposed seeds (the germination in almond is epigeal). Sheikher and Jain (1992a) estimated a loss of around 13% in apple and 12% in peach nursery. The nurseries located near fallow land or some gorge suffer relatively higher damage probably due to the fact that rodents find better refuge and chance to build up their population since human interference is almost negligible here as against the area where well-kept environment prevails. Jain *et al.* (1994) have observed 80% damage in *ber* nursery due to the activity of *Tatera indica* from dusk to dawn, and *Funambulus pennanti* during early morning and late evening hours.

In Hawaii (United States), the rodents incur 5–10% damage annually to macadamia nuts amounting to a net loss of $2–4 million. Damage occurs throughout the crop cycle, from the time when the kernels are small, fleshy, unprotected fruits to when they are fully developed and protected by hard brittle shell and husk (Fellow, 1982; Tobin, 1990). Macadamia orchards provide an ideal habitat for rats. These trees support thick canopy with interlocking branches amongst adjoining trees which provides for safe and easy movement among trees and also the sites for nesting for arboreal rodent species such as roof rat (*Rattus rattus*), Polynesian rat (*R. exulans*) and Norway rat (*R. norvegicus*). These rats typically stay in their burrows on ground or rest in their nests during day and become active emerging from their hiding places to feed in the canopy after sunset (Tobin, 1992, 1993); the macadamia nuts form the principle diet of these rodents throughout the year. Tobin *et al.* (1997) opine that the damage to macadamia nut before maturity is likely to be compensated by the trees. However, if the damage occurs at maturity of the nuts, the trees may not be able to compensate the loss; and the rats undoubtedly tend to forage on trees with relatively mature nuts, as confirmed by radio-telemetry studies (Tobin *et al.*, 1996), may be because the nuts accumulate appreciable amount of high energy oil at maturity.

A variety of rodents damage plantation crops, probably due to their ability to exploit new food sources. Coconut suffers heavily from rodent activity; upto 77% damage in Colombia mainly by *Rattus* and *Sigmodon* species, resulting in total failure of coconut industry was reported by Elias and Fall (1988). *R. rattus rattus* damages coconut by gnawing and penetrating green husk in eastern and western Africa (Fidler, 1988 a). In the Philippines, the rice field rat (*R. rattus mindanensis*) and Polynesian rat (*R. exulans*) are mainly responsible for damage to coconut crop, as these two species were recorded (trapped) from the crown of the plants and possessed coconut meat in their cardiac stomach (Milan, 1990). Rodent damage in this crop starts from nursery till trees are fully grown, and from tender nut stage till storage of nuts. In India, *R. rattus* and *Vandeleuria* sp., *R. rattus wroughtoni* and *R. rattus rufescens* damage coconut on crown, while *B. bengalensis* inflicts damage in nurseries and along with *R. rattus* in storage; the damage varying from 9 to 30% to tender and 8 to 50% to mature nuts, besides 6 to 8% in nursery (Sridhara, 1992; Kapadia, 1995; Rangareddy, 1995). *Funambulus palmarum* and *F. tristriatus* also extensively damage coconut crop (Koshy and Phillip, 1995). Since the fonds/folliage of the adjacent palms often touch each other (even are interlocking), the arboreal species can move from one tree to another without alighting from the tree and thereby avoiding many ground dwelling predators.

Oil palm plantations are severely damaged by rodents in India. Rodens kill young plants (≥ 3 years old) by damaging their collar. In mature and bunch bearing palms, the rats gnaw at the female flower bearing branches and also the tender and ripe fruits (Subiah, 1983a, b). Sridhara (1992), however, reported 5- 57.3% damage to oil palm. Cocoa, likewise is severely damaged by rats and squirrels, the damage being 8-15% (Sridhara, 1992). Valencia *et al.* (1994) reported that 19% of the palms in oil palm plantations in eastern plains of Colombia were damaged annually by rodents.

In Himachal Pradesh, tea plantations are also infested by rodents, especially *B. bengalensis, Golunda ellioti* and *Mus musculus* (Sheikher and Jain, 1990) and since they burrow at the base of the plants, they are likely to affect the quality and quantity of the produce.

**Vegetable crops**

Rodents are also severe on vegetable crops, though quantification of the damage caused in the field in some crops becomes rather difficult owing to the nature of the crop and the damage. Damage is initiated at sowing of the seed itself as the rodents dig and consume the seeds thereby resulting into thinning of the crop, and spreads to early nursery/growth phase, particularly in epigeal crops. However, the damage is not usually alarming at this stage as the crop fields are devoid of adequate cover and the rodents appear not to venture deep in the field owing to the looming danger of predators. Normally in vegetable crops the damage becomes discernible at the maturity or a little earlier at the time of flowering or fruit setting. The usual mode of damage is that the rodents invade the fields and burrow at the base of the plants. Obviously such plants and those nearby are severely damaged while those away from these burrows escape their fury/deprivation.

Rodents inflict heavy damage to tomato crop everywhere, showing special preference for green, unripe and solid berries, consuming their endocarp portion. The extent of damage varies under different conditions and areas. Singh and Saxena (1984) recorded 12.56% damage by a variety of rodent species while Ramesh and Katiyar (1985) observed 8.99% damage to fully mature berries, mainly by *Nesokia indica*. Thomas *et al.* (1986) reported that rats (*Bandicota bengalensis*) not only damaged the big unripe berries on the plants, but also hoarded them in their burrows. Hoarding by *B. bengalensis* has also been observed in Himachal Pradesh. *Milardia meltada, B. bengalensis* and *Tatera indica indica* have been shown to inflict 2.62—35.58% damage to tomato in Gujarat (Katodia *et al.*, 1993). Advani and Mathur (1982) have reported 19% damage in Rajasthan to this crop, chiefly by *T. indica* and *Meriones hurrianae*. In Meghalaya, 4.93% damage is caused due to the activity of *B. bengalensis* and *Rattus nitidus nitidus* (Singh, 1996).

Rodents also inflict severe damage to cauliflower, both curd crop for table purpose and seed crop in Himachal Pradesh due to the activity of *B. bengalensis*, *R. meltada*, *M. musculus and M. booduga* (Sheikher and Jain, 1997). Of these, only *B. bengalensis* and *R. meltada* dig their burrows in the field and at the base of the plants whereas those of mice are restricted to the periphery of the field and away from the plants. Rodent damage varies from 4.44 to 11.37% in curd crop and 6.51 to 8.09% in seed crop. Advani and Mathur (1982) had earlier recorded 5.5% damage to cauliflower and 7.1% to cabbage in Rajasthan. However, Sheikher and Jain (1997) did not observe damage in cabbage in a major cabbage growing area of Himachal Pradesh, though damage was recorded on this crop in Kinnaur district of the state.

It appears that the rodents prefer cereals as food and that they invade the vegetable fields only because they form an alternative food item. Saxena and Sahai (1990) observed that the cereals (viz., wheat and barley) harboured four times more active rodent burrows as compared to those in pea during *Rabi* season. Nevertheless, pea crop suffered rodent damage. Rodent damage to cucumber (8.96%), spongegourd (9.85%), muskmelon (11.4%) and bottlegourd (14.32%) in Haryana (Kumar *et al.*, 1995), to brinjal (4.5%), carrot (15.8%), radish (19.8%), onion (8.8%), garlic (6.6%), spinach (4.7%), bottlegourd (4.1%) and okra (4.4%) in Rajasthan (Adavani and Mathur, 1982), and watermelon (9.94–19.79%) in Punjab (Chopra and Parshad, 1986) has been recorded.

In Africa, vegetables are damaged particularly by the African mole rat, *Tachyorcyctes splendens*, in its preferred moist, upland and non-forested habitat, *Tatera robusta* in dry environment, and ground squirrels (*Xerus rutilus* and *X. erythropus*) in semi-arid marginal farming areas. Other rodent species such as *Meriones shawi* (jird), *Rhabdomys pumalis* (four-striped grass mouse), *Thryonomy swinderianus* (marsh cane rat), *Sciurus carolinensis* (squirrel) and others cause damage to a variety of vegetable crops including pumpkin, egg-plant, pea, beans, etc. (Fiedler, 1988a, b). Marsh rat (*Holochilus brasiliensis*) damages vegetable crops in Argentina (Bruggers and Zaccagnini, 1994). Vegetables including tomato, sweet pepper, egg-plant, radish, carrot, beans, kale, water melon, etc., are damaged in Israel (Moran and Keidar, 1993), and also in Latin American countries, viz., Brazil, Mexico and others (Elias and Fall, 1988).

## MANAGEMENT

The selection of rodent management or control methods is always based upon a multiplicity of considerations including the effectiveness of a control method, the cost to benefit ratio of the control method, its integration with other orchard management strategies/practices, the potential hazards to man, live-stock and other non-target animals, the availability of labour and other inputs such as chemicals, specific traps, bait carrier, bait station, etc., and the climatic/weather conditions. Since the fruit bearing age/life of a tree is quite long, a loss of even a single tree in an orchard tantamounts to a considerable economic loss to the farmer as it will be a perennial loss over the subsequent years. Since a single animal residing near or at the base of the tree, even if it is not arboreal in nature, is capable of severely damaging the roots and also trunk of the tree, the economic threshold level of population of these pests is very low. Further, the value of the tree increases with an increase in the age of plantation, and therefore, the economic threshold decreases. In vegetable crops also, the economic threshold level of population is low as the economic loss depends upon the capability of an individual to inflict damage not only to the plants at the base of which it has dug its burrow, but to a number of plants in its vicinity.

### Cultural control

Rodent population in orchard or a vegetable crop field can be kept under manageable limits through the use of cultural practices including the removal of ground cover or through 'clean culture', physical barrier, traps, etc.

#### *Clean culture*

Ground rodents depend upon ground cover of thick bushy plants and tall grass for safety. Additionally the seeds of these plants and grasses form a part of their diet during scarcity of the main crop in the fields. In absence of such a cover, these are exposed to their predators—raptors as well as ground dwellers (snakes, small carnivores and others). The movement of the rodents, in the absence of shelter and availability of food in such a clean habitat gets restricted and the activity decreases. In apple orchards where grass was regularly mowed and was not allowed to grow tall, 25% decrease in rodent activity was observed in Himachal Pradesh by the author. However, in developing and under developed countries, this grass forms the bulk of the fodder for milch cattle and hence is a prized possession of the farmers, which additionally provides for preservation of ground moisture. Thus maintaining a clean bed around the trees and removal of litter, garbage and bushes should be included in the orchard cultural practices to restrict rodent activity. Treatment of herbicides between the trees is also advisable. Byers (1984) and Godfrey and Askham (1988) advised the maintenance of wide, herbicide treated cultivated strips in tree rows and a regular close mowing of the middle. The grass grown in the orchards should be of a quality which has its tuber/stolons above ground and which forms a close matted cover on the surface. Tillage of the field though destroys the existing burrows, yet can provide the rodents with a loosened soil strata which may enable them to dig burrows in such a manner that they can easily reach the root system of the trees, a condition which may prove fatal to the tree. Therefore, tillage of orchard floor should be avoided as far as possible and use of herbicides should be encouraged.

In vegetable crops, however, one can resort to weeding and hoeing as well as the application of herbicides. However, clean culture is rather an expensive proposition as compared with the use of rodenticides and hence cannot be easily propagated amongst the farming community.

#### *Trapping*

Trapping is generally employed for determining rodent population in an area, but it is also an effective method of rodent control not only in residential premises and godowns but also in the fields. Traps (kill, glue or live traps) should be placed in lines forming a grid. Sheikher and Jain (1992 b) successfully employed live trapping to monitor and manage the rodent activity in vegetable crops. They observed that regular trapping for one year at the rate of three consecutive days trapping per month, reduced the active burrows of rodents (*B. bengalensis*, other rats and mice) by 85% in the fields which supported vegetable crops throughout the year. Regular trapping reduces rodent activity also (expressed in the number of active burrows) in apple orchards (Sheikher and Jain 1992 a). Tobin *et al.* (1993) used trapping in macadamia nut orchards and recorded that the damage to nuts was low in the section where trapping was done as compared to untrapped section. Trapping, however, is an expensive control method since it requires high initial cost for acquiring traps as well as regular labour input in setting the traps and in disposing off of the trapped rodents. Further, traps cannot be exposed for a long period continuously as it may develop trap shyness in the rodent population. However, traps can be used time and again as used traps have no ill-effect on the subsequent trapping of the rodents, rather it may actually increase subsequent trapping as observed by Tobin *et al.* (1994). Byers (1985) opined that trapping could completely remove vole population after 3- 5 days of trapping. Trapping may be poor during extremely cold or hot periods of the year as the surface forays of rodents decrease. During winter the spring loaded wooden traps proved superior over metallic wonder or Sherman traps in Himachal Pradesh. Probably the cold metallic touch/sensory cue prevented the exploratory activity and hence the success rate of trapping.

### *Collars or barriers*

Arboreal rodents can be denied access to fruits by putting trunk guards or collars of aluminium or galvanized iron sheets around the trunk at an appropriate height. Similarly, debarking can also be prevented by erecting guard around tree trunk. However, such guards fail to prevent root damage in trees. The guards should have their upper free edge/margin rolled outward and downward at an angle for better results. However, guards/collars are effective only in orchards with well spaced trees where foliage of adjacent trees does not touch each other. These are also not successful in orchards supporting dwarf varieties and untrained and unpruned trees, especially in hilly terrain, as the drooping shoots come very near to the ground level which enable the arboreal species to climb up the trees.

### Chemical control

Chemical control (use of rodenticides or fumigants) still is the most widely used single method of rodent control. Both acute rodenticides (which kill rodents quickly after consumption of lethal dose, e.g., zinc phosphide, strychnine, etc.) and chronic rodenticides (which kill rodents after a lapse of time on being consumed, e.g., anticoagulants like bromadiolone, brodifacoum, difacinone, etc.) as well as fumigants (e.g., aluminium phosphide) are used for control of rodents in orchards and vegetable crops.

Mode of application of rodenticides is an important aspect, and depends upon various factors, like the distribution and nature of the inhabiting rodent species (whether arboreal or terrestrial), thickness of the ground cover, density and canopy of the crop and topography of the field. Broadcasting of the bait is effective if the crop is very thick, not conducive to man's movement inside the crop, and the rodent infestation is very high. Orchards and vegetable crops generally do not hamper the movement of man due to ample spacing between adjacent plants. Nevertheless, the broadcasting has been attempted with good results in macadamia nut and apple orchards in United States and Canada. Surface baiting would be effective if the ground cover is adequately thick for free movement of rodents. Bait station and burrow baiting are other methods of rodenticide application for ground dwellers. However, in crops where damaging species are arboreal such as in palm plantations and other orchards, crown baiting would be more effective as the rodents stay and spend much of their activity period for forage in the foliage and descend only to rest in their burrows on ground. Hence chances of consumption of lethal doses of the bait are more when the same is placed in the crown.

### *Zinc phosphide ($Zn_3P_2$)*

It is one of the oldest and yet the most preferred and widely used rodenticides in most parts of the world. It is offered as a grain based bait containing normally 2 to 2.5% of active ingredient (though used upto 5% concentration), and has been reported to give upto 87% success under varying conditions in orchards as well as vegetable fields (Advani and Mathur, 1982; Byers and Carbaugh, 1991; Kotadia *et al.*, 1993; Sheikher and Jain, 1996). Sheikher and Jain (1996) opined that in vegetable crops, burrow baiting was far superior over either surface baiting or through bait station. In palm trees crown baiting yielded better results.

Zinc phosphide is effective against a wide range of rodent species but it also induces bait shyness, aversion and even resistance in some rodents (Prakash and Mathur, 1992). Zinc phosphide bait cannot be repeated immediately after it has been used in the field on account of the problem of development of bait shyness and aversion.

### *Anticoagulants*

Anticoagulants are also widely used in rodent management programmes, and are preferred over acute rodenticides by many workers. These anticoagulants have been reported to produce significantly high success in containing rodents (even over 90% success in the field has been reported) in both orchards and vegetable fields (Subiah, 1983 a, b; Milan, 1990; Hoque and Fiedler, 1988; Sheikher and Jain, 1991, 1997; Mathur *et al.*, 1992; Tobin, 1992, 1995). Freshly prepared baits or formulated baits marketed as "ready to use" baits either as loose bait or wax blocks have been used with success. In coconut and other palms, crown baiting produced significant control (Subiah, 1983 a; Hoque and Fiedler, 1988; Milan, 1990; Samiayyan and Chandrasekharan, 1994), while in apple and macadamia nut orchards significant control was achieved with broadcasting, surface baiting and bait station methods of application (Byers and Carbaugh, 1991). In vegetable crops, surface baiting and burrow baiting with anticoagulant baits have been successfully employed to control rodents (Advani and Mathur, 1982; Katodia *et al.*, 1993; Sheikher and Jain, 1997).

### *Fumigants*

Aluminium phosphide and other fumigants can also be used for managing rodents in orchards and vegetable crop fields to control rodent population. It is an effective control measure against the residue population left in the field after carrying out control operation with other rodenticides.

In addition some other chemicals like calciferol/cholecalciferol are also in use in various parts of the world, though their efficacy in the fields is not very encouraging.

### Biological control

Potential of owls, snakes and other carnivores to limit rodent population has often been highlighted (Uchida, 1969; Milan, 1990) and their conservation should be encouraged. However, so far biological control has not yielded any significant success against rodents.

## INTEGRATED RODENT MANAGEMENT

Though many methods of rodent control are available, each with its own merits and demerits, yet each of these methods may not prove effective in every situation. Therefore an integrated approach is advocated.

### Orchards

Proper maintenance of orchards, removal of litter and garbage, and pruning of the drooping shoots as clean culture practice, coupled with banding (fixing collars/guards) of trunk (if arboreal species inhabit the orchard), trapping of rodents and application of rodenticides are required at various stages.

### Vegetable crops

Require special care only at and after the flowering/fruit set stage when the rodents actually invade the fields. Clean culture, trapping and rodenticide application should be used.

### Rodenticides recommended

Zinc phosphide (2–2.5%) and bromadiolone (0.005%). Zinc phosphide should not be repeated at least two months after first treatment to avoid bait shyness and bait aversion in rodents.

## REFERENCES

Advani, R. and Mathur, R.P. 1982. Experimental reduction of rodent damage to vegetable crops in Indian villages. *Agro-Ecosystems,* **8**: 39–45.

Askham, L.R. 1988. A two year study of the physical and economic impact of voles (*Microtus montanus*) on mixed maturity apple (*Malus* spp.) orchards in the Pacific Northwestern United States. *Proceedings of the Thirteenth Vertebrate Pest Conference*, Monterey, California. pp. 151–155.

Bruggers, R.L. and Zaccagnini, M.E. 1994. Vegetable pest problems related to agricultural production and applied research in Argentina. *VIDA Silvestre Neotropical,* **3**: 71–83.

Byers, R.E. 1984. Control and management of vertebrate pests in eastern United States. *Horticultural Reviews,* **6**: 253- 285.

Byers, R.E. 1985. Management and control. *In: Biology of New World Microtus*. R.H. Tamarin (ed).American Society of Mammology, Special Publication No. **8**, pp. 621–646.

Byers, R.F. and Carbaugh, D.H. 1991. Rodenticides for the control of pine and meadow voles in orchards. *Journal of Environment and Horticulture,* **9**: 167–172.

Chopra, G. and Parshad, V.R. 1986. Rodent damage to watermelon. *Indian Farming,* **36** (3): 40–43.

Ciampolini, M., Suss, L., Zerbi, N. and Penna, E. 1985. Voles cause severe damage to the citrus orchards of Calabria. *Informatore Agrario,* **41**: 93–99.

Delany, M.J. 1975. *The rodents of Uganda*. British Museum Natural History, George Press, Kettering, Northamptonshire, England, pp. 165.

Elias, D.J. and Fall, M.W. 1988. The rodent problem in Latin America. *In: Rodent Pest Management*. I. Prakash (ed). CRC Press, Boca Raton. pp. 13–28.

Fellow, D.P. 1982. Rat damage and control in Macadamia. *Proceedings of the Hawaii Macadamia Producers' Association Annual Meeting,* **22**: 94–103.

Ferguson, W.L. 1980. Rodenticide use in apple orchards. *Proceedings of the East Pine and Meadow Vole Symposium,* **4**: 2–8.

Fiedler, L.A. 1988 a. Rodent problem in Africa. *In: Rodent Pest Management*. I. Prakash (ed). CRC Press, Boca Raton. pp. 35–65.

Fiedler, L.A. 1988 b. Rodent pest problems and management in eastern Africa. *FAO Plant Protection Bulletin,* **36**(3): 125–134.

Godfrey, M.E.R. and Askham, L.R. 1988. Non-toxic control technique for *Microtus* spp. in apple orchards. *Bulletin DEPP.* **18**: 265–269.

Goodwin, G.C. 1964. *The Animal Kingdom-Mammalia*. Ward Lock and Co. Ltd., London, Vol. I and II. pp. 1/106- 1/168.

Hoque, M.M. and Fiedler, L.A. 1988. *Rat control in coconut*. Technical Bulletin No. 3/1988. National Crop Protection Centre, University of Philippines, Low Banos. pp. 1–20.

Jain, A.P., Idris, M. and Meena, Y.R. 1994. Rodent damage to ber (*Zizyphus mauritiana*) nursery. *Rodent Newsletter,* **18** (3): 13–14.

Jhala, R.C., Patel, Z.P. and Patel, C.B. 1989. Rodent damage to ber and cashewnut trees. *Rodent Newsletter,* **13**: 4–5.

Kapadia, M.N. 1995. Rodent damage to coconut plantations in Gujarat. *Rodent Newsletter*, **19**: 6–7.

Kotadia, V.S., Virani, V.R., Vyas, H.J. and Mittal, V.P. 1993. Evaluation of some rodenticides in tomato. *Rodent Newsletter*, 17(1–2): 12–14.

Koshy, M.P. and Phillip, V.M. 1995. Rodent pests of horticultural plantation crops in Pathanamthitta district, Kerala. *Rodent Newsletter*, **19**: 6–7.

Kumar, P.; Pasahan, S.C.; Sabhkij, V.P. and Singhal, R.K. 1995. Efficacy of live trapping for the rodent management in vegetable crops. *Rodent Newsletter*, **19** (1): 8–9.

Malhi, C.S., and Sheikher, C. 1989. Rodent damage and control in guava (*Psidium guava*) orchards at Srinagar (Garhwal Himalayas). *Journal of Tree Sciences,* **8**: 27–30.

Manjon, M.C.; Camarasa, L.; Torres, M.P.; Polop, J.J. and Kravetz, F.O. 1983. Estudia de densidad y desplazamiento de reodores en un campo de girasol. *Symposium Zoologia Economica y Vertebrados Como de Plagas de Agricultura IX Congreso Latinoamerican de Zoologia, Ariquipa, Peru*, 9-15 de Octubre, 1983. pp. 156–172.

Mathur, R.P.; Kashyap, N. and Mathur, M. 1992. Bioefficacy of anaticoagulant rodenticides. *In: Rodents in Indian Agriculture*. I. Prakash and P.K. Ghosh (eds). Vol. I. Scientific Publishers, Jodhpur. pp. 517–583.

Milan, P.P. 1990. Evaluation of control methods for rats in Philippine coconut plantation. *Current Mammology,* **2**: 89–17.

Moran, S. And Keidar, H. 1993. Checklist of vertebrate damage to agriculture in Israel. *Crop Protection,* **12**: 173–182.

Parshad, V.R., Ramanna, D.V. and Chopra, G. 1984. Rodent damage in almond nurseries. *Rodent Newsletter,* **8**: 3.

Patel, N.; Tripathi, R.S.; Parveen, F.; Jain, A.P. and Mathur, N. 1993. Vertebrate pest damage to pomegranate in arid horticulture. *Rodent Newsletter*, **17** (1–2): 6–7.

Pearson, K. and Forshey, C.G. 1978. Effect of pine vole damage on tree vigour and fruit yield on New York orchards. *Hort. Science,* **13**: 56–57.

Prakash, I. and Mathur, R.P. 1992. Acute rodenticides. *In: Rodents in Indian Agriculture.* I. Prakash and P.K. Ghosh (eds). Vol.I. Scientific Publishers, Jodhpur, pp. 497– 515.

Ramesh, P. and Katiyar, R.N. 1985. Short tailed bandicoot rat, *Nesokia indica* G. damage in tomato. *Rodent Newsletter,* **9**(1): 6.

Rangareddy, A. 1995. Species composition of rodent pests and their management in coconut in Godavari delta. *Rodent Newsletter,* **19**: 10.

Richmond, M.E.; Forshey, C.G.; Mahaffy, L.A. and Miller, P.N. 1987. Effect of differential pine vole populations on growth and yield of McIntosh apple trees. *Proceedings of the East Wildlife Damage Control Conference,* **3**: 296–304.

Samiayyan, K. And Chandrasekharan, B. 1994. Effect of bromadiolone in rodent control in coconut trees. *Rodent Newsletter,* **18**(4): 1–2.

Saxena, Y. and Sahai, K. 1990. Rodent activity in Rabi crops in Bharatpur region. *Rodent Newsletter,* **14**: 3–4.

Sheikher, C. and Jain, S.D. 1990. Rodents of Himachal Pradesh: A report. *Rodent Newsletter,* **14**: 1.

Sheikher, C. and Jain, S.D. 1991. Rodent damage and control in pecan orchards. *Proceedings of the Indian National Science Academy,* **B57**: 391–396.

Sheikher, C. and Jain, S.D. 1992a. Rodent pests of temperate fruit orchards and their management. *Proceedings of the Golden Jubilee National Seminar*; Sept. 4–5, 1992, Nationa! Horticulture Board, Gurgaon. pp. 291–296.

Sheikher, C. and Jain, S.D. 1992b. Efficacy of live-trapping as a control method in vegetable crops. *Tropical Pest Management,* **38:** 103- 105.

Sheikher, C. and Jain, S.D. 1996. Mode of application and performance of rodenticides in vegetable crops. *Indian Journal of Agricultural Sciences,* **66**: 437–440.

Sheikher, C. and Jain, S.D. 1997. Rodents in cauliflower and cabbage: population, damage and control. *International Journal of Pest Management*, **3**: 63–69.

Singh, R. and Saxena, Y. 1984. A study on rodent losses in tomato crop. *Rodent Newsletter,* **8**: 15.

Singh, Y.P. 1996. Rodent damage in vegetable crops at medium high altitude. *Rodent Newsletter,* **20** (2–4): 5- 6.

Singh, Y.P.; Kumar, D. and Gangwar, S.K. 1994. Status paper on rodents of North Eastern hill region and their management. *Rodent Newsletter*, **18** (4): 2–3.

Sridhara, S. 1992. Rodent pest management in rice, plantation crops, horticultural crops and poultry. *Rodent Newsletter,* **16**(1): 12–14.

Subiah, K.S. 1983 a. Rat damage to oil palm plantation in Little Andaman. *Rodent Newsletter*, 7(1–2): 3- 4.

Subiah, K.S. 1983 b. Rodent problem in oil palm plantation in Hut Bay, Little Andaman and suggested control measures. *Indian Journal of Farm Chemicals,* **1**: 32–40.

Swihart, R.K. and Picone, P.M. 1994. Damage to apple trees associated with woodchuck burrows in orchards. *Journal of Wildlife Management,* **58**: 357–360.

Thomas, M.; Khatri, A.K. and Pachori, R. 1986. Control of bandicoots in tomato field by aluminium phosphide. *Rodent Newsletter,* **10** (1–4): 6.

Tobin, M.E. 1990. Rat control in macadamia orchards. *Proceedings of the 30th Hawaii Macadamia Nut Association, Annual Meeting*. pp. 18–23.

Tobin, M.E. 1992. Rodent damage in Hawaiian macadamia orchards. *In: Proceedings of the 15th Vertebrate Pest Conference*. J. E. Borrecco and R.E. Marsh (eds). University of California, Davis. pp. 272–276.

Tobin, M.E. 1993. Rodent control research in Hawaiian macadamia orchards. *In: Proceedings of the First International Macadamia Research Conference*. H.C. Bittenbender (ed). July 28- 30, 1992. pp. 54–57.

Tobin, M.E. 1995. Rat biology and control of Macadamia orchards. *Proceedings of the Hawaiian Macadamia Nut Association, Annual Meeting,* pp. 54–56.

Tobin, M.E.; Koehler, A.E.; Sugihara, R.T.; Ueunten, G.R. and Yamaguchi, A.M. 1993. Effect of trapping on rat populations and subsequent damage and yields of macadamia nuts. *Crop Protection,* **12**: 243–248.

Tobin, M.E.; Sugihara, R.T. and Engeman, R.M. 1994. Effect of initial rat captures on subsequent capture success of traps. *In: Proceedings of the 16th Vertebrate Pest Conference*. W.S. Halverson and A.C. Crabb (eds). University Of California, Davis, pp. 101–105.

Tobin, M.E.; Sugilhara, R.T. and Koehler, A.E. 1996. Seasonal activity and movement of *Rattus rattus* (Rodentia, Muridae) in a Hawaiian macadamia orchard. *Mammalia,* **60**: 3–13.

Tobin, M.E.; Koehler, A.E. and Sugihara, R.T. 1997. Effect of simulated rat damage on yields of macadamia trees. *Crop Protection,* **16**: 203–208.

Uchida, T.A. 1969. Rat control procedures on the Pacific Islands with special reference to the efficacy of biological control agents. II. Efficacy of Japanese weesel, *Mustela sibirica italitsi* Temminck and Schlegal, as a rat control agent in the Ryukyus. *Journal of the Faculty of Agriculture, Kyushu University,* **15**: 355–385.

Valencia, D., Elias, D.J. and Ospina, J.A. 1994. Rodent pests in Colombian agriculture. *In: Proceedings of the 16th Vertebrate Pest Conference*. W.S. Halverson and A.C. Crabb (eds). University of California, Davis. pp. 92–94.

*Pest Management in Horticulture Crops*
L.R. Verma, A.K. Verma and D.C. Gautam (eds)


## *Chapter 26*

# Birds as Pests of Horticultural Crops

*M.L. Narang*

## INTRODUCTION

BIRDS CONSTITUTE AN important component of most of the ecosystems. In the past when natural ecosystems offered a variety of food to the birds, the damage caused by them to agro-horticultural crops was negligible and was often ignored. But the practice of monoculturing of various crops to feed the rising population has brought about a significant change in the above scenario. At present, the food in the ecosystems (now converted into agroecosystems) is available in abundance. This has created favorable conditions for a few granivorous and frugivorous species of birds whose populations have increased and have attained the status of pests. Such avian species often come in conflict with man and, therefore, require management. In general, the bird community in an agricultural/horticultural area is characterized by overdominance of a few granivorous/frugivorous or omnivorous species while the rest of the species are represented in very small numbers. Some of which are even rare (Dhindsa and Saini, 1994). Dominance by a few common species is typical of disturbed habitats (MacArthur, 1972) and agro/horticultural areas are highly disturbed habitats.

Studies on bird damage to horticultural crops have been undertaken in various parts of India chiefly in Punjab (Ramzan and Toor, 1972; Simwat and Sidhu, 1973; Toor and Ramzan, 1974; Toor and Sandhu, 1981; Sandhu and Dhindsa, 1982), Himachal Pradesh (Chowdhuri and Seam, 1966; Chopra *et al.*, 1972; Kakar *et al.*, 1984; Bhardwaj, 1991), Andhra Pradesh (Shivanarayan, 1982), Karnataka (Prasad and Verghese, 1985), and Uttar Pradesh (Singh and Kumar, 1982; Narang and Lamba, 1984). Most of these studies are qualitative in nature and no proper assessment has been made regarding the economic status of the species involved. However, some of these studies do indicate the damage potential of various species. Long (1985) reported three species of parrots and a cockatoo damaging fruit crops like apple, pear and plum in Australia. In United States, crows alongwith other species of birds were reported damaging fruits like apple and pecan (Dolbeer *et al.*, 1994; Tobin *et al.*, 1989). In another study conducted at Michigan, 12 species of birds were observed feeding on ripening cherries (Guarino *et al.*, 1994). In New

Zealand, gut analysis of mynas, starling, *Sturnus vulgaris*, and silver eve, *Zosterops lateralis*, revealed that these birds fed upon fruits like apple, peach, nightshade, cherry, grape, pear and vegetables like pea and tomato (Moeed, 1975, 1979, 1980).

## PROBLEM BIRDS IN HORTICULTURE

Mehrotra and Bhatnagar (1979) recorded 143 species of birds as minor to serious pests of agricultural and horticultural crops in India. Whereas, Dhindsa and Saini (1994) reported only 25 of about 1200 species of birds found in India, damaging field crops and fruits. In Pakistan, 10 species of birds have been categorized as pests of particular importance (Brooks and Ahmad, 1990).

These reports suggest that parakeets (*Psittacula* spp.), mynas especially Indian myna, *Acridotheres tristis* (L.) and Bank myna, *Acridotheres ginginianus*, Great hill barbet, *Megalaima virens Swinhoe*, Redbilled blue magpie, *Cissa erythroryncha,* crows, *Corvus* spp. and bulbuls, *Pycnonotus* spp., are the important bird pests of horticultural crops in India. Most of the land birds, barring parakeets, play a dual role in the ecosystem. Some bird species cause moderate to extensive damage to agro-horticultural crops during one season and compensate for the same during another season by feeding on insect pests like grasshoppers, bugs, caterpillars and grubs. Parakeets, of which there are several species in India, are purely granivorous/frugivorous in their dietry habits. Even the young ones of parakeets are fed on fruits and grains (Singh and Kumar, 1982) unlike other species of land birds which feed their young ones with soft insects like caterpillars and grubs, etc.

## BIRD AS PESTS OF FRUIT TREES

Parakeets feed on fruits and vegetables like almond, apple, *ber*, guava, mango, peach, pear, pomegranate, tomato, etc., besides feeding and damaging cereal crops like wheat, rice, sorghum, maize, etc., and oilseeds like mustard, sunflower and groundnut. These birds, feeding in large flocks, often inflict far more damage to the crop attacked than they actually consume. Studies carried out in apple orchard in Himachal Pradesh (Narang and Chandel, 1995) revealed that three species of parakeets, viz., slatyheaded parakeet, blossomheaded parakeet and large Indian parakeet caused about 15–29 % damage to early maturing cultivars of apple like Red Gold and Golden Delicious grown mostly as pollinizers. However, the damage was negligible in late maturing and commercially important cultivars like Royal Delicious. In Pakistan, Rose-ringed parakeets are reported to damage soft fruits like guava, mango, orange, loquate, pear, orange, etc. (Brooks and Ahmad, 1990; Khan and Ahmad, 1990). In Australia, three species of parrots and a cockatoo were reported to damage fruits, mainly apple and to a lesser extent pear, plum and nectarines (Long, 1985). More than 1.75 % of the total fruits grown in the orchard were damaged by birds and Golden Delicious and Red Delicious suffered more damage as compared to other varieties. It is because of their depredatory dietry habits that Ali (1949) described.

Parakeets, especially Rose-ringed parakeet and Blossom-headed parakeet, as the most destructive avian pests to cereal crops and fruits of all kinds in India and recommended that their numbers be kept under strict check. In fact, there is a need to declare both these species as 'vermin' as was done long ago in Australia where Port Lincoln parrot, Western rosella and Red-capped parrot were declared 'vermin' during 1920s, 1930s and 1940s, respectively (Long, 1985). Similarly, house crows and jungle crow are omnivorous and as such feed upon fruits like almond, grapes, pear, date palm, kinnow, papaya and jungle jack (Chopra *et al*., 1972; Simwat and Sidhu, 1973; Toor and Ramzan, 1974; Sandhu and Dhindsa, 1982; Prasad and Verghese, 1985; Anonymous, 1996 b) and cereal crops seem to posses no compensatory value.

Mynas on the other hand play a dual role in nature. These birds feed upon a variety of fruits and vegetables (Simwat and Sidhu, 1973; Toor and Ramzan, 1974; Narang and Lamba, 1984) but normally compensate for the same by feeding on insect pests like grasshoppers, bugs, caterpillars and grubs (Narang and Lamba, 1984). In New Zealand nestlings of starling, *Sturnus vulgaris* and myna, *Acridotheres tristis* have been recorded to feed mainly on insects and to lesser extent on fruits like apple, peach, nightshade and cherry (Moeed, 1975). Most of the damage caused by mynas is localised. Bank myna, *A. ginginianus* for instance, is reported to cause extensive damage to grapes in Punjab (Toor and Ramzan, 1974). But a detailed gravimetric and volumetric analysis of food of this species carried out at Dehradun in Uttar Pradesh revealed that these birds mostly fed on insects which were injurious to agricultural crops (Narang and Lamba, 1984). The study further revealed that although mynas were omnivorous in their dietary habits yet they did not depend on fruits and vegetables for their subsistence. Moreover, the economic losses inflicted by various species of myna were not serious enough to label these birds as pests in the real sense.

Three species of bulbuls and *Hypsipetes* sp. have been reported to damage fruit crops like apple, pear, date palm, mulberry and grapes (Table 1). In Pakistan, Redvented bulbul is reported to damage soft fruits like persimmon and after initial damage by parakeets, these birds also feed on citrus fruits (Khan and Ahmad, 1990). Bulbuls can inflict varying degrees of losses to soft fruits but in the case of thick skinned fruits like apple, pear and citrus the birds have to wait for large birds like parakeets and crows to rip open the fruits before they can feed upon the soft pulp of such fruits. The abundance of bulbuls on fruit crops like apple and pear can therefore, be highly misleading. For instance, in a study conducted at Solan, Himachal Pradesh orchards had relatively more abundance of bulbuls than parakeets although their damage potential was negligible as compared to parakeets (Narang and Chandel, 1995).

Additionally, some more species of birds have also been recorded as pests of various horticultural crops. These include redbilled blue magpie, great hill barbet, koel, rosefinch, brownfronted pied woodpecker, house sparrow, common and jungle babblers, etc. The great hill barbet in particular is a big nuisance in apple and pear orchards in Shimla and Kullu districts of Himachal Pradesh.

**Table 1.** Bird damage to horticultural crops in India

| *Fruit/vegetable* | *Bird species* | *Extent of losses (%)* | *Reference* |
|---|---|---|---|
| Almond | Rose-ringed parakeet, Slatyheaded, Parakeet, Himalayan jungle crow | 7 | Sandhu and Dhindsa, 1982; Bhardwaj, 1991 |
| Apple | Slathyheaded parakeet, Blossomheaded parakeet, Large Indian parakeet, Great hill barbet, Whitecheeked bulbul, Redvented bulbul, Redbilled blue magpie, Brownfronted pied woodpecker, Himalayan jungle crow | 15–29 | Narang and Chandel, 1995; Kakar *et al*., 1984; Chouduri and Seam, 1966 |

| | | | |
|---|---|---|---|
| Ber | Rose-ringed parakeet, Brahminy myna, Pied myna | — | Simwat and Sidhu, 1973; Narang and Lamba, 1984 |
| Cherry | Rose-ringed parakeet | — | Simwat and Sidhu, 1973 |
| Date palm | Rose-ringed parakeet, House crow, Whitecheeked bulbul | — | Gupta, 1980; Anonymous, 1996a |
| Grapes | Mynas especially Bank myna, House sparrow, house crow, Rose-ringed parakeet, Koel, Redvented bulbul, Common babbler, Jungle babbler | 12.5 | Chopra *et al.* 1972; Simwat and Sidhu, 1973; Toor and Ramzan, 1974 |
| Guava | Rose-ringed parakeet, Blossomheaded parakeet, House crow | 20–40 | Ramzan and Toor, 1972; Chopra *et al.*, 1972 Singh and Kumar, 1982 Shivanarayan, 1982 Prasad and Verghese, 1985 |
| Jamun | Rose-ringed parakeet, Brahminy myna, Pied myna, Jungle myna | — | Simwat and Sidhu, 1973; Narang and Lamba, 1984 |
| Kinnow | Large Indian parakeet, Jungle crow | — | Anonymous, 1996 b |
| Mango | Rose-ringed parakeet | — | Simwat and Sidhu, 1973 |
| Mulberry | House crow, Rose-ringed parakeet, Black bulbul, Whitecheeked bulbul, Redvented bulbul, Koel, Jungle myna, Rosefinch, Redbilled blue magpie | — | Toor and Ramzan, 1974 |
| Peach | Rose-ringed parakeet, House crow | 32 | Simwat and Sidhu, 1973; Toor and Sandhu, 1981; Kakar *et al.*, 1984 |
| Pear | Slatyheaded parakeet, Blossomheaded parakeet | 15–20 | Kakar *et al.*, 1984 |
| Papaya | Rose-ringed parakeet | | Simwat and Sidhu, 1973; Prasad and Verghese, 1985 |
| Pomegranate | Rose-ringed parakeet | — | Simwat and Sidhu, 1973 |
| Plum | Slatyheaded parakeet | — | Kakar et al., 1984 |
| Chillies | Common myna, bank myna | | Narang and Lamba, 1984 |
| Peas | Blue rock pigeon, Bank myna, Common myna, Brahminy myna | 54 | Narang and Lamba, 1984; Saini and Kaur, 1986; Saini and Toor, 1991 |
| Tomato | Rose-ringed parakeet | — | Simwat and Sidhu, 1973 |

## BIRDS AS PESTS OF VEGETABLE CROPS

Birds like pigeons cause extensive damage to pea (Saini and Kaur, 1986; Saini and Toor, 1991). Different species of myna have been recorded to feed upon pea and chillies (Narang and Lamba, 1984). Redvented bulbul is reported to peck at leaves of crops like cabbage and cauliflower in Pakistan (Khan and Ahmad, 1990). Vegetables like asparagus, pea and tomato have been recorded from the gizzards of starling, *S. vulgaris* in New Zealand (Moeed, 1980). A large lacunae, however, remains in our knowledge on the damage caused to vegetable crops by birds as detailed studies seem to have not been undertaken in this regard in India.

## MANAGEMENT

Human-bird conflict is an age old phenomenon. With the rapid expansion of agriculture, the losses caused by birds to agricultural produce have become more noticeable and economically important. Management of depredatory species of birds has, therefore, become an important aspect of wildlife management. Increased awareness about the environment and wildlife has resulted in stress being laid on non-lethal control of pest birds. Emphasis is also being laid on an integrated approach to bird pest management whereby several control measures are used in combination to minimize the losses.

The use of frightening or scaring devices, toxic baits and repellent chemicals has become common to manage depredatory activities of pest birds. In addition, cultural practices, habitat manipulation and use of nets have also been recommended as bird control measures.

### Frightening or scaring devices

The most widely used frightening device for bird control is the use of automatic bird scaring exploders which produces a loud explosion at timed intervals. Dolbeer (1994) recommended the use of one exploder for a two hectare field. To be more effective the location of this device should be changed after every 2–3 days. Shell crackers, rockets and even gun fire can also be used occasionally to supplement the impact of exploders.

Visual repellents like helium filled balloons with eyespots mimicking predatory birds, black plastic flags hung from a tall pole and use of scare-crows have also been used with varying degrees of success against depredatory birds in the United States and Japan (Nakamura, 1982). In Himachal Pradesh, white polythene bags are tied to apple trees to scare birds. The position of scaring devices should be changed occasionally to prevent habituation by birds.

Another technique employed for scaring birds away from the field is the use of alarm and distress calls of birds broadcast over a public address system. These amplified calls have been found to be effective in repelling birds from the target area (Rao and Jayavardhan, 1987). However, if the calls are played frequently the birds become habituated and tend to ignore them. The most promising and cost effective bird scaring technique that has found wide acceptability in India and abroad is the use of reflective ribbons or bird scaring tape. The elastic, metallic shining tape with one side red and other silver grey reflects sunlight to give flashing effect. To the flying birds it gives the impression as if the crop is on fire and induces fright in them. The fluttering of these stretched ribbons due to wind produces a loud humming sound which supplements the scaring effect. The use of reflective ribbons was found to be quite effective against blackbird population in the United States (Dolbeer, 1994). Reflective tapes have also been found promising against birds in apple orchards in Himachal Pradesh. It is important to remove scaring devices from the field after the crop is harvested to prevent habituation by birds.

### Repellent chemicals

Among the repellent chemicals, methiocarb (Mesurol) has been widely recommended to reduce bird damage in certain fruit trees and to treat seeds of agricultural crops. This chemical causes illness or adverse physiological response in birds upon ingestion which results in avoidance of the treated food. Methiocarb has been used as spray on ripening cherries, blueberries and grapes (Dolbeer, 1994). On ripening cherries, 0.16 % methiocarb sprayed was found to reduce bird damage to the extent of 65.6 % in Michigan, USA (Guarino *et al.*, 1974). Its residues in the pulp averaged 1.23 ppm after 15 days of treatment, well below the tolerance level of 7 ppm. Another repellent chemical, landrin, tested against frugivorous birds (Kakar *et al.*, 1986) was found to leave toxic residues and therefore, could not be recommended. Some fungicides like thiram, copper oxychloride and dithane M-45 are also reported as good repellents for birds like sparrows and parakeets (Vardhani and Rao, 1992). Some neem (*Azadirachta indica* A. Juss.) formulations have also been tried as bird repellents in India. Neem cake extract @ 300 g/litre has been found to repell parakeets and house crows from maize crop (Shivanarayan and Rao, 1988). Despite these studies, our knowledge on non-lethal control of birds is preliminary and more studies are needed in this direction.

### Modification of habitat and cultural practices

Habitat manipulation and changes in cultural practices though labor intensive, have been found to work well in certain situations. Thinning or pruning of trees can cause roosting birds such as parakeets to disperse, thereby reducing damage to fruit crops, the use of lure crops where pest birds are encouraged to feed in order to save commercially important crops, cultivation of bird resistant crop varieties etc. have been recommended as cultural practice (Mehrotra and Bhatnagar, 1979; Dolbeer, 1994). However, these methods should be evaluated under field conditions before they can be recommended as control measures.

### Netting

The use of nylon nets of various mesh sizes is effective in excluding birds from fruit plants such as blueberries and grapes (Fuller Perrine and Tobin, 1993). In Himachal Pradesh, anti-hail nets have been found to afford protection against bird attack in apple crop. But the main drawback of this technique is that nets are cost prohibitive and cannot be employed on a large scale.

## INTEGRATED APPROACH

The methods reported above afford certain degree of protection against harmful birds. However, any one method may not provide complete protection against depredatory birds. Emphases are, therefore, laid on the integrated use of various techniques to achieve better results.

Bird control in agricultural/horticultural crops is a difficult task considering the mobile nature of birds. Adverse public opinion towards lethal control of birds and environmental regulations have made it more complex (Dolbeer *et al.*, 1994). There is, therefore, an urgent need to develop environmentally safe non-lethal techniques which are based on sound economic and ecological principles and implement them as an integral part of overall wildlife management programmes.

## REFERENCES


Ali, S. 1949. Birds, friends and foe of the cultivators. *Indian Farming,* **10** (9): 385–387.

Anonymous 1996 a. *Annual report of All India Network Programme on Agricultural Ornithology*. ICAR, APAU, Rajendranagar, Hyderabad. pp. 6–7.

Anonymous, 1996 b. *Research progress report. Regional Horticultural Research Station.* Dhulakuan, Dr. Y.S. Parmar University of Horticulture and Forestry, Solan.

Bhardwaj, S.P. 1991. Indian house crow damage to almond in Himachal Pradesh, India. *Tropical Pest Management,* **37**(1): 101–102.

Brooks, J.E. and Ahmad, E. 1990. Pest birds of Pakistan: Identification and distribution. *In: Vertebrate Pest Management.* J.E. Brooks, Ezaj Ahmad, Iftikhar Hussain, Shahid Kunir and A.A. Khan (eds). Pakistan Agricultural Research Council, Islamabad, Pakistan. pp. 173–180.

Chopra, S.K., Mahajan, J.R. and Vir, B.S. 1972. Birds are serious menace to horticultural crops. *Indian Farming,* **21** (10): 24–27.

Chowdhuri, A.N. and Seam, J.C. 1966. Hill birds a growing menace to temperate fruits in Simla hills. *Himachal Horticulture,* 13–16.

Dhindsa, M.S. and Saini, H.K. 1994. Agricultural ornithology an Indian perspective. *Journal of Biosci.* 19(4): 391–402.

Dolbeer, A.R., Holler, N.R. and Hawthorne, D.W. 1994. Identification and control of wildlife damage. *In:* Research *and management techniques for wildlife and habitats*. T.A. Bookhout (ed). The Wildlife Society. Beethesda. Maryland. pp. 474–506.

Fuller-Perrine, L.D. and Tobin, M.E. 1993. A method for applying and removing bird-exclusion netting in commercial vinevards. *Wildlife Society Bulletin,* **21**: 47–51.

Guarino, J.L., Shake, W.F. and Schaffer, E.W. 1994. Reducing bird damage to ripening cherries with methiocarb. *Journal of Wildlife Management*, **38** (2): 338–342.

Gupta, M.R. 1980. Resistance of some promising date palm cultivars against rain fall and bird damage. *Punjab Horticulture Journal,* **20**: 74–77.

Kakar, K.L.; Amit Nath and Dogra, G.S. 1984. Studies on the feeding behaviour of birds and their control under mid-hill conditions of Himachal Pradesh. *Proceedings of the National Symposium on Temperate fruits.* Nauni, Solan, H.P. pp.389–395.

Khan, A.A. and Ahmad, E. 1990. Damage pattern of pest birds and assessment methods. *In: Vertebrate Pest Management.* Joe E. Brooks, Ezaj Ahmad, Iftikhar Hussain, Shahid Munir and A. Azij Khan (eds). Pakistan Agricultural Research Council. Islamabad. Pakistan. pp. 181–186.

Long, J.L. 1985. Damage to cultivated fruits by parrots in the south of Western Australia. *Australian Wildlife Research,* **12**: 75–80.

MacArthur, R.H. 1972. *Geographic ecology*. New York: Harper and Row.

Mehrotra, K.N. and Bhatnagar, R.K. 1979. *Status of economic ornithology in India.* ICAR, New Delhi. p 79.

Moeed, A. 1975. Diets of nestling starlings and mynas at Havelock north, Hawke's Bay, NewZealand. *Notornis,* **22** (4): 291–294.

Moeed, A. 1979. Foods of the Silvereye (*Zosterops lateralis*: Aves) near Nelson, New Zealand. *New Zealand Journal of Zoology,* **6**: 475–479.

Moeed, A. 1980. Diets of adult and nestling starlings (*Sturnus vulgaris*) in Hawke's Bay. New Zealand. *New Zealand Journal of Zoology,* **7**: 247–256.

Nakamura, K. 1982. Bird damage in Japan and its control methods. *Farming Japan,* **16** (2): 44–50.

Narang, M.L. and Lamba, B.S. 1984. A contribution to the food habits of some Indian Mynas (Aves). *Records of the Zoological Survey of India*. Misc. Publication. Zoological Survey of India, Calcutta. Occasional Paper No. **44**: 1–76.

Narang, M.L. and Chandel, J.S. 1995. Preliminary studies on bird damage to apple orchards in Himachal Pradesh. *Biannual Conference of Ornithological Society of India.* Srinagar, Garhwal, U.P. Environment and Birds. pp. 54–55.

Prasad, V.G. and Verghese, A. 1985. Birds as pests of horticultural crops. *Bulletin of Entomology,* **26** (1): 94–95.

Ramzan, M. and Toor, H.S. 1972. Studies on damage to guava fruits due to Rose-ringed parakeet, *Psittcula krameri* (Scopoli), at Ludhiana (Pb.). *Punjab Horticulture Journal,* **12**: 144–145.

Rao, P.S. and Jayawardhan, M. 1987. Effect of bioaccoustics in the dispersal of bird pests from maize crop. *Pavo,* **25**: 33–36.

Saini, M.S. and Kaur, H. 1986. Damage by blue rock-pigeon to peas. *Indian Journal of Agricultural Science,* **56**: 486–487.

Saini, H.K. and Toor H.S. 1991. Feeding ecology and damage potential of feral pigeons *Columba livia* in agricultural habitat. *Gerfaut,* **81**: 195–206.

Sandhu, P.S. and Dhindsa, M.S. 1982. Damage by rose-ringed parakeet and some other animal pests to almond at Ludhiana, Punjab. *Indian Journal of Agricultural Sciences,* **52**: 779–781.

Shivanarayan, N. 1982. The rose-ringed parakeet, the problem bird in agriculture and its management. *In: Proc. Seminar on management of problem birds in aviation and agriculture.* RA Aggrawal and RK Bhatnagar (eds). ICAR, New Delhi.

Shivanarayan, N. and Rao, M.A. 1988. Efficacy of neem cake as bird pest repellent on maize. *Pavo,* **26**: 49–52.

Simwat, G.S. and Sidhu, A.S. 1973. Note on the feeding habits of rose-ringed parakeet *Psittacula krameri* (Scopoli). *Indian Journal of Agricultural Sciences,* **43**: 607–609.

Singh, A.K. and Kumar, P. 1982. A study of breeding biology of *Psittacula krameri*. *In: Proc. Seminar on Management of Problem Birds in Aviation and Agriculture.* R.A. Aggrawal and R.K. Bhatnagar (eds.). ICAR, New Delhi. pp. 159–163.

Tobin, M.E.; Dolbeer, R.A. and Woronecki, P.P. 1989. Bird damage to apples in the Mid-Hudson valley of New York. *HortScience,* **24**: 859.

Toor, H.S. and Ramzan, M. 1974. A study on grapes lost to birds. *Punjab Horticulture Journal,* **14**: 46–48.

Toor, H.S. and Sandhu, P.S. 1981. Bird damage to peach *Prunus persica* Batsch. *Indian Journal of Ecology,* **8**: 308–310.

Vardhani, B.P. and Rao, P.S. 1992. Evaluation of certain fungicides as repellent against house sparrow, *Passer domesticus* and rose-ringed parakeet, *Psittacula krameri. Pavo,* **30**: 5–8.

*Pest Management in Horticulture Crops*
L.R. Verma, A.K. Verma and D.C. Gautam (eds)


*Chapter 27*

# Pesticide-Honeybee Interactions

*L.R. Verma*

## INTRODUCTION

THE WIDELY INFLUENTIAL United Nations Conference on environment and Development (UNCED) held in Rio points out that despite outward success of green revolution, crop yields are stagnating or declining in many countries, rural poverty is increasing and the natural resource base is being seriously degraded. This conference concluded: 'major adjustments are needed in agricultural, environmental and macro-economy policy at both national and international level to create the conditions for sustainable agriculture and rural development'. According to Food and Agriculture Organization (FAO) 'sustainable development is the management and conservation of the natural resource base, and the orientation of technological and institutional change in such a manner as to ensure the attainment and continued satisfaction of human needs for present and future generations. Such sustainable development (in agriculture, forestry and fisheries sectors) conserves land, water, plant and animal genetic resources, is environmentally non-degrading, technically appropriate, economically viable and socially acceptable'.

In the context of above, conservation and management of native bees and the maintenance of a thriving beekeeping industry are both essential for sustainable agriculture and rural development. Swaminathan (1986), in his forwarding note in FAO Service Bulletin on Apiculture, stated 'in spite of all global resolutions on food security, several hundred million children, women and men are going to bed hungry everyday, particularly in the countries of the South. Since prospects for global food security system appear to be small at the present moment, it will be for the developing countries, characterized by poverty and under-nutrition, to build their own national food security system. In this context, apiculture can play a useful role. At very little expenditure, honeybees will not only provide food and income but will also enhance the productivity of horticultural and other field crops by their pollination activities'.

Honeybees and other native bees play an indispensable role both in agricultural and natural ecosystems because pollination by bees is vital for seed production not only for the majority of our crop plants but also for an unknown number of wild flowers. In the context of sustainable development of agriculture in terms of resource base availability, resource management practices and production flow, different national and international development agencies have identified beekeeping and pollination ecology as such an income, nutrition and food generating activity which offers comparative advantages and is environmentally most friendly. The pollination activities of honeybees boost the productivity of several mountain crops and help in the conservation of forest and grassland ecosystems (Verma, 1990, 1992; Verma and Partap, 1993)

## PESTICIDE POISONING OF HONEYBEES

Beekeeping and the use of pesticides are both essential inputs for modern techniques of agricultural and horticultural production. Productivity of agricultural crops will be seriously impaired if either of these two is ignored. Since the advent of synthetic pesticides several decades ago, the beekeeping industry, both in developed and developing countries, has been incurring heavy losses. In developed countries, large-scale monoculture cultivation of crops and a high degree of mechanization has greatly amplified the problem of honeybee poisoning by pesticides. On the other hand, in developing countries, the basic problem is the lack of information about the harmful effects of pesticides on honeybees. Conflicts often arise when pesticides are applied to a crop at the inappropriate time or by inappropriate means, or when unsafe pesticides are used.

In recent year, pest control problems are becoming more serious and difficult to manage. This is because large areas of land are being used to cultivate exotic cultivars of agricultural and horticultural crops which, under the new environmental conditions, are more susceptible to pests and diseases. In some cases, pests and diseases affecting exotic cultivars/varieties are also introduced either accidentally or through human error, or lack of proper quarantine facilities. Such pests and diseases multiply and spread more rapidly in their new environment because of the absence of their natural enemies such as predators or parasites. Chemical pest control measures are absolutely essential under such circumstances.

The increasing reliance on chemical method for the control of pests and diseases is also creating serious environmental pollution problems, including health hazards to human beings and decline in other non-target, beneficial insect populations. Amongst the latter, natural insect predators, parasites, and insect pollinators especially honeybees, are the primary victims. These beneficial insects are facing extinction in their natural habitat where they greatly contribute to the conservation of biological diversity and render essential ecological services i.e., pollination and maintaining the population levels of different insects pests below economic thresholds.

In developing countries, the basic problem is lack of information and lack of awareness about the harmful effects of biocides to honeybees along with the indiscriminate prescription of pesticides by agricultural scientists and extension workers and blanket and erratic application of pesticides by farmers due to lack of knowledge regarding what, how much to use, and when.

The practice of using honeybees to enhance the productivity levels of different agricultural and horticultural crops has not become very popular among farmers because indiscriminate use of biocides kills large numbers of bees. However, now other methods are available and these ensure the selective use of biocides at some right time and in the appropriate formulation and concentration. As a result of this, the hazards of bee poisoning by pesticides can be reduced to a minimum (Verma and Partap, 1993).

## HISTORICAL PERSPECTIVE

In California, during 1962–77, there was significant effect on bee colonies due to the use of pesticide. Subsequently, such losses became reduced due to proper use of pesticides. There are only a few reports on pollinator losses due to pesticides in India. However, considering the use of pesticides in this country, these losses should be considerable. Lately, several orchardists in Himachal Pradesh have started using domesticated bees for pollination purpose in areas where earlier natural pollinators population was abundant. This indicates that the natural pollinator population is going down, mainly due to the excessive use of pesticides.

Several reviews on bee toxicity have appeared (Mishra and Sihag, 1987; Verma, 1990; Roubik, 1995; Mishra, 1995). No information is available on toxicity to *Apis dorsata* and only two papers reported toxicity of pesticides to *A. florea* (Sorthia and Chari, 1985; Patel *et al*., 1988). Most of the information pertains to *Apis cerana or A. mellifera* (Anonymous, 1992).

## EFFECTS OF PESTICIDES

Pesticides may be absorbed by bees in one or more of the following ways—oral, respiratory and dermal. Oral intake is likely to occur when nectar and/or pollen are contaminated. For a pesticide to be toxic via this route it must be absorbed and the efficiency of absorption depends upon the characteristics of the pesticides. For example, bees could carry lethal doses of organophospate dimethoate in their honey stomachs without showing signs of intoxication.

Contamination of nectar occurs in a number of plants treated with systemic insecticides. However, there is a little danger of contamination through nectar. But, on the other hand, several scientists have reported that dimethoate applied at the rate of 11kg/ha killed the bees as a result of nectar contamination.

The major cause of bee poisoning is the contamination of pollen by microencapsulated insecticides. For example, when penncap-M $^{R}$ capsules are applied to agricultural crops, the capsules are carried by foraging bees to their hives and stored in the brood frames together with pollen; hive bees feed the contaminated pollen to the developing brood which results in loss of the total colony. Foraging bees are killed while collecting and transporting this pollen, young hive bees are killed while storing and feeding the contaminated pollen and the brood is killed by the poisoned food.

Insecticides such as DDVP, some organic chlorine pesticides such as chlordane, and compounds such a nicotine can be present in sufficient concentration in the air to be toxic to bees by way of their tracheal system. Another possible problem in this regard is the absorption and subsequent release of volatile pesticides with fumigant properties. Beewax has excellent absorptive properties. Combs exposed to DDVP for 48 hours absorb sufficient pesticide to kill bees that are exposed to these combs within two to six minutes.

Direct contact is probably the major way in which pesticides are absorbed by bees. Interception of pesticide droplets in the air during spraying operations and contact with sprayed surfaces are the most likely sources of contamination. The toxicity of the airborne droplets varies according to the method of application and the amount of pesticide available to bees, decreases with increasing absorption of pesticides by the surface.

Bees lose their sense of time when exposed to sublethal doses of pesticides like parathion. However, it is not clear whether this phenomenon is the result of changes in the 'internal clock' of the poisoned bees or in the manner in which they communicate this 'time' to other bees. Disruption in the communication of distance also occurs due to pesticide poisoning.

Outbreaks of European foulbrood and sacbrood virus infections were observed to follow applications of carbaryl insecticides in the foraging area of the affected colonies. The first records of the occurrence of Chalkbrood disease came from colonies that were exposed to fenetrothion spray (Verma and Partap, 1993).

### Symptoms of bee poisoning

One of the obvious signs of poisoning is the presence of a large number of dead or dying bees at the hive entrance. These adult bees are foragers who have been exposed to pesticides sprayed on flowering plants. The mortality figures in table 1 are used as guidelines to assess the extent of bee poisoning by pesticides.

**Table 1.** Comparative hazard of pesticides to honeybee colonies from field applications of pesticides.[1 2]

| *No. of bees killed per colony* | *Comparative hazard to colony* |
|---|---|
| < 150 | L = low |
| 150–350 | ML = moderately low |
| 350–750 | M =moderate (can stand = DDT) |
| 750–1250 | MH = moderately high |
| 1250–1750 | H = high (can stand once per season). This value represents the LD50. |
| 1750–2750 | VH = very high |
| 2750–4000 | EH = extremely high |
| > 4000 | DH= disastrously high |

[1] *Based on field observations of the viability of hives after various spray regimes. The absolute viability of a given hive may vary widely.*
[2] *Source: Atkins (1980)*

As a result of organophosphorous poisoning, dying bees extend their tongues through which nectar is regurgitated and a moist and sticky mass of dead bees is often found at the hive entrance. Fast acting insecticides kill foraging bees in the field itself, and only a small number of such bees manage to return to the hive. Sometimes, the whole bee colony may die instantly. Strong bee colonies suffer greater losses due to pesticide poisoning than weaker ones because the former have a larger number of foraging bees.

Foraging bees often carry residual pesticides in their pollen loads while returning to the hive. The behaviour of bees in the hive changes abruptly as a result, honeybees in such colonies become more agitated or aggressive. As and when the hive with pesticide affected forager bees is opened, they often fly off the top bars of the hive and sometimes straight into the face of the beekeeper handling them. Other symptoms of pesticide poisoning include suffocation, paralysis, and abnormal, jerky, or spinning movements. Carbaryl poisoning causes bees to crawl around at the hive entrance, they lose their ability to fly and ultimately die in two to three days after poisoning.

Nurse bees in pesticide-affected colonies lose their ability to clean dead bees from the hive, as a result of which the hive entrance is completely blocked. (Table 2)

**Table 2.** Symptoms of insecticide poisoning in honeybees.

| *Insecticide* | *Symptoms* |
|---|---|
| Organophospharous | Rugurgitation, (disorientation, lassi-tude, distended abdomen, erratic cleaning attempts, tumbling, wings held away from body but hooked together, more than 50% of the bees die at the colony. |
| Carbamates | Aggressiveness, erratic movements, inability to fly, stupefaction (as though chilled), paralysis, more than 50% of the bees die at the colony. |
| Organochlorines | Erratic movements, abnormal activities, trembling, dragging of hind legs, wings held away from body but hooked together, more than 50% of the bees usually die away from the colony. |
| Uncouplers of oxidative phosphorylation | Similar to the chlorinated hydrocarbon insecticides but regurgitation may occur. More than 50% of the bees usually die at the colony. |
| Botanical insecticides | Regurgitation, erratic movements, inability to fly, stupefaction, paralysis. More than 50% of the bees usually die away from the hive. |

*Source:* Atkins (1975)

Pesticides poisoning also affects the colony strength because there is a break in the brood-rearing cycle and often dead or deserted colonies cease foraging, as a result of which there is a sharp decline in food storage, and incoming foragers are attacked at the hive entrance by other bees. (Atkins, 1980).

### Factors affecting toxicity of insecticide to honeybees

The susceptibility of honeybees to insecticides is governed by several factors:

#### *Environment*

Environment plays an important part in bee-pesticide interaction. Its components such as temperature, humidity, rainfall etc., regulate the bee toxicity, persistence and degradation of insecticides under field conditions.

#### *Relative toxicity*

Different chemicals undergo metabolism by internal detoxification mechanisms such as mixed function oxygenases, glutathione mediated systems or by hydrolysis by esterases. In honeybees, relative toxicity is greatly influenced by the health or stress of the colony.

***Formulation***

The types of formulation of the insecticide regulate incidence of contact and the quantity of the poision picked up or ingested by the insect. Johansen (1977) classified formulation on the basis of the decreasing order of their toxicity as dusts < wettable powder < emulsifiable concentrate < soluble powder < granules.

***Frequency of Exposure***

Levels of foraging activity, number and duration of field visits influence susceptibility of foragers to the toxic residues.

***Herbicides and Fungicides***

Fungicides are comparatively harmless, on the other hand herbicides as such as 2,4,-D adversely effect honey bee foraging due to decrease or cessation of nectar secretion in supporting flora

***Insecticides***

Phytotoxic effects reduce floral nectar, pollen and plant odours, thus reducing the bee activity (Nath, 1997)

## STRATEGIES TO REDUCE BEE LOSSES

It is possible to use simple strategies to reduce bee losses due to pesticides. It involves pest management with non-chemical methods; for example, using cultural methods of control like crop diversification, clean cultivation, resistant crop varieties and biocontrol agents like bacteria (*Bacillus thuringiensis*) and insect viruses are safe to honeybees.

Pesticides, if needed, should have low hazard to bees like chitin inhibitors (Chandel and Gupta, 1992; Gupta and Chandel, 1995), neem preparations (Schmutterer and Holst, 1987; Nauman *et al.*, 1994). The effect of diflubenzuron and penfluron on third and fourth instar larvae and pupae, oviposition and on sealed and unsealed brood of Indian and Italian bees showed medium lethal dose of 1.5–6.0 $\mu$g that compares quite favourably with so called safe conventional insecticides. Adults fed on 50ppm of diflubenzuron and penfluron in sugar syrup laid more eggs than untreated control. The differences in unsealed, sealed broods and effects on honey and pollen stores were not significant. The effects of these chemicals were latent such as reduction in weight and hypopharyngeal gland activity.

Apply high hazard pesticides at a time when there are no bees on the crop. Bees are active during a particular time of the day and if spray on crops is outside this time then the pesticide is less hazardous to bees (Johansen, 1977). All pesticide applications from 2100hr to 0400 hrs are not only safe to bees but give better control of most of the pests that are active at night. Environmental temperature is equally important. Do not spray crop when the night temperature drops below 10–12°C or when there is high humidity (Erickson and Erickson, 1983). Transport bee colony more than a kilometer away from the site of pesticide application. Clear field of all the flowering weeds which may attract bees before spraying the field. Prefer safe formulations such as granules, solutions and emulsifiable concentrates over dust, wettable powders and water dispersible powders. Later formulations are similar to the pollen grains and easily stick on to foraging bees. Ground application of pesticides is safer compared to aerial application. Similarly, fine spray applications are safer than large droplet spray application and combined spray applications are better than individual spray applications as it exposes bees to pesticides less number of times.

Despite best precautions, if bees get poisoned by pesticides, then feed them with pesticide free pollen and sugar syrup and immediately shift hives to a safe place, far away from the treated area. Contaminated pollen is stored in hives, removed by dipping the combs in water and washing by slight shaking.

Roubik (1995) classified pesticides according to their toxicity to bees. Highly toxic chemicals (highly dangerous to bees), moderately toxic chemicals (dangerous to bees, apply at night) and moderately toxic chemicals for bees (use late in the day or when bees are not in flight)

**Protective Measures**

As far as possible, biocides should not be applied during the blooming period. Pesticides that have short residual effects and those that are less hazardous to honeybees should be selected.

- Ignorance and the absence of extension programmes to educate farmers about the harmful effects of pesticides are the biggest problems in developing countries. Both farmers and beekeepers should be properly informed regarding pesticide application schedules and how to reduce poisoning in a particular area.
- Broad-spectrum pesticides should be avoided as they are much more hazardous than selective pesticides which are safer for bees and other beneficial insects.
- Night or early morning application of pesticides is always desirable because foraging honeybees are in the hive at that time and thus out of danger. Night application of pesticides allows adequate time for the pesticide to dissipate or break down into substances that are non-toxic to bees.
- It has been recognized that spray of liquid formulations are safer than dust of wettable powder formulations. The death rate is six times higher if powder formulations are used in comparison to liquid formulations. The addition of solvents of oily substances to sprays significantly reduces bee looses.

Insecticide formulations can be classified in order of their toxicity;

Dust>wettable powder>flowable>emulsifiable concentrate or soluble powder of liquid solution>granular formation.

- It is always advisable to keep bee colonies as far away as possible from the pesticide-treated fields. Even a distance of one kilometer from the sprayed site will reduce honeybee mortality nine fold.
- Remove all the flowering weeds from the fields by either mowing or beating so that they do not become a source of poison to the bees. This practice will force the bees to forage over loner distance, free from the adverse effect of pesticides.
- In the temperate regions, residues will remain toxic for a longer time due to lower temperatures. But, at the same time, low temperatures delay the initiation of the foraging activities

of honeybees in the morning. Keeping this in mind, the time of pesticide application should be shifted accordingly.

- Primary emphasis should be placed on an integrated pest management programme which mainly relies on biological or cultural methods of insect pest control and minimizes the use of poisoning chemicals.

## SAFE PESTICIDE USE FOR HUMAN BEINGS

**General**

- Pesticides should be applied only when it is really necessary.
- Instructions given on the label of the pesticide container should be read carefully and followed accordingly.

**Reducing the incidence of pesticide poisoning in human beings**

- Protective clothing such as a hat, long sleeved shirt, long trousers, face mask, goggles, and rubber or neoprene gloves and shoes should be worn while handling pesticides.
- When handling and applying pesticides, one should not eat, drink, or smoke. After applying pesticides, hands and face should be washed thoroughly with soap.
- Children should never be allowed to apply/use pesticides. Pesticides should not be kept within the reach of children.
- Leaking and defective equipment should never be used.
- Dust or powdered pesticides should not be applied on windy days because these can blow back on to the person applying them.
- Contact with recently sprayed crops should be avoided.
- If pesticide contamination does occur, one should change the clothing and thoroughly wash the contaminated area of the skin with soap as soon as possible.
- Pesticide containers should not be washed in ponds, streams, or other water sources. Waste pesticides and their containers should be carefully disposed of in a way that will not endanger people and other living beings.
- Empty pesticide containers should not be used for any other purpose, such as for food, water, or as cooking pots, because it is not possible to clean them thoroughly enough to render them safe.
- Pesticides should always be kept in their original labeled containers and should never be poured into drinking bottles.
- Before harvesting the crops, the recommended safety intervals should be observed.

## FUTURE RESEARCH STRATEGIES TO REDUCE BEE LOSSES

From toxicity data, it is now possible to calculate doses of pesticides which if sprayed in field will be relatively safe to bees. LD $_{50}$ value in $\mu$g per bee converted into pounds per acre when sprayed on the aerial portion of plants is safe to bees. A pesticide with a slope value of 4 or more is less hazardous when used at slightly reduced dose per hectare. There is a need to develop such models for Indian bees. There is a need to develop insecticide repellent to bees so that bees do not visit the treated area. Allethrin, a synthetic pyrethroid, disulfoton and demeton are bee repellents and are therefore safe to bees under field conditions (Atkins, 1992).

Research on good bee repellents is a need of the day as pesticides sprayed in mixture with bee repellent can be safer to bees. Good bee repellents under laboratory conditions are not good under field conditions (Kumar and Patyal, 1992 and Nauman *et al.*, 1994). Develop pesticide with low toxicity to bees. Such pesticides may have lower interaction with the site of action, faster detoxification and poor penetration and distribution in bees. Neem based pesticide like margosan–O and a pyrethroid, flumethrin not only control crop pests but also find use in hives for mite, *Acaropis woodi* control. Bees treated with syrup containing 3 to 6 ml margosan–O containing 3000 ppm azadirachtin per litre of syrup showed increase in honey and pollen yield and control of adults and eggs of the mite (Liu, 1995).

Special formulation of endosulfan with endosulfan–I and other bee toxic contaminants may be safer to bees than normal endosulfan (Nath *et al.*, 1983; Nath and Sharma, 1995). This formulation will give longer protection from pests as it will have higher endosulfan–II contents and thus is safe not only for bees but also for consumers of produce as it will give sufficient time for dissipation when sprayed on blossom. It can be useful for dipping bees or for treating hives to remove ecto-and endo-parasitic mites without ill effect on bees (Yamada, 1990).

Technology to decontaminate the bees from pesticide residues is feasible and needs refinement (Sharma and Nath, 1996). It is possible to isolate strains of bacteria from honey bees and adapt them to degrade pesticides. Bees when treated with such bacteria become tolerant to pesticides and employed to pollinate crops treated with pesticides.

Proper legislature can save bees from misuse of pesticides. With commercialization of beekeeping, it is now possible to give insurance cover to bee keepers that will go a long way in providing help to beekeeper if there is accidental contamination due to pesticides. Organic farming needs encouragement.

It appears that there is sufficient information on pesticide toxicity to *A. mellifera*, similar information on *A. cerana* is scanty and non-existent on other Indian bee species. It is time that more information be generated on Indian species because it is these which carry out most of the pollination in the country.

## RELATIVE TOXICITY OF SOME PESTICIDES TO THE ASIAN HIVE BEE *APIS CERANA* AND *APIS MELLIFERA*

In several developing countries of southeast Asia, beekeeping with *Apis cerana* has been sustaining heavy losses since the advent of synthetic pesticides several decades ago. Such widespread and careless use of toxic pesticides during the blooming periods of agricultural and horticultural crops not only kills bees but also contaminates hive products.

Several research workers have compared the relative toxicities of commonly used pesticides on *Apis cerana* and these have been classified as highly toxic, moderately toxic, and non-toxic. Application of highly toxic pesticides on blooming crops or weeds may cause severe damage to bees. Beekeepers should be warned in advance by the growers when these insecticides are to be used so that they can temporarily move the colonies to safer locations. Even 10 hours after spraying, these pesticides are still highly toxic to bees. Moderately toxic pesticides should be applied during the late evening when bees are not foraging actively. Beeshive should not be directly exposed to these pesticides. For minimal hazards to honeybees, the dose, the timings, and the methods of application of these moderately toxic pesticides are very important. Relatively non-toxic pesticides cause minimum damage to the bees. These should be applied during late evening, night or early morning. Relative toxicity of some pesticide to Asian hive bee, *Apis cerana* is given in table 3 and list of highly toxic and moderately toxic pesticides to European honeybee *Apis mellifera* is given in tables 4 and 5 alongwith $LD_{50}$ values.

**Table 3.** List of pesticides with high, moderate, and least toxicity to *Apis cerana*

| **Group I. Highly Toxic Pesticides** | Malathion 50 EC |
|---|---|
| Carbaryl 50% WP | Methyl demeton |
| Carbophenothion 20 EC | Monocrotophos 40EC |
| Cypermethrin 10EC | Trichlorfan 50EC |
| Decamethrin 20EC | Diazinon 20EC |
| Dichlorvos 100EC | Ethyl parathion 46% |
| Dimethoate 30EC | Fenitrothion 100EC |
| DDVP 100EC | 50EC |
| Monocrotophos 36WSC | Fenthion 100EC |
| Oxydemeton-methyl 25EC | Formothion 25EC |
| Parathion | Gamma BHC 20% |
| Phosphamidon 100EC | Lindane |
| Phorate | Metasystox 50EC |
| Permethrin 25EC | 25EC |
| Quinalphos 25EC | Mevinphos |
| Sumithion 50EC | Methyl Parathion 50EC |
| Thiometon 25EC | Dithane M-45 75 WP |
| **Group II. Moderately Toxic Pesticides** | Poltaf 80 WP |
| BHC 50 % | Difolitan 50EC |
| Carbyl 50WP | Hexacap 50WP |
| DDT 50% | Bavistin 50WP |
| Dieldrin | **Group. III. Relatively Non-toxic Pesticides** |
| Endrin | Endosulfan 35EC |
| Hinosan 50 EC | Menazon 70 DP |
| Heptachlor 10 WP | Phosalone 35EC |

*Source:* Verma and Partap (1993).

**Table 4.** Highly toxic and moderately toxic pesticides to European honeybee *Apis mellifera*
Group I Highly toxic pesticides
(LD%) = 0.000–1.99 μg/mg

Several losses may be expected if these pesticides are applied when bees are present at treatment time or within a day thereafter, except as indicated by footnotes. (Arranged by toxicity to honeybees–highest to lowest.)

| *Pesticide* | *Laboratory Data: [e] LD50* | *Pesticide* | *Laboratory Data: [e] LD50* |
|---|---|---|---|
| TEPP[1,3] | 0.002 | Methiocarb | 0.372 |
| Bioethanomethrin | 0.035 | Fenvalerate | 0.408 |
| Resmethrin | 0.062 | Famphur | 0.414 |
| Decamethrin | 0.067 | azimphos-methyl[1] | 0.528 |
| Chlorpyrifos | 0.110 | Bendiocarb | 0.428 |
| dieldrin[1] | 0.133 | naled[3] | 0.485 |
| carbofuran[5] | 0.149 | Dichlorvos, DDVP | 0.501 |
| Permethrin | 0.159 | heptachlor[1] | 0.526 |
| Fenitrothion | 0.176 | Isofenphos | 0.606 |
| parathion[1] | 0.177 | malathion[4] | 0.726 |
| Dimethoate | 0.191 | aziupos-ethyl | 0.958 |
| Methidathion | 0.237 | Aminocarb | 1.12 |
| EPN[1] | 0.237 | Phosmet | 1.13 |
| methyl parathion[1, b] | 0.241 | Acephate | 1.20 |
| Etrimfos | 0.264 | methomyl[5] | 1.29 |
| aldicarb [1,5, 7] | 0.272 | propoxur[2] | 1.34 |
| Mexacarbate | 0.302 | Methamidophos | 1.37 |
| dicrotophos[1] | 0.305 | Tetrachlorvinphos | 1.39 |
| Mevinphos[1,3] | 0.305 | Fenamiphos | 1.43 |
| Fenthion | 0.319 | Phosphamidon | 1.45 |
| fensulfothion[5] | 0.337 | Carbaryl | 1.54 |
| Aldrin | 0.352 | Fufencarb | 1.65 |
| monocrotophos[1] | 0.357 | Pyrazophos[2] | 1.85 |
| Diazinon | 0.372 | arsenicals[1] | 27.15 |

*Source:* Modified from University of California, Division of Agricultural Sciences Leaflets Nos. 2286 and 2287 and field tests.

**Table 5.** Moderately toxic pesticides to European honeybee *Apis mellifera*. LD50 = 2.0–10.99 μg/bee).

These can be used in the vicinity of bees if dosage, timing and method of application are correct, but should not be applied directly on bees in the field or at the colonies. (Arranged by toxicity to honeybees –highest to lowest.)

| *Pesticide* | *Laboratory Data:* [e] *LD50* | *Pesticide* | *Laboratory Data:* [e] *LD50* |
|---|---|---|---|
| Tempohos | 1.40 | DDT[1,10] | 6.19 |
| Demeton[1] | 1.71 | ethiofencarb | 6.85 |
| Trichloronate | 2.00 | Larvin | 7.08 |
| endrin[1] | 2.04 | sulprofos | 7.22 |
| Crotoxyphos | 2.31 | endosulfan | 7.81 |
| Pyramat | 2.62 | Fonofos | 8.68 |
| oxydemeton-methyl | 2.86 | chlordane[2] | 8.80 |
| profenfos | 3.46 | phosalone | 8.97 |
| Terbufos | 4.09 | formetanate hydrochloride | 9.21 |
| Ethylan | 4.57 | phorate[1,6] | 10.25 |
| Ethoprop | 5.56 | Oxamyl | 10.26 |
| Runnel | 5.62 | carbophenothion | 12.99 |
| disulfoton[1,2,6] | 6.12 | | |

*Source:* Modified from University of California, Division of Agricultural Sciences Leaflets Nos. 2286 and 2287 and field tests.

## REFERENCES


Anonymous 1992. *International Conference on the Asian Honey Bees and Bee Mites*. Chulalongkorn University, Bangkok, Thailand 9–14 Feburary, 1992. Publishers Research Unit, Department of Biology, Chulalongkorn University and IBRA

Etkins, E.L. 1975. Injury to honeybees by poisoning. *In: The hive and the honeybee*. R.A. Grout (ed). Dadant and Sons, Hamilton, Illionis. pp. 663–696.

Atkin, E.L. 1980. *Personal Communication to the Panel*. Dept. Entomology, University of California, Riverside, Calif.

Atkin, E.L. 1992. Management strategies for protecting honeybees, *Apis mellifera* from pesticides. (cited from Anonymous, 1992).

Chandel, R.S. and Gupta, P.R. 1992. Toxicity of diflubenzuron and penfluron to immature stages of *Apis cerana indica* F. and *Apis mellifera* L. *Apidologie*, **23**: 465–473.

Erickson, E.H. and Erickson, B.J. 1983. Honeybees and pesticides. *Am. Bee. J.* **123**: 797–805.

Gupta, P.R. and Chandel, R.S. 1995. Effect of duflubenzuron and penfluron on workers of *Apis cerana indica* F. and *Apis mellifera* L. *Apidologie,* **26**: 3–10.

Johansen, C.A. 1977. Pesticides and pollinators. *Ann. Rev. Ent.* **22**: 177–191.

Kumar, J. and Patyal, S.K. 1992. Oils from agroforestry plants as protectant for honeybees from insecticidal hazards. IUFRO Workshop Proceedings Agroforestry for Rural Needs. D.K. Khurana and P.K. Khosla (eds). Vol. II. pp. 728–733.

Liu, T.P. 1995. controlling tracheal mites in colonies of honeybees with neem (margosan-O) and flumethrin (Bayvarol). *Am Bee J.* **135:** 562–566

Mishra, R.C. 1995. Pesticidal poisoning to honeybees. *In: Honeybee and their management.* ICAR, New Delhi. pp. 151–163

Mishra, R.C. and R.C. Sihag, (1987). *Apiculture Research in India.* All India Co-Ordinated Research Project on Honey bee Research and Training at Haryana University, Hisar. pp. 120.

Nath, A. (1997). Safe use of pesticides during pollination. *In: Fruit Crop Pollination.* L.R. Verma and K.K. Jindal (eds). Ludhiana: Lalyani Publishers, pp. 363–372.

Nath, A.; Sharma, D.D.; Sharma, I. D. and Kumar, A. 1983. Toxicity of endosulfan to the Indian honey-bee. *A c. indica* 51SP 195–99 (*Les Colleques de INRA* n 21 France).

Nath, A. and Sharma, D.C. 1995. Pharmacodynamics of endosulfan in India honeybee. *Indian J. Exp. Biol.* **23**: 263–265.

Nath, A.; Sharma, D.C. and Sharma, I.D. (1985). Metabolism of endosulfan in the Indian Honey bee. *Experientia*, **41**: 1411–1412.

Nauman, K.; Currie, R.W. and Isman, M.B. (1994). Evaluation of repellent of a neem insecticide on foraging honey bees and other pollinators. *Canadian Entomol.* **126**: 225–230.

Patel, B.H.; Upadhayay, V.R.; Muralidharan, C.M. and Jindal, G.S. 1988. Effect of various insecticides on honey bees, *Apis florae* F. in ber (*Zizyphus mauritiana* Lank). *Curr. Sci.* **57**: 1195–1200.

Roubik David, W. 1995. *Pollination of cultivated plants in the tropics.* FAO Agricultural Services Bulletin 118.

Sharma, I.D. and Nath, A. 1996. Metabolism of 1-Naphthyl-N-methyl carbamate (carbaryl) by bacterial isolates from honeybees and the effect of bacterial inoculations on the carbaryl tolerance in bees. *J. Appl. Bacteriol*, **81:** 235–241.

Schmutterer, H. and Holst, H. 1987. On the effect of enriched and formulated neem seed kernel extract (AZT-VR-K) on the honey bee. *Zeitschrift Angewandte Ent.* **103**: 208–213.

Sorthia, B.K. and Chari, M.S. 1985. Toxicity of some insecticides to honey bee, *Apis florae* F. and *Apis mellifera* L. *J. Ent. Res.* **9**: 195–97.

Swaminathan, M.S. 1996. *Tropical and subtropical apiculture.* FAO Agricultural Services Bulletin No. 68. pp. 283.

Verma, L.R. 1990. *Bee Keeping in Integrated Mountain Development.* Oxford & IBH Publication, New Delhi. pp. 62–67.

Verma, L.R. 1992. *Honeybees in Mountain Agriculture.* Oxford and IBH Publishing Co. Pvt. Ltd. New Delhi.

Verma, L.R. and Partap, U 1993. *The Asian hive bee Apis cerana, as a pollinator in vegetable seed production.* ICIMOD Publication. pp. 35–41.

Yamada, M. 1990. Contorl of chaetodactylus mite, *Chaetodactylus nipponicus* Kurosa and important mortality agent of hornfaced osmia bee, *Osmia cornifons* Radoszkdvski. *Bulletin of the Aomori Apple Experiment Station,* **26**. 39–77.

# Subject Index


Absolute methods 27
*Agrobacterium* mediated transfer of DNA 225
Almond 295, 481
Almond weevil 297
Amla/Aonla 356
Aphid biotypes 265
Aphid forecasting 266
Aphid species as vectors 262
Aphids 259, 296, 301, 331, 394, 430, 449,
Aphids as pests 259
Apical twig gall maker 358
Apple 237, 472
Apple fruit moth 247
Apple leaf roller 242
Apple root borer 289
Approaches of biological control 124
Apricot 279, 481

*Bacillus thuringiensis* 217
Baculoviruses 220
Bael 367
Banana 474
Bark eating caterpillar 287, 354, 357, 361, 363, 369, 377
Bean aphid 414
Bean fly 413
Beans 410
Beetles 460
*Ber* 348
*Ber* fruitfly 348
*Ber* leaf webbers/rollers 355
Biochemical pesticides 222
Biofungicides 228
Biological control 16, 90
Biopesticides 215
Birds as pests of fruit trees 520
Birds as pests of horticultural crops 519
Birds as pests of vegetable crops 522
Black bean bug 411
Black-headed caterpillar 375
Blister beetles 412
Blossom beetles 303
Blossom thrips 248
Borers 249, 305
Botanical pesticides 17
Broad mite 397

Caterpillars 458
Chemical control 17
Cherry 295, 481
Cherry stem borer 300
Chitin synthetase inhibitors 175
Citrus 325, 478
Citrus black/white flies 327
Citrus leaf miner 329
Citrus psylla 325
Clones 264
Coccids 302
Cockchafer beetles 351
Codling moth 246, 285, 307
Concept of injury levels 10
Concepts of pest management 5
Consequences of disregarding indigenous knowledge systems 211
Cotton or melon aphid 442
Cucurbits 437
Cultural control 14, 135
Cut worms 388, 419

Date palm 373
Date palm scale 373
Defoliating and fruit eating beetles 245, 290, 297, 341
Defoliating caterpillars 290, 298, 391

Defoliating pests 290, 391
Defoliators 297, 304
Delta-endotoxins 217
Dispersion 40
Diversity of indigenous knowledge and pest management 204

Ecological factors regulating pest numbers 5
Economic injury level 10
Economic threshold level 11
Effects of insect growth regulators on non-target organisms 178
Electroporation 225
Entomopathogenic nematodes 227
Entomopathogens 111
Environmental economic injury levels 12
*Epilachna* beetles 392
Era of IPM 5
Era of pesticides 4
Era of traditional approaches 3
Estimation of population 70
European red mite 252
Evidence of using indigenous bio-insecticide for crop protection 208

Factors affecting toxicity of insecticides to honeybees 531
Fig 364
Fig leaf rollers 366
Fig midges 366
Fig stem borer 365
Flat-headed peach tree borer 286
Flat-headed stem borer 299
Flea beetles 432
Foliage feeders 242, 303
Frugivorous birds 343
Frugivorous insects 246
Frugivorous pests 284
Fruit borer 342
Fruit eating bat 343
Fruit flies 248, 285, 317, 439
Fruit nematodes 470
Fruit sucking moths 331
Fruits of arid and semi-arid regions 347
Future research strategies to reduce bee losses 534
Future strategies for ITK strengthening 212

Genetic engineering techniques for insect pest management 215
Genetic variations in aphid populations 264
Girdler beetle 341
Grape flea beetle 340
Grapes 339
Grapevine thrips 340
Green semilooper 423

Hadda beetle 441
Hairy caterpillars 352
Heteroecy 259
Holocycly 260
Homopterous pests 368
Host plant resistance 15

Indian gypsy moth 243
Indigenous cultural practices to protect crops from insect pests 205
Indigenous technological knowledge 201
Indirect sampling methods 37
Insecticide resistance 266
Insect growth regulators 171
Insecticide act 169
Insecticide application strategy 167
Integrated approach 524
Integrated rodent management 515
Integration of pest management tactics 13
IPM 18, 26
ITK most threatened source 211

Jamun 369
Jassid 420
Juvenile hormone agonists and antagonists 172

Karonda/corinda 378
Key factor analysis 80

Lac insect 353, 367
Leaf defoliators 303
Leaf eating beetles 304
Leaf eating beetles and weevils 368, 377
Leaf eating caterpillars 305, 320, 378, 433
Leaf folders/rollers 303
Leaf gall psylla 372
Leaf hoppers 396
Leaf roller 332, 342
Leafminers 370, 433, 454
Lemon butterflies 328
Leopard moth 301
Lepidopterous pests 358, 368
Lepidopterous shoot borers 251

Lesion nematodes 499
Life tables 69
Limitation in the use of pesticides 155
*Lincoln index* 30

Main attributes of ITK in relation to pest management 203
Management of aphid pests 267
Management of insect pests of fruit crops with semiochemicals 186
Mango 310
Mango gall midges 314
Mango hoppers 310
Mango leaf webbers 319
Mango mealybug 312
Mango shoot gall psylla 316
Mango stem borer 319
Mango stone weevil 318
Mealybug 340,367, 455
Mechanical control 142
Microbial pesticides 217
Microprojectile bombardment method 225
Miscellaneous pests 359, 362, 364, 367, 369, 372, 376, 378
Mites 333, 461
Mode of acquiring viruses 262
Modern agriculture, agroecological changes and pest problems 7
Moulting hormone agonists and antagonists 172
Mulberry 362
Mulberry bug 364
Mycoherbicides 228
Mycoinsecticides 221

Natural enemies 99
Neem botanical 223
Nematode diseases 471
Nematode management 482, 500
Nematode pests 401
Nematode pests of fruit crops 469, 472
Nematode pests of ornamental plants 499
Nematode pests of vegetable crops 491
Neuropeptides 171

Okra 420
Okra aphid 422
Ornamental plants 449

Papaya 479
Pea 416
Pea blue butterfly 418
Pea leafminer 416
Pea pod borer 418
Peach 279, 480
Peach black aphid 282
Peach leaf curling aphid 280
Peach mealy aphid 281
Peach twig borer 288
Pear 237, 474
Pecannut 482
Pesticide application equipment 163
Pesticide poisoning of bees 528
Pesticide-honey bee interactions 527
Pesticides 154
Phalsa/falsa 377
Pheromones 223
Phytophagous mites 251
Pineapple 479
Plant volatiles 222
Plum 279, 481
Plum fruit moth 284
Plum scale 282
Pomegranate 360
Pomegranate butterfly 360
Population models 55
Population monitoring 27
Population prediction models and pest management 53
Potato 387
Potato cyst nematode 497
Potato tuber moth 397
Potential of ITK in pest management 207
Predatory arthropods 107
Problem birds in horticulture 520
Problems related with pesticide application 168
Problems relating to pesticide residues 158
Pumpkin caterpillar 443

Radish and turnip 427
Red cotton bug 422
Red palm weevil 375
Red pumpkin beetle 437
Red spider mite 415
Red vegetable mite 444
Regulatory control 144
Relative methods 31
Relative toxicity 535
Remote sensing 39
Reniform nematode 494

Rhinoceros beetle 374
Rodent pests of horticultural crops 506
Root borer 249, 301
Root knot nematodes 403, 491, 499

Safe pesticide use for human beings 534
Safe use of pesticides 161
Sampling programme 48
San Jose scale 238, 284
Sap sucking insects 359, 361, 363, 377
Sap sucking pests 238, 296, 301
Sap thripters 279
Sawfly 428
Scale insects 297, 455
Scale insects and mealy bugs 330
Scales 367
Semiochemicals 184
Shoot and fruit borer 421
Shoot and trunk borers 333
Shoot borer 289, 321
Shot hole borers 251, 288, 341
Soft scales 322
Soil pests 388
Stem and root borers 299
Stem and shoot borers 250, 298, 305
Stem borers 289, 362
Stem boring beetle 342
Stink bug 444
Storage pest 397
Strategies to reduce bee losses 532
Strawberry 477
Sucking pests 394
Symptoms of bee poisoning 530

Tent hairy caterpillar 244
Termites 289, 379
Thrips 332, 369, 452
Transgenic bacteria and viruses 221
Transgenic crops 226
Transgenic plant pesticides 224
Traps 32-36
Two spotted mite 253

Uses of semiochemicals 185

Variants 266
Vertebrate pests 343
Virus transmission 261

Walnut 295, 482
Walnut green aphid 304
Walnut weevil 306
White grubs 389
White peach scale 283
Whiteflies 371, 413, 457
Woolly apple aphid 240

Xylophagous pests 286